Acid-Base Interactions:
Relevance to Adhesion Science and Technology, Volume 2

Acid-Base Interactions:

Relevance to Adhesion Science and Technology

Volume 2

Editor: K.L. Mittal

Utrecht • Boston • Tokyo, 2000

VSP BV
P.O. Box 346
3700 AH Zeist
The Netherlands

Tel: +31 30 692 5790
Fax: +31 30 693 2081
vsppub@compuserve.com
www.vsppub.com

First published in 2000

ISBN 90-6764-325-4

Printed in The Netherlands by Ridderprint bv, Ridderkerk.

CONTENTS

Part 3: Applications of Acid-Base Interactions

Acid-Base Interactions, Vol. 2, pp. ix–x
Ed. K.L. Mittal

Preface

This book documents the proceedings of the Second International Symposium on Acid-Base Interactions: Relevance to Adhesion Science and Technology held under the auspices of C4 Technologies, Inc. in Newark, NJ, October 19–21, 1998.

The first symposium on this topic was held on the occasion of the 75th birthday of Professor Frederick M. Fowkes as a part of the 64th Colloid and Surface Science Symposium at Lehigh University, June 18–20, 1990. The proceedings of this symposium were duly documented in a hardbound book [1]. This first book served as a fillip for a flurry of research activity in the arena of acid-base interactions as well as in harnessing such interactions to modulate adhesion in a host of situations.

So it was deemed opportune and necessary to hold the second symposium on this topic. This symposium was organized with the following objectives in mind: (i) to consolidate the R&D activity carried out since the first symposium, (ii) to provide a forum for discussion of latest research results, (iii) to provide an opportunity for cross-pollination of ideas, (iv) to identify topics where there was discordance of opinion or discrepancy, and (v) to highlight areas which needed intensified R&D activities.

The final technical program contained a total of 36 papers by researchers and technologists and they hailed from academia, industry and other organizations. The program was comprised of both invited as well contributed research papers and both overviews and original research contributions were presented. Throughout the symposium there were brisk discussions (which mostly were illuminant, not exothermic), both formally (in the auditorium) and informally (in corridors). During the symposium it became amply clear that there were many putative approaches to acid-base characterization of materials but there was a lack of consistency in these approaches as, at times, it became quite patent that different approaches led to quite divergent results. Thus it was highlighted in this symposium that there was no ecumenical approach to acid-base characterization of materials.

This book contains a total of 32 papers and it should be recorded for posterity that all manuscripts were rigorously peer reviewed and were suitably revised (some two or three times) before inclusion in this book. The book is divided into three parts as follows: Part 1: Fundamental Aspects of Acid-Base Interactions; Part 2: Characterization of the Acid-Base Properties of Materials; and Part 3: Applications of Acid-Base Interactions. The topics covered include: Surface free energy acid-base theory applied to solid surfaces; Good, van Oss and Chaudhury theory; contact angle measurements and interpretation; acid-base theory of contact angles; acid-base strength of solid surfaces; acid-base interactions at solid surfaces; acid-base interactions at the molecular level; characterization of acid-base properties of a host of materials (polymers, wood, glass, ceramics, silica particles, textile fibers, rocks) by XPS, inverse gas chromatography, immersion calorimetry, contact angle titration, and thin layer wicking; and relevance of acid-base interactions to bioadhesion, microbial adhesion, polymer adhesion, and adhesion in reinforced polymer composites.

I certainly hope the bountiful information provided in this compendium will serve as a reference to the veteran researcher as a commentary on contemporary research as well as a guide to the neophyte.

Now comes the pleasant task of thanking those who contributed in many and varied ways. First, I would like to express my sincere thanks to Dr Robert H. Lacombe, dear friend and colleague, for his help during the organization of this symposium. The valuable comments from the unsung heroes (reviewers) were very useful as the comments from the peers are a desideratum to maintain the highest standard of a publication; and my thanks are extended to them for their time and effort. Next I would like to convey my thanks to the staff of VSP (publisher) for the job well done in producing this book. Last, but not least, I am thankful to all the authors for their interest, contribution and enthusiasm without which this book would not have seen the light of day.

K.L. Mittal
P.O. Box 1280
Hopewell Junction, NY 12533, USA

1. K.L. Mittal and H.R. Anderson, Jr. (Eds), *Acid-Base Interactions: Relevance to Adhesion Science and Technology*. VSP, Utrecht, The Netherlands, 1991.

Part 1: Fundamental Aspects of Acid-Base Interactions

Acid-Base Interactions, Vol. 2, pp. 3–53
Ed. K.L. Mittal

"Repulsive acid–base interactions": Fantasy or reality

W. V. CHANG [1,*] and X. QIN [2]

[1] *Department of Chemical Engineering, University of Southern California, Los Angeles, CA 90089-1211*
[2] *Vision-Ease Lens, 16016 Montoya Street, Azusa, CA 91702*

Abstract—About nine years ago, we proposed a phenomenological model for characterizing local thermodynamic interactions at an interface. Despite its many successes, some researchers still show great reluctance to accept it. In this paper a clear discussion on the theoretical basis of interfacial interactions and a critical review of other well-known models is presented. It is pointed out that many of the conflicting conclusions in the literature have resulted from the inconsistent definitions of the 'neutral' materials. By using n-alkanes as the only 'neutral' materials and following a rigorous and logical treatment of literature data, the existence of "negative acid–base contributions" to the work of adhesion is clearly demonstrated. In order to accurately characterize the surface energetics of a material, three parameters are needed: one for the dispersion and two for the acid–base interactions. Methods to determine these parameters are reviewed. Applications of the model to some polymers and common liquids are discussed.

Keywords: Acid-base interactions; work of adhesion; Lifshitz-van der Waals interactions; dispersion interactions; solid-liquid interface.

*To whom correspondence should be addressed. Tel: 213-740-2072, Fax: 213-740-8053.
E-mail: wenji@rcf.usc.edu

NOMENCLATURE

Liquids:

WT	water
GL	glycerol
FA	formamide
MI	methylene iodide
BN	α-bromonaphthalene
TP	tricresyl phosphate
EG	ethylene glycol
A1241	trichlorobiphenyl
A1248	tetrachlorobiphenyl
sTCE	*sym*-tetrachloroethane
TCE	tetrachloroethylene
TBN	*tetra*-butylnaphthalene
CS2	carbon disulfide
BENZ	benzene
TOLU	toluene
$HCCl_3$	chloroform
CCl_4	carbon tetrachloride
ACET	acetone
C3E	propyl ether
C2E	ethyl ether
BPUD	benzyl phenylundecanoate

Polymers:

TFE	Tetrafluoroethylene
CTFE	Chlorotrifluoroethylene
PTFE	Poly(tetrafluoroethylene)
P3FE	Poly(trifluoroethylene)
PVDF	Poly(vinylidene fluoride)
PVF	Poly(vinyl fluoride)
PE	Polyethylene
60/40	Copolymer of TFE and CTFE by weight proportion: 60/40 (mole % of halogen substitution: F/Cl = 90.9/9.1)
80/20	Copolymer of TFE and CTFE by weight proportion: 80/20 (mole % of halogen substitution: F/Cl = 95.6/4.4)
PCTFE	Poly(chlorotrifluoroethylene) (Kel-F) (mole % of halogen substitution: F/Cl = 75/25)
TFE-E	Copolymer of TFE and ethylene (mole % of halogen substitution: F/H = 50/50)
HTA	Hexatriacontane crystal ($C_{36}H_{74}$)
PW	Paraffin wax
PHDFDMS	Poly[(1H,1H,2H,2H–heptadecafluorodecyl)methylsiloxane]

1. INTRODUCTION

The practical strength of adhesive joints, σ_a, has been shown to be directly proportional to the thermodynamic work of adhesion, W_a, at the interface [1, 2]. Correlations between the adhesion strength and the surface thermodynamic properties have been demonstrated using the peel test over a limited range of peel rates [3–7]. Even though σ_a is typically several orders of magnitude higher than W_a, a large thermodynamic work of adhesion is a necessary condition for good joint strength. The work of adhesion, W_a, is equivalent to the Gibbs free energy change for forming a unit interface between two pure substances, and is defined as

$$W_a = \gamma_1 + \gamma_2 - \gamma_{12} \tag{1}$$

where γ_i is the surface free energy of substance i; and γ_{12} is the interfacial free energy. In an immiscible liquid-liquid system, all the free energy terms in Equation (1) can be independently measured; hence, W_a is determined. However, in a solid–liquid system, neither the surface free energy of the solid nor the interfacial free energy can be directly measured. When the liquid forms a finite contact angle on the solid, the work of adhesion can be determined by the Young-Dupré equation:

$$W_a = \gamma_L(1 + \cos\theta) + \pi_e \tag{2}$$

where γ_L is the surface free energy of the liquid, θ is the contact angle of the liquid on the solid, and π_e is the equilibrium spreading pressure of the liquid vapor on the solid surface. On a low energy surface, π_e is usually negligible if the contact angle is larger than ten degrees [8]. If the surface free energy of the solid is so high that the liquid spreads on the solid, the work of adhesion cannot be determined by contact angle. It can be obtained by measuring the free energy change per unit area as a vapor adsorbs to form a liquid film on a powdered solid [9].

It is highly desirable to estimate W_a from the constitutive properties of the two component phases. Girifalco and Good [10] first proposed:

$$W_a = \Phi(W_{11}W_{22})^{1/2} = 2\Phi(\gamma_1\gamma_2)^{1/2} \tag{3}$$

where W_{11} is the cohesive Gibbs free energy and Φ is the interaction parameter. They showed theoretically that the value of Φ was unity if the intermolecular interactions were all due to London-van der Waals force, and the molecules had similar size and shape. In general, Φ is a characteristic of the system.

Researchers have known for a long time that the work of adhesion cannot be uniquely determined from the surface tensions of the component materials. For example, we may have two liquids with similar values of surface tension and yet their respective work of adhesion with a common solid may be very different (see examples in Reference [11]). Fowkes and co-workers [12–16] recognized that there were different intermolecular forces and that they might not interact in the same way. He suggested that apart from the dispersion interaction, the acid–base interaction was the most important contributor to the work of adhesion [14, 17]. Only the dispersion interaction can be expressed as a geometric mean.

Following the Fowkes approach, we proposed a new model for acid–base interactions across an interface [18]. The model has received many criticisms, mainly because the paper challenged two important and well-accepted concepts in this field: 1) the assumption of the commonly used test liquids such as methylene iodide (MI), α-bromonaphthalene (BN), and tricresyl phosphate (TP) as purely dispersive; and 2) the exclusion of acid-acid and base-base repulsions in the traditional Lewis acid–base models. Since the validity of any scientific assumption can only be established through critical experimentation, we have been conducting many critical tests in the last few years. In this paper, we will present ample experimental evidence to support our model. We will show that the assumption of MI, BN and TP as neutral liquids is not consistent with the assumption of n-alkanes as neutral liquids. Consequently, it leads to erratic values for the dispersive component of many solids as reported in the literature. Furthermore, we will show the necessity of allowing repulsive interactions between acid-acid and base-base components. Allowing these interactions neither contradicts the basic Lewis acid–base definition nor violates any fundamental physics. In the next two sections we will critically review the existing theories and models for interfacial interactions. It is followed by a discussion on the need for a new approach. We will then present our model and its use for characterizing the surface acid–base properties. Finally, we will apply the model to contact angle data of solid-liquid systems studied by Zisman and co-workers [19–23], and to some of our and others recent data [24–28].

2. VAN DER WAALS FORCES OF INTERFACIAL INTERACTIONS

2.1. Classical Treatment

One of the important intermolecular interactions results from the van der Waals forces. It was first proposed by J. D. van der Waals in 1873 to explain the deviation from ideality in gases [29]. The van der Waals interactions originate from four sources: the London dispersion force, the Keesom orientation force, the Debye induction force, and a usually negligible electronic-induction contribution.

Let us start with Keesom forces. The interaction potential energy, U_0, between two isolated and position-fixed dipoles can be described by

$$U_0 = -\frac{3[(\boldsymbol{\mu}_1 \cdot \boldsymbol{r})\boldsymbol{r}] \cdot \boldsymbol{\mu}_2}{4\pi\epsilon_0 r^5} + \frac{\boldsymbol{\mu}_1 \cdot \boldsymbol{\mu}_2}{4\pi\epsilon_0 r^3} \tag{4}$$

where $\boldsymbol{\mu}_i$ is the dipole moment, $\boldsymbol{r}$ is the polar coordinate with respect to the center of the first dipole, r is the scalar distance to the dipole center and ϵ_0 is the permittivity of vaccum. In a real situation, the molecules are free to rotate in a gas. A statistical average is required over all spatial orientations of the two interacting dipoles. The average electric field $\langle \boldsymbol{F} \rangle$ generated by the dipole would be zero because there is no preference for a given dipole orientation. The interaction is determined by $\langle \boldsymbol{F}^2 \rangle$.

Thus, the interaction energy between two freely rotating dipoles is given by [30]

$$U_0 = -\frac{2}{3}\frac{1}{(4\pi\epsilon_0)^2}\frac{\mu_1^2\mu_2^2}{kTr^6} \qquad \text{with } kT > \frac{\mu_1^2\mu_2^2}{4\pi\epsilon_0 r^3} \tag{5}$$

Molecules without permanent dipoles are polarizable when placed in an external electric field. The polarization (separation of electropositive and electronegative sites) induces moment, and it is proportional to $\boldsymbol{F}$:

$$\mu_{\text{ind}} = \alpha \boldsymbol{F} \tag{6}$$

where α is called the polarizability. The interactions between dipoles and induced dipoles lead to the Debye induction forces. The induction energy formula worked out by Debye in 1921 [31] is

$$U_\text{I} = -\frac{\mu^2\alpha}{(4\pi\epsilon_0)^2 r^6} \tag{7}$$

Notice that there is no temperature factor in the equation above. The physical basis behind this point is that the induced dipole is instantaneous as compared to the thermal molecular motion.

The dispersion force is the most important van der Waals force. Unlike Keesom and Debye forces, which can be obtained electrostatically, London dispersion force is of quantum mechanical in nature. It could only be derived after the advent of quantum mechanics. The origin of dispersion forces can be understood based on the following physical picture. Molecules (atoms) contain oscillating charges (high frequency, $> 10^{16}$ Hz [29]) due to the circulation of negative electrons around positive nuclei. When two such "oscillators" approach, they start to influence each other, as in the Keesom interaction, with attractive situations having higher probabilities than repulsive ones. London in the 1930s [32] derived the approximate expression for the dispersion energy for the two attracting spherical molecules as

$$U_\text{D} = -\frac{3}{2}\frac{\alpha_1\alpha_2}{(4\pi\epsilon_0)^2 r^6}\frac{h\nu_1\nu_2}{\nu_1+\nu_2} \tag{8}$$

where ν is the zero point (ground state) vibration frequency in Hz, h is the Planck constant, and $3h\nu/2$ is the zero point vibration energy of the molecules for three degrees of freedom. In the above equation, $h\nu$ can be replaced by I, the first ionization energy. Again, due to the instantaneous feature of the "oscillators" there is no temperature factor in the London equation. The London dispersion force differs from the other two in several aspects. First, it is present in all kinds of molecules. Second, it exhibits the retardation phenomenon implying that, for large separation distance (e.g., $r > 10$ nm), the interaction energy decreases more rapidly with distance (factor r^{-7}) than at smaller r. Third, it should be approximately additive, meaning that in case of a large collection of molecules the total dispersion interaction energy is not very different from the sum of pair-interaction energies.

2.1.1. The Berthelot Principle. Noticing the common r^{-6} dependence of three van der Waals forces, we can represent them with

$$U = -\frac{\beta_{12}}{r^6} \tag{9}$$

From the London dispersion expression (equation 8), it is easy to derive the following relations for attraction coefficients:

$$\beta_{12}^{\mathrm{d}} = \frac{2\sqrt{I_1 I_2}}{I_1 + I_2}(\beta_{11}^{\mathrm{d}}\beta_{22}^{\mathrm{d}})^{1/2} \tag{10}$$

$$= \frac{2\beta_{11}^{\mathrm{d}}\beta_{22}^{\mathrm{d}}}{\beta_{11}^{\mathrm{d}}(\alpha_2/\alpha_1) + \beta_{22}^{\mathrm{d}}(\alpha_1/\alpha_2)} \tag{11}$$

For most substances of relevance to the field of adhesion of polymers, ionization potentials lie between 1.28×10^{-18} to 2.08×10^{-18} J (8 to 13 eV). Thus equation (10) can be replaced by a geometric mean with deviations within only a few percent:

$$\beta_{12}^{\mathrm{d}} \cong (\beta_{11}^{\mathrm{d}}\beta_{22}^{\mathrm{d}})^{1/2} \tag{12}$$

The above equation is known as the **Berthelot principle**, i.e., the dispersion interaction energy between two dissimilar molecules can be represented by the geometric mean of their individual interaction energies. On the other hand, if the polarizabilities of interacting pair are approximately the same, i.e. $\alpha_1/\alpha_2 \approx 1$, the dispersion attraction coefficients can be expressed in the harmonic mean:

$$\beta_{12}^{\mathrm{d}} \cong \frac{2\beta_{11}^{\mathrm{d}}\beta_{22}^{\mathrm{d}}}{\beta_{11}^{\mathrm{d}} + \beta_{22}^{\mathrm{d}}} \tag{13}$$

The Keesom orientation interaction (equation 5) automatically satisfies the Berthelot principle:

$$\beta_{12}^{\mathrm{o}} = (\beta_{11}^{\mathrm{o}}\beta_{22}^{\mathrm{o}})^{1/2} \tag{14}$$

Equations 12 to 14 provide the theoretical basis for the geometric mean and harmonic mean models in adhesion of polymers.

2.1.2. Pairwise Approximation of van der Waals Interactions between Macrobodies. All the energy equations discussed so far are only applicable to interactions between isolated atoms or molecules. For interactions of condensed systems the situation will be more complicated. The polar (Keesom and Debye) interactions are considerably reduced due to the highly correlated electric fields produced by dipoles and the vectorial nature of dipole moments [33]. The dispersion force is, however, more or less additive according to London [32]. Hamaker [34] and de Boer [35] further advanced the bimolecular summation approach and derived the approximate treatment in which the total attracton energy was obtained by pair-wise addition of London-van der Waals energies between all molecules of interacting bodies. The original Hamaker-de Boer theory does not account for polar interactions.

Let us start with the case that a single molecule M_1 of medium 1 interacts with a semi-infinite plate of medium 2. If we assume that the molecular forces are pairwise additive and there is no third medium in between media 1 and 2, the total attractive free energy between M_1 and the plate is given by integrating equation (9) over the ring volume v:

$$\Delta G^{\mathrm{d}}_{\mathrm{M,plate}} = -\int_0^{\infty} \frac{\beta_{12}}{r^6} N_2 dv = -\frac{\pi \beta_{12} N_2}{6x^3} \tag{15}$$

where N_2 is the number of molecules of medium 2 per unit volume and x is the distance of M_1 from the plate of medium 2. The interaction energy (in J/m^2) between a column of unit cross section of medium 1 and the entire medium 2 is then obtained by integration of the equation above over x from d (plate-plate distance) to $-\infty$:

$$\Delta G^{\mathrm{d}}_{12} = -W^{\mathrm{d}}_{12} = -\int_d^{-\infty} -\frac{\pi \beta_{12} N_2}{6x^3} N_1 dx = -\frac{A_{12}}{12\pi d_{12}^2} \tag{16}$$

where W^{d}_{12} is the work of adhesion to separate the two phases from van der Waals contact to infinite distance, $N_1 dx$ is the number of molecules in a volume defined by a thin slice of thickness dx and unit cross-sectional area, and A_{12} is the famous **Hamaker constant**:

$$A_{12} = \pi^2 \beta_{12} N_1 N_2 \tag{17}$$

From the previous section, we know that the Berthelot principle is applicable to the London coefficient β_{12}, hence

$$A_{12} \cong (A_{11} A_{22})^{1/2} \tag{18}$$

If the distance d_{12} can also be approximated by the geometric mean of the individual intermolecular distances, that is, $d_{12} \cong (d_{11} d_{22})^{1/2}$, then

$$W^{\mathrm{d}}_{12} \cong \frac{(A_{11} A_{22})^{1/2}}{12\pi d_{11} d_{22}} = (W^{\mathrm{d}}_{11} W^{\mathrm{d}}_{22})^{1/2} = 2(\gamma_1^{\mathrm{d}} \gamma_2^{\mathrm{d}})^{1/2} \tag{19}$$

where, W^{d}_{ii} is the dispersion work of cohesion for medium i. The equation above is the key element in Fowkes treatment of dispersion force contributions to the work of adhesion [13].

2.2. GM and HM Models

Two of the oldest and widely used models for thermodynamic work of adhesion are the geometric mean (GM) model [8, 36] and the harmonic mean (HM) model [8, 37]. The theoretical basis of both models was discussed in the last section. Both assume that there are two distinct components in the surface tension of a material, involving contributions from London dispersive (d) and Keesom polar (p) forces:

$$\gamma = \gamma^{\mathrm{d}} + \gamma^{\mathrm{p}} \tag{20}$$

$$W_a^{GM} = 2[(\gamma_1^d \gamma_2^d)^{1/2} + (\gamma_1^p \gamma_2^p)^{1/2}] \tag{21}$$

$$W_a^{HM} = 4\left[\frac{\gamma_1^d \gamma_2^d}{\gamma_1^d + \gamma_2^d} + \frac{\gamma_1^p \gamma_2^p}{\gamma_1^p + \gamma_2^p}\right] \tag{22}$$

Considering the experimental uncertainty and the small percentage of Debye induction forces contributing to the total van der Waals forces, the dipole-induced component can be neglected.

Although both models have a sound theoretical basis, they are often applied to systems that violate the original assumptions. In most cases, γ^p has been called the specific component of surface free energy. That is, it is considered as the sum of all non-dispersion contributions arising from different types of molecular interactions. This is similar to the treatment of Owens and Wendt [38, 39] who used the geometric mean of hydrogen bonding components to estimate the surface free energies of polymer surfaces. Thus, the "polar" contribution to surface energy γ^p so obtained is not the polarity measure of the molecules, especially for systems with hydrogen-bonds. In hydrogen bonding systems, γ^p is not a measure of the ability of a surface (or interface) to form hydrogen bonds (a subclass of Lewis acid–base bonds), since the complementary nature of attractive hydrogen bonding cannot be expressed in a manner of geometric mean or harmonic mean. The use of GM and HM models has also been criticized by many other researchers, such as Fowkes [14, 15, 40] and Panzer [41].

2.3. The Lifshitz Theory of van der Waals Interactions Between Condensed Media

For many years the theory of van der Waals forces was dominated by the London theory and the Hamaker summation procedure. The London-Hamaker approach, however, was far too limited to adequately deal with those cases of practical interest where macroscopic bodies interact across an intermedium. In such cases, Lifshitz and coworkers [42, 43] showed that the simple pair-wise addition assumption was not valid. Unlike the pair-wise addition approach, the Lifshitz theory considers interacting macrobodies as continuous media. The van der Waals interaction between the media originates from the fluctuation of electromagnetic waves arising from electron motions. The interaction energy is then characterized by macroscopic parameters, i.e., frequency dependent dielectric susceptibilities of the interacting phases under consideration. For two semi-infinite dielectric media 1 and 2 interacting through medium 3, the non-retarded free energy per unit area can be expressed as

$$\Delta G = -\frac{kT}{8\pi d^2} \sum_{n=0}^{\infty'} \sum_{m=1}^{\infty} \Delta_{13}^m \Delta_{23}^m / m^3 \tag{23}$$

$$\approx -\frac{kT}{8\pi d^2} \sum_{n=0}^{\infty'} \Delta_{13} \Delta_{23} \tag{24}$$

where the prime indicates that the zero frequency term is to be multiplied by one-half; Δ_{jk} is defined in terms of the dielectric permeabilities (ϵ) of the media at imaginary frequencies $i\omega_n$:

$$\Delta_{jk} = \frac{\epsilon_j(i\omega_n) - \epsilon_k(i\omega_n)}{\epsilon_j(i\omega_n) + \epsilon_k(i\omega_n)} \tag{25}$$

with $\omega_n = 2\pi nkT/\hbar$ (rad/s), n is the quantum number of the relevant absorption (oscillation), and m is an index to count for the multiple reflections of the electromagnetic field in the gap between the two media [44, 45]. In many cases, the higher order terms ($m = 2, 3, \ldots$) are negligible. In addition, the dielectric permeability of a material can be expressed in terms of the Kramers-Kronig relation [42, 46] as

$$\epsilon(i\omega) = 1 + \frac{2}{\pi}\int_0^{\infty'} \frac{\omega\epsilon''(\omega)}{\omega^2 + \omega_n^2} d\omega \tag{26}$$

where $\epsilon''(\omega)$ is the loss component of the complex dielectric permeability $\epsilon(\omega) = \epsilon'(\omega) - i\epsilon''(\omega)$. Notice that the Lifshitz theory and the pair-wise approximation give the same distance dependency (factor d^{-2}) for the free energy. The effective Hamaker constant for the interaction of two parallel semi-infinite media can then be defined by comparing Equations 24 and 16 as

$$A \cong \frac{3}{2}kT \sum_{n=0}^{\infty} \Delta_{13}\Delta_{23} \tag{27}$$

The Lifshitz equation has two important features. First, it covers all frequencies and thus encompasses dispersion, orientation, and induction energies. At ordinary temperature, the zero-frequency ($n = 0$) contribution to the free energy gives the total energy of orientation and induction interactions, and the finite frequency ($n = 1, 2, \cdots$) contribution corresponds to the dispersion interaction [47]. That is,

$$A_{132} = \frac{3}{2}kT \left\{ \underset{\text{polar}}{\frac{1}{2}(\Delta_{13}\Delta_{23})_{n=0}} + \underset{\text{dispersion}}{\sum_{n=1}^{\infty} \Delta_{13}\Delta_{23}} \right\} \tag{28}$$

Second, it predicts that van der Waals interactions between the same media is always attractive regardless of the inter-media medium. For the case of medium 1 with medium 2 across medium 3, the interaction can be attractive or repulsive depending on the relative scale of the three dielectric permeabilities. When $\epsilon_1 < \epsilon_3$ and $\epsilon_2 > \epsilon_3$, the interaction results in repulsive forces.

2.4. Justification of Geometric Mean Assumption for the Lifshitz (van der Waals) Interactions

More than 30 years ago Fowkes [13] proposed a very powerful concept that the contribution of the London-dispersion interactions to the work of adhesion between two media, W^{d}_{12}, was equal to the geometric mean of the values of the dispersive contributions to the cohesive energy of the media. In this section we will critically evaluate the validity of this assumption based on the Lifshitz theory and experimental data.

2.4.1. Theoretical Justification. The Lifshitz theory has been widely used to calculate the free energy of van der waals interactions [47, 48]. According to the theory, the non-retarded interaction energy between two semi-infinite flat slabs of media 1 and 2 separated by a vacuum (air) layer of thickness l_{12} is,

$$-\Delta G_{12} = \frac{kT}{8\pi l_{12}^2} \sum_{n=0}^{\infty}{}' \sum_{m=1}^{\infty} (\Delta_1 \Delta_2)^m / m^3 \tag{29}$$

with

$$\Delta_j = \frac{\epsilon_j(i\omega_{\mathrm{n}}) - 1}{\epsilon_j(i\omega_{\mathrm{n}}) + 1} \tag{30}$$

and

$$\omega_{\mathrm{n}} = (4\pi^2 kT/h)\, n \tag{31}$$

Where k is the Boltzmann constant, T is the absolute temperature, ϵ_j is the dielectric permittivity of the medium j, h is the Planck constant, and ω_{n} is the frequency. The prime in the summation denotes that the term at $n = 0$ is given half weight. Combining the above equation with the conventional Hamaker constant expression gives:

$$W_{12} = -\Delta G_{12} = \frac{A_{12}}{12\pi l_{12}^2} \tag{32}$$

and one can calculate the Hamaker constant as

$$A_{12} = \frac{3}{2} kT \sum_{n=0}^{\infty}{}' \sum_{m=1}^{\infty} (\Delta_1 \Delta_2)^m / m^3 \tag{33}$$

These equations seem to suggest that $-\Delta G_{12}$ becomes the work of adhesion W_{12} when l_{12} approaches the distance of interatomic separation, L_{c}. However, rigorously speaking, the Lifshitz theory, being a continuum theory, is only applicable for separations greater than interatomic distance. Otherwise, the "graininess" of the material must become important. Therefore, short range interactions such as hydrogen bonding, acid–base interactions and steric forces are not included. In many attempts to estimate the surface energies of materials, L_{c}, in reality, was treated as a fitting parameter. Furthermore, as discussed below, the calculated value of the Hamaker constant is sensitive to the quality of the experimental spectral data of the dielectric permittivity and the accuracy of fitting the data by analytical functions. The dielectric permittivities, $\epsilon_{\mathrm{j}}(i\omega_{\mathrm{n}})$, in the above equations are quantities that can be experimentally determined. The following two representations of Parsegian and Ninham [47] are commonly used:

$$\epsilon(i\omega_n) = 1 + \frac{C_{IR}}{1 + (\omega_n/\omega_{IR})} + \frac{C_{UV}}{1 + (\omega_n/\omega_{UV})} \tag{34}$$

$$\epsilon(i\omega_n) = 1 + \frac{d}{1 + \tau\omega_n} + \sum \frac{f_j}{\omega_j^2 + \omega_n^2 + g_j\omega_n} \tag{35}$$

where ω_{IR} and ω_{UV} are the characteristic infrared and ultraviolet absorption frequencies, and C_{IR} and C_{UV} are constants related to the strength of the respective oscillator. The constants d, τ, f_j, ω_j, and g_j are fitted from experimental dielectric data. The second term in Equation (35) is for microwave contribution, and the third term is the sum of IR and UV contributions.

Equation (34) is used when there is only one single important absorption (relaxation) in each of the IR and the UV regions. In this case,

$$C_{IR} = \epsilon_0 - n_0^2, \quad C_{UV} = n_0^2 - 1 \tag{36}$$

where ϵ_0 is the static dielectric constant, and n_0 is the visible light refractive index. When a material, such as water, has more than one significant absorptions in any of the regions, Equation (35) should be used.

If the geometric mean of Lifshitz interactions is valid, we have

$$W_{12}^L = \sqrt{W_{11}^L W_{22}^L} = 2\sqrt{\gamma_1^L \gamma_2^L} \tag{37}$$

Substitution of the above equation into Equation (32) gives

$$\frac{A_{12}}{l_{12}^2} = \frac{\sqrt{A_{11}A_{22}}}{l_{11}l_{22}} \tag{38}$$

Let us define

$$\Phi_A = \frac{A_{12}}{\sqrt{A_{11}A_{22}}}, \quad \Phi_l = \frac{l_{12}}{\sqrt{l_{11}l_{22}}} \tag{39}$$

Thus,

$$\Phi_A = \Phi_l^2 \tag{40}$$

Table 1 lists some Hamaker constants that were calculated by using Equation (33) and literature spectral data. Since $\Delta_1\Delta_2 < 1$ only the first term (m=1) was kept in our calculations. Equation (33) was then only carried out to n = 2000, after which the dielectric permittivity $\epsilon(i\omega_n)$ converged to nearly unity. The error in truncating the higher order terms is usually a few percent and does not affect our following conclusions.

- The accuracy of computed Hamaker constants is sensitive to the accuracy of the spectral data and the conformity of the mathematical representation to the actual data. For example, the Hamaker constant between two water layers is 12.12 kT

Table 1.
Hamaker Constants calculated from the Lifshitz Theory

	A_{12}, kT			Φ_A	
Material (M)	M-M	WT-M	C14-M	WT-M	C14-M
Water (WT)[a]	12.12				
Water[b]	9.45				
Tetradecane (C14)[b]	12.20	11.98		0.985	
Tetradecane[c]	12.20	10.52		0.980	
PS[b]	19.91	15.33	15.57	0.987	0.999
PS[b]	19.91	13.46[b]		0.981	
PS[c]	16.07	13.63[a]		0.977	
PMMA[c]	17.41	14.38	14.55	0.990	0.998
PVC[c]	19.08	15.03	15.25	0.988	0.999
PTFE[c]	9.37	10.49	10.68	0.984	0.999

[a] Equation 35 and Reference [80]
[b] Equation 35 and Reference [78]
[c] Equation 34 and Reference [48]

if the representation of Roth-Lenhoff was used comparted to 9.45 kT for the representation of Parsegian and Ninham.

- Even though the absolute values of Hamaker constants are sensitive to the accuracy of the dielectric spectra and their mathematical representations, the geometric mean assumption of $\Phi_A = 1$, is quite good for both materials with or without acid–base interactions. Two sets of calculated Φ_A are listed in Table 1. One is for dispersion interactions between water and a second material, and the other is for interactions between tetradecane and a second material. As can be seen, values of Φ_A are always less than or equal to that predicted by the theory and the deviation from unity is less than 2% for those of water and is almost zero for those of tetradecane.
- The Keesom (polar) and the Debye (induced polar) interactions contribute little to the van der Waals interactions in condensed media (liquids and solids). Using the Lifshitz theory, it has been shown that at room temperature the contributions from these polar interactions to the total Lifshitz-van der Waals interactions amount to 8% for water, 6% for acetone and less than 2% for polymers such as polystyrene, poly(vinyl chloride), poly(methyl methacrylate), etc. [49]. Experimentally, dipole-dipole interactions in liquids have been found to be much smaller than that predicted with dipole moment [50].
- Currently, except for a few cases such as normal alkanes, it is not possible to calculate the contribution of the van der Waals interactions to the surface energy of a material with reasonable accuracy. For most systems we do not have accurate dielectric data and more seriously, we do not know how to determine L_c.

- There is really no theoretical justification for the geometric mean assumption for work of adhesion since we do not know whether Φ_l is equal to one. Therefore, we have to justify it by critically reviewing the experimental data.

2.4.2. Experimental Justification. In his pioneering work, Fowkes [12] showed, by using experimental data of interfacial free energies and contact angles, that the dispersion interaction could be reasonably represented by Equation (37), which is the geometric mean (GM) assumption. One reason, in our view, which delayed the further development of a coherent model is that for many years researchers in this field mistakenly took the lack of polarity of a molecule as an indication of the molecule being purely dispersive. Therefore, in addition to n-alkanes, they also treated methylene iodide (MI), tricresyl phosphate (TP), and α-bromonaphthalene (BN) as "neutral" or "purely dispersive" liquids. The Lifshitz components of other materials were then estimated from the interactions with these materials. The electronic states of these compounds are too different as reflected by their differences in physical and chemical properties to place them in the same group. Consequently, the derived values for the dispersive component of materials are not self-consistent. In our model, only n-alkanes are assumed to be neutral, and only the interactions between an n-alkane (nRH) and another material is of purely Lifshitz-van der Waals type. Even this assumption has to be critically justified. Figure 1

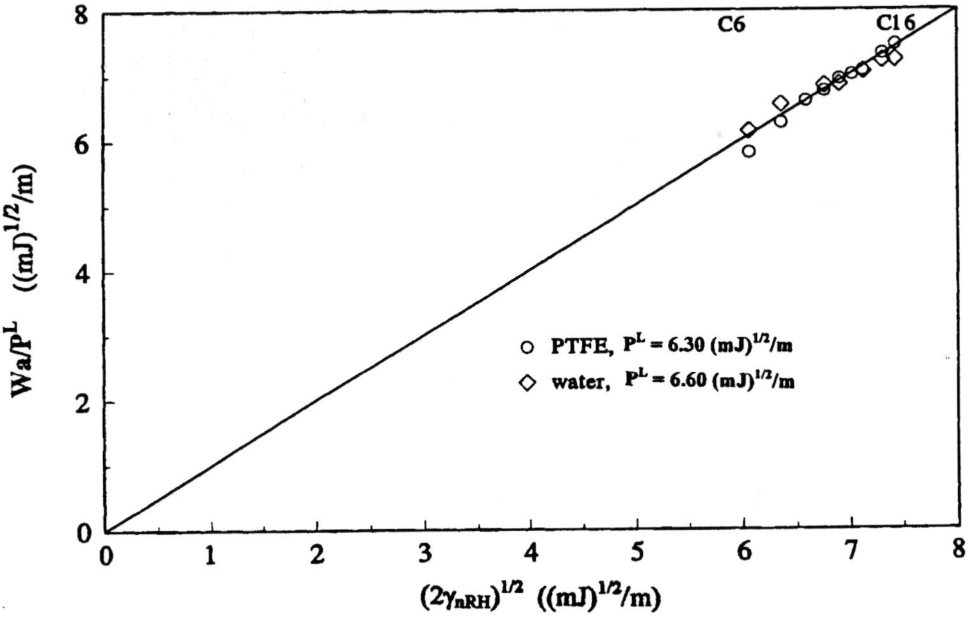

Figure 1. Justification for selecting *n*-alkanes as neutral liquids. The work of adhesion values for *n*-alkanes on PTFE are calculated from the contact angles. The work of adhesion values between water and *n*-alkanes are calculated from the interfacial free energy. *n*-Alkanes are hexane (C6), heptane, octane, nonane, decane, undecane, dodecane, tetradecane, and hexadecane (C16).

uses the works of adhesion between n-alkanes and two other materials, water and PTFE, to illustrate the validity of the GM assumption. In the figure, the work of adhesion between water and an n-alkane is from the experimental interfacial free energy data, and that between PTFE and an n-alkane is from the contact angle data. If the GM assumption is valid W_a should be a linear function of $\sqrt{2\gamma_{nRH}}$ and the slope should be P_S^L or $\sqrt{2\gamma_S^L}$. In this way the Lifshitz-van der Waals component of the substrate can be unambiguously determined. As can be seen, the plots of $W_a/\sqrt{2\gamma_{WT}}$ or $W_a/\sqrt{2\gamma_{PTFE}}$ vs. $\sqrt{2\gamma_{nRH}}$ give excellent straight lines with slope of 1, indicating the validity of the GM assumption. The only exception is the case of n-hexane on PTFE of which the experimental work of adhesion is below the straight line. This deviation is probably due to the high spreading pressure of this compound which was not taken into account in calculating the experimental work of adhesion.

3. MODELS OF INTERFACIAL ACID-BASE INTERACTIONS

The importance of Lewis acid–base interactions in adhesion is reflected by the existence of hydrogen bonds at the interface of two materials. Hydrogen bonding is a a subclass of Lewis acid–base interactions, and has been used to elucidate the importance of acid–base interactions across interfaces. As shown in Pimental and McClellan's 1960 book [51], the strength of a hydrogen bond depends on the acidity of the hydrogen donor and the basicity of the hydrogen acceptor, and is completely independent of the dipole moment. The application of the Lewis acid–base concept in adhesion was promoted to a great deal by the development of acid–base interactions in solution [52]. Research carried out by Fowkes and coworkers [14–16, 53] has shown that the strength of adhesion is directly related to the enthalpy of Lewis acid–base interactions in solution.

Fowkes took the first step to resolve the work of adhesion and the surface free energy into two components corresponding to the above two major contributions [40]:

$$\gamma = \gamma^L + \gamma^{ab} \tag{41}$$

$$W_a = W_a^L + W_a^{ab} \tag{42}$$

And the dispersion (or Lifshitz-van der Waals) contribution to the work of adhesion follows the geometric mean:

$$W_a^L = 2(\gamma_1^L \gamma_2^L)^{1/2} \tag{43}$$

With respect to the acid–base contribution to the work of adhesion, the above geometric mean expression is no longer valid. A typical argument is the acid–base attraction between an acidic liquid (e.g., chloroform) and a basic solid (e.g., polycarbonate) [40].

Fowkes and Mostafa [17] proposed the following expression for W_a^{ab} by making use of the enthalpy of acid–base interactions in solution (or heat of mixing due to acid–base interactions), ΔH^{ab} (J/mol), under infinite dilute conditions.

$$W_{\text{a}}^{\text{ab}} = -fn^{\text{ab}}\Delta H^{\text{ab}} = -fn^{\text{ab}}(C_{\text{A}}C_{\text{B}} + E_{\text{A}}E_{\text{B}}) \quad (44)$$

where f is a fudge factor converting heats of interfacial acid–base interaction to free energies of interfacial acid–base interaction, and n^{ab} the population (mol/m^2) of active acid–base pairs across the interface. ΔH^{ab} follows Drago's model of the acid–base interaction [54].

To be able to predict values of work of adhesion at solid-liquid or solid-solid interfaces, one has to determine f, n^{ab}, and Drago's C and E parameters in addition to knowing the dispersion components of surface free energies. Methods for determining the above variables include the measurement of shifts of IR absorption peaks and NMR chemical shifts, and the measurement of contact angle change due to solute adsorption according to Fowkes *et al.* [16, 53, 55]. However, the following shortcomings have limited the applicability of Equation (44): first, the lack of sufficient data on C and E for potential substrates; Second, n^{ab} converts ΔH from bulk interactions to interfacial interactions. One needs to know how the bonds are formed in the bulk and the interface. This is similar to the attempt of correlating the solubility parameter to the surface energy. Yet another shortcoming is that f and n^{ab} vary from system to system just like Φ in Equation (3). In their study of phenol adsorption from diiodomethane onto poly(methyl methacrylate), Fowkes *et al.* [16] showed that f varied from 0.3 to 0.55 with the increase of phenol concentration.

In addition to the adaption of Drago's model of acid–base interactions in solution by Fowkes to solid surfaces, the idea of donor number and acceptor number introduced by Gutmann [56, 57] was also applied to study the interfacial interactions between polymers and inorganic fibers [58, 59]. The surface energy of polymers and fibers are measured with inverse gas chromatography (IGC). However, as pointed out by Drago *et al.* [54], among many others, the acid–base interaction cannot be properly accounted for by a single parameter ordering such as Gutmann's.

Recently, van Oss, Good, and Chaudhury proposed a two-parameter equation to characterize the acid–base contribution to the surface tension [60–62]

$$\gamma^{\text{ab}} = 2(\gamma^{\oplus}\gamma^{\ominus})^{1/2} \quad (45)$$

$$W_{\text{a}}^{\text{ab}} = 2[(\gamma_1^{\oplus}\gamma_2^{\ominus})^{1/2} + (\gamma_1^{\ominus}\gamma_2^{\oplus})^{1/2}] \quad (46)$$

where $\gamma^{\oplus}$ stands for the electron acceptor (Lewis acid) parameter and $\gamma^{\ominus}$ the electron donor (Lewis base) parameter. They pointed out that a molecule may be apolar ($\gamma^{\oplus} = \gamma^{\ominus} = 0$), mono-polar ($\gamma^{\ominus} = 0$, for mono-acidic; $\gamma^{\oplus} = 0$, for mono-basic), or bipolar (both $\gamma^{\ominus}$ and $\gamma^{\oplus}$ are greater than zero). They emphasized the important differences in the interfacial and cohesive behaviors among the three classes of molecules.

The approach of van Oss *et al.* is consistent with Gutmann's approach, i.e., they also use a single parameter to represent acidity or basicity, except they do not use Gutmann's procedure to determine the value of each parameter, but rather use the values of W_a^{ab} on the arbitrary assumption that for water

$$\gamma_L^{\oplus} = \gamma_L^{\ominus} = 25.5 \text{ mJ/m}^2 \tag{47}$$

The acid–base parameters of other materials reflect the relative measurement of acid–base properties with respect to water. In addition, assumptions such as poly(methyl methacrylate) is Lewis mono-base, and chloroform is Lewis mono-acid, were often made to obtain the relative polarities ($\sqrt{\gamma_{WT}^{\oplus}/\gamma_i^{\oplus}}$ or $\sqrt{\gamma_{WT}^{\ominus}/\gamma_i^{\ominus}}$) of solids or liquids. This can easily cause a large uncertainty. Using their values we have calculated the work of adhesion for their systems of biopolymer-liquid and compared them to the experimental values [18]. The agreement was rather poor, with a standard deviation of 14.8 mJ/m^2. We also calculated the work of adhesion by the GM method, which gave a standard deviation of 8 mJ/m^2. Strong acid–base interactions are expected in some of the biopolymer-liquid pairs. One would expect that their model should give a better result. The poor agreement suggests that the model needs some modification. However, their idea of classifying materials as apolar, monopolar, and bipolar brings fresh thinking to the field although these terms may cause confusion with the traditional polar concept.

4. THE NEED FOR A NEW MODEL

When we started research in this area more than ten years ago we were greatly concerned with the failure of existing models and the confusing arguments in the field. We decided to take a bold approach to challenge some commonly accepted concepts by carefully checking their internal consistencies against the most reliable published data. In particular, we questioned the validity of assuming some of the commonly used test liquids such as MI, BN, and TP as neutral. In addition, we freed ourselves from the restrictions of the traditional concept of Lewis acid–base interaction in solution, that is, if both phases in contact have only basic or acidic sites, they would have no acid–base interactions ($W_a^{ab} = 0$). It has been observed by us that some phenomena in adhesion cannot be explained by this traditional concept. We shall use some experimental data to illustrate it in this section.

Zisman and co-workers at the Naval Research Laboratory have conducted extensive studies on the wetting of low energy polymer surfaces by liquids [63]. For each polymer surface, the contact angles of many liquids have been determined. By examining their data carefully, we found that negative or *repulsive* non-dispersion contribution to the work of adhesion existed in some systems studied by them. Among them, wetting of poly(tetrafluoroethylene) (PTFE) by methylene iodide (MI) is probably the most convincing instance.

A PTFE surface is believed to be neutral by researchers [62, 64, 65]. It is often used to determine the values of γ_L^L for liquids by contact angle measurement [13, 62]. However, experimental results are more likely to indicate that a PTFE surface is not neutral. For example, the work of adhesion for water on PTFE is

Table 2.
Literature Contact Angle Data for Six Liquids on PTFE Surface and Their Values of γ_L^d

	Liquids					
	WT	GL	FA	MI	BN	TP
θ (deg.)	106 [81] 112 [81, 82, 84] 116 [86] 108 [38, 21] 114 [87] 111 [71] 100 [88]	105 [82, 71] 98 [85] 100 [21] 102 [83] 106 [69]	103 [71] 91 [82] 94 [83] 96 [69] 92 [21]	77 [38, 83] 85 [82] 88 [21] 103 [84] 83 [87]	76 [82] 73 [21, 87] 77 [84]	75 [21]
γ_L^L (mJ/m^2)	21.8 [13] 22.6 [8] 10.8 [41] 21.8 (*)	37.4 [13] 34 [62] 20.2 [69] 40.6 [8] 20.5 [41] 32 (*)	39.5 [13] 39 [62] 19.8 [69] 36 [8] 18.1 [41] 26.6 (*)	48.5 [13] 50.8 [62] 50.4 [71] 49.0 [8] 67.3 (*)	47.0 [13] 43.6 [62] 43.7 [71] 44.6 [8] 55.1 (*)	39.2 [13] 39.8 [8] 40.5 (*)

*Values determined in this work.

50.3 mJ/m^2 according to Fox and Zisman [19], which is 20% larger than the value of W_a^L. The latter can be calculated from Equation (43) where the value of γ^L for water is 21.8 mJ/m^2 as determined from the interfacial free energy of water with *n*-alkanes; and the value of γ^L for PTFE is 19.8 mJ/m^2 as determined from the contact angles of *n*-alkanes.

More generally, let us examine the wetting behavior of six liquids: water (WT), glycerol (GL), formamide (FA), methylene iodide (MI), α-bromonaphthalene (BN), and tricresyl phosphate (TP) on PTFE. These liquids are commonly used in contact angle studies of polymer surfaces. Table 2 shows contact angle data from the literature and values of γ_L^L as determined by various research groups with different methods. The values of the work of adhesion can be calculated from Equation (2), omitting the film pressure. They are in turn plotted in Figure 2 versus $(2\gamma_L^L)^{1/2}$. The rectangle associated with each liquid represents the range in the literature data. The dot points in the figure represent the measurement of Fox and Zisman (W_a) and the γ_L^L values determined by us (as discussed later). The solid line passing the origin is the dispersion component of work of adhesion with a slope of $(2\gamma_S^L)^{1/2} = 6.3$ (mJ)^2/m. The deviation of experimental W_a from this line reflects the degree of non-dispersion interactions.

The simple plot shown in Figure 2 indicates that a PTFE surface has significant characteristics that are not attributable to the dispersion force. Among three strong hydrogen-bonding liquids, two (WT and FA) have significant attractive non-dispersion interactions with PTFE. The plot also shows that there exist negative

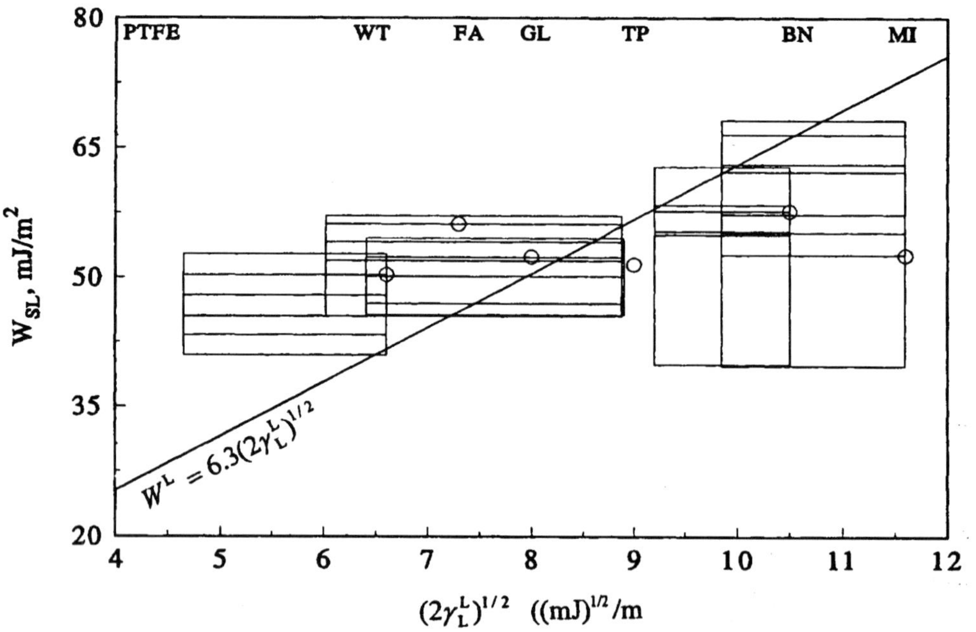

Figure 2. Plot of literature values of work of adhesion for six key liquids on PTFE versus their Lifshitz components, γ_L^L, of surface free energies. Circles are drawn with Zisman's work of adhesion values and their Lifshitz components of surface free energies determined in this work. The line represents the Lifshitz component, W^L, of work of adhesion.

non-dispersion interactions for TP, BN, and MI in most measurements. The negative interaction between MI and PTFE is especially strong. Zisman's data indicate that the value of W_a^L between MI and PTFE is 16% higher than the experimental W_a when taking γ_L^L of MI to be its γ_L. None of the models discussed earlier allows this kind of negative non-dispersion interactions, so they cannot even be qualitatively correct.

Treatment similar to PTFE is also applied to three other fluorinated polymers: poly(trifluoroethylene) (P3FE), poly(vinylidene fluoride) (PVDF), and poly(vinyl fluoride) (PVF). They are compared to PTFE in Figures 3 and 4. The dot points are the data of Ellison and Zisman [23] and the rectangle represents the range of literature values for each liquid. The horizontal side stands for the range of γ_L^L and the vertical side for the experimental error in W_a. The straight line in Figures 3 and 4 for each polymer is constructed from the experimental work of adhesion between n-alkanes and the polymer. Comparing to PTFE, we see that the liquids display increased attractive non-dispersion interactions. However, there seems to be no uniform rate of increase for the six liquids as the H/F ratio in the polymer increases. Therefore, the acid–base contributions cannot be explained by any model using only a single parameter.

MI is commonly considered as a neutral liquid. The above data, however, demonstrate that strong non-dispersion interactions exist between this liquid and

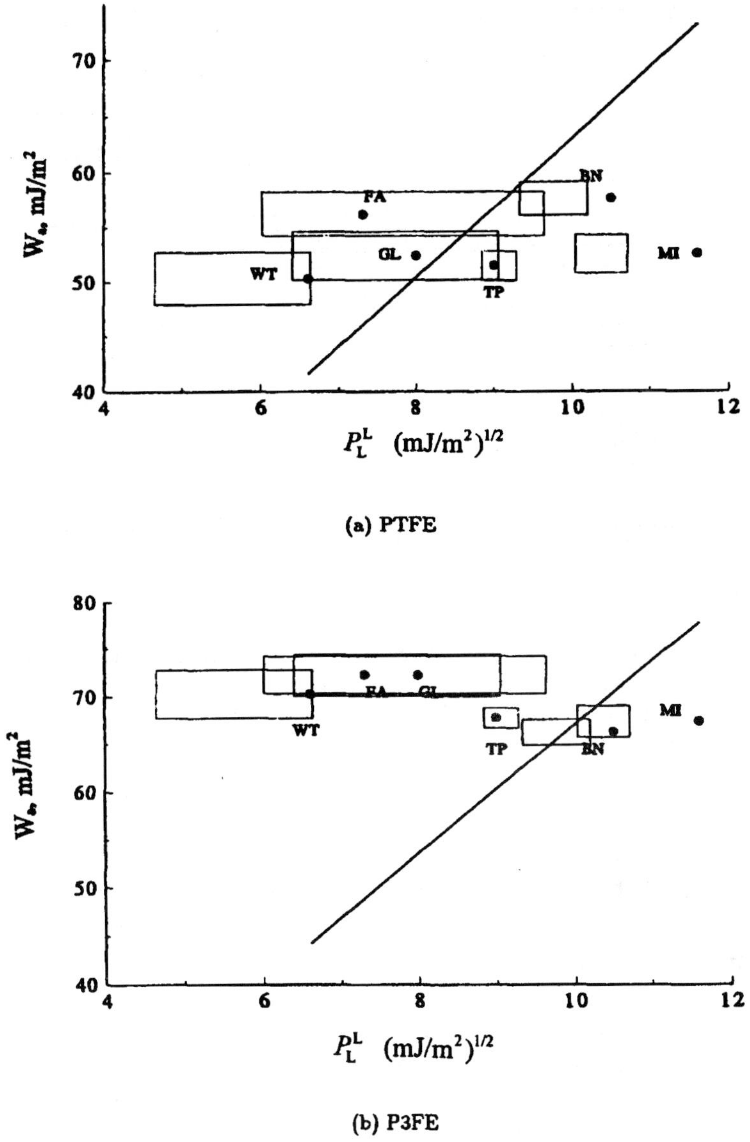

Figure 3. Work of adhesion of six key liquids on (a) PTFE and (b) P3FE.

PTFE. Another simple calculation can further show that it is not a neutral liquid. Water has an interfacial free energy of 48.5 mJ/m^2 with MI. If MI were neutral, there would be only van der Waals interactions between water and MI, and the interfacial free energy would be 57.3 mJ/m^2, i.e., 18% higher than the experimental value.

A more general picture of the surface energetic characteristics of MI can be seen in Figure 5. It is a plot similar to Figure 2, but for MI on various low energy polymer surfaces. The values of W_a were calculated from the literature contact angle data

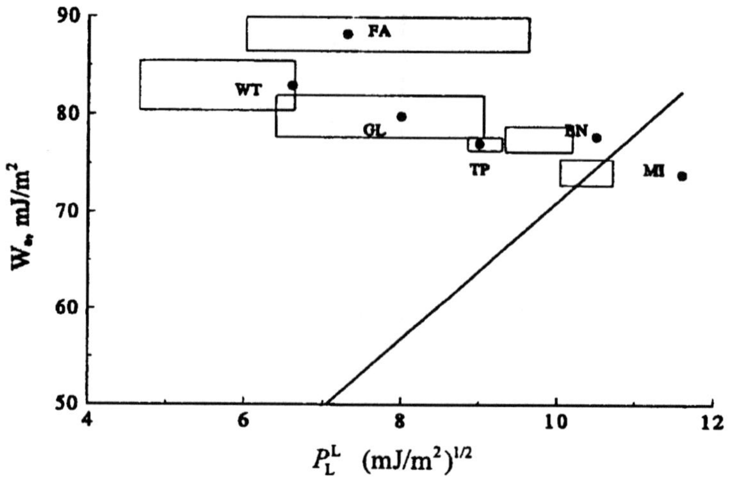

(a) PVDF

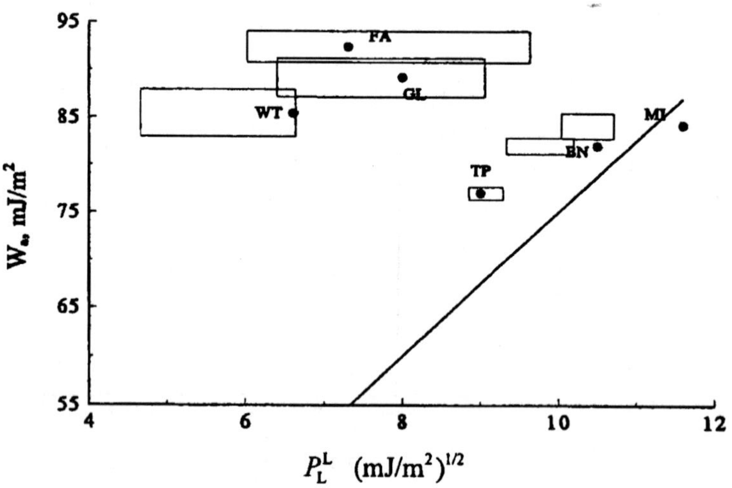

(b) PVF

Figure 4. Work of adhesion of six key liquids on (a) PVDF and (b) PVF.

(Table 5). The error bars were drawn according to the typical experimental error (±2°) in contact angle measurement. Eq. (43) was used to calculate the values of γ_S^L for all the polymer surfaces from the contact angles of n-alkanes. The dashed line is the "neutral" line, that is, it was drawn according to

$$W_a^L = 2\sqrt{\gamma_{MI}\gamma_S^L} = 10.1\sqrt{2\gamma_S^L} \tag{48}$$

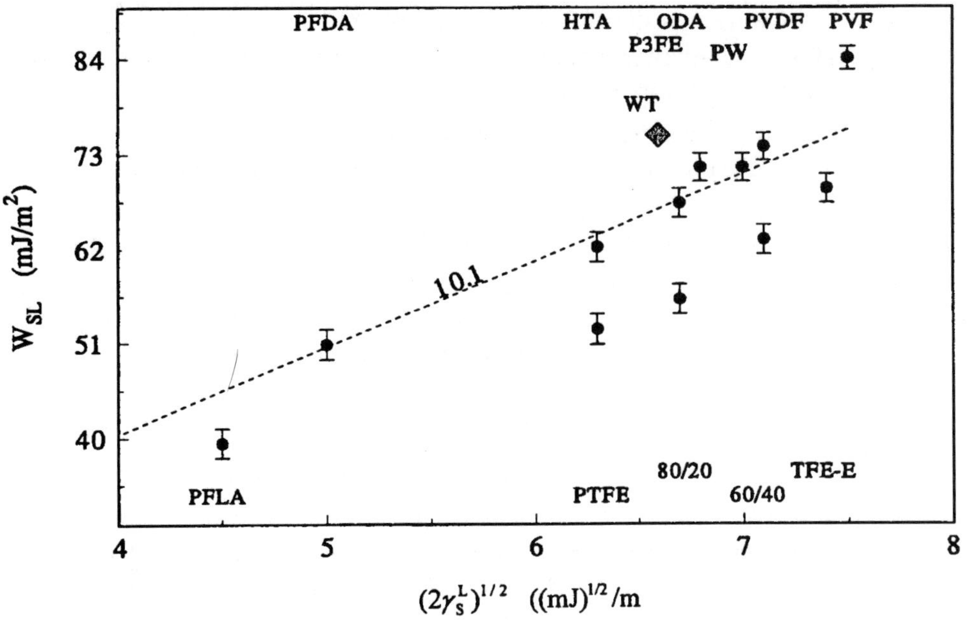

Figure 5. Work of adhesion versus Lifshitz component of surface free energy for MI on low energy polymer surfaces. The line is drawn with the assumption that MI is neutral.

If this liquid could be considered neutral, the experimental values of the work of adhesion should reasonably be on the neutral line regardless of the energetic characteristics of the solid surfaces. In fact, the experimental data points are scattered around so much and it is impossible to find a straight line which can fit them well. Therefore, we cannot ignore the non-dispersion interactions. Furthermore, as can be seen from the figure, it is obvious that W_a values for a group of solid surfaces are well below the values predicted by the above equation. This indicates that there must exist a kind of *repulsive* interaction. It is also interesting to note that all surfaces which show repulsive non-dispersion interactions with MI are perfluoro-polymers.

In summary, a new model for describing interfacial interactions is needed. This model should not only recognize the attractive energetic interactions but also allow for the existence of repulsive interactions.

5. A NEW MODEL FOR INTERFACIAL ACID-BASE INTERACTIONS

To develop a new phenomenological model we set for ourselves the following guidelines:

(1) It should not violate the basic laws of physics.

(2) The model should be operationally well defined. It has to provide clear and self-consistent procedures to determine parameters in the model.

(3) The work of adhesion consists of components due to the Lifshitz interactions (W^{L}) and the acid–base interaction (W^{ab}). The geometrical mean rule is only applicable to the Lifshitz interactions.

(4) Only liquid n-alkanes are selected as our neutral reference materials. Interactions between a liquid n-alkane and any other substance are purely Lifshitz type.

(5) Acid-base interactions of a molecule are characterized by two parameters, the same as in the model of van Oss *et al.* As discussed earlier, one needs at least two terms to characterize the interactions. More parameters are not justifiable considering the difficulty of unequivocally determining their values from the experimental data.

(6) The model should allow both *attractive* and *repulsive* interactions.

(7) A molecule may have both acidic and basic characteristics.

Many researchers have great reluctance to accept and some even strongly object to our allowing repulsive interactions in our model. Attractive acid–base interactions are very familiar to chemists, because they are mostly interested in bond formation. Quantum mechanical models that provide theoretical bases for the attractive interactions are well accepted [66]. However, it is not correct to say that quantum electrodynamics rules out any repulsive acid-acid or base-base interactions. The beauty of the Lewis acid–base concept is its generality which includes the traditional ionic acid–base definition as a special case. Since nobody is against the concept of repulsive interactions between ions of same signs and between dipole moments, our model should be consistent with the Lewis acid–base definition.

Part of the repulsive interactions is of entropic origin because our model is defined in terms of the Gibbs free energy. Entropic interactions depend on local packing orders which are sensitive to the shape and size of interacting units, such as molecules or monomeric units of polymers. Since uncrosslinked polymers are one-dimensional molecules, we expect that they would have different interaction parameters from those of their monomers.

During the early stages of this study, the following physical concept helped us to formulate the model. The origin of the acidic or basic character of a molecule is due to the distribution of outer shell electron clouds. Different orbitals would lead to different charge density distributions. Each molecule has its own well-defined geometry which is not necessarily related to the charge distribution. Charges produce an electric field. A local electrical field vector in general is not normal to the local surface element. An electrical stress tensor can be considered to be the ratio of the electrical field vector and the local surface vector. The influence of a charged surface on the surrounding molecules depends on the local electrical stress tensor.

Hence, it is doubtful that scalar quantities are capable of characterizing the acidic and/or basic strength of a molecule. Without further theoretical guidelines we propose the next simplest model. We assume the strength of acid–base interactions can be represented by a second rank symmetrical tensor. A second rank tensor could have up to three independent principal values and three associated principal directions in a three-dimensional space. However, in a surface problem there are only two principal values. Since n-alkanes (n = 6–16) (the reference molecules) have no acidic or basic character, their principal value is used as the reference (zero) point. A particular axis is acidic or basic depending on whether its associated principal stress is higher (positive) or lower (negative) than that of the reference molecule.

The surface of a material is called acidic if both of its principal values are positive. It is called basic if both of its principal values are negative. In case one is positive and the other is negative, the material has a bi-characteristic surface. When two materials approach each other the relative local orientation (or packing sequence) of the molecules is dictated by the minimization of Gibbs free energy. Our model differs from other acid–base models in defining a common reference point for both acids and bases.

Following Item 3 of our guidelines, the thermodynamic work of adhesion for a liquid on a solid surface is the sum of contributions due to dispersion and acid–base interactions, i.e.,

$$W_a = W_a^L + W_a^{ab} \tag{49}$$

Using the rule of geometric mean, the dispersion term can be expressed as

$$W_a^L = 2\sqrt{\gamma_1^L \gamma_2^L} = P_1^L P_2^L \tag{50}$$

where $P_i^L = \sqrt{2\gamma_i^L}$, which we call the dispersion parameter. Consequently, based on the above model the acid–base component of the work of adhesion is expressed by a combining rule,

$$\begin{aligned} W_a^{ab} &= -\Delta G_a^{ab} \\ \Delta G_{12}^{ab} &= P_1^a P_2^b + P_1^b P_2^a \end{aligned} \tag{51}$$

Similar to W_a (Equation (49)), the surface free energy consists of two components (dispersion and acid–base contributions), too, i.e.,

$$\gamma = \gamma^L + \gamma^{ab} = \frac{1}{2}(P^L)^2 - P^a P^b \tag{52}$$

In the equations above, (P_i^a, P_i^b) are two acid–base parameters (principal values) for any liquid or solid surface i. The acid–base property for a given surface can then be identified by (P^a, P^b), as listed in Table 3. In this paper, we assign P^a to be more acidic and P^b to be more basic to ensure that Equation (51) gives the minimum free energy of adhesion, ΔG_a^{ab}, and hence, the maximum work of adhesion.

Equation (51) not only provides for the complementary nature of acid and base functions (i.e., the acid–base attractive interaction) but also permits the

Table 3.
Identification of Surface Acid-Base Property in Terms of (P^{a}, P^{b})

Characteristic	Surface Parameter
Neutral (or van der Waals)	$P^{a} = P^{b} = 0$
Mono-character (acidic or basic)	$P^{a}, P^{b} \neq 0$, but with same sign
acidic	P^{a} and P^{b} both positive
basic	P^{a} and P^{b} both negative
symmetric	$P^{a} = P^{b}$
Bi-character (acidic and basic)	$P^{a}, P^{b} \neq 0$, but with opposite sign
symmetric	$P^{a} = -P^{b}$

existence of repulsive interaction when both components of the interface are mono-characteristic in the same sense (acids or bases). For example, when both substances are bipolar (that is, P^{a} and P^{b} are of opposite signs), the acid–base attractive interaction will give a positive work of adhesion. In cases when both substances are mono-characteristic in the same sense (i.e., acid-acid or base-base), the repulsive interaction will cause W_{a}^{ab} to be negative and the free energy of adhesion to decrease. But the thermodynamic work of adhesion would still be positive if $W_{a}^{L} > W_{a}^{ab}$. When one or both of the substances at the interface are neutral (i.e., $P^{a} = P^{b} = 0$), the acid–base work of adhesion will be zero.

On substituting Equations (50) to (52) into the Dupré equation, we obtain the expression for the interfacial free energy:

$$\begin{aligned}\gamma_{ij} &= \gamma_{ij}^{L} + \gamma_{ij}^{ab} \\ &= \frac{1}{2}(P_i^{L} - P_j^{L})^2 - (P_i^{a} - P_j^{a})(P_i^{b} - P_j^{b})\end{aligned} \tag{53}$$

It is clear from the equation above that γ_{ij} can be negative if γ_{ij}^{ab} is negatively large enough. The condition for negative γ_{ij}^{ab} is: $P_i^{a} > P_j^{a}$ and $P_i^{b} > P_j^{b}$, or *vice versa*. In other words, the acid–base work of *adhesion* must be greater than the average of the acid–base work of *cohesion*, because

$$\begin{aligned}\gamma_{ij}^{ab} &= \gamma_i^{ab} + \gamma_j^{ab} - W_{ij}^{ab} \\ &= -(P_i^{a}P_i^{b} + P_j^{a}P_j^{b}) + (P_i^{a}P_j^{b} + P_i^{b}P_j^{a})\end{aligned} \tag{54}$$

The acid–base attractive interaction is a necessary but not a sufficient condition for negative γ_{ij}^{ab}.

It is worthwhile to point out that the acid–base component, γ_{ij}^{ab}, of interfacial free energy is usually non-zero for an interfacial system even though the interfacial interaction can be considered Lifshitz only (i.e., $W_{ij}^{ab} = 0$), unless both substances of the system are neutral. For instance, when one of the substances is neutral (e.g., *n*-alkanes (RH)) and the other bi-characteristic (e.g., water (WT)), Equation (53) can be reduced to

$$\gamma_{\mathrm{WT/nRH}} = \frac{1}{2}(P^{\mathrm{L}}_{\mathrm{WT}} - P^{\mathrm{L}}_{\mathrm{nRH}})^2 - P^{\mathrm{a}}_{\mathrm{WT}} P^{\mathrm{b}}_{\mathrm{WT}} = \gamma^{\mathrm{L}}_{\mathrm{WT/nRH}} + \gamma^{\mathrm{ab}}_{\mathrm{WT}} \tag{55}$$

and the work of adhesion is due solely to the Lifshitz interactions. Thus

$$W_{\mathrm{WT/nRH}} = P^{\mathrm{L}}_{\mathrm{WT}} P^{\mathrm{L}}_{\mathrm{nRH}} = \gamma_{\mathrm{WT}} + \gamma_{\mathrm{nRH}} - \gamma_{\mathrm{WT/nRH}} \tag{56}$$

We know $P^{\mathrm{L}}_{\mathrm{WT}} = 6.6\ (\mathrm{mJ/m^2})^{1/2}$, $\gamma^{\mathrm{ab}}_{\mathrm{WT}} = 51\ \mathrm{mJ/m^2}$ and $P^{\mathrm{L}}_{\mathrm{nRH}}$ is in the range of 5.7 to 7.3 $(\mathrm{mJ/m^2})^{1/2}$ for n-alkanes from pentane to tetradecane. This leads the dispersion interfacial free energy to be less than 0.4 $\mathrm{mJ/m^2}$. Therefore, the interfacial free energies of the systems are nearly equal to the acid–base component of water surface free energy.

5.1. Special Cases

When the acid–base structures of the two component substances of an interfacial system are both asymmetric, that is, $P^{\mathrm{a}}_i = -P^{\mathrm{b}}_i$, the work of adhesion can be expressed as (see Equations (50) and (51))

$$W_{\mathrm{a}} = 2\sqrt{\gamma^{\mathrm{L}}_1 \gamma^{\mathrm{L}}_2} + 2P^{\mathrm{a}}_1 P^{\mathrm{b}}_2 = 2\sqrt{\gamma^{\mathrm{L}}_1 \gamma^{\mathrm{L}}_2} + 2\sqrt{\gamma^{\mathrm{ab}}_1 \gamma^{\mathrm{ab}}_2} \tag{57}$$

This is the GM equation. On the other hand, if both components are mono-characteristic in the same sense and one of the two acid–base parameters for each component is zero, the work of adhesion and interfacial free energy are solely due to the dispersion interaction (see Equations (51) and (53)). This is in agreement with the treatment of van Oss, Good, and Chaudhury.

6. APPLICATION OF OUR MODEL TO WETTABILITY STUDIES

6.1. Early Source of Contact Angle Data

Zisman and his associates [19–23] have conducted the most extensive study on wettability and adhesion by any single research group, so we rely heavily on their contact angle data in the process of verification of our model and the determination of surface free energy parameters. Six liquids and ten polymer surfaces are selected to form our key solid/liquid pairs. Their surface energetic parameters were calculated during the first stage of our model application. The contact angles of the key liquids on the key polymers are compiled in Table 4. The corresponding values of the work of adhesion are calculated by Equation (2), and they are given in Table 5.

Table 4.
Contact Angle of Key Liquids on Polymer Surfaces[a]

Polymer	WT	GL	FA	MI	BN	TP
PTFE	108.0	100.0	92.0	88.0	73.0	75.0
P3FE	92.0	82.0	76.0	71.0	61.0	49.0
PVDF	82.0	75.0	59.0	63.0	42.0	28.0
PVF	80.0	66.0	54.0	49.0	33.0	28.0
PE	94.0	79.0	77.0	52.0	35.0	34.0
80/20	100.0	96.0	91.0	84.0	67.0	67.0
60/40	94.0	87.0	75.0	76.0	63.0	56.0
TFE-E	93.0	85.0	79.0	69.0	60.0	55.0
HTA	111.0	97.0	92.0	77.0	67.0	72.0
PARAF	108.0	96.0	91.0	66.0	47.0	62.0
PCTFE	72.8	72.2	66.3	73.1	74.4	70.3
PFLA	105.0	102.0		103.0	94.0	
PFDA	102.0	89.0		90.0	84.0	82.0
ODA	102.0	94.0	81.0	66.0	58.0	65.0

[a]Data compiled from Zisman's θ measurements.

Table 5.
Work of Adhesion of Key Liquids on Polymer Surfaces[a]

Polymer	WT	GL	FA	MI	BN	TP
PTFE	50.3	52.4	56.2	52.6	57.6	51.5
P3FE	70.3	72.2	72.3	67.3	66.2	67.7
PVDF	82.9	79.8	88.2	73.9	77.7	77.0
PVF	85.4	89.2	92.4	84.2	82.0	77.0
PE	67.7	75.5	71.3	82.1	81.1	74.8
80/20	60.2	56.8	57.2	56.1	62.0	56.9
60/40	67.7	66.7	73.3	63.1	64.8	63.8
TFE-E	69.0	68.9	69.3	69.0	66.9	64.4
HTA	46.7	55.7	56.2	62.2	62.0	53.5
PARAF	50.3	56.8	57.2	71.5	75.0	60.1
PCTFE	94.3	82.8	81.6	65.6	56.6	54.7
PFLA	54.0	50.2		39.4	41.5	
PFDA	57.7	64.5		50.8	49.3	46.6
ODA	57.7	59.0	67.3	71.5	68.2	58.2

[a]Data is compiled from Zisman's θ measurements. $W_a = \gamma_l(1 + \cos\theta)$.

For low energy surfaces on which at least some n-alkanes form finite contact angles, it is straightforward to determine the dispersive component of the solid surfaces. Eight of the ten key polymers belong to this category and their contact angles with n-alkanes are available. Thus, the P_S^L values for these polymers can be

Table 6.
Parameters of Key Liquids Determined by Our Model

	WT	GL	FA	MI	BN	TP
P^L	6.60±0.01	8.00±0.01	7.30±0.04	11.60±0.04	10.50±0.03	9.00±0.02
P^b	−7.40±0.01	−9.30±0.08	−4.64±0.09	−4.12±0.07	−3.82±0.10	−5.45±0.46
P^a	6.88±0.01	3.40±0.02	6.92±0.11	−4.11±0.07	−2.67±0.07	0.11±0.04
γ_L (exp.)	72.80	63.40	58.20	50.80	44.60	40.90
γ_L (cal.)	72.75	63.64	58.76	50.35	44.93	41.10
$\gamma_{12}^{\dagger}$ (exp.)	0			48.50	42.40	
γ_{12} (cal.)	0.00	−5.62	0.16	48.64	41.89	16.10

Unit: mJ/m^2 for γ and (mJ/m^2)$^{1/2}$ for P.
$^\dagger\gamma_{12}$: interfacial free energy with water.

Table 7.
Parameters of Key Polymers Determined by Our Model

	$n_\theta{}^\dagger$	P^L	P^a	P^b	γ_S(cal.)
PTFE	8	6.30±0.01	-1.11±0.18	-2.60±0.21	17.0
P3FE	4	6.70±0.02	0.80±0.05	-2.96±0.05	24.8
PVDF	2	7.10±0.03	1.64±0.06	-3.52±0.06	31.0
PVF	-	7.50	2.24±0.22	-3.16±0.26	35.2
PE	-	7.90	0.40±0.12	-1.93±0.13	32.0
80F20Cl	6	6.70±0.05	-0.95±0.33	-2.84±0.54	19.8
60F40Cl	4	7.10±0.05	-0.44±0.10	-3.58±0.11	23.6
TFE-E	2	7.40±0.04	-0.38±0.21	-3.12±0.26	26.2
HTA	5	6.30±0.04	-0.34±0.20	-1.60±0.16	19.3
PW	4	7.00±0.03	-0.31±0.02	-1.04±0.04	24.2
PHDFDMS	7	5.35	-2.86	-3.08	5.5

Unit: mJ/m^2 for γ and (mJ/m^2)$^{1/2}$ for P.
†Number of contact angles used in parameter determination.

obtained directly from the following equation:

$$W_a = \gamma_L(1 + \cos\theta) = P_L^L\, P_S^L \tag{58}$$

The values of P_S^L so determined are listed in Column 1 of Table 7.

Liquid n-alkanes spread, at 20 °C, on the surfaces of the other two polymers (PVF and PE) due to their relatively high surface free energies. Equation (58) can

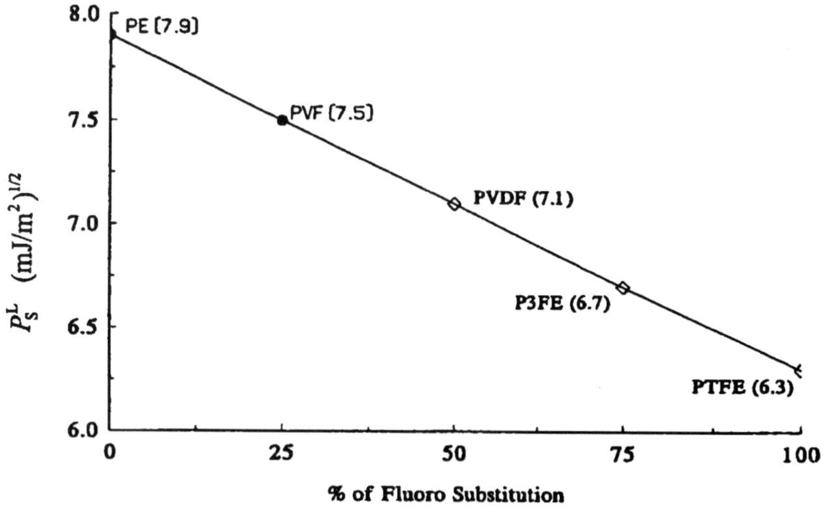

Figure 6. Determination of P_S^L values for PVF and PE from extrapolation. Numbers in parentheses are dispersion parameters.

no longer be used to calculate their P_S^L. Instead, an extrapolation method based on PTFE, P3FE, and PVDF is used. As shown in Figure 6, the relationship between P_S^L and percentage of fluorine substitution to hydrogen is a perfect straight line for PTFE, P3FE, and PVDF. The extrapolated values of P_S^L for PVF and PE are also listed in Column 1 of Table 7.

Water is the only key liquid whose P^L (6.6 (mJ/m^2)$^{1/2}$) is known. It is determined using equation (56), where the interfacial free energy, $\gamma_{WT/RH}$, is available from the literature [13] for a series of n-alkanes. The same method cannot be used to calculate P^L for the other five liquids because they either have significant degree of mutual solubility or are miscible with n-alkanes.

6.2. Numerical Method for Determining Surface Energy Parameters

At the first application stage of our model, the biggest challenge was to assign surface energetic parameters for the key liquids and solids. The only known parameters are the Lifshitz parameters for n-alkanes and water, and those for the polymers in Table 7. It is preferred to have as much experimental data on the work of adhesion as possible, i.e., to have a well overdetermined system, for calculating the surface parameters with

$$W_a = \gamma_L(1 + \cos\theta) = P_S^L P_L^L - (P_S^a P_L^b + P_S^b P_L^a) \tag{59}$$

Also, Equation (52) is incorporated into the calculation of the parameters for each liquid because γ_L is a measurable quantity. Then, the problem can be expressed in the following matrix form, say, for each liquid,

$$\mathbf{F} = \mathbf{f}(\mathbf{P}) + \boldsymbol{\epsilon}_\mathbf{w} \tag{60}$$

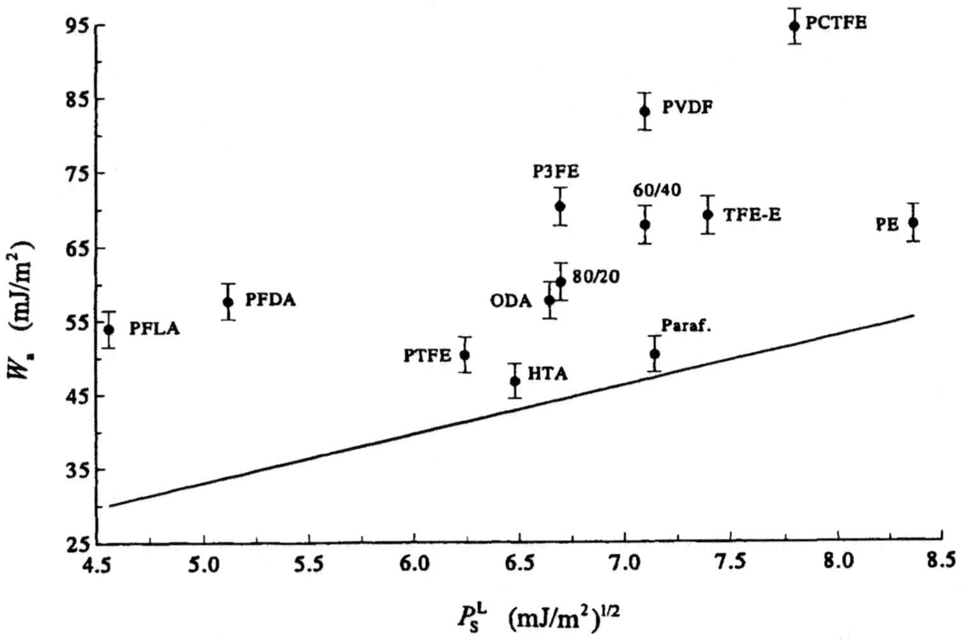

Figure 7. Work of adhesion of water on organic solid surfaces. The deviation of each point from the line reflects the level of non-dispersion contribution.

where $\mathbf{F}$ is a column vector consisting of $(n-1)$ experimental data for W_a and one experimental data for γ_L; ϵ_w is the experimental error. In our case, $(n-1) = 10$ is the number of polymers. The function $\mathbf{f(P)}$ is

$$f_i(P) = \mathbf{P}_{S,i}\,\mathbf{P}_L \qquad (i = 1 \text{ to } n-1) \tag{61}$$

$$f_n(P) = \frac{1}{2}(P_L^L)^2 - P_L^a P_L^b \tag{62}$$

with

$$\mathbf{P}_S = [P_S^L\ P_S^a\ P_S^b], \quad \mathbf{P}_L = \begin{bmatrix} P_L^L \\ -P_L^b \\ -P_L^a \end{bmatrix} \tag{63}$$

The whole parameter calculation process consists of the following steps:

(1) Choosing proper initial values for the Lifshitz parameters of the liquids (except water) and the acid–base parameters of the liquids and the polymers.

(2) Assuming that the values for the acid–base parameters of the polymers are "true", and solving Eq. (60) for each of the liquids.

(3) Solving Eq. (59) for each of the polymer's acid–base parameters with the above determined liquid parameters by the nonlinear least squares method.

(4) Repeat the last two steps, until the average relative deviation of work of adhesion of all pairs is smaller than ϵ_w.

In Step 1, Fowkes' dispersion surface free energy data [13] of the liquids were converted into the initial values of $P_{\mathrm{L}}^{\mathrm{L}}$. The initial acid–base parameters of a liquid were given the same absolute values as derived from its surface free energy and its dispersion parameter value (per Eq. (52)).

In Step 2, Eq. (60) was solved with weighted Gauss-Newton-Raphson nonlinear least squares method [67, 68] to minimize the sum of squares (SS) of residuals (V):

$$SS = \mathbf{V}^{\mathrm{T}}\mathbf{M}^{-1}\mathbf{V} \tag{64}$$

with

$$\mathbf{V} = \mathbf{F}^{\mathrm{exp.}} - \mathbf{F}^{\mathrm{cal.}} \tag{65}$$

where **M** is the variance-covariance (momentum) matrix with its diagonal element being the square of experimental errors (derived from a typical contact angle measurement error of $\pm 2°$). In our calculation, we assume that there is no correlation between different experimental measurements. Thus, the off-diagonal elements of **M** are zero. The same weighting scheme was also used in Step 3 when calculating the acid–base parameters of the polymers.

In addition, for GL, FA, MI, and BN, the interfacial free energy with water was also applied in solving Eq. (60) as an extra constraint.

Throughout the process, $P_{\mathrm{S}}^{\mathrm{L}}$ of all polymers and $P_{\mathrm{WT}}^{\mathrm{L}}$ were fixed; and the dispersion parameter for each liquid was limited in a range defined by examining the literature data (Table 2).

6.3. Selection of Dispersion Parameters for the Key Liquids

6.3.1. Our Calculations. *The determination of the surface parameters cannot be done by just using the mathematical equations. Considerations of the physical meaning of the parameters must be involved.* Before presenting the result, it is necessary to briefly describe some calculations (outputs of the 2-parameter fitting program) that affect the final selection of parameters for the key set of liquids and polymers. The direct outputs include the overall and the individual mean relative errors (OMRE and IMRE) in the work of adhesion, interfacial free energies of water with other key liquids ($\gamma_{\mathrm{WT}/2}$) and surface free energies of liquids (γ_i). The first two outputs are calculated for a set of n liquids and m polymers, according to

$$\mathrm{OMRE} = \left(\sum_{i=1}^{m} \sum_{j=1}^{n} (1 - W_{ij}^{\mathrm{cal.}} / W_{ij}^{\mathrm{exp.}})^2 / (n \times m) \right)^{1/2} \tag{66}$$

$$\mathrm{IMRE} = \left(\sum_{i=1}^{m} (1 - W_{ik}^{\mathrm{cal.}} / W_{ik}^{\mathrm{exp.}})^2 / m \right)^{1/2} \tag{67}$$

for the kth liquid in Step (2) of the determination procedure. γ_i is calculated with Equation (52) and $\gamma_{\mathrm{WT}/2}$ with Equation (53).

During the tuning of parameters, all the outputs are plotted against the Lifshitz parameter P^{L}. The plots give us an intuitive view for locating the range of P^{L}. The initial ranges, as determined from the preliminary runs in which $\gamma_{\mathrm{WT/2}}$ was not considered, were: $7-9, 7-9, > 10, > 10$ and $8-10$ (unit: $(\mathrm{mJ/m^2})^{1/2}$) for GL, FA, MI, BN, and TP, respectively. Example outputs of final runs in those ranges with $\Delta P^{\mathrm{L}} = 0.1$ are shown in Figures 8 to 11 for the first four liquids. In the following, we will use the plots to explain how the parameters were selected.

Glycerol (Figure 8): We can see that OMRE and IMRE vary within only 1% and 2%, respectively, when P^{L} of glycerol varies from 7.5 to 9 $(\mathrm{mJ/m^2})^{1/2}$. These small changes are not meaningful considering the standard deviation in contact angle measurements ($< 2°$, according to Zisman's papers). On the other hand, the calculated interfacial free energies of water with MI and BN are very sensitive to P^{L} of glycerol. For instance, $\gamma_{\mathrm{WT/MI}}$ decreases from 46 $\mathrm{mJ/m^2}$ to 5 $\mathrm{mJ/m^2}$ when P^{L} of glycerol only varies from 7.5 to 8.5 $(\mathrm{mJ/m^2})^{1/2}$. With the known values of $\gamma^{\mathrm{exp.}}_{\mathrm{WT/MI}} = 48.5\ \mathrm{mJ/m^2}$ and $\gamma^{\mathrm{exp.}}_{\mathrm{WT/BN}} = 42.4\ \mathrm{mJ/m^2}$, P^{L} is restricted to between 7.6 to 8 $(\mathrm{mJ/m^2})^{1/2}$. In addition, because glycerol and formamide are spontaneously miscible with water, $\gamma_{\mathrm{WT/GL}}$ and $\gamma_{\mathrm{WT/FA}}$ should be near zero or even negative. Thus, a value of 8.0 was selected for $P^{\mathrm{L}}_{\mathrm{GL}}$.

Formamide (Figure 9): Similar to glycerol, OMRE and IMRE are not the deciding factors for the selection of $P^{\mathrm{L}}_{\mathrm{FA}}$. On the other hand, calculated values of γ_{L}, $\gamma_{\mathrm{WT/FA}}$ and $\gamma_{\mathrm{WT/GL}}$ required the value of $P^{\mathrm{L}}_{\mathrm{FA}}$ to be either less than 7.3 $(\mathrm{mJ/m^2})^{1/2}$ or larger than 8.0 $(\mathrm{mJ/m^2})^{1/2}$. The final value was selected as 7.3 since the correspondingly calculated $\gamma_{\mathrm{WT/BN}}$ (48.0 $\mathrm{mJ/m^2}$) and $\gamma_{\mathrm{WT/MI}}$ (42.0 $\mathrm{mJ/m^2}$) were close to their experimental values within only 1%.

Methylene iodide (Figure 10): The variations in OMRE, IMRE and γ_L with P^{L} are more significant here. Though $\gamma^{\mathrm{L}}_{\mathrm{MI}} = \gamma_{\mathrm{MI}}$ (i.e., $P^{\mathrm{L}}_{\mathrm{MI}} = 10.1\ (\mathrm{mJ/m^2})^{1/2}$) has been widely accepted by other researchers, the selection of this value would cause the IMRE for the work of adhesion to be as high as 9%; the relative error in γ_{MI} to be 3%; and the interfacial free energy between water and α-Bromonaphthalene to be 16 $\mathrm{mJ/m^2}$ well below the experimental value of 42.4 $\mathrm{mJ/m^2}$. In addition, the fact that methylene iodide is miscible with hexane and cyclohexane [74] suggests that $P^{\mathrm{L}}_{\mathrm{MI}}$ should be higher than 11.0 $(\mathrm{mJ/m^2})^{1/2}$; and calculated $\gamma_{\mathrm{WT/MI}}$ prevents $P^{\mathrm{L}}_{\mathrm{MI}}$ to be lower than 11.6 $(\mathrm{mJ/m^2})^{1/2}$. After carefully reviewing the figures, the value of 11.6 was selected for $P^{\mathrm{L}}_{\mathrm{MI}}$.

α-Bromonaphthalene (Figure 11): It seems that the selection for $P^{\mathrm{L}}_{\mathrm{BN}}$ value is the easiest one. All the interfacial free energy outputs and γ_{L} point to $P^{\mathrm{L}}_{\mathrm{BN}} = 10.5\ (\mathrm{mJ/m^2})^{1/2}$, while OMRE and IMRE for the work of adhesion are only 1% and 2% worse than their respective minimum values.

6.3.2. Comparison with the Literature Values. Ever since the monumental work done by Zisman and coworkers, the liquids in Table 6 have been employed in contact angle studies by a number of researchers. However, there are large discrepancies in

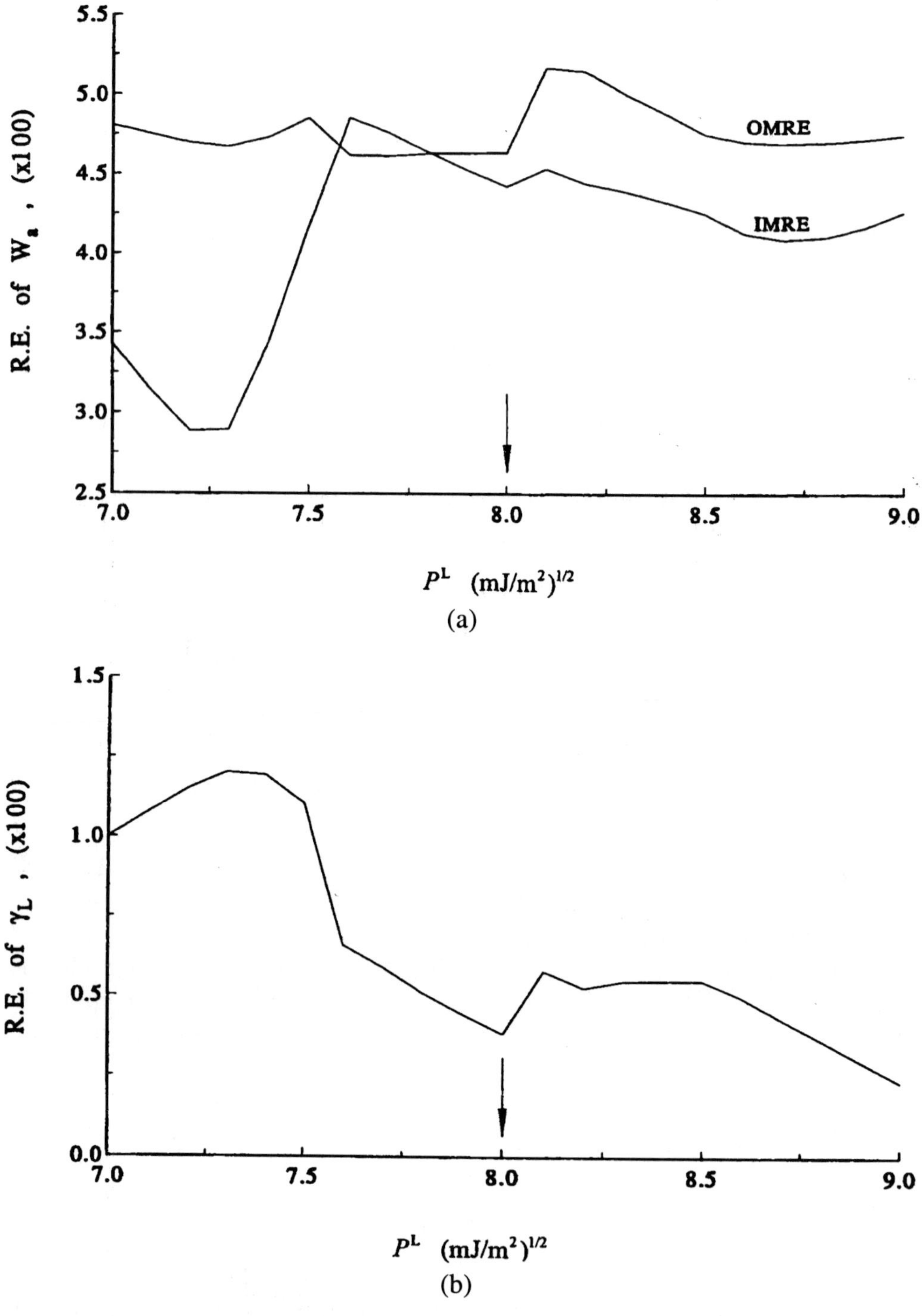

Figure 8. Calculated mean relative errors in W_a (a), γ_L (b) as functions of assigned P^L value of glycerol.

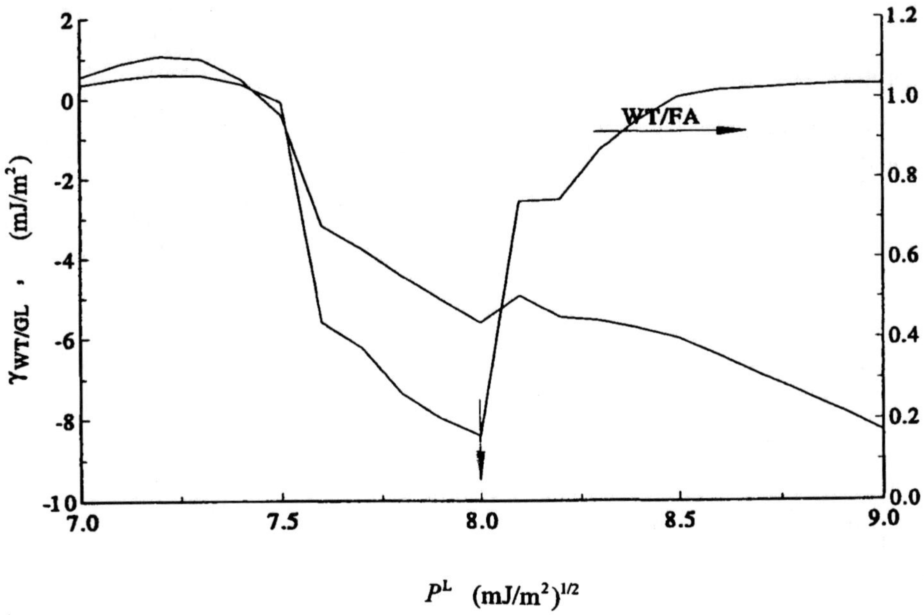

(c)

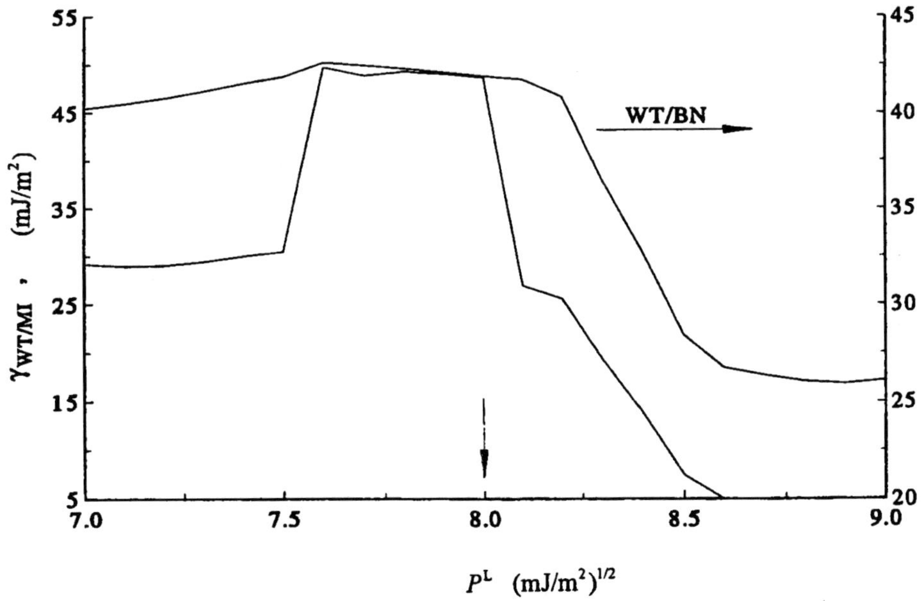

(d)

Figure 8. Calculated interfacial free energies, $\gamma_{WT/GL}$ and $\gamma_{WT/FA}$ (c), $\gamma_{WT/MI}$ and $\gamma_{WT/BN}$ (d) as functions of assigned P^L value of glycerol.

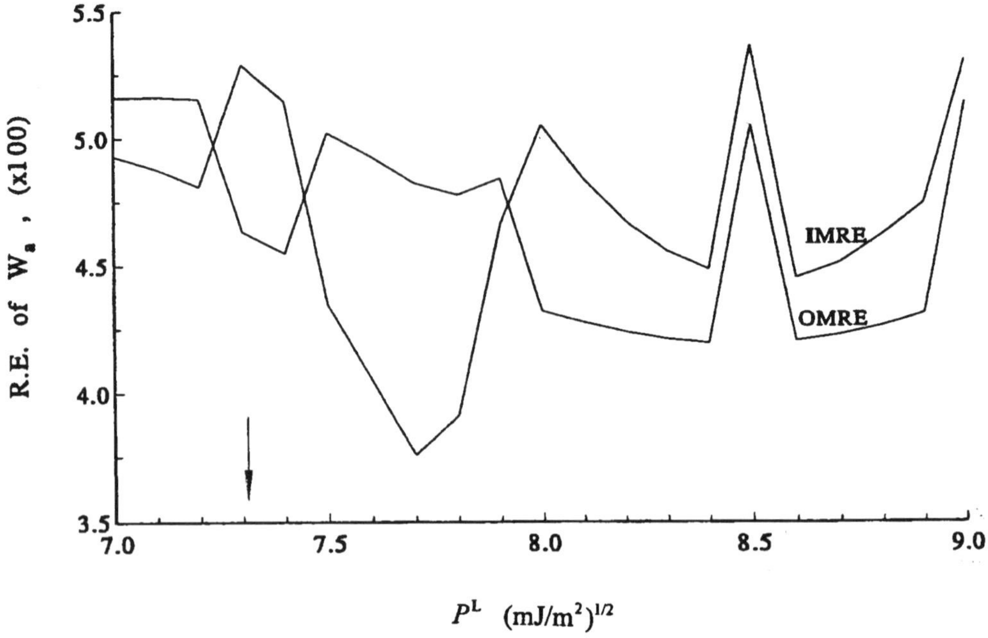

(a)

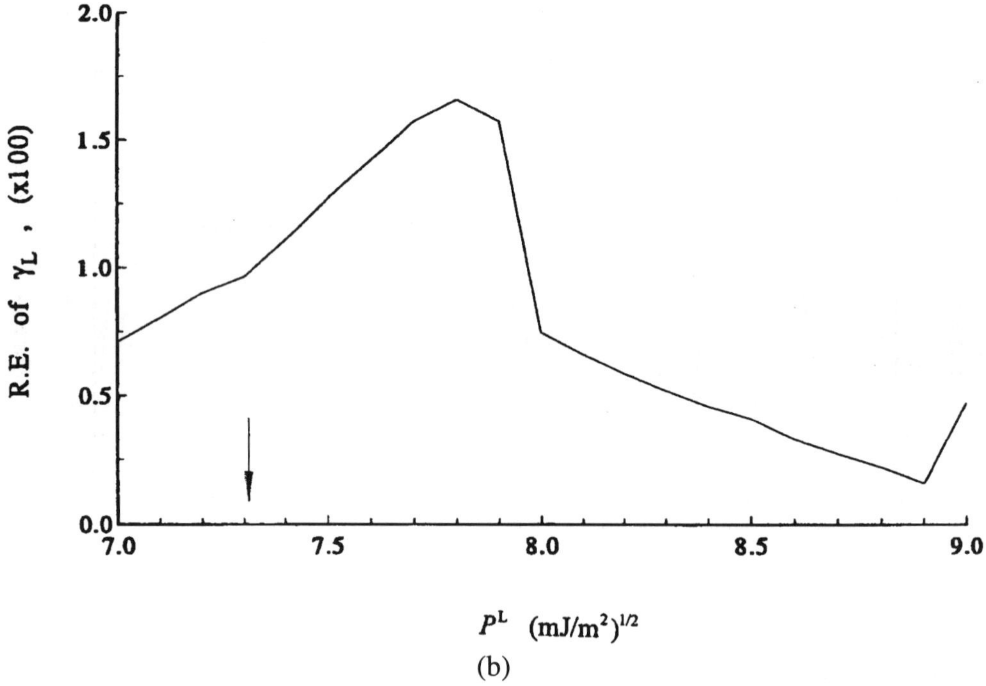

(b)

Figure 9. Calculated mean relative errors in W_a (a), γ_L (b) as functions of assigned P^L value of formamide.

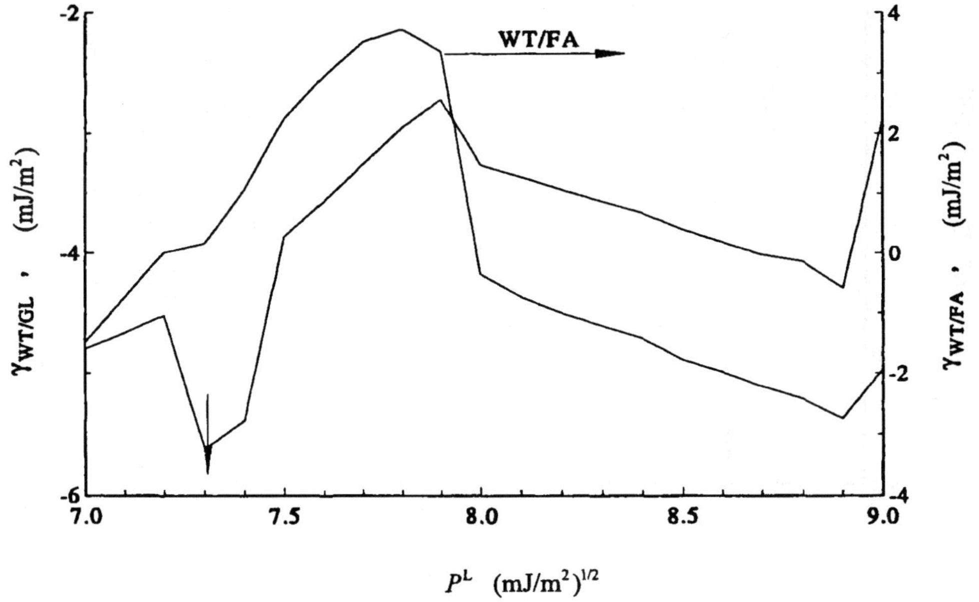

(c)

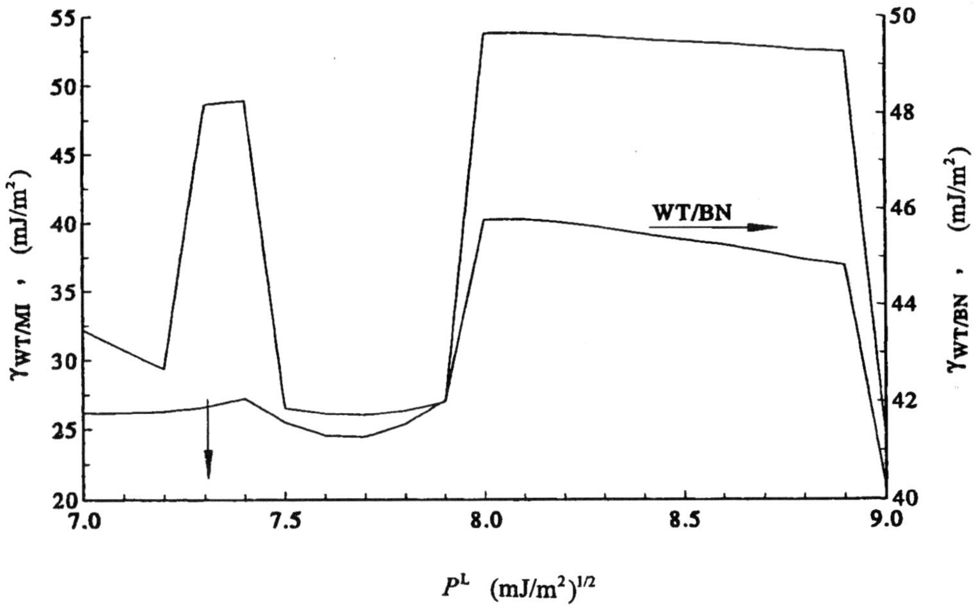

(d)

Figure 9. Calculated interfacial free energies, $\gamma_{WT/GL}$ and $\gamma_{WT/FA}$ (c), $\gamma_{WT/MI}$ and $\gamma_{WT/BN}$ (d) as functions of assigned P^L value of formamide.

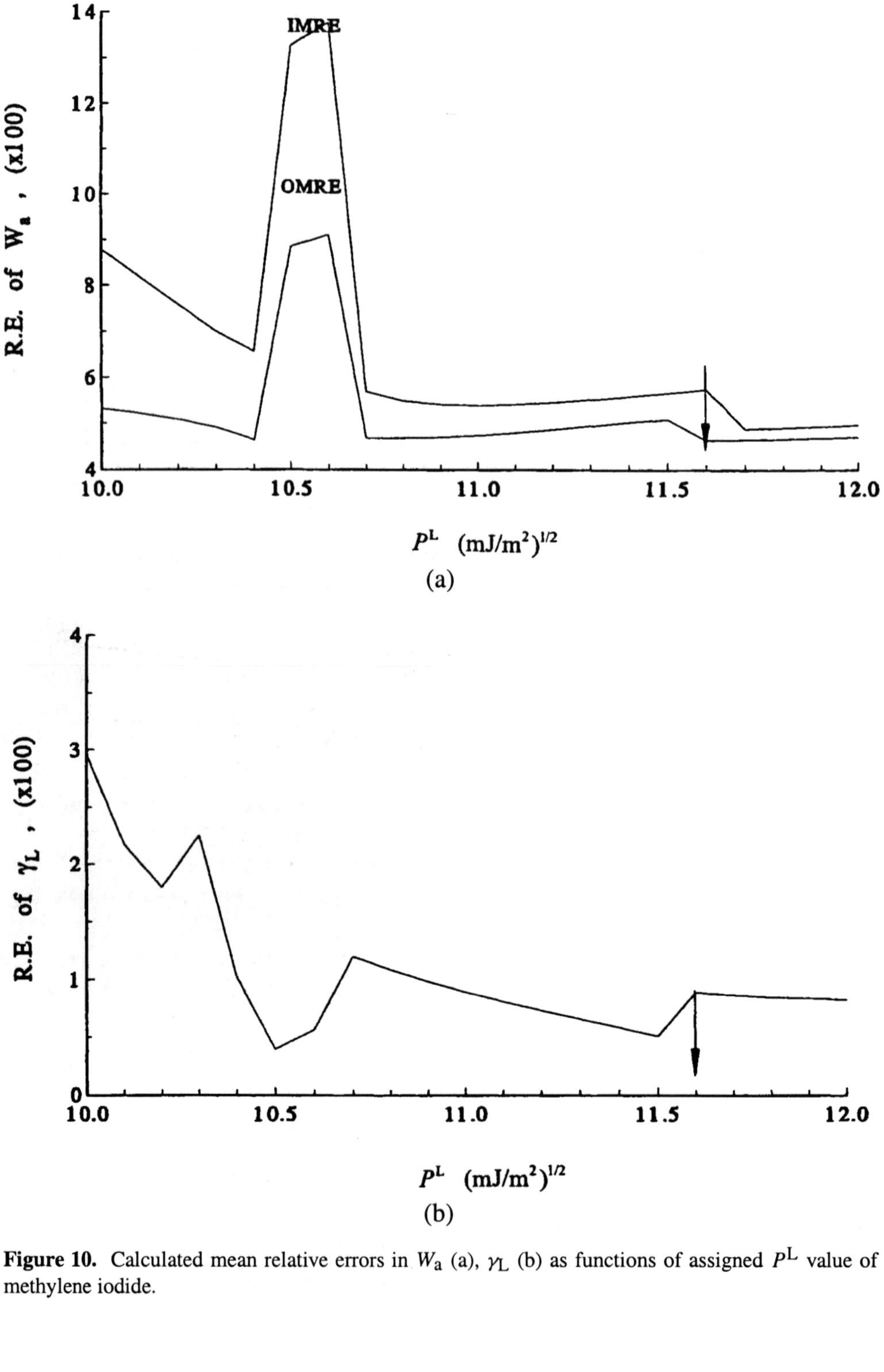

Figure 10. Calculated mean relative errors in W_a (a), γ_L (b) as functions of assigned P^L value of methylene iodide.

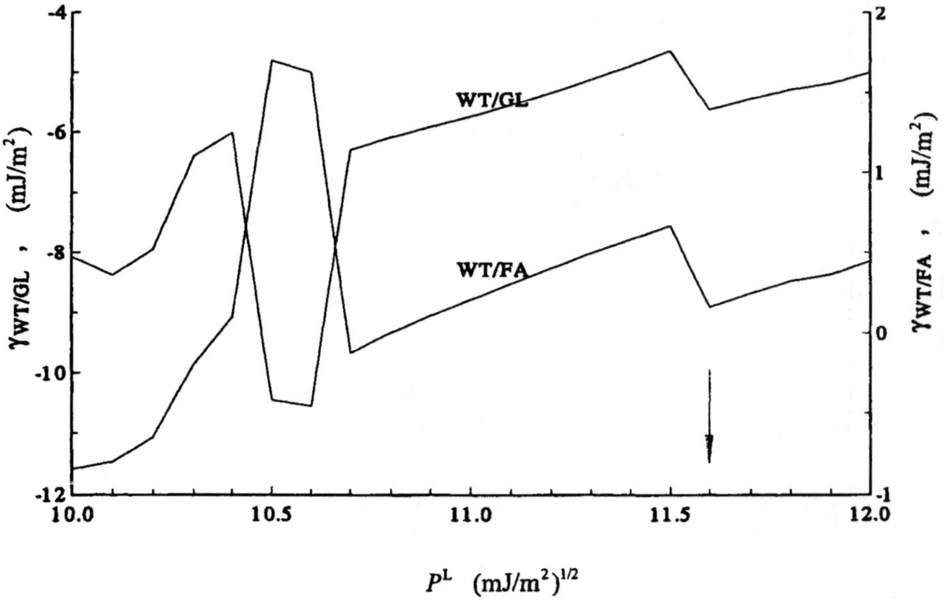

(c)

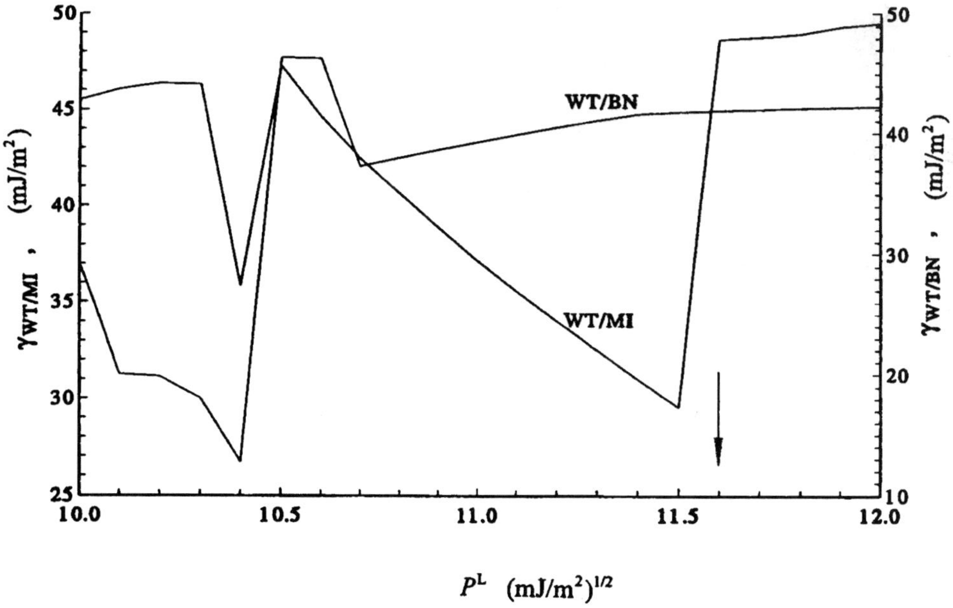

(d)

Figure 10. Calculated interfacial free energies, $\gamma_{WT/GL}$ and $\gamma_{WT/FA}$ (c), $\gamma_{WT/MI}$ and $\gamma_{WT/BN}$ (d) as functions of assigned P^L value of methylene iodide.

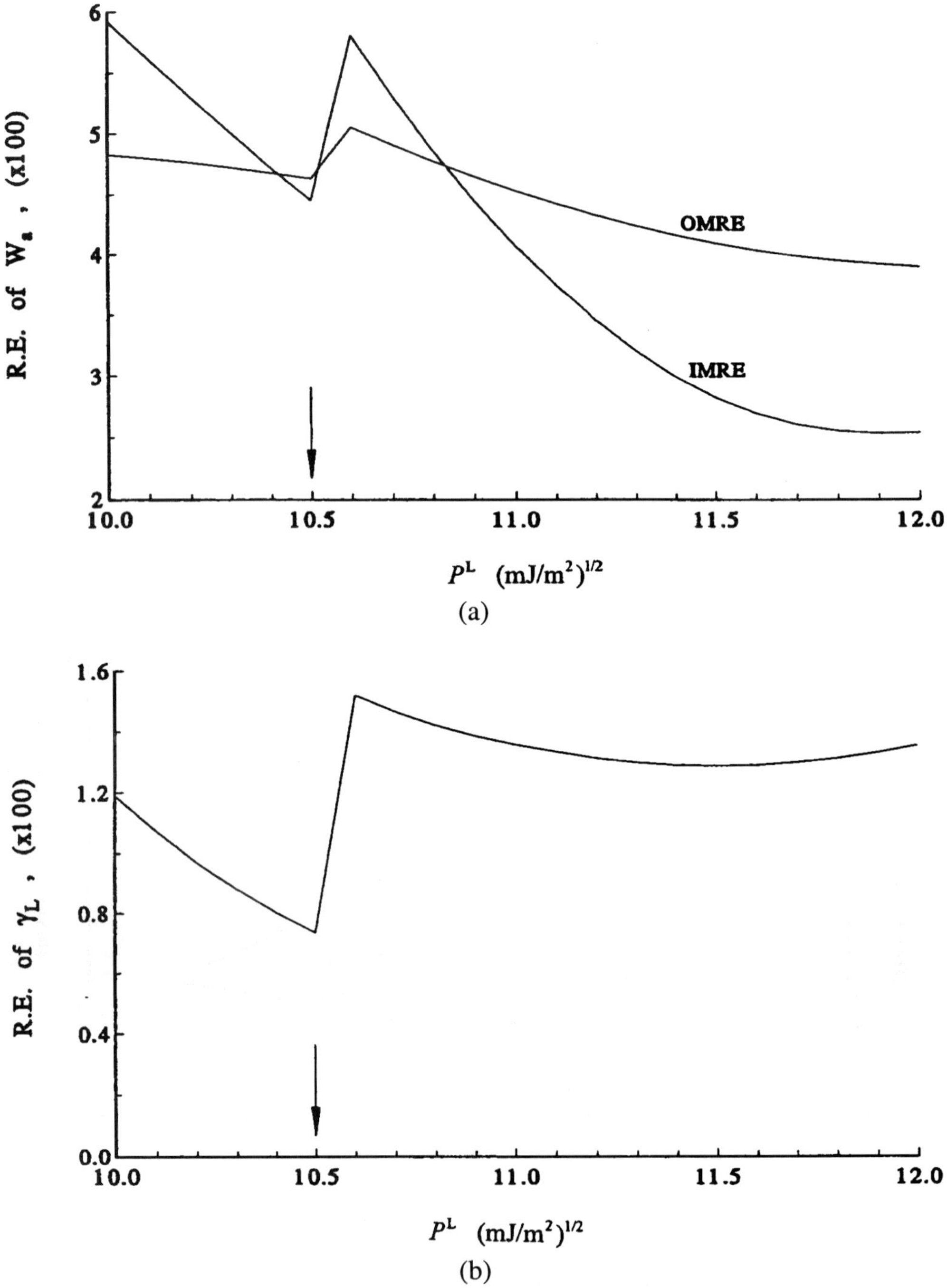

Figure 11. Calculated mean relative errors in W_a (a), γ_L (b) as functions of assigned P^L value of α-bromonaphthalene.

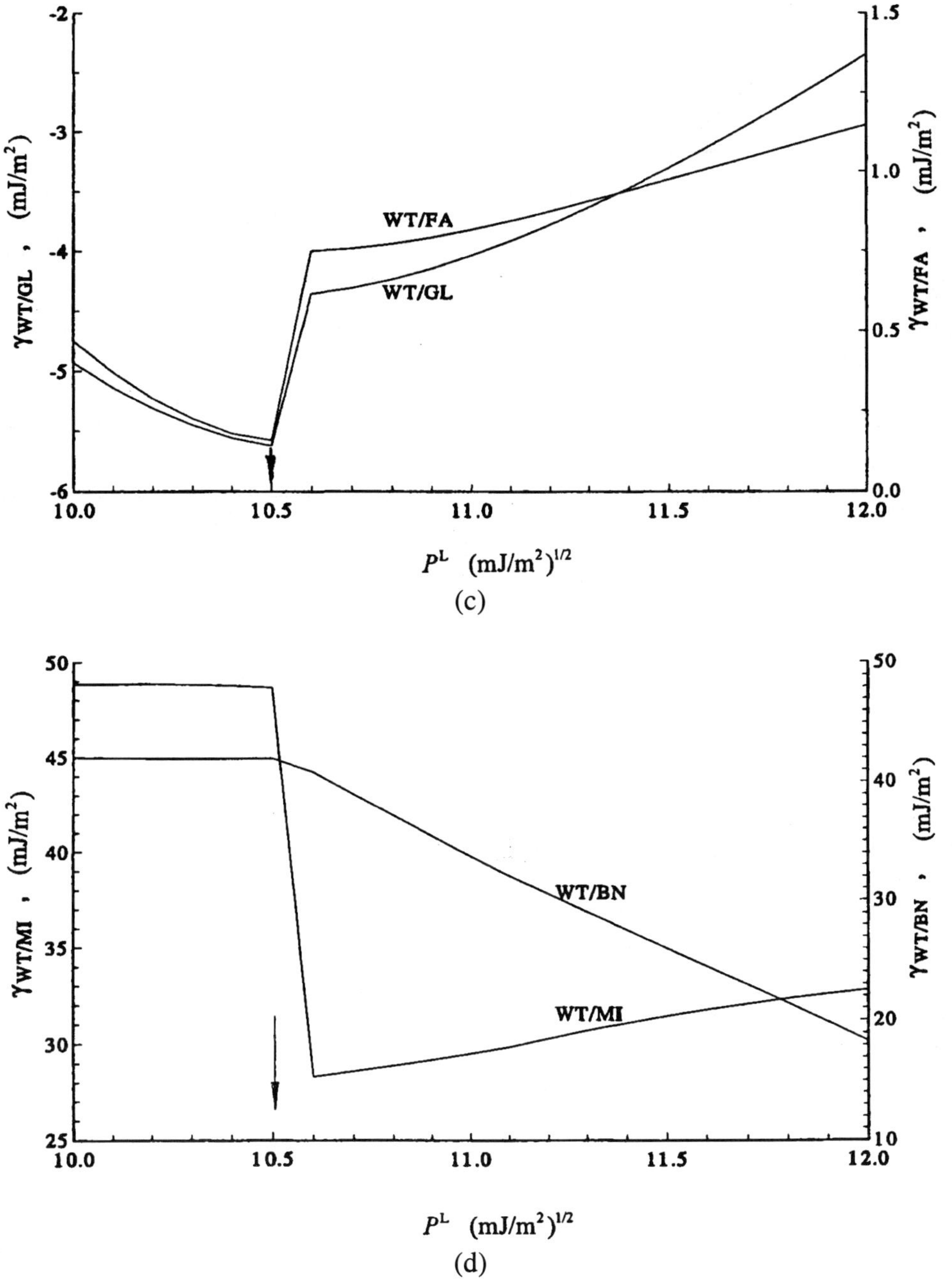

Figure 11. Calculated interfacial free energies, $\gamma_{WT/GL}$ and $\gamma_{WT/FA}$ (c), $\gamma_{WT/MI}$ and $\gamma_{WT/BN}$ (d) as functions of assigned P^L value of α-bromonaphthalene.

the values of γ^{L} estimated by different researchers for these liquids. Values of γ^{L} for the six liquids obtained with different methods are compared in Table 10.

6.3.2.1. Interfacial Free Energy Method. The interfacial free energy between a non-neutral liquid and an n-alkane gives the most direct measure of γ^{L} of the liquid. However, in the case of liquid–liquid contact, it is difficult to avoid the mutual solubility, and to measure the instantaneous interfacial free energy, $\gamma_{i/\mathrm{RH}}(t = 0)$. This is true even for organic hydrogen-bonding liquids like glycerol and formamide. Thus, the available literature data [41, 69, 70] for the interfacial free energy for glycerol and formamide in contact with n-alkanes cannot be confidently used to determine γ^{L} for these two liquids. In addition, a large discrepancy also exists in the literature data on the interfacial free energy. For example, when in contact with dodecane under mutual phase saturation, γ_{12}'s for glycerol and formamide are 43.3 mJ/m^2 and 38.9 mJ/m^2 as reported by Janczuk and Bialopiotrowicz [71], while the corresponding values are only 25.4 mJ/m^2 and 23.7 mJ/m^2 by Panzer [41]. It may be expected that the interfacial free energy between two mutually saturated phases is smaller than that of two pure liquids. The Lifshitz component of a non-neutral liquid so determined is overestimated. As a special case, only γ^{L} of water can be confidently determined to be about 22 mJ/m^2 due to the extremely low solubility between water and n-alkanes.

6.3.2.2. Contact Angle Method. Another method commonly used to estimate γ^{L} of non-neutral liquids is to measure their contact angles on solids which are Lifshitz in nature. Thus, we have

$$W_{\mathrm{a}} = 2\sqrt{\gamma_{\mathrm{S}}^{\mathrm{L}}\gamma_{\mathrm{L}}^{\mathrm{L}}} = \gamma_{\mathrm{L}}(1 + \cos\theta) \tag{68}$$

The problem involved here is that it is difficult to know whether a solid surface is Lifshitz in nature or not. Most often, it is necessary to pre-assume the neutral character of a solid surface. Such assumptions have been widely applied to polymers such as PTFE and PE. In Section 4 we have shown that a PTFE surface is not neutral at all. If one assumes that there are only Lifshitz interactions across an interface of a hydrogen-bonding liquid and a solid surface like PTFE, erroneous results could be obtained. In column 4 of Table 10 (θ method) we present $\gamma_{\mathrm{L}}^{\mathrm{L}}$, the regression result of Equation (68). The "multi-" column lists the result obtained using contact angles of the liquids on eight polymers: PTFE, P3FE, PVDF, 80F20Cl, 60F40Cl, TFE-E, HTA, and PW (Table 4), assuming they were all Lifshitz in nature. In parentheses are Fowkes regression results of the same equation for the same parameters [13]. The "single-" column presents the result of using only one solid, PW. As can be seen, γ^{L} of water is higher than the value determined from its interfacial free energies with n-alkanes. Obviously, the omission of acid–base interactions between water and the solid surfaces results in the overestimation of γ^{L}. By the same reason, the true values of γ^{L} for glycerol and formamide should be lower than the ones calculated by the above methods because these two liquids are known to be strong hydrogen-bonding liquids too.

6.3.2.3. Hansen Plot Method. Both surface free energy and cohesive energy density (C.E.D.) are measures of intermolecular interactions. Hence, many researchers have attempted to correlate them. Work done by Hansen and Beerbower [72] is representative. They have found that it is impossible to find a universal correlation between γ and C.E.D. for all liquids. A correlation established by Beerbower [73] is

$$\gamma = KV^{1/3}(\delta_d^2 + \alpha\delta_p^2 + \beta\delta_h^2) \qquad (69)$$

where δ is the solubility parameter ($(\text{cal/cm}^3)^{1/2}$); V is the molar volume; and 'd', 'p', and 'h' refer to the contributions from dispersion, polar, and hydrogen-bonding. $K = 0.0715$ erg/cal, $\alpha = 0.643$, and $\beta = 0.620$ for common organic liquids [73]. In general, the effect of Keesom dipole-dipole interactions in condensed phases is over-addressed in the Hansen-Beerbower plot because δ_p has been directly estimated from the dipole moment of single molecules. Based on this consideration, the values of γ^L calculated from the Hansen-Beerbower plot should be higher than those shown in Table 10 if a correct δ_p is used. For example, γ^L of water would be 18.2 mJ/m^2, instead of 10.8 mJ/m^2 as given in the table.

Usually, high γ^L values are assigned to liquids of bi-character from either interfacial energy, γ_{12} or contact angle method, and lower values are given by the Hansen-Beerbower plot. Our Lifshitz parameters calculated from the work of adhesion data, with the model under certain physical constraints (like γ_L, $\gamma_{WT/i}$ γ_L^L values), are systematically lower than those from the contact angle method and higher than those from the Hansen plot, for the bi-characteristic liquids.

6.3.2.4. MI and BN. Probably the most striking characteristic of our results (Table 6) is that γ^L of both methylene iodide and α-bromonaphthalene is larger than their respective γ. This results in negative values of γ^{ab} for these two liquids. In Section 4, we have shown that MI should not be considered as a Lifshitz liquid. Instead, there are evidences for the existence of repulsive non-dispersion interactions between MI and other basic materials.

Our results are also confirmed by the following solubility consideration. It is known that methylene iodide is miscible with n-hexane ($\gamma_L^L = 18.4$ mJ/m^2) and cyclohexane [74]. The interfacial free energy of MI with them should be near zero. If we follow the common idea and let $P_{MI}^L = 10.1$ (mJ/m^2)$^{1/2}$, similar to Equation (55) we can write

$$\gamma_{MI/hexane} = \gamma_{MI/hexane}^L - P_{MI}^a P_{MI}^b \qquad (70)$$

with $\gamma_{MI/hexane}^L = (P_{MI}^L - P_{hexane}^L)^2/2 = 8$ mJ/m^2. It is clear that two acid–base parameters of methylene iodide must be of the same sign in order to compensate for the appreciable positive contribution of dispersion force to $\gamma_{MI/hexane}$. That is, there would be a negative acid–base contribution (repulsion) to the surface free energy of methylene iodide. Assuming $\gamma_{MI/hexane}$ is less than 0, the dispersion parameter of

MI can be estimated by (cf. Equation (56))

$$P_{\mathrm{MI}}^{\mathrm{L}} \geqslant (\gamma_{\mathrm{MI}} + \gamma_{\mathrm{hexane}})/P_{\mathrm{hexane}}^{\mathrm{L}} \tag{71}$$

which indicates that $P_{\mathrm{MI}}^{\mathrm{L}}$ is no less than 11.4 $(\mathrm{mJ/m^2})^{1/2}$, or $\gamma_{\mathrm{MI}}^{\mathrm{L}} \geqslant 65.0$ mJ/m^2. Similarly, when the miscibility of α-bromonaphthalene and hexadecane is considered, one concludes that $P_{\mathrm{BN}}^{\mathrm{L}}$ should be larger than 9.7 $(\mathrm{mJ/m^2})^{1/2}$, or $\gamma_{\mathrm{BN}}^{\mathrm{L}} \geqslant 47.2$ mJ/m^2.

6.3.2.5. Lifshitz/Israelachvili method. In the previous paragraphs, we have focused on the experimental side. Theoretically, the famous Lifshitz theory makes it possible to estimate the Lifshitz interactions across an interface from available dielectric spectral data [75, 33]. In brief, the dispersion component of surface free energy can be evaluated by

$$\gamma^{\mathrm{L}} = \frac{A}{24\pi d_{\mathrm{e}}^2} \tag{72}$$

where A is the Hamaker constant, and d_{e} is the van der Waals intermolecular separation distance at equilibrium. It was found that $d_{\mathrm{e}} \approx 1.65$Å[1] [75]. If the absorption spectrum of a material follows a one-frequency dispersion formula and the major part of the dispersion interaction originates from electronic excitation in the ultraviolet region, A is approximately written, according to Israelachvili [77], as:

$$A = \frac{3(n^2-1)^2}{16\sqrt{2}(n^2+1)^{3/2}}\hbar\omega_{\mathrm{UV}} \tag{73}$$

where ω_{UV} is the frequency ($\sim 3 \times 10^{15}$, for organic compounds [77]), n is the refractive index in the visible region and $\hbar$ is the Planck constant.

The calculated results from Equation (72) are shown in the last column of Table 10. We can see that, except for water, the agreement with our result is remarkable. The lower value of γ^{L} for water estimated from Equations (72) and (73) is expected. Different from the other five liquids, water has significant dielectric absorption in infrared and microwave regions [78]. If the experimental data over the full spectrum range are used, the Hamaker constant of water, calculated by Parsegian and Weiss [78], is 4.61×10^{-20} J. Equation (72) will then result in a γ^{L} of 22.4 mJ/m^2 for water.

6.4. Acid-Base Parameters of Liquids

After the values of $P_{\mathrm{L}}^{\mathrm{L}}$ were selected, the other two surface parameters, P^{a} and P^{b}, for the key liquids and polymers could be determined immediately by the program.

[1] By assuming that d_{e} is determined by the balance between the Born repulsion and the van der Waals attraction, van Oss and Good [76] found that $\gamma^{\mathrm{L}} = A/(36\pi d_{\mathrm{e}}^2)$, where $d_{\mathrm{e}} = 1.36$Å. Quantitatively, the γ^{L} value estimated from this expression is only 2% less than that estimated from eq. (72) where d_{e} is assigned as 1.65 Å because $(1.65/1.36)^2 \approx 36/24$.

Table 8.
Absolute Error in Work of Adhesion (mJ/m^2) for Six Key Liquids on Polymers†

	WT	GL	FA	MI	BN	TP
PTFE	-0.98	3.47	-2.67	-5.23	2.69	0.55
P3FE	-0.28	1.08	-0.83	-1.52	0.71	2.73
PVDF	-0.29	-4.20	4.39	-0.76	6.35	3.81
PVF	-2.44	-2.44	5.37	0.89	3.14	-3.05
PE	-0.64	2.04	-1.56	-3.27	1.82	1.33
80-20	3.44	2.36	-6.93	-6.04	2.88	1.44
60-40	-0.54	1.80	-1.28	-2.76	1.52	1.86
TFE-E	1.50	2.67	-4.52	-2.42	-1.00	-0.50
HTA	-3.32	3.03	0.73	-2.88	1.45	-1.47
PW	-0.75	0.11	0.35	-4.22	5.45	-1.34

†$(W^{exp.} - W^{cal.})$: See Table 5 for $W^{exp.}$; $W^{cal.}$ is calculated using eqs. (50) and (51).

The determined parameters for liquids are presented in Table 6, and polymers in Table 7. Also, the calculated values of the work of adhesion are listed in Table 8 for comparing with Table 5. The overall root-mean-square relative deviation between the experimental work of adhesion and those calculated using Equation (59) is about 4%, an error close to that caused by a typical experimental error in contact angle measurement.

6.4.0.6. Calculation of Parameters for Other Liquids. Once the parameters for our key set of liquids and polymers are determined, we can use these parameters and the same calculation method, combining with the experimental work of adhesion, surface free energy, and interfacial free energy if available, to calculate the parameters for other liquids by Step 2 in the above procedure. The surface parameters calculated for some other liquids are summarized in Table 9.

6.4.1. Accuracy of the Parameters. Included in Tables 6, 7, and 9 are deviations for each parameter and the number of experimental data points used to calculate the parameter. The deviations for the Lifshitz parameters of the polymers result from linear regression of Equation (58) in which the contact angles of n-alkanes are used. The deviations for all other parameters are calculated with the common least squares error analysis method [67]. In brief, the variance-covariance matrix of the parameters is estimated with

$$\text{cov}(P) = \frac{[\mathbf{V}^{\mathrm{T}}\mathbf{M}^{-1}\mathbf{V}]}{n-m}[\mathbf{J}^{\mathrm{T}}\mathbf{M}^{-1}\mathbf{J}] \tag{74}$$

where J is the Jacobian of Equation (60); n is the number of experimental work of adhesion data points; and m is the number of variables. The marginal standard deviation of a variable (parameter) is simply the square root of its variance (its corresponding diagonal element of cov(P)). It was found that, from the covariances

Table 9.
Surface Parameters of Some Liquids

Liquid†	n_θ	γ_L mJ/m^2	$\gamma_L^{cal.}$ mJ/m^2	P_L^L (mJ)$^{1/2}$/m	P_L^a (mJ)$^{1/2}$/m	P_L^b (mJ)$^{1/2}$/m
$\gamma_{WT/RH}$						
Pentane		15.8	15.8	5.62	0	0
Hexane		18.4	18.4	6.07	0	0
Heptane		20.1	20.1	6.34	0	0
Decane		23.8	23.8	6.90	0	0
Dodecane		25.4	25.4	7.13	0	0
Hexadecane		27.5	27.5	7.42	0	0
c-C6		25.5	25.6	7.15	0	0
EG	5	47.7	48.2	7.50±0.06	3.69±0.16	-5.44±0.21
$HCCl_3$	3	27.1	27.1	7.10±0.01	2.04±0.01	-0.92±0.01
sTCE	6	36.3	36.5	8.30±0.05	2.69±0.31	-0.77±0.17
ACET	5	26.0	26.0	6.90±0.01	0.32±0.01	-6.95±0.14
C3E	6	17.8	17.8	6.00±0.01	-0.05±0.02	-4.70±0.19
C2E	4	17.0	16.9	6.10±0.01	-0.28±0.01	-5.94±0.01
TBN	13	33.7	34.0	8.20±0.01	0.14±0.01	-2.83±0.08
A1242	5	45.3	45.0	10.20±0.04	-2.15±0.12	-3.26±0.20
A1248	6	44.2	44.2	10.00±0.01	-2.35±0.02	-2.45±0.02
BPUD	13	37.7	37.6	9.50±0.01	-1.76±0.03	-4.30±0.07
BENZ	11	28.9	28.6	7.40±0.03	1.00±0.09	-1.26±0.22
TOLU	5	28.5	28.3	7.40±0.02	1.90±0.05	-0.50±0.04
CCl_4	4	26.8	26.8	7.30±0.01	0.73±0.01	-0.17±0.01
TCE	3	31.7	33.1	8.20±0.01	-0.70±0.02	-0.71±0.02
CS_2	4	31.4	31.3	7.90±0.03	0.51±0.12	-0.11±0.14

†See nomenclature for liquid symbols.

(off-diagonal elements of cov(P)), most correlation coefficients were either near 1 or -1, that is, the errors are highly correlated to each other.

Because of the high correlation of errors, it is more appropriate to use an independent presentation of errors: conditional standard deviation. Equation (74) can again be used to calculate the conditional standard deviations, but assuming for each parameter that all the other parameters have their true values. The conditional standard deviation for one parameter is independent of the values of the other parameters [67]. Thus, the deviations reported in Tables 6 and 7 are conditional deviations for all parameters, except the Lifshitz parameters of the polymers.

6.4.2. Acid-Base Parameters of the Key Liquids. Table 6 shows that, as expected, the three hydrogen bonding liquids all show strong acidic and basic strengths. In their pure state they are able to self associate through their own strong acid–base attractions. The acid–base component of γ of water is appreciably larger than those of glycerol and formamide, which are of similar magnitude. In addition, the acidic and basic parameters of water are comparable (near symmetric), but those of glycerol and formamide are not. The basic character of glycerol is stronger than its acidic character. In contrast, formamide has a smaller basic parameter than its acidic one.

The acid–base parameters of methylene iodide and α-bromonaphthalene indicate that their surfaces are mono-basic by our definition. The conceptual difference between our model and the model of Good, van Oss, and Chaudhury is clearly demonstrated by these two liquids. The interfacial free energies of water with methylene iodide and α-bromonaphthalene can be found in the literature [10]. The calculated values with the surface parameters in Table 6 agree well with the experimental ones.

In the case of glycerol and formamide, it is impossible to directly measure the values of interfacial energies of water-glycerol and water-formamide because of the high miscibility of the systems. van Oss *et al.* [79] have proposed a solvent-gel method to estimate the interfacial free energies between miscible polar liquids. Basically, they measured the contact angle of one liquid on the gel substrate of the second liquid as a function of gel concentration and extrapolated the angles to zero gel concentration, and thus obtained the 'true' contact angle or $\cos\theta$ between the two liquids. Using this method, they obtained negative values for water with other strong polar organic solvents, such as glycerol, formamide, DMSO etc. Our calculations show $\gamma_{\mathrm{WT/GL}} = -5$ mJ/m^2 and $\gamma_{\mathrm{WT/FA}} = 0.16$ mJ/m^2. The former is somewhat in agreement with the value of -8 mJ/m^2 obtained by van Oss *et al.* For water and formamide, the near zero interfacial free energy is also consistent with the observed miscibility. Finally, the calculated water-tricresyl phosphate interfacial free energy is 16.1 mJ/m^2. By examining the table of the Girifalco-Good interaction parameter for water-organic liquid interfaces, we could have a reasonable guess of $\Phi_{\mathrm{WT/TP}} = 0.85 \sim 0.90$, which gives $\gamma_{\mathrm{WT/TP}}$ in the range of $15 \sim 20$ mJ/m^2.

6.4.3. Calculated Surface Parameters of Other Liquids. Table 9 presents the calculated surface parameters of other liquids studied in this work. The top part lists the neutral liquids. In the bottom part are the non-neutral liquids. We can roughly classify the non-neutral liquids, including the key liquids, as strongly bi-characteristic, weakly bi-characteristic, dominantly acidic, dominantly basic, mono-basic, and near-neutral.

Water, glycerol, formamide and ethylene glycol are liquids with known strong hydrogen bond. In term of our model, they also have both strong acid parameter and strong base parameter, and thus they are strong bi-characteristic liquids. Compared to the above liquids, benzene is a weak bi-characteristic liquid.

Chloroform and *sym*-tetrachloroethane, commonly recognized as Lewis acids, are classified as dominantly acidic liquids. Their acidity results from the presence of hydrogen atom(s) and the electron withdrawing ability of -CCl_x group. In addition, our results indicate that they also have a certain degree of basic character. It is interesting to note that toluene belongs to this group too. The acidic parameter of toluene is even comparable to that of chloroform.

The dominantly basic liquids are mainly carbonyl containing liquids, such as tricresyl phosphate, acetone and ethers.

In this work, mono-basicity is defined by two negative non-dispersion parameters. This definition of mono-basicity is different from that of mono-pole in the model of van Oss *et al.* Liquids that match their definition are the dominantly basic liquids discussed above. We found five strong mono-basic liquids: methylene iodide, α-bromonaphthalene, trichlorobiphenyl, tetrachlorobiphenyl, and benzyl phenylundecanoate. Except methylene iodide, the other four liquids all contain multi-aromatic rings in a planar structure.

Finally, we have three near-neutral liquids: carbon tetrachloride, tetrachloroethylene, and carbon disulfide. Their molecules are symmetric, and have zero dipole moment. Both acid and base parameters of these liquids are very weak, compared to other liquids.

6.5. Surface Properties of the Key Polymers

Our calculations show, in Table 7, that none of the polymer surfaces is truly neutral. PE, HTA, and PW are commonly considered as 'apolar'. According to our model, they are also the most 'neutral' ones among all the polymers examined. Their acid and base components of surface free energy are less than 1 mJ/m^2. However, these three polymers still have appreciable basic character that cannot be ignored when interacting with surfaces of strong acidic and/or basic characteristics.

The non-neutral behavior of the polymer surfaces can be better appreciated from Figure 7. The figure plots the values of W_a (points) and W_a^L (straight line) for water on the key polymer surfaces. The values of W_a were already presented in Table 5; and the values of W_a^L are calculated from the corresponding dispersion parameters. As we can see, all polymers exhibit attractive acid–base interactions with water, regardless of the surface constitution. Therefore, none could be truly 'neutral'.

The second striking feature in the table is provided by fluorine and chlorine substituted polyethylenes and PHDFDMS. All hydrogen-absent surfaces (PTFE, 80F20Cl, 60F40Cl, and PHDFDMS) are monobasic, while the polymers containing both fluorine and hydrogen atoms (P3FE, PVDF and PVF) are bi-characteristic.

It is believed that fluorine atoms in these polymers can be electron-donors (base sites) in Lewis acid–base interactions, such as hydrogen-bonding $\overset{\delta-}{A}-\overset{\delta+}{H} \cdots \overset{\delta-}{B}$ [20, 23]. When the contact liquid is a hydrogen-bonding liquid (e.g., water), B is a fluorine atom and A–H is the hydrogen-donating group of the liquid (e.g., O–H). The significant basic parameters calculated from our model for all fluoropolymers support the role of electron-donor played by the fluorine atoms. On the other hand, a hydrogen atom attached to a CF_x is somewhat acidic and it is possible to form a weak hydrogen bond through it [20, 23]. The hydrogen atoms are the source of their acidic characters for the hydrogen-containing fluorinated polymers: no hydrogen atom, no acidity. Thus, PTFE, 80F20Cl, and 60F40Cl have only basic character. Repulsive base-base interactions occur when they are in contact with mono-basic liquids, such as methylene iodide (cf. Figure 5). From this point of view, we can predict that PCTFE is monobasic.

From zero atom percent of fluorine substitution (PE) to 100% (PTFE), the wettability by the key liquids first increases and then decreases (cf. Table 5). Zisman has attributed this phenomenon to the ability of fluorine atom to form hydrogen-bond (i.e., strength of basicity) [23]. According to him, the electronegativity of a fluorine atom should decrease with increasing fluorine substitution. The determined basic parameters of the polymers, as shown in Table 7, agree with this explanation. In addition, the acidic character of the polymers also follows the same pattern. Thus, it is clear that the wettability trend is governed by the non-Lifshitz interactions.

TFE-E, a 50-50 copolymer of ethylene and tetrafluoroethylene, shows a different surface acid–base character from the above hydrogen-containing fluoropolymers. Percentage-wise, it has the same chemical composition as PVDF. The apparent acidic character of PVDF is, however, lacking in TFE-E. Instead, the mono-basic character makes TFE-E more like perfluoropolymers. This characteristic difference is directly reflected by the experimental observation that the wettability of PVDF by non-neutral liquids is higher than that of TFE-E. It is most likely that CH_2 groups in TFE-E are "buried" inside. Just as Zisman pointed out, in the fluorinated polymers the fluorine atoms must be located as much as possible in the surface because the surface energy is thereby reduced to a minimum.

6.6. Surface Parameters of Other Polymers

Our model provides a self-consistent and unambiguous way to characterize the surface parameters of any surface. However, very careful work is needed to accurately determine the three parameters. From our experience we know that it is usually difficult to determine unambiguously their values based only on computer fitting of limited number of wettability measurements. In most cases it is necessary to determine the Lifshitz parameter by an independent method, and then use computer fitting to calculate the acid and base parameters. For low energy surfaces on which some of the n-alkanes form finite contact angles, the Lifshitz parameter can be readily determined as discussed before. However, for most polymer surfaces one cannot find an n-alkane that will not spread on it. For these polymers, a modified two-liquid-contact-angle method was developed to obtain the Lifshitz parameter [25, 26]. Using this approach, the surface parameters for polystyrene, poly(vinyl chloride) and poly(methylmethacrylate) were determined. Using these values, the interfacial free energies between these polymers and various liquids were determined. It was shown that the calculated interfacial free energy correlated very well with both the solubility and solvent crazing resistance of polymers in the liquids.

The model has recently been applied to the wettability data of lignocellulosic surfaces [28] and ethylene-acrylic acid copolymers [27]. It was clearly shown in

Table 10.
Comparison of Dispersion Component of Surface Free Energy (mJ/m^2) of Liquids Obtained from Several Methods

	This work	γ_i/nRH method	θ method[c] multi-	single-	Hansen plot	Israelachvili eqn.(72)
WT	21.8	21.8 24.7(47.5)[a]	37.4(–)	25.8	10.8	16.7
GL	32.0	39.7(25.4)[a] 20.2(43.3)[b]	43.7(37.0)	32.9	21.6	31.3
FA	26.6	35.5(23.7)[a] 19.8(38.9)[b]	41.0(39.5)	33.4	17.2	28.2
MI	67.3		48.0(48.5)	52.1	23.4	65.6
BN	55.1		48.5(47.0)	57.4	36.4	54.1
TP	40.5		39.2(39.2)	36.9	42.1	41.1

[a] $\gamma_{i/n\text{C}12}$ from Reference [41].
[b] $\gamma_{i/n\text{C}12}$ from Reference [71].
[c] See text for details.

both cases that the repulsive (negative) acid–base interactions were real. Without a proper account for the negative interactions the surface thermodynamic behavior of some materials cannot be correctly described. This is the major difference between our approach and that of the van Oss, Good and Chaudhury. The works of Lloyd and Coleman [27] and Gardner *et al.* [28] provide independent supports to this important concept of our model [24–26].

7. SUMMARY

A critical review of the theoretical basis and some well-known models for the work of adhesion and surface energy is presented. It is pointed out that many of the conflicting results and interpretations in this field could be traced to the equivocal and inconsistent definition of a 'neutral compound'. Ample evidence is given to show the need for a new acid–base model, which would allow the repulsive interactions between acid-acid and base-base components. A new model was developed by us to meet this challenge. The model is thoroughly discussed and has been successfully applied to the literature contact angle data. The acid–base parameters calculated from the model for the fluoropolymers are related to their surface compositions. The negative acid–base components of surface free energy for methylene iodide and α-bromonaphthalene, and some fluoropolymers have been calculated and their causes discussed. Many common liquids are categorized based on their surface energetic parameters defined by our model. Our categorization is in agreement with the traditional Lewis acid–base interpretation of the chemical composition of the polymers and the liquids. In addition, methods for determining the dispersion parameter of polymers are discussed.

Acknowledgments

Financial support for this project was provided by the Parker-Hannifin Foundation, the Strategic Highway Research Program, and the J. Warren Chow China Scholar Fund.

REFERENCES

1. K. L. Mittal, in: *Adhesion Science and Technology*, L. H. Lee, (Ed.), Part A, pp.129–168. Plenum Press, New York (1975).
2. A. J. Kinloch, *J. Mater. Sci.* **15**, 2141–2166 (1980).
3. A. N. Gent and J. Schultz, *J. Adhesion*, **3**, 281 (1972).
4. E. H. Andrews and A. J. Kinloch, *Proc. R. Soc. Lond.*, A. **332**, 385–399 (1973).
5. A. D. Roberts, *Langmuir*, **8**, 1479–1486 (1992).
6. D. J. David and A. Misra, *J. Colloid Interface Sci.*, **108**, 371–376 (1985).
7. D. W. Dwight and W. M. Riggs, *J. Colloid Interface Sci.*, **47**, 650 (1974).
8. S. Wu, *Polymer Interface and Adhesion.* Marcel Dekker, New York (1982).
9. A. W. Adamson, *Physical Chemistry of Surfaces.* Chapter VII, Wiley-Interscience, New York, 4th edition (1982).
10. L. A. Girifalco and R. J. Good, *J. Phys. Chem.*, **61**, 904–909 (1957).
11. C. J. van Oss, R. J. Good, and M. K. Chaudhury, *Langmuir*, **4**, 884–891 (1988).
12. F. M. Fowkes, *J. Phys. Chem.*, **66**, 382 (1962).
13. F. M. Fowkes, *Ind. Eng. Chem.*, **56** (12), 40–52 (1964).
14. F. M. Fowkes, in *Physicochemical Aspects of Polymer Surfaces, Vol. 2*, K. L. Mittal, (Ed.), pp. 583–603. Plenum Press, New York (1983).
15. F. M. Fowkes, *J. Adhesion Sci. Technol.*, **1**, 7 (1987).
16. F. M. Fowkes, M. B. Kaczinski, and D. W. Dwight, *Langmuir*, **7**, 2464–2470 (1991).
17. F. M. Fowkes and M. A. Mostafa, *Ind. Eng. Chem. Prod. Res. Dev.*, **17**, 3 (1978).
18. W. V. Chang and F. Chen, *Polym. Mater. Sci. Eng.*, **61**, 607 (1989).
19. H. W. Fox and W. A. Zisman, *J. Colloid Sci.*, **5**, 514–531 (1950).
20. H. W. Fox and W. A. Zisman, *J. Colloid Sci.*, **7**, 109–121 (1952).
21. H. W. Fox and W. A. Zisman, *J. Colloid Sci.*, **7**, 428–442 (1952).
22. E. G. Shafrin and W. A. Zisman, *J. Colloid Sci.*, **7**, 166–177 (1952).
23. A. H. Ellison and W. A. Zisman, *J. Phys. Chem.*, **58**, 260–265 (1954).
24. F. Chen and W. V. Chang, *Langmuir*, **7**, 2401–2404 (1990).
25. X. Qin and W. V. Chang, *J. Adhesion Sci. Technol.*, **9**, 823–841 (1995).
26. X. Qin and W. V. Chang, *J. Adhesion Sci. Technol.*, **10**, 963–987 (1996).
27. T. B. Lloyd and J. A. Coleman, These proceedings.
28. D. J. Gardner, S. Q. Shi, and W. T. Tze, These proceedings.
29. J. Lyklema, *Foundamentals of Interface and Colloidal Science, Vol. 1*, Academic Press, London (1991).
30. W. H. Keesom, *Physik Z.*, **23**, 225 (1922).
31. P. Debye, *Physik Z.*, **22**, 302 (1921).
32. F. London, *Trans. Faraday Soc.*, **33**, 8 (1937).
33. R. J. Good and M. K. Chaudhury, In *Fundamentals of Adhesion*, L.-H. Lee, (Ed.), Chapter 3. Plenum Press, New York (1991).
34. H. C. Hamaker, *Physica*, **4**, 1058 (1937).
35. J. H. de Boer, *Trans. Faraday Soc.*, **32**, 11 (1936).
36. D. H. Kaelble, *Physical Chemistry of Adhesion.* Wiley-Interscience, New York (1971).

37. S. Wu, *J. Polym. Sci., Part C,* **34** (1971).
38. D. K. Owens and R. C. Wendt, *J. Appl. Polym. Sci.*, **13**, 1741–1747 (1969).
39. D. K. Owens, *J. Appl. Polym. Sci.*, **14**, 1725–1730 (1970).
40. F. M. Fowkes, *J. Adhesion*, **4**, 155–159 (1972).
41. J. Panzer, *J. Colloid Interface Sci.*, **44**, 142–161 (1973).
42. E. M. Lifshitz, *Soviet Phys. JETP*, **2**, 73–83 (1956).
43. I. E. Dzyaloshinskii, E. M. Lifshitz, and L. P. Pitaevski, *Adv. Phys.*, **10**, 165–209 (1961).
44. A. D. McLachlan, *Proc. Roy. Soc. Lond*, A **274**, 80–90 (1964).
45. J. N. Israelachvili, *Proc. Roy. Soc. Lond.*, A **331**, 39–55 (1972).
46. L. D. Landau and E. M. Lifshitz, (Eds.), *Electrodynamics of Continuous Media.* Addison–Wesley (1960).
47. V. A. Parsegian and B. W. Ninham, *Biophy. J.*, **10**, 664–674 (1970).
48. D. B. Hough and L. R. White, *Adv. Colloid Interface Sci.*, **14**, 3 (1980).
49. X. Qin, *Adhesion Properties of Polymeric Materials.* Ph.D. Thesis, University of Southern California, Los Angeles (1995).
50. E. F. Meyer and R. E. Wagner, *J. Phys. Chem.*, **70**, 3162 (1966).
51. G. C. Pimentel and A. L. McClellan. *The Hydrogen Bond,* Freeman & Co., San Francisco (1960).
52. W. B. Jensen, *Chem. Rev.*, **78**, 1–22 (1978).
53. F. M. Fowkes, D. O. Tischler, J. A. Wolfe, L. A. Lannigan, C. M. Ademu-John and J. M. Halliwell, *J. Polym. Sci., Polym. Chem. Ed.*, **22**, 547–566 (1984).
54. R. S. Drago, N. Wong, C. Bilgrien, and G. C. Vogel, *Inorg. Chem.*, **26**, 9 (1971).
55. F. M. Fowkes, *J. Adhesion Sci. Technol.*, **4**, 669–691 (1990).
56. V. Gutmann, *The Donor-Acceptor Approach to Molecular Interactions.* Plenum Press, New York (1978).
57. V. Gutmann, *Electrochimica Acta*, **21**, 661 (1976).
58. C. Saint-Flour and E. Papirer, *J. Colloid Interface Sci.*, **91**, 69 (1983).
59. M. Y. Boluk and H. P. Schreiber, *Polym. Composites*, **7**, 295 (1986).
60. C. J. van Oss, M. K. Chaudhury, and R. J. Good, *Adv. Colloid Interface Sci.*, **28**, 35 (1987).
61. C. J. van Oss, M. K. Chaudhury, and R. J. Good, *Separ. Sci. Technol.*, **24** (1&2), 15–30 (1989).
62. R. J. Good, M. K. Chaudhury, and C. J. van Oss, in *Fundamentals of Adhesion*, L.-H. Lee, (Ed.), Chapter 4. Plenum Press, New York (1991).
63. W. A. Zisman, *Ind. Eng. Chem.*, **55**, 19–38 (1963).
64. Y. Tamai, *Progr. Colloid Polym. Sci.*, **61**, 93 (1976).
65. F. M. Fowkes, D. C. McCarthy, and M. A. Mostafa, *J. Colloid Interface Sci.*, **78**, 200–206 (1980).
66. W. B. Jensen, *The Lewis Acid-Base Concepts.* Wiley, New York (1980).
67. W. C. Hamilton, *Statistics in Physical Science.* Ronald Press, New York (1964).
68. D. A. Ratkowsky, *Nonlinear Regession Modeling – A Unified Practical Approach.* Marcel Dekker, New York (1983).
69. B. Jańczuk, T. Bialopiotrowicz, and W. Wójcik, *J. Colloid Interface Sci.*, **127**, 59 (1989).
70. Y. H. Mori, N. Tsui, and M. Kiyomiya, *J. Chem Eng. Data*, **29**, 407–412 (1984).
71. B. Jańczuk and T. Bialopiotrowicz, *J. Colloid Interface Sci.*, **127**, 189–204 (1989).
72. C. Hansen and A. Beerbower, in *Kirk-Othmer Encyclopedia of Chemical Technology*, Vol. 1. Interscience, New York (1971).
73. A. Beerbower, *J. Colloid Interface Sci.*, **35**, 126–132 (1971).
74. M. Windholz, S. Budavari, L. Y. Stroumtsos and M. N. Fertig, (Eds.), *The Merck Index.* Merck & Co., Rahway, NJ, USA, 10th edition (1985).
75. J. N. Israelachvili, *Intermolecular and Surface Forces.* Academic Press, New York (1985).
76. C. J. van Oss and R. J. Good, *Colloids Surfaces*, **8**, 373 (1984).
77. J. N. Israelachvili, *J. Chem. Soc., Faraday Trans.*, **69**, 1729–1738 (1973).

78. V. A. Parsegian and G. H. Weiss, *J. Colloid Interface Sci.*, **81**, 285–289 (1981).
79. C. J. van Oss, M. K. Chaudhury, and R. J. Good, *J. Colloid Interface Sci.*, **128**, 313–319 (1989).
80. C. M. Roth and A. M. Lennoff, *J. Colloid Interface Sci.*, **179**, 637 (1996).
81. A. El-Shimi and E. D. Goaddard, *J. Colloid Interface Sci.*, **48**, 242 (1974).
82. J. R. Dann, *J. Colloid Interface Sci.*, **32**, 302–320 (1970).
83. H. J. Busscher, A. W. J. van Pelt, P. de Beer, H. P. de Jong, and J. Arends, *Colloids Surfaces*, **9**, 319 (1984).
84. L. S. Penn and B. Miller, *J. Colloid Interface Sci.*, **78**, 238 (1980).
85. D. Bargeman, *J. Colloid Interface Sci.*, **40**, 344 (1972).
86. H. J. Busscher, A. W. J. van Pelt, H. P. de Jong, and J. Arends, *J. Colloid Interface Sci.*, **95**, 23 (1983).
87. Y. Tamai, T. Matsunaga, and K. Horiuchi, *J. Colloid Interface Sci.*, **60**, 112–116 (1977).
88. T. Yasuda, T. Okuno, and H. Yasuda, *Langmuir*, **10**, 2435–2439 (1994).
89. H. Kobayashi and M. Owen, *Makromol. Chem.*, **194**, 259–267 (1993).

Acid-Base Interactions, Vol. 2, pp. 55–90
Ed. K.L. Mittal

Troubleshooting of surface free energy acid-base theory applied to solid surfaces: The case of Good, van Oss and Chaudhury theory

C. DELLA VOLPE * and S. SIBONI

Department of Materials Engineering, University of Trento, Via Mesiano 77, 35080 Trento, Italy

Abstract—The Good-van Oss-Chaudhury acid-base theory applied to the calculation of solid surface free energy components has a mathematical nucleus whose correct use has important effects on its results; in particular, one needs to develop an appropriate reference scale to compare the acidic and basic components.

In the present paper the mathematical properties of the equation system expressing the theory are fully analysed and their consequences from both theoretical and practical points of view are shown.

A critical analysis of the most common literature misunderstandings about the theory is performed and some proposals for the choice of a new and proper scale of acid-base strength are made.

Keywords: Contact angle; acid-base surface free energy; Good-van Oss-Chaudhury theory.

1. INTRODUCTION

In the last ten years a significant number of books and papers have dealt with the problem of the evaluation of the surface free energy of solids; the dominant paradigm [1] in this field appears to be the "multicomponent" theory, according to which the surface free energy of a solid or a liquid can be divided in at least two parts, depending on the importance of the interactions which take place at the interface.

These interactions can be summarized as dispersive, or Lifshitz-van der Waals (LW), and Lewis acid-base (AB) interactions [2], terms repeatedly and commonly defined in the literature.

It is worth noting that not all the researchers agree with this multicomponent approach and its implications, and point out that this analysis is not correct; their position is essentially reducible to the so-called equation-of-state (EOS) theory [3].

---

*To whom correspondence should be addressed: Tel. +39-0461-882409, Fax +39-0461-881977. E-mail: devol@devolmac.ing.unitn.it

However, both acid-base and EOS theory share the same basic idea, i.e., the attempt to express in an *explicit* form the work of adhesion (or, equivalently, the interfacial solid-liquid energy γ_{sl}) defined as

$$W_{adh} = \gamma_{lv} + \gamma_{sv} - \gamma_{sl} \tag{1}$$

where the subscript lv, sv and sl refer to the liquid-solid, solid-vapour and solid-liquid interfaces, respectively.

The Young equation permits to easily rewrite this equality in terms of experimental quantities, the liquid-vapour surface tension γ_{lv} and the equilibrium contact angle θ of the liquid on the solid, i.e.,

$$W_{adh} = \gamma_{lv} + \gamma_{sv} - \gamma_{sl} = \gamma_{lv}(1 + \cos\theta) \tag{2}$$

Both theories proposed express the work of adhesion as a function of some quantities particular to the separated surfaces; the EOS theory assumes the three parameter relationship

$$W_{adh} = 2\sqrt{\gamma_{lv}\gamma_{sv}}\,\mathrm{e}^{-\beta(\gamma_{lv}-\gamma_{sv})^2} \tag{3}$$

with an appropriate constant β, while the acid-base model offers two different formulations, depending on the choices proposed by different authors. Fowkes [4, 5] writes

$$W_{adh} = 2\sqrt{\gamma_{lv}^{LW}\gamma_{sv}^{LW}} + fN(-\Delta H^{AB}) \tag{4}$$

where the LW superscript indicates the dispersive component, f is a parameter, generally chosen as 1 and expressing the role of the entropic component and able to convert heats of interfacial acid-base interaction into free energies, N the number of the interfacial acid-base bonds involved (in moles/m^2) and ΔH^{AB} the enthalpy calculated by the Drago theory [6] (which brings the total number of parameters to 8); In the case of Good-van Oss-Chaudhury (GvOC) [7, 8] theory we have instead the six parameters function:

$$W_{adh} = 2\left(\sqrt{\gamma_{lv}^{LW}\gamma_{sv}^{LW}} + \sqrt{\gamma_{lv}^{+}\gamma_{sv}^{-}} + \sqrt{\gamma_{lv}^{-}\gamma_{sv}^{+}}\right) \tag{5a}$$

The definition of the parameters entering the right-hand side of Eq. (5a) will be fully introduced in the following.

The increasing complexity of these equations and the increasing number of variables allow to describe the interactions in terms of quantities similar to those currently used for the bulk materials, particularly in terms of acid-base theory.

The EOS theory has some defenders, but it has been strongly criticized from a thermodynamic point of view [9, 10].

Independent of any evaluation about the formal correctness of this theory, our opinion is that a theory is not simply a device-to-fit-numbers but firstly a lens to look at Nature. The ability to fit numerical data cannot be considered the only requirement to accept or reject a theory. From this point of view the EOS theory does not provide any "microscopic" insight into the understanding of surfaces; while the acid-base theories are able to apply one of the most important paradigms of chemistry, the concept of acid and base, also to this field.

In recent papers the EOS theory has been extended to the analysis of polar solids [11–13], using in all cases axisymmetric drop shape analysis-profile (ADSA-P) technique; so it has become more difficult to evaluate the theory without a detailed analysis of this experimental technique, its use and limitations. These considerations will be developed in a future work together with the analysis of the mathematical form of Eq. (3). Notice, however, that the authors of this theory regard whether β is a universal constant or not as an unresolved problem; and if β is not really constant the EOS theory can suffer some problems.

It is hard to accept the conclusion of ADSA-P users that the results of the standard sessile drop technique are generally meaningless [12]; however, we agree with the evaluation expressed by some defenders of the EOS theory that many published contact angles data suffer from low-quality experimental procedures and that an effort is necessary to obtain better and well-reproducible results.

Our personal opinion is that a round robin on this subject, in which the largest possible number of laboratories in the world should participate, represents the optimum strategy.

The Fowkes theory is complex to apply because of the necessity to evaluate the parameters f and N, along with the various approximations involved. As a consequence, only a small number of experimental results have been analysed using this theory. In particular, as stressed by different authors [14, 15, 16a], the assumption, sometimes made, that the parameter f is 1 does not seem acceptable. We claim that such an approach is compatible with the GvOC theory if some simple hypotheses are made, as illustrated in [16b].

The experimental difficulties in the application of the Fowkes model are partially explained by the eventual measurement of contact angles at different temperatures needed to evaluate the role and the importance of the entropic component of the interaction. In contrast, the GvOC approach is based on contact angles measurements made at only one temperature. This is one of the reasons for its success in terms of numbers of published papers and researchers who tried to apply it to practical situations. This wide range of applications has produced many troubles and problems, which will be fully analysed in this paper.

The considerations below are a short illustration of these problems:

- The analysed surfaces appear to have a basic component systematically greater than the acidic one, so they generally "seem" much more basic than acidic; van Oss has repeatedly supported the idea [7, 17] that this result reveals a universal property of solid surfaces, a conclusion very difficult to accept.
- The results strongly depend on the choice of the liquids employed for the contact angle measurements; this is probably the most important and common criticism raised regarding the theory. It should be noted that an analogous criticism was

levelled at the two-component theory and more than 10 years ago an answer [18] was given to this question in terms very similar to those we are now presenting;

- The unknowns of the calculations are, in practice, the square roots of surface energy components and in some cases they turn out to have negative values. Is this negativity a reproducible fact and if yes, what is its meaning?
- The experimental methods adopted to evaluate the acid-base components of the test liquids are experimentally questionable and, to our knowledge, they have not been fully tested in different laboratories. Actually, in our opinion such methods could be replaced by different and more rational procedures.
- In the majority of papers the contact angles used to compute the acid-base components are the so-called "static" or the advancing angles, although there is the common acceptance that the receding angles have the same scientific relevance and that they are mostly significant with respect to the higher energy portion of a surface [19–20].

In the last two-three years an increasing number of researchers investigated these problems [21–30], showing thus the great interest for a more precise theory of solid interfaces.

Nevertheless, as the measurement of the contact angle, although apparently simple as an experimental technique, is actually difficult to be correctly performed, as well the application of the GvOC theory seems straightforward, but hides some insidious mathematical aspects which must be fully revealed to eliminate errors and misunderstandings.

The removal of these difficulties is the only aim of the present paper.

It is very important to note that we will use here the GvOC theory in the same form which was presented by its authors. We introduce here no new concept; e.g., in the GvOC theory the role of the spreading pressure π_e was not considered, so we have not considered it again. We point out, however, that no well-accepted experimental evaluations of spreading pressure are available even for common liquid-solid pairs and that the few available data suggest that it becomes important only when the contact angle is very low, near zero. When acceptable values will be measured it will eventually be possible to introduce them in the general equations of the theory.

Up to now in the surface energetics theories no author has used explicitly the experimental values of the spreading pressure (as part of the known terms of the equations used) or has tried to develop an acid-base formulation of the term π_e.

As a general rule, it is impossible to introduce in the *present* formulation of the GvOC theory a new *ad-hoc* term for each solid-liquid pair, because this will introduce out of control number of variables; one should use the *already* existing variables, introducing, so, a function $\pi_e = \pi_e(\gamma^+, \gamma^-, \gamma^{LW})$, whose formulation is out of the scope of the present paper.

We agree that the existence of spreading pressure can be invoked as a reason for the problems of the acid-base theory, but we are waiting for more solid experimental measurements. This paper is devoted to the analysis of problems of acid-base theories which are well documented and *do not depend* on the relevance of the spreading pressure. When proper values of spreading pressure will be measured they will be introduced, without problems, in the present proposed *correct* mathematical formulation of the acid-base theory.

This paper does not present a new theory, but only points out some misuses of the GvOC theory itself.

To criticize the GvOC theory is obviously acceptable, if one does not accept some starting point of the theory, which is again acceptable. *We do not accept that people use the theory in a wrong manner, obtaining wrong results and give to these errors the justification that the theory is wrong.*

The reader of this paper should not criticize us for the concepts introduced by the authors of GvOC theory, but only for *our* errors.

2. THE GvOC THEORY: BACKGROUND

2.1. The Mathematical Background of the GvOC Theory: Some Definitions

In two recent papers [31, 32] it has been emphasized that from a mathematical point of view the GvOC theory can be classified in the realm of Linear Free Energy or Solvation Energy Relationships (LFER or LSER) [33, 34], where a thermodynamic quantity F_{th}, inherent to Lewis acid-base properties of two materials X and Y, is expressed as a sum of products of n different coefficients taken two by two, i.e.,

$$F_{th} = \sum_{i,j} X_i Y_j \tag{6}$$

F_{th} can be a free energy or an enthalpy, but this is not mandatory. The index i or j refers to the class of the coefficient: dispersive, acidic, basic, etc. (In an alternative formulation, the X and Y refer to the class of the coefficient, while the index indicates the material).

These kinds of relations are widely employed in physical organic chemistry [33, 34], as is done in the acid-base theory by Drago [6]. It is very important to clarify some aspects of Eq. (6).

The theory developed by Drago considers the reaction enthalpy between acids and bases to which covalent and electrostatic components are attributed, i.e.,

$$-\Delta H = E_A E_B + C_A C_B \tag{7}$$

(The coefficients E and C refer to the electrostatic and covalent components, respectively, and the indexes A and B to the interacting materials); but independently of the particular symbols used, the components pertinent to the two materials A and B play an equivalent role in the relationship. In the GvOC equation this corresponds to considering the X and Y coefficients as the square roots of the surface free energy components.

Notice, however, that the Drago equation exhibits no "complementary" relation between different kinds of coefficients; electrostatic and covalent terms are multiplied by the similar term, as happens in GvOC theory for the dispersive components.

On page 75 of a review published in 1993 [34], in Equation (13) Abraham clearly defines the application of linear solvation energy relationships (LSER) to the cases of gas/condensed-phase and condensed-phase/condensed-phase experiments:

$$\log SP = c + rR_2 + s\pi_2^H = a\alpha_2^H + b\beta_2^H + l\log L \tag{8}$$

in which the dependent variable $\log SP$ refers to some property of a series of solutes in a fixed phase (or phases), c is a fitting constant, and each product on the right-hand side represents specific solute-condensed phase interactions, i.e. dispersion, dipole-dipole, solute acid-solvent base and solute base-solvent acid interactions, respectively; r measures the tendency of the phase to interact through π- and n-electron pairs and R_2 is a solute excess molar refraction, s is the phase dipolarity-polarizability and π_2^H is the solute dipolarity-polarizability, a is the phase hydrogen-bond basicity and α_2^H is the solute hydrogen-bond acidity, b is the phase hydrogen-bond acidity and β_2^H is the solute hydrogen-bond basicity, l is a measure of the ability of the phase to distinguish between or to separate homologues in any homologous series and L is the solute gas-liquid partition coefficient. For processes within a condensed phase the various constants will reflect differences between the phases. (The reader should note that in the case of contact angles three phases co-exist, and, generally, one is gaseous and two are condensed-phases, or all three are condensed, so this is the most complex case.)

In particular, the acid-base quantities follow the same combination rules as GvOC theory: the acidic component of a material, α_2^H, combines with the basic one a and the basic component β_2^H of the same material with the acidic one b of the other one. Moreover the coefficients of the linear combination reflect the properties of the two phases. If one of the "phases" is really a complex system (e.g. in the case of partition coefficient experiments, one has two condensed-phases between which a third component is divided) then the coefficients are inherent to this system. However, there is no apparent reason to assign different physical dimensions to the a, b and α_2^H, β_2^H, parameters; it is obvious that both of them have physical dimensions equal to the square root of the dimension of their product, i.e., the left-hand side quantity. This equivalence is explicitly confirmed by Abraham in the paper mentioned above.

In our case this concept should be used by considering that the interaction is between *two phases*; as a consequence the LSER symbols a and α_2^H and b and β_2^H refer to the properties of these two phases, with an equivalent role in the expression as happens in the case of the gas/condensed-phase interaction. In conclusion, our parameters have as physical dimensions the square roots of an energy per unit surface.

It is worthy of note that in many applications and papers about LFER and LSER, their authors consider a series of materials in a specific experiment, in which a "reference" material interacts with different ones. In these cases the coefficients pertaining to the "reference" material can apparently play a different role with respect to those of the analysed material. As a consequence, a and b quantities are considered as having physical dimensions different from α_2^H and β_2^H. But this is not proven.

Our final conclusion is that with reference to the LSER or LFER theories, the thermodynamic quantity in the GvOC theory is the work of adhesion and the quantities X and Y are the square roots of the surface free energy components; the index i or j denotes the class of the component (acidic, basic, dispersive).

This suggests that any attempt to establish a correlation between an acid-base set of GvOC components and the Abraham solvatochromic scale (or any other scale defined in terms of an LFER) must compare the $\sqrt{\gamma^+}/\sqrt{\gamma^-}$ and α_2^H/β_2^H ratios for the reference liquid (e.g. water). From this point of view the choice made by Lee [31] of considering the ratio between the acidic and basic components and not between their square roots is not justified and may probably be incorrect, regardless of the eventually good results obtained in the calculations.

While the adoption of a simple geometric mean rule for "dispersive" components has a long history, it seems a more original idea of the GvOC theory to introduce the geometric mean of "complementary" components for acid-base coefficients. Note, however, that such a choice already occurs in the Abraham equation (8) [34].

It is easy to convince oneself that an equation like (6) is genuinely nonlinear, when both X_i and Y_i are unknown. In contrast, if the parameters X_i or Y_i are available the expression reduces to a linear form.

In the latter case, i.e. if we know the acid-base coefficients of the test liquids, we can consider the GvOC theory as a good example of LFER or LSER applied to the work of adhesion.

The situation changes if we examine the nonlinear occurrence where the coefficients of both the liquids and the solids are not known. In this case, indeed, while the standard LFER equation set does not contain any further equation, the application of the GvOC theory must take into due account also the definitions of the surface free energies

$$\gamma_{l,i}^{Tot} = \gamma_{l,i}^{LW} + 2\sqrt{\gamma_{l,i}^{+}\gamma_{l,i}^{-}} \qquad \gamma_{s,j}^{Tot} = \gamma_{s,j}^{LW} + 2\sqrt{\gamma_{s,j}^{+}\gamma_{s,j}^{-}} \tag{5b}$$

which appear to play the role of an independent set of equations, not deducible directly from the definition of the work of adhesion and the Young equation. Thus a more complete equation set must be considered [32].

In a nonlinear application of the GvOC theory, it does not appear legitimate to use a reduced set of equations like the so-called "grand equations" adopted by Good in Ref. [28b]. In fact they contain only the work of adhesion relationships and, in an implicit form, the definition of the surface tensions of the liquids. The same importance must be given to the definition of the surface free energy of the solids and its eventual neglect should require a justification. If no information is available on the surface free energy of the solids, a simple argument shows that actually the surface free energy relationships for the solids do not affect the best-fit solutions and may, therefore, be ignored. Good's opinion that the results obtained in [28b] are unsatisfactory could be justified also from this arbitrary simplification.

It is, however, still more important to criticise the basic idea of Ref. [28b] that it is possible to solve with some mathematical trick the equation system of the GvOC theory under conditions which eliminate the necessity of setting a scale of acid-base strength. This is not possible and the argument will be further analysed in the following.

2.2. The Mathematical Background of the GvOC Theory: Some Problems

We have, therefore, a set of Equations (5a–b) which can be considered as linear only if we know or can measure the LW, acidic and basic components of liquids. The original proposal of the GvOC theory is that it is possible to measure experimentally the liquid components, on a scale where water is chosen as a reference with equal acidic and basic components. This assumption is crucial.

The adoption of a reference scale of acid-base strength where water is perfectly neutral pushes all the results towards a certain direction; all of them will crucially depend on this choice, and thus the GvOC scale can be used only as a relative scale. It will be meaningful to compare the relative strength of electron-donicity (or the electron-accepting ability) of two materials, or of the same material after a surface modification, *but a direct comparison of the acidic and basic parameters of the same material makes no sense* and it is not possible to attribute to this comparison any chemical meaning. *This is equivalent to saying that a value of the acidic component greater than the basic one ($\gamma^+ > \gamma^-$) does not necessarily characterise a surface as acidic, and, of course, a similar consideration also applies to the case of a predominant basic component $\gamma^- > \gamma^+$. The calculated values of acidic and basic components are relative to those of water.*

It seems now natural to ask what is the actual acid-base character of common water, in order to introduce a more appropriate ratio γ^+/γ^-, with the aim to obtain an absolute scale, which allows to compare the acidic and basic components of the same material.

In terms of Brønsted theory pure water is neutral for stoichiometric reasons, but with respect to Lewis theory the situation is different. The question allows no simple answer and unless suitable experimental measurements or ab-initio calculations are performed, there is no unquestionable criterion to establish whether the ability of water to accept electron clouds or hydrogen bonds prevails over its electron-donicity. For ultrapure water, a liquid very difficult to obtain and handle, the answer is certainly negative: i.e. we cannot decide on the basis of simple and straightforward chemical arguments if it is a stronger electron donor than acceptor, or a stronger hydrogen bond donor than acceptor.

Moreover, in the case of common distilled or bidistilled water, carbon dioxide, present in the Earth atmosphere, due to its high Henry constant is able to dissolve in water 30 times more than oxygen, with the result that common distilled water is much more acidic than basic, with a pH of 5.5–6, which corresponds to a $[H_3O^+]$ about 900 times greater than $[OH^-]$.

Therefore, in the light of the Lewis theory, ultrapure water is a material whose prevalent acidity or basicity is uncertain; moreover, imposing common water, which is an acid, as a neutral liquid in the GvOC scale actually corresponds to "compressing" and reducing in a very significant way the acidic strength of water and, consequently, of any other material based on this scale.

This is exactly what happens; the typical finding of the GvOC theory is that polymer surfaces are strongly basic. Such a result *does not correspond to the physical reality*, it is simply an artifact, not due to intrinsic errors in the GvOC theory, but to its incorrect use.

However, it is also true that the original application of the GvOC theory involved the use of liquids whose acidic strength was never greater than their basic one; only for water the two coefficients were assumed to be equal. This point will be discussed in greater detail in the following.

The presence of acidic functional groups on the polymer surfaces is not only suggested by the chemical common-sense, *but is also evidenced by the results of direct vacuum analyses, and by measurements of Inverse Gas-Chromatography (IGC), ζ-potential, and so forth* [4, 21, 27, 29]. It is not possible to defend the idea that "all" materials have electron-donor surfaces. Moreover, in some cases, as corona [28a] or flame treated [30] polymers, the acidic component may be as high as the basic one or even higher by using the original GvOC scale of acid-base strength, with water $\gamma^- = \gamma^+$.

In our opinion the hypothesis, proposed by van Oss, that the prevalence of surface electron-donicity is a kind of "general law of Nature" is false.

In a recent paper [17] van Oss and coworkers develop the idea that each polar solid must have a monopolar surface, due to the effect of the so-called "residual" polar property of the solid, which is considered as the strongest one. Assuming that such a reasoning is correct, it provides no explanation of the electron donicity of polar surfaces, but only for their (eventual) monopolar character. Therefore, the predominance of the electron donicity cannot be justified in this way.

A second proposed argument is the importance of the oxygen percentage in air and in the Earth crust material. If this reasoning were literally applied to the case of materials as oxyacids, the conclusion would be that they cannot be considered acidic; e.g., $HClO_4$ contains about 64% of oxygen, but this is not sufficient to remove its acidic nature. Moreover, although in [17] all the oxides are considered basic, this is not really true; as any chemist knows, oxides may have basic, amphoteric or acidic character, depending on the atoms which react with oxygen. For instance, Al_2O_3 has an amphoteric character, because Al^{+3} is a strong Lewis acid.

The arguments invoked about sugars are suspect, or at least questionable: van Oss and coworkers report contact angles on sugar surfaces which are either soluble or at least able to be easily swelled by water [7].

It is certainly true, however, that in many cases simple treatments such as the moulding of a sample, realized near the melting temperature in air, can also oxidize the polymer surfaces, introducing basic functional groups. But this is not sufficient to conclude that surface electron-donicity on Earth is a law of Nature.

Many arguments could be invoked to justify the "electron donicity law"; e.g. the simultaneous presence of carbon dioxide and water can make the Earth atmosphere behave as a predominantly dispersive or weakly Lewis-acid environment. Thus as a polymer surface exposes to the atmosphere its dispersive groups, it could eventually expose also its basic functional groups, due to the polymer surface dynamics.

*But all these reasonings would work **IF** the occurrence of predominant electron-donicity were a real phenomenon, but as previously shown **this is not the case**.* The putative overwhelming electron-donicity is simply an artifact due to the incorrect use of the GvOC reference scale.

The previous considerations are valid in the field of "non-biological" surfaces. Bio-surfaces are actively modified and the pH control is an important factor for biological homeostasis, e.g., human skin is acidic and this factor controls in an active way the bacterial population, so in biological cases the dominance of basic components is surely false.

We analyse in a greater detail the mathematical trick which allowed van Oss [7] to perform calculations by ignoring the exact value of the ratio between the acidic and basic components of water.

It is expressed by an algebraic rearrangement of Eq. (5a): by multiplying and dividing the acidic and basic terms under square roots by the complementary term (basic and acidic, respectively) we obtain an equation of the form

$$W_{adh} = 2\left(\sqrt{\gamma_{lv}^{LW}\gamma_{sv}^{LW}} + \sqrt{\gamma_{lv}^{+}\gamma_{lv}^{-}\frac{\gamma_{sv}^{-}}{\gamma_{lv}^{-}}} + \sqrt{\gamma_{lv}^{-}\gamma_{lv}^{+}\frac{\gamma_{sv}^{+}}{\gamma_{lv}^{+}}}\right) \tag{9}$$

The factor $\sqrt{\gamma_{lv}^{+}\gamma_{lv}^{-}}$ is constant and its value is equal to $(\gamma_{lv} - \gamma_{lv}^{LW})/2$, independent of the particular value assigned to their ratio. Moreover, if we hypothesize that the ratio between the acidic and the basic components of a solid s and those of a certain liquid l is constant, the values of the work of adhesion can be obtained independently on the knowledge of this actual ratio.

In this way, we can calculate for each liquid or solid a quantity $\gamma_{sv}^{-}/\gamma_{lv}^{-}$ or $\gamma_{sv}^{+}/\gamma_{lv}^{+}$ which expresses a ratio with respect to the components of the reference material; obviously also the absolute value of the component can be obtained using a numerical choice, as for example, the equality of acidic and basic components. But this trick cannot overcome the fact that we do not know the actual ratio; all the numerical values we obtain depend on the choice made for the reference material, i.e. for water. *The ratio of the calculated components is constant, the values of some thermodynamic quantities (as the work of adhesion) can remain constant, but the agreement of the coefficients with the real properties of the material is not known, also for the reference one.* We only have a relative scale, and if the choice made for water is not realistic, this unrealistic situation occurs for all other liquids and polymers. And this is, unfortunately, the case. The values of the work of adhesion are correct, but the values used for each acidic or basic component vary with the ratio of water components. Their comparison to establish the acidic or basic character of a surface has no meaning.

Some other scales have been proposed [23–26, 31], also based on interesting mathematical considerations, although questionable [16a–b]. As elsewhere fully discussed [32], infinite scales are possible. All the scales are relative, but only using a scale in which the ratio between the acidic and basic characters of the reference material is appropriately chosen one can hope to correctly compare the acidic and basic characters of the same or of different materials.

The dependence of the calculated components on the test liquids employed is a problem already noted about the two-component theory. Some authors [18, 35] observed that the choice was very important. One of them [18] was able to show that this situation could be dealt with by using a wide set of liquids, some dispersive, some polar. The calculated best fit results, obtained by considering an overdetermined equation set, appeared to be better than those achieved from any single pair of test liquids.

The reason for such a behaviour is a property of the equation set expressing both the two-component as well as the acid-base theory (a three-component theory) of surface free energy: ill-conditioning.

Generally speaking, the ill-conditioning of a matrix system can be referred to each particular operation we are able to do on it, e.g. inversion [36, 37]. We refer here to the case of ill-conditioning in the solution of the corresponding equation set. In this particular case, ill-conditioning means that a slight change in the coefficients of the equations gives rise to a large modification of the solutions. In other terms, the ill-conditioned linear equation set is sensitively dependent on the choice of its coefficients.

The classic mathematical instrument for this analysis is the condition number, corresponding to the operator cond(). Given an equation system $\mathbf{Ax} = \mathbf{b}$ and remembering that the norm of a matrix is a non-negative scalar function of the matrix (or of the vector) satisfying a certain set of axioms, cond() is defined as the product of the norm of the matrix $\mathbf{A}$, $\|\mathbf{A}\|$, times the norm of its inverse $\|\mathbf{A}^{-1}\|$, i.e. $\mathrm{cond}(\mathbf{A}) = \|\mathbf{A}\|\,\|\mathbf{A}^{-1}\|$. Solving the system $\mathbf{Ax} = \mathbf{b}$ with a procedure which yields, for some reason (inaccuracy in calculation, round-off, errors in the quantities, etc.) a residual vector $\mathbf{r} = \mathbf{Ax} - \mathbf{b}$, the relative error in the solution $\mathbf{x}$ is [37]

$$\frac{\|\Delta\mathbf{x}\|}{\|\mathbf{x}\|} \leqslant \|\mathbf{A}\|\,\|\mathbf{A}^{-1}\|\,\frac{\|\mathbf{r}\|}{\|\mathbf{b}\|} \tag{10}$$

The condition number appears to be the parameter able to give us a correct and full information about ill-conditioning.

To show these concepts practically at work we can analyse the case of the use of three liquids in the GvOC theory. Three is the minimum number of liquids required

Table 1a.
Condition numbers for different triplets of liquids calculated from two different norms

liquid 1	liquid 2	liquid 3	C_n from norm 1	C_n from norm s
glycerol	methylene iodide	hexadecane	0.0000	0.0000
glycerol	methylene iodide	α-bromonaphthalene	0.0000	0.0000
methylene iodide	pyrrole	hexadecane	0.0000	0.0000
methylene iodide	DMSO	hexadecane	0.0000	0.0000
methylene iodide	bromoform	hexadecane	0.0000	0.0000
α-bromonaphthalene	bromoform	hexadecane	0.0000	0.0000
water	methylene iodide	hexadecane	0.0000	0.0000
water	methylene iodide	ethylene glycol	6.2821	33.179
water	methylene iodide	DMSO	7.1909	33.312
water	ethylene glycol	α-bromonaphthalene	6.1124	33.404
water	α-bromonaphthalene	DMSO	7.0033	33.599
water	ethylene glycol	hexadecane	5.6486	34.746
water	DMSO	hexadecane	6.8258	35.203
water	glycerol	methylene iodide	6.2996	35.999
water	glycerol	α-bromonaphthalene	6.1337	36.243
glycerol	ethylene glycol	DMSO	118.30	474.04
water	formamide	DMSO	119.84	612.22
formamide	ethylene glycol	pyrrole	133.07	658.90
glycerol	formamide	hexadecane	161.93	689.17
glycerol	formamide	α-bromonaphthalene	175.69	695.56
glycerol	formamide	methylene iodide	180.04	699.18
DMSO	bromoform	pyrrole	210.82	801.56
water	glycerol	ethylene glycol	193.27	910.67

to obtain in the linear case the components of a solid, and in the literature this way of applying the GvOC theory is the most common one, generally based on the liquid components originally proposed by Good and van Oss.

So for each triplet of liquids we can calculate the condition number and easily evaluate which triplet gives rise to the least ill-conditioned set of equations.

Some results are shown in Table 1a and in a graphical form in Fig. 1. What is more interesting is to extract the chemical rationale for these results.

The use of two dispersive liquids is forbidden and corresponds to a singular matrix; this sounds chemically obvious because one polar liquid only is not sufficient to determine *two* acid-base components (this resembles to the case of one equation with two unknowns).

If one does not use any dispersive liquids, this gives the worst results, because it is more difficult to evaluate the LW component. A triplet must contain one prevalently dispersive and two prevalently polar liquids.

The choice of the two polar liquids should be directed to liquids with opposite and complementary character: a pair of liquids with one prevalently acidic and the other prevalently basic in character is better than two liquids of the same polarity. A

Table 1b.
Some condition numbers of different quintuplets, obtained by adding proper liquids to improper liquid triplets

quintuplet	C_n from norm 1
Water, Glycerol, Ethylene glycol + bromoform	improper triplet 193.27
and pyrrole	corrected quintuplet 38.2
Glycerol, Formamide, Hexadecane + Water	improper triplet 161.93
and α-Bromonaphthalene	corrected quintuplet 39.7
DMSO, Bromoform, Pyrrole + Water and	improper triplet 210.82
α-Bromonaphthalene	corrected quintuplet 40.3

lower condition number is the result of a "complementary" choice. The commonly used triplet water (W), methylene iodide (MI) and glycerol (G), is one of the best choices. (An important warning is, however, that the chemical character of the liquids depends on the reference scale used; water and glycerol appear sufficiently unlike or "complementary" in the arbitrary scale chosen by the GvOC, but the situation can change in a different scale.)

In the graphic representation, using the triangular diagram, (as for three-component systems), on each of the axes one of the three properties, dispersive, acidic or basic component is reported in a relative scale (0,1), calculating its co-ordinates by dividing each component by the sum of the components. Each black dot corresponds to a test liquid and, some intervals of values for the condition number being assigned, the center of mass of each liquid triplet is marked by a different symbol according to the interval where the condition number of the given triplet lies. The best choices correspond to that part of the diagram where the acid-base character of the liquids is well balanced.

This kind of graphical representation has no fundamental meaning and it is only a graphic support to show more clearly as the scale commonly used in the GvOC theory is groundlessly shifted towards basic values. We will reconsider this point in the following.

It seems worth noting (and also wonderful for a non-mathematician) that while the function norm of a vector or of a matrix can be defined in many different ways providing different values of the condition numbers, the adoption of a certain norm instead of another does not alter the ordering of the results. The linear correlation among the values of the condition numbers obtained from two different norms is shown in Fig. 2 as an illustration.

In the light of the previous considerations one can better evaluate the significance and the limitations of the choice of a certain liquid triplet.

In a recent chapter by Neumann [3] the criticism towards the acid-base theory is based on the analysis of this aspect of the theory; the results obtained from different triplets are shown to be extremely different. Neumann and coworkers use exactly those triplets which have the largest condition numbers, i.e., choosing triplets in which no dispersive liquid is present. So it is quite obvious that the final results are

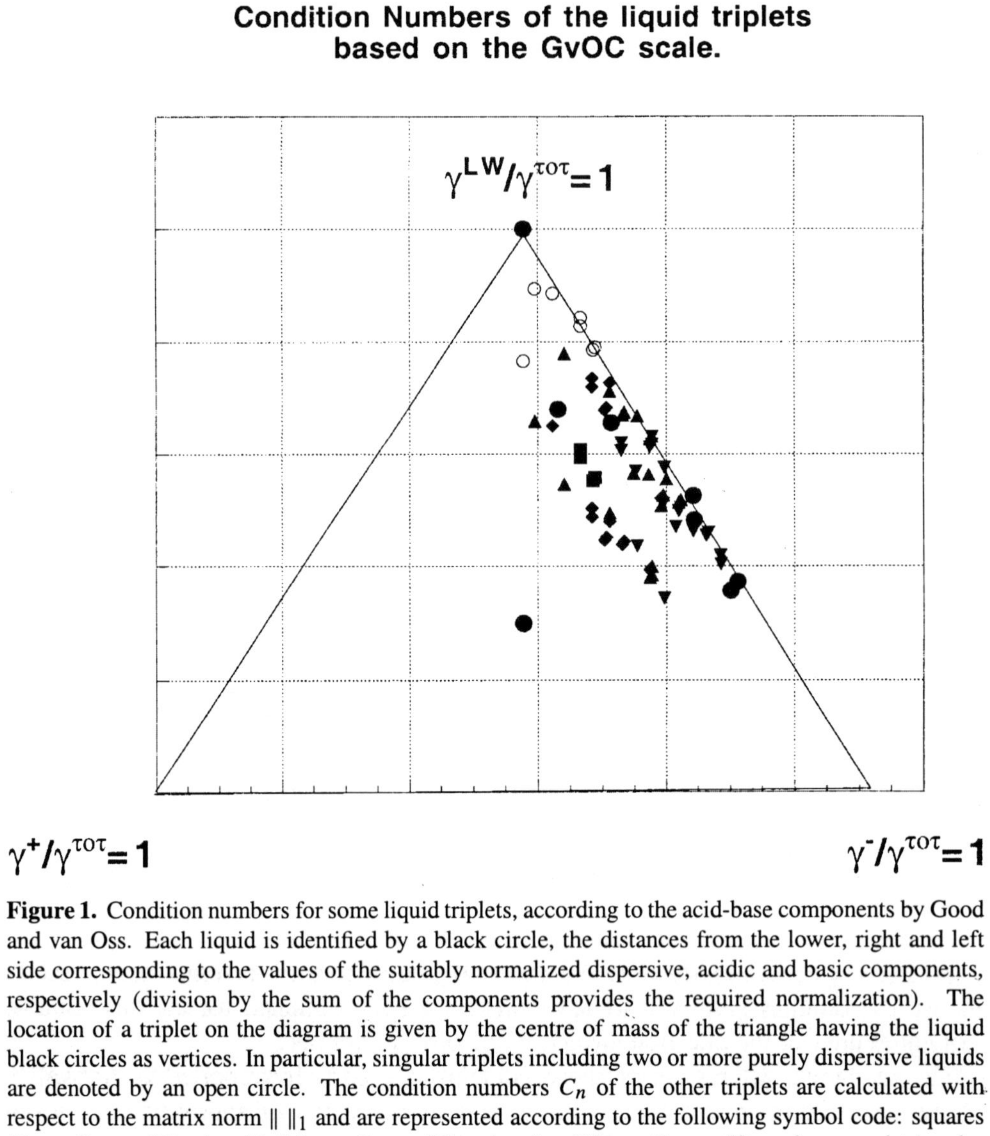

Figure 1. Condition numbers for some liquid triplets, according to the acid-base components by Good and van Oss. Each liquid is identified by a black circle, the distances from the lower, right and left side corresponding to the values of the suitably normalized dispersive, acidic and basic components, respectively (division by the sum of the components provides the required normalization). The location of a triplet on the diagram is given by the centre of mass of the triangle having the liquid black circles as vertices. In particular, singular triplets including two or more purely dispersive liquids are denoted by an open circle. The condition numbers C_n of the other triplets are calculated with respect to the matrix norm $\| \; \|_1$ and are represented according to the following symbol code: squares ($5 < C_n < 10$), rhombi ($10 < C_n < 20$), triangles ($20 < C_n < 50$) and reversed triangles ($50 < C_n < 500$). The coordinates of the graph should be considered as in each other triangular diagram.

the worst possible ones. Unfortunately (for Neumann), this simply means that he used the worst possible liquid triplets, making incorrect choices, and not that the acid-base theory is wrong.

These considerations can be extended to multiliquid analysis.

In this case the concept of condition number is not strictly applicable in the same way as before; the equation system corresponds to a rectangular matrix **A** and its

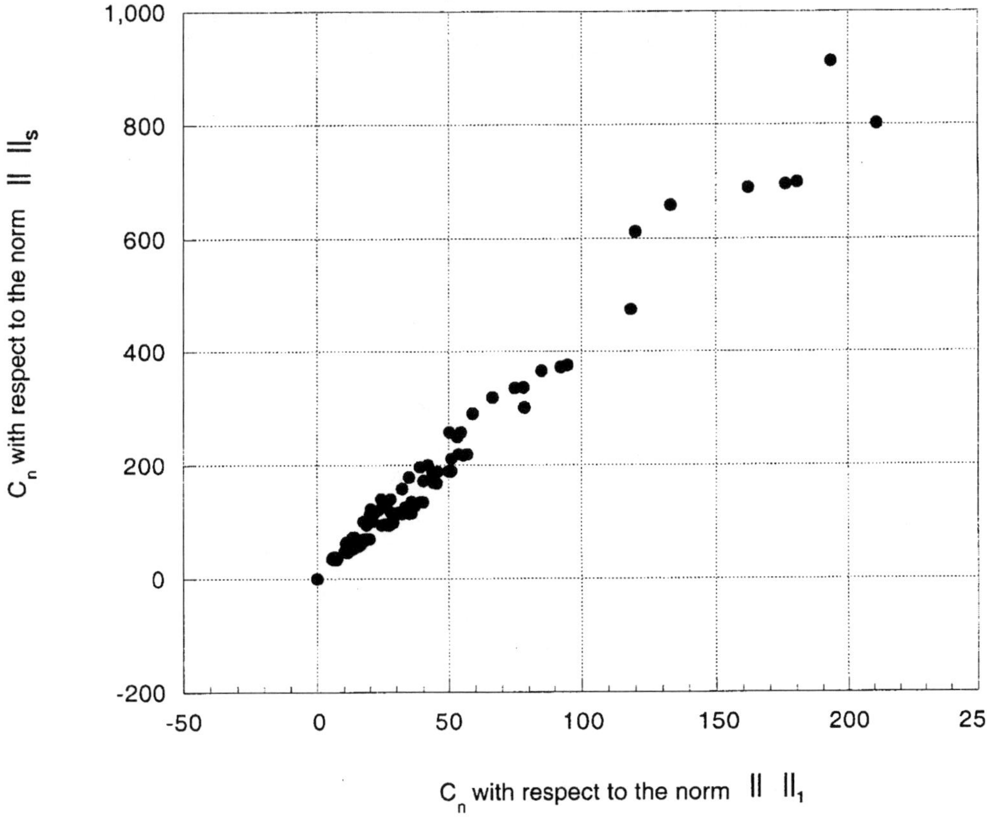

Figure 2. The linear correlation between the condition number calculated according to the matrix norm $\| \|_1$ and that obtained from the $\| \|_s$ norm. Liquid triplets with Good and van Oss values of the acid-base components.

inversion is no more unique. However, as fully explained in Section 4, it is possible to use the product $\|\mathbf{A}^{\mathrm{T}}\mathbf{A}\|$; some results have been obtained for the case of five liquids to answer the question: which is the best quintuplet?

The mathematical results of these calculations, partially reported in Table 1b, show that starting from a certain triplet and adding a pair of appropriately chosen liquids, the multiplets can be "equilibrated" to obtain lower condition numbers. However, the best quintuplets do not have lower condition numbers than the best triplets; so is there a reason to use a larger number of liquids, particularly if properly chosen?

The answer is, yes, if the new measurements are performed with a greater precision. In fact, as previously shown, the relative error in the final results (the calculated coefficients) depends on the product of the condition number times the ratio $\|\mathbf{r}\|/\|\mathbf{b}\|$ (see Eq. (10)), which depends on the accuracy of the experimental data. Generally speaking, the increase in the number of measurements corresponds to a decrease in their uncertainty; so it appears generally convenient to use a larger number of liquids. Consider, however, that only after the measurements it will be possible to answer the question: was it really convenient to increase the number of test liquids? Note that also in the paper of Dalal [18] the comparison between the results obtained from single pairs of liquids and from a complete set was performed after the measurements, as a post-factum conclusion. On the contrary, in the case of triplets our conclusion has been reached on a general basis, independent of the actual measurements, and thus assuming that the weight of the term $\|\mathbf{r}\|/\|\mathbf{b}\|$ was equal in each case. The condition number gives us the possibility of an evaluation of the potential ill-conditioning of a certain liquid triplet choice but, obviously, the experimental accuracy remains extremely important. If one performs measurements with high standard deviation even using a low condition number liquid triplet, his incorrect data will unfavourably affect the results.

Based on the previous discussion about the eventual ill-conditioning of the acid-base equations, the occurrence of occasionally negative estimates for the square roots of the acid-base components, and in particular of the acidic one, can be considered as a numerical artifact, in agreement with the considerations by van Oss and coworkers [17]. It is noticeable that by applying the previous mathematical methods and using an overdetermined set of equations, it is easily possible to evaluate the standard deviation of the calculated components. In the analysed cases, whenever a square root is negative *its standard deviation is always greater than the value itself*; this means that the sign of the quantity is not certain and that the quantity can be considered as zero.

In principle, however, we cannot exclude that the square root of a certain acid-base component can be negative. Thermodynamics requires only the positive sign of the total value of the surface free energy for stability reasons, but this does not prevent acid-base components from being negative, which will be the direct consequence of the existence of a negative square root of a single component.

As a pure mathematical speculation, one can also suppose that the square roots of both the acidic and basic components are negative; the physical meaning of these negative square roots is not clear at the moment and we prefer to consider it only when its experimental existence has been shown with a greater data accuracy.

A great cleverness has been applied in developing the following two experimental methods necessary to the evaluation of the acid-base components of liquids by the GvOC theory. It is our opinion, however, that they are experimentally difficult to perform and their interpretation is questionable.

The "encased gel method" consists in the analysis of a surface of agarose gel encasing the liquid of interest. As an example, in Ref. [38] the gel is produced by water and agarose at three concentrations of water (from 9.65 to 6.85%), and it is then immersed in the liquid of interest for a long time; along the experiment the liquid is systematically replaced and controlled to detect the possible presence of exchanged water, until all traces of water have disappeared.

No analysis, except contact angle measurements, has been performed to characterize the surface; the presence of residual water in the gel cannot be excluded by measurements done in the solution. Finally, only three experimental points, collected at a significant concentration of material, have been used and then fitted with a straight line, by assuming a linear trend of the phenomenon until very low concentration value: this is a "strong" and not justified hypothesis.

The "polar solid method" consists in the measurements of contact angles on polar solids considered as monopolar. Among these solids we find poly-(ethyleneoxide) (PEO), which is soluble or swellable by many polar liquids, protein surfaces and nylon 6.6, materials which intrinsically contain both acidic and basic sites. All these materials have been assumed to be monopolar basic by the GvOC theory.

Many doubts can be raised about the effective monopolarity of these solids chosen as monopolar references. We are not aware of other laboratories which tried to reproduce these results.

In our opinion these methods represent an ingenious example but cannot give a solid foundation to the acid-base theory.

The question is: Are these experimental measurements necessary?

In the general application of the LFER or of Drago equations the criterion has been completely different and the answer to the question is: *no, these experiments are not necessary*.

If one applies the equation of the GvOC theory in a nonlinear form, i.e. considering as unknowns the components of both solids and of liquids, and, if the number of test liquids is sufficiently large, the equation system becomes overdetermined. The situation is essentially equivalent to the application of the least squares method. We cannot obtain an exact solution, but we can have a very good one.

For a system composed of L liquid and S solids, we can have L + S + LS independent equations and 3L+4S unknowns (this assertion can be easily justified).

For each solid or liquid we write down Eq. (5b) defining its surface free energy or its surface tension; then we write down Eq. (5a) expressing the work of adhesion of each pair, for a total of LS equations. The number of unknowns is three for each liquid, because the total value of its surface tension is easily measurable. In the case of the solids, we have no well accepted possibility to perform independent measurements of the total surface free energy, which is, therefore, an unknown as the others. This conclusion can be difficult to accept, but consider that if the equation system allows no exact solution, then each equation is not equal to zero; correspondingly, the definition of the solid surface free energy, Eq. (5b), becomes

$$\gamma_{s,j}^{Tot} - \gamma_{s,j}^{LW} + 2\sqrt{\gamma_{s,j}^{+}\gamma_{s,j}^{-}} = C \tag{11}$$

where C is an arbitrary value, minimised through the application of the least squares criterion.

Although these concepts have been fully developed in previous papers [16a–b], it will be useful to summarize them here and give some preliminary results in the following section.

It is worth noting that some researchers have tried to find some reduction process and some appropriate choice with respect to the number of equations and unknowns to obtain a "non-overdetermined" equation set.

How many liquids and solids are necessary to achieve such a goal?

In the following table we show the number of equations and unknowns involved when one decides to adopt

(a) our point of view, the most general one, which assigns 4 unknowns to each solid;

(b) a second point of view which assigns to each solid only 3 unknowns, and

(c) the reduction method, attempted by Good, to calculate independently the LW components of solids and liquids and using then only the equations for the work of adhesion in which the basic components of the liquids have been substituted by the ratio of the total acid-base component times the acidic one, i.e.,

$$W_{adh} - 2\sqrt{\gamma_{lv}^{LW}\gamma_{sv}^{LW}} = 2\sqrt{\gamma_{lv}^{+}\gamma_{sv}^{-}} + \gamma_{lv}^{AB}\sqrt{\frac{\gamma_{sv}^{+}}{\gamma_{lv}^{+}}} \qquad (12)$$

For each choice, there are some "magic" numbers of solids and liquids which lead to a determined equation set. Can this choice eliminate the necessity to set a scale? The answer is: no.

For L liquids and S solids we will have in the three cases,

No. of equations	No. of unknowns	"Magic" numbers necessary to have a determined set L	S	Notes
LS + L + S	3L + 4S	4	8	case (a)
		5	5	
		6	4	
		9	3	
LS + L	3L + 3S	4	8	case (b)
		5	5	
		6	4	
		9	3	
LS	L + 2S	3	3*	case (c) L + S supplementary equations are used to calculate the L + S LW components
		4	2	

* this set corresponds to the "grand equations" chosen by Good and Hawa [28b]

All the equations are nonlinear, so the number of solutions is not easy to evaluate; but it is possible to show that if one solution exists, then the equation set actually allows infinite solutions. This can be shown by rigorous mathematical statements [32] but also in an easier way. Indeed, due to the "complementarity rule" each acidic component is always multiplied by the basic one; Good and Hawa "grand equations" depend also on the ratio of different acidic components, but this does not affect the final conclusions.

If one has a solution constituted by the set of γ_i^+, γ_j^- for solids and liquids, one can always multiply all the acidic components by an arbitrary nonzero constant, k, and divide all the basic components by the same one, obtaining a new set $k\gamma_i^+$, γ_j^-/k, which will again be a solution of the equation system, because their products maintain the same values. k is an arbitrary positive quantity and thus infinite solutions exist. In other words, the symmetries introduced by the "complementarity rule" valid for both acidic and basic properties cannot be eluded.

We make here two final remarks.

The first one: the importance of using both the receding and the advancing contact angles in the evaluations of acid-base components. This consideration has been recently developed by Good and coworkers [28a] and by one of us [30] in the analysis of corona-treated or flame-treated polypropylene surfaces. In both cases it is evident that the use of the receding angles offers the possibility of a more correct evaluation of the acid-base components. Johnson and Dettre [20] showed about 30 years ago that the receding angles were mainly indicative of the higher energy portion of a surface, and we know that acid-base functionalities are enhanced exactly by those processes which increase the total surface free energy. The processes which are able to eventually increase the surface free energy of a material (chemical oxidation, flame treatment, corona treatment, RF-plasma treatment, etc.) introduce in it acid-base chemical functionalities, which express themselves in stronger acid-base components. Some authors [27] have empirically found that the receding contact angles data are a very good indication of the acid-base functionalities, but have not consistently considered to use this experimental observation, i.e., doing calculations not only by using advancing contant angles, but also by receding ones. This is a common mistake and it is time to correct it.

The second one: some similarities can be found between the conceptual structure of the acid-base theory and the electrochemical definition of the mean-ionic activity coefficient. This consideration can help to answer recent criticisms of some EOS theory defenders about the theoretical impossibility to measure the acidic or basic components of surface free energy [39]. Also in the case of electrochemistry, in fact, such quantities as the activities or the activity coefficients of a single ion are not susceptible to a direct measurement; it is conceptually and/or experimentally impossible to measure them directly. On the contrary, their product is an experimentally accessible quantity.

Also in the case of electrochemistry, however, some reference conditions are necessary to develop a scale of ionic activity or activity coefficients.

This situation is parallel to that of acid-base evaluations of solid surface properties.

So in the electrochemistry it is not possible to prove on a purely thermodynamic basis that the chemical potential of an ionic solution is the sum of a portion depending on the positive ions and a portion depending on the negative ones; it is considered as a postulate, which can only be "proven" by its exact predictions.

In the case of the acid-base theory, the relation between the acid-base components and the work of adhesion or the surface free energy are working hypotheses; on their basis, it is possible to calculate a consistent set of acid-base parameters.

3. SOME SUGGESTIONS FOR A MORE CORRECT ACID-BASE REFERENCE SCALE

The complete set of equations we have to consider is the following

$$\begin{aligned}
&\gamma_{l,i}^{Tot} = \gamma_{l,i}^{LW} + 2\sqrt{\gamma_{l,i}^{+}\gamma_{l,i}^{-}} \qquad \forall i = 1, 2, \ldots, L \\
&\gamma_{s,j}^{Tot} = \gamma_{s,j}^{LW} + 2\sqrt{\gamma_{s,j}^{+}\gamma_{s,j}^{-}} \qquad \forall j = 1, 2, \ldots, S \\
&\gamma_{l,i}^{Tot}(1+\cos\theta_{ij}) = 2\left[\sqrt{\gamma_{l,i}^{LW}\gamma_{s,j}^{LW}} + \sqrt{\gamma_{l,i}^{+}\gamma_{s,j}^{-}} + \sqrt{\gamma_{l,i}^{-}\gamma_{s,j}^{+}}\right] \qquad \forall i, j
\end{aligned} \tag{13}$$

where L and S are the number of liquids and solids respectively, and $\gamma_{l,i}^{Tot}$, $\gamma_{s,j}^{Tot}$, θ_{ij} are assumed to be known experimentally. Owing to the intrinsic approximate validity of the model and to the inaccuracy of experimental data, there is actually no hope to solve the nonlinear system (13) exactly. We shall use instead some numerical best fit algorithm in order to globally minimise the differences, i.e.,

$$\begin{aligned}
&\Delta_{l,i} = \gamma_{l,i}^{LW} + 2\sqrt{\gamma_{l,i}^{+}\gamma_{l,i}^{-}} - \gamma_{l,i}^{Tot} \qquad \forall i = 1, 2, \ldots, L \\
&\Delta_{s,j} = \gamma_{s,j}^{LW} + 2\sqrt{\gamma_{s,j}^{+}\gamma_{s,j}^{-}} - \gamma_{s,j}^{Tot} \qquad \forall j = 1, 2, \ldots, S \\
&\Delta_{ij} = 2\left[\sqrt{\gamma_{l,i}^{LW}\gamma_{s,j}^{LW}} + \sqrt{\gamma_{l,i}^{+}\gamma_{s,j}^{-}} + \sqrt{\gamma_{l,i}^{-}\gamma_{s,j}^{+}}\right] - \gamma_{l,i}^{Tot}(1+\cos\theta_{ij}) \qquad \forall i, j
\end{aligned} \tag{14}$$

It is useful to rewrite the previous relations (14) in a more compact and symmetric form by introducing the matrix notations below

$$X_i = \begin{pmatrix} \sqrt{\gamma_{l,i}^{LW}} \\ \sqrt{\gamma_{l,i}^{+}} \\ \sqrt{\gamma_{l,i}^{-}} \end{pmatrix} \qquad Y_j = \begin{pmatrix} \sqrt{\gamma_{s,j}^{LW}} \\ \sqrt{\gamma_{s,j}^{+}} \\ \sqrt{\gamma_{s,j}^{-}} \end{pmatrix} \qquad \begin{aligned} &\forall i = 1, \ldots, L \\ &\forall j = 1, \ldots, S \end{aligned} \tag{15}$$

which leads to the following equivalent set

$$\begin{aligned} \Delta_{l,i} &= X_i^T R X_i - \gamma_{l,i}^{Tot} \qquad \Delta_{s,j} = Y_j^T R Y_j - \gamma_{s,j}^{Tot} \\ \Delta_{ij} &= X_i^T R Y_j - \frac{1}{2}(1 + \cos\theta_{ij})\gamma_{l,i}^{Tot} \end{aligned} \tag{16}$$

by denoting with the apex ^{T} the transpose of any matrix and with R the 3×3 symmetric orthogonal matrix

$$R = \begin{pmatrix} 1 & 0 & 0 \\ 0 & 0 & 1 \\ 0 & 1 & 0 \end{pmatrix}. \tag{17}$$

The problem of scales arises because of the invariant properties revealed by Eq. (16) under linear transformations of the form

$$X \longrightarrow CX \qquad Y \longrightarrow CY \tag{18}$$

provided that the nonsingular matrix C satisfies the condition

$$C^T R C = R. \tag{19}$$

A very simple example of such matrices is, of course, the following

$$C = \begin{pmatrix} 1 & 0 & 0 \\ 0 & k & 0 \\ 0 & 0 & 1/k \end{pmatrix} \tag{20}$$

with any $k \in \mathbb{R}^+$, which corresponds to the example shown in the previous section, i.e., choosing to multiply each acidic component and to divide each basic component by the same arbitrary constant k. But it is easy to show that there actually exists an uncountable group G_3 of real 3×3 matrices C which obey Eq. (19) and can also be explicitly computed; in particular, G_3 turns out to be isomorphic to the real orthogonal group $O(2,1;\mathbb{R})$ [32].

This means that if X_i, $i = 1, 2, \ldots, L$, Y_j, $j = 1, 2, \ldots, S$ is a set of column vectors for which the differences $\Delta_{l,i}$, $\Delta_{s,j}$ and Δ_{ij} in expressions (14) take some assigned values, then there is a wide choice of matrices C such that the vectors CX_i, $i = 1, 2, \ldots, L$, CY_j, $j = 1, 2, \ldots, S$ leave each of the differences $\Delta_{l,i}$, $\Delta_{s,j}$, Δ_{ij} invariant when replaced with X_i, Y_j. The immediate consequence is that for any given best-fit solution X_i, $i = 1, 2, \ldots, L$, Y_j, $j = 1, 2, \ldots, S$ of expressions (14) there exists a "(triple) infinity" of unlike best-fit solutions CX_i, $i = 1, 2, \ldots, L$, CY_j, $j = 1, 2, \ldots, S$ formally as good as the first one, since invariance of the differences in expressions (14) implies invariance of any merit function one may adopt to apply the best-fit algorithm. The multiplicity of possible scales is in some sense unsatisfactory for practical purposes and must be removed. It is then necessary to establish some criterion which allows a particular scale to be uniquely selected among the possible ones. According to the previous discussion, this is equivalent to specifying the additional requirements on the acid-base components which must be introduced to completely remove the invariance properties of the set (13).

Let us assume to have solved the overdetermined equation system in a satisfactory way for a certain set of donors and acceptors, or liquids and solids. We wish to fix the components of one or more materials in order to eliminate any invariance property and define a specific scale. The task is quite delicate.

If we assume the components of a particular liquid have been calculated, any other set of γ^{LW}, γ^{+}, γ^{-} for the same material must be obtained through a transformation of the invariance group. As $O(2,1;\mathbb{R})$ essentially depends on a triplet of parameters, we can conclude that whenever the components of a reference liquid are assigned, the global best-fit solution is uniquely determined and thus the scale is selected. However, we not only have the problem to choose values in such a way that the three imposed conditions are independent of each other, but we also need to check that the transformation matrix C actually exists in $O(2,1;\mathbb{R})$, or equivalently that it satisfies Eq. (19). The invariance of $X^T RX$ obviously provides a necessary condition to fulfill the last requirement.

Considering our previous example among the infinite solutions we choose that particular one in which the reference liquid (or solid, if we were able) has the measured LW component and a correct acid to base ratio.

In any case we emphasize that the reference liquid and its components must be chosen "judiciously", because a wrong choice could be inconsistent with the experimental data.

Regarding the possibility to estimate the errors in the coefficients we can state that the location of minimum for the nonlinear best-fit function is obviously influenced by our knowledge of constant parameters, which stem from experiment. Errors in the estimate of parameters result in a certain degree of incertitude about minimum position. Owing to nonlinear coupling of variables, a change in even a single parameter will yield a global rearrangement of minima, which will generally be all modified, and so will therefore be the estimated values of acid-base components for all the chemical species involved.

This problem will be considered in detail in the following.

Any liquid can be chosen as a reference liquid, but water is a very good candidate, due to reproducibility of its properties, the purity of which can be obtained, its availability in all the laboratories of the world, etc. There is only one problem, which is particularly important with respect to our analysis: as previously indicated water easily dissolves carbon dioxide, thus becoming an acidic solution of carbonic acid, with a pH of 5.5–6. This phenomenon should be taken into account, when we choose the reference parameters of water.

In previous papers [16a–b, 32] we analysed the dependence of the acidic and basic components of solids and liquids on the numerical value of the chosen reference components for water by using a wide set of contact angles data taken from the literature. We found that it was generally true that the acidic components of solids and liquids increased with the increase of the γ^{+}/γ^{-} ratio for water; on the other hand the basic components became lower, but the variations were not so large as one would expect.

In fact the final results shown in the final tables of our first paper on the argument, [32], indicated that, while it was effectively possible to obtain chemically coherent results for the liquids, the components of solids remained predominantly basic.

For this reason we explored more extensively the domain of the variables by considering also the value of γ^{LW} for water and we have obtained a new and unexpected result. For a value of the dispersive water component of about 23–27 mJ/m^2 there was a strong variation of the calculated acid-base components of the liquids and solids we considered.

In agreement with this result, the global trend of the merit function, whose minimum gave us the acid-base components of liquids and solids, had the minimum at a value of water dispersive component of 26.25 mJ/m^2, completely different from that commonly used value of 21.8 mJ/m^2. This suggests to take a closer look at the origin of this widely accepted value.

The value of 21.8 mJ/m^2 was proposed by Fowkes in 1964 [40]; in that paper he reported measurements of the interfacial energy of water at the interface with completely dispersive liquids, with a very low scatter in the data. In the same paper he discussed also the interfacial energy of water on completely dispersive solids, but in this case the numerical values were not explicitly reported, but only plotted on a graph (Fig. 6 of Ref. [40]) at a grand scale showing that they occupied a wide interval, between 15 and 30 mJ/m^2. It would have been very difficult to recalculate the original values from the figure, so we tried to use other data from the literature to calculate the LW component of water from interactions of water with a purely dispersive solid, whose surface free energy was obtained independently from the contact angle of a purely dispersive liquid on it.

We used Equation (5a) in two modified forms, shown below: the first one to calculate the LW component of a purely dispersive solid from the contact angles of purely dispersive liquids on the solid, and the second one to compute the LW component of water from its contact angle on the same solid

$$\gamma_s^{LW} = \frac{\gamma_l(1+\cos\theta)^2}{4} \qquad \gamma_l^{LW} = \frac{\gamma_l^2(1+\cos\theta)^2}{4\gamma_s^{LW}} \tag{21}$$

It is possible to find many useful collections of data in the literature; we found 10 such sets, whose sources are cited in the "Reference" section [30, 41–44]. The results are shown in Fig. 3 and Table 2. One can see that the estimates are very different, with a large scatter but with a mean value greater than that reported for the liquid-liquid interaction, about 27.1 ± 5.7 mJ/m^2. In contrast, the value obtained by Fowkes for liquid-liquid interaction was 21.8 ± 0.2 mJ/m^2.

Is it conceivable that a liquid-liquid interaction is different from a liquid-solid one, even in the case of purely dispersive interactions, namely that the surface free energy components of a liquid (water) may take different values when it interacts with another liquid or with a solid? The answer is not easy but, in principle, could be

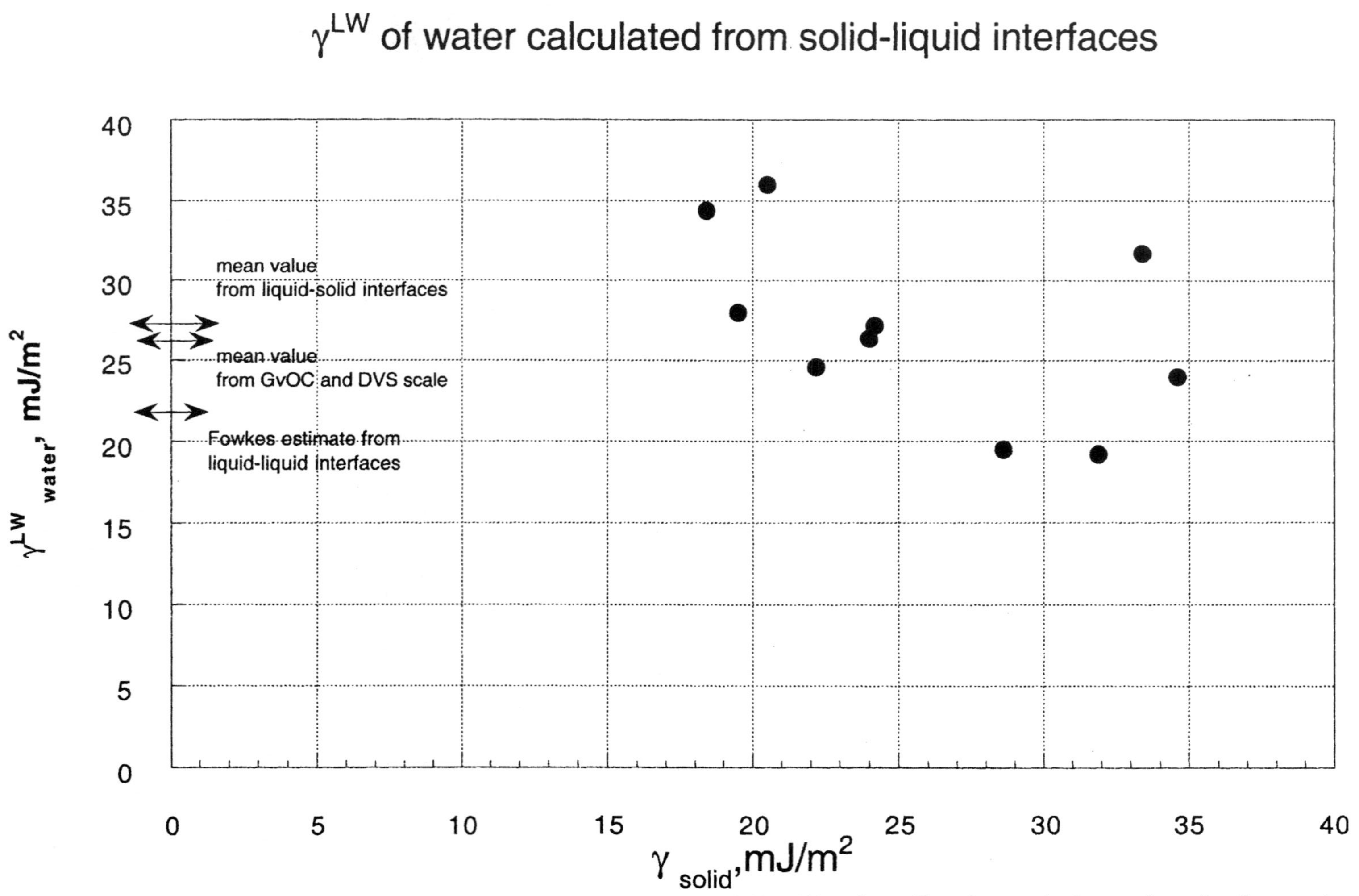

Figure 3. Estimates of the dispersive component of water from data measured on solid-liquid interfaces. The values on abscissa are the surface free energies of the solids employed for the calculation. The data sets were collected from Ref. [30, 41–44]. The mean values by Fowkes, the GvOC and DVS [16b] are

Table 2.
Water dispersive surface free energy component re-calculated using literature data obtained from solid-liquid interfaces

solid	surface free energy of the solid, mJ/m^2	estimated value of γ^{LW} of water, mJ/m^2	references
PTFE	18.4	34.4	41
hexatriacontane	19.5	28.0	41
hexatriacontane	20.5	36.0	42
PTFE	22.2	24.6	43
paraffin	24.0	26.4	41
silicone	24.2	27.2	44
paraffin	28.6	19.5	43
PP	31.9	19.2	30
PE	33.4	31.7	43
PE	34.6	24.0	41

positive; e.g. one can consider that for a solid polymer at a temperature higher than its Tg the molecular motions required to adapt its structure to the force field of water molecules are much more obstructed than in a liquid. Moreover, the geometrical shape of the interface is different, since even the flattest polymer surface cannot be as flat as a liquid surface can be. Finally one should also consider that in a liquid-liquid measurement, even for immiscible liquids, it is conceivable that diffusion and/or spreading can take place at the liquid surface, thus modifying the structure of the interface with respect to the initial one.

A similar result, with a dispersive component for water higher than that commonly accepted, has been obtained by Erbil and Meric [45] using a nonlinear approach to the calculation of the components in a two-component models. Unfortunately Erbil's conclusion, i.e. the presence of a specific contribution for each liquid-solid couple, introduces another *ad hoc* coefficient, and is disastrous for three-component models, because the corresponding set of equations depends on too many parameters and the number of possible solutions proliferates without control.

These coefficients of water can be reasonably used in the definition of a scale. We can assume a value 26.25 mJ/m^2 for the LW component of water, and 48.5 and 11.2 mJ/m^2 for the acidic and basic components, respectively. In a previous version of the scale [32] we used a higher acid-base ratio; in the light of the previous considerations it is possible to reduce it to lower value. The γ^+/γ^- ratio is now 4.35, exactly in the middle point of the interval 3.2–5.5 considered in different versions of LFER parameters calculations and also in our previous scale proposal. If these new coefficients are employed we obtain the best-fit results reported in Table 1a–b of Ref. [16b]. A graphical form of these data for liquids is shown in Fig. 4 using the same method as of Fig. 1. This particular scale is indicated hereafter as the Della Volpe-Siboni (DVS) scale.

The results were relatively acceptable; liquids and solids were found to be acidic or basic, the acid components were significant and greater than basic ones, with a

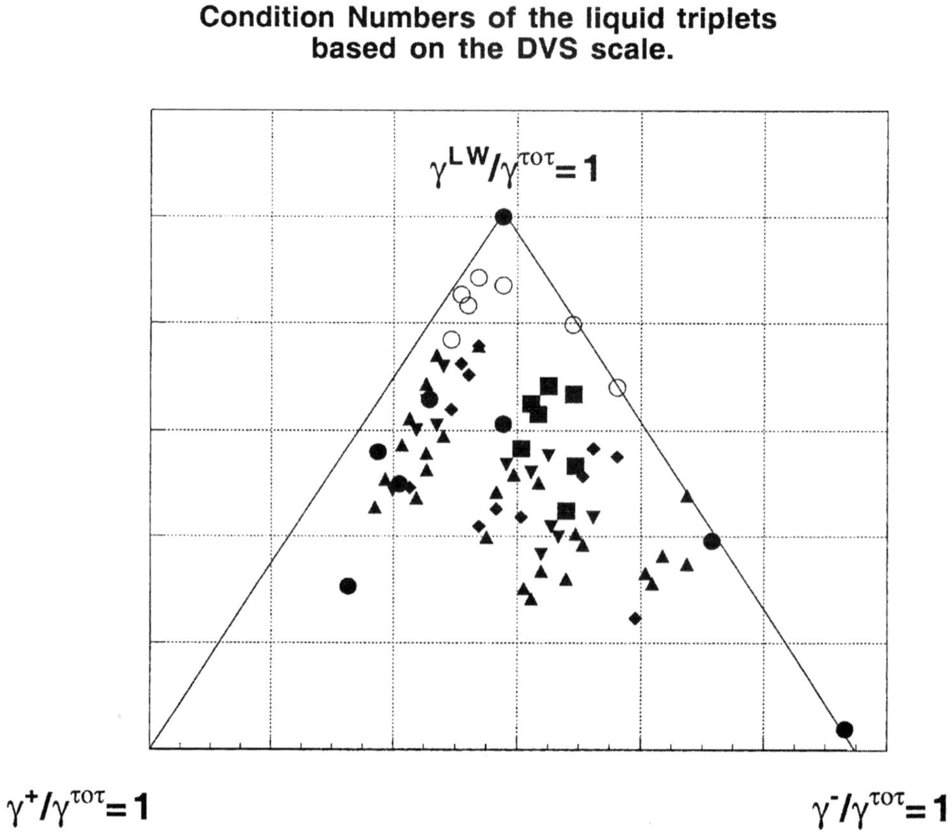

Figure 4. The same as in Fig. 1, but using DVS values of acid-base components for liquids. The basic component of the material surface free energy is generally lower and the acidic one is higher; the more uniform distribution of all the symbols which occupy, now, the whole diagram, corresponds so to a more equilibrated and realistic scale of acid-base components.

considerable agreement with the chemical intuition. Poly(vinyl fluoride) (PVF) or nylon 66 are effectively acidic surfaces, while poly(methyl methacrylate) (PMMA) remains strongly basic. Among liquids dimethyl sulphoxide (DMSO) is the most basic one, while the bromoform, pyrrole, water and glycerol have strong acid content. It is worth noting that an important difference is present between glycerol and ethylene glycol; in fact glycerol becomes prevalently acidic, while ethylene glycol remains prevalently basic. Notice that glycerol, considered as a Brønsted acid, is stronger than ethylene glycol, but weaker than water; this strength order is observed also in the original GvOC scale.

It is interesting to remark that whenever the square root of the acid component is negative the standard deviation is greater than it, except in one case. The odd behaviour of poly(butylene terephthalate) (PBT) was analysed and found to depend on experimental problems, which introduced a systematic error.

In conclusion, the modification of the reference scale for water, including not only the acid-base ratio, but also the dispersive component is able to produce a set of chemically acceptable components for common liquids and solids.

4. SOME INSIGHTS IN THE CONDITIONING

The problem of conditioning arises whenever an equation set containing some uncertain parameters (obtained, for instance, from experimental measurements) must be solved.

In the present case, the calculation of acid-base components of a solid by using the contact angle data on $L \geqslant 3$ liquids implies the solution in $\mathbf{x}$ of the algebraic linear set of the work of adhesion equations

$$\mathbf{Ax} = \mathbf{b} \tag{22}$$

where

$$\mathbf{A} = \begin{pmatrix} \sqrt{\gamma_1^{LW}} & \sqrt{\gamma_1^{+}} & \sqrt{\gamma_1^{-}} \\ \sqrt{\gamma_2^{LW}} & \sqrt{\gamma_2^{+}} & \sqrt{\gamma_2^{-}} \\ \cdots & \cdots & \cdots \\ \sqrt{\gamma_L^{LW}} & \sqrt{\gamma_L^{+}} & \sqrt{\gamma_L^{-}} \end{pmatrix} \quad \mathbf{b} = \begin{pmatrix} \gamma_1(1+\cos\theta_1)/2 \\ \gamma_2(1+\cos\theta_2)/2 \\ \cdots\cdots \\ \gamma_L(1+\cos\theta_L)/2 \end{pmatrix} \quad \mathbf{x} = \begin{pmatrix} \sqrt{\gamma_S^{LW}} \\ \sqrt{\gamma_S^{+}} \\ \sqrt{\gamma_S^{-}} \end{pmatrix} \tag{23}$$

on having denoted with γ_i^{LW}, γ_i^{+}, γ_i^{-} the dispersive and acid-base components of the i-th liquid, $i = 1, 2, \ldots, L$, and with γ_S^{LW}, γ_S^{+}, γ_S^{-} those of the solid, whereas γ_i and θ_i are the surface tension of the i-th liquid and the equilibrium contact angle of the same liquid on the solid, respectively. Owing to the experimental errors in the parameters of the liquids and in the contact angles, both data matrices $\mathbf{A}$ and $\mathbf{b}$ contain a certain amount of uncertainty, which is reflected in an analogous uncertainty in the solution $\mathbf{x}$.

In the classical application of the GvOC method, $L = 3$ and the set (22) allows a unique solution provided that $\mathbf{A}$ is nonsingular. If the errors in $\mathbf{A}$ and $\mathbf{b}$ are expressed by means of the suitably defined "small" matrices $\Delta\mathbf{A}$ and $\Delta\mathbf{b}$ (so that each element ΔA_{ij} or Δb_{ij} of the matrix is the error in the corresponding element of the matrix $\mathbf{A}$ or $\mathbf{b}$), the relative uncertainty in the solution can be generally estimated as

$$\frac{\|\Delta\mathbf{x}\|}{\|\mathbf{x}\|} \leqslant \operatorname{cond}(\mathbf{A}) \left(\frac{\|\Delta\mathbf{A}\|}{\|\mathbf{A}\|} + \frac{\|\Delta\mathbf{b}\|}{\|\mathbf{b}\|} \right) \tag{24}$$

where $\operatorname{cond}(\mathbf{A}) = \|\mathbf{A}\|\,\|\mathbf{A}^{-1}\|$ is the condition number of the matrix $\mathbf{A}$ and $\|\cdot\|$ stands for both vector norm and the corresponding subordinated matrix norm (a "subordinated" matrix norm satisfies the property $\|\mathbf{Ax}\| \leqslant \|\mathbf{A}\|\,\|\mathbf{x}\|$). It is worthy of note that, according to (24), the conditioning of the system (22) can be reduced to a property of the matrix $\mathbf{A}$ only, although conditioning could depend, in principle, on both $\mathbf{A}$ and $\mathbf{b}$.

An important consequence of this observation, crucial for applications, is that the possible ill-conditioning of the GvOC equations related to a certain triplet of liquids can be **a priori** detected, and hence the triplet itself excluded from contact angle measurements and subsequent calculations of solid components. In other words, whenever the acid-base components of the three liquids are known, the appropriateness of using them for contact angle measurements to reliably compute the acid-base components of the given solid can be evaluated **before** effectively using them.

When the number of liquids employed is $L > 3$, the matrix $\mathbf{A}$, whose rows are the square roots of the acid-base components of each liquid, becomes a rectangular $L \times 3$ matrix and the equation set of the work of adhesion is generally expected to allow no solution. It is then necessary to search for the best-fit solutions in the sense of least squares. A well-known result states that whenever $\mathbf{A}$ has the largest rank 3, the least squares best-fit solution $\mathbf{x}$ is unique and is given by the linear set

$$\mathbf{A}^T\mathbf{A}\mathbf{x} = \mathbf{A}^T\mathbf{b} \tag{25}$$

where $\mathbf{A}^T\mathbf{A}$ is a positively defined 3×3 real symmetric matrix. The case of $\text{rank}(\mathbf{A}) < 3$ is analogous, with the only difference that $\mathbf{A}^T\mathbf{A}$ is semi-definite positive and hence singular; the best-fit solution may or may not exist, but anyway it is never unique. In the hypothesis of a nonsingular $\mathbf{A}^T\mathbf{A}$, the solution takes the form

$$\mathbf{x} = (\mathbf{A}^T\mathbf{A})^{-1}\mathbf{A}^T\mathbf{b} \tag{26}$$

for which the problem of conditioning must be considered more carefully. The naive idea of applying estimate (24) to Equation (25), which is formally similar to (22), leads to the upper bound

$$\frac{\|\Delta\mathbf{x}\|}{\|\mathbf{x}\|} \leqslant \text{cond}(\mathbf{A}^T\mathbf{A})\left(\frac{\|\Delta(\mathbf{A}^T\mathbf{A})\|}{\|\mathbf{A}^T\mathbf{A}\|} + \frac{\|\mathbf{A}^T\Delta\mathbf{b}\|}{\|\mathbf{A}^T\mathbf{b}\|}\right) \tag{27}$$

but now the second term in brackets poses some problems. One would wish to provide a further bound to it in terms of the relative error in $\mathbf{b}$

$$\frac{\|\mathbf{A}^T\Delta\mathbf{b}\|}{\|\mathbf{A}^T\mathbf{b}\|} \leqslant \frac{\|\mathbf{A}^T\|\|\Delta\mathbf{b}\|}{\|\mathbf{A}^T\mathbf{b}\|} \leqslant k\frac{\|\Delta\mathbf{b}\|}{\|\mathbf{b}\|} \tag{28}$$

for some suitable constant k independent of $\mathbf{A}$ and $\mathbf{b}$. Due to the noninvertibility of $\mathbf{A}$ it is easily seen that such a constant k does not actually exist. The unit sphere $S = \{\mathbf{b} : \|\mathbf{b}\| = 1\}$ contains vectors belonging to the kernel of $\mathbf{A}^T$

$$\text{Ker}(\mathbf{A}^T) = \left\{\mathbf{b} : \mathbf{A}^T\mathbf{b} = 0\right\} \tag{29}$$

which satisfies, therefore, $\|\mathbf{A}^T\mathbf{b}\| = 0$. A succession $\{\mathbf{b}_h\}_{h\in N}$ in S can be determined such that $\|\mathbf{b}_h\| = 1\ \forall h \in \mathbb{N}$ and $\mathbf{b}_h \underset{h\to+\infty}{\longrightarrow} \mathbf{b}^* \in \text{Ker}(\mathbf{A}^T)$, so that by continuity $\|\mathbf{A}^T\mathbf{b}_h\| \underset{h\to+\infty}{\longrightarrow} 0$.

At a fixed $\Delta\mathbf{b}$, with $\|\mathbf{A}^T\Delta\mathbf{b}\| \neq 0$, this would imply

$$\frac{\|\mathbf{A}^T\Delta\mathbf{b}\|}{\|\mathbf{A}^T\mathbf{b}_h\|} \underset{h\to+\infty}{\longrightarrow} +\infty \tag{30}$$

whereas

$$k\frac{\|\Delta\mathbf{b}\|}{\|\mathbf{b}_h\|} = k\|\Delta\mathbf{b}\| \qquad \forall h \in \mathbb{N}, \tag{31}$$

which is a contradiction.

We conclude that (28) cannot be written as shown and there is no upper bound of the form (24), which is only dependent on the relative errors in **x**, **A** and **b**. The only available estimate is

$$\frac{\|\Delta\mathbf{x}\|}{\|\mathbf{x}\|} \leqslant \mathrm{cond}(\mathbf{A}^T\mathbf{A})\left(\frac{\|\Delta(\mathbf{A}^T\mathbf{A})\|}{\|\mathbf{A}^T\mathbf{A}\|} + \|\mathbf{A}^T\|\frac{\|\Delta\mathbf{b}\|}{\|\mathbf{A}^T\mathbf{b}\|}\right) \tag{32}$$

where the relative uncertainty $\|\Delta\mathbf{b}\|/\|\mathbf{b}\|$ does not appear explicitly. Due to the denominator $\|\mathbf{A}^T\mathbf{b}\|$, the definition of conditioning introduced in the nonsingular case for 3×3 matrices is no more applicable in a proper sense and the "sensitivity" to the data errors must be taken as essentially dependent on <u>both</u> **A** <u>and</u> **b**. The experimental implications are evident, since the data vector is related to the contact angles of the test liquids on the solid; *no meaningful estimate of the sensitivity to data errors can be performed unless contact angle data are available.* Before contact angle measurements, the condition number $\mathrm{cond}(\mathbf{A}^T\mathbf{A})$ provides only a partial information about this sensitivity. A direct bound on the error propagation may be useful

$$\begin{aligned}\|\Delta\mathbf{x}\| \leqslant{}& \left\|\left[(\mathbf{A}+\Delta\mathbf{A})^T(\mathbf{A}+\Delta\mathbf{A})\right]^{-1}(\mathbf{A}+\Delta\mathbf{A})^T - (\mathbf{A}^T\mathbf{A})^{-1}\mathbf{A}^T\right\|\|\mathbf{b}\| + \\ &+\left\|(\mathbf{A}^T\mathbf{A})^{-1}\mathbf{A}^T\right\|\|\Delta\mathbf{b}\|\end{aligned} \tag{33}$$

Particularly if $\|\Delta\mathbf{A}\| \ll \|\Delta\mathbf{b}\|$, then the previous relationship reduces to the simpler approximate form

$$\|\Delta\mathbf{x}\| \leqslant \left\|(\mathbf{A}^T\mathbf{A})^{-1}\mathbf{A}^T\right\|\|\Delta\mathbf{b}\| \tag{34}$$

It is obvious that the experimental data errors are typically much larger than the truncation and round-off ones introduced by the numerical solution of the equation sets. The question about ill- or good-conditioning arises essentially because of the experimental uncertainty in the data, instead of the "noise" due to numerical finite-digits computations. This remark suggests an alternative approach to the problem of error propagation in equations of the form (22) or (26) which may be very useful for applications, although somewhat naive.

Since the experimental data which appear in Equations (22)–(26) are actually represented by confidence intervals, a Monte-Carlo strategy may be adopted. A specific value of each experimental parameter, such as γ_i, γ_i^{LW}, γ_i^+, γ_i^- and θ_i, can be chosen at random inside the appropriate confidence interval and the resulting set of equations can be solved, provided that no singular case occurs. By iterating the procedure a large number of times, a satisfactory approximation of the confidence interval for the solution **x** may be obtained.

Moreover, whenever a statistical distribution of the experimental parameters is available (which in reality seldom happens), the same approach can also yield an accurate estimate of the statistical distribution of the solutions.

The possible occurrence of singular matrices deserves a little more discussion. It is obvious that a nonsingular matrix **A** remains nonsingular when subjected to a sufficiently small perturbation, since its determinant det(**A**) depends continuously on the elements of **A**. Nevertheless, it may happen that a nonsingular matrix is very sensitive to perturbations, in the sense that a very slight modification could be sufficient to yield a singular matrix. Matrices of this kind are known as critically ill-conditioned and should be avoided, if possible, in the computations.

We can naively expect that ill-conditioned matrices, according to the usual definition, are also critically ill-conditioned or that even usual and critical ill-conditionings are equivalent properties, to a certain extent. The physical intuition actually suggests a true correlation, mathematically expressed by the following relationship, due to Kahan [46]

$$\inf_{S:\ \dim[\mathrm{Ker}(S)]>0} \|\mathbf{A}-\mathbf{S}\| = \min_{S:\ \dim[\mathrm{Ker}(S)]>0} \|\mathbf{A}-\mathbf{S}\| = \left\|\mathbf{A}^{-1}\right\|^{-1} \tag{35}$$

and valid for any nonsingular $n \times n$ matrix **A**. The geometrical meaning of (35) is clear; the distance of **A** (in terms of the matrix norm) from the (closed) set of singular $n \times n$ matrices

$$\left\{\mathbf{S}:\ \dim[\mathrm{Ker}(\mathbf{S})] > 0\right\} \tag{36}$$

is exactly given by the inverse of the norm of $\mathbf{A}^{-1}$. As an immediate consequence, whenever the perturbation $\Delta\mathbf{A}$ of **A** has a norm lower than $\|\mathbf{A}^{-1}\|^{-1}$, then the perturbed matrix $\mathbf{A} + \Delta\mathbf{A}$ is certainly still nonsingular. In other words, critical ill-conditioning of **A** occurs for small $\|\mathbf{A}^{-1}\|^{-1}$ and is equivalent to a large condition number, as

$$\min_{S:\ \dim[\mathrm{Ker}(S)]>0} \frac{\|\mathbf{A}-\mathbf{S}\|}{\|\mathbf{A}\|} = \left\|\mathbf{A}^{-1}\right\|^{-1}\|\mathbf{A}\|^{-1} = \frac{1}{\mathrm{cond}(\mathbf{A})} \tag{37}$$

We conclude that the condition number also provides a very useful tool to test critical ill-conditioning in Monte-Carlo simulations.

5. AN EXAMPLE OF APPLICATION OF MONTE-CARLO APPROACH

As an example of the results obtainable with Monte-Carlo approach in the calculation of the acid-base components errors, we report here some calculations on two different triplets, with different values of the condition number. The two triplets are water-formamide-α-bromonaphthalene and water-glycerol-α-bromonaphthalene, used to analyse the PMMA polymer.

The contact angles used were chosen from the literature (see cit. in Ref. [32]): water, 80°, glycerol, 69°, formamide, 64° and α-bromonaphthalene, 16°.

In both cases we have used the scale determined in a recent paper [16b] using the following components for water: $\gamma^{LW} = 26.2$ mJ/m^2, $\gamma^+ = 48.5$ mJ/m^2 and $\gamma^- = 11.2$ mJ/m^2.

These results are only examples of this kind of application; similar results can be obtained using the original GvOC scale ($\gamma^{LW} = 21.8$, $\gamma^+ = \gamma^- = 25.5$ mJ/m^2), or that proposed by Lee [31].

Consider that in the DVS scale glycerol is considerably acidic so the first triplet appears to be constituted of two prevalently acidic and one dispersive liquids, while the second one appears to be more equilibrated, formamide being more basic. Using the GvOC scale these particular triplets show different properties; consider that this is only an example (chosen on the DVS scale) just to show the effect of the condition numbers and not to assert the goodness of a particular triplet.

The condition numbers of the two triplets are shown in the following table.

Water, Formamide, α-Bromonaphthalene			*Water, Glycerol, α-Bromonaphthalene*		
Cond. numb. norm infinity	=	17.7	Cond. numb. norm infinity	=	183
Cond. numb. norm one	=	15.0	Cond. numb. norm one	=	151
Cond. numb. norm two	=	10.0	Cond. numb. norm two	=	109

In both cases the error in the values of contact angles is $\pm 1°$ and that in the values of the surface free energies components is ± 0.2 mJ/m^2. The results are shown in Fig. 5a–c and 6a–c.

As one can see in the first case the values and the errors are in an acceptable range; in the second one the results and their errors are completely out of the range.

This is an extreme example of the importance of the choice of triplets with the lowest norm, or of the relevance of the concepts presented in previous Sections. It is also a good example of the usefulness of Monte-Carlo calculations applied to the evaluation of errors calculation in the present context.

6. CONCLUSIONS

The use of the acid-base theory to calculate the surface free energy components of a surface needs some warnings to eliminate the most common troubles.

1) The equation system obtained from the GvOC theory is strongly ill-conditioned and very precise contact angles are necessary to calculate acceptable values of the acid-base components. To obtain good results one needs to use sets of liquids fulfilling some general criteria, i.e., the presence of at least one dispersive liquid and an equilibrated choice of liquids with complementary characters, some acidic and some basic. The calculation of the condition numbers of the corresponding

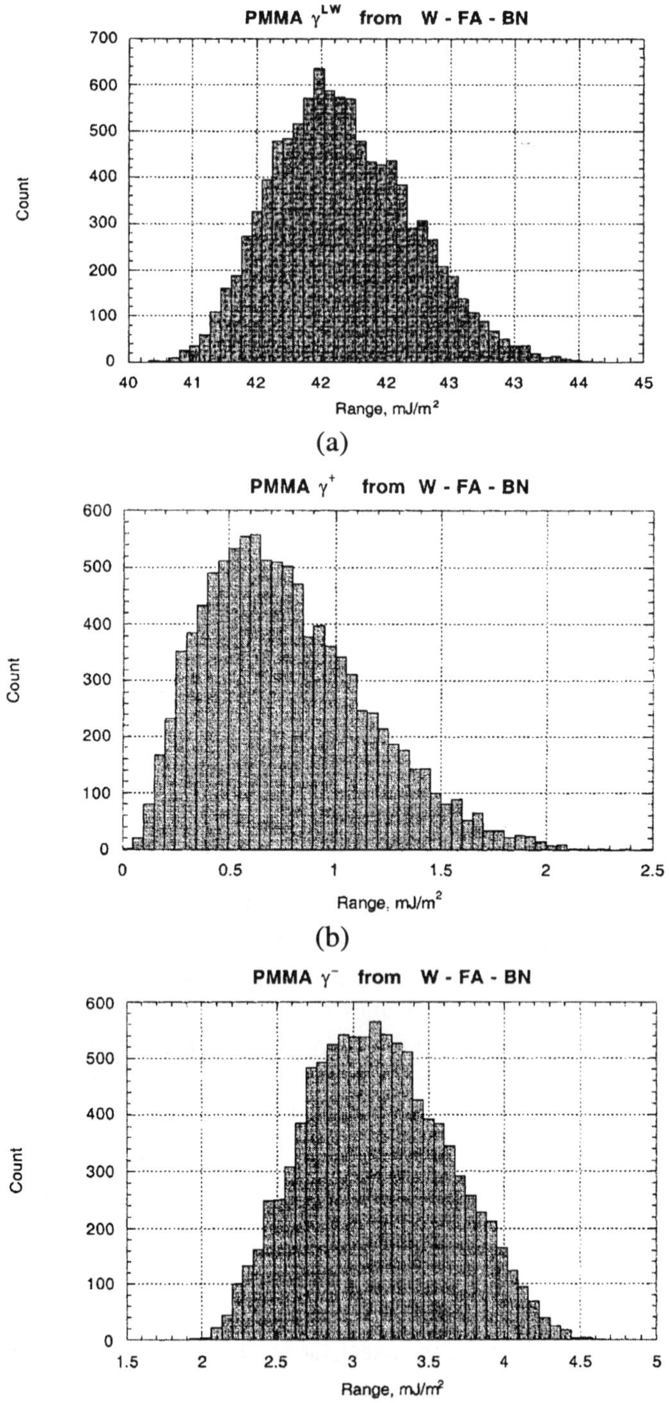

Figure 5. Statistical distribution of the values of the dispersive, acidic and basic components of PMMA evaluated with the Monte-Carlo method; the width of the curve is proportional to standard deviation of the results. The triplet used is Water, Formamide and α-Bromonaphthalene, with components calculated on the DVS scale.

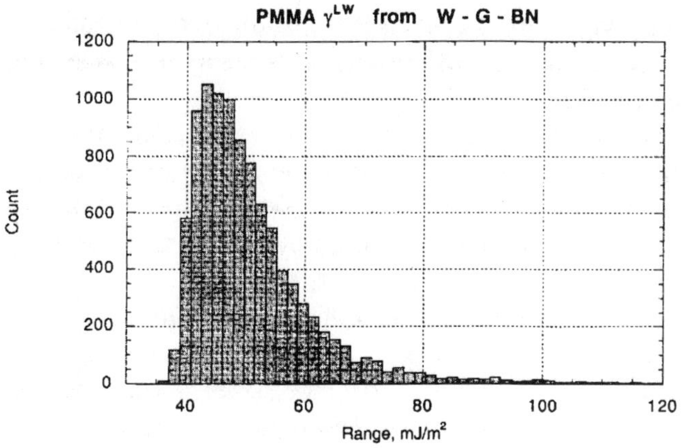

(a)

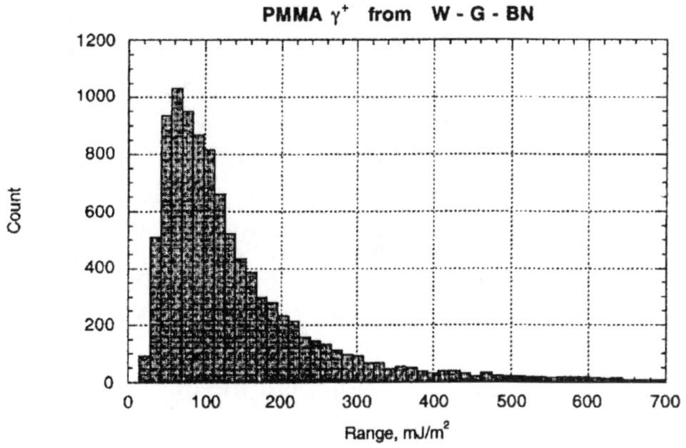

(b)

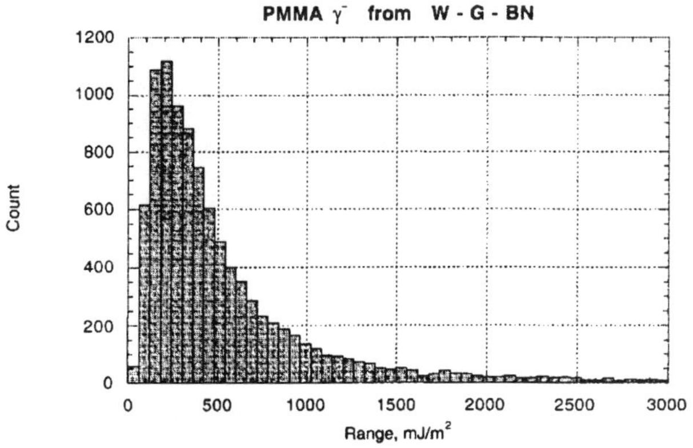

(c)

Figure 6. Statistical distribution of the values of the dispersive, acidic and basic components of PMMA evaluated with the Monte-Carlo method; the width of the curve is proportional to standard deviation of the results. The triplet used is Water, Glycerol and α-Bromonaphthalene, with components calculated on the DVS scale. Due to the more acidic character of glycerol in the DVS scale this triplet is now more ill-conditioned than in the case of original GvOC scale and produces unacceptable results.

equation sets can play a very useful role, indicating the best choice of liquids, able to reduce the ill-conditioning. Moreover, wider sets of proper liquids can assure more satisfactory results.

2) The properties of the test liquids can be calculated by a best-fit procedure together with the solids properties, using a very wide set of liquids and solids and applying the theory in a nonlinear manner. Direct experimental tests of the liquid components, made in the literature, appeared to be complex and questionable, and are not necessary.

3) The solution of the nonlinear equation sets is not unique, due to the symmetry properties of equations themselves; no mathematical trick can eliminate this problem. This corresponds to the existence of an infinite number of equivalent solutions for the acid-base components of liquids and solids.

4) To have a unique solution one must set a *scale*, i.e. to choose a set of properties of some reference material; due to its particular properties water is a good candidate.

5) The choice of the reference material properties is very important; the *most common used scale*, in which the water acid-base properties are considered equal, does not allow to compare the acidic and basic properties of the same or of different materials, i.e. this scale *is a relative one*. Using it in an absolute manner introduces a strong reduction of the acidic components of all the materials. The apparent predominance of the basic component does not correspond to a "law of nature" but to an artifact due to an incorrect application of the theory.

6) Some values or better *some criteria* to choose values for water have been presented. The proposed values are not "the best ones", but simply the result of trials. The mathematical method is reliable, but the numerical results, owing to the large experimental errors in contact angles, are not satisfactory; much work will be necessary to obtain set of contact angle values which would produce acceptable acid-base component values.

A common effort of "surface science community", e.g. an international round robin under the auspices of an international team, is necessary to develop a handbook-level database of contact angles of pure liquids on common polymers from which to calculate their dispersive, acidic and basic components.

Only if this global work does not give acceptable results the "surface science community" will have to reconsider completely its ideas about solid surfaces and the usefulness of contact angles.

Acknowledgements

This work has been supported by the Italian CNR project MSTA II.

REFERENCES

1. T. Kuhn, in *The Structure of Scientific Revolutions*, Harvard University Press, 1962.
2. W. B. Jensen, in: *Acid-Base Interactions: Relevance to Adhesion Science and Technology*, K. L. Mittal and H. R. Anderson, Jr. (Eds.), pp. 3–23, VSP, Utrecht, 1991.

3. A. W. Neumann, in: *Applied Surface Thermodynamics*, A. W. Neumann and J. K. Spelt (Eds.), pp. 293–333, Marcel Dekker, New York, 1996.
4. F. M. Fowkes, in: *Acid-Base Interactions: Relevance to Adhesion Science and Technology*, K. L. Mittal and H. R. Anderson, Jr. (Eds.), pp. 93-115, VSP, Utrecht, 1991.
5. F. M. Fowkes, *J. Adhesion Sci. Technol.*, **1**, 7–27 (1987).
6. R. S. Drago, *Struct. Bond.* **15**, 73–139 (1973).
7. C. J. van Oss, *Interfacial Forces in Aqueous Media*, Marcel Dekker, New York, 1994.
8. R. J. Good and C. J. van Oss, in: *Modern Approaches to Wettability Theory and Applications*, M. E. Schrader and G. I. Loeb (Eds.), pp. 1–28. Plenum Press, New York, 1992.
9. R. E. Johnson, Jr. and R. H. Dettre, *Langmuir*, **5**, 293–295 (1989).
10. I. D. Morrison, *Langmuir*, **5**, 540–543 (1989).
11. D. Y. Kwok, A. Leung, A. Li, C. N. C. Lam, R. Wu and A. W. Neumann, *Colloid Polym. Sci.*, **276**, 459–469 (1998).
12. D. Y. Kwok and A. W. Neumann, *Prog. Colloid Polym. Sci.*, **109**, 170–184 (1998).
13. D. Y. Kwok, A. Leung, A. Li, C. N. C. Lam, R. Wu and A. W. Neumann, *J. Colloid Interface Sci.*, **204**, 000–000 (1998).
14. J. M. Douillard, *J. Colloid Interface Sci.*, **188**, 511–514 (1997).
15. M. D. Vrbanac and J. C. Berg, *J. Adhesion Sci. Technol.*, **4**, 255–266 (1990).
16a. C. Della Volpe and S. Siboni, *J. Adhesion Sci. Technol.*, in press.
16b. C. Della Volpe, S. Siboni and M. Morra, in: *Polymer Interface and Emulsions*, K. Esumi (Ed.), Marcel Dekker, New York, in press.
17. C. J. van Oss, R. F. Giese and W. Wu, *J. Adhesion* **63**, 71–88 (1997).
18. E. N. Dalal, *Langmuir*, **3**, 1009–1015 (1987).
19. R. E. Johnson, Jr. and R. H. Dettre, *J. Phys. Chem.* **68**, 1744 (1964).
20. R. E. Johnson, Jr. and R. H. Dettre, in: *Surface and Colloid Science* Vol. 2, E. Matijevic (Ed.), John Wiley, New York, 1969.
21. H. J. Jacobasch, K. Grundke, S. Schneider and F. Simon. *J. Adhesion*, **48**, 57–73 (1995).
22. M. H. Adao, B. Saramago and A. Fernandes, *Langmuir*, **14**, 4198–4203 (1998).
23. W. V. Chang and X. Qin, Paper presented at the Amer. Chem. Soc. Meeting, San Francisco, California, April, 1992.
24. X. Qin and W. V. Chang, *J. Adhesion Sci. Technol.*, **9**, 823–841 (1995).
25. X. Qin and W. V. Chang, *J. Adhesion Sci. Technol.*, **10**, 963–989 (1996).
26. W. V. Chang and X. Qin, unpublished results.
27. K. Grundke, H. J. Jacobasch, F. Simon and S. Schneider, *J. Adhesion Sci. Technol.*, **9**, 327–350 (1995).
28a. R. J. Good, Li Kuang Shu, Hung-Chang Chiu and C. K. Yeung, *J. Adhesion*, **59**, 25–37 (1996).
28b. R. J. Good and A. K. Hawa, *J. Adhesion* **63**, 5–13 (1997).
29. E. Chibowski, M. Espinosa-Jimenez, A. Ontiveros-Ortega and E. Gimenez-Martin, *Langmuir*, **14**, 5237–5244 (1998).
30. C. Della Volpe, A. Deimichei and T. Riccò, *J. Adhesion Sci. Technol.*, **12**, 1141–1180 (1998).
31. L. H. Lee, *Langmuir*, **12**, 1681–1687 (1996).
32. C. Della Volpe, S. Siboni, *J. Colloid Interface Sci.*, **195**, 121–136 (1997).
33. M. J. Kamlet, J. M. Abboud, M. H. Abraham and R. W. Taft, *J. Org. Chem.*, **48**, 2877–2891 (1983).
34. M. H. Abraham, *Chem. Soc. Rev.* **22**, 73–83 (1993).
35. M. J. Sheriff, *J. Adhesion*, **7**, 257 (1976).
36. A. N. Tikhonov and A. V. Goncharsky (Eds.), *Ill-Posed Problems in the Natural Sciences*. MIR, Moscow (1987).
37. A. Deif, in: *Sensitivity Analysis in Linear Systems*, Springer-Verlag, Berlin, 1986.
38. C. J. van Oss, L. Ju, M. K. Chaudhury and R. J. Good, *J. Colloid Interface Sci.*, **128**, 313–319 (1989).

39. J. K. Spelt, E. Moy, D. Y. Kwok and A. W. Neumann, in: *Applied Surface Thermodynamics*, A. W. Neumann and J. K. Spelt (Eds.), Marcel Dekker, New York, 1996.
40. F. M. Fowkes, *Ind. Eng. Chem.*, **56** (12), 48 (1964).
41. S. Wu, *Polymer Interface and Adhesion*, Marcel Dekker, New York, 1982.
42. G. E. H. Hellwig and A. W. Neumann, *Proc. V. Intern. Congress on Surface Active Substances*, Sec. B, p. 687, Barcelona (1968).
43. B. Janczuk and T. Bialopiotrowicz, *J. Colloid Interface Sci.*, **140**, 362–372 (1990).
44. A. W. Neumann and D. Renzow, *Z. Phys. Chem.* (Frankfurt) **68**, 11 (1969).
45. H. Y. Erbil and R. A. Meriç, *Colloids Surfaces*, **33**, 85–97 (1988).
46. W. Kahan, *Canad. Math Bull.*, **9**, 757–801 (1966).

Acid-Base Interactions, Vol. 2, pp. 91–166
Ed. K.L. Mittal

Contact angle measurements and contact angle interpretation: Relevance to the thermodynamics of adhesion

D. Y. KWOK [1,*] and A. W. NEUMANN [2,†]

[1] *Department of Chemical Engineering, Massachusetts Institute of Technology 77 Massachusetts Ave., Cambridge, Massachusetts 02139*
[2] *Department of Mechanical and Industrial Engineering, University of Toronto 5 King's College Road, Toronto, Ontario M5S 3G8, Canada*

Abstract—Recent progress in the correlation of contact angles with solid surface tensions is summarized. The measurements of meaningful contact angles in terms of surface energetics are also discussed. It is shown that the controversy with respect to measurement and interpretation of contact angles is due to the fact that some (or all) of the assumptions made in all energetic approaches are frequently violated when contact angles are measured and processed. For a large number of polar and non-polar liquids on different solid surfaces, the liquid-vapor surface tension times cosine of the contact angle, $\gamma_{lv}\cos\theta$, is shown to depend only on the liquid-vapor surface tension γ_{lv}, and the solid-vapor surface tension γ_{sv} when the appropriate experimental techniques and procedures are used. Equations which follow these experimental patterns and which allow the determination of solid surface tensions from contact angles are discussed.

Universality of these experimental contact angle patterns is illustrated; other reasons which may cause data to deviate slightly are discussed. It is found that surface tension component approaches do not reflect physical reality. Assuming the fact that solid surface tension is constant for one and the same solid surface, experimental contact angle patterns are employed to deduce a functional relationship to be used in conjunction with Young's equation for determining solid surface tensions. An explicit form of such a relation is obtained by modifying Berthelot's rule together with experimental data; essentially constant solid surface tension values are obtained, independent of liquid surface tension and molecular structure. A new combining rule is also derived based on an expression similar to the one used in molecular theory; such a combining rule should allow a better understanding of the molecular interactions between unlike solid-liquid pairs.

The free energy of adhesion for a solid-liquid pair is calculated from experimental contact angles. Existing static contact angles for 34 different solid surfaces from Zisman and coworkers are evaluated in terms of their solid surface tensions using experimental contact angle patterns.It is found that literature contact angles do not have to be discarded completely; they can be used to determine solid

*This paper represents, in part, the Ph.D. thesis of D. Y. Kwok. E-mail: dykwok@mit.edu

†To whom correspondence should be addressed. Fax: (416) 978-7753.
E-mail: neumann@mie.utoronto.ca

surface tensions and the free energy of adhesion, with caution. The surface tensions for the 34 solid surfaces from Zisman *et al.* are also reported.

Keywords: Contact angle, dynamic; surface tension, equation of state; contact angle, complexity; surface tension of solid; surface tension, from contact angle; surface tension component approach, testing of; adhesion; intermolecular potential.

1. INTRODUCTION

The determination of solid-vapor (γ_{sv}) and solid-liquid (γ_{sl}) interfacial tensions is of importance in a wide range of problems in pure and applied science. For example, the process of particle adhesion can be modeled by the net free-energy change ΔF^{adh} of the system during the adhesion process, which depends explicitly on the solid (particle) surface tensions: the net free-energy change per unit surface area for the adhesion process of a square particle is given by

$$\Delta F^{adh} = \gamma_{pv} - \gamma_{pl} - \gamma_{sl} \tag{1}$$

where γ_{pv} and γ_{pl} are, respectively, the particle-vapor and particle-liquid interfacial tensions. If $\Delta F^{adh} < 0$, the adhesion process is thermodynamically favorable. The free energy of adhesion W^{adh} is simply

$$W^{adh} = -\Delta F^{adh} \tag{2}$$

Because of the difficulties involved in measuring directly the surface tension involving a solid phase, indirect approaches are called for: Several independent approaches have been used to estimate solid surface tensions, including direct force measurements [1]–[9], contact angles [10]–[17], capillary penetration into columns of particle powder [18]–[21], sedimentation of particles [22]–[25], solidification front interaction with particles [26]–[33], film flotation [34]–[38], gradient theory [39]–[42], Lifshitz theory of van der Waals forces [42]–[45], and theory of molecular interactions [46]–[49]. Among these methods, contact angle measurements are believed to be the simplest.

Contact angle measurement is easily performed by establishing the tangent (angle) of a liquid drop with a solid surface at the base. The attractiveness of using contact angles θ to estimate the solid-vapor and solid-liquid interfacial tensions is due to the relative ease with which contact angles can be measured on suitably prepared solid surfaces. It will become apparent later that this seeming simplicity is very misleading.

The possibility of estimating solid surface tensions from contact angles relies on a relation which was recognized by Young [50] in 1805. The contact angle of a liquid drop on an ideal solid surface is defined by the mechanical equilibrium of the drop under the action of three interfacial tensions: solid-vapor, γ_{sv}, solid-liquid, γ_{sl}, and liquid-vapor, γ_{lv}. This equilibrium relation is known as Young's equation:

$$\gamma_{lv} \cos\theta_Y = \gamma_{sv} - \gamma_{sl} \tag{3}$$

where θ_Y is the Young contact angle, i.e. a contact angle which can be inserted into Young's equation. It will become apparent later that the experimentally accessible contact angles may or may not be equal to θ_Y: there exist many metastable contact angles which are not equal to θ_Y.

Young's equation (3) contains only two measurable quantities, the contact angle θ and the liquid-vapor surface tension, γ_{lv}. In order to determine γ_{sv} and γ_{sl}, an additional relation relating these quantities must be sought. Nevertheless, Eq. (3) suggests that the observation of the equilibrium contact angles of liquids on solids may be a starting point for investigating the solid surface tensions, γ_{sv} and γ_{sl}. This has inspired many studies which attempt to develop methodologies for determining solid surface tensions. A common feature of these approaches is the assumption that contact angle measurement is a trivial task.

Since γ_{lv}, γ_{sv} and γ_{sl} are thermodynamic properties of the liquid and solid, Eq. (3) implies a single, unique contact angle; in practice, however, contact angle phenomena are complicated [51]–[53]. In particular, the contact angle made by an advancing liquid (θ_a) and that made by a receding liquid (θ_r) are not identical; nearly all solid surfaces exhibit contact angle hysteresis, H (the difference between θ_a and θ_r):

$$H = \theta_a - \theta_r \tag{4}$$

Contact angle hysteresis can be due to roughness and/or heterogeneity of a solid surface. If roughness is the primary cause, then the measured contact angles are meaningless in terms of Young's equation. On very rough surfaces, contact angles are larger than on chemically identical, smooth surfaces [21]. Obviously, interpreting such angles in terms of Eq. (3) would lead to erroneous results because the contact angle would inevitably reflect surface topography, rather than exclusively surface energetics.

Because of these various complexities, models have been employed to gain a deeper understanding of the thermodynamic status of contact angles. In general, it has been found that the experimentally observed apparent contact angle, θ, may or may not be equal to the Young contact angle, θ_Y [51, 52]:

(1) On ideal solid surfaces, there is no contact angle hysteresis and the experimentally observed contact angle is equal to θ_Y.

(2) On smooth, but chemically heterogeneous solid surfaces, the experimentally observed θ is not necessarily equal to the thermodynamic equilibrium angle. Nevertheless, the experimental advancing contact angle, θ_a, can be expected to be a good approximation of θ_Y. This has been illustrated using a model of a vertical surface consisting of heterogeneous and smooth strips [51, 52]. Therefore, care must be exercised to ensure that the experimental apparent contact angle, θ, is the advancing contact angle in order to be inserted into the Young equation. While the receding contact angle on a heterogeneous and smooth surface can also be a Young angle, it is usually found to be non-

reproducible often because of sorption of the liquid into the solid and swelling of the solid by the liquid [54].

(3) On rough solid surfaces, no such equality between θ_a and θ_Y exists: all contact angles on rough surfaces are meaningless in terms of Young's equation [51, 52].

The thermodynamic equilibrium angles on rough and heterogeneous surfaces are the so-called Wenzel [55] and Cassie [56]–[58] angles, respectively. They are not equal to θ_Y [51, 52]; furthermore, they are not experimentally accessible quantities.

There are as yet no general criteria to answer the question of how smooth a solid surface has to be for surface roughness not to have an effect on the contact angle. This and similar problems can be linked to line tension, which has its own complexities [59]. It is, therefore, of utmost importance to prepare solid surfaces as smooth as possible so that the experimental advancing angles can be a good approximation of θ_Y. In addition to these complexities, penetration of the liquid into the solid, swelling of the solid by the liquid, and physical/chemical reactions can all play a role. For example, swelling of a solid by a liquid [54] can change the chemistry of the solid in an unknown manner and hence affect the values of γ_{sl}, and/or γ_{sv}. Therefore, it is also important to ensure that the solid surfaces are as inert as possible in order to minimize such effects, by appropriate choice of the liquid.

Several contact angle approaches [10]–[17], of current interest, were largely inspired by the idea of using Young's equation for the determination of surface energetics. While these approaches are, logically and conceptually, mutually exclusive, they share, nevertheless, the following basic assumptions:

(1) All approaches rely on the validity and applicability of Young's equation for surface energetics from experimental contact angles.

(2) Pure liquids are always used; surfactant solutions or mixtures of liquids should not be used, since they would introduce complications due to preferential adsorption.

(3) The values of γ_{lv}, γ_{sv}, (and γ_{sl}) are assumed to be constant during the experiment, i.e. there should be no physical/chemical reaction between the solid and the liquid.

(4) The liquid surface tensions of the test liquids should be higher than the anticipated solid surface tension.

(5) The values of γ_{sv} going from liquid to liquid are also assumed to be constant, i.e. independent of the liquids used.

With respect to the first assumption, one requires the solid surfaces to be rigid, smooth and homogeneous so that Young's equation is the appropriate equilibrium condition; the experimentally observed contact angles should also be advancing contact angles. However, many attempts have been made in the literature to interpret surface tensions of solid surfaces, which are not rigid (e.g. gels [60]) and not smooth (e.g. biological surfaces [61]), in conjunction with Young's equation.

Clearly, these results are open to question, since Young's equation may not be valid or applicable in these situations. With respect to the other assumptions, the solid surfaces should be as inert as possible so that effects such as swelling and physical/chemical reactions are minimized.

In order to assure that the experimentally measured contact angles do not violate any of the above assumptions, one requires careful experimentation and suitable methodology. However, contact angles are typically measured simply by depositing a drop of liquid on a given solid surface, and manually placing a tangent to the drop at its base using a so-called goniometer-sessile drop technique [62, 63]. Apart from the subjectivity of the technique, it normally yields contact angle accuracy of no better than $\pm 2°$. More important, the technique cannot be expected to reflect the complexities of solid-liquid interactions. It will become apparent in this paper that much of the controversy with respect to the interpretation of contact angles in terms of surface energetics lies in the fact that not enough attention is given to the above generally accepted assumptions.

In this paper, the correlation of contact angles with solid surface tensions will be discussed. Section 2.1 illustrates the experimental contact angle patterns on inert solid surfaces using an automated axisymmetric drop shape analysis; these results will be contrasted with the patterns obtained on non-inert solid surfaces using a conventional goniometer-sessile drop technique in Section 2.2. By using the automated axisymmetric drop shape analysis — profile (ADSA-P) and the appropriate experimental procedures described in Section 2.3, the discrepancy between the two contact angle patterns is shown to arise from the fact that some (or all) of the above widely accepted assumptions are violated, because the goniometer procedure does not have a built-in quality control mechanism. Section 2.3 illustrates the low-rate dynamic contact angle results for various solid surfaces, as obtained by ADSA-P. The universality of the contact angle patterns will be discussed in Section 2.4. Section 2.5 presents a detailed discussion of the criteria for calculations of surface energetics. Equations which follow these experimental patterns and which allow the determination of solid surface tensions from contact angles will be sought in Section 3; surface tension component approaches are also briefly discussed. In Section 4, the free energy of adhesion of a solid-liquid pair will be calculated from the experimental contact angle data obtained in Section 2.3. Applications of the equation developed in Section 3.2 will be discussed in Section 5; contact angle data from Zisman and coworkers [10] on several solid surfaces will be evaluated for the calculations of surface energetics. A summary, perspective, and future development will be given, respectively, in Sections 6, 7, and 8.

2. CONTACT ANGLE MEASUREMENT

2.1. Experimental Contact Angle Patterns

On carefully prepared solid surfaces, Li *et al.* [64, 65] have performed static (advancing) contact angle experiments using an automated axisymmetric drop shape analysis — profile (ADSA-P). ADSA-P [66] is a technique to determine liquid-fluid interfacial tensions and contact angles from the shape of axisymmetric menisci, i.e., from sessile as well as pendant drops. Assuming that the experimental drop is Laplacian and axisymmetric, ADSA-P finds the theoretical profile that best matches the drop profile extracted from the image of a real drop, from which the surface tension, contact angle, drop volume, surface area and three-phase contact radius can be computed. The strategy employed is to fit the shape of an experimental drop to a theoretical drop profile according to the Laplace equation of capillarity, using surface/interfacial tension as an adjustable parameter. The best fit identifies the correct surface/interfacial tension from which the contact angle can be determined by a numerical integration of the Laplace equation. Details can be found elsewhere [67]–[69]. It has been found [64, 65] that a contact angle accuracy of better than $\pm 0.3°$ can be obtained on well-prepared solid surfaces.

In the experiments of Li *et al.* [64, 65], static (advancing) contact angles were measured by supplying test liquids from below the surface into the sessile drop, using a motor-driven syringe device. A hole of about 1 mm in the center of each solid surface was required to facilitate such procedures. Liquid was pumped slowly into the drop from below the surface until the three-phase contact radius was about 0.4 cm. After the motor was stopped, the sessile drop was allowed to relax for approximately 30 sec. to reach equilibrium. Then 3 pictures of this sessile drop were taken successively at intervals of 30 sec. More liquid was then pumped into the drop until it reached another desired size, and the above procedure was repeated [64]. These procedures ensure that the measured static contact angles are indeed the advancing contact angles.

The three carefully prepared solid surfaces are FC-721-coated mica, Teflon FEP (fluorinated ethylene propylene) heat-pressed against quartz glass slides and poly(ethylene terephthalate) PET. FC-721 is a 3M "Fluorad" brand antimigration coating designed to prevent the creep of lubricating oils out of bearings. The FC-721-coated mica was prepared by a dip-coating technique. Teflon FEP (fluorinated ethylene propylene) surfaces were prepared by a heat-pressing method. The material was cut to 2×4 cm^2, sandwiched between two glass slides, and heat-pressed by a jig in an oven. Poly(ethylene terephthalate) PET is the condensation product of ethylene glycol and terephthalic acid. The surfaces of PET films were exceedingly smooth as received and were cleaned before measurement. Details of the solid surface preparation can be found elsewhere [64].

Figure 1 shows these contact angle results in a plot of $\gamma_{lv} \cos\theta$ vs. γ_{lv}, for a large number of pure liquids with different molecular properties: they are, in the order of increasing surface tension, pentane, hexane, methanol, decane, methyl acetate, dodecane, tetradecane, hexadecane, *trans*-decalin, *cis*-decalin, dimethylformamide, ethyl cinnamate, dibenzylamine, dimethyl sulfoxide, 1-bromonaphthalene, diethylene glycol, ethylene glycol, 2,2′-thiodiethanol, formamide, glycol, and water. The curves are best-fit of experimental data to a simple quadratic equation. It can be seen that, for a given solid surface, $\gamma_{lv} \cos\theta$ changes smoothly and systematically with γ_{lv}. Since the solid surface tension, γ_{sv}, of a given solid surface is expected to be constant, i.e. independent of the choice of the test liquid used, Figure 1 implies that $\gamma_{lv}\cos\theta$ depends only on γ_{lv} at constant γ_{sv}. Replacing the solid surface from the hydrophobic FC-721 surface to the hydrophilic PET surface (and hence the γ_{sv}) shifts the curve in a very regular manner. These results suggest that the values of $\gamma_{lv}\cos\theta$ depend only on γ_{lv} and γ_{sv}, independent of any specific intermolecular forces of the liquids and solid surfaces [64, 65, 70]:

$$\gamma_{lv}\cos\theta = f(\gamma_{lv}, \gamma_{sv}) \tag{5}$$

where f is as yet an unknown function. Because of Young's equation, the experimental contact angle patterns in Figure 1 imply, in light of Eq. (5), that the solid-liquid interfacial tension γ_{sl} depends only on the liquid-vapor γ_{lv} and solid-vapor γ_{sv} surface tensions: combining Young's equation with Eq. (5) yields

$$\gamma_{sv} - \gamma_{sl} = f(\gamma_{lv}, \gamma_{sv}) \tag{6}$$

and hence

$$\gamma_{sl} = \gamma_{sv} - f(\gamma_{lv}, \gamma_{sv}) = F(\gamma_{lv}, \gamma_{sv}) \tag{7}$$

where F is as yet another unknown function. This is indeed in agreement with thermodynamics [71] and the thermodynamic phase rule for capillary systems [72]–[75] which states that there are only two degrees of freedom for such solid-liquid systems. Thus, the contact angle can be changed simply by changing either γ_{lv} or γ_{sv}. While the specific intermolecular forces (e.g., dipole–dipole moments and hydrogen bondings) determine the primary surface tensions of liquids and solids, they do not appear to have any additional and independent effects on the contact angles, in the context of Young's equation. In principle, a plot of $\cos\theta$ vs. γ_{lv}, rather than $\gamma_{lv}\cos\theta$ vs. γ_{lv}, can also be used to deduce the functional dependence of Eq. (7).

We shall point out that the above findings have been controversial [70], [76]–[82]. It is to be noted that the experimental contact angle patterns shown in Figure 1 do not always appear in the literature: curves far less smooth or no unique curves at all are very frequently reported (see later). Such patterns can have a variety of causes. Accurate contact angle measurements require extreme experimental care. Even very minor vibrations can cause advancing contact angles to decrease, resulting in errors of several degrees. Surface roughness can affect the contact angles and make Young's equation inapplicable. Swelling of a solid by a liquid [54] can change the chemistry of the solid and hence the values of γ_{sl} and θ in an unpredictable manner. Non-constancy of γ_{lv}, γ_{sv} and γ_{sl} during the experiment, and non-constancy of γ_{sv} from liquid to liquid can produce scatter in plots of the type of Figure 1 easily.

The interpretation of contact angles has been controversial. It will be illustrated in subsequent sections that such controversy with respect to the experimental contact angle patterns arises from the fact that contact angle phenomena are often complex, and cannot be unraveled by the simple goniometer techniques. In addition to the

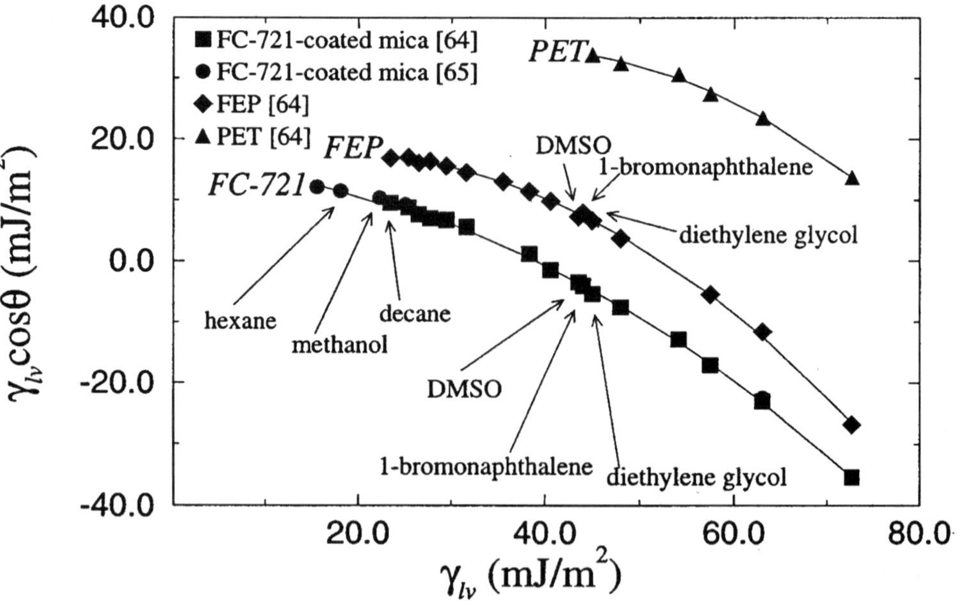

Figure 1. $\gamma_{lv}\cos\theta$ vs. γ_{lv} for three well-prepared inert solid surfaces: FC-721-coated mica, heat-pressed Teflon FEP (fluorinated ethylene propylene), and poly(ethylene terephthalate) PET [64, 65]. The smoothness suggests that $\gamma_{lv}\cos\theta$ depends only on γ_{lv} and γ_{sv}.

solid surface tensions given by Young's equation (3), experimental contact angles often contain other information about a given solid surface, such as molecular orientation at the surface or surface topography. Substantial experimental efforts are required to measure contact angles which represent solid surface tensions exclusively as given by Young's equation and which do not violate the widely accepted assumptions for the interpretation of contact angles. An example of the experimental patterns, which are very different from those of Figure 1, will be illustrated, simply by measuring advancing contact angles by a goniometer technique.

2.2. *Experimental Contact Angle Patterns from a Goniometer*

In order to compare the experimental contact angle patterns obtained by a goniometer with those by an automated ADSA-P technique shown in Figure 1, two well-defined copolymers were selected. These copolymers were selected purposely to be not as inert as the FC-721-coated mica, Teflon FEP, and PET surfaces used by Li *et al.* [64, 65]. Poly(propene-*alt*-*N*-(*n*-propyl)maleimide) and poly(propene-*alt*-*N*-(*n*-hexyl)maleimide) are copolymers with different side chains [83] and were used for contact angle measurements. For each copolymer, a 2% solution was prepared using tetrahydrofuran as the solvent. Si wafers <100> were selected as the substrate for the copolymer coating. The wafers were obtained as circular discs of about 10 cm diameter and were cut into rectangular shapes of about 2 × 3 cm^2. Each surface was first cleaned with ethanol, acetone, and then soaked in chromic acid for at least 24 hrs. The cleaned surfaces were rinsed with doubly-distilled water, and dried under a heat lamp before the copolymer coating. A few drops of the 2% copolymer/tetrahydrofuran solution were deposited on the dried wafers inside petri dishes overnight; the solution spread and a thin layer of the copolymer formed on the wafer surface after tetrahydrofuran evaporated. This preparation produced good quality coated surfaces, as manifested by light fringes, due to refraction at these surfaces, suggesting that roughness was of the order of nanometers or less. It should be noted that if preparation of the solid surfaces and liquids for contact angle measurements is less careful, impurities can easily contaminate the test liquids and solids. This would inevitably result in the contact angle measurements of contaminated liquids and solids, rather than the presumed pure liquids and solids.

Thirteen liquids were chosen in this study. Selection of these liquids was based on the following criteria: (1) liquids should include a wide range of intermolecular forces; (2) liquids should be non-toxic; and (3) the liquid surface tension should be higher than the anticipated solid surface tension [10, 13, 21]. They are, in the order of increasing surface tension, *cis*-decalin, 2,5-dichlorotoluene, ethyl cinnamate, dibenzylamine, dimethyl sulfoxide (DMSO), 1-bromonaphthalene, diethylene glycol, ethylene glycol, diiodomethane, 2,2′-thiodiethanol, formamide, glycerol and water.

The procedures to measure the advancing contact angles using a goniometer-sessile drop technique were the same as those typically used in the literature: a sessile drop of about 0.4–0.5 cm radius was formed from above through a Teflon capillary. The three-phase contact line of the drop was then slowly advanced by supplying more liquid from above through the capillary which was always kept in contact with the drop. The maximum (advancing) contact angles were measured carefully from the left and right sides of the drop and subsequently averaged. The above procedures were repeated for 5 drops on 5 new surfaces. All readings were then averaged.

Figure 2 shows the contact angle results for the two copolymers from the goniometer-sessile drop measurements in a plot of $\gamma_{lv} \cos\theta$ vs. γ_{lv}. Contrary to the contact angle patterns shown in Figure 1, considerable scatter is apparent. On a given solid, say the poly(propene-*alt*-*N*-(*n*-hexyl)maleimide) copolymer, γ_{sv} is expected to be constant; since the values of $\gamma_{lv} \cos\theta$ here do not appear to give a smooth and systematic change with γ_{lv}, one might argue that the contact angle (or $\gamma_{lv} \cos\theta$) cannot be a simple function g of only γ_{lv} and γ_{sv}, but has to depend also on the various specific intermolecular forces of the liquids and solids:

$$\gamma_{lv} \cos\theta = g(\gamma_{lv}, \gamma_{sv}, \textit{dipole–dipole, hydrogen bonding, etc.}) \quad (8)$$

Because of Young's equation, the apparent scatter would seem to favor the stipulation of the surface tension component approaches [11], [15]–[17], [84]–[86] that γ_{sl} depends on various surface tension components due to the specific intermolecular forces, in addition to γ_{lv} and γ_{sv}:

$$\gamma_{sv} - \gamma_{sl} = g(\gamma_{lv}, \gamma_{sv}, \textit{dipole–dipole, hydrogen bonding, etc.}) \quad (9)$$

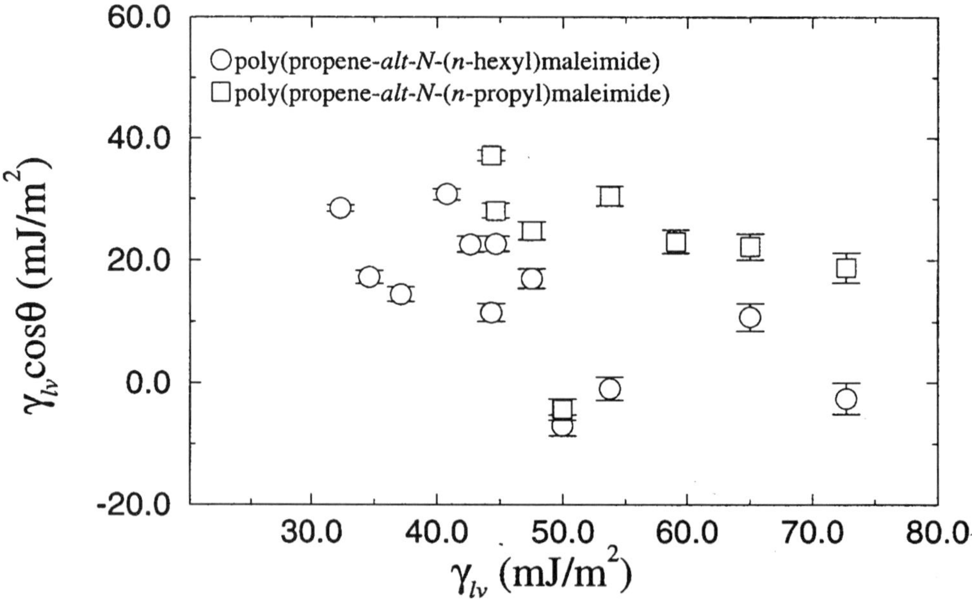

Figure 2. $\gamma_{lv}\cos\theta$ vs. γ_{lv} for poly(propene-*alt*-*N*-(*n*-propyl)maleimide) and poly(propene-*alt*-*N*-(*n*-hexyl)maleimide) copolymers, from advancing contact angles measured by a conventional goniometer technique [83]. Due to the scatter, one might argue that $\gamma_{lv}\cos\theta$ cannot be a simple function of only γ_{lv} and γ_{sv}.

and hence

$$\begin{aligned}\gamma_{sl} &= \gamma_{sv} - g(\gamma_{lv}, \gamma_{sv}, \textit{dipole–dipole, hydrogen bonding, etc.}) \\ &= G(\gamma_{lv}, \gamma_{sv}, \textit{dipole–dipole, hydrogen bonding, etc.})\end{aligned} \tag{10}$$

where G is an unknown function. While the contact angle patterns in Figure 2 can be easily found in the literature, they do not really support the above stipulation of the surface tension component approaches [11], [15]–[17], [84]–[86]. In the next section, it will be shown that this scatter indeed comes from the fact that many of the experimental contact angles in Figure 2 have violated some (or all) of the assumptions usually made by all present contact angle approaches, as summarized in the Introduction. Thus, the apparent additional degrees of freedom (inferred from the scatter) do not come from the putative independent effect of intermolecular forces on the contact angles: $\gamma_{lv}\cos\theta$ can be shown to change smoothly and systematically with γ_{lv} if suitable experimental techniques and procedures are employed, such as those described below, to delete measurements which violate any of the above assumptions.

2.3. Low-Rate Dynamic (Advancing) Contact Angles by ADSA-P

2.3.1. Experimental Procedures. Sessile drop contact angle measurements using ADSA-P were performed dynamically, by using a motor-driven syringe to pump liquid steadily into the sessile drop from below the surface [87]. A quartz cuvette ($5 \times 5 \times 5$ cm^3) was used to isolate the drop from its environment. It has been found that there are virtually no differences between the measured contact angles with or without a cuvette under laboratory conditions, for the liquids used. The dynamic advancing and receding contact angle measurements can be performed, respectively, by pushing or pulling the syringe plunger of a motorized syringe mechanism, leading to an increase or decrease in drop volume. Normally, at least 5 and up to 10 dynamic contact angle measurements were performed on a new solid surface each time, at velocities of the three-phase contact line in the range from 0.1 to 1.5 mm/min. It will become apparent later that low-rate dynamic contact angles in this velocity range are essentially identical to the static contact angles for relatively smooth surfaces.

In these dynamic procedures, liquid is supplied into the sessile drop from below the solid surface using a motorized-syringe device [87]. In order to facilitate such an experimental procedure, a hole of about 1 mm diameter in the solid surface is required. The strategy of pumping liquid from below the surface was pioneered by Oliver *et al.* [88, 89] because of its potential for avoiding drop vibrations and for measuring true advancing contact angles without disturbing the drop profile. In order to avoid leakage between a stainless steel needle and the hole (in the surface), Teflon tape was wrapped around the end of the needle before insertion into the hole. In the literature, it is customary to first deposit a drop of liquid on a given solid surface using a syringe or a Teflon needle; the drop is then made to advance by supplying more liquid from above using a syringe or a needle in contact with the drop. Such experimental procedures cannot be used for ADSA-P since ADSA determines the contact angles and surface tensions based on the complete and undisturbed drop profile.

In actual experiments, an initial liquid drop of about 0.3 cm radius was carefully deposited from above using a gas-tight Hamilton syringe with a stainless steel needle, covering the hole in the surface. This is to ensure that the drop will increase axisymmetrically in the center of the image field when liquid is supplied from the bottom of the surface and will not hinge on the lip of the hole. The motor in the motorized-syringe mechanism was then set to a specific speed, by adjusting the voltage from a voltage controller. Such a syringe mechanism pushes the syringe plunger, leading to an increase in drop volume and hence the three-phase contact radius. A sequence of pictures of the growing drop was then recorded by the computer typically at a rate of 1 picture every 2 sec., until the three-phase contact radius was about 0.5 cm or larger. For each low-rate dynamic contact angle experiment, at least 50 and up to 500 images were normally taken. Since ADSA-P determines the contact angle and the three-phase contact radius simultaneously for each image, the advancing dynamic contact angles as a function of the three-phase contact radius (i.e. location on the surface) can be obtained. In addition, the change in the contact angle, drop volume, drop surface area, and the three-phase contact radius can also be studied as a function of time. The actual rate of advancing can be determined by linear regression, by plotting the three-phase contact radius over time. For each liquid, different rates of advancing were studied, by adjusting the speed of the pumping mechanism [87].

Measuring contact angles as a function of the three-phase contact radius has an additional advantage: the quality of the surface is observed indirectly in the measured contact angles. If a solid surface is not very smooth, irregular and inconsistent contact angle values will be seen as a function of the three-phase contact radius. When the measured contact angles are essentially constant as a function of surface location, the mean contact angle for a specific rate of advancing can be obtained by averaging the contact angles, after the three-phase contact radius reaches 0.3 to 0.5 cm (see later). The purpose of choosing these relatively large drops is to avoid any possible line tension effects on the measured contact angles [53, 90, 91].

2.3.2. Inert (Non-Polar) Surface: FC-722-Coated Mica Surface. A fluorocarbon, FC-722, dip-coated onto mica surfaces was chosen as the substrate for the dynamic contact angle experiments. FC-722 is a fluorochemical coating available from 3M and is chemically very similar to the FC-721 used in earlier studies. The samples were prepared by the same dip-coating procedures described elsewhere [64, 65], using freshly cleaved mica surfaces, originally received as sheets.

Before dip-coating, the mica surfaces were prepared using the following procedures: (1) mica sheets were cut into small mica plates of about 2×3 cm^2; (2) a hole of about 1 mm in diameter was made, by drilling, in the center of each surface; (3) each mica surface was then cleaved by a sharp knife, cleaned with ethanol, and acetone, and dried in the air before dip-coating.

Seventeen liquids were chosen for the contact angle measurements [87]. The surface tensions of these liquids were determined independently by applying ADSA-P to pendant drops at room temperature, 23.0 ± 0.5°C. Selection of these liquids was based on the criteria described in the preceding section.

Figure 3 shows a typical example of a low-rate dynamic contact angle experiment: water on a FC-722 surface. As can be seen in this figure, increasing the drop volume V linearly from 0.18 cm^3 to 0.22 cm^3, by the motorized-syringe mechanism, increases the contact angle θ from about 108° to 119° at essentially constant three-phase contact radius R. This is due to the fact that even carefully placing an initial drop from above on a solid surface can result in a contact angle somewhere between advancing and receding. Therefore, it takes addition of a certain amount of liquid for the initial drop front to start advancing. Further increase in the drop volume causes the three-phase contact line to advance, with θ essentially constant as R increases. Increasing the drop volume in this manner ensures the measured θ to be an advancing contact angle. The rate of advancing for this experiment can be determined by linear regression of the three-phase contact radius over time: the drop periphery was being advanced at a rate of 0.14 mm/min., in the specific example given in Figure 3.

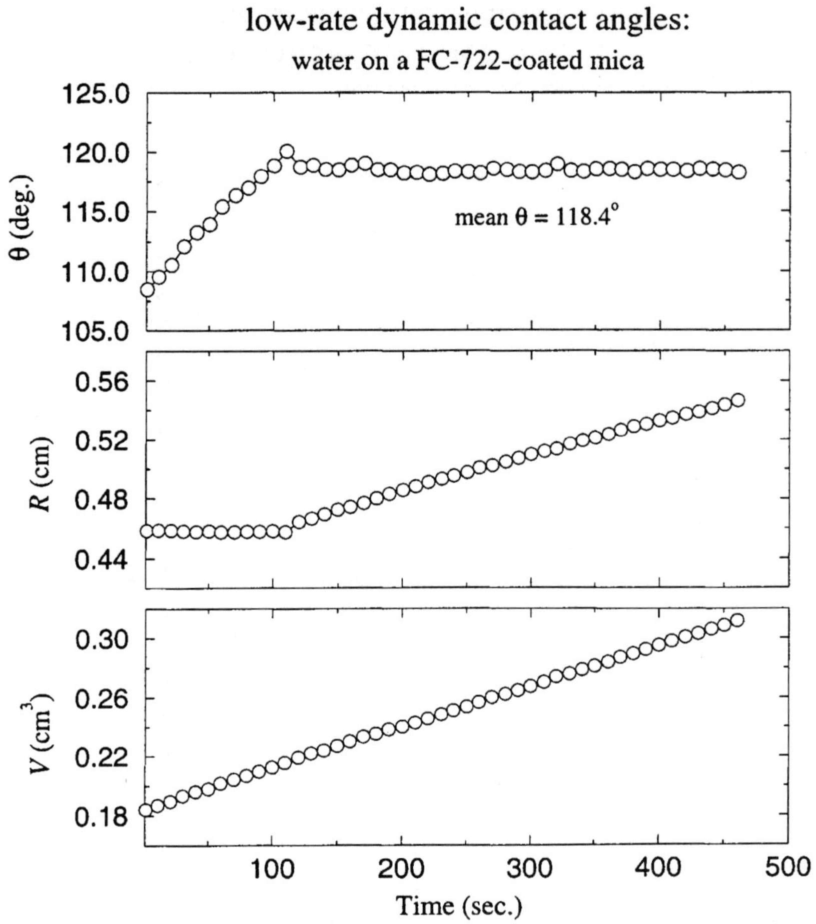

Figure 3. Low-rate dynamic contact angles θ of water on a FC-722-coated mica [87]. R and V are radius and volume of the sessile drop, respectively.

In Figure 3, the measured contact angles are essentially constant as R increases, indicating good surface quality of the solid surface used. It turns out that averaging the measured contact angles after R reaches 0.50 cm is convenient, since the drop is guaranteed to be in the advancing mode and that possible line tension effects are negligible [59, 90, 91]. Averaging the measured contact angles, after R reaches 0.50 cm, yields a mean contact angle of 118.4° for water. While a three-phase contact radius of 0.50 cm may seem to be an arbitrary value, it turns out that there is virtually no difference between average θ for R between 0.48 cm and 0.54 cm; the contact angles are essentially constant after $R = 0.50$ cm. Similar results were also obtained for other liquids [87].

Dynamic/static contact angle experiments have also been performed: A liquid drop is first selected to advance at a specific rate of advancing and then stopped, while a sequence of images is recorded. A typical experiment is shown in Figure 4

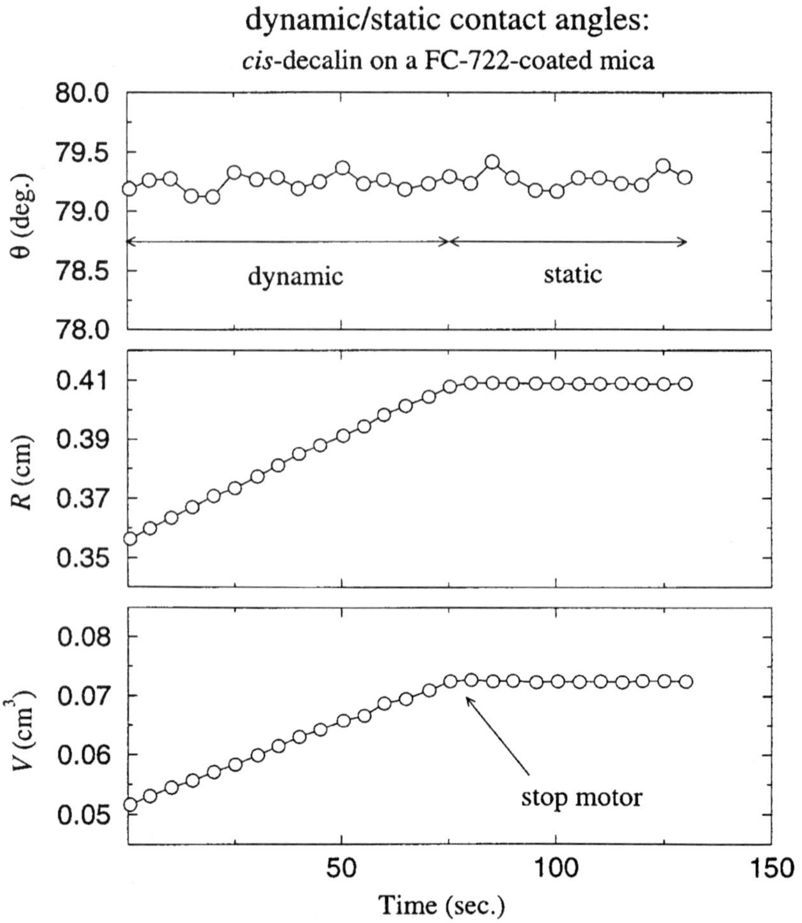

Figure 4. Dynamic and static contact angles of *cis*-decalin on a FC-722-coated mica [87]. This result suggests that the low-rate dynamic contact angle is identical to the static angle. R and V are radius and volume of the sessile drop, respectively.

for *cis*-decalin. As drop volume increases from 0.05 to 0.07 cm^3, the three-phase contact line advances from about 0.36 to 0.41 cm at a rate of 0.41 mm/min. A sequence of drop images was acquired after the motor was stopped at $R = 0.41$ cm. In Figure 4, the contact angle is independent of the low rate of advancing, suggesting that the low-rate dynamic contact angle θ_{dyn} is identical to the static contact angle θ_{stat}. This result re-confirms validity of the experimental protocol used by Li *et al.* [64, 65] to measure static contact angles and is also in good agreement with recent work to determine low-rate dynamic contact angles by the automated capillary rise technique [92, 93].

With each liquid, ten different measurements (i.e. ten different rates of advancing on ten new surfaces) were performed [87]. These dynamic contact angles were found to be essentially independent of the velocity of the three-phase contact line, as is, in principle, obvious from Figure 4. Since the low-rate dynamic contact angles

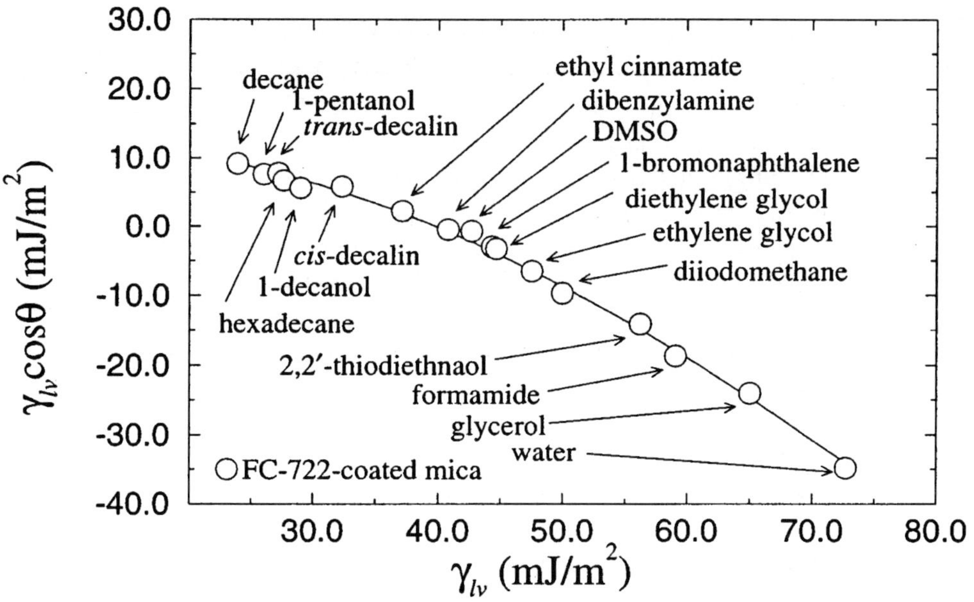

Figure 5. $\gamma_{lv}\cos\theta$ vs. γ_{lv} for the dynamic contact angles of the 17 liquids on a FC-722-coated mica [87]. This result reconfirms the finding of Li *et al.* [64, 65] shown in Figure 1 that at constant γ_{sv}, $\gamma_{lv}\cos\theta$ changes smoothly with γ_{lv}, independent of other liquid properties.

are essentially independent of the velocity of the three-phase contact line, a mean dynamic contact angle for each pure liquid was determined by averaging the contact angles at the ten different rates of advancing.

Figure 5 shows all experimental contact angle results at very slow motion of the three-phase contact line in a plot of $\gamma_{lv}\cos\theta$ vs. γ_{lv}. This result reconfirms the finding of Li *et al.* [64, 65] shown in Figure 1 that, at constant γ_{sv}, $\gamma_{lv}\cos\theta$ changes smoothly with γ_{lv}, independent of the liquid properties. Clearly, if there were any additional and independent effects of intermolecular forces on the contact angles, the data points for the polar liquids (1-pentanol, 1-decanol, DMSO, 2,2′-thiodiethanol, diethylene glycol, ethylene glycol, glycerol, and water) in Figure 5 would not fall completely on a smooth curve along with the non-polar liquids. It can be seen that the data point for 1-pentanol is slightly below the curve; 1-decanol is again slightly below; DMSO is above and ethylene glycol is on the curve: there is no evidence of any systematic deviation of the polar liquids away from such a curve. This question will also be addressed in Section 2.4.1.

In these low-rate dynamic contact angle experiments, images of an advancing drop (and hence information such as surface tension and contact angle) are recorded continuously as drop volume is steadily increased from below the surface. The procedures used here are different from those by Li *et al.* [64, 65] in that the contact angles measured by Li *et al.* were static angles, i.e. contact angles at zero velocity of the three-phase contact line.

*2.3.3. Non-Inert (Polar) Surfaces: Poly(propene-*alt-N-*(*n*-propyl)-maleimide) and Poly(propene-*alt-N-*(*n*-hexyl)maleimide).* Experience has shown that non-polar surfaces, such as Teflon and fluorocarbons, often are quite inert with respect to many liquids; however, polar surfaces often are less inert and hence may show different contact angle patterns, due to such causes as physical/chemical reaction and/or swelling and dissolution of the solid by the liquid. Since low-rate dynamic contact angle experiments when interpreted by ADSA-P have many advantages over the conventional way of manually placing tangents to the sessile drops, ADSA-P is employed here to measure low-rate dynamic contact angles on the copolymers, poly(propene-*alt*-*N*-(*n*-propyl)maleimide) and poly(propene-*alt*-*N*-(*n*-hexyl)maleimide), in order to elucidate the discrepancies between the results in Figures 1 and 5 on the one hand, and those in Figure 2 by a goniometer on the other. It will become apparent that a goniometer technique is liable to produce a mixture of meaningful and meaningless angles, with no criterion to distinguish between the two: if one disregards the meaningless angles to be identified by dynamic ADSA measurements, the results are in harmony with those patterns shown in Figures 1 and 5.

Eight liquids were selected for the contact angle measurements on poly(propene-*alt*-*N*-(*n*-propyl)maleimide) and thirteen liquids on poly(propene-*alt*-*N*-(*n*-hexyl)maleimide). The chemical structure of these copolymers has been described elsewhere [83]. Si wafers <100> were again selected as the substrate. They were obtained as circular discs of about 10 cm diameter and were cut into rectangular size of about 2×3 cm^2. Before soaking the rectangular wafer surfaces into chromic acid, a hole of about 1 mm diameter was made by drilling, using a diamond drill bit from Lunzer (New York; SMS-0.027). The general procedures to prepare the polymer-coated solid surfaces were similar to those described in Section 2.2. Such surfaces have been studied by Atomic Force Microscopy [94] and found that surface roughness is of the order of nanometer.

Figure 6 shows the contact angle result for glycerol on the poly(propene-*alt*-*N*-(*n*-propyl)maleimide) surface. Since the contact angles are essentially constant for all experiments, they can be averaged: in this specific example, averaging yields a mean contact angle of $70.8 \pm 0.1°$. Similar results were also obtained for water, 2,2′-thiodiethanol and 1-bromonaphthalene [83]. For each liquid, at least 5 and up to 10 different experiments (for different rates of advancing) were performed, on each newly prepared surface.

Unfortunately, not all liquids yield essentially constant advancing contact angles. Figure 7 shows the results of formamide on the same copolymer. It can be seen that as drop volume increases initially, contact angle increases from 60° to 63° at essentially constant three-phase radius. As V continues to increase, θ suddenly decreases to 60° and the three-phase contact line starts to move. The contact angle then decreases slowly from 60° to 54°. The surface tension-time plot indicates that the surface tension of formamide decreases with time. This suggests that dissolution of the copolymer occurs, causing γ_{lv} to change from that of the pure liquid. Similar

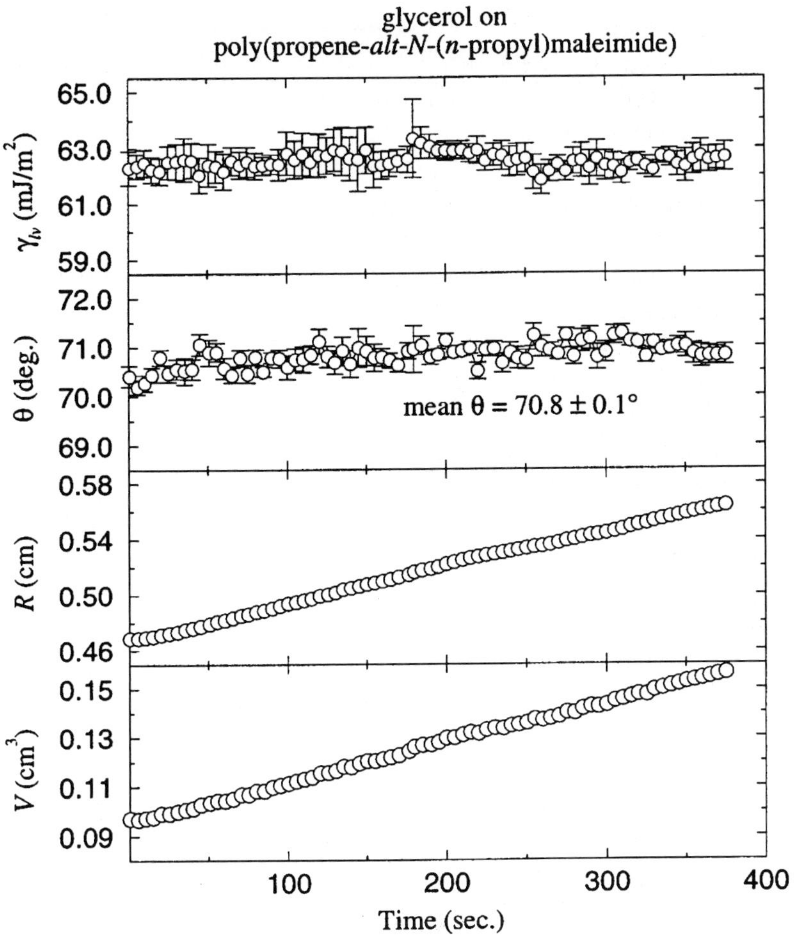

Figure 6. Low-rate dynamic contact angles of glycerol on a poly(propene-*alt*-*N*-(*n*-propyl)-maleimide) copolymer [83]. γ_{lv}, *R* and *V* are liquid-vapor surface tension, radius and volume of the sessile drop, respectively.

behavior can also be observed in other experiments [83]. It is an important question to ask which contact angles one should use for the interpretation in terms of surface energetics. Since physical/chemical reactions such as polymer dissolution alter the liquid-vapor, solid-liquid and solid-vapor interface (interfacial tensions) in an unknown manner, the contact angle data in Figure 7 should be disregarded for the interpretation in terms of surface energetics. The criteria for rejecting contact angles for calculations of solid surface tensions will be discussed in detail in Section 2.5. Obviously, it is virtually impossible for goniometer measurements to detect the complexities shown in Figure 7, e.g. the decrease in γ_{lv}. Thus, the contact angle obtained from a goniometer for this and similar solid-liquid systems cannot be meaningful.

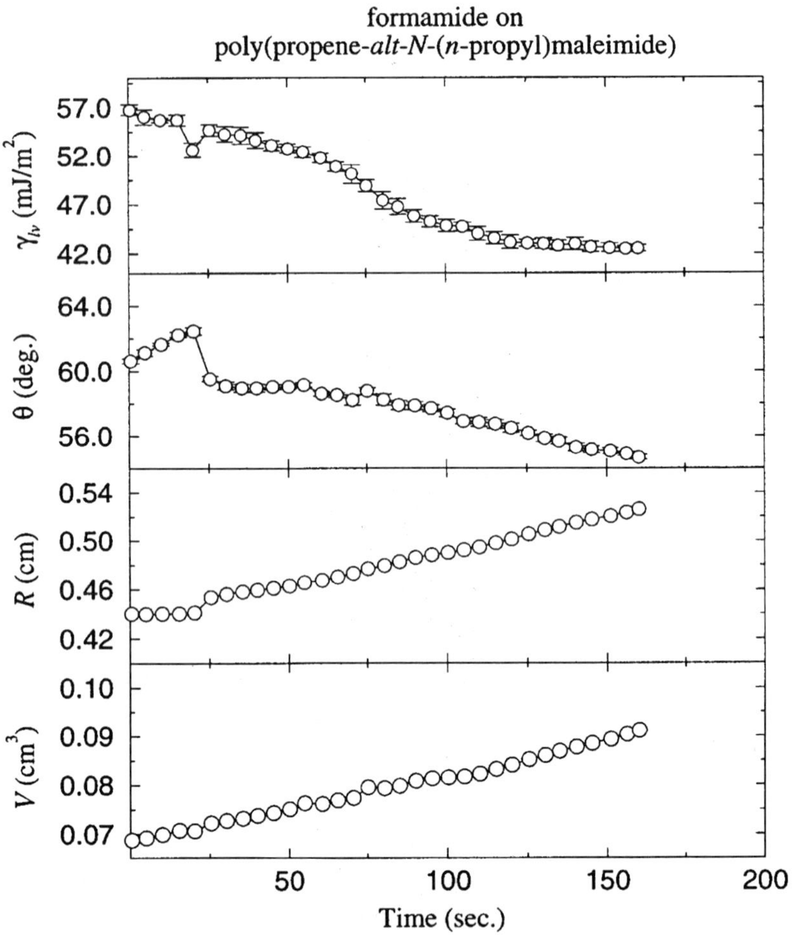

Figure 7. Low-rate dynamic contact angles of formamide on a poly(propene-*alt*-*N*-(*n*-propyl)-maleimide) copolymer [83]. γ_{lv}, R and V are liquid-vapor surface tension, radius and volume of the sessile drop, respectively.

The results of ethylene glycol are shown in Figure 8. It can be seen that the contact angle increases slowly from 54° to 57° as the three-phase contact line advances from 0.46 to 0.54 cm. While the cause of this increase in the contact angle is unclear, it is suspected that the operative solid-liquid interfacial tension is changed slowly due to physico-chemical reaction: according to Young's equation, if the values of γ_{lv} and γ_{sv} are constant, a change in the contact angle must be a consequence of a change in γ_{sl}. The observed trends in the contact angle may well start immediately after drop formation, not only after the measurement procedure was set in motion. Also, there is no reason to suspect that the change of contact angle with time (or radius) would cease once the measurement was terminated. More likely, such trends would continue. Because there is no unique apparent contact angle and it is unclear whether or not γ_{sl} will remain constant and whether Young's equation

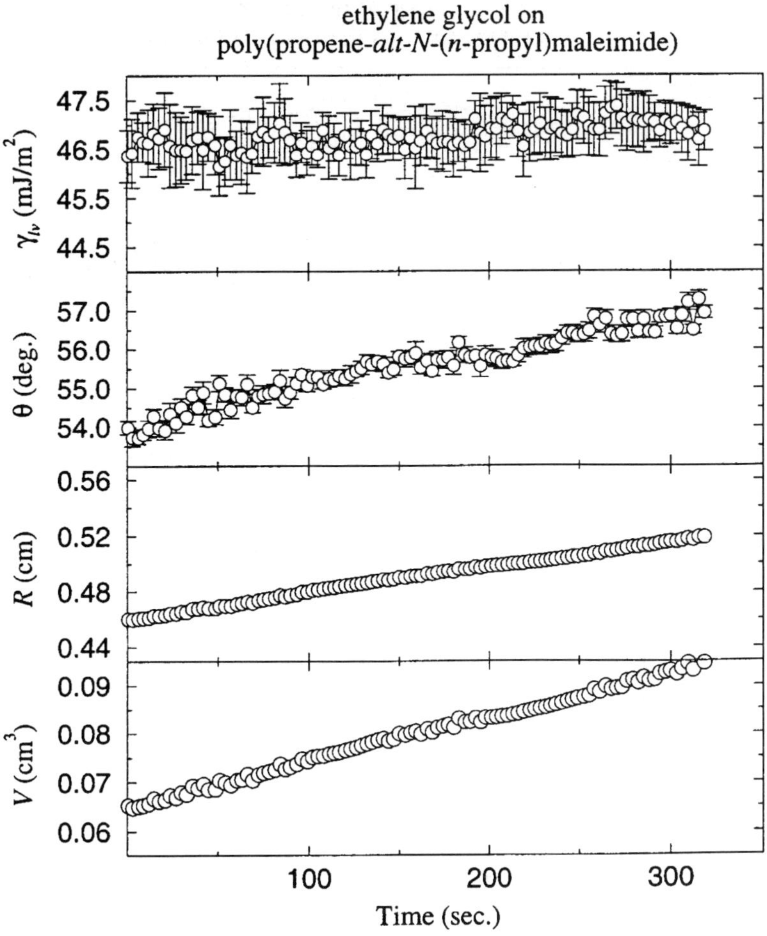

Figure 8. Low-rate dynamic contact angles of ethylene glycol on a poly(propene-*alt*-*N*-(*n*-propyl)-maleimide) copolymer [83]. γ_{lv}, *R* and *V* are liquid-vapor surface tension, radius and volume of the sessile drop, respectively.

is applicable, these angles should be excluded from the interpretation in terms of surface energetics. However, one might consider averaging the contact angles for *R* larger than, e.g., 0.48 cm, since γ_{lv} seems to be constant and since the contact angle error after averaging would not be very large, i.e. $\theta = 55.7 \pm 0.6°$. Such averaging is not allowed: apart from the experimental reasons given above, the question of whether averaging over some data is allowed or not has firm quantitative answers based on the laws of statistics. The main fact is that the statistical analysis rejects "averaging" of the above contact angle data over time by a very large margin [83].

While the contact angle data of formamide, diiodomethane, ethylene glycol, and diethylene glycol for poly(propene-*alt*-*N*-(*n*-propyl)maleimide) should be disregarded, the contact angles of water, glycerol, 2,2′-thiodiethanol, and 1-bromonaphthalene can be used for the interpretation in terms of surface energetics.

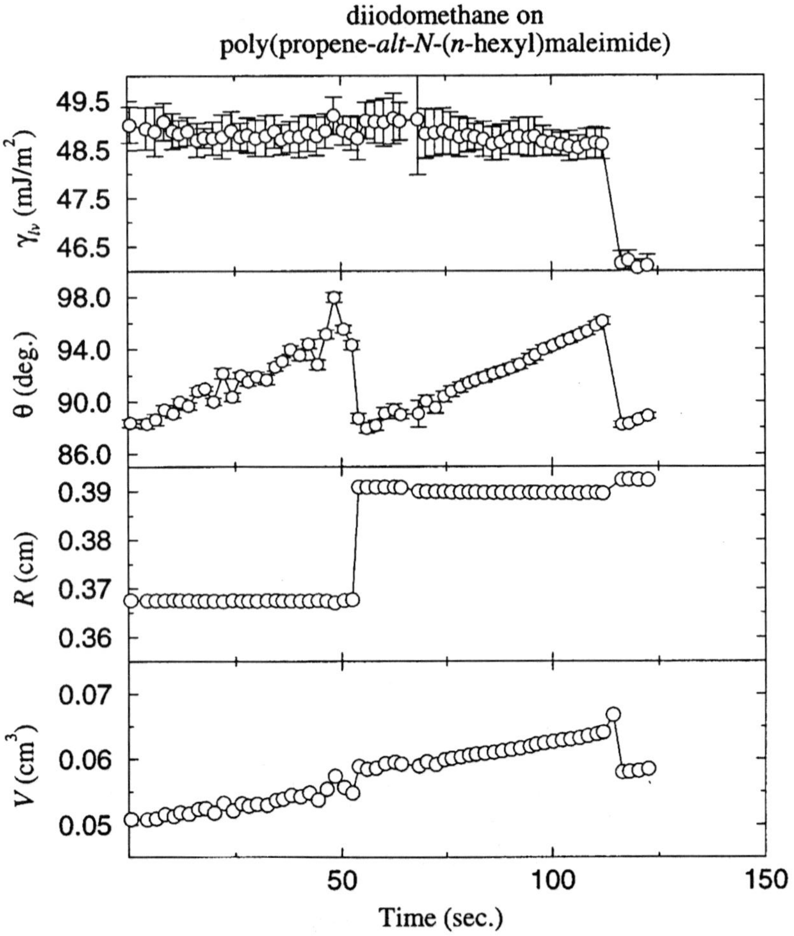

Figure 9. Low-rate dynamic contact angles of diiodomethane on a poly(propene-*alt*-*N*-(*n*-hexyl)-maleimide) copolymer [83]. γ_{lv}, R and V are liquid-vapor surface tension, radius and volume of the sessile drop, respectively.

A similar study was conducted for the second copolymer, poly(propene-*alt*-*N*-(*n*-hexyl)maleimide). It was found that only the advancing contact angles of water, glycerol, diethylene glycol, and *cis*-decalin are essentially constant [83]. The remaining liquids, formamide, 2,2′-thiodiethanol, diiodomethane, ethylene glycol, 1-bromonaphthalene, dimethyl sulfoxide (DMSO), dibenzylamine, ethyl cinnamate, and 2,5-dichlorotoluene all showed very complex contact angle patterns which had to be excluded from the interpretation in terms of surface energetics and testing of approaches for interfacial tensions.

A different contact angle experiment is shown in Figure 9 for diiodomethane: initially the apparent drop volume, as perceived by ADSA-P, increases linearly, and the contact angle increases from 88° to 96° at essentially constant three-phase radius. Suddenly, the drop front jumps to a new location as more liquid is supplied into the sessile drop. The resulting contact angle decreases sharply from 96° to 88°. The contact angle increases again as more liquid is supplied into the drop. Note that the change in the contact angles happened for R moved in only 0.2 mm. Obviously, this mechanism cannot be easily detected by the goniometer technique and procedures. Such slip/stick behavior could be due to non-inertness of the surface. Phenomenologically, an energy barrier for the drop front exists, resulting in sticking, which causes θ to increase at constant R. However, as more liquid is supplied into the sessile drop, the drop front possesses enough energy to overcome the energy barrier, resulting in slipping, which causes θ to decrease suddenly. It should be noted that as the drop front jumps from one location to the next, it is unlikely that the drop will remain axisymmetric. Such a non-axisymmetric drop will obviously not meet the basic assumptions underlying ADSA-P, causing possible errors, e.g., in the apparent surface tension and drop volume. This can be seen from the discontinuity of the apparent drop volume and apparent surface tension with time as the drop front sticks and slips. Similar behavior can also be observed in other experiments [83, 95]. Obviously, the observed contact angles cannot all be Young contact angles, since γ_{lv}, γ_{sv} (and γ_{sl}) are constants, so that because of Young's equation, θ ought to be a constant. In addition, it is difficult to decide unambiguously at this moment whether or not Young's equation is applicable at all. Therefore, these contact angles should not be used for the interpretation in terms of surface energetics.

While pronounced cases of slip/stick behavior can indeed be observed by the goniometer, it is virtually impossible to record the entire slip/stick behavior manually. In this case, the goniometer contact angle can be very subjective, depending on the skill of the experimentalist. It is expected that a contact angle thus recorded by the goniometer should agree with the maximum angle obtained by ADSA-P. Indeed, a contact angle of 98° was observed, in reasonable agreement with the maxima in the entire slip/stick pattern of the ADSA-P results ($\theta \approx 96°$ in Figure 9). In cases where the liquid-vapor surface tension of the sessile drop decreases due to, e.g., dissolution of the surface, only ADSA can detect changes in the liquid-vapor surface tension. The distinctions and differentiations made by ADSA-P are not possible in a goniometer study. Thus, circumspection is necessary in the decision whether or not experimental contact angles can be used in conjunction with Young's equation; contact angles from a conventional goniometer-sessile drop technique may produce contact angles which do not fulfill the basic assumptions made in all surface energetic approaches [10]–[17], e.g. constancy of γ_{lv} and applicability of Young's equation already mentioned in the Introduction. These various assumptions will be discussed in detail in Section 2.5.

A comparison of the angles measured here and those by the goniometer technique had been made [83]. It was found that the goniometer angles correspond very well with those of ADSA-P only in situations where the contact angles are constant; in cases where complexities of contact angles arise, the goniometer contact angles correspond only to the maxima of the angles from ADSA-P. The picture emerging in

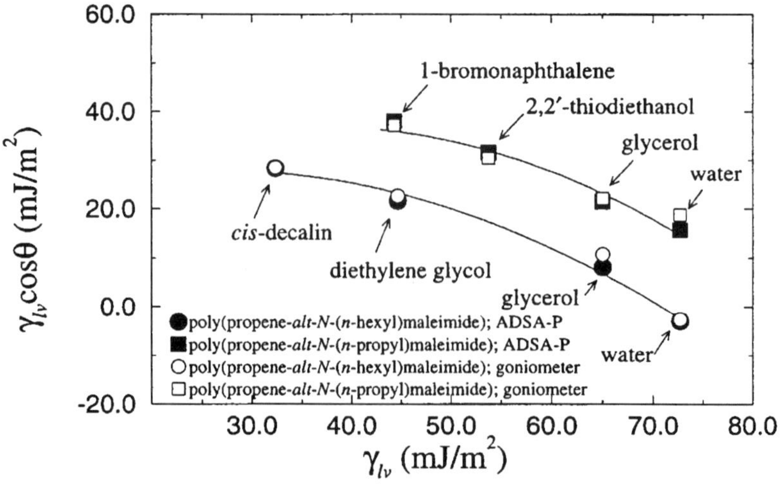

Figure 10. $\gamma_{lv}\cos\theta$ vs. γ_{lv} for the poly(propene-*alt*-*N*-(*n*-propyl)maleimide) and poly(propene-*alt*-*N*-(*n*-hexyl)maleimide) copolymers [83]. Figure 2 changes drastically upon elimination of the angles shown to be meaningless using the dynamic contact angle procedures by ADSA-P.

Figure 2 changes drastically upon elimination of the angles shown to be meaningless in the ADSA-P study, see Figure 10. The curves in Figure 10 are in harmony with the results obtained for more inert polar and non-polar surfaces in Figures 1 and 5. Again, it can be concluded that $\gamma_{lv}\cos\theta$ changes smoothly and systematically with γ_{lv} at constant γ_{sv}. Because of Young's equation, a relation of the form of Eq. (7) can be deduced:

$$\gamma_{sl} = f(\gamma_{lv}, \gamma_{sv})$$

Thus, the experimental procedures and techniques used are crucial in the collection of contact angle data for the determination of surface energetics; conventional goniometer techniques may produce contact angles which violate the basic assumptions made in all surface energetic approaches. In other words, the most serious shortcoming of goniometer studies is not subjectivity and lack of accuracy, but its inability to distinguish between meaningful and meaningless contact angle measurements. A recent study [96] has shown that using the above dynamic procedures with a simple polynomial fit to sessile drop images may also allow one to identify meaningless and meaningful contact angles.

2.3.4. Other Non-Polar and Polar Surfaces. Similarly, low-rate dynamic contact angles of a large number of liquids were studied extensively on various solid surfaces. They are FC-722-coated Si wafer [96], FC-725-coated Si wafer [97], poly(*n*-butyl methacrylate) P*n*BMA [98], polystyrene PS [99], poly(styrene-(hexyl/10-carboxydecyl 90:10)maleimide) [100], poly(methyl methacrylate/*n*-butyl methacrylate) P(MMA/*n*-BMA) [101], poly-(methyl methacrylate) PMMA [102], and poly(propene-*alt*-*N*-methylmaleimide) [96]. Details with respect to solid surface preparation and experimental results have been given elsewhere [96]–[102].

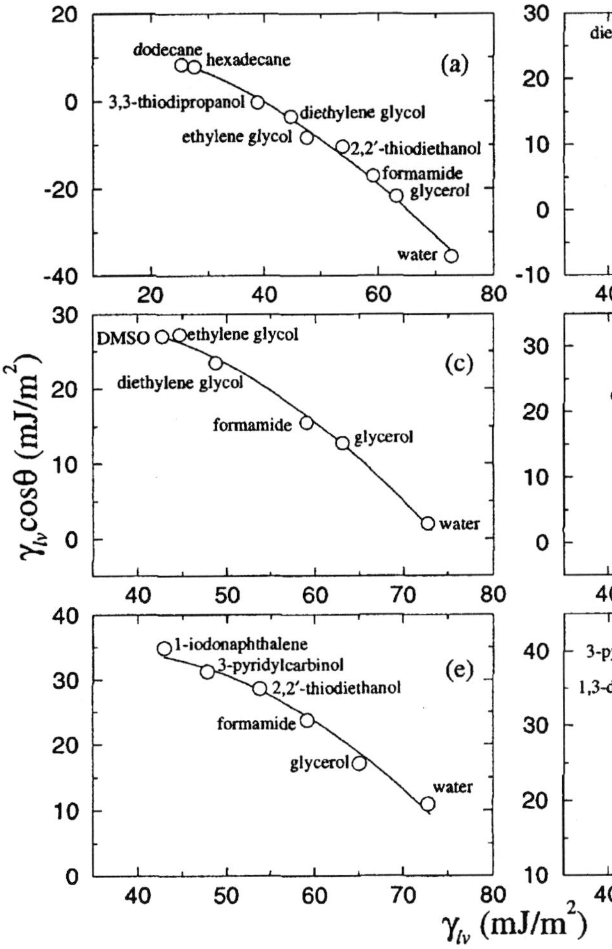

Figure 11. $\gamma_{lv}\cos\theta$ vs. γ_{lv} for (a) FC-725-coated Si wafer [97], (b) poly(*n*-butyl methacrylate) P*n*BMA [98], (c) polystyrene PS [99], (d) poly(styrene-(hexyl/10-carboxydecyl 90:10)maleimide) [100], (e) poly(methyl methacrylate/*n*-butyl methacrylate) P(MMA/*n*BMA) [101], and (f) poly(methyl methacrylate) PMMA [102].

Again, it was found that not all contact angles could be used for surface energetics purposes. For example, slip/stick of the three-phase contact line, γ_{lv} decrease, and θ decrease or increase as the drop front advances, and dissolution of the polymers by the liquids were observed. The meaningful angles for these polymers (copolymers) are plotted in Figure 11, in plots of $\gamma_{lv}\cos\theta$ vs. γ_{lv}. These curves are in harmony with those patterns shown in Figures 1, 5, and 10. The universality of these contact angle patterns will be discussed in the next section.

2.4. Universality of Contact Angle Patterns

In addition to ADSA-P, the capillary rise at a vertical plate technique [92, 93, 103] is also suitable for the determination of low-rate dynamic contact angles. Since the capillary rise at a vertical plate technique has been automated [103] to perform various dynamic advancing and receding contact angle measurements at immersion speeds ranging from 0.008 to 0.9 mm/min., measurements can be and have been performed [92, 93] under dynamic conditions closely resembling those in the ADSA-P experiments. The capillary rise experiments are performed by immersing a vertical (plate) solid surface into the liquid, at a constant (slow) speed; during the immersion, motion of the three-phase contact line along the vertical solid surface is always monitored. The task of measuring a contact angle is reduced to the measurement of a length (capillary rise), which can be performed optically with a high degree of accuracy. In several instances, the capillary rise technique has been employed on relatively inert and well-prepared surfaces: FC-721-coated mica [93], heat-pressed Teflon FEP [93], hexatriacontane [51, 104], and cholesteryl acetate [51, 105]. The FC-721 and FEP surfaces were prepared, respectively, by a dip-coating technique and a heat-pressing technique [64]. Hexatriacontane and cholesteryl acetate were produced by vapor deposition in a vacuum [51, 104, 105]; the surface quality was found to be so good that there was no contact angle hysteresis when water was used. It has been found that the contact angles determined from ADSA-P and the capillary rise at a vertical plate technique are virtually identical for the same solid-liquid systems [92, 93].

In order to illustrate the universality of the contact angle patterns, Figure 12 shows recent dynamic contact angles for various solid surfaces measured by the capillary rise technique and those by ADSA-P: FC-721-coated mica [93], FC-722-coated mica [87] and -coated Si wafer [96] FC-725-coated Si wafer [97], heat-pressed Teflon FEP [93], hexatriacontane [51, 104], cholesteryl acetate [51, 105], poly(propene-*alt*-*N*-(*n*-hexyl)maleimide) [83, 95], poly(*n*-butyl methacrylate) P*n*BMA [98], polystyrene PS [99], poly(styrene-(hexyl/10-carboxydecyl 90:10)-maleimide) [100], poly(methyl methacrylate/*n*-butyl methacrylate) P(MMA/*n*BMA) [101], poly(propene-*alt*-*N*-(*n*-propyl)maleimide) [83, 95], poly(methyl methacrylate) PMMA [102] and poly(propene-*alt*-*N*-methylmaleimide) [96]. Details of the solid surface preparation have been given elsewhere [51, 83, 87, 93], [95]–[105]. The ADSA-P (low-rate) dynamic contact angles were established also by the procedures developed in the preceding section. Again, the results in Figure 12 suggest that the values of $\gamma_{lv} \cos\theta$ change systematically with γ_{lv} in a very regular fashion, from the hydrophobic hydrocarbon surfaces to the hydrophilic poly(propene-*alt*-*N*-methylmaleimide) surface, and that the patterns are independent of the experimental technique on the one hand and liquid structure on the other. The regularity of the contact angle patterns is remarkable.

For a given solid surface, γ_{sv} is assumed — reasonably — to be constant; the results suggest that $\gamma_{lv} \cos\theta$ depends only on γ_{lv}. Changing the solid surface (and hence γ_{sv}) shifts the curve in a very regular manner. Overall, these experimental results suggest again that $\gamma_{lv} \cos\theta$ depends only on γ_{lv} and γ_{sv}:

$$\gamma_{lv} \cos\theta = f(\gamma_{lv}, \gamma_{sv})$$

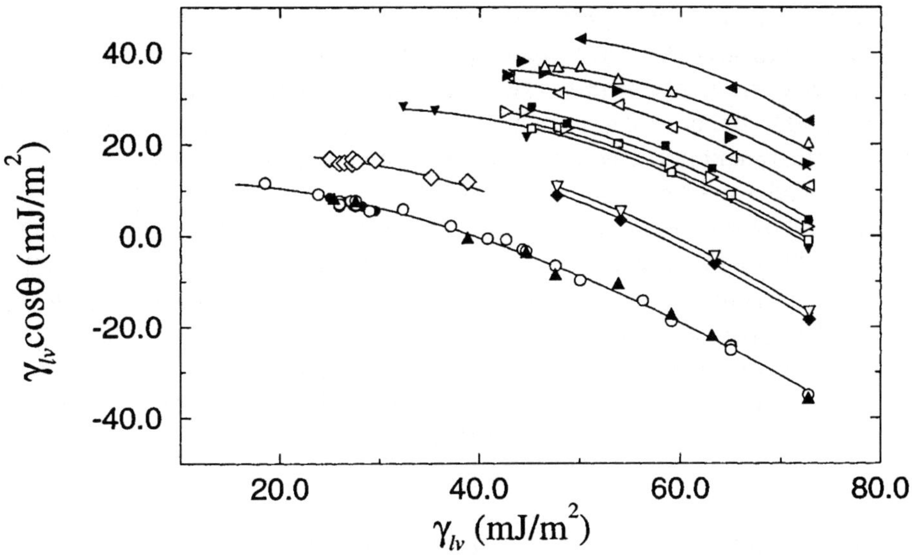

● FC-721-coated mica [93]
○ FC-722-coated mica [87] and Si wafer [96]
▲ FC-725-Si coated wafer [97]
◇ Teflon FEP [93]
◆ hexatriacontane [51,104]
▽ cholesteryl acetate [51,105]
▼ poly(propene-*alt*-*N*-(*n*-hexyl)maleimide) [83,95]
□ poly(*n*-butyl methacrylate) [98]
▷ polystyrene [99]
■ poly(styrene-(hexyl/10-carboxydecyl 90:10)maleimide) [100]
◁ poly(methyl methacrylate/*n*-butyl methacrylate) [101]
▶ poly(propene-*alt*-*N*-(*n*-propyl)maleimide) [83,95]
△ poly(methyl methacrylate) [102]
◀ poly(propene-*alt*-*N*-methylmaleimide) [96]

Figure 12. $\gamma_{lv}\cos\theta$ vs. γ_{lv} for various solid surfaces.

where f is as yet an unknown function. The specific intermolecular forces of the liquids and solids, which give rise to the surface tensions, do not have additional and independent effects on the contact angles. Thus, one can change the contact angle, and because of Young's equation, the solid-liquid interfacial tension, by the simple mechanism of changing either the liquid or the solid. Combining Eq. (5) with Young's equation, we can express γ_{sl} as another unknown function F of only γ_{lv} and γ_{sv}, allowing a search for an equation (7) in the form

$$\gamma_{sl} = \gamma_{sv} - f(\gamma_{lv}, \gamma_{sv}) = F(\gamma_{lv}, \gamma_{sv})$$

This is called an equation-of-state relation, as a relation involving intensive properties [106]. Indeed, equations of this form have been in the literature for a long time, e.g., Antonow's [107] and Berthelot's [108] rules, see later.

2.4.1. Reasons for Deviation from Smoothness. As Figure 12 stands, on one and the same solid surface, there are minor deviations of some contact angle data from the curves and one might wish to argue that intermolecular forces (e.g. hydrogen bonding) could still have some independent effects on the contact angles. This question will be addressed here, by focusing on the contact angle data of three chemically similar methacrylate polymers.

Figure 13 shows a plot of $\gamma_{lv} \cos\theta$ vs. γ_{lv} curves for the PnBMA, P(MMA/n-BMA), and PMMA polymers. As can be seen in this figure, there are slight deviations of the data points from the curves; however, it is not apparent that these deviations are systematic and are caused by any independent effects of intermolecular forces on the contact angles. For example, in the case of 3-pyridylcarbinol with γ_{lv} of about 48.0 mJ/m^2, the data point on the PMMA curve appears to be slightly higher, while the point on the P(MMA/nBMA) is slightly lower and the one for PnBMA slightly higher again. There is no evidence for any systematic variation, for this and other liquids. Clearly, if there were a deviation due to specific intermolecular forces, one would, at the most, expect a monotonous change of the deviation when going from PMMA to P(MMA/nBMA) to PnBMA. Obviously, this does not mean that intermolecular forces are irrelevant; they determine the primary surface tensions of the liquids and solids.

To keep matters in perspective, it has to be realized that the curves in Figure 13 would have to be considered completely smooth if a conventional goniometer technique with $\pm 2°$ contact angle accuracy had been used. Further, the fact that an equilibrium spreading pressure π_e

$$\pi_e = \gamma_s - \gamma_{sv} \tag{11}$$

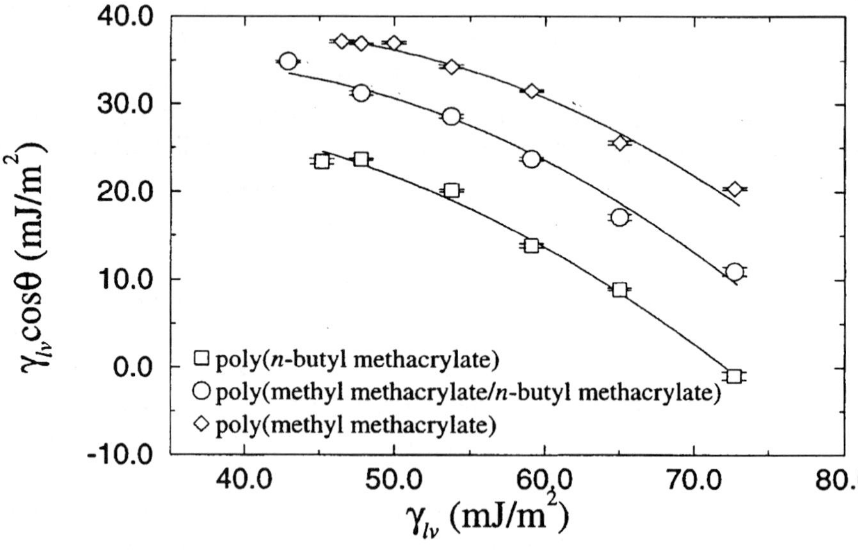

Figure 13. $\gamma_{lv} \cos\theta$ vs. γ_{lv} for three methacrylate polymer-coated Si wafer surfaces.

of as low as 1 mJ/m^2 would easily contribute to such a variation should not be overlooked; γ_s is the surface tension of a solid in a vacuum. On one and the same solid surface, γ_{sv} is expected to be constant when vapor adsorption is negligible; however, if vapor adsorption of liquids does play a role, then γ_{sv} can be different from liquid to liquid even on one and the same type of solid surface. The answer to the question of whether or not vapor adsorption has an effect on the contact angle patterns in Figure 13 (and similar figures) would require general criteria to distinguish this from all other effects, such as swelling of a solid surface. As Figure 13 stands, it does not appear that vapor adsorption plays a significant role in the contact angles presented here; otherwise, one would expect such an effect to manifest itself as random variations of the contact angles, rather than the remarkable regularity. Nevertheless, minor adsorption which causes the equilibrium spreading pressure π_e to vary by as much as 1 mJ/m^2 has been considered possible for low-energy solid surfaces [109]. A recent study has examined the effect of adsorption on the contact angles for some high-energy ceramic/metal systems [110].

Although the liquids and solids (polymers) selected are of high purity, e.g. >99%, minute impurities present in either the liquids or the solids are unavoidable, and finding liquids and solids (polymers) which contain absolutely no impurities is unrealistic. There is no guarantee that such matters could not have caused the apparent minor deviations of the points in Figure 13 (and similar figures). Even very minor swelling of the polymer [54] or creeping of the liquid could easily introduce a slight deviation of some points from such curves. Therefore, only after considering all such possibilities would the need or justification arise for explanations in terms of direct effects of intermolecular forces [45]. There is no reason to suppose that intermolecular forces have any additional and independent effects on the contact angles beyond the fact that intermolecular forces determine the primary interfacial tensions, γ_{lv} and γ_{sv}; the two interfacial tensions determine γ_{sl} through Eq. (7); the three interfacial tensions then determine the contact angle, as given by Young's equation. This fact is most easily understood in terms of the experimental contact angle patterns shown in Figures 1, 5, and 10–13. Recent studies of dynamic contact angles on various copolymer/polymer surfaces revealed similar findings [111]–[113].

2.5. Criteria for Calculations of Surface Energetics

In Section 2.3.3, we have shown that there exist a large number of contact angle complexities which prevent use of the measurements for surface energetic calculations. To arrive at the curves shown in Figure 12, several assumptions already mentioned before were used to eliminate the meaningless contact angles. However, there remains one last assumption which has not been used and discussed: the constancy of γ_{sv} in going from liquid to liquid. It will be apparent in this section that such an assumption is indeed needed for surface energetics calculations; the importance of the various underlying assumptions in the determination of solid surface tensions will be discussed in detail. They will be addressed further in

Section 3.1, in conjunction with the surface tension component approaches [11], [15]–[17].

2.5.1. Accepted Assumptions for Calculations of Surface Energetics.

(1) It has to be realized that Young's equation (3)

$$\gamma_{lv} \cos \theta_Y = \gamma_{sv} - \gamma_{sl}$$

is the only thermodynamic relation to interrelate the three surface tensions, γ_{lv}, γ_{sv}, and γ_{sl}, with the experimental contact angle θ. Therefore, the expectation of the applicability of Young's equation has to be fulfilled. As in the case of slip/stick of the three-phase contact line shown in Figure 9, Young's equation cannot be applicable: Since γ_{lv}, γ_{sv}, and γ_{sl} are material properties and are expected to be constant, Young's equation implies a unique contact angle. Thus, contact angles from slip/stick of the three-phase contact line have to be discarded.

(2) Obviously, contact angle interpretation of surfactant solutions or mixtures of liquids is expected to be more complicated than that of a pure liquid. It has been found that if one measures contact angles of mixture solutions on one and the same solid surface, scatter, or patterns different from those in plots of the type in Figure 12 would arise [114]. Thermodynamically, such systems have three degrees of freedom [72]–[75]. The additional degree of freedom comes from the effect of an additional liquid component. While Young's equation may still be applicable in this case, no contact angle approach as yet allows the determination of solid surface tensions from such angles. Therefore, pure liquids are always used in contact angle measurements. However, even if this is the case, one has to ensure that γ_{lv} remains constant during the experiment. In an example shown earlier in Figure 7, polymer dissolution occurs, causing the liquid surface tension to differ from that of the pure liquid. This reflects the fact that the presumed pure liquid has been changed to a mixture of liquid/polymer solution. If a conventional goniometer technique had been used instead, the change in the operative γ_{lv} might not be detected. Such goniometer measurements would inevitably reflect a solid-liquid system with a changed γ_{lv}, rather than the γ_{lv} of the presumed pure liquid. Thus, deductions and conclusions from such angles would be in error. A recent study has attempted to interpret contact angles of binary liquids in terms of surface energetics; the results obtained from a thermodynamic approach similar to that used in the equation-of-state approach (see later) appears to be promising [115].

(3) If a chemical/physical reaction takes place, any of γ_{lv}, γ_{sv}, and γ_{sl} could change during the experiment, and because of Young's equation (3), the contact angle θ would also change. Therefore, changes in θ suggest that at least one of γ_{lv}, γ_{sv}, and γ_{sl} is changing. In an example shown in Figure 8, the contact angle increases as the drop front advances with essentially constant γ_{lv}. While there

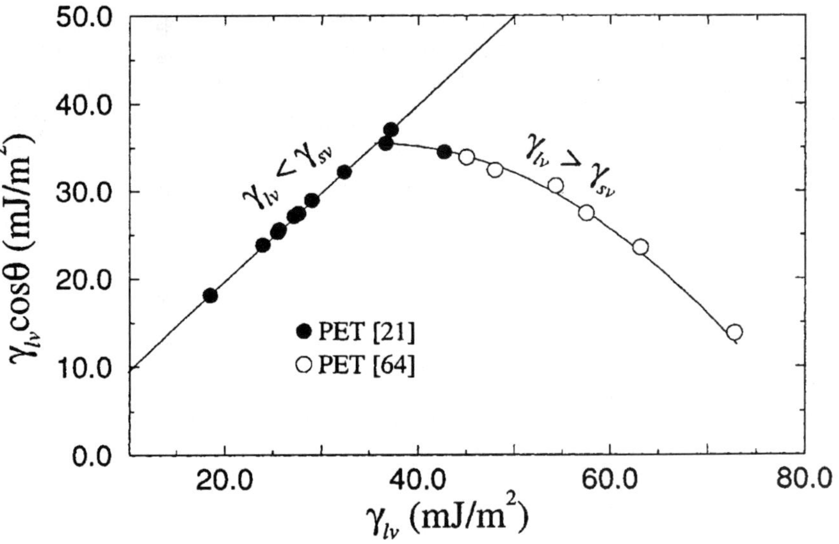

Figure 14. $\gamma_{lv}\cos\theta$ vs. γ_{lv} for a PET surface. Contact angle data for $\gamma_{lv} < \gamma_{sv}$ and $\gamma_{lv} > \gamma_{sv}$ are shown.

is no reason to question the applicability of Young's equation, γ_{sl} is suspected to change: from Young's equation, if γ_{lv} and γ_{sv} are constant, changes in θ must be a consequence of a change in γ_{sl}. While such cases might be very interesting in a different context, they have to be discarded because our goal is to deduce solid surface tension from the simplest possible situations.

(4) If $\gamma_{lv} \approx \gamma_{sv}$ or $\gamma_{lv} < \gamma_{sv}$, complete wetting occurs; a change in γ_{sv} induced by the solid/liquid contact (autophobicity) may also occur. In order to illustrate this, a plot of $\gamma_{lv}\cos\theta$ vs. γ_{lv} is shown in Figure 14 for a PET surface. In this figure, liquids having surface tensions less than the anticipated PET surface tension are included, i.e. $\gamma_{lv} < \gamma_{sv}$. As can be seen, the values of $\gamma_{lv}\cos\theta$ increase as γ_{lv} decreases, reaching a global maximum. Further decrease in γ_{lv} causes the data point to fall on a straight line. Thus, the liquid surface tensions of the test liquids should be higher than that of the anticipated solid surface tension, by the appropriate choice of the liquids. Another possible effect of $\gamma_{lv} < \gamma_{sv}$ is liquid adsorption, which could cause γ_{sv} to be different from liquid to liquid. Therefore, the test liquids used in this study were selected to fulfill the condition $\gamma_{lv} > \gamma_{sv}$. A recent contact angle study has reconfirmed that contact angles should be measured for $\gamma_{lv} > \gamma_{sv}$ and that the measurements with $\gamma_{lv} < \gamma_{sv}$ contain no information about γ_{sv} [116].

(5) One last assumption which has not been used to identify meaningful contact angles in Section 2.3 is the assumption of the constancy of γ_{sv} going from liquid to liquid. This assumption is needed for reasons of procedure in deducing solid surface tensions. Between the three variables, γ_{lv}, γ_{sv}, and γ_{sl}, only γ_{lv} is measurable; an additional measurable quantity is, however, the contact

angle θ which can be interrelated to these quantities only by Young's equation. Therefore, it is impossible to study the direct effect of changing γ_{lv} on a second (non-measurable) quantity, γ_{sl}, through θ, unless the third (non-measurable) quantity, γ_{sv}, is kept constant. Otherwise, interpretation of the data showing implicitly the effect of changes of γ_{lv} on γ_{sl} would not be possible by any of the contact angle approaches of current interest. The assumption of constant γ_{sv} will be used to deduce solid surface tensions from contact angles in Section 3.

2.5.2. Experimental Criteria. The experimental results in Section 2.3.3 illustrate that there are three apparent contact angle complexities:

(1) slip/stick of the three-phase contact line;

(2) contact angle increase/decrease as the drop front advances; and

(3) liquid surface tension changes as the drop front advances.

These contact angles should not be used for the determination of solid surface tensions. With respect to the first point, slip/stick of the three-phase contact line indicates that Young's equation is not applicable. Increase/decrease in the contact angle and change in the liquid surface tension as the drop front advances violate the expectation of no physical/chemical reaction. Therefore, when the experimental contact angles and liquid surface tensions are not constant, they should be disregarded. However, the question arises as to whether the reverse is true, i.e. whether constancy of θ and γ_{lv} in the dynamic measurements described in Section 2.3 will always guarantee that the above contact angle assumptions (in the preceding section) are fulfilled. This question will be explored below.

In practice, many solids are not truly inert with respect to many liquids; even swelling of fluorocarbon surfaces by alkanes has been reported [54]. For example, if swelling of a solid surface occurs quickly upon the contact with a liquid as the drop front advances, θ will change to reflect the changed surface energetics. If the time scale of such an effect is much less than that of the contact angle measurements, θ could reflect the changed γ_{sl}. This might result in a rather constant contact angle, although the energetics would have changed. If this mechanism would not affect the liquid, γ_{lv} would remain constant, i.e. be independent of time. However, such constant contact angles would reflect only the energetics of the already swollen solid surface in contact with the liquid. Further, one and the same solid surface could swell differently by different test liquids in an unknown manner. This effect would manifest itself in scatter of the plots of $\gamma_{lv} \cos\theta$ vs. γ_{lv} because

$$\gamma_{sl} \neq F(\gamma_{lv}, \gamma_{sv}) \tag{12}$$

In this case, a systematic study of the effect of surface energetics is not easily possible since the energetics have changed in an unknown, more complex manner. Obviously, such contact angles would have to be disregarded even if the measured liquid surface tensions and contact angles are constant, because the assumption of no physical/chemical reaction would have been violated. Unfortunately, the strategies employed here will not allow identification of such matters, and it remains a possibility that such effects contribute to the minor deviations of some data points from a smooth curve.

An example for this more complex pattern is self-assembled monolayers (SAMs), which have been widely used to produce monolayer surfaces of different chemical compositions and wettabilities [117]–[120]. For this type of surfaces, it is expected that penetration of liquids into the SAMs is inevitable; different liquids would have different effects on such surfaces. Because penetration of liquid is expected to occur almost instantly as the liquid contacts the solid, the observed advancing angle might reflect a γ_{sl} value caused by a modification of the solid surface.

Low-rate dynamic contact angles of water on a octadecanethiol ($HS(CH_2)_{17}$-CH_3) SAM on a gold substrate are shown in Figure 15. It can be seen that the advancing

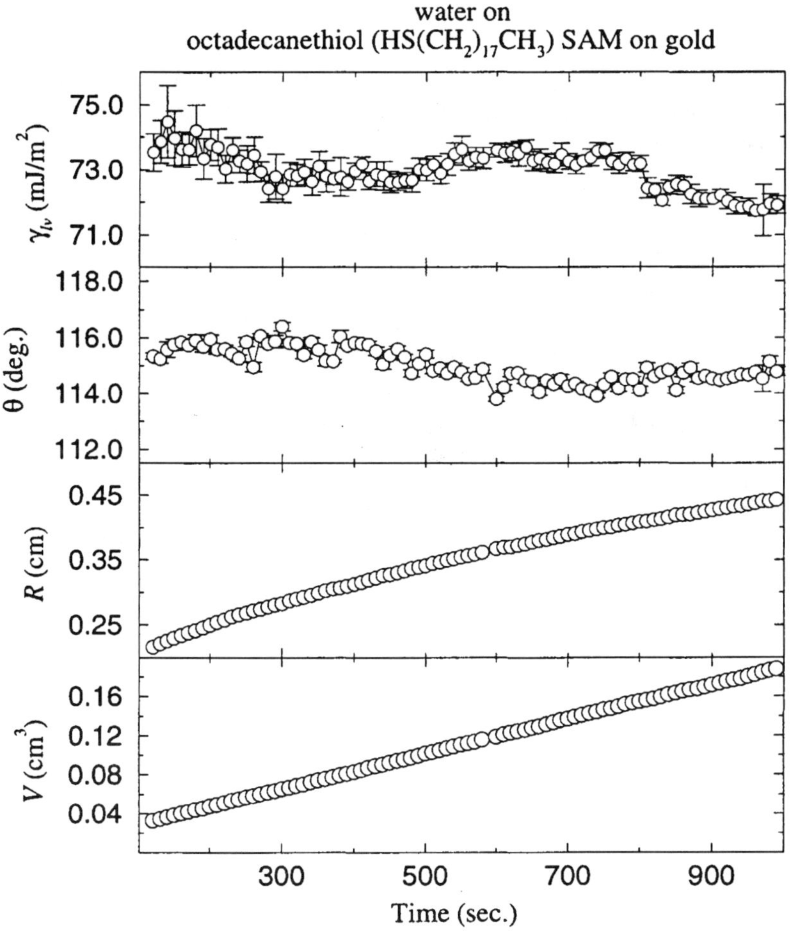

Figure 15. Low-rate dynamic contact angles of water on a octadecanethiol ($HS(CH_2)_{17}CH_3$) SAM on a gold substrate [91]. γ_{lv}, R and V are liquid-vapor surface tension, radius and volume of the sessile drop, respectively.

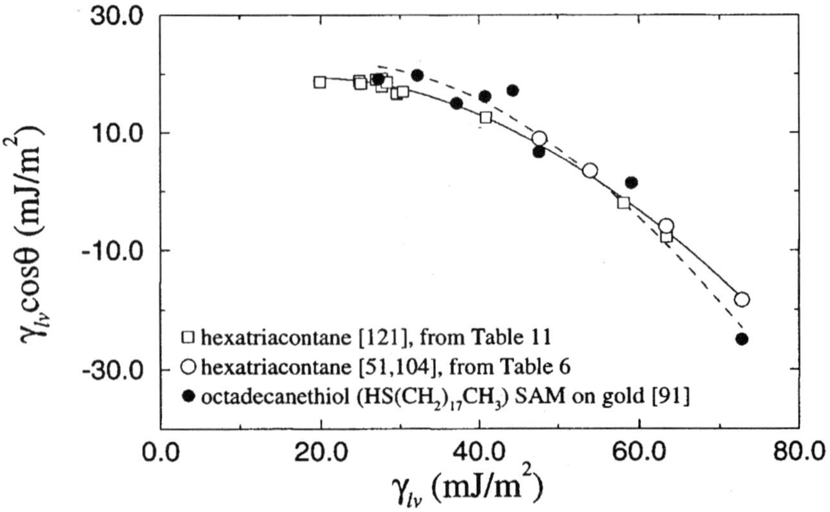

Figure 16. $\gamma_{lv}\cos\theta$ vs. γ_{lv} for a octadecanethiol ($HS(CH_2)_{17}CH_3$) SAM on a gold substrate, and for a hexatriacontane surface. Both curves are best-fit of Eq. (46) to experimental data from the octadecanethiol ($HS(CH_2)_{17}CH_3$) SAM (dashed line) and the hexatriacontane surface (solid line).

contact angles and liquid surface tension are quite constant. A plot of $\gamma_{lv}\cos\theta$ vs. γ_{lv} of various liquids for this surface is shown in Figure 16 (solid symbols). These contact angles were determined dynamically by ADSA-P in another study [91]. Contrary to the patterns shown in Figures 1, 5, and 10–13, no smooth curve results in Figure 16: both curves are best-fit of Eq. (46) to experimental data from the octadecanethiol ($HS(CH_2)_{17}CH_3$) SAM (dashed line) and a hexatriacontane surface (solid line). The reason for the difference in pattern is believed to be the effect of liquid penetration into the SAMs. Such mechanism is believed to be absent or negligible for the relatively thick polymer films used in Section 2.

It is instructive to compare this contact angle pattern with that obtained from the hexatriacontane surface [51, 104, 105, 121] (also plotted in Figure 16). Since both surfaces are expected to consist entirely of CH_3 groups, their solid surface tensions should be similar; one would then expect the two contact angle patterns to be essentially the same. However, the hexatriacontane data follow a smooth curve. From the point of view of surface energetics, the only difference for the two surfaces is that the SAM is a monolayer, whereas the hexatriacontane is a relatively thick crystalline layer. Obviously, it is the effect of liquid penetration and possible contact with the substrate of the SAMs which causes γ_{sl} to change in a pattern different from that which would prevail in the absence of liquid penetration. In this case, there is no reason to question the validity and applicability of Young's equation; however, different liquids are expected to penetrate differently even on one and the same solid surface (in an unknown manner) and hence the surface energetics could be very different from liquid to liquid. Since nothing is known about the changed surface energetics, use of these contact angles in any contact angle approach naively could be meaningless [122]. Although the interpretation of contact angles on SAMs in terms of solid surface tensions could be misleading, wettability studies of SAMs can be very interesting [120, 123].

3. CONTACT ANGLE INTERPRETATION

The calculation of solid surface tension γ_{sv} from the contact angle of a liquid of surface tension γ_{lv} starts with Young's equation, Eq. (3). Of the four quantities in Young's equation, only γ_{lv} and θ are readily measurable. Thus, in order to determine γ_{sv}, further information is necessary. Conceptually, an obvious approach is to seek one more relation among the parameters of Eq. (3), such as an equation-of-state relation, of the form of Eq. (7) which is now known to exist from experimental facts. The simultaneous solution of Eqs. (3) and (7) would solve the problem.

Historically, the interpretation of contact angles in terms of solid surface energetics started with the pioneering work of Zisman and co-workers [10]; the key observation they made was that for a given solid the measured contact angles did not vary randomly as the liquid was varied; rather, $\cos\theta$ changed smoothly with the liquid surface tension γ_{lv} within a band in a fashion that may suggest a straight-line relationship.

Subsequent to Zisman's work, two major schools of thought evolved: one is the surface tension component approach [11], [15]–[17]; the other is the equation-of-state approach [12]–[14] which can be understood as a further development of Zisman's approach. The former approach stipulates that γ_{sl} depends not only on γ_{lv} and γ_{sv}, but also on surface tension components due to the various specific intermolecular forces; the latter approach is based on an equation-of-state relation which has been shown to exist from thermodynamics [71]–[75] and the experimental results in Figures 1, 5, and 10–13.

From the experimental results shown in Figures 1, 5, and 10–13, the curves for all solid surfaces with very different chemical properties are very smooth, independent of intermolecular forces and liquid structures. This point has been addressed elsewhere [70, 83, 87], [95]–[102]. Within the framework of all surface tension component approaches [11], [15]–[17], it stands to reason that, for a given solid surface, the values of $\gamma_{lv}\cos\theta$ would have to depend not only on γ_{lv}, but also on the various intermolecular forces of the liquids. This expectation, however, clashes directly with the experimental results shown in Figures 1, 5, and 10–13. The only possible conclusion is that surface tension component approaches [11], [15]–[17] contradict physical reality [70, 83, 87], [95]–[102]. Only equation of state type approaches can be used: two well-known equation-of-state relations are those of Antonow [107] and Berthelot [108]. Nevertheless, surface tension component approaches will be briefly discussed and tested in the next section.

3.1. Surface Tension Component Approaches

3.1.1. Fowkes Approach. The approach of surface tension components was pioneered by Fowkes [11]. He postulated that the total surface tension could be expressed as a sum of different surface tension components, each of which arises due to a specific type of intermolecular forces:

$$\gamma = \gamma^d + \gamma^h + \gamma^{di} + \dots \tag{13}$$

where γ, γ^d, γ^h, and γ^{di} are, respectively, the total surface tension, dispersive surface tension component, and surface tension components due to hydrogen and dipole-dipole bonding. Equation (13) is often rearranged into

$$\gamma = \gamma^d + \gamma^n \tag{14}$$

i.e. the total surface tension γ is a sum of only the dispersive γ^d and non-dispersive γ^n surface tension components. The former is claimed to result from molecular interaction due to London forces, the latter from all other interactions due to non-London forces. A geometric mean relationship was postulated both of the solid-liquid and liquid-liquid interfacial tensions:

$$\gamma_{12} = \gamma_1 + \gamma_2 - 2\sqrt{\gamma_1^d \gamma_2^d} \tag{15}$$

For solid-liquid systems, combining Eq. (15) with Young's equation (3) yields

$$\gamma_l \cos\theta_Y = -\gamma_l + 2\sqrt{\gamma_s^d \gamma_l^d} \tag{16}$$

Typically, experimental contact angles of different liquids with known γ_l^d on a dispersive solid surface ($\gamma_s = \gamma_s^d$) are employed to determine the surface tension of a solid. In these procedures, the applicability of Young's equation and the constancy of solid surface tension from liquid to liquid stated in the Section 2.5.1 are, obviously, implied.

The Fowkes approach, Eq. (15) and hence Eq. (16), can be easily tested using the contact angle results in Section 2. For simplicity, two liquids of nearly the same surface tension but very different intermolecular forces are selected on the FC-722-coated mica surface; the polar (non-dispersive) liquid is 1-pentanol (γ_l = 26.0 mJ/m^2) and the dispersive one is *trans*-decalin ($\gamma_l = \gamma_l^d$ = 27.2 mJ/m^2). From Eq. (16), the γ_s (= γ_s^d) value determined using the contact angle (73.4°) of *trans*-decalin is 11.2 mJ/m^2. Since 1-pentanol is a polar liquid, γ_l^d is expected to be very different (smaller) than the total surface tension γ_l; assuming for the moment that dispersion force contributes 75% to the total surface tension of 1-pentanol. This leads to γ_l^d = 19.5 mJ/m^2 for γ_l = 26.0 mJ/m^2. From the experimental contact angle (73.0°) of 1-pentanol, Eq. (16) predicts the γ_s value to be 14.5 mJ/m^2. This value is almost 30% larger than that calculated using *trans*-decalin; a γ_s value of 11.2 mJ/m^2 would imply γ_l^d of 1-pentanol to be 25.2 mJ/m^2. This is clearly absurd: 1-pentanol would have to be classified as a non-polar liquid, i.e. $\gamma_l \approx \gamma_l^d$. Similar calculations can be performed for other dispersive/non-dispersive liquid pairs, with similar results. This result simply illustrates the fact that the perceived "surface tension components" in Fowkes' sense do not reflect physical reality. In principle, this is obvious from the experimental results that $\gamma_{lv}\cos\theta$ changes smoothly with γ_{lv}, independent of intermolecular forces and liquid structures. Thus, the basic postulate of the Fowkes approach is false and generalization of this approach must suffer from the same deficiency.

3.1.2. Owens-Wendt-Kaelble Approach. Owens and Wendt [15] extended Fowkes' concept to cases where both dispersion and hydrogen bonding forces may operate. They regarded the surface tension as being composed of two components such that

$$\gamma = \gamma^d + \gamma^h \tag{17}$$

where γ^h denotes the component of surface tension due to both hydrogen bonding and dipole-dipole interactions. They postulated

$$\gamma_{sl} = \gamma_s + \gamma_l - 2\sqrt{\gamma_s^d\gamma_l^d} - 2\sqrt{\gamma_s^h\gamma_l^h} \tag{18}$$

Combining this equation with Young's equation (3) yields

$$\gamma_l(1 + \cos\theta_Y) = 2\sqrt{\gamma_s^d\gamma_l^d} + 2\sqrt{\gamma_s^h\gamma_l^h} \tag{19}$$

Nearly at the same time, Kaelble [84] also published a very similar equation in terms of dispersion and polar forces. Thus, Eq. (18) is often called the Owens-Wendt-Kaelble equation.

Clearly, the applicability of Young's equation discussed in Section 2.5.1 is implied in Eq. (19). Since Eq. (19) contains two unknowns (γ_s^d and γ_s^h) of the solid, it is suggested to use contact angle measurements of at least two different liquids on one and the same solid surface, by solving two simultaneous equations. Such procedures also imply constancy of the solid surface tension from liquid to liquid; if the operative solid surface tension is not constant from one liquid to the next, simultaneous solution of different equations (from contact angles of different liquids) would be meaningless.

Equation (19) can be easily tested using the contact angles of *trans*-decalin and 1-pentanol on the FC-722 surface in Figure 5. Assuming the polarity of the (dispersive) FC-722 surface to be unknown and that γ_l^d of 1-pentanol to be again 19.5 mJ/m^2, two simultaneous equations can be used to determine the two unknowns (γ_s^d and γ_s^h) from the experimental contact angles of 1-pentanol and *trans*-decalin. Equation (19) predicts γ_s^d and γ_s^h to be, respectively, 11.2 mJ/m^2 and 0.6 mJ/m^2. Since diiodomethane is claimed to be a dispersive liquid [16, 17] ($\gamma_l = \gamma_l^d = 50.0$ mJ/m^2) and water is a non-dispersive liquid [11] ($\gamma_l = 72.7$ mJ/m^2 and $\gamma_l^d = 21.8$ mJ/m^2), two additional simultaneous equations can be obtained by insertion of the experimental contact angles from Figure 5. This procedure yields γ_s^d and γ_s^h to be 8.1 mJ/m^2 and 2.8 mJ/m^2, respectively. Clearly, different choices of the liquid pairs yield different surface tension components of the solid surface: if nothing were known about the dispersive property of the FC-722 surface, Eq. (19) would have predicted this dispersive surface to be non-dispersive, with over 25% contribution from, e.g., dipole-dipole interactions and hydrogen bonding (polarity) in the total surface tension; the relative magnitudes would depend on the choice of the liquid pairs. Similar calculations can be performed for other pairs of liquids, with similar results.

3.1.3. Lifshitz–van der Waals/Acid-Base (van Oss and Good) Approach. The Lifshitz–van der Waals/acid-base (van Oss and Good) approach [16, 17] was claimed to be a generalization of the Fowkes approach, by considering perceived acid-base interactions at an interface. van Oss and coworkers divided the surface tension into different perceived components, i.e. the so-called Lifshitz–van der Waals (LW), acid (+), and base (−) components, such that the total surface tension is given by

$$\gamma_i = \gamma_i^{LW} + 2\sqrt{\gamma_i^+ \gamma_i^-} \tag{20}$$

where i denotes either the solid or liquid phase. The interfacial tension was postulated both of solid-liquid and liquid-liquid systems as

$$\gamma_{12} = \gamma_1 + \gamma_2 - 2\sqrt{\gamma_1^{LW}\gamma_2^{LW}} - 2\sqrt{\gamma_1^+\gamma_2^-} - 2\sqrt{\gamma_1^-\gamma_2^+} \tag{21}$$

For solid-liquid systems, combining Eq. (21) with Young's equation yields

$$\gamma_l(1 + \cos\theta_Y) = 2\sqrt{\gamma_l^{LW}\gamma_s^{LW}} + 2\sqrt{\gamma_l^+\gamma_s^-} + 2\sqrt{\gamma_l^-\gamma_s^+} \tag{22}$$

Equation (22) is often used to determine the solid surface tension components (γ_s^{LW}, γ_s^+, and γ_s^-) from contact angles, using three simultaneous equations by inserting properties of calibration liquids. These procedures also imply the applicability of Young's equation and constancy of the solid surface tension from one liquid to the next, as discussed in Section 2.5.1. The failure of this approach can be easily illustrated, by employing the postulated liquid surface tension components from van Oss *et al.* [16, 17] and experimental contact angles of liquid triplets; three simultaneous equations of Eq. (22) (from contact angles of three different polar liquids) will yield the three unknown solid properties. Table 1 illustrates the solid surface tension components of the FC-721-coated mica surface [124]. It can be seen that the calculated γ_s value varies from −30.0 to 107.0 mJ/m^2, depending on the choice of the liquid triplets; the resulting "surface tension components" vary by a very large margin. However, it was recently suggested [125] that at least one non-polar liquid should be used for the above calculations, such as 1-bromonaphthalene. It is the contention of the proponents of this acid-base approach that 1-bromonaphthalene is a non-polar compound. No reason was given for the necessity of including a non-polar liquid. Nevertheless, the

Table 1.
Calculated solid surface tension components (mJ/m^2) for FC-721 using three simultaneous equations of the acid-base approach, Eq. (22), from contact angles of three polar liquids

liquid triplets	$\sqrt{\gamma_s^{LW}}$	$\sqrt{\gamma_s^+}$	$\sqrt{\gamma_s^-}$	γ_s
formamide/glycerol/water	2.6	0.5	0.9	7.7
formamide/glycerol/dmso	2.3	1.3	−1.6	1.1
formamide/glycerol/ethylene glycol	2.1	2.2	−4.2	−14.1
formamide/water/dmso	7.1	−4.1	1.2	40.6
formamide/water/ethylene glycol	1.1	1.9	0.8	4.3
formamide/dmso/ethylene glycol	1.7	2.1	−2.0	−5.5
glycerol/water/dmso	3.4	−0.1	0.7	11.4
glycerol/water/ethylene glycol	10.0	−4.7	−0.8	107.5
glycerol/dmso/ethylene glycol	0.9	3.2	−4.8	−29.9
water/dmso/ethylene glycol	2.4	1.0	0.5	6.8

Table 2.
Calculated solid surface tension components (mJ/m^2) for FC-721 using three simultaneous equations of the acid-base approach, Eq. (22), from contact angles of one non-polar (1-bromonaphthalene) and two polar liquids

liquid triplets	$\sqrt{\gamma_s^{LW}}$	$\sqrt{\gamma_s^+}$	$\sqrt{\gamma_s^-}$	γ_s
formamide/glycerol/1-bromonaphthalene	3.0	−1.1	5.6	−3.2
formamide/water/1-bromonaphthalene	3.0	0.0	0.9	9.0
formamide/dmso/1-bromonaphthalene	3.0	0.5	−1.2	7.8
formamide/ethylene glycol/1-bromonaphthalene	3.0	2.4	−9.2	−35.2
glycerol/water/1-bromonaphthalene	3.0	0.2	0.8	9.3
glycerol/dmso/1-bromonaphthalene	3.0	0.4	−0.1	8.9
glycerol/ethylene glycol/1-bromonaphthalene	3.0	1.4	−3.8	−1.6
water/dmso/1-bromonaphthalene	3.0	0.3	0.6	9.4
water/ethylene glycol/1-bromonaphthalene	3.0	0.5	0.4	9.4
dmso/ethylene glycol/1-bromonaphthalene	3.0	0.0	2.7	9.0

solid surface tension components of the same surface were recalculated using 1-bromonaphthalene and shown in Table 2. Again, the calculated γ_s values are not consistent, varying from −36.0 to 9.5 mJ/m^2. If nothing were known about the properties of the non-polar FC-721 surface, and if only a small number of measurements were available, this approach would predict all kinds of different molecular properties: non-polar, monopolar, bipolar acidic and basic as well as "negative" solid and solid-liquid interfacial tensions. Thus, the ability of the approach to explain, e.g., "negative" interfacial tensions does not imply its validity. The reason that the scatter of the results is not as bad as in Table 1 is due to the use of Berthelot's rule, see later. Clearly, this approach is just as erroneous as the earlier surface tension component approaches.

Table 3.
Calculated solid surface tension components (mJ/m^2) for polystyrene using three simultaneous equations of the acid-base approach, Eq. (22), from contact angles of three liquids

liquid triplets	$\sqrt{\gamma_s^{LW}}$	$\sqrt{\gamma_s^+}$	$\sqrt{\gamma_s^-}$	γ_s
formamide/glycerol/water	4.0	1.4	2.4	22.7
formamide/glycerol/dmso	3.7	2.5	−0.9	9.2
formamide/glycerol/ethylene glycol	3.6	2.9	−2.2	0.2
formamide/water/dmso	19.7	−14.9	4.1	265.9
formamide/water/ethylene glycol	2.8	2.6	2.2	19.3
formamide/dmso/ethylene glycol	3.3	2.8	−1.0	5.3
glycerol/water/dmso	5.1	0.6	2.1	28.5
glycerol/water/ethylene glycol	9.4	−2.4	1.2	82.6
glycerol/dmso/ethylene glycol	2.8	3.5	−2.6	−10.4
water/dmso/ethylene glycol	4.3	1.5	2.0	24.5

Table 4.
Calculated solid surface tension components (mJ/m^2) for poly(methyl methacrylate) PMMA using three simultaneous equations of the acid-base approach, Eq. (22), from contact angles of three liquids

liquid triplets	$\sqrt{\gamma_s^{LW}}$	$\sqrt{\gamma_s^+}$	$\sqrt{\gamma_s^-}$	γ_s
formamide/glycerol/diiodomethane	6.2	−0.8	7.4	26.6
formamide/water/diiodomethane	6.2	0.2	3.3	39.8
glycerol/water/diiodomethane	6.2	0.3	3.2	40.4
glycerol/water/formamide	5.8	0.6	3.3	37.6

It should be realized that the failure of this approach is not limited to the non-polar FC-721 surface. As an illustration, the low-rate dynamic contact angles (from Section 2.3.4) of various liquids (with very different liquid structures) on the polystyrene PS and poly(methyl methacrylate) PMMA surfaces are employed to test the approach; the calculated solid surface tensions and component values for PS and PMMA are shown in Tables 3 and 4, respectively. Clearly, the calculated $\sqrt{\gamma_s^{LW}}$, $\sqrt{\gamma_s^+}$, and $\sqrt{\gamma_s^-}$ and hence γ_s^{LW}, γ_s^+, and γ_s^- are not consistent. In particular, the calculated γ_s values for PS and PMMA vary, respectively, from -10.4 to 266.0 mJ/m^2 and from 26.6 to 40.4 mJ/m^2. The calculated values show a strong dependence on the choice of the liquid triplets.

We wish to point out that testing of the van Oss and Good approach requires *a priori* the liquid surface tension components that, in turn, were postulated by the proponents. The exact values of such postulated components has been a concern [126]–[128]. Since the availability of the liquid surface tension components is limited, the contact angle results from some solid surfaces do not provide enough equations to be able to solve for the solid surface tension components: the approach is limited due to lack of the postulated liquid surface tension components. The failure of the van Oss and Good approach to predict the correct acidic (electron-acceptor) character of PVC has been discussed elsewhere [129, 130]. The limitation of the approach is broad and is not restricted only to the solid surfaces in question. For example, van Oss *et al.* [17] attached "surprising" when several acidic carbohydrates were found to be basic. The predominantly basic (electron-donor) character predicted by the approach for virtually all kinds of solid surfaces is indeed well-known [17, 129, 130]. It should be realized that the equation-of-state approach, which will be discussed in the next section, yields essentially constant γ_{sv} values using the same experimental set of γ_{lv} and θ pairs, independent of the choice of the liquids used; no *a priori* knowledge of, e.g., liquid polarities is required. Testing of the above van Oss and Good approach has recently been a hot subject [124], [126]–[138].

Since solid surface tension and surface tension components are expected to be unique material properties, these results suggest that the acid-base approach is not capable of predicting correct solid surface tensions and "surface tension components". Thus, question arises as to whether or not one should attempt to infer intermolecular forces from the calculated "solid surface tension components", for these and other solid surfaces using Eq. (22), and hence the underlying equation (21).

Obvious as it may be, we would like to comment on the name of the approach which we have been criticizing: It is not the pioneering work of van der Waals and of Lifshitz which is questioned here but only an empirical, postulatory approach with no direct link to the work of these great scientists.

3.2. Equation-of-state Approach

In this and the following sections, an equation-of-state approach will be discussed, as such a relation has been inferred from the experimental results in Section 2.3. In the literature, two well-known equation-of-state relations are Antonow's [107] and Berthelot's [108] rules.

3.2.1. Antonow's Rule. An old equation-of-state is that of Antonow [107]. Antonow's rule relates γ_{sl} with γ_{lv} and γ_{sv} for solid-liquid systems in the following simple manner:

$$\gamma_{sl} = |\gamma_{lv} - \gamma_{sv}| \tag{23}$$

This equation, although never having been derived in any fashion, appears in the literature from time to time. Combining this relation with Young's equation gives

$$\cos\theta_Y = -1 + 2\frac{\gamma_{sv}}{\gamma_{lv}} \tag{24}$$

Thus, assuming, for the sake of the argument, the validity of Eq. (24), γ_{sv} can be determined when γ_{lv} and θ_Y are known. Once γ_{sv} is known, γ_{sl} can be found either from Young's equation (3) or from Eq. (23).

3.2.2. Berthelot's (Geometric Mean) Combining Rule. Another equation-of-state relation can be obtained from the Berthelot combining rule [108]. Unlike Antonow's rule, Berthelot's rule has a theoretical basis, based on molecular interactions of like pairs: From the London theory of dispersion [139, 140], the long range dispersion energy function for two identical molecules is given by

$$U_{disp} = \frac{C_6}{r^6} \tag{25}$$

where C_6 is a negative constant, so that the energy contribution is attractive; r is the radius of a hard sphere. A more general form [140] of the long range dispersion energy function including electrostatic and induction interactions can be expressed as

$$U_{disp} = \frac{C_6}{r^6} + \frac{C_8}{r^8} + \frac{C_{10}}{r^{10}} + \dots \tag{26}$$

where C_8 and C_{10} are all negative constants. It has been shown [139, 140] that the energy coefficient C_6 for two identical molecules can be expressed as

$$C_6 = -\frac{3}{4}\frac{E_I \alpha^2(0)}{(4\pi\epsilon_0)^2} \tag{27}$$

where ϵ_0 is the permittivity of free space, $\alpha(0)$ is the static polarizability, and E_I is the ionization energy.

From Eq. (27), the dispersion energy coefficient of a pair of molecules of species i is given by

$$C_6^{ii} = -\frac{3}{4}\frac{E_{I_i} \alpha_i^2(0)}{(4\pi\epsilon_0)^2} \tag{28}$$

and similarly, for a pair of molecules of species j,

$$C_6^{jj} = -\frac{3}{4}\frac{E_{I_j} \alpha_j^2(0)}{(4\pi\epsilon_0)^2} \tag{29}$$

It has been shown [140] that the interaction C_6^{ij} between two dissimilar molecules i and j can be related by the following expression:

$$C_6^{ij} = -\frac{3}{2}\frac{E_{I_i} E_{I_j}}{E_{I_i} + E_{I_j}}\frac{\alpha_i(0)\alpha_j(0)}{(4\pi\epsilon_0)^2} \tag{30}$$

Because the ionization energies vary only slightly from molecule to molecule, i.e. $E_{I_i} \approx E_{I_j}$, the dispersion energy coefficient C_6^{ij} of Eq. (30) can be written in terms of C_6^{ii} and C_6^{jj} as

$$C_6^{ij} = \sqrt{C_6^{ii} C_6^{jj}} \tag{31}$$

This relation indeed forms a basis of the Berthelot (geometric mean) combining rule [108]:

$$\epsilon_{ij} = \sqrt{\epsilon_{ii}\epsilon_{jj}} \tag{32}$$

where ϵ_{ij} is the potential energy parameter (well depth) of unlike-pair interactions; ϵ_{ii} and ϵ_{jj} are the potential energy parameters (well depth) of like-pair interactions.

Thermodynamically, a relation of the free energy of adhesion per unit area of a solid-liquid pair is equal to the work required to separate unit area of the solid-liquid interface [141]:

$$W_{sl} = \gamma_{lv} + \gamma_{sv} - \gamma_{sl} \tag{33}$$

From the geometric mean combining rule (i.e. the Berthelot rule, Eq. (32)), the free energy of adhesion W_{sl} can be approximated in terms of the free energy of cohesion of the solid, W_{ss}, and the free energy of cohesion of the liquid, W_{ll} [108, 140, 142, 143], i.e.

$$W_{sl} = \sqrt{W_{ll} W_{ss}} \tag{34}$$

By the definitions $W_{ll} = 2\gamma_{lv}$ and $W_{ss} = 2\gamma_{sv}$, Eq. (34) becomes

$$W_{sl} = 2\sqrt{\gamma_{lv}\gamma_{sv}} \tag{35}$$

By combining Eq. (35) with Eq. (33), the solid-liquid interfacial tension γ_{sl} can be written as

$$\gamma_{sl} = \gamma_{lv} + \gamma_{sv} - 2\sqrt{\gamma_{lv}\gamma_{sv}} = (\sqrt{\gamma_{lv}} - \sqrt{\gamma_{sv}})^2 \tag{36}$$

Combining this equation with Young's equation (3) yields

$$\cos\theta_Y = -1 + 2\sqrt{\frac{\gamma_{sv}}{\gamma_{lv}}} \tag{37}$$

Thus, the solid-vapor surface tension can be determined when experimental (Young) contact angle and liquid-vapor surface tension are known.

The question then arises as to what to expect when such an equation-of-state relation, in conjunction with Young's equation, is used to calculate solid surface tensions. There is one immediate criterion that the results obtained with this equation (or, for that matter, any other equation) must satisfy; i.e., when measuring contact angles with a number of liquids on a low-energy solid, the solid-vapor surface tension γ_{sv} is expected to be constant, independent of the liquid surface tension γ_{lv}. In other words, different pairs of θ and γ_{lv} for one and the same solid should yield sensibly constant values of γ_{sv}. The γ_{sv} values obtained from Eqs. (37) and (24) are given in Table 5 for the FC-722-coated mica from Figure 5. It can be seen that the γ_{sv} values obtained from Antonow's and Berthelot's rule are not

Table 5.
Calculated γ_{sv} values (mJ/m^2) of the FC-722-coated mica surface from Antonow's and Berthelot's rules

liquid	γ_{lv} (mJ/m^2)	θ (deg.)	γ_{sv} Antonow's rule, Eq. (24)	γ_{sv} Berthelot's rule, Eq. (37)
decane	23.88	67.36	16.5	11.5
1-pentanol	26.01	72.95	16.8	10.9
trans-decalin	27.19	73.38	17.5	11.2
hexadecane	27.62	75.94	17.2	10.7
1-decanol	28.99	78.84	17.3	10.3
cis-decalin	32.32	79.56	19.1	11.3
ethyl cinnamate	37.17	86.54	19.7	10.5
dibenzylamine	40.80	90.70	20.1	10.0
dimethyl sulfoxide	42.68	90.95	21.0	10.3
1-bromonaphthalene	44.31	93.81	20.7	9.7
diethylene glycol	44.68	94.22	20.7	9.6
ethylene glycol	47.55	97.87	20.5	8.9
diiodomethane	49.98	101.18	20.1	8.1
2,2′-thiodiethanol	56.26	104.56	21.1	7.9
formamide	59.08	108.49	20.2	6.9
glycerol	65.02	111.73	20.5	6.5
water	72.70	118.69	18.9	4.9

constant; rather, the former tends to increase as γ_{lv} increases; the latter tends to decrease as γ_{lv} increases.

As mentioned above, Antonow's rule does not have a theoretical basis. But, in the theory of intermolecular interactions and the theory of mixtures, combining rules are used to evaluate the parameters of unlike-pair interactions in terms of those of the like interactions [144, 145]. However, as for many other combining rules, the Berthelot rule [108], i.e. Eq. (36), is only a useful approximation and does not provide a secure basis for the understanding of unlike-pair interactions. For the interactions between two very dissimilar types of molecules or materials, where there is an apparent difference between ϵ_{ii} and ϵ_{jj}, it has been demonstrated [44, 146] that the geometric mean combining rule generally overestimates the strength of the unlike-pair interaction, i.e. the geometric mean value is too large an estimate. Clearly, this is why the geometric mean combining rule does not work for situations of large differences $|W_{ll} - W_{ss}|$ or $|\gamma_{lv} - \gamma_{sv}|$, as can be seen in Table 5. This fact is indeed well-known [46]–[49], [140].

In general, ϵ_{sl}, the minimum of the solid-liquid interaction potential, is often expressed in the following manner [45]–[49]:

$$\epsilon_{sl} = g(\sigma_s/\sigma_l)\sqrt{\epsilon_{ss}\epsilon_{ll}} \tag{38}$$

where ϵ_{ss} and ϵ_{ll} are, respectively, the minima in the solid-solid and liquid-liquid potentials; σ_s and σ_l are the appropriate core diameters for the molecules of the solid and liquid, respectively. It should be noted that many explicit forms of $g(\sigma_s/\sigma_l)$ have been suggested. For example, Hudson and McCoubrey [147] and Matyushov and Schmid [49] employed

$$g(\sigma_s/\sigma_l) = \left(\frac{4\sigma_s/\sigma_l}{(1+\sigma_s/\sigma_l)^2}\right)^3 \tag{39}$$

and Steele [148] and others [149] set

$$g(\sigma_s/\sigma_l) = \frac{1}{4}\left(1+\frac{\sigma_s}{\sigma_l}\right)^2 \tag{40}$$

in attempts to give a better representation of solid-liquid interactions from solid-solid and liquid-liquid interactions. In general, these functions are normalized such that $g(\sigma_s/\sigma_l) = 1$ when $\sigma_s = \sigma_l$. In this case, Eq. (38) reverts to Berthelot's geometric mean combining rule, Eq. (32). Therefore, a modification of Berthelot's rule is called for, leading to an equation-of-state approach for solid-liquid interfacial tensions.

3.2.3. Equation-of-state Approach: Modified Berthelot's Rule. In the study of mixtures [144, 145], it has become common practice to introduce a factor $(1 - K_{ij})$ to the geometric mean combining rule,

$$\epsilon_{ij} = (1 - K_{ij})\sqrt{\epsilon_{ii}\epsilon_{jj}} \tag{41}$$

where K_{ij} is an empirical parameter quantifying deviations from the geometric mean combining rule. Since the geometric mean combining rule overestimates the strength of the unlike-pair interactions, the modifying factor $(1 - K_{ij})$ should be a decreasing function of the difference $\epsilon_{ii} - \epsilon_{jj}$ and be equal to unity when the difference $\epsilon_{ii} - \epsilon_{jj}$ is zero.

On the basis of this thought, Li and Neumann [150] have considered a modified combining rule of the form

$$\epsilon_{ij} = \sqrt{\epsilon_{ii}\epsilon_{jj}}\exp\left(-\alpha\left(\epsilon_{ii} - \epsilon_{jj}\right)^2\right) \tag{42}$$

where α is an empirical constant; the square of the difference $(\epsilon_{ii} - \epsilon_{jj})$, rather than the difference itself, reflects the symmetry of this combining rule, and hence the anticipated symmetry of the equation-of-state. Correspondingly, for the cases of large differences $|W_{ll} - W_{ss}|$ or $|\gamma_{lv} - \gamma_{sv}|$, the combining rule for the free energy of adhesion of a solid-liquid pair can be written as

$$W_{sl} = \sqrt{W_{ll}W_{ss}}\exp\left(-\alpha(W_{ll} - W_{ss})^2\right) \tag{43}$$

or, more explicitly, by using $W_{ll} = 2\gamma_{lv}$ and $W_{ss} = 2\gamma_{sv}$,

$$W_{sl} = 2\sqrt{\gamma_{lv}\gamma_{sv}}\exp\left(-\beta(\gamma_{lv} - \gamma_{sv})^2\right) \tag{44}$$

In the above, α and β are as yet unknown constants. Clearly, when the values of γ_{lv} and γ_{sv} are close to each other, Eq. (44) reverts to Eq. (35), the geometric mean combining rule. By combining Eq. (44) with Eq. (33), an equation-of-state for solid-liquid interfacial tensions can be written as

$$\gamma_{sl} = \gamma_{lv} + \gamma_{sv} - 2\sqrt{\gamma_{lv}\gamma_{sv}}\,\exp\left(-\beta(\gamma_{lv} - \gamma_{sv})^2\right) \tag{45}$$

Since the Berthelot's rule was derived based on the dispersion energy coefficient C_6, the above correction factor, $\exp\left(-\beta(\gamma_{lv} - \gamma_{sv})^2\right)$, is indeed an attempt to account for systems which deviate from the predictions of the London theory of dispersion. Combining Eq. (45) with Young's equation (3) yields

$$\cos\theta_Y = -1 + 2\sqrt{\frac{\gamma_{sv}}{\gamma_{lv}}}\,\exp\left(-\beta(\gamma_{lv} - \gamma_{sv})^2\right) \tag{46}$$

Thus, the solid surface tensions can be determined from experimental (Young) contact angles and liquid surface tensions when β is known, e.g., by the Newton method [151]. Obviously, for a given set of γ_{lv} and θ data measured on one and the same type of solid surface, the constant β and γ_{sv} values can be determined by a least-squares analysis technique [64, 152, 153]. Starting out with arbitrary values for γ_{sv} and β, iterative procedures can be used to identify that pair of γ_{sv} and β values which provides the best fit of the experimental data to the set of experimental γ_{lv} and θ_Y pairs belonging to one and the same solid surface. From the experimental contact angles on the FC-721-coated mica, heat-pressed Teflon FEP, and poly(ethylene terephthalate) PET surfaces (not used here), an averaged β value of 0.0001247 $(m^2/mJ)^2$ was obtained [14, 64].

The γ_{sv} values calculated from Eq. (46) with the above β value are given in Table 6, for all solid surfaces shown in Figure 12. In contrast to the results obtained the surface tension component approaches described in Section 3.1 (cf. Tables 1–4), the γ_{sv} values calculated from the equation-of-state approach [14, 64] are quite constant, essentially independent of the choice of the liquids. This reconfirms the validity of the approach to determine solid surface tensions from contact angles.

It should be noted that the above calculations were based on a constant β value of 0.0001247 $(m^2/mJ)^2$. Alternatively, the γ_{sv} values of various solids can be determined by the two-variable least-squares analysis [64, 152, 153], by assuming γ_{sv} and β in Eq. (46) to be constant, as described above. This leads to a β value and a γ_{sv} value for every given solid surface. The results, together with those obtained using the β value of 0.0001247 $(m^2/mJ)^2$, are given in Table 7. It is evident that there is good agreement between the γ_{sv} values determined from the two strategies, and the slight difference in the β values have very little effect on the calculated γ_{sv} values. Since the β values do not appear to change systematically with the solid surface (and hence γ_{sv}), a weighted β value can be calculated. Alternatively, a statistical analysis can be used to decide whether or not averaging of β over γ_{sv} is allowed [83], [152]–[155]. This procedure yields a correlation coefficient r_{cal} of 0.477 for the 14 observations in Table 7. A tabulated correlation coefficient r_{tab} at

Table 6.
Summary of the dynamic contact angles on various solids by ADSA-P and capillary rise techniques. The γ_{sv} values were calculated from the equation-of-state approach, Eq. (46), using $\beta = 0.0001247$ $(m^2/mJ)^2$

solid surface/technique	liquid	γ_{lv} (mJ/m^2)	θ (deg.)	γ_{sv} equation-of-state Eq. (46)
FC-721-coated mica/capillary rise [93]	dodecane	25.03	70.4	11.7
	2-octanol	26.00	73.5	11.3
	tetradecane	26.50	73.5	11.6
	1-octanol	27.28	75.1	11.5
	hexadecane	27.31	75.6	11.3
	1-hexadecene	27.75	74.0	12.0
	1-decanol	28.29	76.6	11.5
	1-dodecanol	29.53	79.2	11.3
FC-722-coated mica/ADSA-P [87]	decane	23.88	67.36	11.9
	1-pentanol	26.01	72.95	11.5
	trans-decalin	27.19	73.38	11.9
	hexadecane	27.62	75.94	11.4
	1-decanol	28.99	78.84	11.2
	cis-decalin	32.32	79.56	12.4
	ethyl cinnamate	37.17	86.54	12.2
	dibenzylamine	40.80	90.70	12.2
	dimethyl sulfoxide	42.68	90.95	12.9
	1-bromonaphthalene	44.31	93.81	12.4
	diethylene glycol	44.68	94.22	12.4
	ethylene glycol	47.55	97.87	12.1
	diiodomethane	49.98	101.18	11.7
	2,2′-thiodiethanol	56.26	104.56	12.7
	formamide	59.08	108.49	12.0
	glycerol	65.02	111.73	12.8
	water	72.70	118.69	12.2
FC-722-coated Si wafer/ ADSA-P [96]	hexane	18.50	50.83	12.4
	2-octanol	26.42	74.74	11.2
	hexadecane	27.62	75.64	11.5
	glycerol	65.02	111.89	12.7
FC-725-coated Si wafer/ ADSA-P [97]	dodecane	25.64	71.02	11.8
	hexadecane	27.62	73.41	12.1
	3,3-thiodipropanol	39.83	90.48	11.9
	diethylene glycol	45.16	94.47	12.5
	ethylene glycol	48.66	100.05	11.6
	2,2′-thiodiethanol	53.77	101.07	13.2
	formamide	59.08	106.89	12.7
	glycerol	63.13	110.21	12.7
	water	72.70	119.31	11.9

Table 6.
(continued)

solid surface/technique	liquid	γ_{lv} (mJ/m^2)	θ (deg.)	γ_{sv} equation-of-state Eq. (46)
Teflon (FEP)/capillary rise [93]	dodecane	25.03	47.8	17.7
	2-octanol	26.00	52.3	17.2
	tetradecane	26.50	52.6	17.5
	1-octanol	27.28	54.4	17.5
	hexadecane	27.31	53.9	17.7
	1-hexadecene	27.75	54.2	17.9
	1-dodecanol	29.53	55.7	18.6
	dimethylformamide	35.21	68.6	17.7
	methyl salicylate	38.85	72.7	18.4
hexatriacontane/capillary rise [51, 104]	ethylene glycol	47.7	79.2	20.3
	2,2′-thiodiethanol	54.0	86.3	20.3
	glycerol	63.4	95.4	20.6
	water	72.8	104.6	20.3
cholesteryl acetate/ capillary rise [51, 105]	ethylene glycol	47.7	77.0	21.3
	2,2′-thiodiethanol	54.0	84.3	21.3
	glycerol	63.4	94.0	21.3
	water	72.8	103.3	21.1
poly(propene-*alt*-*N*-(*n*-hexyl)maleimide)/ ADSA-P [83, 95]	*cis*-decalin	32.32	28.81	28.5
	triacetin	35.52	39.45	28.3
	diethylene glycol	44.68	61.04	26.7
	glycerol	65.02	82.83	28.6
	water	72.70	92.26	27.8
poly(*n*-butyl methacrylate)/ ADSA-P [98]	diethylene glycerol	45.16	58.73	28.0
	3-pyridylcarbinol	47.81	60.30	29.2
	2,2′-thiodiethanol	53.77	68.00	29.4
	formamide	59.08	76.41	28.5
	glycerol	65.02	82.11	29.0
	water	72.70	90.73	28.7
polystyrene/ADSA-P [99]	dimethyl sulfoxide	42.68	50.67	29.7
	diethylene glycol	44.68	52.41	30.5
	ethylene glycol	48.66	61.20	29.3
	formamide	59.08	74.76	29.4
	glycerol	63.11	78.38	30.0
	water	72.70	88.42	30.2
poly(styrene-(hexyl/10-carboxydecyl 90:10) maleimide)/ADSA-P [100]	diethylene glycol	45.16	51.32	31.3
	ethylene glycol	48.66	59.72	30.2
	formamide	59.08	70.28	31.8
	glycerol	63.13	76.51	31.0
	water	72.70	87.13	31.0

Table 6.
(continued)

solid surface/technique	liquid	γ_{lv} (mJ/m^2)	θ (deg.)	γ_{sv} equation-of-state Eq. (46)
poly(methyl methacrylate/ *n*-butyl methacrylate)/ ADSA-P [101]	1-iodonaphthalene	42.92	35.67	35.7
	3-pyridylcarbinol	47.81	49.22	34.2
	2,2′-thiodiethanol	53.77	57.84	34.6
	formamide	59.08	66.33	34.0
	glycerol	65.02	74.72	33.3
	water	72.70	81.33	34.6
poly(propene-*alt*-*N*-(*n*-propyl)maleimide)/ADSA-P [83, 95]	1-iodonaphthalene	42.92	35.19	35.9
	1-bromonaphthalene	44.31	30.75	38.6
	1,3-diiodopropane	46.51	39.98	37.1
	2,2′-thiodiethanol	53.77	54.04	36.5
	glycerol	65.02	70.67	35.7
	water	72.70	77.51	37.0
poly(methyl methacrylate)/ ADSA-P [102]	1,3-diiodopropane	46.51	36.95	38.3
	3-pyridylcarbinol	47.81	39.47	38.4
	diiodomethane	49.98	42.25	39.0
	2,2′-thiodiethanol	53.77	50.35	38.3
	formamide	59.08	57.73	38.6
	glycerol	65.02	66.84	37.9
	water	72.70	73.72	39.3
poly(propene-*alt*-*N*-methyl-maleimide)/ADSA-P [96]	diiodomethane	49.98	30.71	43.7
	glycerol	65.02	60.25	41.7
	water	72.70	69.81	41.8

99% confidence is 0.623. Since, $r_{cal} < r_{tab}$, there is no correlation between β and γ_{sv} at over 99% confidence, for the results given in Table 7. Thus, averaging of β is allowed. Remarkably, this procedure together with the new experimental data yield a weighted β value of 0.0001234 (m^2/mJ)^2, which is in excellent agreement with that (β = 0.0001247 (m^2/mJ)^2) obtained by Li and Neumann [64] from contact angles on only three solid surfaces not considered here.

It should be noted that other forms of the modified Berthelot's rule (or any other equation in the form of Eq. (7)) can also be used, as long as they are capable of describing the experimental contact angle patterns in Figures 1, 5, and 10–13. A different form of the modified Berthelot's rule will be proposed and discussed in the next section, which yields essentially the same γ_{sv} values as Eq. (46) does.

3.2.4. Equation-of-state Approach: Alternative Formulation. Following the arguments leading to Eq. (41), we propose an alternative formulation [154, 155] of the equation-of-state approach, i.e. a modified Berthelot's rule with a different modify-

Table 7.
Averaged γ_{sv} values from Table 6 using $\beta = 0.0001247$ $(m^2/mJ)^2$; the γ_{sv} (mJ/m^2) and β $(m^2/mJ)^2$ values determined by a two-variable least-squares fit from Eq. (46) are also given

solid surface/technique	# of system	Table 6[†] γ_{sv}	least-squares-fit β	least-squares-fit γ_{sv}
FC-721-coated mica/capillary rise [93]	8	11.5 ± 0.2	0.000124	11.7
FC-722-coated mica and Si wafer/ ADSA-P [87, 96]	21	12.1 ± 0.3	0.000111	11.8
FC-725-coated Si wafer/ADSA-P [97]	9	12.3 ± 0.4	0.000114	11.9
Teflon FEP/capillary rise [93]	9	17.8 ± 0.3	0.000142	18.0
hexatriacontane/capillary rise [51, 104]	4	20.4 ± 0.2	0.000124	20.3
cholesteryl acetate/capillary rise [51, 105]	4	21.3 ± 0.2	0.000128	21.5
poly(propene-*alt*-*N*-(*n*-hexyl)-maleimide/ADSA-P [83, 95]	5	28.0 ± 1.0	0.000122	27.9
poly(*n*-butyl methacrylate)/ADSA-P [98]	6	28.8 ± 0.5	0.000124	28.8
polystyrene/ADSA-P [99]	6	29.9 ± 0.5	0.000120	29.7
poly(styrene-(hexyl/10-carboxydecyl 90:10)maleimide)/ADSA-P [100]	5	31.0 ± 0.6	0.000120	30.8
poly(methyl methacrylate/*n*-butyl methacrylate)/ADSA-P [101]	6	34.4 ± 0.8	0.000136	34.7
poly(propene-*alt*-*N*-(*n*-propyl)-maleimide)/ADSA-P [83, 95]	6	36.8 ± 1.1	0.000133	36.9
poly(methyl methacrylate)/ADSA-P [102]	7	38.5 ± 0.4	0.000113	38.3
poly(propene-*alt*-*N*-methylmaleimide/ ADSA-P [96]	3	42.4 ± 2.8	0.000167	43.4

[†] error bars are 95% confidence limits.

ing factor similar to the $(1 - K_{ij})$ used in Eq. (41),

$$1 - \kappa_1(\epsilon_{ii} - \epsilon_{jj})^2 \tag{47}$$

so that $(1 - \kappa_1(\epsilon_{ii} - \epsilon_{jj})^2)$ is also a decreasing function of the difference $(\epsilon_{ii} - \epsilon_{jj})$ and is equal to unity when $\epsilon_{ii} = \epsilon_{jj}$; κ_1 is an unknown constant. Thus, Eq. (41) can then be written as

$$\epsilon_{ij} = \left(1 - \kappa_1(\epsilon_{ii} - \epsilon_{jj})^2\right)\sqrt{\epsilon_{ii}\epsilon_{jj}} \tag{48}$$

The square of the difference $(\epsilon_{ii} - \epsilon_{jj})$, rather than the difference itself, reflects again the symmetry of this combining rule, and hence the anticipated symmetry of the equation-of-state. Because the free energy is directly proportional to the energy parameter ϵ [142, 143], i.e. $W \propto \epsilon$, and for the cases of large differences $|W_{ll} - W_{ss}|$ or $|\gamma_{lv} - \gamma_{sv}|$, the combining rule for the free energy of adhesion of a solid-liquid pair can be written as

$$W_{sl} = \left(1 - \alpha_1(W_{ll} - W_{ss})^2\right)\sqrt{W_{ll}W_{ss}} \tag{49}$$

Table 8.
γ_{sv} (mJ/m^2) and β (m^2/mJ)^2 values determined by a two-variable least-squares fit, for the two equation-of-state formulations

solid surface/technique	# of system	Eq. (46) β	Eq. (46) γ_{sv}	Eq. (52) β	Eq. (52) γ_{sv}
FC-721-coated mica/capillary rise [93]	8	0.000124	11.7	0.000097	11.7
FC-722-coated mica and Si wafer/ ADSA-P [87, 96]	21	0.000111	11.8	0.000092	11.6
FC-725-coated Si wafer/ADSA-P [97]	9	0.000114	11.9	0.000092	11.6
Teflon FEP/capillary rise [93]	9	0.000142	18.0	0.000118	17.8
hexatriacontane/capillary rise [51, 104]	4	0.000124	20.3	0.000100	19.7
cholesteryl acetate/capillary rise [51, 105]	4	0.000128	21.5	0.000104	20.9
poly(propene-*alt*-*N*-(*n*-hexyl)-maleimide/ADSA-P [83, 95]	5	0.000122	27.9	0.000109	27.8
poly(*n*-butyl methacrylate)/ADSA-P [98]	6	0.000124	28.8	0.000108	28.5
polystyrene/ADSA-P [99]	6	0.000120	29.7	0.000107	29.5
poly(styrene-(hexyl/10-carboxydecyl 90:10)maleimide)/ADSA-P [100]	5	0.000120	30.8	0.000106	30.6
poly(methyl methacrylate/*n*-butyl methacrylate)/ADSA-P [101]	6	0.000136	34.7	0.000124	34.6
poly(propene-*alt*-*N*-(*n*-propyl)-maleimide)/ADSA-P [83, 95]	6	0.000133	36.9	0.000123	36.9
poly(methyl methacrylate)/ADSA-P [102]	7	0.000113	38.3	0.000106	38.2
poly(propene-*alt*-*N*-methylmaleimide/ ADSA-P [96]	3	0.000167	43.4	0.000155	43.3

where α_1 is an empirical constant which is in general different from the κ_1 in Eq. (48); more explicitly, by using $W_{ll} = 2\gamma_{lv}$ and $W_{ss} = 2\gamma_{sv}$,

$$W_{sl} = 2\left(1 - \beta_1(\gamma_{lv} - \gamma_{sv})^2\right)\sqrt{\gamma_{lv}\gamma_{sv}} \tag{50}$$

In the above, α_1 and β_1 ($= 4\alpha_1$) are as yet unknown constants. Clearly, when the values of γ_{lv} and γ_{sv} are close to each other, Eq. (50) reverts to the geometric mean combining rule. By combining Eq. (50) with Eq. (33), an alternative equation-of-state for solid-liquid interfacial tensions [154, 155] can be written as

$$\gamma_{sl} = \gamma_{lv} + \gamma_{sv} - 2\sqrt{\gamma_{lv}\gamma_{sv}}\left(1 - \beta_1(\gamma_{lv} - \gamma_{sv})^2\right) \tag{51}$$

Combining Eq. (51) with Young's equation (3) yields

$$\cos\theta_Y = -1 + 2\sqrt{\frac{\gamma_{sv}}{\gamma_{lv}}}\left(1 - \beta_1(\gamma_{lv} - \gamma_{sv})^2\right) \tag{52}$$

The question then becomes whether this combining rule fits the experimental contact angle patterns shown earlier. Obviously, for a given set of γ_{lv} and θ data measured on one and the same type of solid surface, the constant β_1 and γ_{sv} values can be determined by a two-variable least-squares analysis technique [64, 152, 153]. Starting out with arbitrary values for γ_{sv} and β_1, iterative procedures can be used to

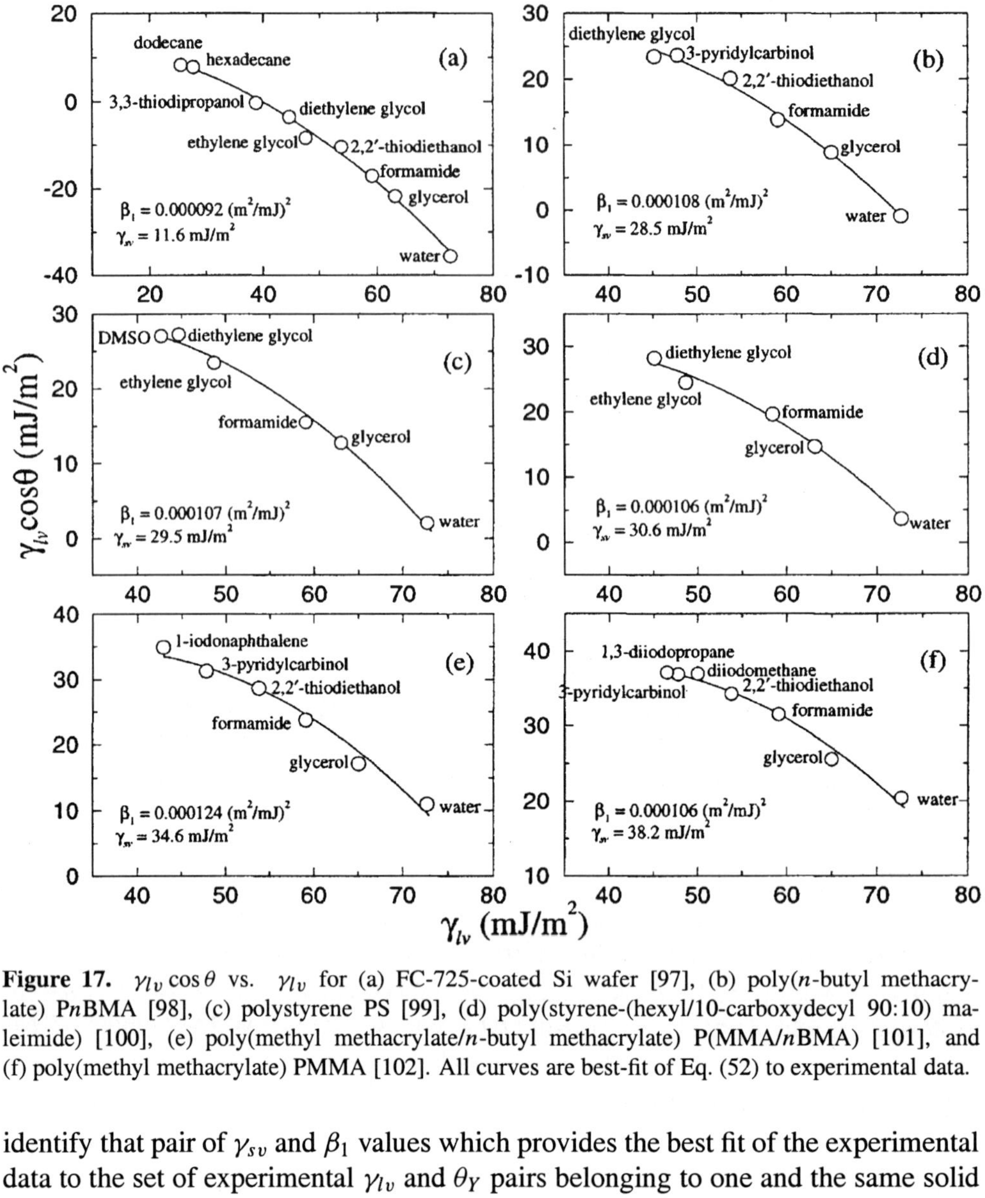

Figure 17. $\gamma_{lv}\cos\theta$ vs. γ_{lv} for (a) FC-725-coated Si wafer [97], (b) poly(n-butyl methacrylate) PnBMA [98], (c) polystyrene PS [99], (d) poly(styrene-(hexyl/10-carboxydecyl 90:10) maleimide) [100], (e) poly(methyl methacrylate/n-butyl methacrylate) P(MMA/nBMA) [101], and (f) poly(methyl methacrylate) PMMA [102]. All curves are best-fit of Eq. (52) to experimental data.

identify that pair of γ_{sv} and β_1 values which provides the best fit of the experimental data to the set of experimental γ_{lv} and θ_Y pairs belonging to one and the same solid surface. Figure 17 shows the best-fitted curves of Eq. (52) onto the experimental contact angle data for several solid surfaces previously considered. It can be seen that Eq. (52) fits these experimental data very well. The results of such a least-squares analysis are summarized in Table 8. Since the β_1 values do not appear to show any dependence on the solid surfaces [154, 155], a weighted mean β_1 was calculated. This yields $\beta_1 = 0.0001057$ (m^2/mJ)^2. A comparison between the γ_{sv} values for various solid surfaces obtained from the two equation-of-state formulations is given in Table 8. It can be seen that the γ_{sv} values obtained from the two formulations (equations) are virtually identical. Although the two relations are very different, Eq. (47) can indeed be obtained from a first-order approximation of a Taylor-series expansion of the adjusting parameter in Eq. (42).

Since Eq. (48) was formulated here as an expression similar to those of Eqs. (38) and (41), insight about molecular interactions of unlike pairs can be gained; the rationale is as follows. Knowing $\beta_1 = 0.0001057$ $(m^2/mJ)^2$, the present form of the combining rule for the free energy of adhesion, Eq. (49), can be written as

$$W_{sl} = \left(1 - 0.0000264(W_{ll} - W_{ss})^2\right)\sqrt{W_{ll}W_{ss}} \tag{53}$$

where $\alpha_1 = \beta_1/4 = 0.0000264$ $(m^2/mJ)^2$. Since $W \propto \epsilon$ and if $W \approx k\epsilon$, a combining rule for the energy parameter can be derived as:

$$\epsilon_{sl} = \left(1 - 0.0000264k^2(\epsilon_{ll} - \epsilon_{ss})^2\right)\sqrt{\epsilon_{ll}\epsilon_{ss}} \tag{54}$$

where k is a proportionality constant.

Although the proportionality constant k is unknown at the moment, Eq. (54) should, in principle, be able to tell us something about the molecular interactions between unlike solid-liquid pairs. In the study of molecular interactions between solids and liquids through, e.g., the Lennard-Jones solid-fluid potential [45, 48, 49, 140, 147], a combining rule is always required to represent the interactions of unlike pairs of molecules ϵ_{ij} from those of like pairs, ϵ_{ii} and ϵ_{jj}. However, the exact form of the combining rule for solid-liquid energy potential is often uncertain owing to the fact that estimating solid-liquid interfacial tensions directly is a difficult task and hence any specific (proposed) form of the combining rule cannot be easily tested and verified. As Eq. (54) stands, information about unlike molecular interactions can be inferred when the proportionality constant k is known. It should be realized that Eq. (49) can be formulated for k to drop out, e.g. by normalization. This procedure should allow the determination of a combining rule for the specific solid-liquid interactions. A recent work has employed the contact angles reported here to infer a combining rule for energy parameters and intermolecular potentials [156].

3.2.5. Comparison with the Original Equation-of-state Formulation. The formulation of an equation-of-state approach is essentially an empirical curve fit to contact angle data. As such there are a variety of ways in which one might proceed. It was decided in the original formulation [12, 13] to correlate the data in terms of Good's interaction parameter Φ:

$$\Phi = \frac{\gamma_{sv} + \gamma_{lv} - \gamma_{sl}}{2\sqrt{\gamma_{lv}\gamma_{sv}}} \tag{55}$$

The explicit form of such an equation-of-state relation was formulated as

$$\gamma_{sl} = \frac{(\sqrt{\gamma_{lv}} - \sqrt{\gamma_{sv}})^2}{1 - 0.015\sqrt{\gamma_{lv}\gamma_{sv}}} \tag{56}$$

Combining Eq. (56) with Young's equation yields

$$\cos\theta_Y = \frac{(0.015\gamma_{sv} - 2.00)\sqrt{\gamma_{lv}\gamma_{sv}} + \gamma_{lv}}{\gamma_{lv}(0.015\sqrt{\gamma_{lv}\gamma_{sv}} - 1)} \tag{57}$$

It should be noted that Eq. (56) (and hence Eq. (57)) was derived based on extensive contact angle measurements on eight polymer surfaces from Zisman and coworkers almost 50 years ago. This formulation was first attempted in the 1960s [12] and completed in the 1970s [13]. Note that difficulties may arise for liquids with relatively large liquid surface tensions γ_{lv}, since the denominator of Eq. (57) can become zero. This limitation of the equation-of-state formulation is due to the use of Good's interaction parameter Φ and is purely mathematical. Physical reasoning has been used to "mend" Eq. (57), and in practice Eq. (57) is implemented as a computer program [13], a set of tables [157] or a hand-held calculator [158]. However, as mentioned, the use of Good's interaction parameter Φ is not the only way to find an explicit expression for the equation-of-state; two different formulations have been discussed in the previous sections.

It is instructive to compare Eq. (57) with that proposed by Li *et al.* [14, 64], Eq. (46), and the new formulation [154, 155], Eq. (52), by calculating γ_{sv} with hypothetical values of γ_{lv} and θ. Examples of these results are given in Table 9. It is evident that the three equations yield essentially the same γ_{sv} values. We wish to point out that these results are not trivial. The above three formulations were based on entirely different sets of experimental contact angle data: Eq. (57) was obtained using Zisman's contact angles on eight polymer surfaces [13]; the β in Eq. (46) was calibrated by the contact angles of Li *et al.* [14, 64] on three different solid surfaces; the β_1 in Eq. (52) was obtained from the low-rate dynamic contact angles on the 14 solid surfaces presented in Sections 2.3.2, 2.3.3, and 2.3.4 [154, 155]. It is remarkable that, despite the different formulations and experimental contact angles, Eqs. (46), (52) and (57) yield virtually the same γ_{sv} values, as given in Table 9. For practical purposes, any of the three equations can be used. Of interest is a recent work to infer a combining rule for intermolecular potentials based on experimental contact angle/adhesion patterns, allowing the formulation of a different equation-of-state relation [156].

3.2.6. Predictive Power of the Equation-of-state Approach. In this section, the predictive power of the equation-of-state approach will be illustrated simply by predicting, from a single contact angle measurement, all the other contact angles measured on one and the same solid surface. Since the γ_{sv} values obtained from the three formulations are virtually the same, the choice of any specific formulation becomes irrelevant.

For example, if the surface tension (γ_{lv} = 23.9 mJ/m^2) and contact angle (θ = 67.4°) of decane on a FC-722-coated mica surface are chosen (from Figure 5), the equation-of-state approach [14, 64], Eq. (46), yields a γ_{sv} value of 11.9 mJ/m^2. From this γ_{sv} value, Eq. (46) can generate a series of contact angles for different

Table 9.
Comparison between the three equation-of-state formulations by calculating the γ_{sv} values for hypothetical values of γ_{lv} (mJ/m^2) and θ (deg.)

γ_{lv}	θ	γ_{sv} (mJ/m^2)		
		Li and Neumann [64], Eq. (46)	Kwok and Neumann [154], Eq. (52)	Neumann *et al.* [13], Eq. (57)
70.0	20.0	66.1	66.1	66.0
70.0	30.0	61.9	61.8	61.7
70.0	40.0	56.9	56.7	56.5
70.0	50.0	51.5	51.0	50.8
70.0	60.0	45.7	45.1	45.0
70.0	70.0	39.7	39.0	39.1
70.0	80.0	33.6	33.0	33.2
70.0	90.0	27.5	27.0	27.1
70.0	100.0	21.5	21.3	21.1
70.0	110.0	15.8	15.9	15.2
50.0	20.0	47.1	47.1	47.1
50.0	30.0	43.9	43.9	44.0
50.0	40.0	40.0	39.8	40.0
50.0	50.0	35.4	35.3	35.1
50.0	60.0	30.8	30.5	30.2
50.0	70.0	26.0	25.6	25.4
50.0	80.0	21.2	20.8	20.6
50.0	90.0	16.5	16.2	16.2
50.0	100.0	12.2	11.9	12.0
30.0	20.0	28.2	28.2	28.2
30.0	30.0	26.2	26.2	26.2
30.0	40.0	23.6	23.6	23.6
30.0	50.0	20.7	20.6	20.6
30.0	60.0	17.5	17.5	17.5
30.0	70.0	14.4	14.3	14.3
30.0	80.0	11.3	11.2	11.3

values of γ_{lv}. These procedures are employed to produce a set of hypothetical contact angles corresponding to the surface tensions of the other liquids on the FC-722 surface. The predicted contact angle results are given in Table 10. In all cases, the predicted and measured contact angles agree well, generally within the usual error limit given in the literature as $\pm 2°$.

Conversely, a series of hypothetical γ_{lv} values for different experimental contact angle values can also be generated from a single decane contact angle measurement. The predicted γ_{lv} values are also shown in Table 10 for the FC-722 surface; it can be seen that the deviations between the predicted and measured values vary from ± 0.1 mJ/m^2 to ± 2.0 mJ/m^2. While these deviations are somewhat outside the experimental error limits, there is no evidence of any systematic variation. Similar calculations can be performed for other solid surfaces, with similar results. From a single contact angle measurement on one and the same solid surface, the equation-

Table 10.
Comparison between the measured and predicted contact angles and liquid surface tensions on a FC-722-coated mica

liquid	θ (deg.)		γ_{lv} (mJ/m^2)	
	measured in Table 5	predicted† from Eq. (46)	measured in Table 5	predicted† from Eq. (46)
decane	67.36	—	23.88	—
1-pentanol	72.95	71.4	26.01	26.8
trans-decalin	73.38	73.4	27.19	27.2
hexadecane	75.94	74.2	27.62	28.6
1-decanol	78.84	76.4	28.99	30.4
cis-decalin	79.56	81.3	32.32	31.3
ethyl cinnamate	86.54	87.4	37.17	36.6
dibenzylamine	90.70	91.5	40.80	40.2
dimethyl sulfoxide	90.95	93.5	42.68	40.4
1-bromonaphthalene	93.81	95.2	44.31	43.3
diethylene glycol	94.22	95.6	44.68	43.5
ethylene glycol	97.87	98.4	47.55	47.0
diiodomethane	101.18	100.7	49.98	50.5
2,2′-thiodiethanol	104.56	106.3	56.26	54.4
formamide	108.49	108.7	59.08	59.0
glycerol	111.73	113.5	65.02	62.9
water	118.69	119.3	72.70	71.9

† based on a γ_{sv} value of 11.9 mJ/m^2 calculated from the contact angle of decane.

of-state approach [14, 64] is capable of predicting the contact angles of different liquids as well as the liquid surface tensions.

4. CALCULATION OF FREE ENERGY OF SOLID-LIQUID ADHESION

Thermodynamically, the free energy of adhesion W^{adh} for a solid-liquid pair is given by Eq. (33) as

$$W^{adh} = \gamma_{lv} + \gamma_{sv} - \gamma_{sl}$$

Because of Young's equation (3), W^{adh} may be written in terms of the two directly measurable quantities, γ_{lv} and θ:

$$W^{adh} = \gamma_{lv}(1 + \cos\theta_Y) \tag{58}$$

θ_Y is the Young contact angle, i.e. a contact angle which can be inserted into Young's equation. Equation (58) was obtained by assuming again that Young's equation is applicable, allowing us to calculate the free energy of solid-liquid adhesion W^{adh} from a single pair of experimental data, γ_{lv} and θ. However, as is — in principle — obvious from the discussion in Section 2.5.1, one additional assumption is needed here for reasons for procedures in interpreting W^{adh}: γ_{sv} of a given solid surface

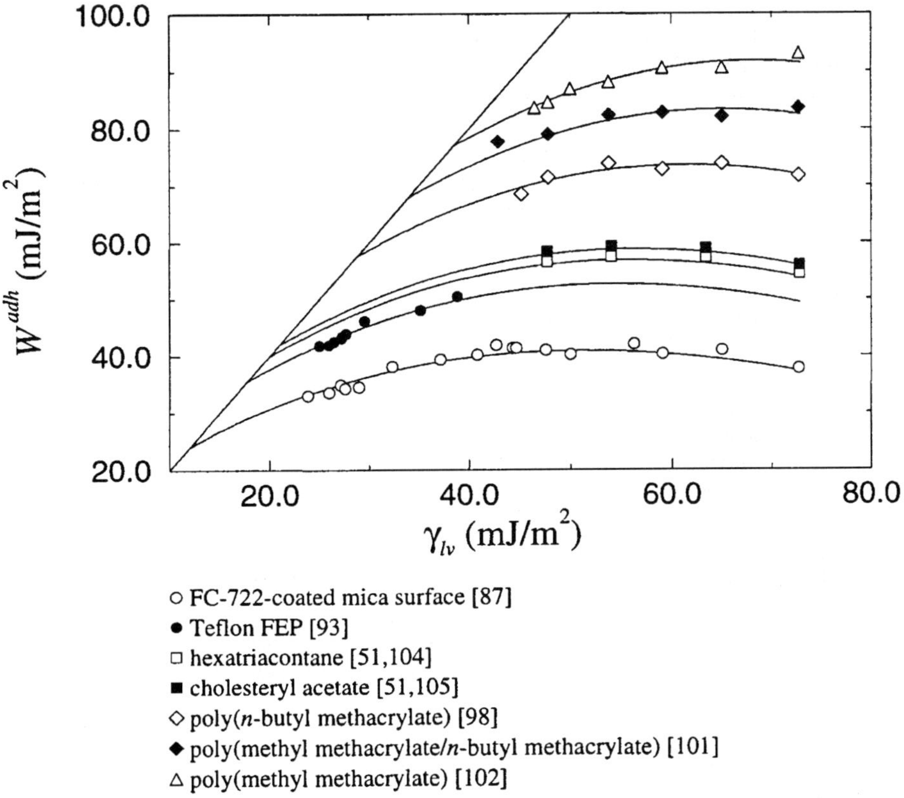

Figure 18. Free energy of adhesion W^{adh} vs. liquid surface tension γ_{lv} for several solid surfaces shown in Figure 12. The curves are hypothetical constant γ_{sv} curves generated from Eq. (45), by calculating γ_{sl} for different γ_{lv} values; Eq. (33) was then used to calculated W^{adh}.

has to be assumed constant from liquid to liquid; otherwise, a systematic study of the direct effect of changing γ_{lv} on W^{adh} through θ by the second (non-measurable quantity) γ_{sl} cannot be possible, unless the third (non-measurable quantity) γ_{sv} is kept constant.

The low-rate dynamic contact angles obtained in Section 2.3 can be employed for the calculations of W^{adh}. Figure 18 shows the calculated free energy of adhesion vs. the liquid-vapor surface tension γ_{lv} for the several solid surfaces from Figure 13; for reasons of clarity, only the data from certain solid surfaces are plotted. In Figure 18, for one and the same solid surface (and hence constant γ_{sv}), W^{adh} increases as γ_{lv} increases, reaching a global maximum; further increase in γ_{lv} causes W^{adh} to decrease.

Interpretating these experimental results in terms of Eq. (33) suggests that any changes in W^{adh} with changing γ_{lv} must be a consequence of changes in the difference between γ_{sv} and γ_{sl}. Since γ_{sv} is expected to be constant for one and the same solid surface, such variations in W^{adh} must result from the effect of changing γ_{sl} as a consequence of changing γ_{lv}. The results in Figure 18 suggest that γ_{sl}

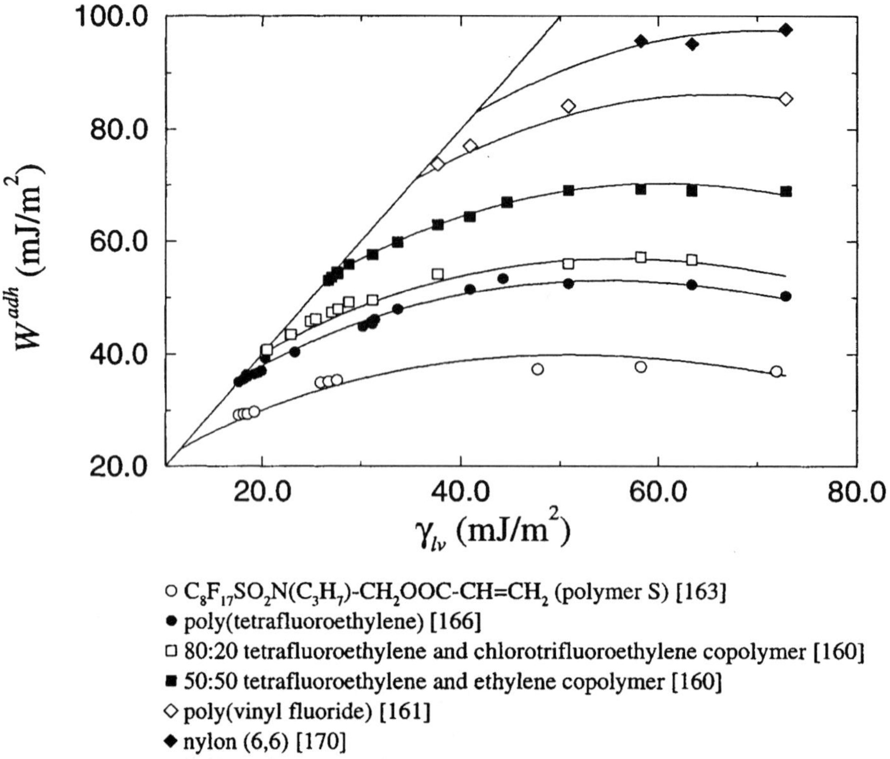

Figure 19. Free energy of adhesion W^{adh} vs. liquid surface tension γ_{lv} for Zisman's contact angle data. The curves are hypothetical constant γ_{sv} curves generated from Eq. (45), by calculating γ_{sl} for different γ_{lv} values; Eq. (33) was then used to calculated W^{adh}.

increases with γ_{lv}, causing W^{adh} to increase; as γ_{sl} increases and equals γ_{sv}, W^{adh} reaches a global maximum. Further increase in γ_{lv} causes γ_{sl} to increase faster than γ_{lv}, causing W^{adh} to decrease. Similar results can also be found in Figure 19, from Zisman's contact angles which will be discussed in the next section. The straight lines in Figures 18 and 19 are lines of zero contact angle calculated from Eq. (58); the curves are theoretical constant γ_{sv} curves generated from the equation-of-state approach, i.e. Eq. (45), by calculating γ_{sl} from γ_{lv} at constant (hypothetical) γ_{sv} values; Eq. (33) was then used to deduce W^{adh}. The good fit between the experimental free energy of adhesion and theoretical prediction again illustrates the applicability of the equation-of-state approach. The adhesion patterns showed in Figure 18 has been used to derive a combining rule for intermolecular potentials [156].

5. EVALUATION OF EXISTING CONTACT ANGLE DATA

In Sections 2.2 and 2.3.3, we have shown that static contact angles measured by conventional goniometer techniques could be meaningless in the context of

Young's equation. By measuring dynamic contact angles at very slow motion of the three-phase contact line using an automated drop shape technique, we obtained a variety of contact angle patterns, which cannot be easily detected by conventional goniometer contact angle techniques and procedures. For example, slip/stick of the three-phase contact line was obtained, where the contact angle increased steadily by as much as 35° at constant three-phase radius. Although such patterns are very interesting, they have to be discarded if the aim is to deduce solid surface tensions: since γ_{lv}, γ_{sv}, and γ_{sl} are material properties and are expected to be constant, Young's equation implies a unique contact angle. Thus, the interpretation of contact angles for solid surface tensions from slip/stick of the three-phase contact line would be meaningless.

Apart from the applicability of Young's equation, several other widely made assumptions are often overlooked. In an example (Figure 8) shown in Section 2.3.3, the contact angle increases as the drop front advances with essentially constant γ_{lv}. While there is no reason to question the applicability of Young's equation, γ_{sl} is suspected to change: from Young's equation, if γ_{lv} and γ_{sv} are constant, changes in θ must be a consequence of a change in γ_{sl}. While such cases might be very interesting in a different context, they have to be discarded in the present context.

The above findings are not readily attainable for static contact angle measurements using conventional goniometer techniques and procedures: goniometer data may contain a mixture of meaningful and meaningless angles for solid surface tensions. If one neglects the goniometer angles shown to be meaningless in the low-rate dynamic contact angle study by, e.g., axisymmetric drop shape analysis — profile (ADSA-P), smooth curves emerge in plots of $\gamma_{lv}\cos\theta$ vs. γ_{lv}. This pattern was used to deduce a functional relationship to be used in conjunction with Young's equation for determining solid surface tensions.

Nevertheless, there is an abundance of contact angle data in the literature which were obtained from static measurements, and are hence of unknown status. The question then arises as to whether one should neglect completely the literature contact angle data and repeat all measurements using a dynamic procedure such as ADSA-P, or whether some of the data can be salvaged. This question will be explored in this section [159], using Zisman's contact angle data on 34 different types of solid surfaces.

The rationale of evaluating literature contact angle data is as follows. After discarding all those measurements which violate the basic assumptions made in all contact angle approaches of current interest [11]–[17], the low-rate dynamic contact angle measurements yielded contact angles which fell on smooth curves in plots of $\gamma_{lv}\cos\theta$ vs. γ_{lv}, for one and the same solid surface [83, 87], [95]–[102]. Assuming that this would also have happened in the systems studied in the laboratory of Zisman [10], we propose to discard those contact angle data which do not sensibly fall on a smooth curve.

The equation-of-state approach for solid-liquid interfacial tensions [14], i.e. Eq. (46), described in Section 3.2.3 can be used to evaluate these data by generating

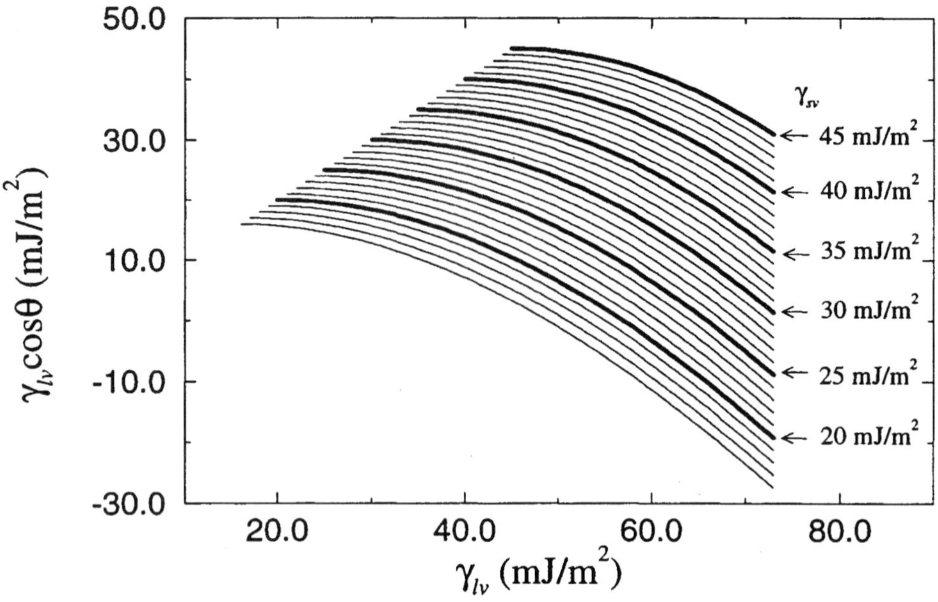

Figure 20. $\gamma_{lv}\cos\theta$ vs. γ_{lv} generated from the equation-of-state approach, Eq. (46), from γ_{sv} = 16.0 mJ/m^2 to 45.0 mJ/m^2, in 1.0 mJ/m^2 increments.

a series of theoretical constant γ_{sv} curves, in plots of $\gamma_{lv}\cos\theta$ vs. γ_{sv}. The experimental (parallel) curves of those shown in Figures 1, 5, and 10–13 could in principle be used; however, a quantitative comparison would be difficult. Figure 20 shows the theoretical curves generated by the equation-of-state approach, i.e. Eq. (46), from γ_{sv} = 16.0 mJ/m^2 to γ_{sv} = 45.0 mJ/m^2 in 1.0 mJ/m^2 increments; each curve represents a constant γ_{sv} value and hence a hypothetical solid surface. Thus, if goniometer contact angle data are plotted in this graph, it can be decided how well data for any one solid fall onto an individual curve.

Figure 21 shows the contact angle data of Zisman and coworkers for a copolymer of 80:20 tetrafluoroethylene and chlorotrifluoroethylene [160], copolymer of 50:50 tetrafluoroethylene and ethylene [160], and poly(vinyl fluoride) [161] surfaces, as well as the theoretical curves from Figure 19. In Figure 20, the data points for the 50:50 tetrafluoroethylene and ethylene surface fall consistently close to the theoretical constant γ_{sv} curve of 27.0 mJ/m^2. However, considerable scatter is apparent for the other surfaces. As we have discussed in Section 2.4.1, vapor adsorption of liquids which could cause γ_{sv} to change from one liquid to the next by as low as 1 mJ/m^2 has been considered possible [109]. If $\pm$1.0 mJ/m^2 is selected as the allowable error, contact angle data which deviate from a theoretical γ_{sv} curve by more than $\pm$1.0 mJ/m^2 can be disregarded. It should be noted that this criterion is not an overestimate: the widely-accepted goniometer accuracy of $\pm 2°$ could easily cause the data points in Figure 20 to vary along the $\gamma_{lv}\cos\theta$ axis by as much as $\pm$2.0 mJ/m^2. Following the above procedures, it was found that 7 of the 22 data points for the 80:20 tetrafluoroethylene and chlorotrifluoroethylene surface were

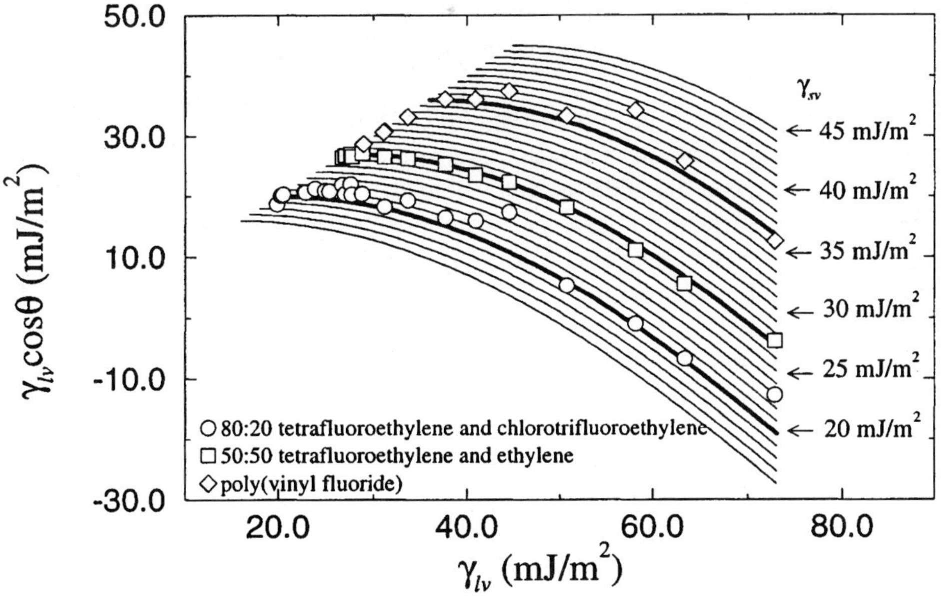

Figure 21. $\gamma_{lv}\cos\theta$ vs. γ_{lv} for Zisman's data on a copolymer of 80:20 tetrafluoroethylene and chlorotrifluoroethylene [160], copolymer of 50:50 tetrafluoroethylene and ethylene [160], and poly(vinyl fluoride) [161], as well as hypothetical constant γ_{sv} curves from Figure 20.

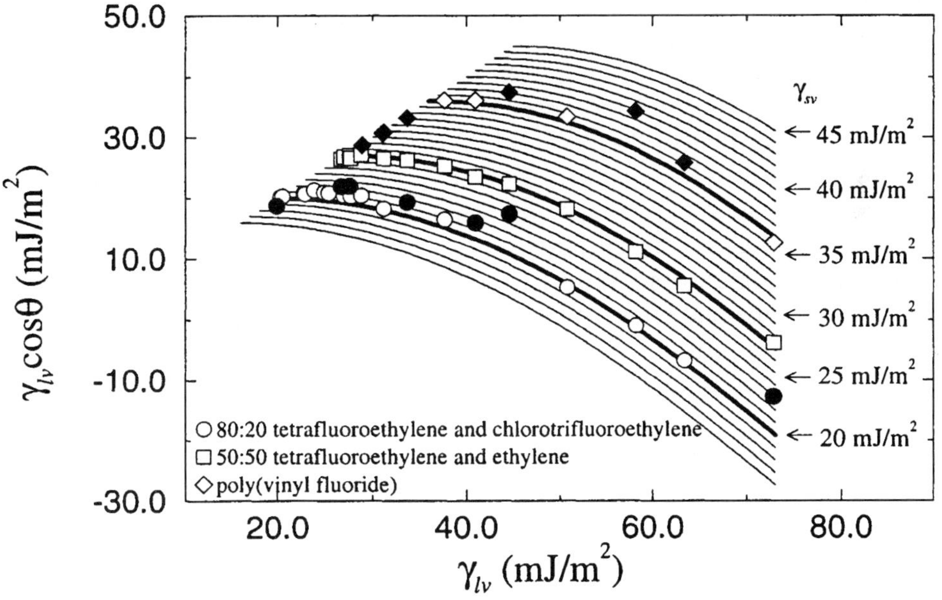

Figure 22. $\gamma_{lv}\cos\theta$ vs. γ_{lv} for Zisman's data on a copolymer of 80:20 tetrafluoroethylene and chlorotrifluoroethylene [160], copolymer of 50:50 tetrafluoroethylene and ethylene [160], and poly(vinyl fluoride) [161]. Contact angle data which are clearly outside a constant γ_{sv} curve by more than ± 1.0 mJ/m^2 are discarded (solid symbols).

clearly outside the error band of ±1.0 mJ/m² and, similarly, 6 out of 11 data for the poly(vinyl fluoride) surface; these data are rejected, as shown in Figure 22 in solid symbols.

In principle, the above procedures can be employed for all other solid surfaces. However, if a large amount of contact angle data and solid surfaces are available, these procedures can be tedious and subjective. For this reason, they were automated by constructing a least-squares objective function [152] to calculate the deviation of all contact angle data from a theoretical curve generated by a given γ_{sv} value. The best-fitted constant γ_{sv} curve can then be used to discriminate against data which are clearly above or below the curve within a band, e.g., ±1.0 mJ/m². A FORTRAN computer program which automates these procedures is available [159].

This computer program was applied to the goniometer data for the following 34 solid surfaces from Zisman and coworkers: perfluorocapric acid [162], perfluorolauric acid [162], $C_8F_{17}SO_2N(CH_3H_7)-CH_2CH_2OOC-CH=CH_2$ (polymer S) [163], $C_7F_{15}CH_2OOC-C(CH_3)=CH_2$ (polymer A) [163], perfluorocaprylic acid [162], perfluorocaproic acid [162], perfluorovaleric acid [162], perfluorobutyric acid [162], 17-(perfluoroheptyl)-heptadecanoic acid [164], 17-(perfluoropentyl)-heptadecanoic acid [164], perfluorodecanoic acid [165], polytetrafluoroethylene [166], 17-(perfluoropropyl)-heptadecanoic acid [164], paraffin [121], hexatriacontane [121], 17-(perfluoroethyl)-heptadecanoic acid [164], copolymer of 80:20 tetrafluoroethylene and chlorotrifluoroethylene [160], stearic acid [167], *n*-octadecylamine [168], ω-monohydroperfluoroundecanoic acid [169], copolymer of 60:40 tetrafluoroethylene and chlorotrifluoroethylene [160], copolymer of 50:50 tetrafluoroethylene and ethylene [160], polytrifluoroethylene [161], polychlorotrifluoroethylene [160], polystyrene [170], 2-ethylhexanoic acid [171], 2-ethylhexylamine [171], 2-ethylhexanol [171], poly(vinylidene fluoride) [161], poly(ethylene terephthalate) [170], poly(vinylidene chloride) [161], poly(vinyl fluoride) [161], nylon (6,6) [170], and perchloropentadienoic acid [161]. Figure 22 shows the contact angle data for three solid surfaces, together with the fitted γ_{sv} curves; the rejected angles are given in solid symbols. It can be seen that, disregarding the rejected angles, the experimental data points fall closely to a distinct theoretical γ_{sv} curve for each solid within ±1.0 mJ/m². The contact angles for those pairs falling to ±1.0 mJ/m² on smooth theoretical γ_{sv} curves are given in Table 11, together with the calculated γ_{sv} values from the equation-of-state approach [14, 64], Eq. (46). It can be seen that the calculated γ_{sv} values for these surfaces are indeed quite constant. The γ_{sv} values for the 34 solid surfaces obtained from the above least-square fit are summarized in Table 12; the number of contact angles available and being rejected are also shown.

For several solid surfaces in Table 12, γ_{sv} values had been obtained from more accurate contact angles. It is, therefore, instructive to compare these values with those determined from the above fitting procedures using Zisman's goniometer data. For example, γ_{sv} values of 28.3 mJ/m² and 29.9 ± 0.5 mJ/m² (in Table 7) had been reported for two different types of polystyrene-coated Si wafer surfaces [99, 172];

Table 11.
Static contact angles from Zisman *et al.* which fall within ±1.0 mJ/m^2 on a fitted γ_{sv} curve

solid surface	liquid	γ_{lv} (mJ/m^2)	θ (deg.)	γ_{sv} equation-of-state Eq. (46)
perfluorocapric acid† [162]	octane	21.8	64.5	11.5
	decane	23.9	69.5	11.3
	dodecane	25.4	73.0	11.1
	tetradecane	26.7	75.0	11.2
	hexadecane	27.6	77.0	11.1
perfluorolauric acid [162]	pentamer	18.1	60	10.3
	octane	21.8	65	11.3
	decane	23.9	70	11.2
	di-(*n*-amyl) ether	24.9	68	12.2
	dodecane	25.4	74	10.9
	tetradecane	26.7	78	10.4
	hexadecane	27.6	78	10.8
	polyethylene V-120	27.8	78	10.9
	di-(*n*-decyl) ether	28.4	79	10.9
	benzene	28.9	75	12.3
	1,6-hexamethylene glycol bis-2-ethylhexanoate	30.2	77	12.3
	pentaerythritol tetracaproate	30.4	80	11.5
	polyethylene SS903	30.4	81	11.2
	8-*p*-tolylnonadecane	30.7	83	10.7
	bis-(2-ethylhexyl) sebacate	31.1	79	12.1
	n-decylbenzene	31.2	83	10.9
	9-(*α*-(cis-0,3,3-bicycl-oöctyl)-methyl)-heptadecane	31.2	84	10.6
	4-cyclohexyleicosane	31.6	83	11.1
	1-cyclohexyl-2-(cyclohexyl-methyl)-pentadecane	32.7	84	11.2
	1-*α*-decalylhendecane	32.7	84	11.2
	9-*n*-dodecylperhydrophenanthrene	34.2	86	11.2
	1,7-dicyclopentyl-4-(3-cyclopentylpropyl)-heptane	34.6	86	11.3
	1,3-dicyclopentylcyclopentane	34.6	85	11.7
	1,1-di-(*α*-decalyl)-hendecane	35.1	88	10.9
	benzylphenyl hendecanoate	37.7	92	10.6
	tetrachlorobiphenyl	44.2	96	11.6
	methylene iodide	50.8	103	11.3
$C_8F_{17}SO_2N(C_3H_7)$–CH_2CH_2OO–$CH{=}CH_2$ (polymer S)† [163]	tetramer	17.6	48.5	12.2
	pentamer	18.1	51.5	12.0
	hexamer	18.5	54.0	11.8
	nonamer	19.2	56.5	11.7
	tridecane	25.9	69.5	12.4
	tetradecane	26.7	71.5	12.2

Table 11.
(continued)

solid surface	liquid	γ_{lv} (mJ/m^2)	θ (deg.)	γ_{sv} equation-of-state Eq. (46)
	hexadecane	27.6	73.5	12.1
	ethylene glycol	47.7	102.5	10.4
	formamide	58.2	110.5	10.8
	water	71.9	119.0	11.8
$C_7F_{15}CH_2OOC$–$C(CH_3)$= CH_2 (polymer A)† [163]	pentamer	18.1	48.5	12.6
	hexamer	18.5	51.5	12.3
	nonamer	19.2	53.0	12.5
	dodecane	25.4	67.0	12.8
	tridecane	25.9	68.0	12.8
	tetradecane	26.7	71.5	12.2
	hexadecane	27.6	73.0	12.2
	formamide	58.2	109.5	11.2
	water	71.9	118.5	12.0
perfluorocaprylic acid† [162]	octane	21.8	60.0	12.5
	decane	23.9	65.8	12.3
	dodecane	25.4	68.0	12.5
	tetradecane	26.7	71.0	12.3
	hexadecane	27.6	74.0	11.9
perfluorocaproic acid† [162]	octane	21.8	59.0	12.8
	decane	23.9	64.5	12.6
	tetradecane	26.7	68.8	13.0
	hexadecane	27.6	71.8	12.6
perfluorovaleric acid† [162]	decane	23.9	70.5	13.3
	dodecane	25.4	67.5	13.2
	tetradecane	26.7	65.5	13.3
	hexadecane	27.6	70.5	13.0
perfluorobutyric acid† [162]	decane	23.9	61.0	13.5
	dodecane	25.4	64.5	13.5
	tetradecane	26.7	67.0	13.5
17-(perfluoroheptyl)-heptadecanoic acid [164]	heptane	20.3	52	13.4
	octane	21.8	57	13.3
	decane	23.9	64	12.8
	undecane	24.7	64	13.2
	dodecane	25.4	66	13.1
	tridecane	25.9	68	12.8
	tetradecane	26.7	70	12.6
	2-octanol	26.7	70	12.6
	1-octanol	27.8	68	13.8
	bis-(2-ethylhexyl) phthalate	31.3	78	12.5
	dicyclohexyl	32.8	75	14.2

Table 11.
(continued)

solid surface	liquid	γ_{lv} (mJ/m^2)	θ (deg.)	γ_{sv} equation-of-state Eq. (46)
	t-butylnaphthalene	33.7	78	13.6
	tricresyl phosphate	40.9	90	12.5
	tetrachlorobiphenyl	44.2	93	12.7
	formamide	58.2	105	13.2
	glycerol	63.4	107	14.4
	water	72.8	115	14.3
17-(perfluoropentyl)-heptadecanoic acid [164]	perfluoroalkane (FCD-330)	20.2	33	17.1
	dicyclohexyl	32.8	72	15.2
	t-butylnaphthalene	33.7	73	15.3
	tricresyl phosphate	40.9	81	16.0
	methylene iodide	50.8	94	15.1
	formamide	58.2	100	15.6
	water	72.8	110	17.1
perfluorodecanoic acid [165]	perfluorokerosene	18.1	26	16.3
	di(*n*-butyl) ether	22.8	47	16.3
	di(isoamyl) ether	23.0	49	16.0
	$H(CF_2)_2CH_2OH$	24.5	55	15.5
	di(*n*-amyl) ether	24.9	54	16.0
	carbon tetrachloride	26.8	60	15.6
	$CFCl_2–CF_2–CCl_3$	27.8	56	17.4
	sec-butylbenzene	28.7	65	15.2
	propylbenzene	29.0	59	17.2
	butylbenzene	29.2	63	16.1
	tris-(2-ethylhexyl)tricarballylate	29.6	65	15.7
	1,6-hexamethylene glycol bis-2-ethylhexanoate	30.2	66	15.7
	t-butylnaphthalene	33.7	71	16.0
	perchlorocyclopentadiene	37.5	73	17.3
	butylphenylhendecanoate	38.0	76	16.5
	tricresyl phosphate	40.9	82	15.6
	pentamethylene glycol	43.4	82	16.8
	methylene iodide	50.8	90	16.9
poly(tetrafluoro-ethylene) [166]	tetramer	17.6	8	17.4
	pentamer	18.1	15	17.5
	hexane	18.4	12	18.0
	hexamer	18.5	19	17.5
	nonamer	19.2	26	17.4
	dodecamer	19.6	29	17.3
	heptadecamer	19.9	30	17.3
	heptane	20.3	21	19.1
	polyethylsiloxane	23.3	43	17.6

Table 11.
(continued)

solid surface	liquid	γ_{lv} (mJ/m^2)	θ (deg.)	γ_{sv} equation-of-state Eq. (46)
	di-(2-ethylhexyl) adipate	30.2	61	17.3
	di-(2-ethylhexyl) sebacate	31.1	62	17.6
	di-(2-ethylhexyl) phthalate	31.2	63	17.3
	carbon disulfide	31.4	62	17.8
	t-butylnaphthalene	33.7	65	18.2
	tricresyl phosphate	40.9	75	18.4
	tetraachlorodiphenyl	44.2	78	18.9
	methylene iodide	50.8	88	17.3
	glycerol	63.4	100	18.1
	water	72.8	108	18.3
17-(perfluoropropyl)-heptadecanoic acid [164]	perfluoroalkane (FCD-330)	20.2	23	18.6
	bis-(2-ethylhexyl) phthalate	31.3	60	18.4
	dicyclohexyl	32.8	62	18.6
	tricresyl phosphate	40.9	72	19.6
	ethylene glycol	47.7	84	18.4
	methylene iodide	50.8	86	18.8
	water	72.8	106	19.4
paraffin [121]	std. fluorolube	25.1	38	20.2
	formamide	58.2	91	20.1
	glycerol	63.4	96	20.2
	water	72.8	108	18.2
hexatriacontane [121]	polymethylsiloxane	19.9	20	18.7
	di(*n*-amyl) ether	24.9	41	19.3
	std. fluorolube	25.1	43	19.0
	di(*n*-heptyl) ether	27.0	45	19.9
	hexadecane	27.6	46	20.1
	di(*n*-decyl) ether	27.7	50	19.0
	n-heptylic acid	28.3	49	19.8
	tri(2-ethylhexyl)tricarballylate	29.6	56	18.6
	pentaerythtitol tetraeaproate	30.4	56	19.1
	trieresyl phosphate	40.9	72	19.6
	formamide	58.2	92	19.6
	glycerol	63.4	97	19.7
17-(perfluoroethyl)-heptadecanoic acid [164]	dicyclohexyl	32.8	61	19.0
	tricresyl phosphate	40.9	72	19.6
	ethylene glycol	47.7	80	19.9
	methylene iodide	50.8	83	20.2
	formamide	58.2	94	18.6
	water	72.8	105	20.0

Table 11.
(continued)

solid surface	liquid	γ_{lv} (mJ/m^2)	θ (deg.)	γ_{sv} equation-of-state Eq. (46)
80:20 tetrafluoroethylene and chlorotrifluoroethylene copolymer [160]	heptane	20.3	8	20.1
	di(*n*-propyl) ether	20.5	8	20.3
	di(*n*-butyl) ether	22.8	25	20.8
	nonane	22.9	26	20.7
	di(*n*-amyl) ether	24.9	33	21.1
	dodecane	25.4	35	21.1
	di(*n*-heptyl) ether	27.0	41	21.0
	di(*n*-octyl) ether	27.7	43	21.0
	benzene	28.8	45	21.3
	di(2-ethylhexyl)-phthalene	31.2	54	20.2
	benzyl phenylun-decanoate	37.7	64	20.9
	methylene iodide	50.8	84	19.7
	formamide	58.2	91	20.2
	glycerol	63.4	96	20.2
stearic acid [167]	octanoic acid	29.2	40	23.0
	nonanoic acid	29.5	43	22.4
	tricresyl phosphate	40.9	65	22.5
	tetrachlorobiphenyl	44.2	70	22.4
	thiodiglycol	54.0	81	23.0
	glycerol	63.4	92	22.4
n-octadecylamine [168]	di(*n*-butyl) ether	22.8	9	22.5
	nonane	22.9	13	22.3
	decane	23.9	18	22.7
	di(*n*-amyl) ether	24.9	22	23.2
	dodecane	25.4	30	22.2
	tetradecane	26.7	34	22.4
	di(*n*-heptyl) ether	27.0	34	22.7
	hexadecane	27.6	38	22.2
	di(*n*-octyl) ether	27.7	36	22.8
	di(*n*-decyl) ether	28.4	37	23.1
	tri(2-ethylhexyl)-tricarballylate	29.6	45	21.9
	hexylbenzene	30.0	39	22.3
	1,6-hexamethylene glycol di-2-ethylhexanoate	30.2	44	22.6
	pentaerythritol-tetracaproate	30.4	43	23.1
	di(2-ethylhexyl) sebacate	31.1	46	22.7
	di(2-ethylhexyl) phthalate	31.2	49	21.9

Table 11.
(continued)

solid surface	liquid	γ_{lv} (mJ/m^2)	θ (deg.)	γ_{sv} equation-of-state Eq. (46)
	decylbenzene	31.2	44	23.4
	α-methylnaphthalene	36.4	55	23.5
	benzylphenylundecanoate	37.7	58	23.2
	tricresyl phosphate	40.9	65	22.5
	glycerol	63.4	90	23.6
	water	72.8	102	21.9
ω-monohydroperfluoro-undecanoic acid [169]	n-amyl amine	25.2	9	24.9
	1,6-hexamethylene glycol			
	di-2-ethyl hexanoate	30.2	39	24.1
	di-(2-ethylhexyl) phthalate	31.2	40	24.6
	glycerol	63.4	89	24.1
	water	72.8	97	24.9
60:40 tetrafluoroethyl-ene and chlorotrifluoro-ethylene copolymer [160]	di(n-heptyl) ether	27.0	22	25.1
	hexadecane	27.6	24	25.2
	benzene	28.8	25	26.2
	t-butylnaphthalene	33.7	45	25.0
	tricresyl phosphate	40.9	56	26.3
	α-bromonaphthalene	44.6	63	25.8
	glycerol	63.4	87	25.2
	water	72.8	94	26.7
50:50 tetrafluoroethyl-ene and ethylene copolymer [160]	tetradecane	26.7	9	26.4
	di(n-heptyl) ether	27.0	9	26.7
	hexadecane	27.6	12	27.0
	di(n-octyl) ether	27.7	16	26.6
	benzene	28.8	20	27.1
	di(2-ethylhexyl) phthalate	31.2	32	26.8
	t-butylnaphthalene	33.7	39	26.9
	benzyl phenylundecanoate	37.7	48	27.0
	tricresyl phosphate	40.9	55	26.7
	α-bromonaphthalene	44.6	60	27.1
	methylene iodide	50.8	69	27.0
	formamide	58.2	79	26.5
	glycerol	63.4	85	26.4
	water	72.8	93	27.4
poly(trifluoroethylene) [161]	bis-(2-ethylhexyl)-phthalate	31.2	22	29.0
	benzyl phenylundecanoate	37.7	44	28.5
	tricresyl phosphate	40.9	49	29.0
	formamide	58.2	76	28.1
	glycerol	63.4	82	28.1
	water	72.8	92	28.0

Table 11.
(continued)

solid surface	liquid	γ_{lv} (mJ/m^2)	θ (deg.)	γ_{sv} equation-of-state Eq. (46)
poly(chlorotrifluoro-ethylene) [160]	methylene iodide	50.8	64	29.5
	glycerol	63.4	82	28.1
	water	72.8	90	29.2
polystyrene [170]	formamide	58.2	74	29.2
	glycerol	63.4	80	29.2
	water	72.8	91	28.6
2-ethylhexanoic acid [171]	decyl acetate	28.5	2	28.5
	bis-(n-butyl)-protartrate	29.3	5	29.2
	bis-(2-ethylhexyl)-β-methyl adipate	30.2	14	29.3
	1,6-hexanediol bis-(2-ethylhexanoate)	30.2	10	29.7
	bis-(2-ethylhexyl) tetrahydrophthalate	30.7	14	29.8
	polyethylene SS 906	30.7	23	28.3
	bis-(2-ethylhexyl)-sebacate	31.1	15	30.0
	1-cyclohexyl-2-(cyclo-hexylmethyl)-pentadecane	32.7	27	29.3
	9-n-dodecylperhydro-phenanthrene	34.2	32	29.4
	1,1-bis-(α-decalyl)-hendecane	35.1	34	29.6
2-ethylhexylamine [171]	α-bromonaphthalene	44.6	53	30.1
	methylene iodide	50.8	63	29.9
2-ethylhexanol [171]	tricresyl phosphate	40.9	37	33.5
	bis-(2-phenylethyl)-β-methyl adipate	41.3	42	32.0
	α-bromonaphthalene	44.6	45	33.5
poly(vinylidene fluoride) [161]	benzyl phenylundecanoate	37.7	26	34.1
	α-bromonaphthalene	44.6	42	34.7
	water	72.8	82	34.2
poly(ethylene terephthalate) [170]	formamide	58.2	61	36.2
	water	72.8	81	34.9
poly(vinylidene chloride) [161]	formamide	58.2	61	36.2
	water	72.8	80	35.5
poly(vinyl fluoride) [161]	benzyl phenylendecanoate	37.7	17	36.1
	tricresyl phosphate	40.9	28	36.4
	methylene iodide	50.8	49	36.6
	water	72.8	80	35.5

Table 11.
(continued)

solid surface	liquid	γ_{lv} (mJ/m^2)	θ (deg.)	γ_{sv} equation-of-state Eq. (46)
nylon (6,6) [170]	formamide	58.2	50	41.9
	glycerol	63.4	60	40.6
	water	72.8	70	41.7
perchloropentadienoic acid [161]	methylene iodide	50.8	34	43.1
	glycerol	63.4	56	42.8
	water	72.8	66	44.2

† contact angles were estimated from graphs to the nearest 0.5°.

they were determined from recent low-rate dynamic contact angles. In view of the difference in polymer properties and molecular weights, a value of 28.9 mJ/m^2 for the polystyrene given in Table 12 is indeed plausible. Further, $\gamma_{sv} = 35.6 \pm 0.4$ mJ/m^2 for poly(ethylene terephthalate) reported elsewhere [64] is again in good agreement with the value of 35.2 mJ/m^2 in Table 12. These findings give confidence in our procedures in deducing solid surface tension values from goniometer contact angles, so that dynamic contact angle procedures by, e.g., ADSA-P do not need to be repeated for the solid surfaces studied by Zisman *et al.* Thus, contact angles can indeed be used for the determination of solid surface tensions and literature contact angles do not have to be discarded completely.

6. SUMMARY

This study illustrates clearly that obtaining meaningful contact angles for the determination of solid surface tensions depends heavily on how contact angles are measured and whether or not the widely made assumptions have been violated. Experimental procedures and criteria for measuring and interpreting meaningful contact angles have been developed. The discrepancy in the literature in terms of the contact angle patterns have been explained. A large amount of meaningful contact angle data was also generated on various solid surfaces. Unique experimental curves of $\gamma_{lv} \cos\theta$ vs. γ_{lv} were obtained. The widely used surface tension component approaches [11], [15]–[17] were shown to contradict these experimental contact angle patterns. The new contact angle data established here provide an experimental foundation for future development of the equation-of-state approach for solid surface tensions; a new formulation of such an approach which allows the determination of solid surface tensions has been developed. The contact angle data presented here also allow the determination and interpretation of the free energy of adhesion. Literature static contact angles do not have to be discarded completely and can be used for determining solid surface tensions, with caution.

Table 12.
Summary of the γ_{sv} (mJ/m^2) values from curve fitting of Eq. (46) for 34 solid surfaces, using the contact angle data from Zisman and coworkers

solid surface	# of data		γ_{sv}
	available	rejected	
perfluorocapric acid† [162]	5	0	11.2
perfluorolauric acid [162]	45	18	11.2
$C_8F_{17}SO_2N(C_3H_7)$–CH_2CH_2OO–CH=CH_2 (polymer S)† [163]	19	9	11.4
$C_7F_{15}CH_2OOC$–$C(CH_3)$=CH_2 (polymer A)† [163]	19	10	11.9
perfluorocaprylic acid† [162]	5	0	12.3
perfluorocaproic acid† [162]	5	1	12.7
perfluorovaleric acid† [162]	4	0	13.2
perfluorobutyric acid† [162]	3	0	13.5
17-(perfluoroheptyl)-heptadecanoic acid [164]	30	13	13.7
17-(perfluoropentyl)-heptadecanoic acid [164]	17	10	16.2
perfluorodecanoic acid [165]	88	70	16.4
polytetrafluoroethylene [166]	56	37	18.0
17-(perfluoropropyl)-heptadecanoic acid [164]	17	10	19.1
paraffin [121]	20	16	19.3
hexatriacontane [121]	35	23	19.6
17-(perfluoroethyl)-heptadecanoic acid [164]	19	13	19.7
80:20 tetrafluoroethylene and chlorotrifluoroethylene copolymer [160]	22	8	20.2
stearic acid [167]	18	12	22.6
n-octadecylamine [168]	38	16	22.6
ω-monohydroperfluoroundecanoic acid [169]	24	19	24.6
60:40 tetrafluoroethylene and chlorotrifluoroethylene copolymer [160]	17	9	26.1
50:50 tetrafluoroethylene and ethylene copolymer [160]	14	0	26.9
polytrifluoroethylene [161]	15	9	28.1
polychlorotrifluoroethylene [160]	9	6	28.9
polystyrene [170]	7	4	28.9
2-ethylhexanoic acid [171]	18	8	29.3
2-ethylhexylamine [171]	13	10	30.0
2-ethylhexanol [171]	13	11	33.1
poly(vinylidene fluoride) [161]	13	10	34.3
poly(ethylene terephthalate) [170]	7	5	35.2
poly(vinylidene chloride) [161]	6	4	35.7
poly(vinyl fluoride) [161]	11	7	35.7
nylon (6,6) [170]	7	4	41.5
perchloropentadienoic acid [161]	3	0	43.7

† contact angles were estimated from graphs to the nearest 0.5°.

7. PERSPECTIVE

A look at the likely future of research on contact angles and surface energetics will benefit from a view back, particularly at persistent obstacles to progress. There are three key attitudes which have hampered progress immensely:

(1) Contact angles are simple quantities which can be readily measured and interpreted by anybody, requiring no particular skill, methodology, or knowledge.

(2) Contact angle measurements show hysteresis. Hence, experimental contact angles are not equilibrium values, and therefore nothing useful can be learned from contact angle studies.

(3) Contact angles contain readily accessible information about intermolecular forces.

All of these propositions are false. With respect to the first proposition, the present study shows how misleading conventional contact angle measurements on sessile drops can be.

The second proposition, while false, is the most difficult to refute. Operationally, the key question is not whether a contact angle is an equilibrium contact angle, but whether it can be used in conjunction with Young's equation. For instance, contact angle hysteresis caused by a modest degree of surface heterogeneity does not make Young's equation inapplicable. Furthermore, it is clear from the systems examined here that contact with a liquid will often affect the solid surface. This does not preclude the possibility that the advancing contact angle is the equilibrium angle; rather this equilibrium cannot be reached from "the other side", i.e. under receding conditions, because the solid surface is now different. Furthermore, it is difficult to see how the smooth curves of, e.g., Figures 1, 5, and 10–13 could arise if the contact angle data did not contain surface energetics meaning.

Finally, the widely held view expressed in the third proposition is demonstrably false, from curves like those in Figures 1, 5, and 10–13. For a given solid, the contact angle depends only on the liquid surface tension, not directly on the intermolecular forces which give rise to these surface tensions. Referring to the curve for FC-721 in Figure 1, the fact that methanol falls on the same smooth curve as hexane and decane would imply that the polar component of surface tension of methanol is zero, i.e. methanol would have to be classified as non-polar, which is clearly absurd.

While the first and third propositions have been dealt with extensively in this study, the answer to the question of contact angle hysteresis in the second proposition would require further study.

8. FUTURE DEVELOPMENT

The present state of knowledge is well summarized in the equation-of-state approach outlined above. The approach can be used to good advantage in numerous applications, including interfacial phenomena of powders [25, 27]. Nevertheless, there remain many unsolved questions relating to the measurement and interpretation of contact angles.

(1) While vapor adsorption does not appear to play a significant role in the contact angle data presented here, it is quite possible that it could enter the picture for more hydrophilic surfaces, particularly if $\gamma_{lv} \approx \gamma_{sv}$ or $\gamma_{lv} < \gamma_{sv}$, i.e. cases which were excluded in the present study. While adsorption may produce contact angle patterns different from those shown in Figures 1, 5, and 10–13, this would not necessarily imply that Young's equation is not applicable or that contact angle approaches might not be feasible. The answer to these questions would require general criteria to distinguish the effect of vapor adsorption on the contact angles from all other effects.

(2) Recently, self-assembled monolayers (SAMs) have been widely used to produce surfaces of different chemical compositions and wettabilities. However, it is believed that penetration of liquid into the SAMs is inevitable, depending on crystallinity and packing density. Given that two solids having the same solid surface tensions and that a specific liquid may penetrate into one solid but not the other, the contact angles on the two chemically identical surfaces can be different: one would be due to pure energetic effects, and the other to the effects of the changed energetics and liquid penetration. In the latter case, attempts [122] to interpret surface energetics from contact angle approaches naively, e.g. by means of the above equation-of-state approach, could be misleading (cf. Figure 16).

(3) It is well-known that roughness affects the experimentally observed contact angles. However, there are as yet no general criteria to quantify roughness and at what level of smoothness surface topography has no longer an effect on the contact angle. The answer to such questions will have to involve considerations of line tension [59] and the contortions of three-phase lines.

(4) It is still an open question of whether or not β in Eq. (46) and β_1 in Eq. (52) are "universal" constants, i.e. independent of the solid surface. Such a question can be addressed only after an even larger body of accurate contact angle data on various solids has been generated.

(5) While the slip/stick contact angle patterns, e.g., that shown in Figure 9 and similar patterns cannot be used to determine surface energetics, they are still very interesting and worth investigating.

(6) Routine production of systems which are free of contact angle hysteresis might be considered an important goal. It is firmly believed that all that would be produced is further confirmation of the contact angle story presented here. It is expected that more benefits can be obtained from the much simpler systematic study of receding contact angles of available systems, using concepts and procedures described in Section 2.3.

Acknowledgments

This research was supported by the Natural Science and Engineering Research Council (NSERC) of Canada under grants no. A8278 and no. EQP173469. Financial support from a NSERC Postdoctoral Fellowship (D.Y.K.) held at M.I.T. is gratefully acknowledged. D.Y.K. gratefully acknowledges Professor Paul E. Laibinis at M.I.T. for financial support and general discussion.

REFERENCES

1. B. V. Derjaguin, V. M. Muller, and Y. P. Toporov. *J. Colloid Interface Sci.*, **73**, 293 (1980).
2. K. L. Johnson, K. Kendall, and A. D. Roberts. *Proc. Roy. Soc. (London)*, **A324**, 301 (1971).
3. V. M. Muller, V. S. Yushchenko, and B. V. Derjaguin. *J. Colloid Interface Sci.*, **92**, 92 (1983).
4. A. Fogden and L.R. White. *J. Colloid Interface Sci.*, **138**, 414 (1990).
5. R. M. Pashley, P. M. McGuiggan, R. G. Horn, and B. W. Ninham. *J. Colloid Interface Sci.*, **126**, 569 (1988).
6. H. K. Christenson. *J. Phys. Chem.*, **90**, 4 (1986).
7. P. M. Claesson, C. E. Blom, P. C. Horn, and B. W. Ninham. *J. Colloid Interface Sci.*, **114**, 234 (1986).
8. P. M. Pashley, P. M. McGuiggan, and R. M. Pashley. *Colloids Surfaces*, **27**, 277 (1987).
9. R. M. Pashley, P. M. McGuiggan, B. W. Ninham, and D. F. Evans. *Science*, **229**, 1088 (1985).
10. W. A. Zisman. In *Contact Angle, Wettability and Adhesion. Advances in Chemistry Series*, volume 43. American Chemical Society, Washington, D.C. (1964).
11. F. M. Fowkes. *Ind. Eng. Chem.*, **56** (12), 40 (1964).
12. O. Driedger, A. W. Neumann, and P. J. Sell. *Kolloid – Z. Z. Polym.*, **201**, 52 (1965).
13. A. W. Neumann, R. J. Good, C. J. Hope, and M. Sejpal. *J. Colloid Interface Sci.*, **49**, 291 (1974).
14. J. K. Spelt and D. Li. In *Applied Surface Thermodynamics*, J. K. Spelt and A. W. Neumann (Eds), pp. 239–292. Marcel Dekker, New York (1996).
15. D. K. Owens and R. C. Wendt. *J. Appl. Polym. Sci.*, **13**, 1741 (1969).
16. C. J. van Oss, M. K. Chaudhury, and R. J. Good. *Chem. Rev.*, **88**, 927 (1988).
17. R. J. Good and C. J. van Oss. In *Modern Approaches to Wettability: Theory and Applications*, M. Schrader and G. Loeb (Eds), pp. 1–27. Plenum Press, New York (1992).
18. H. G. Bruil. *Colloid Polym. Sci.*, **252**, 32 (1974).
19. G. D. Cheever and J. C. Ulcing. *J. Coating Technol.*, **55**(679), 53 (1983).
20. H. W. Kilau. *Colloids Surfaces*, **26**, 217 (1983).
21. K. Grundke, T. Bogumil, T. Gietzelt, H.-J. Jacobasch, D. Y. Kwok, and A. W. Neumann. *Progr. Colloid Polym. Sci.*, **101**, 58 (1996).
22. E. I. Vargha-Butler, T. K. Zubovits, D. R. Absolom, and A. W. Neumann. *J. Dispersion Sci. Technol.*, **6**, 357 (1985).
23. E. I. Vargha-Butler, E. Moy, and A. W. Neumann. *Colloids Surfaces*, **24**, 315 (1987).
24. E. I. Vargha-Butler, T. K. Zubovits, D. R. Absolom, and A. W. Neumann. *Chem. Eng. Commun.*, **33**, 255 (1985).
25. D. Li and A. W. Neumann. In *Applied Surface Thermodynamics*, J. K. Spelt and A. W. Neumann (Eds), pp. 509–556. Marcel Dekker, New York (1996).
26. S. N. Omenyi and A. W. Neumann. *J. Appl. Phys.*, **47**, 3956 (1976).
27. D. Li and A. W. Neumann. In *Applied Surface Thermodynamics*, J. K. Spelt and A. W. Neumann (Eds), pp. 557–628. Marcel Dekker, New York (1996).
28. A. E. Corte. *J. Geophys. Res.*, **67**, 1085 (1962).

29. P. Hoekstra and R. D. Miller. *J. Colloid Interface Sci.*, **25**, 166 (1967).
30. J. Cissé and G. F. Bolling. *J. Crystal Growth*, **10**, 67 (1971).
31. J. Cissé and G. F. Bolling. *J. Crystal Growth*, **11**, 25 (1971).
32. A. M. Zubko, V. G. Lobonov, and V. V. Nikonova. *Sov. Phys. Crystallogr.*, **18**, 239 (1973).
33. K. H. Chen and W. R. Wilcox. *J. Crystal Growth*, **40**, 214 (1977).
34. D. W. Fuerstenau and M. C. Williams. *Colloids Surfaces*, **22**, 87 (1987).
35. D. W. Fuerstenau and M. C. Williams. *Particle Characterization*, **4**, 7 (1987).
36. D. W. Fuerstenau and M. C. Williams. *Int. J. Mineral Process*, **20**, 153 (1987).
37. D. W. Fuerstenau, M. C. Williams, K. S. Narayanan, J. L. Diao, and R. H. Urbina. *Energy and Fuels*, **2**, 237 (1988).
38. D. W. Fuerstenau, J. Diao, and J. Hanson. *Energy and Fuels*, **4**, 34 (1990).
39. S. J. Hemingway, J. R. Henderson, and J. R. Rowlinson. *Faraday Symp. Chem. Soc.*, **16**, 33 (1981).
40. R. Guermeur, F. Biquard, and C. Jacolin. *J. Chem. Phys.*, **82**, 2040 (1985).
41. B. S. Carey, L. E. Scriven, and H. T. Davis. *AIChE J.*, **26**, 705 (1980).
42. E. Moy and A. W. Neumann. In *Applied Surface Thermodynamics*, J. K. Spelt and A. W. Neumann (Eds), pp. 333–378. Marcel Dekker, New York (1996).
43. H. C. Hamaker. *Physica*, **4**, 1058 (1937).
44. J. N. Israelachvili. *Proc. R. Soc. London A.*, **331**, 39 (1972).
45. A. E. van Giessen, D. J. Bukman, and B. Widom. *J. Colloid Interface Sci.*, **192**, 257 (1997).
46. T. M. Reed. *J. Phys. Chem.*, **55**, 425 (1955).
47. B. E. F. Fender and G. D. Halsey, Jr. *J. Chem. Phys.*, **36**, 1881 (1962).
48. D. E. Sullivan. *J. Chem. Phys.*, **74**, 2604 (1981).
49. D. V. Matyushov and R. Schmid. *J. Chem. Phys.*, **104**, 8627 (1996).
50. T. Young. *Philos. Trans. R. Soc. London*, **95**, 65 (1805).
51. A. W. Neumann. *Adv. Colloid Interface Sci.*, **4**, 105 (1974).
52. D. Li and A. W. Neumann. In *Applied Surface Thermodynamics*, J. K. Spelt and A. W. Neumann (Eds), pp. 109–168. Marcel Dekker, New York (1996).
53. A. Marmur. *Colloids Surfaces A*, **116**, 25 (1996).
54. R. V. Sedev, J. G. Petrov, and A. W. Neumann. *J. Colloid Interface Sci.*, **180**, 36 (1996).
55. R. N. Wenzel. *Ind. Eng. Chem.*, **28**, 988 (1936).
56. A. B. D. Cassie. *Discuss. Faraday Soc.*, **3**, 11 (1948).
57. S. Baxer and A. B. D. Cassie. *J. Text. Inst.*, **36**, 67 (1945).
58. A. B. D. Cassie and S. Baxter. *Trans. Faraday Soc.*, **40**, 546 (1944).
59. J. Gaydos and A. W. Neumann. In *Applied Surface Thermodynamics*, J. K. Spelt and A. W. Neumann (Eds), pp. 169–238. Marcel Dekker, New York (1996).
60. C. J. van Oss, L. Ju, M. K. Chaudhury, and R. J. Good. *J. Colloid Interface Sci.*, **128**, 313 (1989).
61. C. J. van Oss and R. J. Good. *J. Macromol. Sci-Chem.*, **A26** (8), 1183 (1989).
62. A. W. Neumann and R. J. Good. In *Surface and Colloid Surface*, volume 11, R. J. Good and R. R. Stomberg (Eds), pp. 31–91. Plenum Press, New York (1979).
63. J. K. Spelt and E. I. Vargha-Butler. In *Applied Surface Thermodynamics*, J. K. Spelt and A. W. Neumann (Eds), pp. 379–411. Marcel Dekker, New York (1996).
64. D. Li and A. W. Neumann. *J. Colloid Interface Sci.*, **148**, 190 (1992).
65. D. Li, M. Xie, and A. W. Neumann. *Colloid Polym. Sci.*, **271**, 573 (1993).
66. Y. Rotenberg, L. Boruvka, and A. W. Neumann. *J. Colloid Interface Sci.*, **93**, 169 (1983).
67. P. Cheng, D. Li, L. Boruvka, Y. Rotenberg, and A. W. Neumann. *Colloids Surfaces*, **93**, 169 (1983).
68. P. Cheng. PhD thesis, University of Toronto (1990).
69. S. Lahooti, O. I. del Río, P. Cheng, and A. W. Neumann. In *Applied Surface Thermodynamics*, J. K. Spelt and A. W. Neumann (Eds), pp. 441–507. Marcel Dekker, New York (1996).

70. D. Y. Kwok, D. Li, and A. W. Neumann. *Colloids Surfaces A*, **89**, 181 (1994).
71. C. A. Ward and A. W. Neumann. *J. Colloid Interface Sci.*, **49**, 286 (1974).
72. R. Defay. *Etude Thermodynamique de la Tension Superficielle*. Gauthier Villars, Paris (1934).
73. R. Defay and I. Prigogine. *Surface Tension and Adsorption*. Longmans, London (1954).
74. D. Li, J. Gaydos, and A. W. Neumann. *Langmuir*, **5**, 293 (1989).
75. D. Li and A. W. Neumann. *Adv. Colloids Interface Sci.*, **49**, 147 (1994).
76. R. E. Johnson, Jr. and R. H. Dettre. *Langmuir*, **5**, 293 (1989).
77. I. D. Morrison. *Langmuir*, **5**, 540 (1989).
78. I. D. Morrison. *Langmuir*, **5**, 1833 (1989).
79. D. Li, E. Moy, and A. W. Neumann. *Langmuir*, **6**, 885 (1990).
80. J. Gaydos, E. Moy, and A. W. Neumann. *Langmuir*, **6**, 888 (1990).
81. J. Gaydos and A. W. Neumann. *Langmuir*, **9**, 3327 (1993).
82. D. Li and A. W. Neumann. *Langmuir*, **9**, 3728 (1993).
83. D. Y. Kwok, T. Gietzelt, K. Grundke, H.-J. Jacobasch, and A. W. Neumann. *Langmuir*, **13**, 2880 (1997).
84. D. H. Kaelble. *J. Adhesion*, **2**, 66 (1970).
85. S. Wu. In *Adhesion and Adsorption of Polymers*, L. H. Lee (Ed.), pp. 53–65. Plenum, New York (1980).
86. C. J. van Oss, R. J. Good, and M. K. Chaudhury. *Langmuir*, **4**, 884 (1988).
87. D. Y. Kwok, R. Lin, M. Mui, and A. W. Neumann. *Colloids Surfaces A*, **116**, 63 (1996).
88. J. F. Oliver, C. Huh, and S. G. Mason. *J. Colloid Interface Sci.*, **93**, 169 (1983).
89. J. F. Oliver, C. Huh, and S. G. Mason. *Colloids Surfaces*, **1**, 79 (1980).
90. D. Duncan, D. Li, J. Gaydos, and A. W. Neumann. *J. Colloid Interface Sci.*, **169**, 256 (1995).
91. A. Amirfazli, D. Y. Kwok, J. Gaydos, and A. W. Neumann. *J. Colloid Interface Sci.*, **205**, 1 (1998).
92. D. Y. Kwok, D. Li, and A. W. Neumann. In *Applied Surface Thermodynamics*, J. K. Spelt and A. W. Neumann (Eds), pp. 413–440. Marcel Dekker, New York (1996).
93. D. Y. Kwok, C. J. Budziak, and A. W. Neumann. *J. Colloid Interface Sci.*, **173**, 143 (1995).
94. K. Grundke, P. Weidenhammer, P. Werner, R. Trinowitz, A. Janke, D. Y. Kwok, K. Pöschel, H.-J. Jacobasch, and A. W. Neumann, in preparation.
95. D. Y. Kwok, C. N. C. Lam, A. Li, A. Leung, and A. W. Neumann. *Langmuir*, **14**, 2221 (1998).
96. O. I. del Río, D. Y. Kwok, R. Wu, J. M. Alvarez, and A. W. Neumann. *Colloids Surfaces A*, **143**, 197 (1998).
97. D. Y. Kwok, C. N. C. Lam, A. Li, A. Leung, R. Wu, E. Mok, and A. W. Neumann. *Colloids Surfaces A*, **142**, 219 (1998).
98. D. Y. Kwok, A. Leung, A. Li, C. N. C. Lam, R. Wu, and A. W. Neumann. *Colloid Polym. Sci.*, **276**, 459 (1998).
99. D. Y. Kwok, C. N. C. Lam, A. Li, K. Zhu, R. Wu, and A. W. Neumann. *Polym. Eng. Sci.*, **38**, 1675 (1998).
100. D. Y. Kwok, A. Li, C. N. C. Lam, R. Wu, S. Zschoche, K. Pöschel, T. Gietzelt, K. Grundke, H.-J. Jacobasch, and A. W. Neumann. *Macromol. Chem. Phys.*, **200**, 1121 (1999).
101. D. Y. Kwok, C. N. C. Lam, A. Li, and A. W. Neumann. *J. Adhesion*, **68**, 229 (1998).
102. D. Y. Kwok, A. Leung, C. N. C. Lam, A. Li, R. Wu, and A. W. Neumann. *J. Colloid Interface Sci.*, **206**, 44 (1998).
103. C. J. Budziak and A. W. Neumann. *Colloids Surfaces A*, **43**, 279 (1990).
104. G. H. E. Hellwig and A. W. Neumann. Proc. 5th Int. Congr. on Surface Activity, Section b. p. 687 (1968).
105. G. H. E. Hellwig and A. W. Neumann. *Kolloid-Z. Z. Polym.*, **40**, 229 (1969).
106. H. B. Callen. *Thermodynamics and an Introduction to Thermostatistics*, second edition. John Wiley, New York (1985).
107. G. Antonow. *J. Chim. Phys.*, **5**, 372 (1907).

108. D. Berthelot. *Compt. Rend.*, **126**, 1857 (1898).
109. R. J. Good. In *Adsorption at Interfaces*, K. L. Mittal (Ed.), ACS Symp. Ser. No. 8. American Chemical Society, Washington, D.C. (1975).
110. R. Asthana. *J. Colloid Interface Sci.*, **165**, 256 (1994).
111. D. Y. Kwok, A. Li, and A. W. Neumann. *J. Polym. Sci. B.*, **37**, 2039 (1999).
112. D. Y. Kwok, R. Wu, A. Li, and A. W. Neumann. *J. Adhesion Sci. Technol.*, **14** (5), 1021.
113. D. Y. Kwok, A. Li, and A. W. Neumann. *Russian J. Colloid Interface Sci.*, in press.
114. D. Li, C. Ng, and A. W. Neumann. *J. Adhesion Sci. Technol.*, **6**, 601 (1992).
115. E. Tronel-Peyroz and A. Lhassani. *J. Colloid Interface Sci.*, **171**, 552 (1995).
116. D. Y. Kwok and A. W. Neumann. *J. Colloid Interface Sci.*, in press.
117. C. D. Bain, E. B. Troughton, Y. Tao, J. Eval, G. M. Whitesides, and R. G. Nuzzo. *J. Am. Chem. Soc.*, **111**, 321 (1989).
118. C. D. Bain and G. M. Whitesides. *Angew. Chem. Int. Ed. Engl.*, **28**, 506 (1989).
119. P. E. Laibinis and G. M. Whitesides. *J. Am. Chem. Soc.*, **113**, 7152 (1991).
120. P. E. Laibinis. PhD thesis, Harvard University (1991).
121. H. W. Fox and W. A. Zisman. *J. Colloid Sci.*, **7**, 428 (1952).
122. J. Drelich and J. D. Miller. *J. Colloid Interface Sci.*, **167**, 217 (1994).
123. G. S. Ferguson and G. M. Whitesides. In *Modern Approaches to Wettability: Theory and Applications*, M. Schrader and G. Loeb (Eds), pp. 143–177. Plenum Press, New York (1992).
124. D. Y. Kwok, D. Li, and A. W. Neumann. *Langmuir*, **10**, 1323 (1994).
125. W. Wu, R. F. Giese, Jr., and C. J. van Oss. *Langmuir*, **11**, 379 (1995).
126. L. H. Lee. *Langmuir*, **12**, 1681 (1996).
127. C. Della Volpe and S. Siboni. *J. Colloid Interface Sci.*, **195**, 121 (1997).
128. B. Jańazuk, T. Bialopiotrowicz, and A. Zdziennicka. *J. Colloid Interface Sci.*, **211**, 96 (1999).
129. J. C. Berg. In *Wettability*, J. C. Berg (Ed.), p. 75. Marcel Dekker, New York (1993).
130. M. Morra. *J. Colloid Interface Sci.*, **182**, 312 (1996).
131. J. K. Spelt, E. Moy, D. Y. Kwok, and A. W. Neumann. In *Applied Surface Thermodynamics*, J. K. Spelt and A. W. Neumann (Eds), pp. 293–332. Marcel Dekker, New York (1996).
132. D. Y. Kwok and A. W. Neumann. *Can. J. Chem. Eng.*, **74**, 551 (1996).
133. D. Y. Kwok, Y. Lee, and A. W. Neumann. *Langmuir*, **14**, 2584 (1998).
134. D. Y. Kwok. *Colloids Surfaces A*, in press.
135. J. M. Douillard. *J. Colloid Interface Sci.*, **188**, 511 (1997).
136. M. Michalski, J. Hardy, and B. J. V. Saramago. *J. Colloid Interface Sci.*, **208**, 319 (1998).
137. B. Bouali, F. Ganachaud, J. Chapel, C. Pichot, and P. Lanteri. *J. Colloid Interface Sci.*, **208**, 81 (1998).
138. M. Greiveldinger and M. E. R. Shanahan. *J. Colloid Interface Sci.*, **215**, 170 (1999).
139. F. London. *Z. Phys. Chem. (B)*, **11**, 222 (1930).
140. G. C. Maitland, M. Rigby, E. B. Smith, and W. A. Wakeham. *Intermolecular Forces: Their Origin and Determination.* Clarendon Press, Oxford (1981).
141. A. Dupré. *Théorie Mécanique de la Chaleur.* Gauthier-Villars, Paris (1868).
142. L. A. Girifalco and R. J. Good. *J. Phys. Chem.*, **61**, 904 (1957).
143. R. J. Good and E. Elbing. *Ind. Eng. Chem.*, **62**(3), 72 (1970).
144. K. C. Chao and R. L. Robinson, Jr. *Equation of State: Theories and Applications.* American Chemical Society, Washington, D.C. (1986).
145. J. S. Rowlinson and F. L. Swinton. *Liquids and Liquid Mixtures.* Butterworth Scientific, London (1981).
146. J. Kestin and E. A. Mason. *AIP Conf. Proc.*, **11**, 137 (1973).
147. G. H. Hudson and J. C. McCoubrey. *Trans. Faraday Soc.*, **56**, 761 (1960).
148. W. A. Steele. *Surf. Sci.*, **36**, 317 (1973).
149. J. E. Lane and T. H. Spurling. *Aust. J. Chem.*, **29**, 8627 (1976).
150. D. Li and A. W. Neumann. *J. Colloid Interface Sci.*, **137**, 304 (1990).

151. E. W. Swokowski. *Calculus with Analytic Geometry*. Prindle, Weber and Schmidt, Boston (1979).
152. C. Lipson and N. J. Sheth. *Statistical Design and Analysis of Engineering Experiments*. McGraw-Hill, New York (1973).
153. J. A. Mann, Jr. In *Surface and Colloid Science*, E. Matijević and R. J. Good (Eds), p. 213. Plenum Press, New York (1984).
154. D. Y. Kwok and A. W. Neumann. *Colloids Surfaces A* (1999).
155. D. Y. Kwok. *Contact Angles and Surface Energetics*. PhD thesis, University of Toronto (1998).
156. D. Y. Kwok and A. W. Neumann. *J. Phys. Chem.*, in press.
157. A. W. Neumann, D. R. Absolom, D. W. Francis, and C. J. van Oss. *Sep. Purification Methods*, **9**, 69 (1980).
158. C. P. S. Taylor. *J. Colloid Interface Sci.*, **100**, 589 (1984).
159. D. Y. Kwok and A. W. Neumann. *Colloids Surfaces A*, in press.
160. H. W. Fox and W. A. Zisman. *J. Phys. Chem.*, **7**, 109 (1952).
161. A. H. Ellison and W. A. Zisman. *J. Phys. Chem.*, **58**, 260 (1954).
162. E. F. Hare, E. G. Shafrin, and W. A. Zisman. *J. Phys. Chem.*, **58**, 236 (1952).
163. M. K. Bernett and W. A. Zisman. *J. Phys. Chem.*, **66**, 1207 (1962).
164. E. G. Shafrin and W. A. Zisman. *J. Phys. Chem.*, **66**, 740 (1962).
165. F. Schulman and W. A. Zisman. *J. Colloid Sci.*, **7**, 465 (1952).
166. H. W. Fox and W. A. Zisman. *J. Colloid Sci.*, **5**, 514 (1950).
167. E. G. Shafrin and W. A. Zisman. *J. Phys. Chem.*, **61**, 1046 (1957).
168. E. G. Shafrin and W. A. Zisman. *J. Colloid Sci.*, **7**, 166 (1952).
169. A. H. Ellison, H. W. Fox, and W. A. Zisman. *J. Phys. Chem.*, **57**, 622 (1953).
170. A. H. Ellison and W. A. Zisman. *J. Phys. Chem.*, **58**, 503 (1954).
171. H. F. Hare and W. A. Zisman. *J. Phys. Chem.*, **59**, 335 (1955).
172. A. Augsburg, K. Grundke, H.-J. Jacobasch, K. Pöschel, and A. W. Neumann. *Acta Polymerica*, **49**(8), 417 (1998).

Acid-Base Interactions, Vol. 2, pp. 167–171
Ed. K.L. Mittal

On the acid/base theory of contact angles

ROBERT J. GOOD, PROFESSOR EMERITUS *

Department of Chemical Engineering, State University of New York at Buffalo, Buffalo, NY 14260

Abstract—The need to distinguish between two "polar" parameters, γ^{P} and γ^{AB}, is analyzed mathematically. Incorrect and misleading results will be obtained, in treating contact angle data, if a dipolar term which follows a geometric mean law is used. The complete combining rule, due to Good, Chaudhury and van Oss, is explained.

Keywords: Acid-base; polar parameters; contact angle; Good, Chaudhury and van Oss theory.

1. INTRODUCTION

In the field of acid/base interaction and contact angles, there is a need for some clarification and development of the theory. Specifically, the difference between the dipolar (γ^{P}) and acid/base (γ^{AB}) parameters has not been recognized by some authors. This had led to misinterpretation of some contact angle measurements.

2. THE ACID/BASE (γ^{AB}) vs. THE DIPOLAR (γ^{P}) COMPONENTS OF INTERFACIAL FREE ENERGY

The so-called Good–Girifalco theory [1] first arose when the analogy between the Scatchard–Hildebrand theory of the energy of mixing of nonelectrolyte solutions and the theory of the interfacial free energy was recognized. The use of the "regular solution" concept [2], together with the London [3] theory of the dispersion force, led to the "geometric mean" theory [4] relating the free energy of adhesion ($\Delta G^{\mathrm{a}}_{12}$) and the free energy of cohesion (ΔG^{c}) of each of the separate phases. Thus, if molecules of substances 1 and 2 interact only through the dispersion force (see refs [4] and [5]), then the surface free energy of each pure phase is

$$\gamma_1^{\mathrm{LW}} = \frac{3\alpha_1^2 I_1}{4r_{\mathrm{o}1}^2}, \qquad \gamma_2^{\mathrm{LW}} = \frac{3\alpha_2^2 I_2}{4r_{\mathrm{o}2}^2} \tag{1}$$

*Telephone number of Dr. R. Good is: (716) 6452911; fax: (716) 6453822.

where α is polarizability, I is ionization potential, and r_o is equilibrium intermolecular distance. For two immiscible substances 1 and 2,

$$\gamma_{12}^{\mathrm{LW}} = \frac{3\alpha_1\alpha_2 I_1 I_2}{4(I_1 + I_2) r_{\mathrm{o},12}^2} \tag{2}$$

Combining Eqs. (1) and (2), we obtain,

$$\gamma_{12}^{\mathrm{LW}} = \gamma_1^{\mathrm{LW}} + \gamma_2^{\mathrm{LW}} - 2\sqrt{\gamma_1^{\mathrm{LW}}\gamma_2^{\mathrm{LW}}} \tag{3}$$

$$= \left(\sqrt{\gamma_1^{\mathrm{LW}}} - \sqrt{\gamma_2^{\mathrm{LW}}}\right)^2 \tag{4}$$

A corresponding treatment can be given to the Keesom [6] dipole–dipole component of intermolecular forces for freely rotating dipoles. If μ is dipole moment,

$$\gamma_1^{\mathrm{P}} = 2\mu_1/3kTr_{\mathrm{o}1}^2 \tag{5}$$

$$\gamma_{12}^{\mathrm{P}} = 2\mu_1^2\mu_2^2/3kTr_{\mathrm{o},12}^2 \tag{6}$$

$$\gamma_{12}^{\mathrm{P}} = \gamma_1^{\mathrm{P}} + \gamma_2^{\mathrm{P}} - 2\sqrt{\gamma_1^{\mathrm{P}}\gamma_2^{\mathrm{P}}} \tag{7}$$

$$= \left(\sqrt{\gamma_1^{\mathrm{P}}} - \sqrt{\gamma_2^{\mathrm{P}}}\right)^2 \tag{8}$$

The dispersion and dipole–dipole forces are not fully independent of each other. There exists a "cross-term," commonly called the Debye (induction) force, or dipole-induced dipole force [7], signified by the superscript i. Then an induction free energy term can be written [4]:

$$\gamma_1^{\mathrm{i}} = \frac{\alpha_1\mu_1^2}{r_{\mathrm{o}}^6}, \qquad \gamma_2^{\mathrm{i}} = \frac{\alpha_2\mu_2^2}{r_{\mathrm{o}}^6} \tag{9}$$

The cross-term, then, is [4, 7]

$$\gamma_{12}^{\mathrm{i}} = \frac{\left(\alpha_1\mu_2^2 + \alpha_2\mu_1^2\right)}{2r_{\mathrm{o},12}^2} \tag{10}$$

This interfacial term turns out to be small compared to the dispersion force or the dipole–dipole force, whichever is larger.

Fowkes [8] has pointed out that the components of surface free energy should be additive. He called the hydrogen bonding component γ^{h}. Neglecting γ^{i},

$$\gamma = \gamma^{\mathrm{LW}} + \gamma^{\mathrm{P}} + \gamma^{\mathrm{h}} + \cdots \tag{11}$$

Fowkes later identified γ^{h} with acid/base interaction, γ^{AB}.

In estimating the relative importance of the acid/base term, Fowkes computed the dispersion force energy of interaction for pairs of molecules, using Eqs (1) and (2). He also computed the dipole–dipole energies using Eqs (5) and (6), and found $\gamma_{12}^{\mathrm{P}} \ll \gamma^{\mathrm{LW}}$ for a large number of cases. So Fowkes concluded that γ^{P} in Eq. (11)

was small; and, by elimination, he arrived at the fact that γ_{12}^{h} (which he recognized as an acid/base term) must be comparable to $\gamma_{12}^{\mathrm{LW}}$.

van Oss and colleagues [5, 9, 10] have proposed the following combining rule for the hydrogen bond component, treated as an acid/base interaction.

$$\gamma_{12}^{\mathrm{AB}} = 2\left(\sqrt{\gamma_1^+} - \sqrt{\gamma_2^+}\right)\left(\sqrt{\gamma_1^-} - \sqrt{\gamma_2^-}\right) \tag{12}$$

For substance i,

$\gamma_{\mathrm{i}}^+ \equiv$ the Lewis acidic, or proton–donor (electron acceptor) parameter of i,

$\gamma_{\mathrm{i}}^- \equiv$ the Lewis basic, or proton–acceptor (electron donor) parameter of i.

Unlike dispersion force or dipole–dipole force, an acid–base interaction is one in which the components are complementary. The combining rule, Eq. (12), is based on the quantum theory of hydrogen bonding due to Kollman *et al.* [11, 12] and Hobza and Zaharadnik [13].

Some substances (e.g., ketones) act soley as bases in hydrogen bonding. Such compounds are referred to as monopolar, basic. Others are exclusively acids (e.g., chloroform) and are referred to as monopolar, acidic. And many are bipolar, e.g., water. For a pure, *bipolar* compound

$$\gamma_{\mathrm{i}}^{\mathrm{AB}} = 2\sqrt{\gamma_{\mathrm{i}}^+\gamma_{\mathrm{i}}^-} \tag{13}$$

For a strictly *monopolar* compound, $\gamma^+ = 0$ or $\gamma^- = 0$, and

$$\gamma_{\mathrm{i}}^{\mathrm{AB}} = 0 \tag{14}$$

Note the difference between dipole–dipole surface energy of a pure liquid, which is always positive (Eq. (5)), and the acid–base surface energy of a strictly monopolar liquid, which is zero. Thus,

$$\gamma_{\mathrm{i}}^{\mathrm{P}}(\mu_{\mathrm{i}} > 0) \neq \gamma^{\mathrm{AB}} \qquad (\gamma^+ \text{ or } \gamma^- = 0) \tag{15}$$

For an interface between two polar liquids, the acid-base component of surface energy may be negative, according to Eq. (12).

We have been disappointed to see, in publication as recently as September, 1999 [14], the expression that is based on setting $\gamma^{\mathrm{P}} = \gamma^{\mathrm{AB}}$, used namely

$$\gamma^{\mathrm{P}} = 2\sqrt{\gamma^+\gamma^-} \tag{a}$$

This form, historically, is attributed to Owens and Wendt [16]. Fowkes *et al.* [17] criticized Eq. (a) very strongly; however, Fowkes did not at that time propose an alternative combining rule.

Good *et al.* [15] have carried out a test of Eq. (a) such as Fowkes recommended (ref. [17, p. 381]): The interfacial tension of CCl_4 vs. water is 45.0 mJ/m^2, which is much larger than that of $CHCl_3$ vs. water, which is 31.6 mJ/m^2. But, surprisingly, the surface tensions of the two organic liquids are nearly equal: 26.95 for CCl_4 and

27.31 for $CHCl_3$. The dipole moment of $CHCl_3$ is 1.01 Debye units; and $CHCl_3$ is known to form hydrogen bonds with water. CCl_4, of course, is nonpolar and does not form hydrogen bonds with water.

Therefore Eq. (a) is wrong; and

$$\gamma_{12}^{\mathrm{P}} = 0 \quad \text{if} \quad \gamma^{+} \text{ or } \gamma^{-} = 0 \tag{16}$$

It is recommended that the term, γ^{P}, not be used at all.

Equation (17) is the complete form, for an interface between two polar phases.

$$\gamma_{12} = \left(\sqrt{\gamma_1^{\mathrm{LW}}} - \sqrt{\gamma_2^{\mathrm{LW}}}\right)^2 + 2\left(\sqrt{\gamma_1^{+}} - \sqrt{\gamma_2^{+}}\right)\left(\sqrt{\gamma_1^{-}} - \sqrt{\gamma_2^{-}}\right) \tag{17}$$

3. CONCLUDING REMARKS

This discussion, arranging the historical argument in a logical progression culminating in Eq. (17), shows why the Lewis acid and base terms can be expected to account, to a good approximation, for the excess of interfacial free energy over that predicted by the simple geometric mean approximation, i.e., over the first term on the right of Eq. (17).

Owens and Wendt [16] reported a correct intuition, in introducing an explicit polar component, γ^{P}, in the interfacial free energy. Fowkes made an important revision to the Owens–Wendt model, by identifying this interracial polar effect as an acid-base interaction. This superseded the "γ^{P}" approach. The two approaches cannot be used simultaneously.

REFERENCES

1. L. A. Girifalco and R. J. Good, *J. Phys. Chem.* **91**, 904 (1957).
2. J. H. Hildebrand and R. L. Scott, *Solubility of Nonelectrolytes*, 3rd ed. Reinhold, New York, 1950.
3. F. H. London, *Trans. Faraday Soc.* **33**, 8 (1937).
4. R. J. Good and E. Elbing, *Ind. Eng. Chem.* **62** (3), 54 (1970); In: *Chemistry and Physics at Interfaces*, S. Ross (Ed.), p. 72. Am. Chem. Soc., Washington, DC (1971).
5. C. J. van Oss, R. J. Good and M. K. Chaudhury, *Chem. Rev.* **88**, 927 (1988).
6. W. H. Keesom, *Phys. Z.* **22**, 126 (1921); **23**, 225 (1922).
7. P. Debye, *Phys. Z.* **21**, 178 (1920); **22**, 302 (1921).
8. F. M. Fowkes, *Ind. Eng. Chem.* **56** (12), 440 (1964).
9. R. J. Good, *J. Adhesion Sci. Technol.* **6**, 1269–1302 (1992).
10. C. J. van Oss, *Interfacial Forces in Aqueous Media*. Marcel Dekker, New York, 1994.
11. P. Kollman, J. McKelvey, A. Johansson and S. Rothenberg, *J. Am. Chem. Soc.* **97**, 955 (1975).
12. P. Kollman, *J. Am. Chem. Soc.* **99**, 4875 (1977).
13. P. Hobza and R. Zaharadnik, *Weak Intermolecular Interactions*, pp. 171–176. Elsevier, New York, 1980.
14. Y. Cai, D. Gardner and G. T. Caneba, *J. Adhesion Sci. Technol.* **13**, 1017 (1999).
15. R. J. Good, M. K. Chaudhury and C. J. van Oss, in: *Fundamentals of Adhesion*, L-H. Lee (Ed.), pp. 153–172. Plenum, New York, 1991.

16. D. K. Owens and R. C. Wendt, *J. Appl. Polym. Sci.* **13**, 1741 (1969).
17. F. M. Fowkes, F. L. Riddle, Jr., W. E. Pastore and A. A. Weber, *Colloids Surfaces* **43**, 367–387 (1990).

Acid-Base Interactions, Vol. 2, pp. 173–179
Ed. K.L. Mittal

Acid–base interactions as the driving force for both hydrophobic attraction and hydrophilic repulsion

C. J. VAN OSS*
Departments of Microbiology, Chemical Engineering and Geology, SUNY at Buffalo, Buffalo, NY 14214-3000, USA

Abstract—The 'hydrophobic effect', i.e. the attraction between apolar molecules or particles (i), immersed in water (w), is caused by the Lewis acid–base (AB) component of the free energy of cohesion of water: $\Delta G_{iwi}^{hydrophobic} = \Delta G_{ww}^{AB}(coh) = -102$ mJ/m^2, at 20°C. Therefore, hydrophilic repulsion in water between hydrophilic molecules or particles (i) must achieve a free energy of repulsion, $\Delta G_{iwi}^{hydrophilic} >> 102$ mJ/m^2, before actual repulsion can prevail. Methods of measurement of the subcomponents and parameters of these free energies are outlined, as well as methods for calculating solubilities, or critical micelle concentrations of partly polar compounds and surfactants. Similarly methods are given for predicting stability or otherwise of aqueous particle suspensions. Finally, the linkage between the hydrophilicity *vs.* hydrophobicity of particles, and their surface (ς) potential, is outlined, with indications how to modulate these parameters.

Keywords: Lewis acid–base interactions; hydrophobic attraction forces; hydrophilic repulsion forces; apolar molecules; apolar particles; polar molecules; polar particles; interaction free energies in water; hydrophobic effect.

1. INTRODUCTION

Apolar molecules or particles (i), immersed in water (w) attract each other hydrophobically with a free energy $\Delta G_{iwi}^{hydrophilic} = -102$ mJ/m^2, at 20°C. In contrast with this strong hydrophobic energy of attraction, the Lifshitz–van der Waals (LW) attraction between such molecules or particles immersed in water only amounts to ΔG_{iwi}^{LW} values between –5 and –7 mJ/m^2. Under the same conditions the electrostatic (EL) repulsive energy, ΔG_{iwi}^{EL} tends to have a (repulsive, i.e. positive) value in the same range as the negative value of ΔG_{iwi}^{LW} for most biological (macro) molecules, particles or cells. The hydrophobic attraction on the other hand is driven by the Lewis acid–base (AB) free energy of cohesion between the water molecules of the liquid medium [1, 2]:

* To whom correspondence should be addressed. E-mail: cjvanoss@acsu.buffalo.edu

$$\Delta G_{iwi}^{hydrophobic} = \Delta G_{ww}^{AB} \text{ (coh)} \tag{1}$$

Equation (1) also signifies that the hydrophobic effect, which is caused by the AB free energy of cohesion between the surrounding water molecules, and which therefore is always present in liquid water, plays an inexorable role in all interactions taking place in aqueous media. A net hydrophilic repulsion between particles or molecules immersed in water must, therefore, comprise a gross repulsive free energy (i.e. $\Delta G_{iwi}^{hydrophilic} >> 0$) which significantly exceeds the value of $\Delta G_{ww}^{AB} = -102$ mJ/m^2 (at 20°C). In practice very hydrophilic organic as well as inorganic materials (i) tend to be monopolar electron donors with a Lewis basic surface tension parameter $\gamma_i^- \geq 28.5$ mJ/m^2, so that [3]:

$$\Delta G_{iwi}^{hydrophilic} = 4\,(\gamma_i^- \gamma_w^+)^{1/2} >> 102 \text{ mJ/m}^2 \tag{2}$$

where γ_w^+ is the Lewis acid (electron accepting) surface tension parameter of water, equal to 25.5 mJ/m^2 at 20°C, in order to overcome not only the omnipresent underlying hydrophobic attraction, but also to compensate for the small, but often non-negligible, Lifshitz–van der Waals attraction ΔG_{iwi}^{LW}.

The criterion for hydrophobicity is $\Delta G_{iwi}^{interfacial} < 0$, and for hydrophilicity:

$$\Delta G_{iwi}^{interfacial} > 0 \text{ [4]}$$

where:

$$\Delta G_{iwi}^{interfacial} = \Delta G_{iwi}^{LW} + \Delta G_{iwi}^{AB} \tag{3}$$

Hydrophobic particles, for which $\Delta G_{iwi}^{interfacial} < 0$, clump together when immersed in water and under the same conditions hydrophobic molecules are insoluble in water [5]. Hydrophilic particles, for which $\Delta G_{iwi}^{interfacial} > 0$, form stable suspensions in water (without requiring an electrostatic repulsion) and under the same conditions of $\Delta G_{iwi}^{interfacial} > 0$, hydrophilic molecules and macromolecules are soluble in water.

2. DETERMINATION OF INTERFACIAL FREE ENERGIES OF INTERACTION IN WATER

The free energy of interaction between two particles or molecules (i) immersed in water (w) can be expressed as [3]:

$$\Delta G_{iwi}^{interfacial} = -2\gamma_{iw} \tag{4}$$

where γ_{iw} is the interfacial tension between substance, i, and water, expressed as [3]:

$$\gamma_{iw} = [(\gamma_i^{LW})^{1/2} - (\gamma_w^{LW})^{1/2}]^2 + 2[(\gamma_i^+\gamma_i^-)^{1/2} + (\gamma_w^+\gamma_w^-)^{1/2} - (\gamma_i^+\gamma_w^-)^{1/2} - (\gamma_i^-\gamma_w^+)^{1/2}] \quad (5)$$

Now, γ_{iw} can be measured directly when i is a liquid, immiscible with water [6] (mainly applicable to alkanes; see also Section 3, below). However, γ_{iw} between water and a miscible liquid or between water and a solid surface cannot be measured directly. Indirectly, however, γ_{iw} can be determined using Equation (5), once the different γ-components are known. These can be determined by means of contact angle (θ) measurements, using the following version of the Young–Dupré Equation [3]:

$$(1+\cos\theta)\gamma_L = 2[(\gamma_S^{LW}\gamma_L^{LW})^{1/2} + (\gamma_S^+\gamma_L^-)^{1/2} + (\gamma_S^-\gamma_L^+)^{1/2}] \quad (6)$$

where L and S stand for liquid and solid, respectively. Because for a given solid, S, there are three unknowns ($\gamma_S^{LW}, \gamma_S^+$ and γ_S^-), Equation (6) has to be used three times, with three different well-characterized liquids, L, of which at least two must be polar [3]. Contact angles (θ) can be measured directly [3], or, when S is a powder, cosθ can be determined by thin layer wicking [7].

Knowing γ_{iw}, (see Equation (5)) the interfacial free energy of interaction ($\Delta G_{iwi}^{interfacial}$) between similar molecules or particles (i), immersed in water, then follows (cf. Equations (4) and (5)). Similarly one can also determine the free energy of interfacial interaction between two different particles or molecules, i and j, immersed in water, via [3]:

$$\begin{aligned}\Delta G_{iwj}^{interfacial} &= \gamma_{ij} - \gamma_{iw} - \gamma_{jw} \\ &= 2\,[(\gamma_i^{LW}\gamma_w^{LW})^{1/2} + (\gamma_j^{LW}\gamma_w^{LW})^{1/2} \\ &\quad - (\gamma_i^{LW}\gamma_j^{LW})^{1/2} - \gamma_w^{LW}] \\ &\quad + 2\{(\gamma_w^+)^{1/2}[(\gamma_i^-)^{1/2} + (\gamma_j^-)^{1/2} - (\gamma_w^-)^{1/2}] \\ &\quad + (\gamma_w^-)^{1/2}[(\gamma_i^+)^{1/2} + (\gamma_j^+)^{1/2} - (\gamma_w^+)^{1/2}] \\ &\quad - (\gamma_i^+\gamma_j^-)^{1/2} - (\gamma_i^-\gamma_j^+)^{1/2}\}\end{aligned} \quad (7)$$

Contrary to general belief there usually is a sizable hydrophobic attraction, even between a hydrophobic and a hydrophilic site, immersed in water [1–3], in which case $\Delta G_{iwj}^{interfacial} < 0$.

3. HYDROPHOBIC ATTRACTION, THE LOW AQUEOUS SOLUBILITY OF APOLAR SOLUTES, AND THE INSTABILITY OF HYDROPHOBIC PARTICLE SUSPENSIONS

Totally apolar solute molecules, i, immersed in water, undergo a mutual hydrophobic attraction (at 20°C), ΔG^{AB}_{iwi} = –102 mJ/m^2. (The total free energy of attraction, $\Delta G^{interfacial}_{iwi}$, of alkanes, with the exception of octane, can be slightly stronger than 102 mJ/m^2, due to an additional small van der Waals contribution, which however for octane is zero.)

Now the aqueous solubility of apolar compounds is function of two parameters only (at constant temperature and pressure): the interfacial tension with water, γ_{iw} and the contactable surface area (S_c) between two of the apolar molecules, i [5]:

$$-2\gamma_{iw}\, S_c = kT.\ 1n\ s \tag{8}$$

where k is Boltzmann's constant, T the temperature in degrees Kelvin and s the solubility in mole fractions. For linear alkanes which can be taken to have a cylindrical shape, S_c corresponds to 1/6 of the curved part of the cylinder's surface area, measured at the van der Waals, or Born-Kihara surface of the molecules [5].

For octane, γ_{iw} = 50.8 mJ/m^2 and S_c = 0.518 nm^2. Equation (8) then yields a value for the aqueous solubility, s, for octane of 2.2 × 10^{-6} mole fractions = 0.122 mM = 1.4 mg/100 g water, which is its solubility at 20°C. It should be noted, however, that this pathway (Equation (8)) cannot be followed for compounds which are less than totally apolar. This is because with even mildly amphipathic compounds, such as for example, octanol, the γ_{iw} value, when measured by e.g. drop-shape analysis [6], is much too low. The reason for this is that at the octanol–water interface one measures mainly the OH-water interface, because water most strongly attracts the hydrophilic part of the organic molecule. It is nonetheless possible to utilize Equation (8), provided one separate the γ^{apolar}_{iw} from the γ^{polar}_{iw} values for the two different moieties of a polar molecule, convert them to units kT, and then add them together, for insertion into Equation (8). For instance for octanol, [5] the octyl group, like octane, has a γ_{iw} = 50.8 mJ/m^2 which corresponds to 6.51 kT at 20°C. The OH group [5] has a value for γ_{iw} = –1.80 kT. Thus for octanol γ^{tot}_{iw} = 6.51 – 1.80 = 4.71 kT. Using Equation (8) this yields a solubility in water of 8.1 x 10^{-5} mole fractions, which corresponds to a solubility for octanol = 58.6 mg per 100 g H_2O, which is correct. Should one however use the total <u>measured</u> γ_{iw} -value for octanol = 8.5 mJ/m^2, which corresponds to 1.3 kT, one would find an apparent solubility [5] (using Equation (8)) of s=0.074 mole fractions, which is more than 900 times too high.

The same procedure as outlined here for octanol can be applied to determine the critical micelle concentrations (cmc) of surfactants. It thus is possible to calculate the cmc for nonionic as well as for anionic surfactants, with an accuracy better than ± 10%, from the interfacial tensions (γ^{TOT}_{iw}) of the recombined but separately determined hydrophobic and hydrophilic moieties of these surfactants and their respective contactable surface areas, (Ref. [3], pp. 290–294). Surfactants remain, of course, soluble in water at concentrations greater than their cmc (s = cmc), but this only occurs after their monomeric molecules organize into complexes (i.e. micelles), which orient their hydrophilic moieties toward their interface with the water.

Suspensions of hydrophobic particles, i, where $\Delta G_{iwi}^{interfacial} < 0$, flocculate because such particles attract each other hydrophobically when immersed in water. This is analogous to the aqueous insolubility (or more precisely the very low solubility) of apolar, or slightly polar, organic compounds discussed above. However, nonionic as well as ionic surfactants can confer stability to aqueous hydrophobic particle suspensions by attaching their hydrophobic 'tails' to the particles' hydrophobic surface, which allows them to orient their hydrophilic 'heads' toward the water interface. This results in a multi kT repulsive energy per particle pair, in close similarity to the stability or solubility of surfactant micelles.

4. LINKAGE BETWEEN ELECTROSTATIC AND LEWIS ACID–BASE INTERACTIONS

Molecules or particles which are strong electron-donors exert a sizeable hydrophilic repulsion on each other (e.g. dextran, polyethylene oxide) in the absence of any electrical surface potential. However, molecules or particles that do have, for example, a negative surface-, or ζ-potential, exhibit at the same time a strong electron-donicity which is linked directly to the ζ-potential [8].

When one decreases the ζ-potential, e.g. through the admixture of plurivalent counterions, or by changing the pH, the value of γ_i^- decreases concomitantly. This causes the molecules or particles to change from hydrophilic to hydrophobic, so that precipitation or flocculation ensues.

Thus the real cause of the flocculation of charged particles upon the admixture of plurivalent conterions (Schulze-Hardy effect) is not the fact that the Lifshitz–van der Waals attraction energy (ΔG_{iwi}^{LW}) prevails over the electrostatic repulsion which has decreased as a consequence of the drop in ζ-potential. Actually, flocculation occurs [8] when $|\Delta G_{iwi}^{EL}|$, although diminished, is still larger than $|\Delta G_{iwi}^{LW}|$. Flocculation takes place because as a result of the decrease in ΔG_{iwi}^{EL}, ΔG_{iwi}^{AB} changes sign and becomes much more negative than ΔG_{iwi}^{LW} [8]. Upon hydrophobization of the particles caused by a decrease in ζ-potential and thus in ΔG_{iwi}^{EL}:

$$|\Delta G_{iwi}^{AB}| >> |\Delta G_{iwi}^{EL}| > |\Delta G_{iwi}^{LW}| \tag{9}$$

5. CONCLUSIONS

(1) Although not intuitively obvious, the hydrophobic effect is a simple concept and its energy is easily measured and readily expressed in S.I. units. The driving force for the hydrophobic effect is the Lewis acid–base (AB) free energy of cohesion between the water molecules of the liquid medium, which acts upon all solutes or particles immersed in water. In water, the hydrophobic attraction energy generally is about an order magnitude greater than the two other commonly occurring noncovalent interactions: van der Waals attraction and electrostatic repulsion.

(2) Contrary to a widespread belief, hydrophobic attraction between a hydrophobic molecule or particle and a hydrophilic molecule or particle, readily occurs.

(3) As a consequence of the permanent presence of the AB energy of cohesion of water in all aqueousmedia, the total free energy of hydrophilic repulsion (which also is of Lewis acid–base origin) has to be larger than the omnipresent underlying hydrophobic attraction of 102 mJ/m^2, to achieve a net repulsion between hydrophilic molecules or particles, immersed in water.

(4) All organic solutes that are not exceedingly polar are, therefore, sparsely soluble or almost totally insoluble in water. From an organic compound's molecular dimensions (yielding S_c) and its interfacial tension with water (γ_{iw}) its aqueous solubility can be accurately qualitatively predicted. For an amphipathic organic compound, S_c and γ_{iw} values have to be separately determined for its apolar and polar moieties, expressed in units kT and then added together.

(5) There is a strong one-way linkage between electrostatic (EL) and Lewis acid–base (AB) interactions. For instance, strongly negatively charged molecules or particles are hydrophilic, i.e. they have a surface electron-donicity, $\gamma_i^- > 28.5$ mJ/m^2. When ΔG^{EL} decreases, the value of γ_i^- also decreases to the point where ΔG_{iwi}^{AB} changes from strongly positive (hydrophilic repulsion) to strongly negative (hydrophobic attraction). Thus, changes in pH or the admixture of plurivalent counterions that cause flocculation of charged particles operate by hydrophobization of the particles and not by making the electrostatic repulsion weaker than the van der Waals attraction.

(6) Compounds or particles that are electrostatically neutral can still be quite hydrophilic (e.g. dextran, polyethylene oxide) and have a γ_i^- value that is significantly higher than 28.5 mJ/m^2. They differ from charged compounds or particles, however, in that manipulations that cause a decrease in the ζ-potentials of charged materials and make these more hydrophobic have no influence at all on electrically neutral hydrophilic materials.

REFERENCES

1. C. J. van Oss, *Colloids Surfaces* **B5**, 91–110 (1995).
2. C. J. van Oss, *Curr. Opin. Colloid Interface Sci.* **2**, 503–512 (1997).
3. C. J. van Oss, *Interfacial Forces in Aqueous Media*, p. 44. Marcel Dekker, New York (1994); *Forces Interfaciales en Milieux Aqueux*, p. 47. Masson, Paris (1996).
4. C. J. van Oss and R. F. Giese, *Clays Clay Miner.* **43**, 474–477 (1995).
5. C. J. van Oss and R. J. Good, *J. Dispersion Sci. Technol.* **17**, 443–449 (1996).
6. A. W. Adamson, *Physical Chemistry of Surfaces*, pp. 28–36. Wiley/Interscience, New York (1990).
7. C. J. van Oss, R. F. Giese, Z. Li, K. Murphy, J. Norris, M. K. Chaudhury and R. J. Good, *J. Adhesion Sci. Technol.* **6**, 413–428 (1992).
8. W. Wu, R. F. Giese and C. J. van Oss, *Colloids Surfaces* **A89**, 241–252 (1994).

Acid-Base Interactions, Vol. 2, pp. 181–186
Ed. K.L. Mittal

Irrelevance of the ratio of the electron-accepticity to the electron-donicity of water with respect to the determination of polar surface tension components, interfacial tensions and free energies of interaction of liquids and/or solids

C. J. VAN OSS*
Departments of Microbiology, Chemical Engineering and Geology, State University of New York at Buffalo, Buffalo, NY 14214-3000, USA

Abstract—It is shown that the simple assumption of a fixed value for the hitherto unknown and to date unknowable electron-accepticity/electron-donicity ratio (x) for water at 20°C, does not influence the outcome of the determination of the polar surface tension components (γ_i^{AB}) interfacial tensions (γ_{12}^{AB}) and free energies of interaction (ΔG^{AB}) of polar liquids and/or solids. What does remain linked to the assumption of a reference value, x, for the above ratio for water is the outcome of the electron-accepticity (γ^+) and electron-donicity (γ^-) surface tension parameters for any polar compound, as determined with a system based upon the assumed reference value of that ratio, x.

A more cumbersome system, using dimensionless ratios between the electron-accepticity of polar compound i and that of water, and between the electron-donicity of polar compound i and that of water, can also be used, to obviate any assumptions about the polar properties of water. This system is used here to demonstrate the mathematical correctness of the above contentions, with an example based on contact angle measurements on poly (methyl methacrytate), but its use is not advocated, as it is less user-friendly than the system based on a simple assumption about the electron-accepticity/electron-donicity ratio for water, which assumption cancels out in the calculation of all polar (acid–base) components, interfacial tensions and free energies of adhesion.

Keywords: Acid–base forces; electron-accepticity; electron-donicity; electron-accepticity/electron donicity ratio of water.

1. INTRODUCTION

A certain unrest (see, e.g. [1]) seems to have developed concerning the validity of the contention [2, 3] that assumptions about the ratio of the γ_w^+ to γ_w^- values (i.e. of the electron-acceptor to the electron-donor parameters of the Lewis acid–base component of the surface tension [γ_w^{AB}] of water) ***have no influence on the outcome of the de-***

* To whom correspondence should be addressed: E-mail: cjvanoss@acsu.buffalo.edu

termination of the polar surface tension components[1]***, interfacial tensions and free energies of interaction of polar liquids and/or solids*** (i.e. γ_i^{AB}, γ_{12}^{AB}, ΔG_{121}^{AB}, ΔG_{132}^{AB}, and ΔG_{12}^{AB} values).

It is true that the values found for γ_i^+ and γ_i^-, using contact angles based on the assumed ratio γ_w^+/γ_w^-, will be found to be altogether different when a system is used based on a value for γ_w^+/γ_w^- that deviates from unity and indeed γ_i^+ and γ_i^- values will vary according to each different γ_w^+/γ_w^- value adopted for a given system. Although this has been quite clearly stated in refs. [2, 3], the different results obtained for γ_i^+ and γ_i^- values, when based on different assumed γ_w^+/γ_w^- ratios, may well have been the source of recent papers about the flaws in the system, by authors who may have too superficially read our various publications on the acid–base theory of surface thermodynamics concerning polar systems [1, 4].

Early in 1987 [5] we published a paper in which we mentioned that a γ_w^+/γ_w^- ratio equal to 2.0, might be closer to reality than a value of 1.0, taking some of Gutmann's data into consideration [6]. However, upon growing realization that no real hard facts were known about the quantitative values of γ_w^+ and γ_w^- for water [ref. 3, pp. 27–28], and that the values for both were quite sizable, an arbitrary value of γ_w^+/γ_w^- of unity was decided upon, for water at 20°C, as the basic reference system for expressing all the other values for γ_w^+/γ_w^- of condensed-phase materials [2, 3]. At the same time our designation of SR (for short-range) for Lewis acid–base surface tension components [5], was changed to AB [2], in view of the realization that AB forces were really not as short-range as thought initially. In the same publication [2] we demonstrated how the whole problem of having to adopt a system based on a fixed γ_w^+/γ_w^- ratio for water could be entirely circumvented, at the cost however of having to get used to a completely new and somewhat complicated symbolism in the guise of:

$$\delta_{iw}^+ = (\gamma_i^+/\gamma_w^+)^{1/2} \tag{1A}$$

and

$$\delta_{iw}^- = (\gamma_i^-/\gamma_w^-)^{1/2} \tag{1B}$$

by means of such dimensionless numbers the need for assuming any fixed γ_w^+/γ_w^- ratio[2] could be avoided. Meanwhile, however, at the same time the dimensionless δ^+ and δ^- approach can furnish proof of the fact that the resulting γ_i^{AB}, γ_{12}^{AB}, ΔG_{121}^{AB}, and ΔG_{132}^{AB} values do not depend on any assumptions about the γ_w^+/γ_w^- ratio. An explanation of the δ_{iw}^+ and δ_{iw}^- approach is reiterated below.

---

[1] It should be stressed that γ_i^{AB} is a surface tension component of condensed material, i, similar to γ_i^{LW}, such that $\gamma_i = \gamma_i^{LW} + \gamma_i^{AB}$. One must not confound surface tension components with the polar (AB) surface tension parameters, γ_i^+ and γ_i^-. These are factors of γ_i^{AB}, such that $\gamma_i^{AB} = 2(\gamma_i^+\gamma_i^-)^{1/2}$. The values of the surface tension components (e.g. γ_i^{AB}) are not influenced by assumptions about the value of the γ_w^+/γ_w^- ratio, but the values of the surface tension parameters γ_i^+ and γ_i^- are influenced by such assumptions.

[2] A γ_w^+/γ_w^- ratio is also, of course, dimensionless, but the dimensionless δ_{iw}^+ and δ_{iw}^- numbers can be readily determined, whereas the value of the 'real' γ_w^+/γ_w^- ratio remains, so far, unknown.

2. USE OF THE MEASURABLE γ_i^+/γ_w^+ AND γ_i^-/γ_w^- RATIOS INSTEAD OF THE ESSENTIALLY UNKNOWN γ_w^+/γ_w^- RATIO

Using the ***relative*** acid and base polarities (defined above in Equations (1A) and (1B)), of substance, i, with respect to water, the Young–Dupré equation [2, 3] may be written for water as the contact angle liquid, as:

$$(1 + \cos\theta_w)\gamma = 2[(\gamma_i^{LW}\gamma_w^{LW})^{1/2} + (\gamma_i^+\gamma_w^-)^{1/2} + (\gamma_i^-\gamma_w^+)^{1/2}] \qquad (2)$$

It is known that for water at 20°C, $2(\gamma_w^+\gamma_w^-)^{1/2} = \gamma_w^{AB} = 51$ mJ/m^2 [7, 8].

Therefore:
$(\gamma_w^+\gamma_w^-)^{1/2} = 25.5$ mJ/m^2, so that $(\gamma_w^+)^{1/2} = 25.5/(\gamma_w^-)^{1/2}$ and $(\gamma_w^-)^{1/2} = 25.5/(\gamma_w^+)^{1/2}$ [2, 3]. The Young–Dupré equation for water as contact angle liquid, (Equation (2)) can then be written as:

$$(1 + \cos\theta_w)\,\gamma_w = 2\,[(\gamma_i^{LW}\gamma_w^{LW})^{1/2} + 25.5\,(\gamma_i^+/\gamma_w^-)^{1/2} + 25.5\,(\gamma_i^-/\gamma_w^+)^{1/2}] \qquad (3)$$

or, taking Equations (1A) and (1B) into account, as:

$$\gamma_w\,(1 + \cos\theta_w) = 2\,(\gamma_i^{LW}\gamma_w^{LW})^{1/2} + 51\;\delta_{iw}^+ + 51\;\delta_{iw}^- \qquad (4)$$

Equation (4) is now a form of the Young–Dupré equation, for water as contact angle liquid, from which all assumptions about the γ_w^+/γ_w^- ratio have been eliminated. In other words, Equation (4) is true even if a γ_w^+/γ_w^- ratio is adopted that differs from unity, and all terms of Equation (4) can be determined quantitatively without the need for knowing the ratio γ_w^+/γ_w^-. The dimensionless polarity ratios δ_{iw}^+ and δ_{iw}^- can be determined for any condensed-phase material, i [2, 3].

Equations of the type of Equation (4) can be readily devised for other polar test liquids. Taking glycerol (g) as an example of another high-surface energy test liquid, for which it is known that $\gamma_g^{AB} = 30$ mJ/m^2 [8], the Young–Dupré equation for glycerol as a contact angle liquid then is:

$$(1 + \cos\theta_g)\,\gamma_g = 2\,(\gamma_i^{LW}\gamma_g^{LW})^{1/2} + 30\;\delta_{ig}^+ + 30\;\delta_{ig}^- \qquad (5)$$

As;

$$\delta_{ig}^+ = \delta_{iw}^+/\delta_{gw}^+ \qquad (6A)$$

and:

$$\delta_{ig}^- = \delta_{iw}^-/\delta_{gw}^- \qquad (6B)$$

Equation (5) can then be expressed as:

$$(1 + \cos\theta_g)\,\gamma_g = 2\,(\gamma_i^{LW}\gamma_g^{LW})^{1/2} + 30\;\delta_{iw}^+/\delta_{gw}^+ + 30\;\delta_{iw}^-/\delta_{gw}^- \qquad (7)$$

By using water as well as glycerol for measuring contact angles on a number of different monopolar solid surfaces, it follows from the results given in ref. [9] that $\delta_{gw}^+ = 0.392$ and $\delta_{gw}^- = 1.50$. Using these values, Equation (7) can be rewritten for use of glycerol as a contact angle liquid, as [2, 3]:

$$(1 + \cos\theta_g)\,\gamma_g = 2\,(\gamma_i^{LW}\gamma_g^{LW})^{1/2} + 76.53\;\delta_{iw}^+ + 20.0\;\delta_{iw}^- \qquad (8)$$

When for any given condensed-phase substance, i, γ^{LW} is known or has been determined by contact angle measurement with an apolar liquid, and contact angles have been measured with water as well as with glycerol, using Equations (4) and (8), one can solve for δ^{+}_{iw} and δ^{-}_{iw}. The γ_i^{AB} component of the surface tension of substance, i, then is:

$$\gamma_i^{AB} = 51\ \delta^{+}_{iw}\delta^{-}_{iw} \tag{9}$$

The polar (AB) component of the interfacial tension between condensed-phase substances 1 and 2 becomes:

$$\gamma_{12}^{AB} = 51(\delta^{+}_{1w}\delta^{-}_{1w} + \delta^{+}_{2w}\delta^{-}_{2w} - \delta^{+}_{1w}\delta^{-}_{2w} - \delta^{+}_{2w}\delta^{-}_{1w}) \tag{10}$$

The polar (AB) component of the free energy of interaction between molecules or particles of substance, 1, immersed in liquid, 2, is obtained via:

$$\Delta G_{121}^{AB} = -2\gamma_{12}^{AB} \tag{11}$$

Finally, using Dupré's equation:

$$\Delta G_{132} = \gamma_{12} - \gamma_{13} - \gamma_{23} \tag{12}$$

the polar (AB) component of the free energy of interaction between molecules or particles 1 and 2, immersed in liquid 3, is:

$$\Delta G_{132}^{AB} = 51[\delta^{+}_{3w}(\delta^{-}_{1w} + \delta^{-}_{2w} - \delta^{-}_{3w}) + \delta^{-}_{3w}(\delta^{+}_{1w} + \delta^{+}_{2w} - \delta^{+}_{3w}) - \delta^{+}_{1w}\delta^{-}_{2w} - \delta^{-}_{1w}\delta^{+}_{2w}] \tag{13}$$

As far as its polar (AB) component is concerned, one can express the polar (AB) part of the free energy of adhesion between condensed-phase polar particles or molecules 1 and 2, as:

$$\Delta G_{12}^{AB} = -51(\delta^{+}_{1w}\delta^{-}_{2w} + \delta^{-}_{1w}\delta^{+}_{2w}) \tag{14}$$

by using a combination of Equations (9) and (10), and inserting the appropriate terms into the following form of Dupré's equation:

$$\Delta G^{AB} = \gamma_{12} - \gamma_1 - \gamma_2 \tag{15}$$

3. VERIFICATION OF THE IRRELEVANCE OF THE RATIO $\gamma_w^{+}/\gamma_w^{-}$ IN OBTAINING γ_i^{ab} OF A POLAR MATERIAL

For flat surfaces of poly (methyl methacrylate) (PMMA) the following contact angles were found at 20°C with the liquids [9]: 1-bromonaphthalene: θ_{BN} = 13°; diiodomethane; θ_{DIM} = 40°; water: θ_w = 72°; glycerol: θ_g = 65.5°. Averaging the values for γ^{LW}_{PMMA} found with θ_{BN} and θ_{DIM}, the value adopted is γ^{LW}_{PMMA} = 41.4 mJ/m^2. Then from θ_w and θ_g, using a value for the reference ratio $\gamma_w^{+}/\gamma_w^{-}$ = 1.0, if follows that using this reference system, γ^{+}_{PMMA} = 0.023 mJ/m^2 and γ^{-}_{PMMA} = 11.12 mJ/m^2, yielding γ^{AB}_{PMMA} = 2 $(\gamma^{+}_{PMMA}\gamma^{-}_{PMMA})^{1/2}$ = 1.01 mJ/m^2.

Now, using instead of the γ_w^+/γ_w^- equals unity system, the dimensionless δ_{iw}^+ and δ_{iw}^- values and the Young–Dupré equations (4) and (8) for θ_w and θ_g, respectively, values for $\delta_{iw}^+ = 0.030$ and $\delta_{iw}^- = 0.660$ are found. Applying these values to Euquation (9), it is found that, again $\gamma_{PMMA}^{AB} = 1.01$ mJ/m^2, Q.E.D.

4. DISCUSSION

The actual γ_w^+/γ_w^- ratio for water at 20°C remains inaccessible at present with the data and equations now at our disposal. To paraphrase our second paper on the subject, published in 1987 [2]: "It may not be impossible for novel types of physical measurements to allow the precise determination of the γ_i^+ and/or γ_i^- values for given polar compounds, at some future time." Pending this eventuality, however, my collaborators and I have opted to continue using $\gamma_w^+ = \gamma_w^- = 25.5$ mJ/m^2, at 20°C, as the preferred reference system, because of its easy recognizability and its close connection to the familiar surface tension symbolism. We felt that using the $\gamma_w^+/\gamma_w^- = 1.0$ ratio would be the best approach, if repeatedly accompanied by careful reiteration of its limitations, on account of the fact that its assumption cancels out when calculating, γ_i^{AB}, γ_{12}^{AB}, ΔG_{121}^{AB}, ΔG_{132}^{AB} and ΔG_{12}^{AB} values in S.I. units [2, 3, 10].

We did not advocate the use of the dimensionless approach utilizing δ_{iw}^+ and δ_{iw}^- values (Equations (1A) and (1B)) because of the (probably well-founded) fear that very few readers would be sufficiently motivated to master its unfamiliar and much more cumbersome methodology. In addition, expressing the polar properties of a condensed-phase material in terms of γ^+ and γ^-, immediately shows whether such a material is an electron-donor or (more rarely) an electron-acceptor and whether it is bipolar (mainly in the case of liquids) or monopolar (in the case of dry solids [11]). It is, of course, necessary that the reader take the trouble to understand the limitations as well as the non-limitations of using a system based on an assumption regarding the ratio of γ_w^+/γ_w^- for water at 20°C.

4.1. Limitations

The limitations of using a given fixed γ_w^+/γ_w^- ratio for water at 20°C as a reference point are only that the γ_i^+ and γ_i^- values found while using this reference system are linked to this reference system alone and will be different when a different γ_w^+/γ_w^- ratio is used as the basis of a reference system, e.g. at 20°C.

4.2. Non-limitations

The non-limitations are that irrespective of the γ^+/γ^- ratio chosen as the basis of the reference system (at 20°C), the values of γ_i^{AB}, γ_{12}^{AB}, ΔG_{121}^{AB}, ΔG_{12}^{AB} and ΔG_{132}^{AB} ***remain unchanged***.

4.3. Temperature

There is a reason for my continuous reiteration that the standard reference value for the γ_w^+/γ_w^- ratio of 1.0 is restricted to determinations done at 20°C. This is because the "real" γ_w^+/γ_w^- ratio although unknown for the time being, ***increases*** as the temperature is raised above 20°C. For instance, in comparison with γ_w^+/γ_w^- = 1.0 at 20°C, at 38°C, this ratio becomes 1.75 [ref. 3, p. 301].

5. CONCLUSIONS

(1) Using an assumed reference value for the γ_w^+/γ_w^- ratio for water allows the determinations of γ_i^{AB}, γ_{12}^{AB}, ΔG_{121}^{AB}, ΔG_{132}^{AB} and ΔG_{12}^{AB} values which are independent of the choice of the γ_w^+/γ_w^- ratio. In the absence of any known or knowable γ_w^+/γ_w^- ratio, the assumption of unity for that ratio, at 20°C is still the most practical one.

(2) Only the γ_i^+ and/or γ_i^- values obtained for a polar condensed-phase material, i, are dependent on the chosen γ_w^+/γ_w^- reference ratio.

(3) The need for choosing a given γ_w^+/γ_w^- reference ratio can be eliminated by determining the dimensionless $\delta_{iw}^+ = (\gamma_i^+/\gamma_w^+)^{1/2}$ and $\delta_{iw}^- = (\gamma_i^-/\gamma_w^-)^{1/2}$ values for any polar condensed-phase material, i.

(4) An example is given of determinations of the γ_{PMMA}^{AB} values, using either a given γ_w^+/γ_w^- ratio or the dimensionless δ_{PMMA}^+ and δ_{PMMA}^- values, based on contact angles measured with water and glycerol. With both approaches the same γ_{PMMA}^{AB} value is found.

(5) In spite of the advantage, by means of the dimensionless value (δ_{iw}^+ and δ_{iw}^-) approach, of being able to avoid any assumption about the (so far unknown) γ_w^+/γ_w^- ratio for water, the use of an assumed fixed reference value γ_w^+/γ_w^- for water is much more user-friendly than the more intricate and laborious manipulation of δ_{iw}^+ and δ_{iw}^- values.

(6) It should be noted, however, that any given assumption of a γ_w^+/γ_w^- value for water is valid only for one given temperature (e.g. 20°C) as the Lewis acidity of water increases with the temperature. The δ_{iw}^+ and δ_{iw}^- values can also, to some extent, be temperature-dependent, when i is a polar liquid.

REFERENCES

1. C. Della Volpe and S. Siboni, these proceedings.
2. C. J. van Oss, M. K. Chaudhury and R. J. Good, *Adv. Colloid Interface Sci.* **28**, 35 (1987).
3. C. J. van Oss, *Interfacial Forces in Aqueous Media*. Marcel Dekker, New York (1994); *Forces Interfaciales en Milieux Aqueux*. Masson, Paris (1996).
4. M. Greiveldinger and M. E. R. Shanahan, *J. Colloid Interface Sci.* **215**, 170 (1999).
5. C. J. van Oss, M. K. Chaudhury and R. J. Good, *Separ. Sci. Technol.* **22**, 1515 (1987).
6. V. Gutmann, *The Donor-Acceptor Approach to Molecular Interactions*. Plenum Press, New York (1978).
7. F. M. Fowkes, in: *Chemistry and Physics at Interfaces*, S. Ross (Ed.), p. 1. Amer. Chem. Soc. Publ., Washington, D.C. (1965); *Ind. Eng. Chem.* **56**(12), 40 (1964).
8. M. K. Chaudhury, Short-range and Long-range Forces in Colloidal and Macroscopic Systems, Ph.D. Dissertation, SUNY at Buffalo (1984).
9. C. J. van Oss, R. J. Good and H. J. Busscher, *J. Dispersion Sci. Technol.* **11**, 75 (1990).
10. C. J. van Oss, *Colloids Surfaces-A*. **78**, 1 (1993).
11. C. J. van Oss, R. F. Giese and W. Wu, *J. Adhesion* **63**, 71 (1997).

Acid-Base Interactions, Vol. 2, pp. 187–204
Ed. K.L. Mittal

Electronic properties and modifications of acid-base strengths at non-stoichiometric and polar oxide surfaces

FABIO FINOCCHI and CLAUDINE NOGUERA
Laboratoire de Physique des Solides, UMR CNRS 8502, Université Paris-Sud, 91405 Orsay, France

Abstract—The electronic structure of oxygen-deficient and polar MgO surfaces, obtained by a combination of ab initio density-functional and semi-empirical Hartree–Fock calculations, is analysed with special emphasis on the acid-base properties of the surface sites. On these surfaces, the character of the highest occupied molecular orbital displays a progressive transformation from a state localised on surface oxygens to a an extended metallic state. This transformation may be driven by the sub-stoichiometry or by the polarity, and it induces profound changes in the reactivity of the surfaces with resect to ad-molecule dissociation or metal adsorption.

Keywords: Electronegativity (atom-in-the-solid); acid-base; oxide surfaces; reactivity; vacancies; polar surfaces.

1. INTRODUCTION

Although oxides have been the subject of active research for many years, they have attracted a renewed interest in the last decade. Oxides are good catalysts, which are largely used, for instance, in petrochemistry. Specific experimental techniques have been developed to study the connection between their catalytic activity and their surface geometry. In layered materials produced nowadays, oxides can be used as supports for thin metallic films or grafted polymeric layers, as intercalation layers in electronic devices, etc. They are often present, although in an uncontrolled way, whenever a material is in contact with the ambient atmosphere, playing a fundamental role in corrosion processes.

The control of the physical and chemical properties of oxide surfaces, such as the change in the electronic and magnetic properties as a function of the surface morphology, still remains a challenge [1, 2]. Among these open issues, the understanding of how surface orientation or surface defects—stoichiometric defects such as oxygen vacancies, or extended geometric defects such as steps, kinks, etc—drive the reactivity of an oxide catalyst or the adhesion force between an oxide and a deposit are of crucial importance. It is interesting to note that, in this context, the concept of acid-base (AB) strength, largely used in molecular chemistry, may offer a valuable help. At the same time, the AB concept should be considered in order to account for the long range effects which take place in a solid, and for the existence of multiple surface sites which can have, in principle, various AB strengths [2, 3].

It this framework we analyze, in the following, the electronic characteristics of exotic surface orientations (i.e. other than those obtained by cleavage) or highly non-ideal oxide surfaces. We focus on oxygen-deficient surfaces as well as polar surfaces, and we show how the electron redistributions which take place on such surfaces can lead to important modifications of the basicity of surface oxygens or the acidity of surface cations, making them useful as AB-tunable materials.

Oxygen deficiency in the outer layers of an oxide often occurs as a result of surface preparation. It has been shown that annealing the surfaces, after cutting and polishing, results in a desorption of oxygen atoms, which is favored by the increase of the annealing temperature. Such oxygen deficiency may promote an ordering of the oxygen vacancies, leading to surface reconstructions, as in the case of $TiO_2(100)$ [4–7], $TiO_2(110)$ [8, 9], $SnO_2(110)$ [10] $SrTiO_3(001)$, [11–13], and α-$Al_2O_3(0001)$ [14–19]. The resulting surface morphologies are nowadays considered as substrates potentially useful for the fabrication of artificial structures such as self assembled clusters, metallic nanowires, etc.

Polar surfaces, on the other hand, show distinctive electronic features because of the presence of a net dipole moment along the normal to the surface [20]. They are generally much less stable than non-polar faces and, for this reason, they have not been studied as thoroughly as the latter. It is usually thought that the electronic structure of pure polar surfaces should be very different from their non-polar counterparts, yielding a higher surface reactivity. This is exemplified by the (111) face of periclasite (native magnesia MgO), which is found in nature as a result of its enhanced ability to dissociate water, compared to the cleavage (100) face [21]. In many oxides, however, planar defect-free polar orientations cannot be produced. The observed configurations are rather non-stoichiometric and reconstructed.

In this paper, we discuss the connection between the acid-base strength and the electronic structure of insulating oxide surfaces (Section 2). Then, we analyze the results of quantum-mechanical calculations of oxygen-deficient MgO(100) surfaces (Section 3) and of the polar (111) orientation of magnesium oxide (Section 4), focusing on the particular features of the electronic structure, due to the oxygen vacancies or to the (111) stacking sequence.

2. ACID-BASE STRENGTH OF SURFACE SITES: DISCUSSION IN TERMS OF THE LOCAL ELECTRONIC STRUCTURE

2.1. Acid-base Concept

While the importance of acids and bases in chemistry has been recognized for centuries, it is only in the last few decades that the same concept has been applied to interactions involving solids. Solid catalysts are interesting since they show generally high activity and selectivity. Moreover, they are generally less polluting

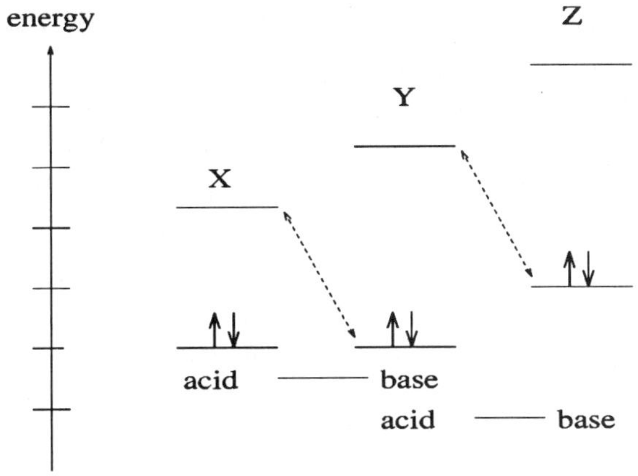

Figure 1. Frontier orbital diagram for an acid-base reaction: the HOMOs are filled with an electron pair; the LUMOs are empty. The Y compound presents an amphoteric character: it is a base in the XY conjugate pair, and an acid in the YZ pair. When the energy difference between the base HOMO and the acid LUMO is small, the reaction is controlled by the frontier orbitals; otherwise, it is charge-controlled.

than metals. In adhesion science, the interfacial interactions between the adhesive and the substrate, including dispersion and acid-base forces, are responsible for the adhesion strength: the AB interaction arises from the formation of mixed iono-covalent bonds between the surface and the adhesive (e.g. a polar polymer), one of them behaving as the acid and the other as the base [22–24]. Acidity or basicity is thus revealed as a tendency for the exchange of either a proton or electrons [25]. In the first case, Brønsted's definition applies. It is suited to processes that take place in solutions or at surfaces in contact with the ambient atmosphere, on which water molecules generally adsorb dissociatively. More generally, when a mixed iono-covalent bond forms between a surface and an adsorbate, Lewis-type of acidity may be invoked: according to Lewis, an acid A is able to accept a pair of electrons, while a base B is able to donate one. On insulating oxide surfaces, cations are Lewis acid sites and oxygens have Lewis base properties.

In order to take into account the iono-covalent character of the bonding in AB reactions, quantum calculations are required. Mulliken [26, 27] and Hudson and Klopman [28, 29] rephrased Lewis' electronic theory with the tools of quantum mechanics. A key role is played by the electronic *frontier orbitals*: the highest occupied molecular orbital (HOMO), acting as a base, and the lowest unoccupied molecular orbital (LUMO), acting as an acid. The energy levels of the HOMO and the LUMO are related to the first ionization potential of the base and with the electron affinity of the acid, respectively. A typical diagram of frontier orbitals is shown in Fig. 1.

However, the prediction of the acid strength on surfaces is complicated by the large number of degrees of freedom, such as surface orientation, roughness, structural or stoichiometry defects, which do not permit to define a *unique* acid strength. Various types of reactive sites have to be discriminated and a careful analysis of the electron localization in the HOMO and LUMO has to be performed. This has been exemplified by various studies of water dissociation on oxide surfaces, which show that perfect cleavage surfaces are usually inert, while more under-coordinated surface sites, such as step edges, or corners, present a reactivity which increases as a function of their decreasing coordination number [2, 3]. The study of reactivity can thus benefit from the use of concepts usually employed in surface science.

2.2. Electronegativity of an Atom in the Gas Phase and in a Solid

In Lewis' viewpoint, the strength of a surface acidic or basic site is directly related to the value of the electronegativity or the electron affinity of this site. Mulliken proposed an electronegativity scale which is easily related to the quantum chemical approach to bonding. According to his definition, the electronegativity χ of an atom is equal to minus the first partial derivative of its total energy E with respect to the electron number N:

$$\chi = -\frac{\partial E}{\partial N}. \tag{1}$$

As a consequence, the Mulliken electronegativity is the opposite of the chemical potential, which, itself, is equal to either minus the ionisation potential I or minus the electron affinity A, depending upon whether subtraction ($dN < 0$) or addition ($dN > 0$) of electrons is considered. According to Koopman's and Janak's theorems—the latter may be considered as a generalization of the former in the framework of density functional theory (DFT)—I and A are simply equal to minus the HOMO or LUMO levels respectively [30], under the assumption that the orbitals are little affected by the variation of the number of electrons.

In the case of stoichiometric insulating oxides, the HOMO level is the top of the valence band. It is usually strongly localized on the oxygens. On the other hand, the LUMO level is the bottom of the conduction band; it has a prevalent cation character.

To emphasize the importance of the charge distribution in semiconductors and insulators, and its implication in the electronegativity, we consider the *effective* atomic orbital energies in the framework of Hartree theory, as the diagonal matrix elements of the self consistent Hartree operator H, on the atomic orbital basis set $|i\rangle$. Under simplifying assumptions, they read (atomic units are used) [2]:

$$\epsilon_i = \langle i|H|i\rangle = \epsilon_i^{(0)} - U_i Q_i - V_i. \tag{2}$$

In this expression, $\epsilon_i^{(0)}$ is the atomic orbital energy in the isolated neutral atoms i; U_i the intra-atomic electron–electron repulsion integral, Q_i is the ionic charge, and V_i the electrostatic potential (Madelung potential) exerted by the surrounding charges. These charges result from the electron redistribution which takes place in the solid when atoms are put together [31]. In the Hartree–Fock approximation, the

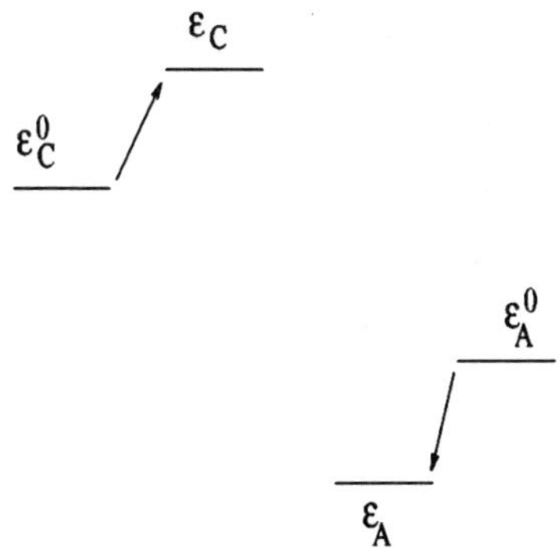

Figure 2. Shifts of the atomic energy levels induced by the Madelung potential.

diagonal terms of H have a slightly more complex expression but they behave, as a function of the charge, in a qualitatively similar way.

The diagonal matrix elements $\langle i|H|i\rangle$ have two corrections, with respect to ϵ_i^0. The first one, $U_i Q_i$ is an intra-atomic correction. It is related to the excess (the loss) of electron–electron repulsion on the anion (the cation) under consideration, with respect to the neutral atoms. The second correction is due to the electrostatic potential V_i which lowers the anion (A) levels and raises the cation (C) levels (Fig. 2), while the intra-atomic term acts in the opposite direction. The two corrections are thus in competition, and their balance is controlled by the value of the optical dielectric constant.

When the HOMO or LUMO has a well-defined atomic character, using Equation 2, we can write the Mulliken electronegativity under the following form:

$$\chi_i = \chi_i^0 + U_i Q_i + V_i. \quad (3)$$

which ascertains that the electronegativity χ_i of the *atom in the solid* is different from that of the isolated neutral atom χ_i^0.

2.3. Application to Oxide Surfaces

Because of the discontinuity caused by the surface, the value of the electrostatic potential exerted on an ion depends on the actual location of the latter with respect to vacuum. The Madelung potential on atoms close to the surface is usually smaller than in the bulk. Generally, the lower the coordination number Z_s of the atom under consideration, the lower the Madelung potential. For instance, on the rocksalt (100), (110) and (211) faces, on which the atoms are, respectively, five-fold, four-fold and three-fold coordinated ($Z_s = 5$, 4 and 3)—six-fold in the bulk—the surface to bulk Madelung potential ratios are equal to 0.96, 0.88 and 0.65, respectively.

As a result, the effective levels of the surface cations are lower than in the bulk. The reverse is true for the surface anions. The importance of this effect is closely related to the ionicity of the compound and, for a given compound, to the surface orientation. Fig. 3 schematically shows the shift of the surface atomic levels from their bulk values.

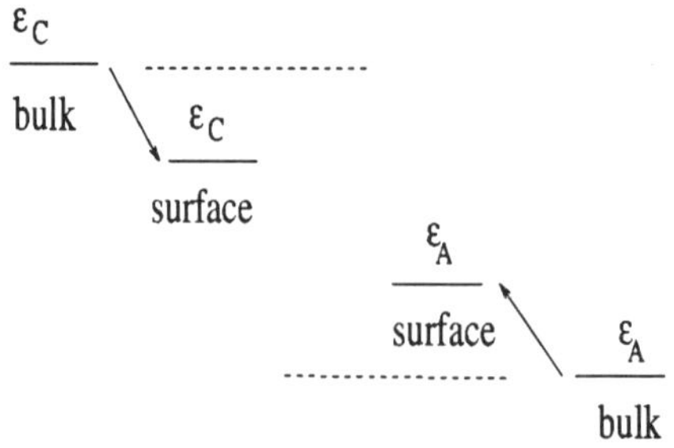

Figure 3. Shifts of the surface atomic levels due to the reduction of the Madelung potential at the surface.

At a surface, considering the chemical shifts discussed above, it is thus expected that the acidic strength of a surface cation and the basic strength of a surface anion increase when the coordination numbers of the ions become lower. This is a well recognized property, sometimes expressed in short by 'the more undercoordinated surface sites the more reactive'. A thorough study of proton and hydroxyl group adsorption on various stoichiometric oxide surfaces has confirmed this conclusion quantitatively [32]. In the following sections, we examine how the change of stoichiometry due to the creation of oxygen vacancies at the MgO(100) surface influences the surface electronic structure and thus the AB strength of surface sites. The case of the polar (111) orientation of magnesium oxide will then be discussed.

3. OXYGEN-DEFICIENT MgO(100) SURFACES

We present in this section an analysis of the electronic structure of oxygen-deficient MgO(100) surfaces, with a special emphasis on the electron redistribution associated with vacancies and on the nature of the HOMOs.

We have performed total energy calculations in the framework of DFT within the Local Density Approximation (LDA) for the exchange and correlation energies. The atomic geometries are optimized, and the electronic structure is fully consistent with the surface morphology. Further computational details are given in Reference [33].

We consider periodic arrangements of vacancies within a $(2\sqrt{2} \times 2\sqrt{2})$ surface supercell and several vacancy concentrations n_F, namely $n_F = 1/8$, $1/4$, $1/2$ and 1. This corresponds to removing one, two, four or eight neutral oxygen atoms per unit cell in the surface top layer (Fig. 4), respectively. For the two intermediate concentrations (1/4 and 1/2), we simulate two distributions of vacancies: the first one, which is homogeneous, and another configuration in which the vacancies are close-packed in the unit cell (denoted as (h) and (cp) in Fig. 4 and in the following, respectively).

We do not describe here the details of the atomic relaxation which takes place around the vacancies, nor comment on the variation of the formation energy of vacancies as a function of the vacancy density. We focus on the electronic structure

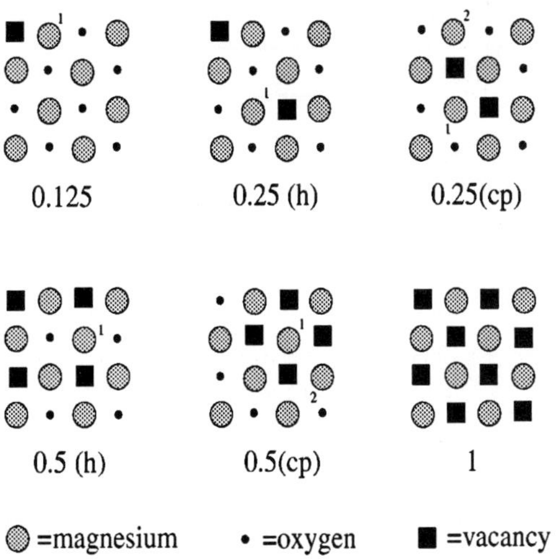

Figure 4. Schematic top view of the surface configurations with 1/8, 1/4, 1/2 and 1 concentrations of neutral oxygen vacancies. *h* and *cp* refer to homogeneous or close packed configurations of vacancies, respectively.

of those defective surfaces for the purpose of understanding the evolution of the HOMO/LUMO levels as a function of the concentration and the arrangement of the O vacancies. Other details can be found in Reference [33].

3.1. Electronic Structure

The electronic structure of stoichiometric MgO consists of a filled valence band, of mainly oxygen character, and an empty conduction band mainly built from magnesium states. Upon removal of neutral oxygens, states disappear in the valence band, and the two electrons left by each missing oxygen are accommodated in quantum states whose energy generally lies in the bulk band gap. We report in Fig. 5 the energy of these F_s states at the center of the two-dimensional Brillouin zone (Γ point), with respect to the the valence band maximum (VBM), up to $n_F = 1/2$. The use of such a reference to specify the energy of the defect states avoids the well-known drawback of DFT-LDA calculations, in which the HOMO-LUMO gap is systematically underestimated. One can note the increase in the F_s band width with n_F, the shift towards higher energies of the HOMO level, of the order of few eV, and its sensitivity to the distribution of oxygen vacancies.

The degree of localization of the excess electrons in the jth defect state is better visualized by plotting the surfaces of equal electron density $|\psi_j(\vec{r})|^2 = \text{const}$. In the case of a single vacancy we find a high degree of localization of the electrons in the vacancy site, in agreement with most authors [34, 35]. The two electrons are trapped by the positive electrostatic potential exerted by the neighboring magnesium atoms, which is high in such an ionic material. The F_s center may thus be thought of as a kind of 'vacancy orbital' localized at the vacancy site.

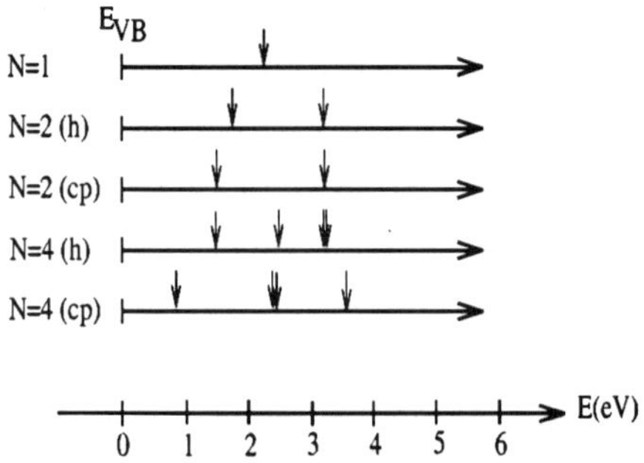

Figure 5. Energy levels of the filled defect states in the gap, for the different surface configurations. The tops of the bulk valence bands E_{VB} have been aligned.

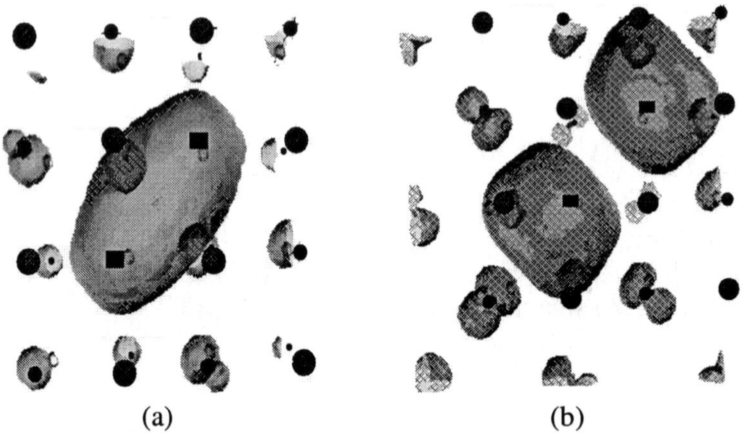

Figure 6. Iso-density surfaces (about 15% of maximum density) for the two gap states associated with a $1/4$ concentration of vacancies in a close-packed configuration. (a) on the left: the bonding state; (b) on the right, the antibonding state (see text). A top view of the surface is presented with oxygen atoms as black dots and magnesium atoms as light spheres. The location of the vacancies is indicated by black boxes.

Figures 6a and 6b represent iso-density plots for the two gap states associated with a $1/4$ concentration of vacancies in a close-packed configuration. For the lower energy state (Fig. 6a), the electronic density is enhanced between the two vacancies. For the state at higher energy (Fig. 6b), a density showing two lobes is found, with a nodal surface separating the vacancy sites. These are the signatures of bonding and antibonding states formed by hybridization between the two vacancy orbitals.

A similar hybridization is actually found on all the defective surfaces that we consider. The hybridization between vacancy orbitals is one of the key factors determining the overall character (either insulating or metallic) of the electronic structure of the defective surfaces.

However, the fully defective ($n_{\mathrm{F}} = 1$) surface presents some features which are *qualitatively* different from those occurring at low vacancy concentration. The analysis of the charge density maps reveals that only seven out of the eight filled gap states are hybridized vacancy orbitals, which, remarkably, maintain a strong identity even on the fully defective surface. However, the eighth gap state, just below the Fermi level E_F, shows density maxima above the surface magnesium atoms and its symmetry is inconsistent with that of a totally anti-bonding state built from vacancy orbitals. The totally antibonding F_s state is actually found at higher energies, about 2.3 eV above E_F.

A full band structure calculation assesses this point. Two bands may be recognized within the gap: the most dispersive one results from the hybridization of vacancy orbitals (the F_s band). The second one, which is clearly hybridized with the conduction band, involves states which are localized on the surface magnesiums. The latter band may be called a surface conduction band (Mg_s band): it is shifted towards lower energies, with respect to the bulk conduction band. The Fermi level intercepts the two bands, conferring a metallic character to the surface. A similar picture applies to the surface with $n_F = 1/2$ in a close-packed configuration (cp).

3.2. Transition Towards Metallicity

The analysis of the electronic structure of defective MgO surfaces thus shows that there are two kinds of quantum states, which are in competition to accommodate the excess electrons left by the missing oxygens. At very low concentration of vacancies ($n_F = 1/8$), the electrons are trapped into the vacancy orbitals, whose energy lies in the gap. These orbitals are only slightly hybridized, because the distance between the vacancies is much larger than the typical radius of the vacancy state. A very narrow band is formed. This band is entirely located within the surface band gap, so that it is completely filled and the surface conduction band is empty. The surface is insulating. When the density of vacancy increases ($n_F = 1/4$(h), $n_F = 1/4$(cp), and $n_F = 1/2$(h)), the vacancy F_s band(s) and the surface conduction band(s) of Mg_s do not overlap despite the increase of the vacancy orbital band width W. The surface remains insulating, since the vacancy orbital bands can accommodate exactly all the excess electrons left by the missing neutral oxygen atoms.

In our calculation, the surface with $n_F = 1/2$(cp) is the critical configuration at which the two bands start to overlap, since the energy difference between the top of the valence band and the bottom of the Mg_s band is of the order of W. For this configuration, and at higher vacancy concentrations, the F_s band and the Mg_s band overlap and get partially filled, conferring a metallic character to the surface. The filled gap states belong either to the F_s or to the Mg_s band, and their degree of localization varies accordingly.

In conclusion, non-stoichiometric MgO(100) surfaces are systems in which a progressive transition from an insulator to a metal takes place. On the non-defective surface, the basic and acidic sites are the surface oxygens and magnesiums, respectively, and their AB strength is determined by the position of the top of the valence band (HOMO) and the bottom of the conduction band (LUMO). When small concentrations of vacancies are produced, the HOMO changes character, and its energy shifts towards higher values. It is associated with a quantum state trapped at the site of the vacancy by the Madelung potential exerted by neighbouring cations. Finally, at high vacancy concentrations, the surface becomes metallic, with close resemblance to a pure magnesium surface.

4. POLAR MgO(111) SURFACE

For materials involving at least two types of atoms, such as oxides, a classification of surfaces into three families has been established on the basis of simple electrostatic criterion [36] (Fig. 7). Type 1 and Type 2 surfaces involve atomic layers which can be either neutral or charged but for which the stacking sequence is such that the total dipole moment of the repeat unit vanishes. In the third family, named *polar* surfaces, the layers are not electrically neutral and the repeat unit bears a non-zero dipole moment. The (111) face of rocksalt crystals, the zinc blende (111) and (100) faces, the wurtzite (0001) face, the fluorite (110) face, etc., are polar. It is important to stress the difference between type 1 and type 2 surfaces, which can have a *surface* dipole (due, e.g. to a different relaxation of anions and cations in the outer layers), and type 3 surfaces, whose dipole is *macroscopic*, that is, proportional to the slab thickness. Because of the presence of a net dipole moment in the repeat unit, polar surfaces generally have a very low stability and, for this reason, they have not been studied as thorougly as non-polar surfaces.

A semi-infinite crystal cut along a polar direction has a diverging electrostatic energy, because the electrostatic field has a non-zero mean value in the material. If one considers a slab of equidistant layers, bearing positive and negative charge densities equal to $\pm\sigma$, the repeat unit has a dipole moment density equal to σR. As a result, the electrostatic potential increases across the slab by an amount $\delta V = 4\pi\sigma R$ per double layer. δV is large, typically of the order of several tens eV. The total dipole moment of a slab involving N bilayers is equal to $\mu = N\sigma R$

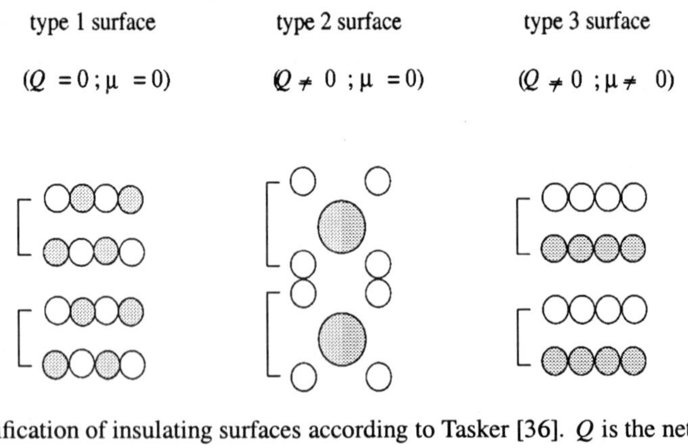

Figure 7. Classification of insulating surfaces according to Tasker [36]. Q is the net charge borne by the atomic layers, and μ is the dipole moment of the repeat sequence in the direction perpendicular to the surface, indicated by brackets.

and the macroscopic electrostatic energy amounts to: $E = 4\pi N R\sigma^2$. It is very large, even for thin films. This is the origin of the surface instability.

In order to cancel the macroscopic component of the dipole moment, a modification of the charge density on the outer layers of the slab is required. It can be shown [2, 20, 37] that a reduction of the surface charge by a factor of two with respect to the bulk: $\sigma' = \sigma/2$, suppresses the monotonic increase of the electrostatic potential. The resulting dipole moment $\mu = \sigma R/2$ is no longer proportional to the slab thickness and the electrostatic energy decreases by about a factor of two: $E = 2\pi R\sigma^2(N - 1/2)$. The surface charges σ' provide the charge compensation necessary to suppress the polarity. More generally, the condition for the cancellation of the macroscopic dipole moment reads:

$$\left|\sum_{j=1}^{m} \sigma_j\right| = \frac{\sigma}{2}, \tag{4}$$

in which m is the number of outer layers affected by the modification of charges.

The stabilizing charges may be induced by charged defects, such as oxygen vacancies, impurities or pairs of charged defects. The energy gain obtained through adsorption and/or non-stoichiometric reconstructions may be rather large, explaining the high reactivity of polar surfaces. In the case of oxides, for instance, the saturation of surface sites by adsorbed protons and hydroxyl groups is an effective mechanism for compensation.

All the mechanisms that have been hencefore proposed in the literature for the stabilization of a polar surface are fully consistent with equation 4. They are based either on internal electron redistribution, on the presence of charged adsorbates or impurities [38, 39], or on substoichiometry, as in compound semiconductors [40, 41]. The first of these mechanisms was discussed in a cluster calculation of MgO(111) [42], and later generalized to periodic slabs [43]. A metallization of the surface layers was predicted. We will discuss in the following these three mechanisms in the case of MgO(111), which is the lowest index polar face of the rocksalt structure. Equally spaced anion and cation planes alternate along the normal to the surface. This face has a high surface energy and cannot be produced defect-free. In nearly all experiments performed until now, it turned out to be either faceted or non-stoichiometric.

4.1. The Stoichiometric MgO(111) Surface

We present results obtained by a total energy Hartree–Fock semi-empirical approach, which has the advantages of (i) being fast enough to allow the consideration of large systems, nevertheless (ii) being able to account for self-consistent screening effects. Further computational details can be found in Reference [37]. Preliminary results obtained in our group through DFT on the unreconstructed surface confirm the following description.

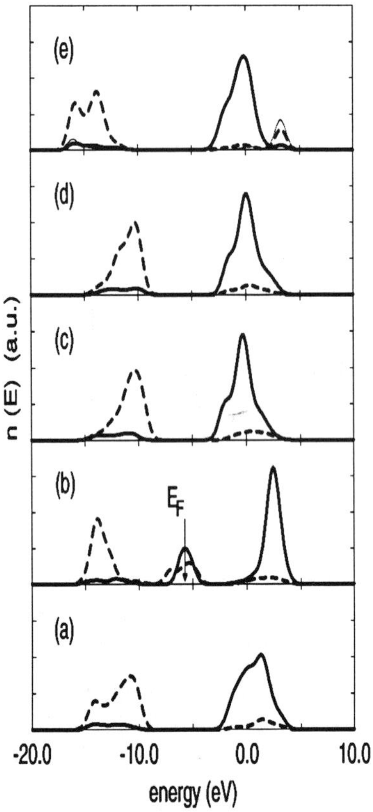

Figure 8. Local density of states (LDOS) on MgO{111}: (a): the bulk LDOS; (b): the ideal (1 × 1) surface; (c): the reconstructed (2 × 1) surface; (d): the reconstructed (2 × 2) surface and (e): the hydroxylated (1 × 1) surface. The LDOS' on oxygens, magnesiums and protons are represented by dashed, bold plain and light plain lines, respectively. The calculated LDOS' have been convoluted by a 0.7 eV wide Gaussian function.

The unreconstructed (1 × 1) MgO(111) surface has hexagonal symmetry. The outer layer is composed of either O or Mg atoms, which are three-fold coordinated with the atoms in the underlying layer. Figures 8 a and b show the local density of states (LDOS) on bulk and surface atoms, respectively. At the (111) surface, the bulk valence (VB) and conduction bands (CB) are, each, split into two peaks. The higher VB peak and the lower CB peak overlap in the energy region of the bulk band gap. The Fermi level E_F intersects these two surface bands, which gives a metallic character to the two faces of the slab.

As a result of metallicity, a strong electron redistribution takes place. In the surface layers, the charges are found equal to $Q_S = \pm 0.59|e|$ (e is the electron charge). They amount to $Q_{SS} = \pm 1.19|e|$ in the sub-surface layers, while in all the inner layers, they are close to their bulk values $Q = \pm 1.22|e|$. The charge reduction on the surface oxygens and magnesiums is assigned to the incomplete filling of the surface O valence band and to the partial filling of the surface Mg conduction band, respectively. We do not wish to emphasize too much on the precise charge values, but rather on the fact that $|\sigma_S + \sigma_{SS}|$ is close to one half the bulk charge density, as required from Equation 4.

An electron redistribution on the surface layers resulting in an open-shell electronic structure can be shown to be a necessary condition for the cancellation of the macroscopic electric and dipole on the pure (1×1) surface. However, this is achieved at the expense of a very high surface energy. We find that the average of the surface energies of the oxygen and a magnesium terminations is equal to 7.3 J/m^2, about a factor 6 larger than the surface energy of the stable (100) face. This result is obtained by excluding any reconstruction, that is, by imposing a (1×1) periodicity in the direction parallel to the surface. When this constraint is removed, a spontaneous desorption of half the number of surface atoms occurs, confirming the unstable nature of the (1×1) pure surface and suggesting that non-stoichiometric reconstructions may be an effective mechanism to stabilize the (111) face of MgO.

4.2. Reconstructed MgO(111) Surface

We consider in the following two surface configurations in which a macroscopic number of atoms are missing in the outer layer(s). The atom concentrations have been chosen in such a way that the surface charge density is the one required for suppressing the polarity (Equation 4), assuming that the charges keep their bulk values.

Starting from the stoichiometric MgO(111) surface and its hexagonal (1×1) unit cell (Fig. 9a), a non stoichiometric surface with half of the atoms missing in the outer layer and a (2×1) unit cell may be imagined as shown in Fig. 9b. The surface displays parallel streaks, with {100} and {110} facets. The surface atoms are three-fold coordinated ($Z = 3$).

Indeed, we find that the surface electronic structure is not very different from the bulk one (Fig. 8c): the oxygen and magnesium bands are weakly shifted towards higher and lower energies, respectively. No band splitting takes place, and the surface retains its insulating character. A finite gap is present between the surface valence and conduction bands, whose width is equal to 6.9 eV. This value is close to that found on the MgO(211) surface (6.5 eV) [44], a non-polar surface on which the atoms have a similar environment.

A weak charge redistribution takes place on the two outer planes on each side of the slab. We find atomic charges equal to $Q_S = \pm 1.14$ on the surface layers and $Q_{SS} = \pm 1.19$ on the sub-surface layers. After normalization to the surface unit cell of the unreconstructed surface, one obtains charge densities equal to $\sigma_S = \pm 0.57|e|$, and $\sigma_{SS} = \pm 1.19|e|$. The value of $|\sigma_S + \sigma_{SS}|$, is close to $\sigma/2$, and thus fulfills the electrostatic criterion (Equation 4) for surface stability.

Figure 9c represents another non-stoichiometric surface structure in which three atoms out of four have been removed in the outer layer, and one out of four in the sub-surface layer. Square facets of {100} orientations involving four atoms, and pyramid-like structures around the three-fold coordinated surface atoms ($Z = 3$)

(a)

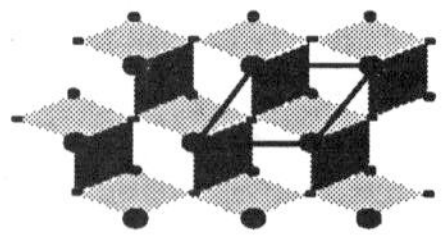

(b)

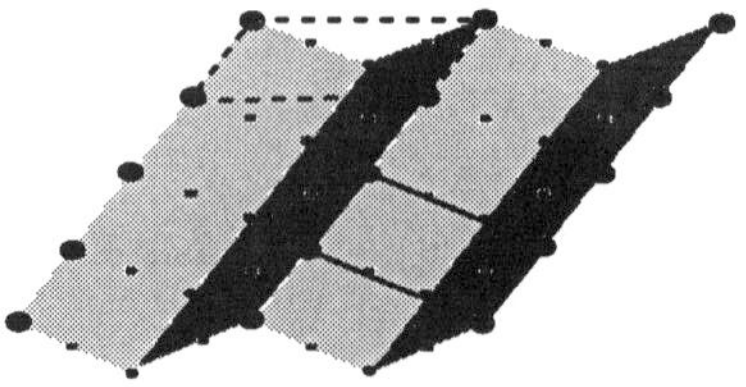

(c)

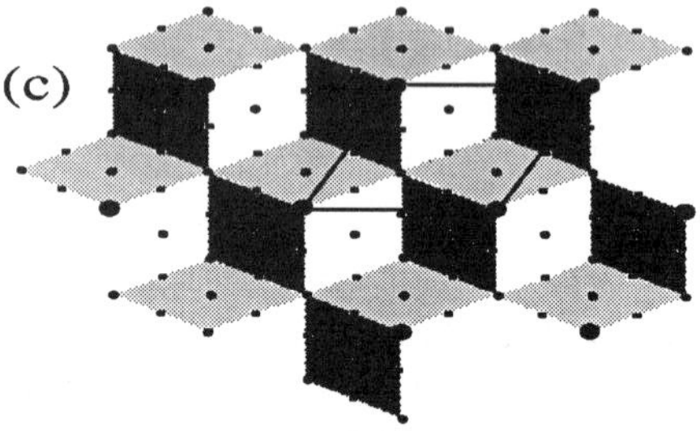

Figure 9. Relief views of the MgO{111} surfaces. (a): the unreconstructed (1 × 1) surface; (b): the (2 × 1) surface; (c): the (2 × 2). The atoms in the outer layer are marked by large black dots, and the surface unit cells are superimposed on the structures. On the (2 × 1) surface, the {100} and {110} facets have been colored in dark and light grey, respectively. On the (2 × 1) surface, the (2 × 1) and the rectangular cells are both indicated.

are produced, a geometry already proposed by several authors [45, 46]. The surface cell is hexagonal and its area is four times larger than that of the unreconstructed (1 × 1) face. The surface is p(2 × 2) reconstructed.

The electronic structure found for this surface configuration (Fig. 8d) is remarkably similar to that of the (2 × 1) surface. The shift of the surface valence and conduction bands with respect to the bulk bands is slightly smaller, so that the resulting gap (7.5 eV) is slightly wider than that of the (2 × 1) surface. The surface charges on the outer planes of the (2 × 2) slab are close to the bulk values. We find $Q_S = \pm 1.20$ on the surface layers, $Q_{SS} = \pm 1.21$ on the sub-surface layers and $Q_{SSS} = \pm 1.21$, $Q'_{SSS} = \pm 1.25$ respectively on the five- and six-fold coordinated atoms of the third layers. These values, once normalized to the unit cell of the unreconstructed surface, correspond to charge densities equal to $\sigma_S = \pm 0.3|e|$, $\sigma_{SS} = \pm 0.91|e|$ and $\sigma_{SSS} = \pm 1.22|e|$, which fulfill the condition (4.1).

The energies required to form the (2×1) and (2×2) surfaces amount, respectively, to $\Sigma_{(2\times1)} = 2.6\ \mathrm{J/m^2}$ and $\Sigma_{(2\times2)} = 2.3\ \mathrm{J/m^2}$. They are much lower than the energy to form the unreconstructed stoichiometric surface. It is possible to obtain a good estimate of these energies by noting that the surface is actually made of {100} and {110} facets linked by edges. By counting the number of atoms involved in the facets, edges and corners per unit cell, and by considering the formation energies calculated for each of these surface defects [47], surface energies in agreement with the numerical values to within 10% may be computed for the facetted structures. This model can also be used to predict that even larger facets than those simulated here may be formed.

It can thus be concluded that reconstructed surfaces might represent the most stable configurations of *pure* magnesium oxide, cut along the (111) orientation, since they have a much lower surface energy than the stoichiometric planar geometry. Their electronic structure is very close to that of non-polar surfaces: the energy of the HOMO, with prevalent oxygen character, is determined to a large extent by the coordination number N_N of surface oxygens. Most of them are undercoordinated with respect to those on the (100) cleavage surface ($N_N = 3$ instead of 5). A strong reactivity of these reconstructed surfaces is therefore expected.

4.3. Reactivity of MgO(111) Surface

In order to exemplify the enhanced reactivity of the MgO(111) surface with respect to the cleavage (100) face, we have modelled the dissociative adsorption of water in the limit of a total saturation of the surface sites by adsorbed OH^- and H^+ groups. The hydroxyl groups are adsorbed in threefold coordinated sites on the magnesium-terminated face, while protons bind on-top a single oxygen on the opposite face of the slab. Constructed in this way, the hydroxylated slab is symmetric, with layers of composition: H, O, Mg, O, Mg...

In a fully ionic picture, in which the ionic charges are equal to ± 2 whatever the local environment of the atoms, the configuration of the hydroxylated surface fulfills the criterion for cancellation of the macroscopic dipole moment. The LDOS' on the outer hydrogen, oxygen and magnesium atoms of the hydroxylated (1×1) surface are shown in Fig. 8e. The adsorption of hydroxyl groups at a site threefold coordinated with Mg atoms restores the octahedral environment of the surface magnesiums. Their LDOS is thus practically identical to the bulk LDOS. The oxygens in the second layer, on the other hand, have three magnesium and one hydrogen neighbours. As far as bands originating from the hydroxyl groups are concerned, bonding and non-bonding states are located in the lower part of the bulk valence band, while anti-bonding states are found above the bulk conduction band. The hydroxylated slab is insulating with a gap width equal to 7.9 eV, in close resemblance to the values found for the hydroxylated MgO(211) surface [48].

The value of the OH charge density, normalized to the (1×1) surface cell, is equal to $\sigma_S = -0.63|e|$ and the charge density on the layer below amounts to $|\sigma_{SS}| = 1.26|e|$. The relation (4) is also fulfilled on this surface.

The energy of hydroxylation of the MgO{111} surface is calculated to be 8.3 eV, per water molecule, as the energy difference between the clean substrate plus the isolated water molecules, and the hydroxylated slab. The positive value indicates that the water dissociation is an exothermic process, so that water may spontaneously dissociate and adsorb, provided that contingent activation barriers could be overcome. The process induces a more pronounced energy lowering than on other MgO surfaces that we have studied previously (−0.9 eV, 1.9 eV and 5.9 eV on MgO(100), (110) and (211) respectively) [48].

It can thus be concluded that the MgO(111) surface is highly reactive with respect to water dissociation, at variance with the (100) planar terraces which are known to be largely inert. This may explain the surprisingly high occurrence of natural (111) faces in MgO minerals, grown in humid atmosphere. Similarly, (111) layers are observed in MgO precipitates in a Cu matrix, suggesting that they are also highly reactive for metal adsorption. Preliminary results obtained in our group confirm this conclusion.

5. CONCLUSION

The electronic structure of oxygen-deficient and polar MgO surfaces, obtained by a combination of ab initio DFT-LDA and semi-empirical Hartree–Fock calculations, has been analysed with special emphasis on the acid-base properties of the surface sites.

The non-stoichiometric MgO(100) surface displays a progressive transition from an insulator to a metal, as a function of the oxygen vacancy distribution. When small vacancy concentrations are produced, the two electrons in the HOMO are trapped at the site of the vacancy by the Madelung potential exerted by neighbouring cations and their energy varies in the bulk band gap region as a function of the vacancy concentration and distribution. At high vacancy concentrations, the surface becomes metallic, with close resemblance to a pure magnesium surface.

The stoichiometric polar MgO(111) surface is open-shell and unstable. It can be stabilized by a spontaneous desorption of a macroscopic number of surface atoms, leading to nanofacetted surface configurations. The (2×2), so-called octopolar, surface displays nano-pyramids with a high concentration of tri-coordinated surface atoms, whose reactivity is expected to be much higher than on the cleavage (100) face. The polar (1×1) surface may also be stabilised upon adsorption: water dissociation and adsorption of metal atoms provide two examples of very efficient mechanisms.

Acknowledgements

This work benefits from interesting discussion with A. Pojani and J. Goniakowski. The major calculations were performed on the CRAY C98 computer at IDRIS/CNRS, under the project numbers 970109 and 980732.

REFERENCES

1. V. E. Henrich and P. A. Cox, *The Surface Science of Metal Oxides*. Cambridge University Press (1994).
2. C. Noguera, *Physics and Chemistry at Oxide Surfaces*. Cambridge University Press (1996).
3. C. Noguera, in: *Organic Coatings*, P. C. Lacaze (Ed.), American Institute of Physics Conference Proceeding **354**. Woodbury, New York (1996).
4. P. Zschack, J. B. Cohen and Y. W. Chung, *Surf. Sci.* **262**, 395 (1992).
5. P. J. Hardman, N. S. Prakash, C. A. Muryn, G. N. Raikar, A. G. Thomas, A. F. Prime and G. Thornton, *Phys. Rev. B* **47**, 16056 (1993).
6. P. W. Murray, F. M. Leibsle, C. A. Muryn, H. J. Fischer, C. F. J. Flipse and G. Thornton, *Surf. Sci.* **321**, 217 (1994).
7. P. W. Murray, F. M. Leibsle, C. A. Muryn, H. J. Fischer, C. F. J. Flipse and G. Thornton, *Phys. Rev. Lett.* **72**, 689 (1994).
8. M. Sander and T. Engel, *Surf. Sci.* **302**, L263 (1994).
9. H. Onishi and Y. Iwasawa, *Surf. Sci.* **313** , L783 (1994).
10. D. F. Cox, T. B. Fryberger and S. Semancik, *Surf. Sci.* **224**, 121 (1989).
11. T. Matsumoto, H. Tanaka, T. Kawai and S. Kawai, *Surf. Sci.* **278**, L153 (1992).
12. Y. Liang and D. A. Bonnell, *Surf. Sci.* **285**, L510 (1993).
13. H. Tanaka, T. Matsumoto, T. Kawai and S. Kawai, *Surf. Sci.* **318**, 29 (1994).
14. C. C. Chang, *J. Appl. Phys.* **19**, 5570 (1968).
15. T. M. French and G. A. Somorjai, *J. Phys. Chem.* **74**, 2489 (1970).
16. S. Baik, D. E. Fowler, J. M. Blakely and R. Raj, *J. Am. Ceram. Soc.* **68**, 281 (1985).
17. M. Gautier, J. P. Duraud, L. Pham Van and M. J. Guittet, *Surf. Sci.* **250**, 71 (1991).
18. M. Gautier, G. Renaud, L. Pham Van, B. Villette, M. Pollak, N. Thromat, F. Jollet and J. P. Duraud, *J. Am. Ceram. Soc.* **77**, 323 (1994).
19. G. Renaud, B. Villette, I. Vilfan and A. Bourret, *Phys. Rev. Lett.* **73**, 1825 (1994).
20. M. Lannoo and P. Friedel *Atomic and Electronic Structure of Surfaces*, Springer Series in Surface Science, Vol. 16. Springer Verlag, Berlin (1991).
21. K. Refson, R. A. Wogelius, D. G. Fraser, M. C. Payne, M. H. Lee and V. Milman, *Phys. Rev.* **B52** 10823 (1995).
22. F. M. Fowkes, *J. Adhesion Sci. Technol.* **1**, 7 (1987).
23. F. M. Fowkes, *J. Adhesion Sci. Technol.* **4**, 669 (1990).
24. K. L. Mittal and H. R. Anderson, Jr. (Eds), *Acid-Base Interactions: Relevance to Adhesion Science and Technology*. VSP, Utrecht, The Netherlands (1991).
25. H. L. Finston and A. C. Rychtman, *A New View of Current Acid-Base Theories*. John Wiley, New York (1982).
26. R. S. Mulliken, *J. Chem. Phys.* **19**, 514 (1951).
27. R. S. Mulliken, *J. Am. Chem. Soc.* **74**, 811 (1952).
28. R. F. Hudson and G. Klopman, *Tetrahedron Lett.* **12**, 1103 (1967).
29. G. Klopman and R. F. Hudson, *Theor. Chim. Acta* **8**, 165 (1967).
30. J. F. Janak, *Phys. Rev. B* **18**, 7165 (1978); J. P. Perdew, R. G. Parr, M. Levy and J. L. Balduz, Jr., *Phys. Rev. Lett.* **23**, 1691 (1982). For a more comprehensive discussion, see also R. G. Parr and W. Yang, *Density Functional Theory of Atoms and Molecules*. Oxford University Press, Oxford (1989).
31. Given an atom X (X $=$ A, C) the charge Q_X is defined as the difference between the number of electrons $N^{(0)}$ of the neutral atom and the number of electrons N in the solid. The precise value of the latter cannot in general be defined unambiguosly, due to the delocalised nature of the electronic density. Nevertheless, it can be considered as an indication of the charge state of the atom X in the solid. Moreover, the computed values of the physical quantities which depend

on charges, such as the ionization potentials, the electronic structure, etc., are generally in good agreement with experiments.
32. J. Goniakowski, S. Bouette-Russo and C. Noguera, *Surf. Sci.* **284**, 315 (1993).
33. F. Finocchi, J. Goniakowski and C. Noguera, *Phys. Rev. B* **59**, 5178 (1999).
34. G. Pacchioni, A. M. Ferrari and G. Ieranò, *Faraday Discuss.* **106**, 155 (1997).
35. L. N. Kantorovich, J. M. Holender, M. J. Gillan, *Surf. Sci.* **343**, 221 (1995).
36. P. W. Tasker, *J. Phys. C* **12**, 4977 (1979).
37. A. Pojani, F. Finocchi, J. Goniakowski and C. Noguera, *Surf. Sci.* **387**, 354 (1997).
38. N. Floquet and L. C. Dufour, *Surf. Sci.* **126**, 543 (1983).
39. P. A. Cox and A. A. William, *Surf. Sci.* **152/153**, 791 (1985).
40. S. Y. Tong, G. Xu and W. N. Mei, *Phys. Rev. Lett.* **52**, 1963 (1984).
41. M. Lannoo, *Mater. Sci. Ing. B* **22**, 1 (1993).
42. M. Tsukada and T. Hoshino, *J. Phys. Soc. Jpn.* **51**, 2562 (1982).
43. C. Noguera, A. Pojani, F. Finocchi, J. Goniakowski, in: *Chemisorption and Reactivity on Supported Clusters and Thin Films*, R. M. Lambert and G. Pacchioni (Eds), NATO ASI series E: Applied Sciences. Kluwer, Dordrecht, The Netherlands (1997).
44. J. Goniakowski, C. Noguera, *Surf. Sci.* **319**, 68 (1994); *ibid.* **319**, 81 (1994).
45. R. Lacmann, *Colloq. Int. CNRS* **152**, 195 (1965).
46. D. Wolf, *Phys. Rev. Lett.* **68**, 3315 (1992).
47. J. Goniakowski and C. Noguera, *Surf. Sci.* **340**, 191 (1995).
48. J. Goniakowski and C. Noguera, *Surf. Sci.* **330**, 337 (1995).

Acid-Base Interactions, Vol. 2, pp. 205–222
Ed. K.L. Mittal

Accurate calculations of core-electron binding energies (CEBEs) via density functional theory: Towards an evidencing of acid-base effects at real interfaces via XPS

J. CHARLIER,[1] C. BUREAU,[1,*] D. P. CHONG,[2] S. KRANIAS[1] and G. LÉCAYON[1]
[1]*CEA-Saclay, DSM-DRECAM-SRSIM, bat. 466, F-91191 Gif-sur-Yvette Cedex (France)*
[2]*Department of Chemistry, University of British Columbia, 2036 Main Mall, Vancouver, BC, Canada*

Abstract—A recent procedure for computing core-electron binding energies (CEBEs) which makes use of the Density Functional Theory (DFT) is presented. It is based on a generalization of the Transition-State concept of Slater, and affords calculated CEBEs with a systematic average absolute deviation (aad) from the experimental values of the order of 0.2 eV on the 200 benchmark systems on which it was tested. The favorable CPU-time (Computation per Unit Time) scaling of DFT allows this procedure to be applied to fairly large systems such as polymers and interfaces. By obtaining a systematic agreement of the order of 0.2 to 0.3 eV between theory and experiment, one may use this procedure as a powerful tool to identify unexpectedly weak effects in XPS spectra. It will be shown that measurable shifts can also be obtained in the presence of several intermolecular effects such as ion-pairing, dipolar interactions, and hydrogen bonding within polymer phases. Possible extensions of this method to the calculation of CEBEs of molecules chemisorbed on metallic surfaces will also be discussed.

Keywords: XPS; core levels; computation; density functional theory; nitriles; γ-aminopropylethoxysilane.

1. INTRODUCTION

A thorough characterization of the interfacial region between two materials (polymer/polymer, organic molecule/metallic substrate, ...) has long been a goal in surface science. This stems from the peculiar role played by the interface region in complex phenomena of technological interest such as adhesion or lubrication [1].

Towards this aim, a need for accurate analytical techniques focusing on the very interface region has emerged, in particular when the analysis can be connected with predictive models.

In the field of adhesion, the concept of acids and bases brought about by Lewis at the beginning of the century has raised many hopes [2–4]. This concept states that significant predictive information on the nature of molecular interactions can be gained from a knowledge of isolated systems, and isolated systems only (i.e. prior to

* To whom correspondence should be addressed. E-mail: christophe.bureau@cea.fr

their actual interactions). This approach is very appealing as it opens the possibilities towards an *a priori* estimation of the physicochemical behavior of complex assemblies. As regards the tentative theoretical modeling of interface interactions, the very recent point of view brought about by the Density Functional Theory (DFT) on the Lewis acid-base concept is very attractive [5–12]. Let us recall briefly that from certain parameters (such as hardness (η), chemical DFT potential (μ) and electronic density ($f(r)$) having a very physical significance and correlation to experimental values), it becomes possible to determine in a relative way the overall predominant nature of the interactions (electrostatic interactions or charge transfer), the sign of the electron transfer when the charge transfer is the major component and the regio-selectivity of the interaction in the case of an interaction driven by the electron transfer. However, it is important to notice that the DFT point of view on the Lewis acid-base concept is restricted to the initial steps of the interaction (the so-called non-crossing assumption); i.e. the attempt is clearly to "feel" the reaction path as far as possible with the electronic and geometric characteristics of the reactants forced to remain in their unperturbed situation [10].

One way to circumvent this problem is to admit—once and for all—the non-crossing assumption, and merely say that, up to a semi-empirical parametrization, one is seeking only a functional relationship between measurable quantities (e.g. spectroscopic) and heats of adsorption. This point of view is very well described in [4], along with the various empirical scales (Drago, Gutmann… etc) that have emerged. Within the framework of a given scale, i.e. of a given working equation (Drago's constants (E&C), acceptor/donor number of Gutmann (AN/DN)...etc), several types of experimental measurements have been used as probes of the interacting region: NMR shifts of a given reference system ($SbCl_5$...), infra-red shifts...etc. In most circumstances, the idea is to determine how a given "probe" molecule, supposedly non- reactive, may give information on its surroundings, i.e. (may render spectroscopically the intermolecular interactions it undergoes). The same type of approach, developed for molecule/surface interactions, greatly benefits from the use of X-ray Photoelectron Spectroscopy (XPS), which may directly give indications on the electronic rearrangements within a molecule upon interaction. XPS is among the very powerful techniques used to gain information on the chemical composition, bonding and homogeneity of surfaces and interfaces [11–13]. Core electron binding energies (CEBEs) constitute the key spectral information obtained by this technique as they are related to the physicochemical environment a nucleus feels in a molecule. The usual rule-of-thumb is that the CEBE of an atom is raised when it is in a more electronegative surrounding, and that it is lowered when experiencing the effects of a more electropositive environment [14]. Despite the invaluable usefulness of this simple rule, it is sometimes difficult to relate precisely the various CEBEs to the underlying molecular structure, especially when the CEBEs differ by a few electron volts (eV) or fractions of an eV [13, 14]. In a number of cases, theoretical calculations have proven quite useful or even indispensable to a correct interpretation [15, 16]. The gain of a theory vs. experiment comparison is a cross-check enabling one to validate plausible molecular structures for spectrum interpretation.

Moreover, this technique is essentially surface-sensitive, which makes it well suited to follow interactions at surfaces and interfaces [17–20]. For example, Figure 1 reports the correlation between molar enthalpies for the formation of Lewis acid-base complexes (deduced from IR measurements), versus the chemical shift observed in Cl $2p^{3/2}$ line for the Lewis acid chloroform probe adsorbed on the surface of various polymers [17]. The value of the slope of this curve is 0.9 eV/(kcal/mol) (in absolute value): this shows that chemical shifts of a fraction of an eV are likely to result in interactions of the order of a kcal/mol. This result is pretty much surprising at first glance, as 1 eV, strictly speaking, corresponds to about 25 kcal/mol. The slope obtained in Fig. 1 may be interpreted in terms of that the XPS probes also density distortions (polarization, relaxation, correlation...) and not directly the ground state energies of a system. One may thus expect a sensitivity, which is larger than the very energies involved in the photoelectronic mechanism of XPS. However, interpreting or even ascertaining chemical shifts as low as a fraction of one eV is difficult. Thus, molecular probes which involve chemical shifts as large as possible will be required. In addition, the quality of its predictions depends much on the type of molecule chosen to probe a given surface.

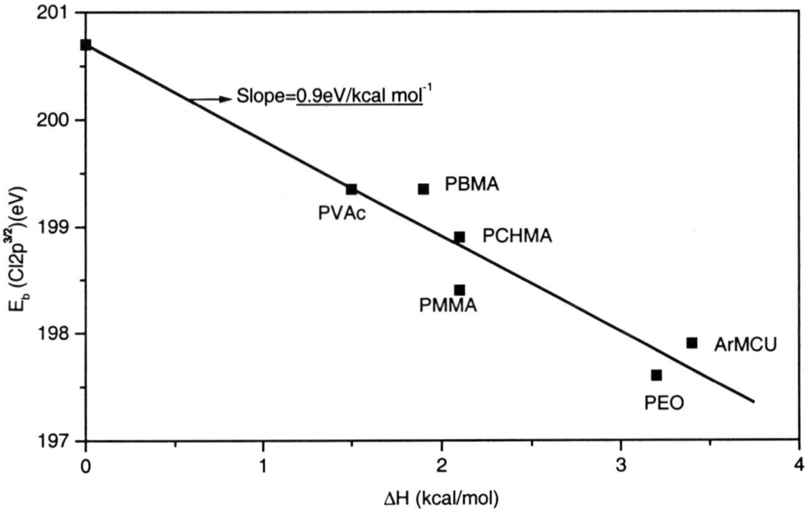

Figure 1. Correlation between molar enthalpies for the formation of Lewis acid-base complexes (deduced from IR measurements), versus the chemical shift observed in chlorine $2p$ line for the Lewis acid chloroform probe adsorbed on the surface of various polymers [17]: Poly(vinyl acetate) (PVAc), Poly(butyl methacrylate) (PBMA), Poly(cyclohexyl methacrylate) (PCHMA), Poly(methyl methacrylate) (PMMA), Poly(ethylene oxyde) (PEO), polyurethane resin (ArMCU).

One purpose of the present paper is to show that one may nowadays compute core-electron binding energies (CEBEs) of a given molecular system using quantum mechanics, with a precision compatible with the resolution of the best XPS experiments, i.e. of the order of 0.2 eV. This result is obtained by employing a recent procedure, which makes use of the Density Functional Theory [21]. The prospects of such a powerful tool are threefold: (i) one may obtain the reference ("gas phase") XPS spectrum of a given molecule, even when this molecule cannot be obtained in the gas phase, and can only be obtained by coating on a substrate, thus already interacting with other molecules; (ii) relying on the accuracy of the method, one may find out what are the various effects that may contribute to shifts in the CEBEs when the molecule is now in the condensed phase or adsorbed; and (iii) with a systematic precision of 0.2 eV, one may even be able to identify various intermolecular effects, which is quite surprising for a core-level probing method.

In the following sections, we will briefly review the available theoretical methods enabling the practical computation of XPS spectra. Keeping in mind necessary applications to large systems (polymers, surfaces, ...), our purpose will be to emphasize computationally simple yet effective methods. Special attention will be paid to some recent methods, which make use of the Density Functional Theory. Then, we will illustrate the validity of this method for the cases of some real systems ranging from small molecules in the gas-phase to polymers. We will also mention some recent results obtained on γ-aminopropyltriethoxyxysilane (γ-APS), a molecule which is used as an adhesion promoter in the microelectronics for coating oxidized silicon wafers with polyimide films. Possible extensions of this method to the calculation of CEBEs of molecules chemisorbed on metallic surfaces will also be discussed.

2. THEORETICAL APPROACH: A NON-EXHAUSTIVE REVIEW

At the simplest level, Koopmans' theorem [22] provides a straightforward method to make a rough evaluation of the CEBEs (I_k) by equating these with $-\varepsilon_k$ ($1 \leq k \leq N/2$, for a closed shell system, where N is the number of electrons):

$$I_k = -\varepsilon_k$$

Which is the opposite of the orbital energies resulting (ε_k) from a Hartree-Fock (HF) calculation on a neutral parent molecule.

This method gives rather poor estimates of the CEBEs with differences as large as 10 to 20 eV from experimental values [23]. This large difference is mainly due to the significant relaxation effects, which are not taken into account since the calculation is done on the neutral parent only, and which are only partly compensated by the simultaneous neglect of correlation effects [14].

At the other end in the hierarchy of techniques used to compute CEBEs, one finds various ΔSCF approaches. These consist in making the explicit computation of the electronic structure of both the neutral parent molecule M and the core-ionized species M_k^+, and then calculating the difference between the two:

$$I_k = E^{SCF}(M_k^+) - E^{SCF}(M).$$

where $E^{SCF}(M_k^+)$ and $E^{SCF}(M)$ are the SCF computed orbital energies of the core-ionized species and the neutral parent molecule, respectively.

The quality of the results then depends on the theoretical treatment used in both SCF calculations. At the HF level, the approach explicitly takes relaxation effects into account—as opposed to Koopmans' theorem, i.e. it describes the way the valence levels rearrange when a core electron is removed. The correlation effects, which are not taken into account at the HF level, are thus not compensated anymore, and need to be included to deliver results of acceptability quality. Various post-HF methods enable a partial account of correlation effects, and pretty good results have been obtained with this approach [16], mainly when correlation effects are included pertubatively up to the second order via the Moller-Plesser method (ΔMP2). Even though the calculated CEBEs are still 0.5 to 1.0 eV apart from experimental values, i.e. the precision of this method is lower than the best experimental resolutions [16, 24].

Recently, Chong has proposed a method called the unrestricted generalized transition-state (uGTS) using Density Functional Theory calculations [21] based on the ideas of Slater's transition-state [25] later generalized by Williams *et al.* [26]. As with ΔSCF approaches, this method involves the explicit SCF calculation on the two systems: the neutral parent molecule and a partially ionized system. Because of the use of DFT, the procedure scales as N^3 (where N is the number of electrons) only, which offers much hope towards its applicability to large systems. This procedure has been tested on about two hundred molecular systems ranging from small molecules (CO, CH_4, ... etc.) to larger models (n-unit models of polymer, n=1–3) [24, 27, 28]. It has been shown to predict CEBEs in excellent agreement with experiment, with a systematic average absolute deviation (aad) from experimental results of only 0.2 eV for CEBEs. The precision of this technique is thus compatible with the best experimental resolutions (apart from synchrotron experiments).

3. ON THE UGTS METHOD IN DFT

Over twenty years ago, Slater introduced the transition-state (TS) method [25]. Consider an electronic process such as ionization or excitation at a fixed molecular geometry. Let us define:

$$E(\lambda) = E_0 + \lambda E_1 + \lambda^2 E_2 + \lambda^3 E_3 + \lambda^4 E_4 + \cdots$$

where $E(0)$ and $E(1)$ correspond to the initial and final states respectively, and λ, assumed to be a continuous variable, is the amount of electron, which is taken out from the system. From this equation, the difference ΔE in energy between the neutral parent molecule and the ionized species, which is the experimental CEBE, is equal to:

$$\Delta E = E(1) - E(0) = E_1 + E_2 + E_3 + E_4 + \cdots.$$

Now if we define the first derivative:

$$F(\lambda) = \left(\frac{\partial E(\lambda)}{\partial \lambda}\right) = E_1 + 2\lambda E_2 + 3\lambda^2 E_3 + 4\lambda^3 E_4 + \cdots.$$

Then, in the generalized Slater's transition-state (GTS) method, ΔE is approximated as:

$$\Delta E = E(\lambda = 0) - E(\lambda = 1) = \frac{1}{4}\left\{F(\lambda = 0) + 3F\left(\lambda = \frac{2}{3}\right)\right\}$$

Several attempts showed that the best way to carry out the calculation with λ=2/3 was to perform an unrestricted computation (uGTS), i.e. removing 2/3 electron in the kth orbital of the α spin manifold while leaving an occupation of 1 in the kth orbital of the β spin manifold [18]. $F(0)$ and $F(2/3)$ are obtained via the molecular orbital (MO) energies of the partially emptied Kohn-Sham orbitals. Hence, at least (M+1) SCF calculations have to be carried out to obtain M CEBEs: the one on the parent neutral molecule, and at least M SCF calculations with a partial core-hole localized on the M desired atomic centers. The detailed mathematical procedure is described elsewhere [21, 29].

4. TEST CALCULATIONS ON GAS-PHASE MOLECULES

Figure 2 gives a plot of the computed vs. experimental C1s CEBEs for a series of 20 small molecules such as CH_4, CO, CO_2, $HCONH_2$, CH_3COOH, etc. for which the gas-phase spectra have been recorded with an accuracy better then 0.3 eV [27]. The results obtained are very good, with an aad of 0.2 eV. Results of comparable accuracy are obtained on CEBEs for elements other than carbon [27–29]. As a comparison with other methods, Fig. 3 shows a plot of computed vs. experimental C1s CEBEs for acetonitrile, propionitrile and acrylonitrile (scheme 1) between the present DFT/uGTS approach and the best ΔSCF method previously available [27c]. At the best ΔMP2 level of theory, the various CEBEs are 0.5 to 1.0 eV lower than the experimental values. This feature prevents a direct unambiguous assignment of the various lines, as the resolution of the best XPS experiments now approaches 0.2 eV. As these molecules are usually used as model systems for the interpretation of Polyacrylonitrile (PAN) XPS spectrum, it is difficult to know, with such computation methods with an accuracy lower than the resolution of the best XPS experiments (0.2 eV), whether the fairly broad high binding energy lines of PAN, acetonitrile, propionitrile and acrylonitrile can be decomposed into several structures on safe grounds, and whether intermolecular effects invoked to account for discrepancy do play a part. On contrary, with our procedure, an aad of only 0.11 eV is reached and that precision is compatible with the best experimental resolutions. The DFT/uGTS method appears more accurate for a smaller computational effort. With such a computational method, we can then hope to progress in the assignment of the several components of the C1s line for PAN.

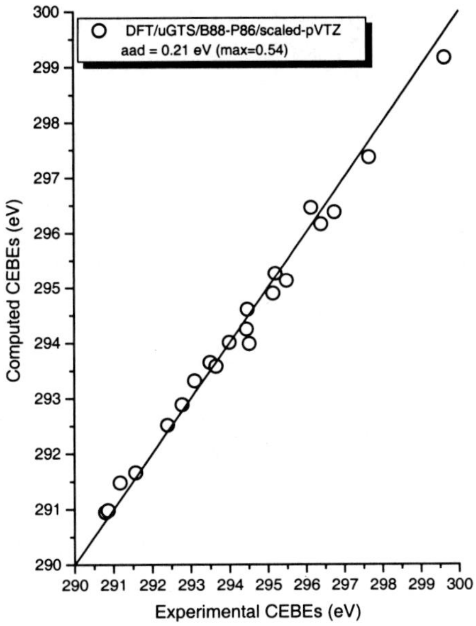

Figure 2. Computed vs experimental C1s CEBEs (in eV) for a series of 20 organic molecules of size ranging from CH_4 to propanoic acid. The solid line is the unit slope line. All values have been computed using the DFT/uGTS/B88-P86 procedure with a so-called-pVTZ basis. The average absolute deviations (aad) are given in eV in the inset.

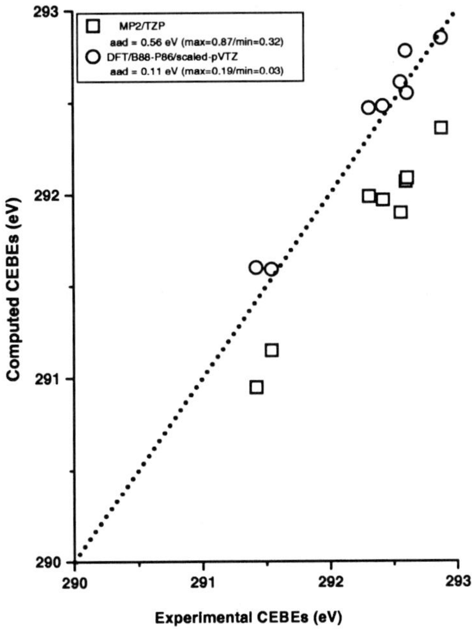

Figure 3. Computed vs experimental C1s CEBEs (in eV) for acetonitrile, propionitrile and acrylonitrile. Circles indicate DFT/uGTS values, while squares indicate ΔMP2/TZP values taken from Ref. [16]. The solid line is the unit slope line. Average maximum and minimum absolute deviations are given (in eV) in the inset.

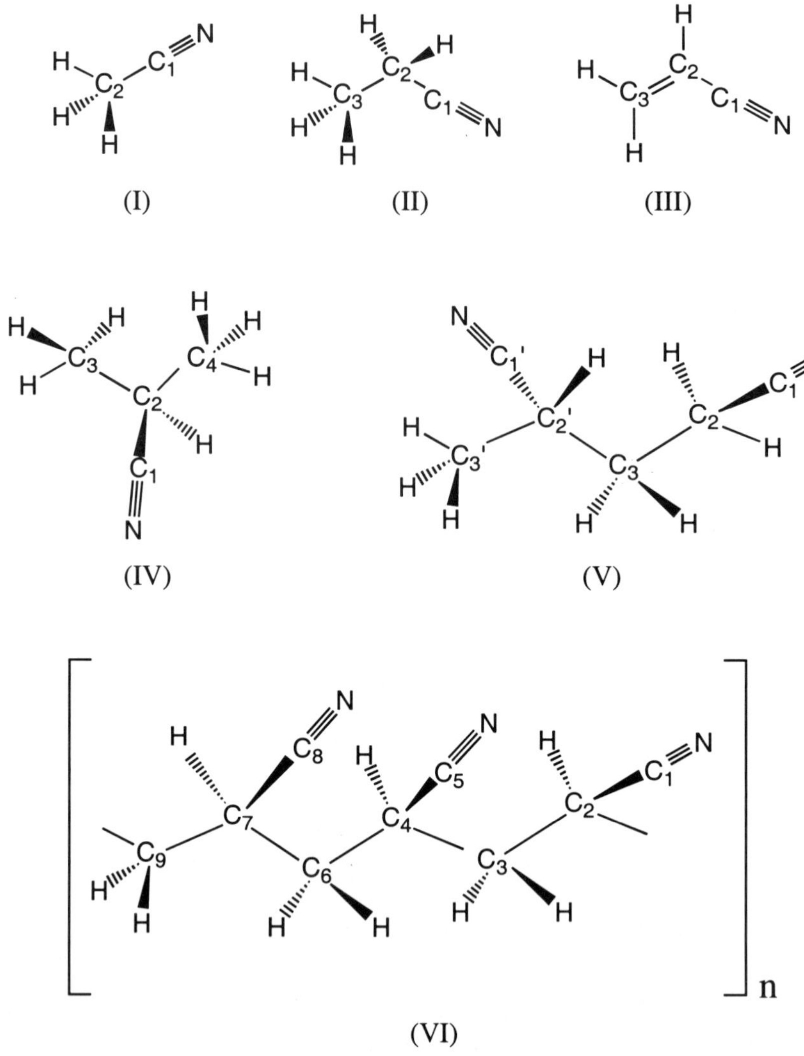

Scheme 1. Molecular structures and atom labeling for Acetonitrile (I), Propionitrile (II), Acrylonitrile (III), 2-cyanopropane (IV) and the lowest energy conformer of 2-methylglutaronitrile (V).

5. APPLICATION TO POLYMERS: POLYACRYLONITRILE (PAN)

We have used our method to help interpret the C1s line in the XPS spectrum of polyacrylonitrile, a technologically important polymer with $[-C^3H_2-C^2H(C^1N)-]_n$ as the repeat unit. The experimental XPS spectrum of the C1s region of the PAN [30] can be decomposed in three lines at 285.48, 286.35 and 286.73 eV respectively as shown in Table 1. Several assignments have been proposed in order to be in agreement with the 2/1 ratio (high/low energy components, respectively). The high energy structure has been assigned either to C^3 and C^2 [31] or to C^1 and C^2 [32]. Theoretical calculations have been carried out on model molecules to help ascertain the assignments. Propionitrile (C^3H_3-C^2H_2-C^1N -scheme 1, II) and 2-cyanopropane (C^3H_3-$C^2H(C^1N)$-C^4H_3 - scheme 1, IV) have been selected as truncated models for PAN (scheme 1, VI). Because calculations for single molecules are compared to experimental results on solid polyacrylonitrile, we must shift each computed CEBE by a quantity WD [33]. The quantity (WD) denotes the sum of the workfunction of the sample and other energy effects, such as the polarization energy, the width of the interchain band formation and the peak broadening in the solid state. The calculated CEBEs and the corresponding experimental values are shown in Table 1. One can see that C 1s CEBEs are in excellent agreement with the available experimental data with an aad from experiment of 0.09 eV with propionitrile and 0.05 eV with 2-cyanopropane. The results of the calculations have confirmed the overall assignment of Beamson and Briggs, namely that the high energy component in the C 1s region of the XPS spectrum of PAN is due to carbons C^1 and C^2, and the low energy line to carbons C^3. We also find that the difference between the high and low energy lines is 0.94 eV from propionitrile and 1.00 eV from 2-cyanopropane. These values are quite close to the experimental 1.06 eV shift of Beamson and Briggs [30].

As it seemed that using larger molecular models improved the calculated difference between the high and low energy components, so we also investigated the CEBEs of the lowest energy conformer of 2-methylglutaronitrile (scheme 1,V), taken as a dimeric model of PAN. These results are also reported in Table 1. Focusing on the three carbon atoms of 2-methylglutaronitrile which most closely resemble the C^1, C^2 and C^3 atoms in PAN (labeled $C^{1'}$, $C^{2'}$ and C^3 in scheme 1), we see that the high energy line is still due to $C^{1'}$ and $C^{2'}$, and the low energy line to C^3. However, we find that CEBE ($C^{1'}$) is lower than the CEBE ($C^{2'}$). The computed CEBEs remain in quite good agreement with experiment, with an aad of 0.15 eV. But, this good result masks, in fact, a failure in the present dimer model of PAN, as the predicted shift between the high and low energy lines is now 0.64 eV, i.e. much worse than that obtained with propionitrile or 2-cyanonitrile. Hence, as suggested by Raynaud *et al.* [34], we find that the interplay of neighboring groups (in particular the cyano groups) is not innocuous in the simulation of the XPS spectrum of PAN via truncated models.

A first conclusion that can be drawn from these results is that the DFT/uGTS is capable of predicting accurate CEBEs (0.2-0.3 eV at worse). Thus, the discrepancies of more than this value upon comparison with experiment can only be due to a wrong (or insufficient) molecular model, and not to the method of calculation itself.

In the case of PAN, we can suggest two directions to improve the quality of the results:

(i) The use of a trimer. Indeed, the C^2 and $C^{3'}$ carbons we have selected in our model do not exactly feel the same chemical environment as in PAN, since the CN group interacts with only one nearest CN group instead of two. This would show that the increase in the size of the system is indeed useful in the thorough understanding of CEBEs of PAN

(ii) One could also try to examine intermolecular effects (involving CN-CN interactions), as was done for a correct interpretation of the infrared spectra of nitriles in condensed media [35].

Table 1.
Calculated CEBEs (in eV) of propionitrile, 2-cyanonitrile and the lowest energy conformer of 2-methylglutaronitrile (scheme 1) as model molecules for polyacrylonitrile, using the uGTS/B88-P86 method with the mixed cc-pVTZ/scaled-pVTZ basis set

	Calc.[a]	Obs.[b]	a.d.[c]
Propionitrile			
WD[d] = 6.03 eV			
C^1	286.58	286.73[b]	0.15
C^2	286.44	286.35[b]	0.09
C^3	285.57	285.48	0.09
N	399.26	399.28	0.02
		aad[c]	0.09
Δ[e]	0.94	1.06	
2-cyanopropane			
WD[d] = 5.85 eV			
C^1	286.66	286.73[b]	0.07
C^2	286.40	286.35[b]	0.05
C^3	285.53	285.48	0.05
N	399.26	399.28	0.02
		aad[c]	0.05
Δ[e]	1.00	1.06	
2-methyl glutaronitrile			
WD[d] = 6.19 eV			
$C^{2'}$	286.47	286.73[b]	0.26
$C^{1'}$	286.32	286.35[b]	0.03
$C^{3'}$	285.76	285.48	0.28
N'	399.27	399.28	0.01
		aad[c]	0.15
Δ[e]	0.64	1.06	

[a]see Ref. [24]; [b]Ref.[30]: the assignment for the high energy C1s line of PAN is uncertain; [c]Absolute deviation between calculated and experimental values (eV). Aad is the average absolute deviation; [d]WD: see text; [e]Δ corresponds to the shift between the high and the low C1s components.

6. INTERMOLECULAR EFFECTS: γ-AMINOPROPYLTRIETHOXYSILANE (γ-APS)

Assuming the same viewpoint as above concerning the accuracy of our uGTS approach, we have tried to shed some light on the interpretation of the XPS spectrum of γ-APS ($H_2N\text{-}CH_2\text{-}CH_2\text{-}CH_2\text{-}Si(OCH_2CH_3)_3$), first when deposited on a silicon surface, and then after thermal treatment. γ-APS is commonly used as an adhesion promoter for polyimide on silicon wafers, in microelectronics. This compound oligomerizes upon heating so that no gas-phase reference XPS spectrum can be obtained [36, 37]. Hence, γ-APS can only be observed as deposited on a surface, i.e. already interacting with others molecules and/or with the surface. This raises some difficulties as to the possibility of safely separating various components in the decomposition of the XPS spectra of γ-APS and its by-products.

The experimental broad C1s peak (Fig. 4a)—centered at 286.10 eV—obviously stems from several contributions, as suggested by the Full Width at Half Maximum (FWHM) of 2.8 eV. Knowing the molecular structure of γ-APS, it is evident to decompose this peak into three components (FWHM ≈ 1.35 eV) corresponding to each

(a)

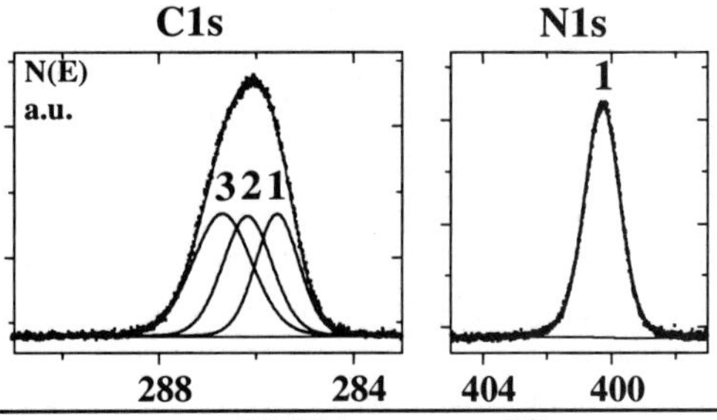

Notation Scheme 3	Energy (eV)	FWHM	G/L (%)	Normalized Area
n1	400.2	1.50	0.40	2360
c3	286.7	1.37	0.35	2560
c2	286.1	1.34	0.35	2525
c1	285.5	1.34	0.35	2530

(b)

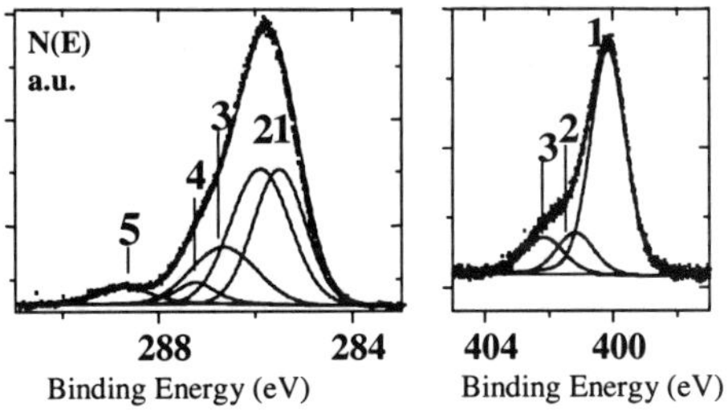

Notation Scheme 3	Energy (eV)	FWHM	G/L (%)	Normalized Area
n3	402.2	1.50	0.40	321
n2	401.2	1.50	0.44	308
n1	400.2	1.50	0.35	1725
c5	288.8	1.20	0.50	285
c4	287.2	1.20	0.47	290
c3	286.7	1.37	0.35	1783
c2	286.0	1.34	0.35	2554
c1	285.5	1.34	0.35	2592

Figure 4. (a, top) C1s and N1s regions of the XPS spectrum of γ-APS spin-coated onto a silicon wafer before thermal treatment. The tentative decompositions are those delivering the value in Table 2. (b, bottom) C1s and N1s lines of the same sample after thermal treatment under H_2O/CO_2 atmosphere. The C1s lines labeled 1–3 have the same binding energy as in (a), while lines 4–5 are new lines. The Nls line labeled 1 has the same binding energy as in (a), while lines 2–3 are new lines.

carbon atom of the molecule. In this case, however, it is the structure of the molecule which serves to interpret the XPS spectrum. This reduces considerably the a priori analytical power of the technique—and, in short, its interest—when trying to follow in situ the physico-chemical modifications of the original system under post-treatment such as thermal curing [38].

One alternative we have proposed is to make a computation of the CEBEs of the γ-APS molecule—modeled by $H_2N\text{-}CH_2\text{-}CH_2\text{-}CH_2\text{-}Si(OH)_3$,—and examine whether the observed chemical shifts are intrinsic to the molecular structure and do not result from intermolecular interactions due to the deposition of γ-APS on the silicon wafer. The results are shown in Table 2. These indicate that the three carbons of γ-APS are separated by 0.6 eV, i.e. by an amount, which is larger than the theoretical aad [39]. This suggests that the C 1s broad line can be safely decomposed on the basis of three peaks corresponding to each and every carbon center of the γ-APS molecular skeleton [39]. The experimental XPS spectrum [38] may thus be taken as a reference spectrum of original γ-APS, and used to follow the modifications upon thermal treatment.

Table 2.
Core-Electron-Binding-Energies (CEBEs, in eV) for the triethoxy-form of γ-APS, compared with a tentative three-peak decomposition of the C1s region of spin-coated γ-APS [38]. All values are referenced to the Fermi level of the spectrometer

	Obs.(3-peak fit)[a]	Obs.(fit)-C_{rel}[b]	DFT (a.d.[d])
$H_2\underline{N}CH_2CH_2CH_2Si(OH)_3$	400.30	400.20	400.29 (0.09)
$H_2N\underline{C}H_2CH_2CH_2Si(OH)_3$	286.70	286.65	286.65 (0.00)
$H_2NCH_2\underline{C}H_2CH_2Si(OH)_3$	286.10	286.05	285.94 (0.11)
$H_2NCH_2CH_2\underline{C}H_2Si(OH)_3$	285.50	285.45	285.45 (0.00)
$H_2NCH_2CH_2CH_2Si(\underline{O}H)_3$	533.50	533.31	533.75 (0.44)
aad[e]			0.13

[a] Ref. [38].
[b] C_{rel} is a relativistic correction [27].
[c] uGTS with B88-P86/cc-pVTZ [39].
[d] Absolute deviation from experimental values, in eV.
[e] Average absolute deviation from experiment, in eV.

When heated under a H_2O/CO_2 atmosphere, H_2CO_3 forms, which is thought to react with the ammonium end of γ-APS via a Brönsted-type reaction (scheme 2), giving an ammonium hydrogen-carbonate ion pair ($-CH_2\text{-}NH_3^+, HCO_3^-$). The XPS spectrum of the thermally treated sample (Fig. 4b) pretty much looks like that of γ-APS, with additional components at 287.20 eV and 288.80 eV in the C 1s line [38]. These were first attributed to $-\underline{C}H_2\text{-}NH_3^+$ and $H\underline{C}O_3^-$, respectively. The N 1s line (Fig. 4b), which was made of a unique peak at 400.10 eV (Fig. 4a), also has two additional component at 401.20 eV and 402.2 eV, attributed to the ammonium bicarbonate ion and to the ammonium silicate ion respectively (scheme 3) [38].

$$H_2N\text{-}CH_2\text{-}CH_2\text{-}CH_2\text{-}Si(OH)_3 + H_2O + CO_2$$
$$\rightleftharpoons HCO_3^- + H_3N^+\text{-}CH_2\text{-}CH_2\text{-}CH_2\text{-}Si(OH)_3$$

Scheme 2. Sketch of the Brönsted acid-base reaction which is thought to occur upon thermally treating a film of γ-APS under H_2O/CO_2 atmosphere.

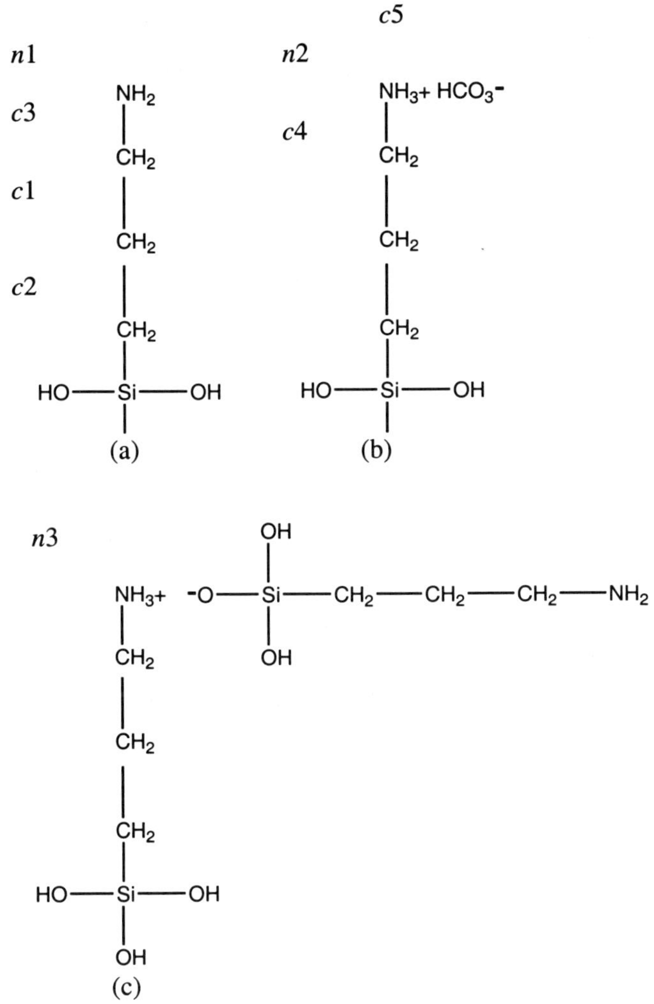

Scheme 3. Forms of γ-APS: (a) neutral, hydrolyzed form; (b) ammonium bicarbonate form; and (c) ammonium silicate form.

However, when we compute the CEBEs of CH_3-NH_3^+ and HCO_3^-, separately (case 1, Fig. 5), the aad's are 7.14 and 4.39 eV for the two additional carbons, and 5.74 eV for the presumed ammonium ion [39]. Assuming that our DFT/uGTS procedure is correct and should deliver better aad's, we improved the molecular model:

(i) First, we let the CH_3-NH_3^+ and HCO_3^- ions to actually interact, i.e. we consider the (CH_3-NH_3^+,HCO_3^-) ion-pair (case 2, Fig. 5). The aad obtained in that case is much better, 0.72 eV. This alludes to these ionic compounds as interesting candidates but for which further insights are needed.

(ii) Second, we examine whether intermolecular effects are capable of delivering favorable aad's using the ion-pair model. Thus we consider the "solvation" of the (CH_3-NH_3^+,HCO_3^-) ion-pair by an increasing higher number of water molecules in the surrounding of this ion-pair (cases 3 to 5, Fig. 5). In each case, the optimal geometry was found by simulated annealing using van der Waals plus dispersion intermolecular potentials [40].

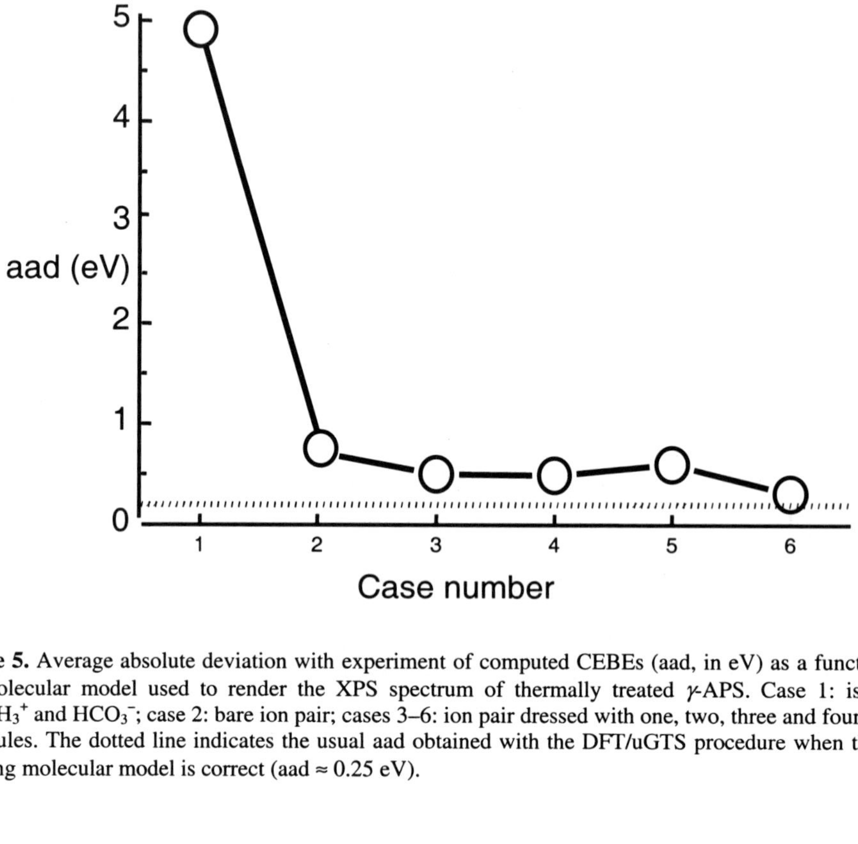

Figure 5. Average absolute deviation with experiment of computed CEBEs (aad, in eV) as a function of the molecular model used to render the XPS spectrum of thermally treated γ-APS. Case 1: isolated $CH_3NH_3^+$ and HCO_3^-; case 2: bare ion pair; cases 3–6: ion pair dressed with one, two, three and four water molecules. The dotted line indicates the usual aad obtained with the DFT/uGTS procedure when the underlying molecular model is correct (aad ≈ 0.25 eV).

The solvation effects illustrated here with water molecules may be achieved by (i) true water molecules which remain in the γ-APS film even after the solvent has been evaporated, (in analogy with crystallization water in crystals); (ii) OH groups from silanols ($Si(OH)_3$) or alcohol (CH_3CH_2OH) functions of vicinal fragments resulting from the hydrolysis of γ-APS, or (iii) siloxane Si-O-Si groups resulting from oligomerization of γ-APS molecules. All these groups have in common that they bear an oxygen atom, the lone pair of which is capable of acting as a Lewis base towards the ammonium end of protonated γ-APS (hence lowering its N 1s CEBE), while acting as a Lewis acid toward HCO_3^-, and hence lowering its C 1s CEBE as well. The results summarized graphically for cases 3 to 5 show that the computed CEBEs get closer and closer to experimental values as the number of water molecules increases (while deviating by more than 0.3 eV even with the $[CH_3\text{-}NH_3^+,HCO_3^-;(H_2O)_4]$ system). The H_2O molecules tend to surround the $CH_3\text{-}NH_3^+$ cation and stay closer to the nitrogen atom than any other atom, while their hydrogen atoms point toward the HCO_3^-anion. Thus, the water molecules are ideally oriented to raise the electron density around the ammonium nitrogen (which lowers its N 1s CEBE), while withdrawing electrons from the HCO_3^- anion due to hydrogen bonds (which raises its C 1s CEBE as compared to the isolated ion-pair).

These results suggest that the resulting characteristics of the core-XPS spectrum of thermally treated γ-APS cannot be explained properly if intermolecular effects are not considered. One alternative is that the solvation effects illustrated here with water molecules may be achieved by -Si(OH)_3 groups of vicinal γ-APS molecules. However, our findings indicate that CEBEs are—in this case—clearly sensitive to intermolecular effects. This rather new aspect is due to the interplay of charged species, for which Coulombic and polarization effects are long-range.

7. EXTENSION TO ORGANIC MOLECULES ADSORBED ON METALLIC SURFACES

We have also proposed a new method of performing the direct calculations of CEBEs of small molecules chemisorbed on metallic surfaces [41]. In this method, the surface chemisorption site is described by a small cluster. The use of cluster models for the surface sites appears to us as a good compromise, as it allows the use of accurate quantum chemistry tools for both the adsorbate and the metal. Practically, we cannot expect reliable metal workfunctions using aggregates unless these reach a size of about 100 atoms [39, 40]. For this reason, we propose a new approach for CEBEs calculations via a formal two-step path. Let us decompose the core-ionization process in the following way:

(i) first step promotes an electron from the desired core-level X to the conduction band (c.b.) of the metal. This step builds an excited state and involves an energy change equal to $\Delta E(X \rightarrow c.b.)$.

(ii) The remaining path involves an extraction of the excited electron from the conduction band of the metal to vacuum, with zero kinetic energy at infinity. This step involves an energy change equal to $\varphi^X\,(c.b \rightarrow vac.)$, where the superscript X indicates that the adsorbate has a core-hole on center X. This term can be replaced by the actual workfunction ϕ^{M} of the metal, in the limit of the low adsorbate coverage [44–47]. Hence, the overall non-relativistic CEBE, $I_{nr}(X)$, for X becomes:

$$I_{nr}(X) = \Delta E(X \rightarrow c.b.) + \phi^M.$$

We have tested the value of this approximation in the case of CO atop- and bridge-chemisorption on Ni, Pd(100) and Pd(110) within the DFT framework [41]. Our predicted CEBEs show an average absolute deviation (aad) from experimental values of

0.75 eV to 0.55 eV for clusters bearing one to four atoms. The first tests we have performed on Pd(100) and Pd(110) using larger clusters (up to four atoms) indicate that the results of our equation are less cluster-size dependent and converge faster than using direct ΔSCF calculations (Table 3). In the present case, aad's as low as 0.23 eV are obtained with Pd_4 clusters modeling Pd(100) and Pd(110), suggesting that the method is adequate for quantitative predictions at an acceptable computational cost.

Table 3.
Theoretical vs. Experimental C1s and O1s binding energies (in eV) for CO adsorbed on Pd(100) and Pd(110). The values in parentheses indicate the absolute deviation (in eV) with respect to the experimental values. All calculated values are corrected for relativistic effect and referenced to the vacuum

		C 1s			O 1s	
Cluster	Exp. [47]	ΔSCF [40]	our DFT/uGTS procedure[a] (a.d[b])	Exp. [47]	ΔSCF [40]	our DFT/uGTS procedure[a] (a.d[b])
CO/Pd (100)	291.20			536.90		
Pd_2-CO		292.89 (1.69)	291.50 (0.30)		538.12 (1.22)	536.88 (0.02)
Pd_4-CO		292.80 (1.60)	291.02 (0.18)		538.60 (1.70)	536.68 (0.22)
CO/Pd (110)	291.10			536.80		
Pd_2-CO		292.89 (1.79)	291.5 (0.30)		538.12 (1.32)	536.88 (0.08)
Pd_4-CO		292.35 (1.25)	291.09 (0.01)		538.20 (1.40)	536.77 (0.03)

[a] see details in ref. [41].
[b] Absolute deviation from experimental values, in eV.

8. CONCLUDING REMARKS

The DFT/uGTS procedure is capable of delivering computed CEBEs of experimental quality on a known structure, with aad's which are of the order of experimental resolution (0.2 eV). By obtaining a systematic agreement of the order of 0.2 to 0.3 eV between theory and experiment we can use this technique as a "probe" to help interpret the molecular structure of real systems, which are not precisely known. So, we decided to rely completely on the calculation procedure to help us scan various plausible structures mimicking the presumed structural changes. Thus the strategy followed is: (i) we propose plausible molecular models and use our procedure to obtain predicted CEBEs, as our procedure is meant to be as precise as experiment, (ii) as soon as we have found the correct underlying molecular structures, we shall consider as nonplausible those molecular models which give too large discrepancy between the theory and experiment.

Using this procedure, we have been able to identify unexpected weak effects in XPS spectra. In particular, we have shown that intermolecular effects such as ion-pairing, dipolar interactions and hydrogen bonding [48] within a polymer phase can be separated from intrinsic structural effects. In this way, the spectral modifications accompanying morphological and/or structural changes can easily be identified. In the case of the γ-APS study, we have shown that the ion-pairing alone cannot deliver aad's lower than about 0.70 eV, and that interactions with at least second nearest neighbors are mandatory to lower the aad to about 0.30 eV (either in the form of clusters of ion-pairs or in the form of "solvated" ion-pairs). This suggests that these intermolecular effects, which should involve purely Coulombic and polarization terms, can be responsible for shifts of as high as 0.4–0.5 eV.

Moreover, the theoretical binding energies can be obtained with moderate computational effort, taking advantage of the intrinsic virtues of DFT as regards the increase of CPU time as a function of the size of the system. This offers much hope as to the usefulness of the method to study large systems such as polymers or molecules adsorbed on metallic surfaces. In this last field, first attempts were in encouraging agreement with the experimental values. The new approach developed for metallic clusters fulfills our need for an efficient, accurate (in relation to experimental resolution) and yet computationally acceptable procedure for determining core electron binding energies for chemisorbed molecules. In short, we can say that this tool should find particular interest in the field of polymers and, in general, solid interfaces.

REFERENCES

1. See for example *Proc. ECASIA'97:* I. Olefjord, L. Nyborg and D. Briggs (Eds). John Wiley & Sons, 1997.
2. F. M. Fowkes, *J. Adhesion* **4**, 155 (1972).
3. F. M. Fowkes, *J. Adhesion Sci. Technol.* **1**, 7 (1987).
4. K. L. Mittal and H. R. Anderson, Jr. (Eds), *Acid-Base Interactions: Relevance to Adhesion Science and Technology.* VSP, Utrecht, The Netherlands (1991).
5. R. G. Parr and W. Yang, *Density Functional Theory of Atoms and Molecules.* Oxford University Press, New York, 1989.
6. W. Yang and R. G. Parr, *Proc. Natl. Acad. Sci. USA* **82**, 6723 (1985).
7. M. Berkowitz and R. G. Parr, *J. Chem. Phys.* **88**, 2554 (1988).
8. L. H. Lee, in: *Acid-Base Interactions: Relevance to Adhesion Science and Technology*, K. L. Mittal and H. R. Anderson, Jr. (Eds), pp. 25–46. VSP, Utrecht, The Netherlands (1991).
9. L. H. Lee, in: *Fundamentals of Adhesion*, L. H. Lee (Ed.). Plenum Press, New York (1990).
10. C. Bureau, G. Deniau, P. Viel, and G. Lécayon, *J. Adhesion* **58**, 101 (1996).
11. J. Charlier, V. Detalle, C. Bureau, F. Valin, and G. Lécayon, *J. Adhesion* **66**, 275 (1998).
12. J. Charlier, V. Detalle, C. Bureau, F. Valin, and G. Lécayon, *J. Vac. Sci. Technol., Part A* **15**, 353 (1997).
13. (a) S. Hüfner, in: *Photoelectron Spectroscopy*, M. Cardona, P. Fulde, K. von Klitzing, and H.-J. Queisser (Eds). Springer Series in Solid-State Physics 82, Springer-Verlag, Berlin, Heidelberg (1995); (b) see for example: "*Practical Surface Analysis: Vol. 1: Auger and X-Ray Photoelectron Spectroscopy*", second edition, D. Briggs and M. P. Seah (Eds). J. Wiley and Sons, Chichester, England (1990).
14. (a) S. W. Gaarenstroom and N. Winigrad, *J. Chem. Phys.* **67**, 3500 (1977); (b) C. Battistoni, J. L. Dormann, D. Fiorani, E. Paparazzo and S. Viticoli, *Solid State Commun.* **39**, 581 (1981); (c) J. C. Klein and D. M. Hercules, *J. Catalysis* **82**, 424 (1983).

15. See for example: C. S. Fadley, in: *Electron Spectroscopy: Theory, Techniques and Applications*, C. R. Brundle and A. D. Baker (Eds), Vol. 2, p. 75. Academic Press, London (1977).
16. (a) U. Gelius, *Phys. Scripta* **9**, 133 (1974); (b) A. Naves de Brito, H. Svensson, H. Agren, and J. Delhalle, *J. Electron Spectrosc. Relat. Phenom.* **63**, 239 (1993) and references therein.
17. M. M. Chehimi, J. F. Watts and W. K. Eldred, *J. Mater. Chem.* **4**, 305 (1994).
18. R. B. Borade, A. Adnot and S. Kaliaguine, *J. Chem. Soc., Faraday Trans.* **86**, 3949 (1992).
19. M. Huang, A. Adnot and S. Kaliaguine, *J. Catalysis* **137**, 322 (1992).
20. J. Xie, M. Huang and S. Kaliaguine, *Appl. Surf. Sci.* **115**, 157 (1997).
21. D. P. Chong, *Phys. Lett.* **232**, 486 (1995).
22. T. Koopmans, *Physica* **1**, 104 (1933).
23. M. E. Schwartz, in: *Application of Electronic Structure Theory*, H. F. Schaeffer (Ed.) Plenum Press, New York(1977).
24. C. Bureau, D. P. Chong, G. Lécayon, and J. Delhalle, *J. Electron Spectrosc. Relat. Phenom.* **83**, 227 (1997).
25. J. C. Slater, *Adv. Quantum Chem.* **6**, 1 (1972).
26. A. R. Williams, R. A. de Groot, and C. B. Sommers, *J. Chem. Phys.* **63**, 628 (1975).
27. (a) D. P. Chong, *Can. J. Chem.* **74**, 1005 (1996); (b) D. P. Chong, C.-H. Hu, and P. Duffy, *Chem. Phys. Lett.* **249**, 491 (1996); (c) C. Bureau, and D. P. Chong, *Chem. Phys. Lett.* **264**, 186 (1997).
28. P. Duffy and D. P. Chong, *Org. Mass. Spectr.* **28**, 321 (1993).
29. C. Bureau and D. P. Chong, in: *Proceedings of the Second International Conference on Polymer-Solid Interfaces: From Model to Real Systems,* J. J. Pireaux, J. Delhalle, and P. Rudolf (Eds). Presse Universitaire de Namur, Namur, Belgium (1996).
30. G. Beamson and D. Briggs, *High Resolution XPS of Organic Polymers.* J. Wiley and Sons, Chichester, UK, 1992.
31. T. Fujikawa, T. Ohta and H. Kuroda, *Bull. Chem. Soc. Jpn.* **49**, 1486 (1976).
32. O. A. Barry, B. Riedl, A. Adnot and S. C. Kaliaguine, *J. Electron Spectrosc. Relat. Phenom.* **57**, 47 (1991).
33. J. Delhalle, J. M. André, S. Delhalle, J. J. Pireaux, R. Caudano, and J. J. Verbist, *J. Chem. Phys.* **60**, 595 (1974).
34. M. Raynaud, J. Riga, C. Reynaud and Y. Ellinger, *J. Electron. Spectrosc. Relat. Phenom.* **53**, 251 (1991).
35. D. Mathieu, PhD Thesis, Université de Grenoble, France (1993).
36. E. P. Plueddemann, in: *Fundamentals of Adhesion*, L.-H. Lee (Ed.), p. 279. Plenum Press, New York (1990).
37. (a) E. P. Plueddemann, *J. Adhesion Sci. Technol.* **5**, 261 (1991); (b) H. Linde and R. T. Gleason, *J. Polym. Sci. Chem.* **22**, 3034 (1984); (c) L. P. Buchwaller, *J. Adhesion Sci. Technol.* **4**, 697 (1990); (d) V. A. Ogarev and S. L. Selector, *Prog. Org. Coat.* **21**, 135 (1992).
38. I. George, P. Viel, C. Bureau, J. Suski, and G. Lécayon, *Surf. Interf. Anal.* **24**, 774 (1996).
39. S. Kranias, C. Bureau, D. P. Chong, V. Brenner, I. George, P. Viel, and G. Lécayon, *J. Phys. Chem. B* **101**, 10 254 (1997).
40. O. Hess, M. Caffarel, J. Langlet, J. Caillet, C. Hiszoon and P. Claverie, in: *Proc. Physical Chemistry in Modeling of Molecular Structures and Properties in Physical Chemistry and Biophysics.* J. L. Rivail (Ed.). Elsevier, Amsterdam, (1990).
41. C. Bureau, *Chem. Phys. Lett.* **269**, 378 (1997).
42. E. K. Parks, T. D. Klots and S. Riley, *J. Chem. Phys.* **92**, 3813 (1990).
43. M. B. Knickelbein, S. Yang and S. Riley, *J. Chem. Phys.* **93**, 94 (1990).
44. V. Russier, D. R. Salahub and C. Mijoule, *Phys. Rev. B* **42**, 5046 (1990).
45. V. Russier and C. Mijoule, *J. Phys. Condens. Matter.* **3**, 3193 (1991).
46. C. Mijoule, M. Filali Baba and V. Russier, *J. Mol. Catal.* **83**, 367 (1993).
47. L. Pedocchini, M. R. Ji, G. Comelli and G. Rovida, *Surf. Rev. Lett.* **2**, 563 (1995).
48. C. Bureau, J. Charlier, J. Cousty and Z. Xi, accepted in *Langmuir.*

Acid-Base Interactions, Vol. 2, pp. 223–234
Ed. K.L. Mittal

Acid–base interactions at the molecular level: Adhesion and friction studies with interfacial force microscopy

A. R. BURNS,* J. E. HOUSTON, R. W. CARPICK and T. A. MICHALSKE

Surface and Interface Sciences Department, Sandia National Laboratories, Albuquerque, NM 87185-1413, USA

Abstract—To examine the forces of acid–base adhesion interactions at a molecular level, we have utilized the scanning probe Interfacial Force Microscope (IFM). Unlike cantilever-based atomic force microscopes, the IFM is a non-compliant, mechanically stable probe that provides a complete adhesion profile without jump-to-contact. In this way, we are able to quantitatively measure the work of adhesion and bond energies at well-defined, nanometer-scale single asperity contacts. In particular, we will discuss the displacement-controlled adhesion forces between self-assembled monolayers of functionalized alkanethiols strongly bound to a gold substrate and a similarly functionalized tip. We also discuss a method utilizing decoupled lateral and normal force sensors to simultaneously observe the onset of both friction and chemical bond formation. Measurements show that friction can be directly attributed to bond formation and rupture well before repulsive contact.

Keywords: Adhesion; friction; scanning probe microscopy; interfacial forces; interfacial force microscopy.

1. INTRODUCTION

In the last ten years, the field of scanning probe force microscopy of surfaces has grown enormously. The primary factors in this growth stem from the realization that attractive and repulsive forces between the probe tip and the sample surface can be utilized to provide not only a "feedback" signal for detailed nanometer-scale topographic imaging, but also to provide a window into those molecular force interactions with nN sensitivity and unprecedented displacement and lateral resolutions. Atomic force microscopes (AFM) [1–3] and interfacial force microscopes (IFM) [4, 5], can detect weak van der Waals forces as well as strong bonding interactions. Fundamental studies of friction and tribology have undergone a considerable rebirth due to the ability of scanning probe microscopes to detect lateral forces simultaneously with normal forces [6–8], although the fundamental relationship between friction and adhesion remains elusive.

* To whom correspondence should be addressed: E-mail: aburns@sandia.gov

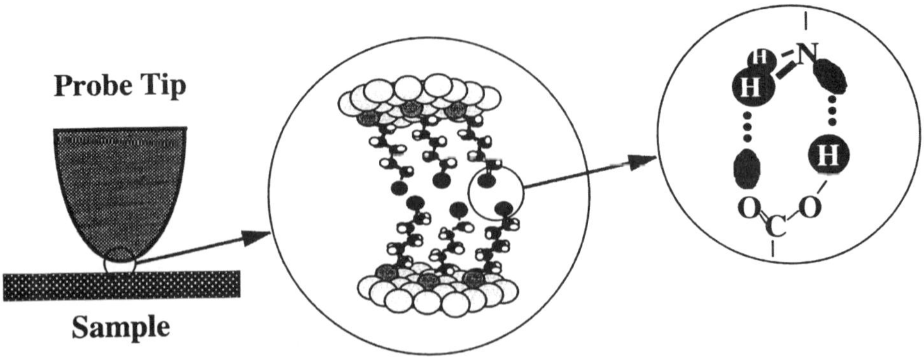

Figure 1. Schematic representation of scanning probe studies of (acid–base) hydrogen bonding between functional groups terminating alkanethiol monolayers. The thiol monolayers self-assemble on gold substrates, thus both tip and sample are initially coated with gold.

Chemical modification of the sample and/or scanning probe tip with hydrophobic or hydrophilic functional groups has been shown to be a sensitive way to probe "chemical forces" and to map the location of functional groups [5, 9–11]. An extremely effective and reproducible way of tailoring functional groups for fundamental studies of adhesion is to use self-assembled monolayers of n-alkanethiols [12, 13] that are terminated with various tail groups such as CH_3, COOH, OH, NH_2. A schematic representation is shown in Fig. 1. Alkanethiols have also been used extensively as model "lubricants" because they can be tailored to examine dissipative losses as a function of "mechanical" parameters such as chain length, packing density [7, 14], and cross-linking [15], as well as "chemical" parameters such as molecular adhesion.

In this paper we will discuss how the IFM can be used to study adhesion interactions at the molecular level. Extensive reviews of cantilever-based AFM studies can be found in Refs. [7, 10]. The IFM was developed [4] to overcome the inherent mechanical instabilities of AFM cantilevers when making ("jump-to-contact") and breaking ("pull-off") interfacial contacts. As shown in Fig. 2, these instabilities arise when the attractive force gradient of the tip-sample interaction exceeds the spring constant of the cantilever. Much of the detail in the adhesion force-distance profile is lost when these mechanical instabilities occur, and the contact itself always has some repulsive character [16]. Another concern is the phenomenon of adhesion hysteresis, which is the reason that we show *two* force *vs*. displacement curves in Fig. 2 instead of a single curve. The causes of hysteresis go beyond the scope of this paper, but can be attributed to irreversible mechanical, chemical, and/or structural effects [17]. On approach, the jump-to-contact can completely obscure the depth of the initial, bond-forming adhesion interaction; whereas on withdrawal, the much stronger adhesion interaction is observed directly as the pull-off force. In fact, it is the pull-off force that is commonly used to determine the work of adhesion (see Equation (1) in next section). Thus we see that in the case of significant adhesion hysteresis, AFM analysis of adhesion energies must be viewed with caution. Furthermore, mechanical impact resulting from jump-to-contact may significantly alter the character of the adhesion interface. It is clearly desirable to have a mechanically stable, displacement-controlled

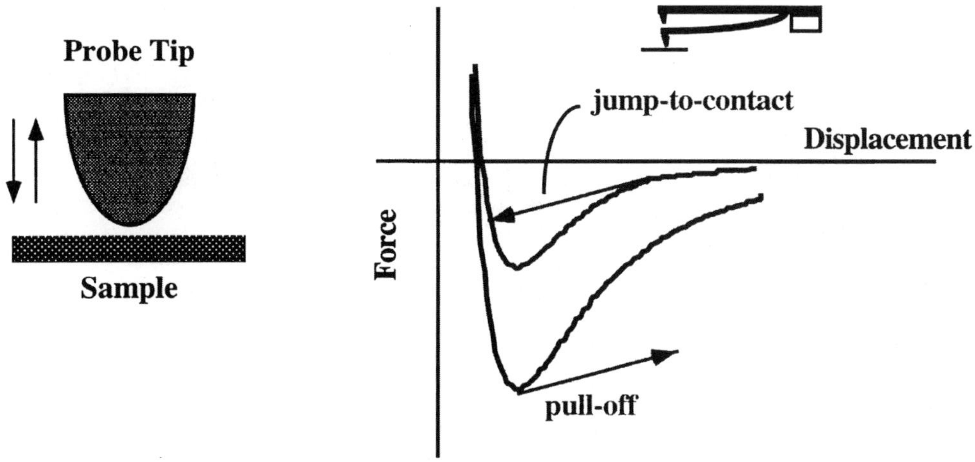

Figure 2. Schematic representation of force profile during tip approach to sample and during withdrawal after repulsive contact. Cantilevered AFM tips shown at top will suffer from "jump-to-contact" if the approach force gradient exceeds the cantilever spring constant (slope of arrow). Similarly, the cantilever will exhibit the "pull-off" instability during withdrawal from contact. Note that the approach and withdrawal curves are not the same if there is significant adhesion hysteresis.

method to measure the tip-sample adhesion interaction. While we exclusively discuss adhesion force vs. displacement experiments here, we still retain the advantages of nanoscale resolution and imaging capabilities of scanning probe microscopy.

2. INTERFACIAL FORCE MICROSCOPY

The basic idea underlying an IFM is shown in Fig. 3. Details concerning operation of the sensor may be found elsewhere [4, 18], and are only briefly summarized here. An IFM sensor consists of a teeter-totter-like common plate of a differential capacitor suspended by torsion bars above two identical gold pads fixed in a glass substrate, creating capacitances C_1 and C_2 in Fig. 3. The position of the common plate is determined by an rf bridge circuit and is initially balanced by static voltages applied to the capacitor pads. The balanced sensor is placed under proportional-integral-derivative feedback control by a circuit that supplies voltages to one pad to counteract any error signal due to normal forces applied by the tip. The normal force can be directly calculated with reasonable accuracy and precision (±20%) from the feedback voltage, the capacitances, and the static voltages, or it can be directly calibrated (with greater accuracy and precision) using a laboratory microbalance [4, 18]. Unlike AFM cantilevers, the sensor is mechanically stable and non-compliant. Scanning and displacement-controlled approaches and withdrawals are performed with standard piezoelectric devices.

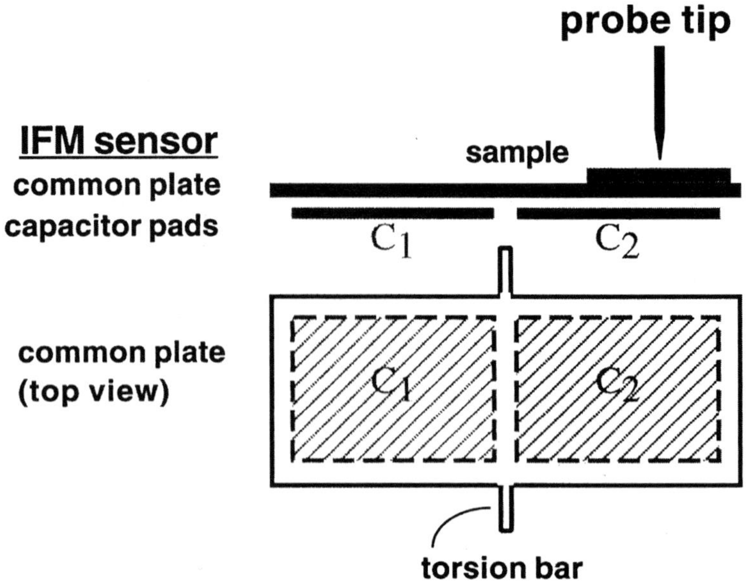

Figure 3. Schematic of interfacial force microscope (IFM). A probe tip is brought into contact with a sample resting on an IFM sensor. The sensor measures the attractive and repulsive normal forces on the tip by maintaining an electrostatic balance of two capacitors C_1 and C_2, formed by the common plate and identical gold pads fixed on a glass substrate (not shown).

Representative force-displacement curves [5] in the tip/sample adhesion interaction are shown in Fig. 4. These experiments were performed in an atmosphere of dry nitrogen. For these data, various pairs of thiol tail groups were prepared on the gold-coated probe tips and samples by 24-hour exposure to 0.5 mM solutions of specific n-alkanethiols $X(CH_2)_{11}SH$, where $X{=}CH_3$, NH_2, COOH. The force profiles were measured at a displacement rate of 8 nm/s, and the zero of displacement was arbitrarily set as the point where the IFM signal (load) goes through zero prior to positive (net repulsive) loading. It was also convenient for further evaluation to normalize the adhesion forces relative to the tip radius (Equation (1)) in order to scale results for the different tip radii measured beforehand by scanning electron microscopy and shown in parentheses. Qualitatively, one can see that a complete range of adhesion interaction between tip and sample was obtained for both approach and withdrawal, with no evidence of mechanical instabilities. The adhesion forces increase considerably in the expected order as one progresses from the CH_3 *vs.* CH_3 van der Waals interaction to the full two-center acid–base bonding in NH_2 *vs.* COOH.

Quantitative analysis of the force-displacement curves is based on the following. From contact mechanics, the work of adhesion W_A between the parabolic tip with radius R and the sample is a function of the critical load L_c (usually negative), or peak attractive force, at pull-off:

$$W_A = \frac{-L_c}{\pi C R}. \tag{1}$$

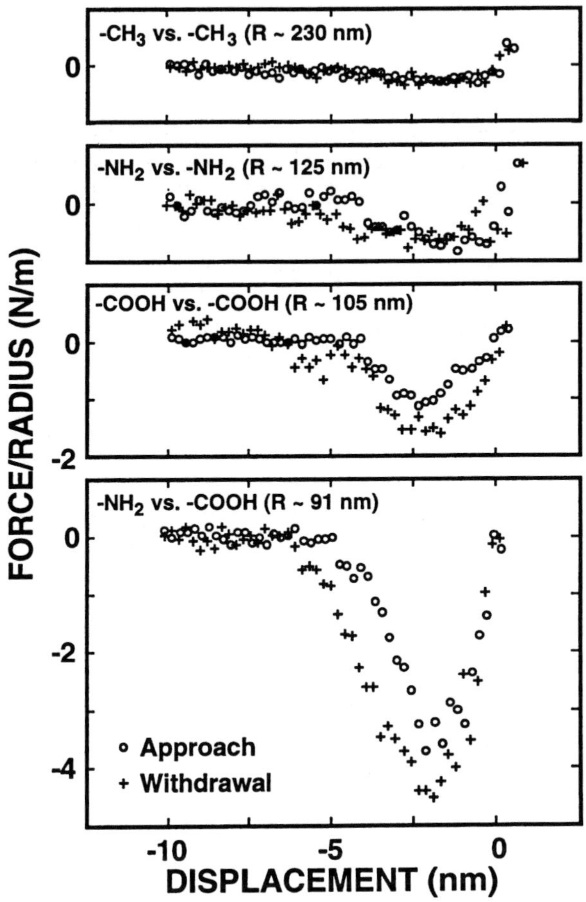

Figure 4. Force-displacement approach (O) and withdrawal (+) curves for various pairings of functional terminal groups of alkanethiols (from Ref. [5]). From top to bottom these are CH_3 *vs.* CH_3, NH_2 *vs.* NH_2, COOH *vs.* COOH, and COOH *vs.* NH_2. The force axis is normalized to the probe radius R, thus proportional to the work of adhesion (Equation (1)). Zero displacement is arbitrarily chosen to be the point where the force goes through zero at contact (far right). Approach is from left to right.

The constant $C = 3/2$ in the Johnson, Kendall, Roberts (JKR) [19] model of adhesion contact mechanics, whereas $C = 2$ in the Derjaguin, Muller, Toporov (DMT) [20] model. Much controversy has surrounded the applicability of one model over the other and scanning probe experiments often require elements of both. We use the JKR model here because it correctly predicts a finite contact area at the critical load L_c, while the DMT model predicts zero contact area at L_c. This becomes important in the next section where we discuss friction *vs.* load. In Fig. 4, (L_c/R) is usually larger for the withdrawal relative to the approach because of adhesion hysteresis. It is well known from the Dupré equation (Equation 2) [17] that the work of adhesion W_A is a function of the surface free energies γ_1 and γ_2 of the contacting surfaces and the interfacial free energy γ_{12}:

$$W_A = \gamma_1 + \gamma_2 - \gamma_{12}. \qquad (2)$$

For identical pairs of interfaces, $\gamma_1 = \gamma_2$ and $\gamma_{12} = 0$; hence, W_A becomes the work of cohesion $2\gamma_1$ and we obtain the surface free energies γ_1 shown in Table 1. The CH_3 *vs.* CH_3 pair is expected to have a purely van der Waals (dispersive) interaction, with each monolayer having a surface free energy of ~ 25 mJ/m^2 [17]. We see that the result is too high (it would be much closer to the expected value if $C = 2$). The other two identical pairs indicate increasing surface free energies due to hydrogen-bonding contributions. Fowkes [21] realized that contributions to the surface free energy could be additive, i.e.

$$\gamma_1 = \gamma_1^d + \gamma_1^h, \qquad (3)$$

where the first term in Equation (3) is the non-directional dispersion forces contribution, and the second term is the directional hydrogen-bonding contribution.

Table 1.
Summary of IFM adhesion experiments

Interface	W_A (mJ/m^2)	γ_1 or γ_{12} (mJ/m^2)	H-bonding (mJ/m^2)	Bond energy (kcal/mole)	Gas phase* (kcal/mole)
CH_3 vs. CH_3	80±43	40±21 (γ_1)			
NH_2 vs. NH_2	133±32	67±16 (γ_1)	53±32	1.6±1.0	3.4
COOH vs. COOH	304±72	152±36 (γ_1)	224±72	6.9±2.2	14.0
COOH vs. CH_3	80±43	0 (γ_{12})			
COOH vs. NH_2	907±83	-688±41 (γ_{12})	688±41	21.1±1.3	

* From Ref. [22].

To estimate the latter for the pairs NH_2 *vs.* NH_2 and COOH *vs.* COOH we simply subtract the 80 mJ/m^2 dispersion contribution measured from the hydrophobic CH_3 *vs.* CH_3 pair (see Table 1) which has no bonding; i.e., $\gamma_1 = \gamma_1^d$ = 40 mJ/m^2. We show below that the dispersive component is the same for the COOH and CH_3 tail groups. The homologous pairs indicate very weak hydrogen bonding in the NH_2 vs. NH_2 case and relatively stronger bonding in COOH *vs.* COOH case. In order to compare the results in Table 1 with gas-phase values, we convert the bond energy per unit area (mJ/m^2) to kcal/mole by assuming a uniform molecular spacing of 21.4 Å^2/molecule. The gas-phase energetics for dimerization of NH_2 *vs.* NH_2 and COOH *vs.* COOH are, respectively, 3.4 kcal/mole and 14 kcal/mole [22], which are roughly twice that estimated here. However, the gas-phase value for the latter represents two bonds per dimer. It is possible that the discrepancy may be due to hydrogen-bonding between NH_2 (or COOH) groups within the same monolayer and/or steric constraints for close-packed monolayers that do not permit optimum bonding orientation.

The work of adhesion of hydrocarbon/polar interfaces such as CH_3 *vs.* COOH can be estimated from the purely dispersive contributions [17, 21] through

$$W_A \approx 2\sqrt{\gamma_1^d \gamma_2^d}\ . \tag{4}$$

Since W_A is the same for the CH_3 *vs.* COOH pair and the CH_3 *vs.* CH_3 pair, the van der Waals components for the two tail groups are indistinguishable and thus no other interaction was significant. Since $W_A = 2\gamma_1^d$ for the CH_3 *vs.* CH_3 pair, we obtain from Equation (4) that $W_A = 2\gamma_1^d$. It is important to note that analogous experiments [9, 10] performed in a liquid such as water or ethanol will have three interfaces and thus very different results.

We finally obtain evidence for a true acid/base interaction with the pair COOH *vs.* NH_2. The large work of adhesion for this pair in Fig. 4 is due to the interfacial free energy γ_{12}. From Equation (2) and the values of γ_1 and γ_2 obtained from Table 1, we estimate γ_{12} = -688 mJ/m^2. We saw from the CH_3 *vs.* COOH pair that van der Waals interactions could be neglected in γ_{12}, thus we attribute the entire amount to hydrogen bonding. This value corresponds to ~21 kcal/mole, which is certainly in the range of strong hydrogen bonding [17], and is consistent with the increased polarity of the two-center interaction (see Fig. 1) relative to the COOH *vs.* COOH dimer.

Thus we have shown that the IFM technique is sensitive to the strength of interfacial bonding as well as being capable of monitoring the entire adhesion interaction. As pointed out elsewhere [10], a distinct advantage of scanning probe adhesion studies is the ability to probe the surface energies of high free energy surfaces. These materials are not amenable to contact angle analysis because they are wet by most liquids. Furthermore, contact angle measurements are macroscopic and thus do not necessarily account for surface roughness [17]. However, an important issue in both AFM and IFM studies is the presence of trace water monolayers condensed from the atmosphere, even when the conditions are "inert". Water layers can potentially reduce surface energies, particularly for strongly hydrophilic surfaces. The results discussed here thus represent a lower bound to the binding energies.

3. FRICTION MEASUREMENTS WITH INTERFACIAL FORCE MICROSCOPY

Scanning probes can also be used in sliding contacts to measure friction on the nanometer scale. The effects of adhesion on the *lateral* forces experienced by scanning probes has been an active area not only because of model lubricant studies, but also because this is the basis for "chemical force" imaging of functional groups [7, 9, 10]. Unfortunately, the mechanical instabilities of cantilever-based measurements tend to obscure the exact relation between friction, interfacial bonding, and applied load. That is because the friction forces are generally measured only in the repulsive regime of the contact, or for a limited range preceding pull-off that has both repulsive and attractive components [16]. This situation is exacerbated when chemical modification of the free tail groups leads to stronger adhesion forces at the interface and hence much greater force gradients [9, 10]. Finally, the finite mechanical coupling between the normal and lateral force sensing modes of a typical AFM cantilever [23] must also be taken into account in examining the friction-load relation of model lubricants.

In this section, we describe a new way of examining frictional forces over the *entire* adhesion interaction between a probe tip and the sample surface that not only avoids the inherent mechanical instabilities of AFM cantilevers, but also completely decouples normal and lateral force sensors. The method is based on bringing together the techniques of shear-force microscopy [24, 25] with the IFM. In this arrangement, frictional forces acting to dampen the tip's lateral motion are monitored independently as a function of both positive (repulsive) and negative (attractive) loads. Thus the adhesion nature of friction at the earliest stages of contact can be clearly characterized. Experimental details have been published elsewhere [26], but the basic design is to mount the probe tip shown in Fig. 3 on a small "dither" piezo tube, which vibrates it laterally. The probe is made of a glass fiber which provides a fairly high figure of merit $Q = m\omega/R \approx 100$, where ω is the natural resonance frequency (25–50 kHz), m is the fiber mass, and R is the damping out of contact. The diameter of the probe tip is reduced to <100 nm by using a commercial fiber puller [27]. Fiber motion induces voltages on the piezo tube, which are detected on the remaining quadrants by a phase-sensitive lock-in amplifier [28]. Attenuation of the fiber amplitude upon interaction of the probe tip with the surface is the basis for our friction measurements. The shear force damping (friction force) is proportional to the quantity $(1-V/V_0)$, where V is the attenuated signal at a given displacement and V_0 is the unattenuated signal due to the free lateral amplitude prior to contact [26]. For an initial 12 nm free amplitude, we estimate that the frictional force at complete damping is 18 nN [26]. By keeping the initial free amplitude constant, we in effect use the *same range of friction force* (0–18 nN) for each sample. (For much higher normal loads and lateral forces, another technique has been recently developed involving lateral force detection by the IFM itself [29].)

We have studied the behavior of two thiols discussed above which have identical chain lengths, but differing tail groups: $CH_3(CH_2)_{11}SH$ (hereafter called "CH_3-thiol"), and $COOH(CH_2)_{11}SH$ (hereafter called "COOH-thiol"). Although the chemistry of the bare glass tip can be modified in principle, we choose here to let the native glass silanol groups interact with the molecular monolayers; therefore, our functional group "pairings" are CH_3 *vs.* OH and COOH *vs.* OH. All of the experiments were conducted in a filtered, dry nitrogen atmosphere where the relative humidity was less than 8%, and no capillary condensation of water was observed.

We show at the top of Fig. 5 the lateral "shear" damping of the tip amplitude as the oscillating fiber approaches the CH_3-thiol monolayer, together with the *simultaneous* normal-force response of the IFM sensor. The tip approaches the substrate at a rate of 0.25 nm/s until the extent of lateral damping exceeds 80% of the undamped amplitude, at which point the tip is withdrawn at the same rate. We estimate the displacement is accurate to within 15%, with independent piezo tube calibration performed against a known 5 nm feature. Unlike Fig. 4, we now arbitrarily set zero in the displacement axis in Fig. 5 to be the point where the damping (friction) begins. (Also, the direction of approach is now right to left, instead of left to right). The initially negative IFM sensor signal indicates a weak attractive interaction that reaches a maximum of ~100 nN, and appears to have a total range of 0.7±0.1 nm. As the tip continues to approach past approximately (-0.7 nm), the attractive interaction gives way to increasingly repulsive forces that compress the CH_3-thiol monolayer and gold substrate. A JKR force-displacement fit in this ~3.2 nm compressive regime provides an effective Young"s modulus of 8±1 GPa that is comparable to a value of 9.3±3.1 GPa previously measured for a supported hydrocarbon film [30]. Thus, in summary, friction in the CH_3-thiol system is due first to weak attractive forces, followed by compression of the film.

Under identical conditions and approach rate, the interaction of the *same tip* with the COOH-thiol sample is predominantly attractive; as shown in the lower part of Fig. 5, the tip motion is completely damped over the same range of lateral forces long before any appreciable repulsive compression occurs. In fact, the steep rise in friction with displacement suggests that a significant fraction of COOH-thiol chains "stand up" from their usual 30° tilt [12] to meet the probe. The ~0.7 nm range of the tip-COOH-thiol attractive interaction is essentially the same as that observed for the CH_3-thiol, although the forces are clearly much stronger. Thus it is possible that some of the CH_3-thiol chains stand up as well, but they are unable to significantly slow the tip down because of the relatively weak attractive interaction. However, by virtue of the stronger attractive interaction, the COOH-thiols bring the tip to a halt. The collective motion of the chains will act as an efficient channel for vibrational energy dissipation in addition to the losses incurred by the making and breaking of adhesion bonds. We can account for ~0.3 nm of the ~0.7 nm attractive interaction range through the motion of the thiol chains from the initial 30° tilt to a 0° upright position. The additional ~0.4 nm could be due to surface roughness over the ±12 nm lateral motion.

If we examine the data in terms of shear damping (friction) *vs.* load (Fig. 6), the adhesion difference between the two thiols is readily apparent. The JKR model predicts the dependence of tip-sample contact area A upon the applied load L. Thus, assuming friction F is proportional to contact area through a constant shear strength τ, we have the relation given in Equation 5 for a parabolic tip,

$$F = \tau A = \tau\pi \left\{ \frac{R}{K} \left[L + 3\pi R W_A + \sqrt{6\pi R W_A L + (3\pi R W_A)^2} \right] \right\}^{2/3} \tag{5}$$

where

$$K = \frac{4}{3} \left(\frac{1-\nu_1^2}{E_1} + \frac{1-\nu_2^2}{E_2} \right)^{-1}. \tag{6}$$

The reduced modulus K of the two materials in contact (Equation 6) is a function of the respective Young's moduli (E_1, E_2) and Poisson ratios (ν_1, ν_2) of the tip and sample. If R and K are known, a plot of friction *vs.* load can be used to determine both constants W_A and τ. Since our glass tips were determined beforehand by scanning electron microscopy to be blunt and non-parabolic, we must apply the extensions to Equation (5) developed in Ref. [31]. As mentioned above, the JKR model also predicts the relationship between load and displacement, which yielded a value of E_1=8 GPa. If we assume ν_1=0.4 as for most materials, and use the (E_2, ν_2 of glass [32], weobtain K=11 GPa. With this value of K in the extensions to Equation (5), we observe the JKR friction *vs.* load fits shown in Fig. 6 for CH_3-thiol and COOH-thiol. Although adhesion hysteresis was observable, approach and withdrawal JKR fits were the same within experimental error.

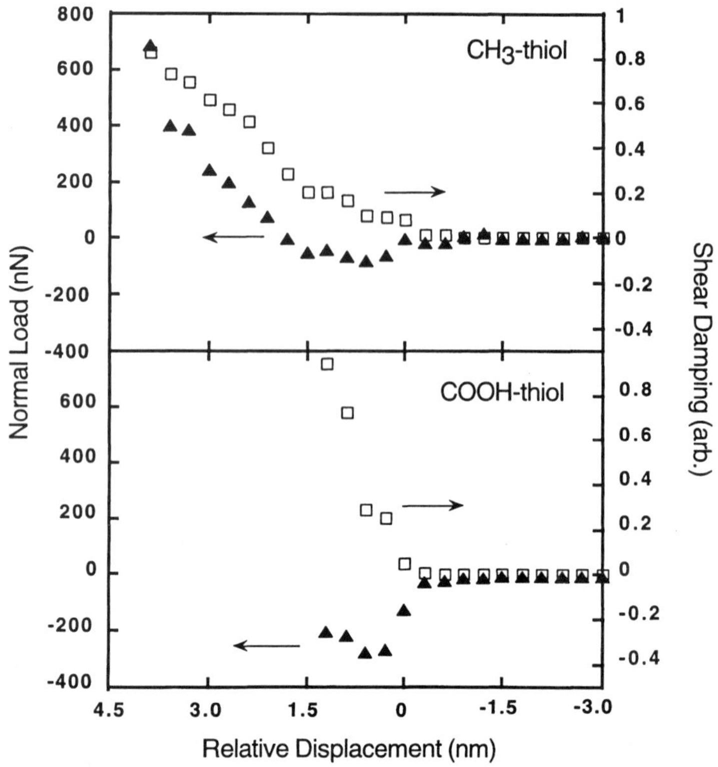

Figure 5. Simultaneous shear damping of tip lateral motion (□) and normal forces (▲) on tip acquired for approach to CH_3-thiol monolayer (top) and to COOH-thiol monolayer (bottom). The same glass tip, terminated by intrinsic silanols, is used for both data sets. The corresponding withdrawal curves are not shown for the sake of clarity. Zero displacement is arbitrarily set at the point of initial contact, and the approach is from right to left.

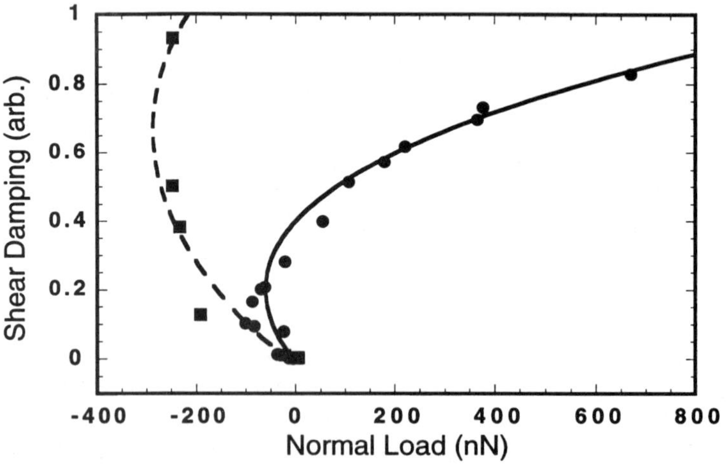

Figure 6. Johnson-Kendall-Roberts (JKR) fits to shear damping (friction) *vs.* normal load for different monolayers: (●-solid curve) CH_3-thiol, and (■-dashed curve) COOH-thiol.

Averaging over many fits for different regions of the CH_3-thiol monolayer, we obtain W_A = 50±21 mJ/m^2, which is in better agreement with purely van der Waals molecular interactions [17] than observed in the previous section. The COOH-thiol monolayer has an average adhesion energy W_A = 449±49 mJ/m^2 with the glass silanol groups, which we see from Table 1 is stronger than the COOH *vs.* COOH pair, but weaker than the COOH vs. NH_2 pair. If we subtract the 50 mJ/m^2 van der Waals contribution, the 400 mJ/m^2 result yields a hydrogen-bond energy of ~12 kcal/mole.

From F = 18 nN at 100% damping and the contact areas given by the fits [33], we obtain average shear strength (τ) values of 13.7±1.4 MPa and 20.0±1.9 MPa for CH_3-thiol and COOH-thiol, respectively. The CH_3-thiol shear strength is similar to those obtained for hydrocarbon films with the surface forces apparatus [34]. It is initially surprising that the difference in shear strengths for the CH_3-thiol and COOH-thiol does not correspond with the large difference in adhesion. However, the JKR fit for the CH_3-thiol includes a significant portion of positive applied loads, whereas the JKR fit for COOH-thiol only has negative loads. The absence of data in the repulsive contact regime for the COOH-thiol could mean that the JKR fit provides a lower bound to the estimated shear stress. More important is the fact that the JKR model assumes that the interfacial forces have zero spatial range. Models other than DMT which allow a small, finite range to these attractive forces [35] produce curves essentially the same as those depicted in Fig. 6, and predict that contact is first made with the surface displaced significantly upward from its equilibrium position, i.e. there is tensile deformation of the surface toward the tip. Indeed, observing finite friction with such negative loads indicates that the materials have made contact under tensile stress and thus the "range" of attractive forces should take account of film deformation as the molecules extend upward in response to the attractive forces. Our observations are unique, since measurements of load, displacement and friction in this tensile, molecularly-deformed regime are not possible with sensors exhibiting jump-to-contact instabilities.

4. SUMMARY AND CONCLUSIONS

Scanning force microscopy is ideally suited for fundamental studies of acid–base interactions at the molecular-level. In particular, we have focused on the interfacial force microscope (IFM), which avoids mechanical instabilities such as "jump-to-contact" that affect cantilever-based atomic force microscopes (AFM). With the non-compliant IFM, one is able to map the entire adhesion interaction *vs.* displacement and also reveal the extent of adhesion hysteresis. Furthermore, we have discussed the sensitivity of IFM to measure the magnitude of adhesion forces between functionalized self-assembled monolayers which range from weak van der Waals dispersion forces to "full strength" acid–base (hydrogen-bonding).

We have shown that the molecular origins of friction can be examined in detail with an instrument which combines the IFM with shear-force microscopy in a way that decouples normal and lateral force sensors. Correlation of normal forces with lateral shear forces reveals both the attractive and repulsive nature of friction between the contacting surfaces. Chemical modification of the surface dramatically changes the relative contributions of adhesion forces to friction. In addition to dissipative processes due to bond breaking and making, strong adhesion forces may result in additional channels of energy loss characterized by tensile deformation and collective chain motion. Studies such as those presented here can help reveal the detailed relationship between friction and adhesion, and identify modes of deformation unique to molecular films.

Acknowledgments

R. W. C. acknowledges support from the Natural Sciences and Engineering Research Council of Canada. Sandia is a multiprogram laboratory operated by Sandia Corporation, a Lockheed Martin Company, for the United States Department of Energy under Contract DE-AC04-94AL85000.

REFERENCES

1. G. S. Blackman, C. M. Mate, and M. R. Philpott, *Phys. Rev. Lett.* **65**, 2270 (1990).
2. N. A. Burnham, D. D. Dominguez, R. L. Mowery, and R. J. Colton, *Phys. Rev. Lett.* **64**, 1931 (1990).
3. N. A. Burnham, R. J. Colton, and H. M. Pollock, *Nanotechnology* **4**, 64 (1993).
4. S. A. Joyce and J. E. Houston, *Rev. Sci. Instrum.* **62**, 710 (1991).
5. R. C. Thomas, J. E. Houston, R. M. Crooks, T. Kim, and T. A. Michalske, *J. Amer. Chem. Soc.* **117**, 3830 (1995).
6. J. Krim, *Sci. Amer.* **275**, 74 (1996).
7. R. W. Carpick and M. Salmeron, Chem. Rev. **97**, 1163 (1997).
8. G. Hähner and N. Spencer, *Phys. Today* **51**, 22 (Sept. 1998).
9. C. D. Frisbie, L. F. Rozsnyai, A. Noy, M. S. Wrighton, and C. M. Lieber, *Science* **265**, 2071 (1994).
10. A. Noy, D. V. Vezenov, and C. M. Lieber, *Annu. Rev. Mater. Sci.* **27**, 381 (1997).
11. J.-B. D. Green, M. T. McDermott, M. D. Porter, and L. M. Siperko, *J. Phys. Chem.* **99**, 10960 (1995).
12. L. H. Dubois and R. G. Nuzzo, Annu. *Rev. Phys. Chem.* **43**, 437 (1992).
13. A. Ulman, *Chem. Rev.* **96**, 1533 (1996).
14. A. Lio, D. H. Charych, and M. Salmeron, *J. Phys. Chem. B* **101**, 3800 (1997).
15. N. D. Shinn, T. Kim, C. Daly, T. Mayer, R. Crooks, and T. Michalske, *Langmuir*, in press.
16. R. G. Horn, J. N. Israelachvili, and F. Pribac, *J. Colloid Interface Sci.* **115**, 480 (1986).
17. J. N. Israelachvili, *Intermolecular and Surface Forces*. Academic Press, London (1992).
18. O. L. Warren, J. F. Graham, and P. R. Norton, *Rev. Sci. Instrum.* **68**, 4124 (1997).
19. K. L. Johnson, K. Kendall, and A. D. Roberts, *Proc. Roy. Soc. Lond. A* **324**, 301 (1971).
20. B. V. Derjaguin, V. M. Muller, and Yu. P. Toporov, *J. Colloid Interface Sci.* 53, 314 (1975).
21. F. M. Fowkes, *J. Phys. Chem.* **66**, 382 (1962).
22. G. C. Pimentel and A. L. McClellan, *The Hydrogen Bond*. W. H. Freeman and Co., San Francisco (1960).
23. D. F. Ogletree, R. W. Carpick, and M. Salmeron, *Rev. Sci. Instrum.* **67**, 3298–3306 (1996).
24. E. Betzig, P. L. Finn, and J. S. Weiner, *Appl. Phys. Lett.* **60**, 2484 (1992).
25. R. Toledo-Crow, P. C. Yang, Y. Chen, and M. Vaez-Iravani, *Appl. Phys. Lett.* **60**, 2957 (1992).
26. A. R. Burns, J. E. Houston, R. W. Carpick, and T. A. Michalske, *Langmuir* **15**, 2922 (1999).
27. Sutter Instruments, Novato, CA.
28. J. Barentz, O. Hollricher, and O. Marti, *Rev. Sci. Instrum.* **67**, 1912 (1996).
29. J. D. Kiely and J. H. Houston, *Langmuir* **15**, 4513 (1999).
30. T. P. Weihs, Z. Nawaz, S. P. Jarvis, and J. B. Pethica, *Appl. Phys. Lett.* **59**, 3536 (1991).
31. R. W. Carpick, N. Agraït, D. F. Ogletree, and M. Salmeron, *J. Vac. Sci. Technol. B* **14**, 1289 (1996).
32. H. Muramatsu, N. Chiba, and M. Fujihira, *Appl. Phys. Lett.* **71**, 2061 (1997).
33. ~400 nm^2 at the minimum load. This area is smaller than the tip shape, due to small asperities and/or surface roughness.
34. J. N. Israelachvili, in: *Handbook of Micro/Nanotribology*, B. Bhushan (Ed.), p. 267. CRC Press, Boca Raton, FL (1995).
35. J. A. Greenwood, *Proc. R. Soc. Lond. A* **453**, 1277 (1997).

Acid-Base Interactions, Vol. 2, pp. 235–244
Ed. K.L. Mittal

Quantifying the impact of surface complexation on acid–base interactions

MICHAEL A. BUTKUS[1,*] and DOMENICO GRASSO[2]

[1]*Department of Geography and Environmental Engineering, United States Military Academy, West Point, NY 10996-1695, USA*
[2]*Department of Civil and Environmental Engineering, The University of Connecticut, Storrs, CT 06269-2037, USA*

Abstract—We have examined the utility of the DLVO model as modified by van Oss and colleagues ($DLVO_{EX}$) for predicting changes in the hydrophilicity of a colloidal hydrous ferric oxide matrix (HFM) derived at a drinking water treatment plant, as a function of phosphate adsorption. Increase in phosphate adsorption, below monolayer coverage, resulted in increased hydrophilicity as indicated by decreased mean HFM aggregate diameter. Results of the $DLVO_{EX}$ model suggested that adsorption of phosphate resulted in an increase in the colloid's Lewis base component (γ^-) of surface free energy. The increase in γ^- resulted in a concomitant increase (and change of sign) in the predicted free energy of interaction (ΔG_{1w1}) between colloidal aggregates. The change in hydrophilicity suggested by ΔG_{1w1} is conceptually congruent with the Gibbs isotherm equation, which predicts a decrease in solid-liquid interfacial energy with increased adsorption. The increase in predicted aggregate hydrophilicity is also consistent with the observed decrease in aggregate size mentioned above. Phosphate adsorption beyond monolayer coverage resulted in an increase in mean HFM aggregate diameter suggesting a decrease in particle hydrophilicity. This decrease in hydrophilicity may be a result of increased ionic strength and/or phosphate bridging.

Keywords: Adhesion; surface complexation; vOCG theory; DLVO theory; phosphate adsorption; Lewis acid–base.

1. INTRODUCTION

Hydrous ferric oxide contains surface functional groups whose charge is a function of the bulk water pH [1]. Due to their ability to hydrogen bond, it is possible that surface functional groups may play a role in governing the stability of colloidal matter in solution [2]. In addition, surface complexation of these groups with ions, such as phosphate, in the bulk solution may result in changes in the Lewis acid–base properties of these particles. Changes in surface energy (which can be caused by changes in Lewis acid–base properties) of particles can significantly influence their behavior in solution [3, 4]. However, other studies have suggested that ions can also have the opposite effect. For example, Anderson and co-workers [5] reported aggregation of geothite via phosphate bridges, which resulted in a decrease in BET surface area. Using transmission electron spectroscopy, He *et al.* [6] reported similar findings with ferric hydrox-

* To whom correspondence should be addressed. E-mail: bm8375@exmail.usma.edu

ide. Models that can predict changes in particle hydrophilicity, as a function of surface complexation (hereafter referred to as adsorption) may provide valuable insight into colloidal interactions in both engineered and natural systems.

Derjaguin–Landau–Verwey–Overbeek *Sensu stricto* ($DLVO_{SS}$) models have been used to describe particle-particle interactions for a good part of the twentieth century [7, 8]. $DLVO_{SS}$ is predicated on a force (or energy) balance approach, which can be used to predict particle stability in aqueous media. These models account for van der Waals (vdW) and electrostatic forces. For two identical particles in aqueous media, the vdW force is always attractive and dominates at both close range and very large distances. The electrostatic force is typically repulsive and dominates at intermediate distances. Due to its simplicity and often serendipitous results [9], $DLVO_{SS}$ has been widely used to explain particle-particle interactions in many applications (e.g. filtration, coagulation, colloid deposition, microbial adhesion, etc.). However, it is well documented that $DLVO_{SS}$ predictions are often quite different from experimental findings [10–12].

Van Oss and co-workers [13] have developed a model ($DLVO_{EX}$), which allows for quantitative determination of ancillary interfacial forces (Lewis acid–base forces) in addition to the well-known $DLVO_{SS}$ forces. Their approach accounts for the electron donor/acceptor (Lewis acid/base, AB) properties of both the solvent and solid surfaces, which are not included in $DLVO_{SS}$ models. Thus, the $DLVO_{EX}$ model is often more consistent with experimental findings than $DLVO_{SS}$ models. A significant advantage of this approach is that interfacial forces can be determined *a priori*. Consequently, the van Oss approach lends itself to a wide range of applications [14–20]. The free energy of interaction (ΔG_{1w1}) between two identical particles, 1, in water, w, can be determined using the $DLVO_{EX}$ model [13]:

$$\Delta G_{1w1} = -2\left(\sqrt{\gamma_1^{LW}} - \sqrt{\gamma_w^{LW}}\right)^2 - 4\left(\sqrt{\gamma_1^+\gamma_1^-} + \sqrt{\gamma_w^+\gamma_w^-} - \sqrt{\gamma_1^+\gamma_w^-} - \sqrt{\gamma_1^-\gamma_w^+}\right) \quad (1)$$

where

γ^{LW} = van der Waals component of surface free energy (mJ/m^2),
γ^+ = Lewis acid component of surface free energy (mJ/m^2), and
γ^- = Lewis base component of surface free energy (mJ/m^2).

A negative value of ΔG_{1w1} indicates attractive energy and positive value indicates repulsive energy. The ΔG_{1w1}, coupled with the electrostatic energy, can be used to predict the total energy between two surfaces as a function of distance. A more detailed explanation of the modeling approach is presented elsewhere [9, 12, 13, 17].

The work reported here examines the hydrophilicity of a hydrous ferric oxide matrix (HFM) as a function of phosphate adsorption. The $DLVO_{EX}$ colloid stability model is utilized for predicting changes in the Lewis acid–base properties and subsequent hydrophilicity of HFM aggregates as a function of phosphate adsorption. We demonstrate the utility and shortcomings of a $DLVO_{EX}$ model for describing the impact of phosphate adsorption on colloidal hydrophilicity. This novel application of the $DLVO_{EX}$ model may be useful in predicting changes in colloidal interactions in numerous natural and engineered systems.

2. MATERIALS AND METHODS

The HFM used in this study was generated from July 1995 to August 1996 at the Lake Gaillard Water Treatment Plant, in New Haven, CT, operated by the South Central Connecticut Regional Water Authority. The primary coagulant used at this direct filtration plant was $Fe_2(SO_4)_3$. The composition (organic and inorganic) of the HFM used in this study has been characterized and is documented elsewhere [21, 22].

Surface thermodynamics of HFM colloids, as a function of adsorbed phosphate and pH, was determined via contact angle measurements [9]. Stock solutions were prepared by combining various quantities of dibasic ammonium phosphate (Fisher Scientific, Pittsburgh, PA) with 75 ml of HFM (1.43% solids) and the pH adjusted to six. The range of total phosphate (adsorbed + aqueous) was 0.05 – mole PO_4/kg HFM. All experiments were conducted in 0.01 *M* NaCl and 0.02% NaN_3 [23]. The samples were allowed to equilibrate for at least 48 hours [21, 24–26]. Following equilibration, pH and conductivity were measured and the suspension was spread on prewashed microslides that were held normal to the gravitational field. The samples were dried in a desiccator for a minimum of 24 hours [12]. Contact angle measurements were conducted on the HFM coated glass slides, using a Ramé–Hart goniometer, following the method outlined by van Oss [9]. Advancing contact angles were measured as they are assumed to represent thermodynamic equilibrium and are therefore extensible to interpretation by the Young equation [9]. A minimum of 20 contact angle measurements were recorded for each phosphate treatment. Surface energy components for the solid surfaces, 1, were determined with three liquids, 2, via the following relationship [13]:

$$(1+\cos\theta)\gamma_2 = 2\left(\sqrt{\gamma_1^{LW}\gamma_2^{LW}} + \sqrt{\gamma_1^{+}\gamma_2^{-}} + \sqrt{\gamma_1^{-}\gamma_2^{+}}\right) \tag{2}$$

where,

θ = contact angle (degrees), and

γ_2 = total surface energy of liquid (mJ/m^2).

Glycerol, diiodomethane, and DI water were used as probe liquids to characterize the HFM surfaces as a function of phosphate concentration. Surface energy data for the liquids were taken from van Oss [9].

Phosphate partitioning was quantified by withdrawing 3 aliquots from each of the stock solutions used for contact angle measurements. Samples were centrifuged and vacuum-filtered through 0.45 μm nominal pore size nylon filters (Micron Separations Inc., No. RO4SP04700, Westboro, MA). Aqueous extracts were analyzed for phosphate using ion chromatography (Dionex Corp., series 4000, Sunnyvale, CA).

Electrophoretic mobility of the HFM-phosphate solutions, described above, was measured by laser diffraction techniques using a Zetasizer IIc (Malvern Instruments Inc., Southborough, MA). All electrophoretic mobility measurements were conducted with HFM and phosphate treated samples that were suspended in a 0.01 *M* NaCl solution. A minimum of 10 measurements were taken and averaged at voltages ranging from 75 to 175 volts.

Jar tests were conducted to measure the hydrophilicity (as determined by changes in particle size distribution) of HFM colloids as a function of phosphate concentration and pH. Aliquots were drawn from the stock solutions discussed above, diluted in demineralized water (25 μl/500 ml) and injected into a particle counter (HIAC ROYCO, model 325, Pacific Scientific) following the protocol outlined by Srivastava [27]. The particle counter channels were set to 0.27, 0.31, 0.36, 0.41, 0.47, 0.54, 0.63, 0.72, 0.83, 0.96, 1.1, and >1.27 μm. Each dilution was counted a minimum of three times in order to minimize the effect of sample heterogeneity. Prior to drawing the sample, stock solutions were sonicated for one minute to break up weakly-held flocs. This approach allowed for thorough mixing of the sample and also minimized the effects of floc shear due to sample passage through the HIAC sensor [28].

3. RESULTS

3.1. Experimental

The HFM studied here consisted of a complex mixture of colloidal material deriving from the raw water source, coagulant and organic polyelectrolytes [21]. Based on the etiology of this material, we hypothesize that its surface is heterogeneous and composed of a mixture of sites. HFM aggregates are likely to be a combination of different heteroflocculated particles bound together at sites with varying energies.

As part of this work, we report on changes in HFM aggregate size as a function of solution chemistry. Figure 1(a) depicts the effect of phosphate adsorption on HFM mean aggregate diameter, at a pH of 6, determined from particle size distributions using a first moment number average. A two variable t-test ($\alpha = 0.05$) indicated that trends in particle size distribution were statistically significant [21]. Figure 1(a) illustrates that aggregate size decreased as a function of apparent phosphate adsorption, and reached a minimum diameter at approximately 100 mg PO_4/g HFM. This finding suggests that phosphate adsorption may result in partial re-peptization of heteroflocculated HFM particles. Decrease in mean aggregate diameter as a function of phosphate adsorption was expected because stabilization of colloidal metal oxides (with surface characteristics similar to that of HFM) by phosphate has been reported elsewhere [3, 29, 30]. Figure 1(b) depicts trends in ionic strength (determined from conductivity measurements [31]) as a function of phosphate adsorption, at a pH of 6. Although average particle diameter decreased from 0 to 100 mg PO_4/g HFM, ionic strength appeared to increase slightly. As phosphate adsorption continued beyond 100 mg PO_4/g HFM, we observed an increase in aggregate diameter and a concomitant increase in ionic strength.

Figure 1 (a)

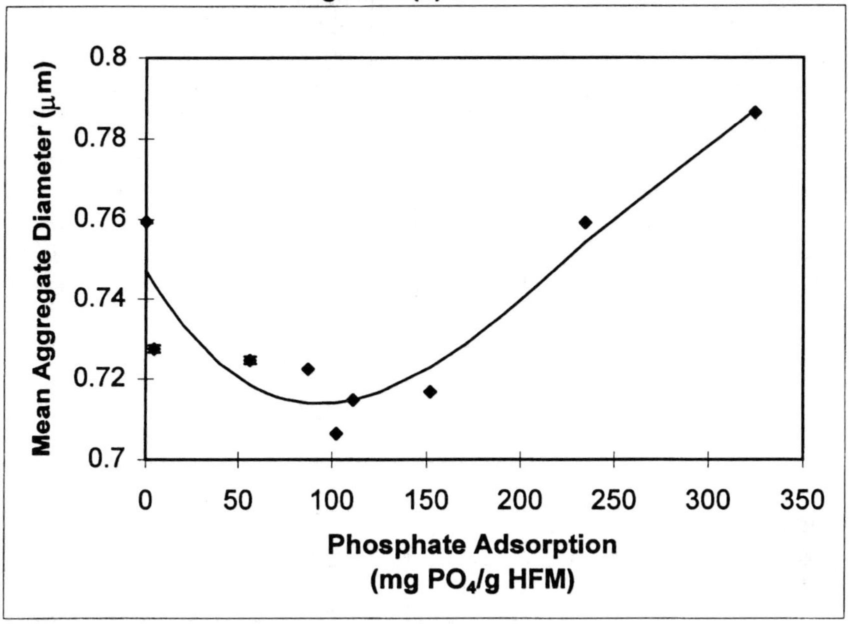

Figure 1 (b)

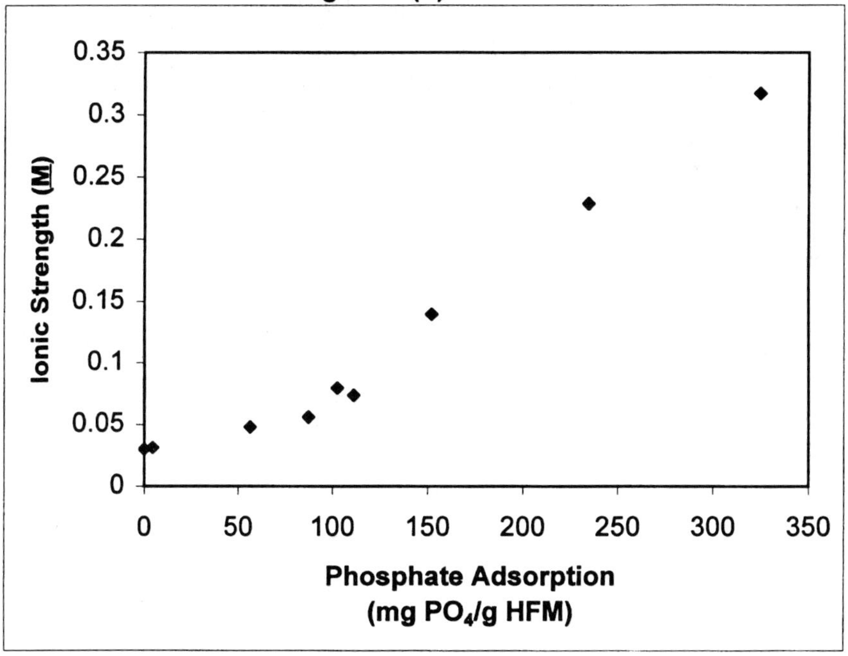

Figure 1. (a) HFM mean (number average) aggregate diameter as a function of phosphate adsorption (0.01 M NaCl, pH = 6). Error bars: 95% confidence interval. A trace has been superimposed on the figure to illustrate data trends. (b) Ionic strength (determined from conductivity measurements) as a function of apparent phosphate adsorption (0.01 M NaCl, pH = 6).

Figure 2 illustrates the effect of adsorbed phosphate on the zeta potential of HFM aggregates. Adsorption of phosphate shifted the HFM isoelectric point (IEP) to lower pH values. This finding is similar to that reported by Breeuwsma and Lyklema [32] for phosphate adsorbed onto iron oxide. The IEP shifted from a pH of approximately 5.6, for control HFM, to a value of approximately 4.2 for HFM that had been treated with a total of 4.5 g PO_4/l (152 mg PO_4/g HFM at pH 6). The zeta potential was observed to be *ca.* –10 mV in the pH range of 6–6.5 for *both* systems. Adsorption of phosphate had a more pronounced influence on surface charge at pH values less than 6. It should be noted that although all samples were suspended in 0.01 M NaCl, HFM samples treated with phosphate had a slightly higher ionic strength than the stock HFM (cf. Figure 1(b) for pH of 6). The difference in ionic strength was most notable at extreme pH values because more acid/base was required due to buffering caused by aqueous phosphate.

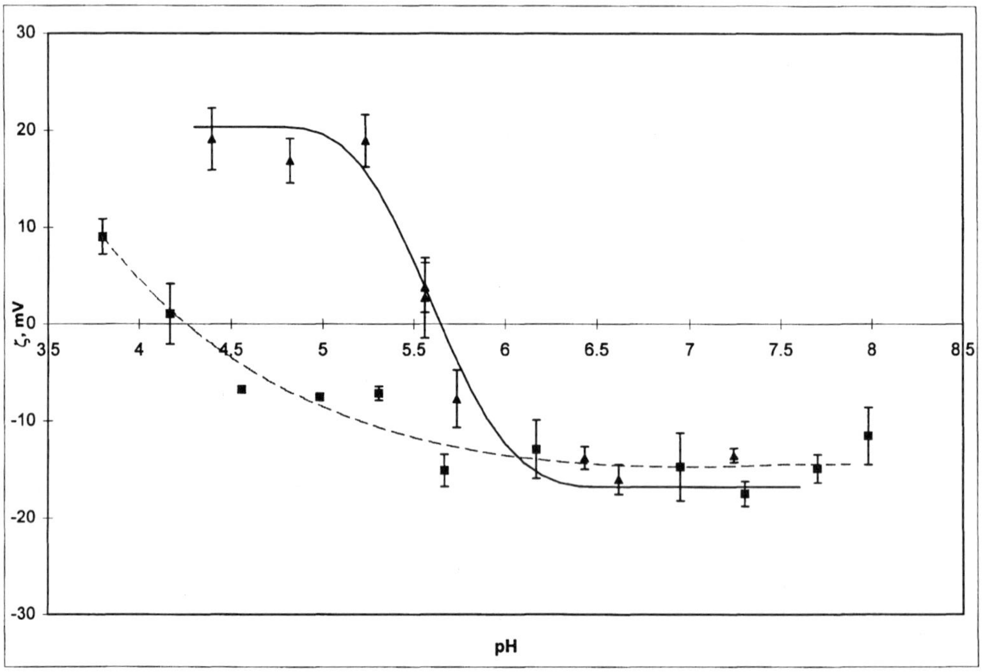

Figure 2. Zeta potential measurements as a function of pH for HFM Control suspended in 0.01 M NaCl (▲) and HFM treated with 4.5 g PO_4/l (■). Error bars: 95% confidence interval. Traces have been superimposed on the figure to illustrate trends in the data.

3.2. Modeling

The results of contact angle measurements; Lewis acid–base components, determined via Equation (2); and, the free energy of interaction determined via Equation (1) for control HFM and phosphate treated HFM are shown in Table 1. We observed a significant decrease in DI water contact angle as a function of increased phosphate adsorption, suggesting an increase in hydrophilicity. The influence of phosphate adsorption on glycerol and diiodomethane contact angles remains unclear. The Lewis

Table 1.
Average contact angle measurements, Lewis acid–base components, and free energy of interaction for control HFM and HFM treated with phosphate. (pH = 6; 0.01 M NaCl)

Phosphate Treatment, mg PO_4/g HFM	Average Contact Angle, degrees (Standard Deviation)			Surface parameters, mJ/m^2			
	DI water	glycerol	Diiodo-methane	γ_s^{LW}	γ_s^+	γ_s^-	ΔG_{lwl}
Control	57 (5)	47 (4)	47 (2)	35.9	1.9	17.7	-15.7
4.41	31 (4)	36 (4)	41 (2)	39.3	1.3	41.6	16.7
56.3	32 (3)	47 (3)	43 (2)	38.1	0.2	50.8	33.4
87.3	22 (3)	36 (3)	44 (2)	37.6	1.1	51.4	29.6
103	19 (2)	47 (2)	49 (2)	34.7	0.2	65.5	52.6
111	17 (3)	45 (2)	45 (2)	37.1	0.2	64.9	51.7
152	16 (2)	37 (5)	50 (2)	34.1	1.2	58.6	38.6
234	10 (1)	35 (1)	50 (1)	34.3	1.4	59.5	38.3
324	8 (2)	33 (3)	50 (2)	34.9	1.4	59.3	37.9

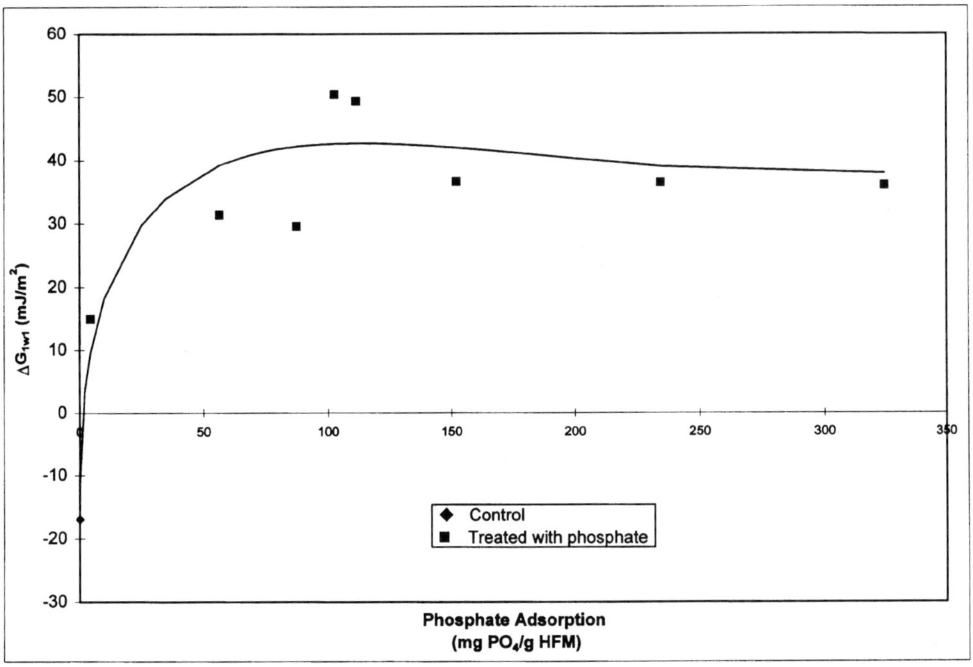

Figure 3. Free Energy of interaction between identical HFM aggregates in water, ΔG_{lwl}, as a function of phosphate amendment (0.01 M NaCl, pH = 6). A trace has been superimposed on the figure to illustrate data trends.

base component of surface free energy (γ^-) increased as a function of increased phosphate adsorption. The increase in γ^- resulted in a concomitant increase (and change of sign) in the predicted free energy of interaction (ΔG_{1w1}) between colloidal aggregates, illustrated in Figure 3, suggesting an increase in the hydrophilicity of the HFM surface. This result is conceptually congruent with the Gibbs isotherm equation, which predicts a decrease in solid-liquid interfacial energy with increased adsorption. The control HFM (suspended in a solution of 0.01 M NaCl without phosphate treatment) gave a ΔG_{1w1}= -16 mJ/m^2. This result suggests that the sum of the vdW and AB energies is attractive between control HFM particles suspended in an electrolyte solution. The increase in hydrophilicity suggested by ΔG_{1w1} is consistent with the observed decrease in aggregate size discussed above. At approximately 100 mg PO_4/g HFM, ΔG_{1w1} appears to reach a maximum. This correlates with the minimum aggregate diameter measurements described earlier. However, beyond this quantity of phosphate adsorption, model predictions and experimental observations diverged somewhat; the increase in mean aggregate diameter did not correspond to the decrease in ΔG_{1w1}.

4. DISCUSSION

Apparent adsorption of phosphate at surface coverages of less than *ca.*100 mg PO_4/g HFM resulted in an increase in the HFM γ^- component of surface energy. The increase in particle hydrophilicity (ΔG_{1w1}) predicted by the model was congruent with the observed changes in particle size distribution (decrease in mean aggregate diameter).

A slight (0.05 units) increase in ionic strength was observed for increased phosphate loadings below *ca.* 100 mg PO_4/g HFM (cf. Figure 1(b)). Traditional colloid modeling theory would predict that increases in ionic strength would compress the double layer and consequently decrease particle stability. We observed the converse; the presence of phosphate increased ionic strength slightly, but particle size decreased. It should also be noted that given the magnitude of the surface charge (*ca.* –10 mV) the electrostatic energy of interaction is relatively minor compared to the vdW and AB energy components at the distance of closest approach [9]. This might explain why the influence of ionic strength on particle size distribution was not apparent below *ca.* 100 mg PO_4/g HFM.

We observed that trends in particle size distribution and surface hydrophilicity changed significantly near *ca.*100 mg PO_4/g HFM. Adsorption data and changes in ionic strength indicate that monolayer coverage of phosphate on HFM occurred at approximately 100 mg PO_4/g HFM. Figure 4 illustrates phosphate adsorption onto HFM as a function of aqueous phosphate concentration. The change in slope near 100 mg PO_4/g HFM may be indicative of saturation of surface adsorption sites [1, 21, 22]. Additional binding of phosphate may be attributed to surface precipitation [1, 22, 33], and/or the formation of soluble metal-phosphate, and/or organic-phosphate complexes [22]. As illustrated in Figure 1(b), ionic strength increased dramatically as a function of phosphate loading above monolayer coverage (approximately 100 mg PO_4/g HFM) which may be indicative of high aqueous phosphate and ammonium concentrations. Surface area measurements also suggest that monolayer coverage occurred at *ca.* 100 mg PO_4/g HFM [21, 34].

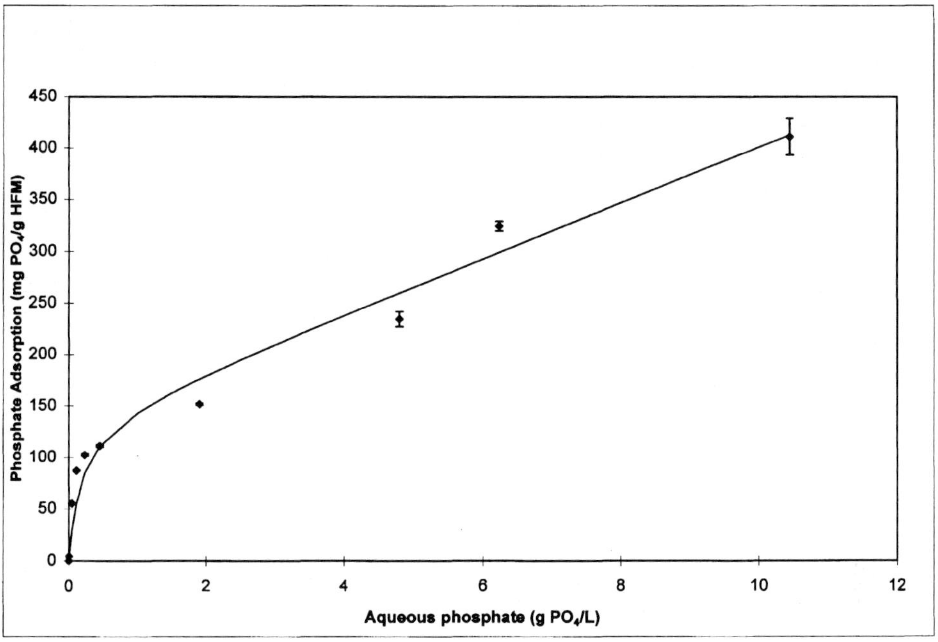

Figure 4. Adsorption of phosphate onto HFM as a function of aqueous phosphate concentration. (0.01 M NaCl, pH = 6).

Apparent adsorption beyond monolayer coverage, due to another mechanism such as surface precipitation, did not result in re-peptization of HFM aggregates. The HFM flocs appeared to grow at higher phosphate concentrations. Increase in aggregate size, as a function of increasing phosphate adsorption beyond monolayer coverage, may be attributed to phosphate bridging between HFM particles to form larger oriented aggregates [5, 6]. An alternative explanation may be related to the impact of ionic strength on the structure (and solvency) of water [35] and/or the electrostatic energy of interaction. High ionic strengths resulting from high aqueous concentrations of phosphate and ammonium may have resulted in a decrease in the free energy of interaction between HFM aggregates and resulted in further heteroflocculation [12, 17, 21, 35]. The dominant mechanism behind the increase in HFM heteroflocculation at high phosphate loading remains unclear.

5. CONCLUSION

We used the $DLVO_{EX}$ model to predict changes in particle hydrophilicity as a function of phosphate adsorption. Because it accounts for changes in ancillary forces, e.g. hydrogen bonding, as a result of surface complexation of selected ions and particles, the $DLVO_{EX}$ model may be of more use than the $DLVO_{SS}$ model. However, beyond monolayer coverage the $DLVO_{EX}$ model does not appropriately address bridging of particles by anions in solution and/or the effects of ionic strength.

Acknowledgements

This work was conducted at the University of Connecticut while M.A. Butkus was a doctoral student. The authors would like to thank Prof. Carel J. van Oss, State University of New York at Buffalo, for his helpful comments during the preparation of this manuscript. The authors gratefully acknowledge the South Central Connecticut Regional Water Authority, New Haven, CT, for funding this work.

REFERENCES

1. D. A. Dzombak and F. M. M. Morel, *Surface Complexation Modeling: Hydrous Ferric Oxide*. John Wiley, New York (1990).
2. S. R. Holmes-Farley, R. H. Reamey, T. J. McCarthy, J. Deutch and G. M. Whitesides, *Langmuir* **1**, 725 (1985).
3. J. Lyklema, in: *The Scientific Basis of Flocculation*, K. J. Ives, (Ed.), p. 36. Sijhoff and Noordhoff, Alphen aan den Rijn, The Netherlands (1978).
4. E. Matijevic, *CHEMTECH*, 176 (March 1991).
5. M. A. Anderson, M. I. Tejedor-Tejedor and R. R Stanforth, *Environ. Sci. Technol.* **19**, 632 (1985).
6. Q. H. He, G. G. Leppard, C. R. Paige and W. J. Snodgrass, *Water Res.* **30**, 1345 (1996).
7. B. V. Derjaguin and L. D. Landau, *Acta Physicochim U.S.S.R.* **14**, 633 (1941).
8. E. J. W. Verwey and J. Th. G. Overbeek, *Theory of the Stability of Lyophobic Colloids*. Elsevier, Amsterdam (1948).
9. C. J. van Oss, *Interfacial Forces in Aqueous Media*. Marcel Dekker, New York (1994).
10. S. W. Swanton, *Adv. Colloid Interface Sci.* **54**, 129 (1995).
11. P. M. Claesson, in: *Surface Forces and Surfactant Systems*, H. G. Kilian and G. Lagaly (Eds), p. 48. Springer-Verlag, Darmstadt (1987).
12. P. Chheda and D. Grasso, *Langmuir* **10**, 1044 (1994).
13. C. J. van Oss, M. K. Chaudhury and R. J. Good, *Chem. Rev.* **88**, 927 (1988).
14. C. J. van Oss, R. F. Giese and P. M. Constanzo, *Clays Clay Minerals* **38**, 151 (1990).
15. C. J. van Oss and R. J. Good, *J. Dispersion Sci. Technol.* **12**, 95 (1991).
16. P. Chheda, D. Grasso and C. J. van Oss, *J. Colloid Interface Sci.* **153**, 226 (1992).
17. D. Grasso, J. C. Carrington, P. Chheda and B. Kim, *Water Res.* **29**, 49 (1995).
18. P. LaFrance and D. Grasso, *Environ. Sci. Technol.* **29**, 1346 (1995).
19. B. Janczuk, J. A. Méndez-Sierra, M. L. González-Martin and J. M. Bruque, *J. Colloid Interface Sci.* **184**, 607 (1997).
20. D. Grasso, B. F. Smets, K. A. Strevett, B. D. Machinist, C. J. van Oss, R. F. Giese and W. Wu, *Environ Sci. Technol.* **30**, 3604 (1996).
21. M. A. Butkus, Ph.D. Dissertation, University of Connecticut, Storrs, CT (1997).
22. M. A. Butkus, D. Grasso, C. P. Schulthess and H. Wijnja, *J. Environ. Qual.* **27**, 1055 (1998).
23. D. M. Dohse and L. W. Lion, *Environ. Sci. Technol.* **28**, 541 (1994).
24. R. E. White and A. W. Taylor, *J. Soil Sci.* **28**, 8 (1977).
25. P. S. Nair, T. J. Logan, A. N. Sharpley, L. E. Sommers, M. A. Tabatabai and T. L. Yuan, *J. Environ. Qual.* **13**, 591 (1984).
26. P. Persson, N. Nilsson and S. Sjöberg, *J. Colloid Interface Sci.* **177**, 263 (1996).
27. R. M. Srivastava, *Water Res.* **27**, 939 (1993).
28. R. J. Gibbs, *Environ. Sci. Technol.* **16**, 298 (1982).
29. L. Liang and J. J. Morgan, *Aquatic Sci.* **52**, 32 (1990).
30. J. P. Boisvert, T. C. To, A. Berrak and C. Jolicoeur, *Water Res.* **31**, 1939 (1997).
31. V. L. Snoeyink and D. Jenkins, *Water Chemistry*. John Wiley, New York (1990).
32. A. Breeuwsma and J. Lyklema, *J. Colloid Interface Sci.* **43**, 437 (1973).
33. K. J. Farley, D. A. Dzombak and F. M. M. Morel, *J. Colloid Interface Sci.* **106**, 226 (1985).
34. M. A. Butkus and D. Grasso, *Environ Eng Sci.* **16**, 117 (1999).
35. M. A. Butkus and D. Grasso, *J. Colloid Interface Sci.* **200**, 172 (1998).

Acid-Base Interactions, Vol. 2, pp. 245–271
Ed. K.L. Mittal

Acid–base interactions on surfaces of wet and dry inorganic oxides

CRISTIAN I. CONTESCU* and JAMES A. SCHWARZ
Department of Chemical Engineering and Materials Sciences, Syracuse University, Syracuse, NY 13244

Abstract—The acid–base chemistry of the surfaces of inorganic solids, both in their wet and (semi) dry state, is reviewed. The results of different experimental methods to probe acidity, namely the Proton Affinity Distribution (PAD) approach and FTIR spectroscopy, are compared. These two methods provide information on the types and acid strength of surface hydroxyl groups on the wet and (semi) dry surface of oxides, respectively. Several examples show that a relationship can be established between the acid–base properties of the wet and (semi) dry surface of the same oxide. This is the result of a common structure of the outermost surface layers. In particular, for oxides, it is proposed that at both the wet and (semi) dry surfaces the proton binding / proton donor sites are surface hydroxyl groups. These groups terminate the oxide layer and may bind a surface layer of water molecules, which are in contact with either an adjacent liquid phase or a gaseous environment. Following an analysis of the factors that determine the acid–base properties of oxide surfaces, several examples are given which illustrate the various principles and practical procedures for the control of the acidity or basicity of several classes of oxide materials.

Keywords: Surface acidity; surface basicity; inorganic oxides; proton affinity distribution; FTIR; surface hydroxyls; wet surfaces; dry surfaces.

1. INTRODUCTION

The whole edifice of pure and applied chemistry was built upon a simple rule: acids and bases should be compared when confined to the same homogeneous phase, either liquid or gaseous. As long as this convention is observed, the reliability of analytical methods in acid–base chemistry or the meaning of their results is unquestionable. For practical purposes, two scales of acidity were selected—the aqueous scale and that based on gas phase measurements—but they do not coincide. In fact, due to the inherent relativity of the measurements, any discussion of acid–base chemistry depends on the standards used, or is specific to experimental conditions. Fortunately, this fact is very well understood.

When the discussion focuses on the acid–base properties of solid surfaces, it is inevitable that the established conventions will require some modifications. It is obvious that solids possess both acidity and basicity. The challenge is to characterize their acid–base behavior. The presence of two phases and the location of acid–base sites at the *interface* between the solid and either a gas, a liquid, or even a solid phase makes

* To whom correspondence should be addressed. E-mail: cjcontes@mailbox.syr.edu

the task difficult. The conflict with the simple rule stated above is apparent. If microporosity is also present, the distinction between "pure" acid–base processes and perturbations due to the confined environment is difficult. Both the solid matrix and the molecules confined to the internal spaces may show altered acid–base behavior in comparison with their "free space" counterparts. Thus, when the acid–base chemistry occurs in micropores or interlayers of nanostructured materials, the rules are broken a second time. Now all references based on the classical approach to acid–base properties of macroscopically homogeneous phases become inconsistent.

In the discussion that follows we attempt to organize the vast amount of information available on the acid–base behavior of oxide surfaces. The focus is on molecular-scale interactions and the influence of structural and chemical factors on the acid–base properties of solid oxides. It is recognized that, under ambient conditions, the surfaces of materials are hydrated; the role of water at the surface of oxides, in contact with either an aqueous environment or a gaseous, semi-dry atmosphere, is discussed first. The atomic structure in the surface layer is viewed as the common denominator to be adopted for it allows for a unified description of the changes in acid–base properties that accompany wet-dry transformations of oxide surfaces. In the last section of the review, several examples are given. They show means and principles that can be used to modify the acidity or basicity of oxide systems of practical importance.

2. ACID–BASE PROPERTIES OF SOLID SURFACES

Brönsted and Lewis acids and bases are frequently encountered in the chemistry of solid surfaces, but in some instances the definition of acid–base processes on solid surfaces is complicated by the "in-between two phases" location of the acidic-basic sites. From the point of view of the Brönsted definition, a solid is acidic if it is able to donate (or at least transfer partially) a proton to a basic probe molecule to which the proton bonds (or at least becomes partially associated). According to the Lewis definition, a solid is acidic if it is able to accept an electron pair from a basic molecule *and* to form a coordinative bond with it.

Molecular interactions across the interface between condensed phases may be split into physical and chemical terms. Although, traditionally, only physical interactions (nonspecific) have been recognized across the interface, modern theories explain the interfacial phenomena on solids, such as adhesion or wetting, in terms of chemical interactions. Because the Lewis definition of acidity is very comprehensive, it can be easily accepted that most of chemical interactions at solid/gas and solid/liquid interfaces may be effectively described as acid–base interactions [1]. Even for the solid/solid interfaces, a continuously growing literature re-interprets surface interactions in terms of acid–base properties [2].

The range of chemical interactions that can possibly occur between a solid surface and molecules of an adjacent fluid phase is very broad. At the solid/liquid interface one can envisage first-order local interactions (such as coordinative bonding or proton exchange) and second-order interactions (such as hydrogen bond formation). All interactions that may be destructive for the solid, such as dissolution and colloid formation, are, of course, excluded from our discussion.

The solid/gas interface was traditionally studied with respect to adsorption and catalysis. A long-term goal of surface studies for the characterization of solid catalysts was to distinguish and quantify the number of Brönsted or Lewis sites with potential catalytic activity for gas phase reactants. Here the assertion that the Brönsted definition of acidity is a particular case of the Lewis definition is neither obvious nor even helpful because many catalytic transformations of hydrocarbons on heterogeneous catalysts require, specifically, the presence of Brönsted acid sites.

The acidity (basicity) of solid surfaces manifests itself only in relation to a base (acid) that must be present, like in the case of homogeneous one-phase systems. The strength of acid–base sites at a given solid surface depends, therefore, on the particular molecular probe involved. A universal scale of solid acidity, comparable to the pK scale for aqueous solutions or to proton affinities for acid–base reactions in the gas phase, is not defined for solids. This fact obscures the quantification of acidity of oxide surfaces. The intrinsic property of interfaces to serve as accumulation zones for adsorbed species further complicates the assessment of acid–base properties of solid surfaces.

3. ORIGIN OF ACID–BASE PROPERTIES OF SOLID OXIDE SURFACES

In this section we examine how the general acid–base concept, either in the limited interpretation of the Brönsted theory or in the more general sense of the Lewis theory, can be applied to solid surfaces [3].

3.1. Role of Water

The chemistry of acid–base interactions on solid oxides is strongly conditioned by the degree of hydration of the surface. Water is present almost invariably at the surface of oxide materials. Due to its high intrinsic reactivity, water may act either as an electron pair donor, or as a proton donor or acceptor, or as a hydrogen bond donor or acceptor. The surface of oxides, with exposed oxygen ions or metal cations, is energetically unstable, and it is easily stabilized by reactions with water from the ambient. Several reaction paths are shown in Scheme 1. They depict only first order interactions, which result in redistribution of bonds, or formation of new chemical bonds at the oxide surface. Second-order interactions that occur through hydrogen bonds are not shown in Scheme 1.

When an ideally perfect oxide surface interacts with water, one can assume that the oxygen atom of water may occupy a vacant anion position at the surface, the same way as to continue the lattice of the oxide. The oxygen from the first layer of adsorbed water is, therefore, indistinguishable from the oxygen in the outermost layer of the bare oxide surface. Only in the case of dissociative adsorption of water, does important differences arise between the "intrinsic" oxygen from the original oxide lattice and the "additional" oxygen of adsorbed water molecules. The differences relate to the magnitude and the sign of localized electrical charges created through water dissociation. A localized surface charge may exist, even on very pure materials. It depends on local crystallographic factors (surface orientation) and on the extent of charge transfer (proton exchange) at the surface. The latter depends on the degree of

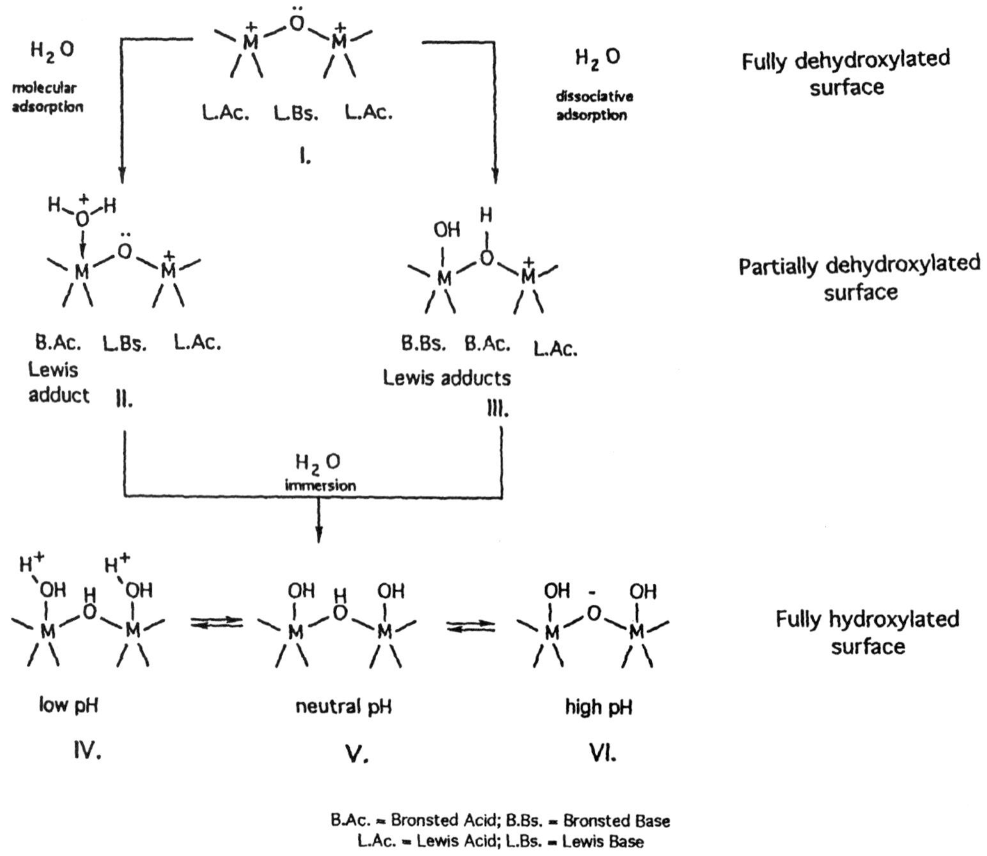

Scheme 1.

hydration of the surface. Whenever the oxide surface is in contact with a large proton buffer, such as an aqueous solution, proton transfer to or from the hydroxylated surface sites will occur to minimize the surface energy.

3.2. Fully Dehydroxylated Surfaces

A fully dehydroxylated oxide surface can be viewed as an extended array of coordinately unsaturated oxygen ions and metal cations; such a "perfectly dry" surface is not stable and does not have physical reality. From the acid–base point of view, surface ions act as both Lewis base and Lewis acid centers (structure **I** in Scheme 1). The net charge on each surface center depends on the stoichiometry, ionicity, and local surface structure which, in turn, vary from one crystal plane to another. If geometrical defects exist on the surface (corners, edges, dislocations) they may enhance the acid–base character, because of their additional chemical unsaturation.

Molecules impinging on the oxide surface may have either a Lewis base or Lewis acid character. For instance, CO with a filled orbital at the carbon end displays Lewis basicity, while CO_2 shows Lewis acidity [4]. On oxides, the adsorption interaction is strictly localized, and surface adducts are formed with the acidic or basic centers of the solid. Adsorption of water may occur nondissociatively on coordinatively unsaturated cations as:

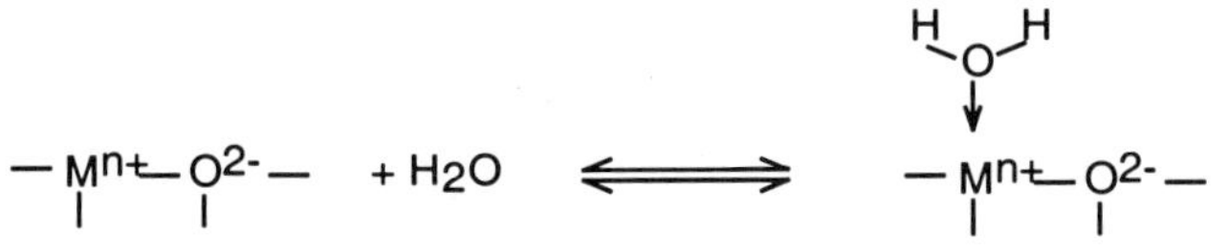

(1)

The newly formed adsorption complex (structure **II** in Scheme 1) is a Lewis adduct and, at the same time, a Brönsted acid site.

A peculiarity of solid surfaces is the occurrence of adjacent Lewis acid and Lewis base sites in repetitive structures. If they interact in a concerted manner with the adsorbed species, the result is dissociative adsorption. The most important example to cite is the dissociative adsorption of water on dehydroxylated oxide surfaces; it leads to the formation of surface hydroxyl groups which have tremendous impact on the reactivity of most oxides under ambient or close to ambient conditions:

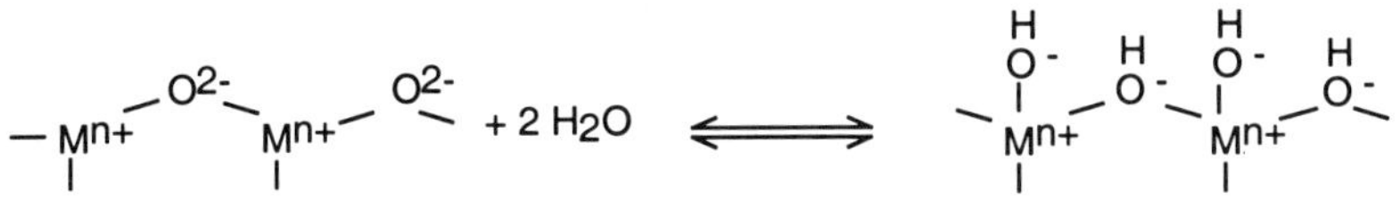

(2)

In Scheme 1, dissociative adsorption of water transforms the Lewis acid–base pair from structure **I** into a pair of Brönsted acid–base sites (structure **III**) that can be regarded, at the same time, as Lewis adducts. Also note the conversion of the Lewis acid into a Brönsted base and that of the Lewis base into a Brönsted acid.

3.3. Partially Dehydroxylated Surfaces

Fully dehydroxylated oxide surfaces, as those considered above, do not exist under normal conditions. Upon exposure of "perfectly dry" surfaces to ambient, Brönsted acidity is generated by chemisorption of water, and the surfaces acquire different degrees of hydroxylation (depicted by structures **II** and **III** from Scheme 1). Longer exposure to ambient moisture results in surfaces with various degrees of hydration. Although at this stage the samples appear to be "dry" according to the common sense, their surface is in reality hydrated, as evidenced from infrared spectroscopy.

Quantum mechanical calculations support the validity of localized models for acid–base sites on oxide surfaces. Brönsted acidity is present only when hydroxyl groups are present. On ionic inorganic oxides, the strength of Brönsted acid sites is controlled by the electrostatics of the system and is mainly governed by three structural parameters: the coordination number of the hydroxyl group, the charge on the cation(s), and the coordination number(s) of cations neighboring the hydroxyl group [5].

Surface hydroxyl groups may differ with respect to their coordination with underlying metal cations, and the magnitude and sign of the uncompensated electron charge. In turn, this results in variations of the strength of O–H bonds at various surface oxygen sites. The acid–base properties of surface oxygen ions are, therefore, determined by surface structural factors. For example, the two Lewis adducts of structure **III** in Scheme 1 differ with respect to their Brönsted properties: the monodentate hydroxyl is basic, and the bidentate hydroxyl is acidic. This fact can be recognized in the fine structure of O–H stretching vibrations in infrared spectra of partially dehydroxylated oxide surfaces.

The literature on infrared spectroscopy abounds with information that reveals the presence of several types of surface hydroxyls with various stretching frequencies, ν_{OH}, on partially dehydroxylated oxides. Structural considerations suggest that the number of all possible configurations for hydroxyl (and oxygen) groups on low index planes of oxides is limited [6].

Figure 1 illustrates the DRIFT spectra of several simple oxides in the region of deuterium-exchanged hydroxyl vibrations [7]. The spectra were collected using a Nicolet Impact 400 spectrometer equipped with a catalytic chamber. The O–D rather than O–H vibrations were investigated in order to isolate the contribution of oxide surface groups from other perturbing influences during the measurements.

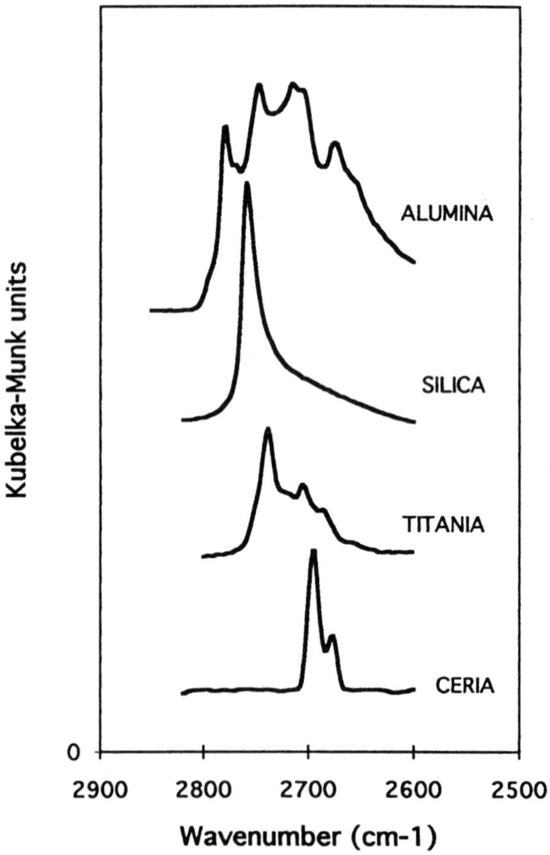

Figure 1. DRIFT spectra of O–D stretching vibrations for several oxides after dehydration at 775 K.

Knözinger and Ratnasamy [8] assigned the ν_{OH} bands on alumina to surface hydroxyl groups in various configurations. They assumed that only single-, double-, and triple- coordinated surface oxygen positions existed, in ratios and geometrical arrangements that differed from one surface plane to another. With some modifications [9] their scheme was generalized by Busca *et al.* [10] for surface sites on several oxides with spinel and corundum structures. In general, the frequency of ν_{OH} bands depends on the nature, coordination state, and number of directly neighboring cations. For OH groups bonded to the same element, the ν_{OH} frequencies decrease in the order: terminal over a tetrahedral cation > terminal over an octahedral cation > bridging > triply bridging; the breadth of the OH stretching band increases in the order terminal < bridging << triply bridging [10b].

The infrared spectra in Figure 2 show the changes in the O–D stretching vibrations during dehydration of three mixed oxide catalysts (SiO_2-TiO_2 with 67 % TiO_2, SiO_2-ZrO_2 with 50 % ZrO_2, and SiO_2-Al_2O_3 with 10 % Al_2O_3) [11]. The samples shown in Figure 2a were mildly dehydrated under vacuum, at 423 K. The spectra show the characteristic vibration of deuterated silanol groups (–Si–OD) and large populations of deuteroxyl (and hydroxyl) groups, strongly perturbed by hydrogen bond interactions. The latter are associated, in part, with chemisorbed or physically bound deuterated water, which resists dehydration at 423 K; the presence of water on the surface is proven by the occurrence of the deformation vibration of D_2O (1420–1460 cm^{-1}) in the same spectra (not shown).

At higher dehydration temperatures, the population of laterally interacting deuteroxyls (and hydroxyls) diminishes, as shown in Figure 2b for dehydration at 575 K. However, the pattern of isolated deuteroxyls (and hydroxyls) other than silanols could only be seen after heating at 875 K (Figure 2c). For comparison, the spectrum of a silica sample subjected to the same treatment is shown in Figure 2c. Although the intense peak of deuterated silanol (2759 cm^{-1}) dominates all spectra, mixed oxide samples reveal additional bands, not seen in pure silica; they were assigned by Contescu *et al.* [11] to bridging deuteroxyls (hydroxyls) in Si-OH(D)-M surface groups. It is significant that the position of these bands (indicated with arrows in Figure 2c) on mixed oxides follows the order $\nu_{Si\text{–}OH\text{–}Ti} > \nu_{Si\text{–}OH\text{–}Zr} > \nu_{Si\text{–}OH\text{–}Al}$ which is also the expected order of increase of acidity based on electronegativity considerations: Si–OH–Ti < Si–OH–Zr < Si–OH–Al. Similar trends were reported for other mixed oxide systems [12].

The results shown in Figure 2 demonstrate that partially dehydrated oxide surfaces exhibit a large inventory of OH groups and water molecules interacting through hydrogen bonds. Only severe dehydroxylation at 875 K produces isolated surface groups, i.e. almost free of lateral interactions. They coexist with coordinatively unsaturated Lewis acid and Lewis base sites (structures **II** and **III** in Scheme 1), which may also be present on the surface after a severe dehydroxylation. For incompletely dehydrated surfaces, close to those depicted by Figures 2a and 2b, the hydroxyl population and the layer of adsorbed water are the sources of protons that cause enhanced surface electrical conductivity and catalytic activity. Significantly, the increase of surface conductivity is paralleled by growth of either the amount of weakly bound protons or of their mobility [13].

Almost all metal oxides are active in catalytic isomerization of alkenes, which is one of the least demanding reactions in terms of the requirements for the acid strength of active sites [14]. Studies on several oxide systems show that the catalytic activity is generated by mild dehydration, when the state of the surface is close to that depicted by spectra of Figure 2a [11]. The activity is lost after extensive dehydration and is partially restored by rehydration [15]. In all these examples the catalytically active sites were proton donors (Brönsted acids) of the oxide surface.

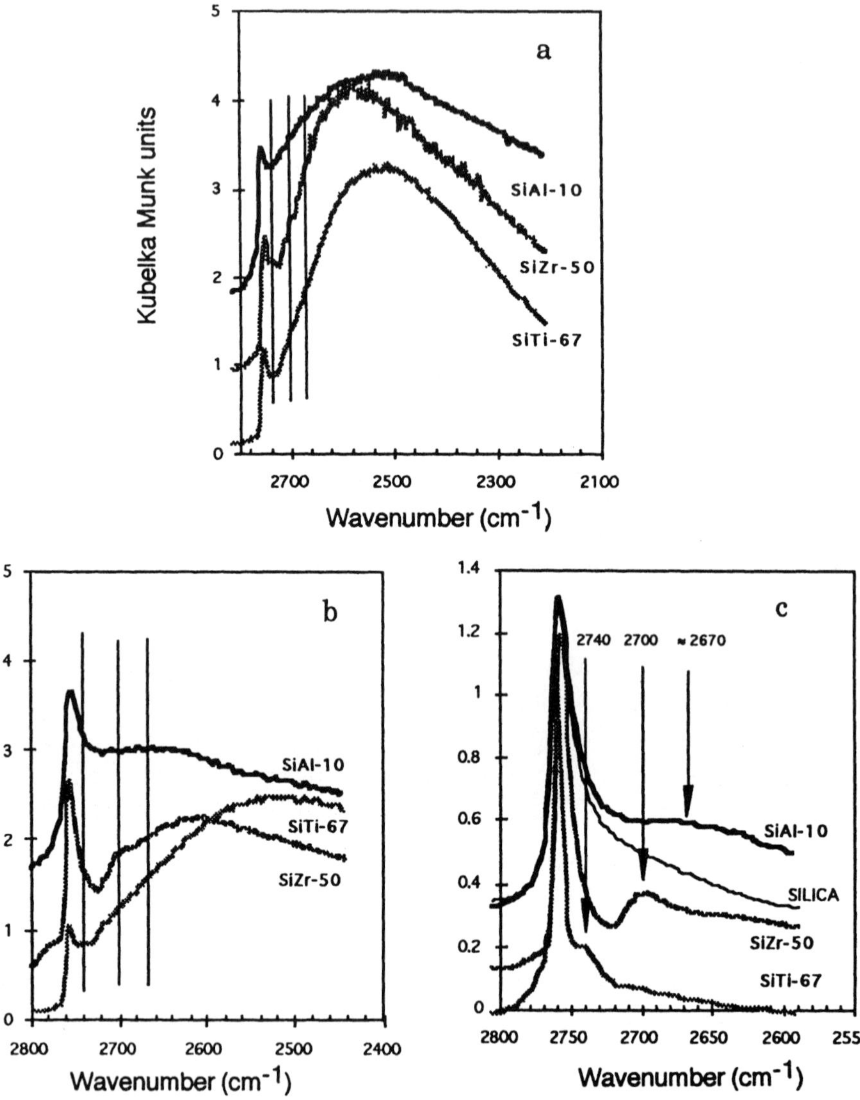

Figure 2. DRIFT spectra of O–D stretching vibrations of three mixed oxides catalysts after various dehydration treatments in the measuring cell of the FTIR spectrometer: (a) vacuum, 425 K, 30 min; (b) Ar flow, 575 K, 30 min; (c) Ar flow, 875 K, 30 min. The following catalysts are shown: SiO_2-TiO_2 with 67 % TiO_2 (SiTi); SiO_2-ZrO_2 with 50 % ZrO_2 (SiZr); SiO_2-Al_2O_3 with 10 % Al_2O_3 (SiAl).

3.4. Fully Hydrated Surfaces

A complete hydration of surface species corresponding to structures **II** and **III** from Scheme 1 results in species depicted by structure *V*. At this stage, the surface density of surface hydroxyls is so high that they enter in strong lateral interactions through hydrogen bonds and yield the typical infrared pattern shown in Figure 2b. The surface is fully hydroxylated and hydrated, although the sample may still appear "dry". According to common sense, only samples that are soaked in water are referred to as "wet". For our discussion, the surface of a wet sample is assumed to be completely hydrated *and* in contact with a liquid phase. If the latter is an aqueous solution, it acts as an infinite proton reservoir that exchanges protons with the hydroxyl groups on the oxide surface. Thus, new species are formed on "wet" oxide surfaces (structures **IV**, **V** and **VI** in Scheme 1).

At the solid/solution boundary, interfacial phenomena include electrostatic adsorption, ion exchange, and ligand binding of species from solution. However, the species that count as charge determining in aqueous media are, in many instances, H^+ and OH^- ions. Because the concentration of these ions in solution is pH-dependent, the predominance of one or another of the surface structures **IV**, **V** and **VI** is also pH-dependent.

From the liquid side of the solid/solution interface, the charge neutralization at the interface is best approached by an electrostatic or ligand chemistry description, or by a combination of both. However, from the solid side of the interface, the charge development is viewed as a combination of proton binding or proton release processes, which occur at the outermost oxygen layer of the oxide lattice. All proton transfer processes are localized on the solid side of the interface. They can be regarded as acid–base interactions between the oxygen ions at the oxide boundary and one or two interface-structured layer(s) of hydrogen-bonded water molecules in the liquid phase.

It is still a matter of controversy which site-binding model is best suited to describe adsorption of H^+ and OH^- ions on the hydrated sites of the oxide/water interface. The classical approach [16] is that all surface oxygen ions are equivalent and the corresponding hydrated surface sites have equal tendency to bind or release protons. To explain the amphoteric acid–base properties of most inorganic oxides, a charge build-up mechanism was postulated with a two-step protonation process ("two p*K*-model"):

$$MO^- + H^+ \Leftrightarrow MOH \qquad pK_{H,1} \tag{3a}$$

$$MOH + H^+ \Leftrightarrow MOH_2^+ \qquad pK_{H,2} \tag{3b}$$

The two proton binding constants are interrelated by the pH of zero charge (PZC), an experimentally measurable quantity:

$$PZC = \frac{pK_{H,1} + pK_{H,2}}{2}$$

Bolt and van Riemsdijk proposed the alternative representation of the "one p*K* model". [17]

$$M - OH^{-1/2} + H^+ \Leftrightarrow M - OH_2^{+1/2} \qquad pK_H \qquad (4)$$

In this model, only one protonation step and a single pK_H value describes the amphoteric character of surface oxygen groups. If Δ pK = log $K_{H,1}$ – log $K_{H,2}$ < 2, the "two pK-model" reduces to the "one pK model" and the unique K_H value obtained corresponds to the point of zero charge: PZC = log K_H.

These two approaches are, of course, idealized. It has been recognized for a long time that the surface of real metal oxides and hydroxides is non-uniform, i.e. several types of oxo- and hydroxo- groups are present. It has also been assumed, as in the case of partially dehydroxylated surfaces, that the surface chemistry of the oxide/solution interface can be related to the structure of the underlying solid using Pauling's concept of bond valence. Recently, Hiemstra *et al.* [18] and others [19] suggested an integrated approach based on the following protonation scheme:

$$M_x - OH^{(\upsilon x-2)} + H^+ \Leftrightarrow M_x - OH^{(\upsilon x-1)} \qquad K_{H(x,1)} \qquad (5a)$$

$$M_x - OH^{(\upsilon x-1)} + H^+ \Leftrightarrow M_x - OH_2^{(\upsilon x)} \qquad K_{H(x,2)} \qquad (5b)$$

This representation is the basis of the multisite complexation model (MUSIC). It is assumed that the number of protons bound to surface oxygen sites on different crystal planes may be 0, 1 or 2. The corresponding surface species are related to each other in distinct proton binding equilibria, each with a particular constant: $K_{H(x,1)}$ (for oxo-groups), and $K_{H(x,2)}$ (for hydroxo- groups). In the above chemical equilibria, v is the bond valence, which depends on the coordination number, x, of surface oxygen by metal cations; the formal charge per site may be any fractional number, positive or negative.

Figure 3 depicts the two-fold variability of surface groups that may serve as sites for proton binding. First, surface oxygen exposed to the solution on different surface planes may have different coordination numbers with respect to metal cations. Second, for each type of surface oxygen (terminal, or type I; bridging, or type II; triple bridging, or type III) both proton binding equilibria shown by equations 5a and 5b above are, in principle, possible. The MUSIC model incorporates, therefore, the "one pK" and "two pK" models as particular cases. However, in the pH range where most oxides are stable (about pH 3 to 11), only one of the two protonation reactions may be "active". Thus, in practical terms, a "one pK – multisite" approach [20] seems more appropriate for description of proton binding equilibria and of corresponding thermodynamic properties [21] at the oxide-solution interface. Based on Pauling's valence rule, it is assumed that the basicity of surface oxygen species increases in the order type III < type II < type I. Therefore, the equilibrium between species **IV** and **V** in Scheme 1 (deprotonation of monodentate complexes) occurs at a lower pH than the equilibrium between species **V** and **VI** (deprotonation of bridging hydroxyl groups). Proton dissociation of terminal hydroxyls of species **VI** is possible at even a higher pH.

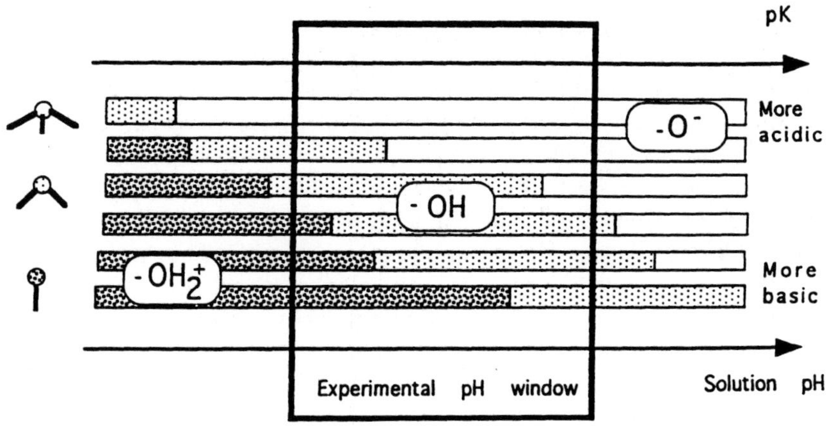

Figure 3. The variability of proton binding sites at the hydrated oxide surface is determined by structural factors, such as the occurrence of surface oxygen ions with one, two or three neighboring cations. Overlapped on this, the inventory of surface species that may be formed by binding of 2, 1 or 0 protons (and of the associated surface charges) depends on solution pH. The assumption of the MUSIC model is that only one proton binding equilibrium is observed for each type of surface sites in the experimentally accessible pH window from about 3 to 11.

An example which illustrates the above crystallochemical considerations is shown in Figure 4. The upper part shows the pK spectrum (or the Proton Affinity Distribution) determined by Contescu *et al.* from potentiometric titration of γ Al_2O_3 [20]; the cartoon at the bottom depicts a protonation scheme of surface oxygen species that was considered most probable. The surface oxygen groups on alumina were previously classified by Knözinger and Ratnasamy [8] according to the number and type (i.e., tetrahedral or octahedral) of neighboring Al atoms. Based on this classification, Contescu *et al.* [20] made the following assignments for the proton exchange equilibria revealed in the Proton Affinity Distribution: $pK_{II,a}$ = 3.8–4 for deprotonation of type II-a hydroxo groups (bridging between a tetrahedral and an octahedral aluminum ions); $pK_{I,a}$ = 6.8–7 for deprotonation of water coordinated to type I-a groups (terminal on tetrahedral aluminum); $pK_{I,b}$ = 9.8–10 for deprotonation of water coordinated to type I-b groups (terminal on octahedral aluminum); $pK_{III} < 3.5$ was tentatively assigned to deprotonation of type III hydroxo groups (in triple bridging positions), but this value is uncertain. It was also assumed that type II-b hydroxyls (bridging between two aluminum ions in octahedral positions) were stable in the experimental pH range (3 to 11) because of their high symmetry which results in charge neutrality [20a].

A consequence of the crystallochemical model with the above assignments is that at almost any solution pH, all crystal planes exhibit surface sites with both positive and negative charges. As shown in Figure 5, the distribution of these sites on various low index planes of alumina obeys specific geometrical arrangements [20b].

The authors of the MUSIC model assumed that the uncompensated surface charge was evenly distributed between metal cations and oxygen ions. They suggested a first-principle method for calculation of intrinsic proton binding constants for surface oxygen sites in various coordination environments [18a]. The model was able to estimate correctly the PZC values of several metal oxides and hydroxides. In a latter development [18c] other factors were taken into account related to structural details around surface oxygen ions. Following a suggestion by Bleam [22], the MUSIC model was modified to account for hydrogen bond formation between the oxide surface sites and neighboring water molecules.

The merit of the original MUSIC model [18] and its subsequent refinements [19] is that proton-binding constants at the solid oxide-solution interface were no longer regarded as free fitting parameters, but as interface constants. They are subjected to the constraints of physical and chemical reality, such as distribution of oxygen ions on various crystallographic planes and polarizing power of metal cations [23]. Even though the model may not be totally quantitative, the tabulated values of proton affinity constants [18a, 19] have their own merit as reference values.

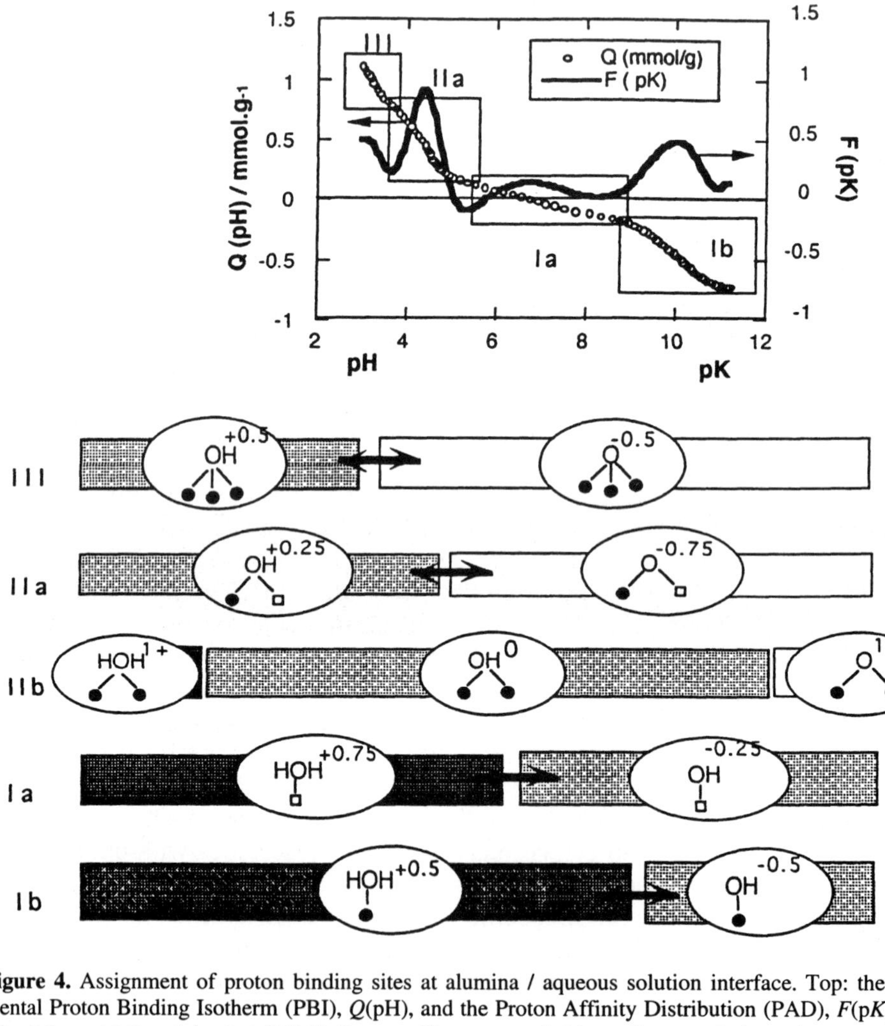

Figure 4. Assignment of proton binding sites at alumina / aqueous solution interface. Top: the experimental Proton Binding Isotherm (PBI), Q(pH), and the Proton Affinity Distribution (PAD), F(pK), determined for γ-Al_2O_3 calcined at 875 K. Bottom: The most probable surface speciation scheme on alumina surface [20] using the Knözinger and Ratnasamy's [8] classification of sites.

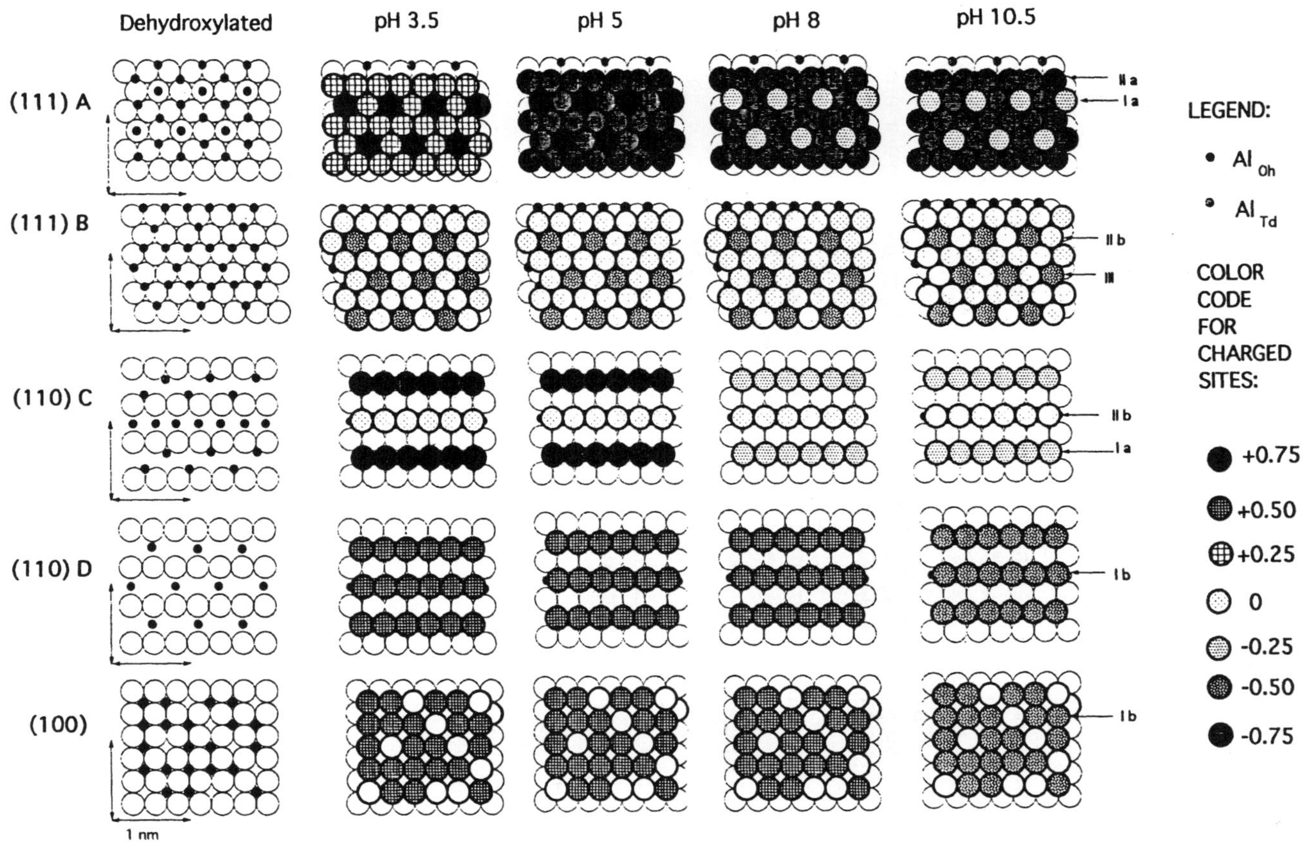

Figure 5. Ideal distribution of surface species on (111), (110) and (100) surface planes of alumina in equilibrium with aqueous solutions of selected pH values; the pH values were selected at the minima in the measured Proton Affinity Distribution shown in Figure 4. The intensity of gray shades (for positive charges) or the density of filled dots (for negative charges) shows the magnitude of electric charge on various surface species.

A method for the derivation of Proton Affinity Distributions (PAD) spectra from experimental potentiometric titration data was developed by Contescu *et al.* [20]. As shown in Figure 6, the potentiometric titration curve (pH vs. volume of added acid or base) is converted into a Proton Binding Isotherm (PBI), which shows the variation in surface excess of protons (relative to the state of the surface at the immersion pH) as a function of solution pH. The overall PBI is attributed to a non-uniform surface with continuously varying Langmuir-type behavior:

$$\Theta\ ([\mathrm{H}]) = \int_{\mathrm{p}K_{\mathrm{min}}}^{\mathrm{p}K_{\mathrm{max}}} \left(1+10^{\ \mathrm{pH\text{-}p}K}\right)^{-1} F\ (\mathrm{p}K)\ \mathrm{dp}K$$

In the above equation, F (pK) is the distribution function of acid sites, which expresses (in moles per unit weight or surface area) the fraction of surface sites with acidity constants in the interval (pK, pK + dp$K$). Several methods are available for solving the above integral equation of adsorption [20, 24]. In general, they give similar or closely related results [25]. The challenge is to correctly derive a molecular picture of surface sites at the hydrated oxide surface, based on the numerical information (number of sites and their acid strength) revealed by the distribution function. Any additional information regarding the nature and types of surface sites that may possibly exist at the interface, if judiciously used, helps to narrow the ambiguity of the mathematical solution to this "ill-posed" problem.

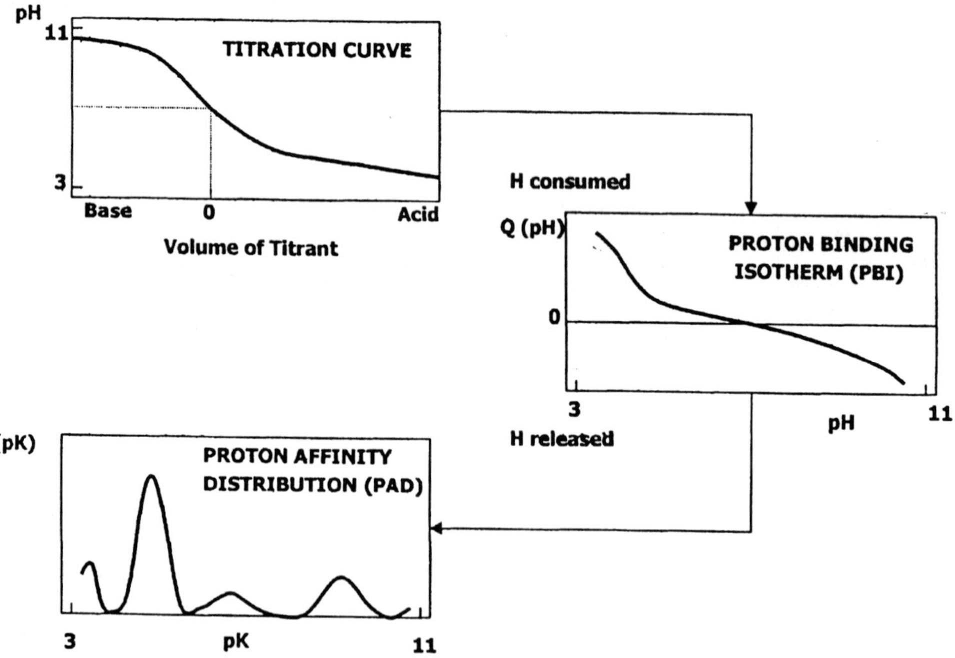

Figure 6. Schematics of data processing steps for the Proton Affinity Distribution approach.

When experimental methods to measure Proton Affinity Distributions (PAD) from potentiometric titration data became available [20], Contescu *et al.* [21, 26] found that the apparent values of proton binding constants measured from PAD were in semi-quantitative agreement with log K_H values for H^+ binding predicted by the MUSIC model (Table 1). The characteristic PAD for several oxides are shown in Figure 7.

Table 1.
Acidity constants (pK_a) for hydroxyl groups with different coordination on fully hydrated surfaces of various oxides (data predicted theoretically by the MUSIC model and measured experimentally by PAD and other techniques)

Oxide	pK calculated from MUSIC model	pK from PAD measurements	Other pK estimates
Al_2O_3	< 3 [20a]	< 3 [20a]	
	4.5 [20a]	4.5 [20a]	
	6.7 [20a]	6.7 [20a]	
	10 [27a]; 9.9 [18c]	9.5 - 9.8 [20a]	
	12.3 [18c]	-	
TiO_2 (rutile)	4.4 [18c]	-	
	5.3 [27a]; 5.8 [18c]	5.6 [26]	
	6.3 [27a]; 6.1 [18c]	6.9 [26]	
	7.5 [18c]	9.2 [26]	
TiO_2 (anatase)	4.7 [18c]	4.4 [26]	
	5.6 [18c]	5.4 [26]	5.4 (CC) [27b]
	6.3 [18c]	6.9 [26]	7.7 (CC) [27b]
	7.2 [18c]	8.7 [26]	
	-	9.7 [26]	9.5 (CC) [27b]
SiO_2	7.5 [27a]; 7.9 [18c]	8.0 [27c]	7.2 (S) [27d]
ZrO_2	4.2 [27c]	4.5 [27c]	
	-	5.4 [27c]	
	6.1 [27c]	6.7 [27c]	6.7 - 7.8 (DL) [27e]
	9.2 [27c]	9.6 [27c]	
CeO_2	-	4.0 [27f]	3.1-3.3 (DL) [27g]
	5.5 [27h]	5.8 [27f]	6.0 - 6.3 (DL) [27g]
	10.4 [27h]	10.2 [27f]	11.1-11.2 (DL) [27g]

Note: DL – double layer model; CC – constant capacitance model; S – spectroscopic methods.

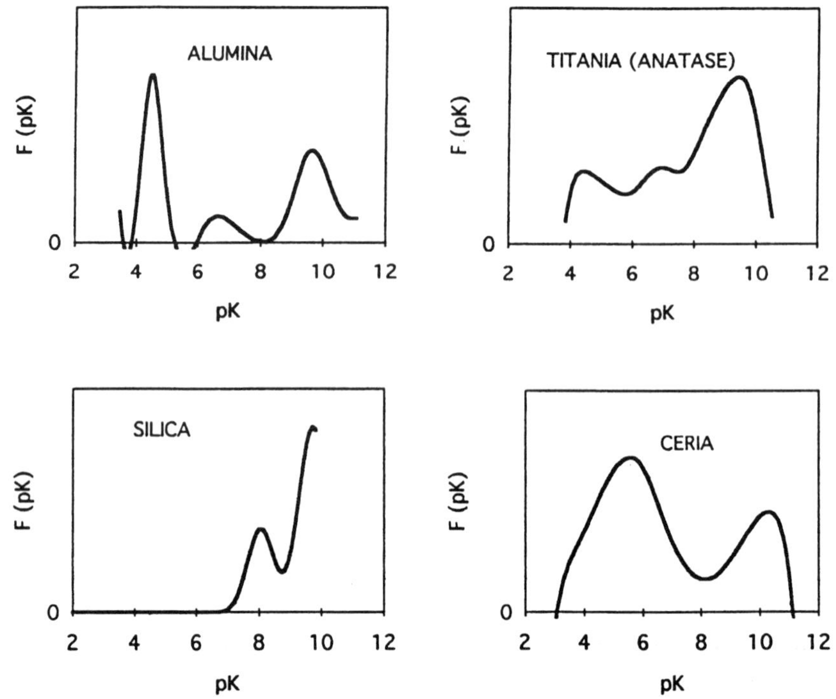

Figure 7. Proton Affinity Distributions (PAD) measured for several oxides in 0.1 N $NaNO_3$ solution.

4. FACTORS DETERMINING ACID–BASE PROPERTIES OF OXIDE SURFACES

On the basis of the above discussion it seems appropriate to relate the appearance of Lewis acidity and disappearance of Brönsted acidity with the increase in the degree of dehydroxylation of oxide surfaces. Indeed, for some oxides (ZnO, MoO_3, Cr_2O_3, WO_3) it was demonstrated that interconversion of Lewis acidity to Brönsted acidity occurred on dehydration and rehydration [28]. But which factors determine the strength of acid sites?

The chemical factor, i.e. the chemical nature of metal cations, must be considered first. The identity of metal cations and their oxidation state as well as other related properties such as ionic radius and electronegativity are primordial in determination of the ionicity of the crystalline lattice of oxides. Also, the symmetry type of the crystalline lattice, the local arrangement of surface ions and the energetic stability (and relative occurrence) of densely packed planes depend on the ratio of ionic radius of cation to that of O^{2-} ion.

In general, a fully dehydroxylated covalent surface acquires a homopolar character, with no sensible charge separation. Adsorption on covalent solids is mainly physical in nature; covalent surfaces tend to be hydrophobic, and extensive hydroxylation requires high temperatures. Hydroxyl groups present on partially dehydroxylated surfaces show very weak acidic character. They act as hydrogen bond donors with most bases, and can protonate only very strong bases. A typical example is silica [4].

The acid–base properties at the surface of oxides with ionic character are controlled mostly by electrostatic factors. The stronger the ionic character, the stronger is the electric field pointing outward from the surface. The electric field of the surface may enhance (or diminish) the local electron donor (or acceptor) properties of surface ions, and thus modify their Lewis acid–base properties; inductive effects akin to those known for small molecules were proposed. Various factors, such as electronegativity [29], effective charge on oxygen [30], and the polarizing power of metal ions [31] have been correlated with the acid–base properties of ionic oxides.

In stable oxide structures, the principle of electroneutrality requires that the charge of the cation be neutralized by the charge of surrounding oxygen ions. The distribution of charge over the surrounding ligands can be expressed per bond, leading to the concept of bond valence. Introduced by Pauling, the bond valence ν is defined as the charge of the cation divided by its coordination number:

$$\nu = z \,/\, \mathrm{CN}$$

Pauling's bond valence rule has been used in numerous attempts to correlate acidic properties of surface hydroxyls with their formal charge, as expressed by the above relationship [8, 18a, 32]. For example, the assignment of stretching vibrations of surface hydroxyls ν_{OH} already mentioned [8] is based on calculations of the formal charge on surface hydroxyls using Pauling's rule. Based on the possible relationship between acidity and stretching frequencies of hydroxyls in similar structures [33], the above reference to Pauling's rule illustrates, in fact, a relationship between structure and properties (acidity) for surface hydroxyls on partially dehydrated oxides. Pauling's valence bond rule served as a basis for the MUSIC model of proton binding to surface hydroxyls on fully hydrated oxides. The results of these semi-empirical approaches are in remarkable agreement with exact quantum mechanical calculations based on local Hartree–Fock models [38] and with many experimental trends.

An examination of Figures 1 and 7 indicates that FTIR spectra of isolated hydroxyls on oxides have their counterpart in PAD of the fully hydroxylated surfaces. The parallelism is seen with respect to both the number and strength of hydroxyl groups present on the surface. A plot of stretching frequencies of -OD groups (from Figure 1) versus the p*K* values revealed by PAD of several oxides (Figure 7) is shown in Figure 8. Despite some scatter in the data, a linear trend is evident for each oxide. At this stage one can only speculate that the slope of the straight line(s) describes the intensity of the electrostatic field(s) at the oxide surface. It apparently depends on the sample's chemical nature (possibly, through the cation's electronegativity or charge-to-radius ratio) and/or its surface structure (anion packing, preferential development of crystal planes, surface imperfections). In any case, the correlation shown in Figure 8 is similar to a relationship found for zeolites, where the increase of Brönsted acidity (from catalytic activity) parallels a decrease in the strength of O–H bond that can be monitored through the corresponding decrease of stretching frequency in FTIR spectra [4].

Besides chemical factors, the role of physical factors in determining the acid–base properties of solid oxides should also be considered. If polymorphic structures are likely, they may have different acidities depending on the packing density of surface O^{2-} ions, the degree of exposure of surface cations, and their preference for defect sites. For example, in η-Al_2O_3 the oxygen lattice is more loosely packed, the frequency of stacking faults is higher, and the fraction of cations in tetrahedral positions is slightly lower than in γ-Al_2O_3, although the local ionic configuration of Al^{3+} cations is the same in both modifications. Because of structural differences between η-Al_2O_3 and γ-Al_2O_3, the former modification has significantly higher acidity, surface reactivity, and catalytic activity than the latter [8].

Surface imperfections of crystallographic structure may occur in small sized metal oxide crystallites. Metal or oxygen vacancies are typical imperfections that cause a local charge imbalance. Therefore, the acid properties may be strongly affected by the crystalline size, in parallel with (and perhaps as a consequence of) the variation of electronic properties. For anatase particles with sizes in the range of 75–100 μm, formation of new acid sites, with higher strength on smaller sized particles, was observed [34].

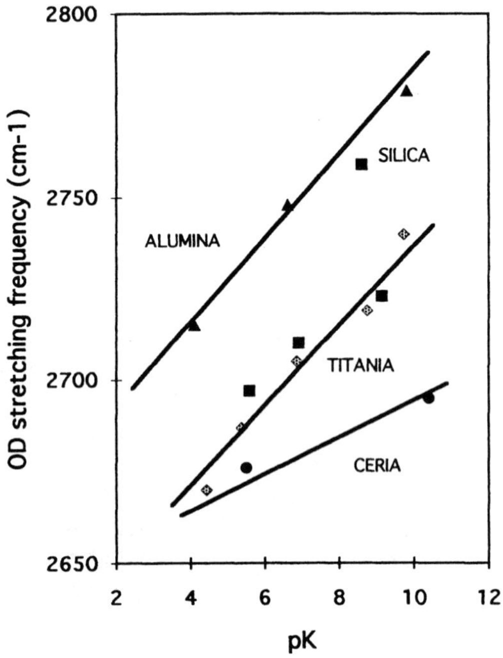

Figure 8. Relationship between stretching frequencies of isolated deuteroxyls on partially dehydrated surfaces (Figure 1) and the acidity constants pK_a determined for several oxides by PAD at the oxide/solution interface (Figure 7).

5. CONTROL OF ACID–BASE PROPERTIES OF OXIDE SURFACES

The acid–base properties of oxide materials can be controlled by mixing oxides with different properties; if phase separation does not occur, the substitution of cations of the minor component oxide in the host oxide lattice results in bulk phases of true mixed oxides.

A number of models have been advanced to predict the generation of new acid sites in dilute solid oxides solutions. From the earliest models [35] it was understood that acid centers in mixed oxide structures owed their existence to localized electrical charges created through isomorphous substitution. The model proposed by Tanabe and coworkers [36, 37] to explain acid site formation is frequently cited. This model attributes the generation of surface acidity to the charge imbalance that develops along the M–O–M' heterolinkages (where M and M' are different cations) due to the differences in cation valence and coordination number at the site of lattice substitution of the minor component cation. Kung [38] proposed an alternate model which accounts for variations in the electrostatic potential experienced by the guest cation from all ions in the oxide matrix. Although based on different assumptions, in most cases Kung's model gives the same prediction as Tanabe's model. However, an important difference exists between the two models: Tanabe's model is localized, while the model by Kung is delocalized. In other words, in Tanabe's model the site of acidity is fixed at the substituted cation, while in Kung's model the newly created acid sites may be located at any position with respect to the substitution site. Arguments supporting the localized model came recently from the works of Ko and coworkers [39] on mixed oxide aerogels. Using preparation techniques of sol-gel chemistry that afford a better control of molecular scale mixing, they demonstrated that well-mixed two component oxides (zirconia-silica, titania-silica) displayed higher total site densities, larger Brönsted populations, and enhanced activity as acidic catalysts than less homogeneously mixed samples of the same composition. These observations were explained in terms of the much higher number of heteroatom M–O–M' linkages accessible in each well-mixed oxide pair.

The acidity of zeolites is easily explained by Tanabe's model. Substitution of a Si atom by a tetrahedral Al atom in a silica-rich host lattice generates AlO_4^- units with a localized negative charge that may be compensated by a cation or a proton. This picture was confirmed by an extensive body of quantum mechanical calculations of acidity from the proton affinity of small clusters [40, 41]. On theoretical grounds it was concluded that the O-H bond weakened and the proton was more easily ionized if the hydroxyl was located in bridging positions between Al and Si atoms. A general discussion on the acidity concept in zeolites can be found in Ref. [42].

The chemical modification of the outermost atomic layers is another method for adjustment of surface acidity of solidoxides. Connel and Dumesic [43] were able to obtain acid centers (both Lewis and Brönsted) by doping small amounts of a second metal cation on the surface of several catalysts, ranging from very basic (MgO) to acidic (Al_2O_3) or inert (SiO_2 and TiO_2). The "impurity" ions were not uniformly substituted in the bulk matrix oxide lattice, but incorporated in defect cation positions at the surface of the host oxide. The generation of acidity on chemically modified catalyst surfaces was explained on the basis of changes in electronegativity and coordination environment experienced by dopant metal ions on the host oxide surface.

Another practical method for control of acid–base properties of oxide surfaces is doping the bulk of the major oxide component with electropositive or electronegative elements. Scokart and Rouxhet [44] used Pauling's valence rule to evaluate the net charge of non-protonic acid sites thus formed, and found agreement with the strength of Lewis acid sites as measured by pyridine adsorption. Doping is frequently used in the practice of catalyst preparation, where modifiers from the group of alkali [45] or alkaline earth [44] metals are used to neutralize acidity, and those selected from electronegative elements [44, 46, 47] are used to promote acidity. Berteau *et al.* [48] summarized a scale of the effects of various modifying ions on acid–base properties of alumina as follows:

$$Na^{+} \leq Mg^{2+} < \text{pure } \gamma\, Al_2O_3 < PO_4^{3-} \leq Cl^{-} < SO_4^{2-} < F^{-}$$

$$\text{increase of acidity } \rightarrow \rightarrow \rightarrow$$

$$\leftarrow \leftarrow \leftarrow \text{ increase of basicity}$$

The discovery of method of chemical modification with sulfate ions was a breakthrough for the development of new non-zeolitic solid acid catalysts. It was found that metal oxides, especially ZrO_2, TiO_2, Fe_2O_3, and zirconia-containing mixed oxides showed unusually strong acidity after modification with sulfate ions. The strong "catalytic" acidity of these materials is evident by their ability to activate paraffins at lower reaction temperatures than conventional acid catalysts. Several authors [49] denoted this property as "superacidity" and defined superacids as materials with an acid strength stronger than 100 % sulfuric acid.

No agreement has been reached about the structure, the strength and the nature of the acid sites in "superacidic" catalysts. In many cases the properties are strongly influenced by the preparation conditions. Brönsted acidity is commonly observed on sulfated zirconia, although strong Lewis acidity has also been implicated in the high reactivity of these materials [50]. Non-sulfated zirconia exhibits two types of isolated OH groups. After modification with sulfuric acid, the terminal (most basic) OH group disappears, the acidity of the bridging OH group increases, and new, more acidic, Brönsted sites are formed [51].

There is spectroscopic evidence that water converts the Lewis acid sites of sulfated zirconia into Brönsted acid sites [52]. At the same time, water promotes isomerization reactions over sulfated zirconia for moderate extent of catalyst dehydration. Similarities were reported between the effect of rehydration on isomerization activity of sulfated zirconia [53] and other oxide catalysts [15] that are consistent with the role of surface H^{+} donor sites in reactions of hydrocarbon isomerization. However, when spectroscopic methods using basic probes were used to compare sulfated zirconia and zeolites in terms of the strength of their acid sites, the results were inconsistent with all catalytic data. In these systems, which have different structures and compositions, the "catalytic" and the "physico-chemical" scales for the strength of acid–base interaction may differ significantly.

The deposition of bidimensional oxide overlayers on oxide supports opens another avenue for the control and modification of acid–base properties of oxides. For example, a strong increase in proton acidity of alumina was found on addition of small amounts (maximum 5 %) of MoO_3, WO_3 and CrO_3 [54]. The systems of bidimensional oxide overlayers have attracted an increased interest in recent years, as they are intermediates in the preparation of hydrotreating catalysts, some of the most widely used industrial catalysts [55].

All oxides that form bidimensional overlayers (CrO_3, MoO_3, WO_3, Re_2O_7, V_2O_5) on a second, high surface area oxide (Al_2O_3, SiO_2, TiO_2) have a tendency to aggregate as polyoxoanions in acidic media and to disperse as monomeric oxoanions in basic media. Hence, the match of acid–base properties between the dispersed and host oxides decisively influences the stability of bidimensional structures in various systems. Supported oxide species follow the same chemistry as they do in aqueous solution as a function of pH. According to Wachs [56], the bidimensional structures formed under ambient conditions correspond to those chemical species that are stable at the point of zero charge (PZC) of the composite oxide. Therefore, oxide supports with basic character favor those molecular species in the surface overlayer that are stable in basic solutions, and vice-versa. Also, the structural conformation of the overlayer oxide depends on the degree of surface coverage [57]. Moreover, when exposed to aqueous solutions of variable pH, the structure of the overlayer changes with the solution's pH. Based on this observation, Adachi *et al.* [58] showed that the PAD of supported MoO_3–Al_2O_3 systems measured by potentiometric titration could be correlated with the catalytic activity of the catalysts obtained after sulfidation in a typical hydrodesulfurization reaction. This is another interesting application in which the properties of the wet and dry surfaces of the same oxide system could be correlated.

6. RELATIONSHIP BETWEEN "WET" AND "DRY" SURFACES

In the discussion above it was shown, based on several examples, that a relationship could be established between the acid–base properties of the wet and (semi) dry surface of the same oxide. In particular, results from two experimental techniques, the Proton Affinity Distribution (PAD) approach and FTIR spectroscopy were compared in Figure 8 for several oxides, both in wet and dry conditions. It is evident that these two techniques characterize different properties of surface hydroxyl groups, namely the apparent acidity constants at the "wet" oxide/solution interface (PAD), and the frequency of stretching vibrations of O–H bonds of non-interacting hydroxyls at the "dry" oxide surface (FTIR). The latter are the measure of the corresponding force constants along the O–H bonds. At the molecular level, the higher the force constant, the larger is the vibration frequency of a particular surface group; on the other hand, higher force constants imply stronger O–H bonds and therefore stronger basicity (lower tendency of proton dissociation) of the corresponding surface groups. A molecular interpretation based on the localized character of chemical bonds is, therefore, able to provide a simple rationale for the correlation shown in Figure 8. Although from a different perspective, these two properties describe the acid–base character of surface hydroxyl species on both the "wet" and "dry" surface of the same oxide.

The reversibility of wet–dry transformations is high, especially for ionic oxides. The common denominator that generates stable oxygen species in both surface conditions, i.e. "wet" (fully hydrated) and "dry" (partially dehydroxylated), is the arrangement of ions in the outermost surface layers of the oxide lattice. It is assumed that this is not influenced by hydration of the oxide surface, except for adsorption of water to coordinatively unsaturated cations and proton binding to surface oxygen sites. More specifically, in the picture presented here it is assumed that the proton binding / proton donor sites at both interfaces ("wet" and "dry") are surface oxo- and hydroxo-groups which terminate the oxide layer. They bind (through secondary interaction) a layer of water molecules, which is in contact with molecules from an adjacent phase, either liquid or gas. The magnitude of uncompensated charge on various surface sites is determined by the underlying structure of the oxide lattice and can be rationalized, within the crystallochemical model, based on the Pauling valence rule.

This picture of acid–base properties of solid surfaces in contact with a "wet" or "dry" environment has received an independent confirmation from measurements of catalytic activity. A correlation of the strength of surface acid sites on oxides with the rates of catalytic reactions in acid-catalyzed reactions on heterogeneous catalysts is a challenge. A relationship is known in homogeneous catalysis between the acid strength (pK_a) of an acid in solution and the rate of a reaction catalyzed by that acid: the stronger the acid, the easier is to activate the reactant and the faster is the catalyzed reaction. The Brönsted-Pedersen relationship relates the effectiveness of a general acid catalyst to its acid–base strength as:

$$k_A = G\,(K_a)^{\beta}$$

where k_A is the specific rate constant for a reaction catalyzed by an acid whose acid dissociation constant is K_a ; G and β are constants for a series of similar catalysts.

Several authors assumed that a similar correlation should exist for heterogeneous catalysis [59], but the difficulty of measuring at the same time the number and the strength of Brönsted acid sites on solid catalysts has hindered any experimental proof of a complementary relationship. Recently, Contescu *et al.* [11, 27c] demonstrated the occurrence of a Brönsted type relationship for isomerization of 1-butene on a series of mixed oxides catalysts (SiO_2–Al_2O_3, SiO_2–ZrO_2, and SiO_2–TiO_2). The number and strength of Brönsted acid sites at the oxide/solution interface were determined from Proton Affinity Distributions (PAD) of solid acid catalysts [27c]. For the same catalysts, the catalytic properties (1-butene isomerization at 423 K) after a mild dehydration (473 K) were also evaluated. In these experiments, the state of the surface during the catalytic tests was close to that depicted by DRIFT spectra already shown in Figure 2a, i.e. the oxide surface was covered by hydrogen bonded hydroxyls and water molecules. For catalysts of various compositions it was found that the reaction rates per unit area were correlated with the *number* of a particular type of acid site (Figure 9a), identified as the bridging hydroxyl in Si–O–M structures. Significantly, in each binary oxide system the active sites were characterized by nearly constant *acid strength* values (expressed on the pK scale of water). This shows, indeed, that the acid strength in Si–O–M structures depends on the electronegativity of M. Moreover, the logarithms of specific reaction rates constants gave linear plots versus the pK values of the active sites identified on catalysts from various binary systems, see Figures 9b and 9c [11, 27c]. This is the equivalent in heterogeneous catalysis of the Brönsted relationship, first developed in homogeneous catalysis.

The fact that a Brönsted type relationship was experimentally found for solid acid catalysts has both practical and theoretical significance. From a practical point of view, this shows that the Proton Affinity Distribution approach affords a direct and quantitative determination of both the number and the strength of Brönsted acid sites on surfaces of oxide catalysts. Also, the fact that potentiometric titration data for the oxide/water interface could be correlated with the catalytic activity of the oxide/gas interface (in systems with similar chemical properties) confirms that a common denominator exists which determines the acid–base properties of the surface in both states, "wet" and "dry". The very same finding re-opens the discussion on the relevance of acidity scales, either that of gases or that of water, in characterization of surface acidity of solids.

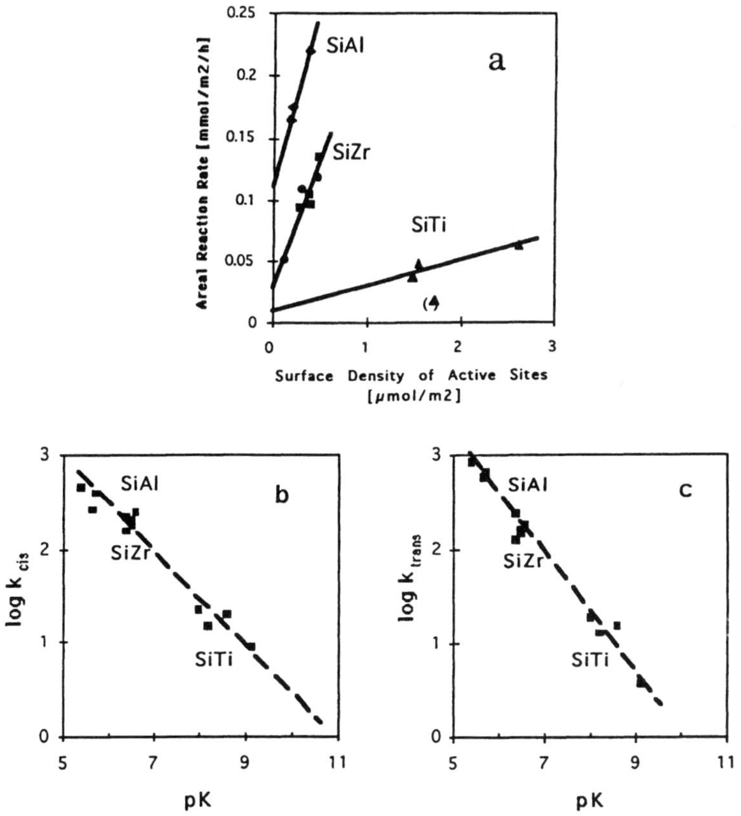

Figure 9. Relationship between catalytic activity (isomerization of 1-butene to 2-*cis* and 2-*trans* butenes) on several mixed oxide catalysts (SiO_2-Al_2O_3, SiO_2-ZrO_2, SiO_2-TiO_2) and their surface acidity on the aqueous pK_a scale: (a) linear dependence between specific reaction rates and the density of active sites from PAD data; (b) linear Brönsted relationship between the reaction rate of 2-*cis*-butene formation and the acid strength of corresponding active sites on different mixed oxides; (c) same relationship for formation of 2-*trans*-butene.

7. SUMMARY AND CONCLUSION

An understanding the acid–base properties of solid surfaces is important in many fields of science and engineering. For example, the solid/gas interface intervenes when solid acids (or bases) are used as heterogeneous catalysts. The processes at the solid/liquid interface between natural waters and rocks, minerals or soils are important in the weathering processes that determine the geohydrochemical circuit of elements.

Similar to homogeneous one-phase systems, the acidity of solid surfaces manifests itself only in relation with a base that must be present; the strength of the acid–base interaction depends, therefore, on the particular base involved. This is true for both Brönsted and Lewis acid sites, and a similar statement can be made for surface basic sites.

At the molecular level, the acid–base properties of oxide surfaces originate in the charge disparity between coordinatively unsaturated ions (metal cations and O^{2-} ions) at the oxide surface. The local charge imbalance at the surface is related to chemical factors (nature of metal cation) and geometrical effects (particular atomic arrangements on various crystal planes). These factors are inherent to a particular oxide and determine the intrinsic properties of the atoms in the outermost surface layer, such as the proton affinity (i.e., basicity) of oxygen ions, and the electron affinity (i.e., Lewis acidity) of exposed cation sites. The Pauling valence rule was successfully used in various crystallochemical and quantum-mechanical models of acid–base properties.

The role of surface-bound water is crucial in determining the type (i.e., Brönsted or Lewis) of prevailing acid–base sites. The wet–dry transformations at the surface of ionic oxide materials are, in most cases, reversible. In contact with the ambient, oxide surfaces become covered with a layer of chemisorbed and/or physisorbed water molecules; the oxygen of the water molecule in the first adsorbed layer is indistinguishable from the surface oxygen of the oxide lattice. Protons attached to the oxygen sites compensate the charge deficit caused by missing cations and create centers of Brönsted acidity. The existence of surface hydroxyls and bound water on oxides, even after strong dehydration, is shown by infrared spectroscopy. The presence of a layer of surface water on oxides is important for the catalytic reactivity of many oxide systems. Even after complete hydration, the local structures in the first molecular layer of water on oxide surfaces are not much different from those found on moderately dehydrated oxides and identified by infrared spectroscopy. In addition, the degree of protonation and the surface charge of oxides dispersed in an aqueous environment are pH-dependent.

The above description of acid–base properties of oxide surfaces is strictly *localized* from an atom- or molecular- size perspective. A localized description is useful for the explanation of molecular events involved in, for example, the mechanism of a surface catalytic reaction at the "dry" oxide surface, or the dissolution process at a mineral surface in contact with the liquid phase. However, many properties of solid surfaces, such as wetting, adhesion, surface energetics, do not require a localized description. In these cases, the observable properties are the result of many local interactions at the atomic scale, averaged over a significantly large surface region. Other structural and morphological factors, associated with the occurrence of surface imperfections, defect sites, surface roughness, may become significant. They may intervene with various contributions in the averaged result. However, it is our belief that a molecular description is needed. Such a localized description may serve to define the baseline for any subsequent analysis of acidity and basicity of solid surfaces. *After all, acidity and basicity are molecular properties of individual chemical species, rather than global descriptors of statistically significant populations.*

REFERENCES

1. J. C. Berg, in: *Wettability*, J. C. Berg (Ed.), p. 76. Marcel Dekker, New York, (1993).
2. (a) K. L. Mittal and H. R. Anderson, Jr. (Eds.), *Acid–Base Interactions: Relevance to Adhesion Science and Technology*, VSP, Utrecht, 1991. (b) M. F. Finlayson and B. A. Shah, *J. Adhesion Sci. Technol.* **4**, 431 (1990); (c) H. F. Webster, and J. P. Wightman, *J. Adhesion Sci. Technol.* **5**, 93 (1991); (d) T. S. Oh, L. P. Buchwalter, and J. Kim, *J. Adhesion Sci. Technol.* **4**, 303 (1990); (e) E. P. Papirer and H. Balard, *J. Adhesion Sci. Technol.* **4**, 357 (1990); (f) A. Larsson and W. E. Johns, *J. Adhesion* **25**, 121 (1988); (g) D. Ma, W. E. Johns, A. K. Dunker, and A. E. Bayoumi, *J. Adhesion Sci. Technol.* **4**, 411 (1990).
3. P. C. Stair, *J. Amer. Chem. Soc.* **104**, 4044 (1982).
4. A. Zechina and C. O. Arean, *Catal. Rev. – Sci. Eng.* **35**, 261 (1993).
5. P. Nortier, A. P. Borosy, and M. Allavena, *J. Phys. Chem. B* **101**, 1347 (1997).
6. A. A. Tsyganenko and V. N. Filimonov, *J. Mol. Structure* **19**, 579 (1973).
7. C. Contescu and J. A. Schwarz, unpublished results (1995).
8. H. Knözinger and P. Ratnasamy, *Catal. Rev. – Sci. Eng.* **17**, 31 (1978).
9. (a) P. Nortier, P. Fourre, A. B. Mohammed Saad, O. Saur, and J. C. Lavalley, *Appl. Catal.* **61**, 141 (1990); (b) A. B. Mohammed Saad, V. A. Ivanov, and J. C. Lavalley, *Appl. Catal. A, General* **94**, 71 (1993).
10. (a) G. Busca, V. Lorenzelli, V. Sanchez Escribano, and R. Guidetti, *J. Catal.* **131**, 167 (1991); (b) G. Busca, V. Lorenzelli, G. Ramis, and R. J. Willey, *Langmuir* **9**, 1492 (1993).
11. C. Contescu, V. T. Popa, J. B. Miller, E. I. Ko, and J. A. Schwarz, *Chem. Eng. J.* **64**, 265 (1996).
12. (a) D. Barthomeuf, in: *Catalysis by Acids and Bases*, B. Imelik (Ed.), p. 75. Elsevier, Amsterdam (1985); (b) D. Barthomeuf, in: *Molecular Sieves II*, A.C.S. Symposium Series No. 40, R. Katzer (Ed.), p. 453. American Chemical Society, Washington D.C. (1977).
13. (a) S. R. Morrison, *The Chemical Physics of Surfaces*, Plenum Press, New York (1977); (b) V. F. Kiselev and O. V. Krylov, *Electronic Phenomena in Adsorption and Catalysis on Semiconductors and Dielectrics*. Nauka, Moscow (1979).
14. H. Kung, in: *Stud. Surf. Sci. Catal.* vol. 45, B. Delmon and J. T. Yates (Eds.). Elsevier, Amsterdam (1989).
15. (a) S. G. Hindin, A. G. Oblad, and G. A. Mills, *J. Am. Chem. Soc.* **77**, 535 (1955); (b) E. I. Ko, J. P. Chen, and J. G. Weissman, *J. Catal.* **105**, 511 (1987); (c) M. R. Gonzales, K. M. Kobe, K. B. Fogash, and J. A. Dumesic, *J. Catal.* **160**, 290 (1996); (d) K. H. Nam, *M. S. Thesis*, Syracuse University (1997).
16. (a) T. W. Healy and R. L.White, *Adv. Colloid Interface Sci.* **9**, 303 (1978); (b) R. D. James and G. A. Parks, in: *Surface and Colloid Science*, vol. 12, E. Matijevic (Ed.), Plenum Press, New York (1990).
17. G. H. Bolt and W. H. van Riemsdijk, in: *Soil Chemistry, Physico-Chemical Models*, G. H. Bolt (Ed.), Elsevier, Amsterdam (1982).
18. (a) T. Hiemstra, J. C. M. De Wit, and W. H. Van Riemsdijk, *J. Colloid Interface Sci.* **133**, 91 (1989); (b) T. Hiemstra and W. H. Van Riemsdijk, *J. Colloid Interface Sci.* **179**, 488 (1996); (c) T. Hiemstra, P. Venema and W. H. Van Riemsdijk, *J. Colloid Interface Sci.* **184**, 680 (1996).
19. L. K. Koopal, *Electrochimica Acta* **41**, 2293 (1996).
20. (a) C. Contescu, J. Jagiello, and J. A. Schwarz, *Langmuir* **9**, 1754 (1993); (b) C. Contescu, J. Hu, and J. A. Schwarz, *J. Chem. Soc. Faraday Trans.* **89**, 4091 (1993).
21. C. Contescu, A. Contescu, and J. A. Schwarz, *J. Phys. Chem.* **98**, 4327 (1994).

22. W. F. Bleam, *J. Colloid Interface Sci.* **159**, 312 (1993).
23. G. M. S. El Shafei, *J. Colloid Interface Sci.* **182**, 249 (1996).
24. (a) J. Jagiello, *Langmuir* **10**, 2778 (1994); (b) J. Jagiello, T. Bandosz, K. Putyera, and J. A. Schwarz, J. *Colloid Interface Sci.* **172**, 341 (1995); (c) M. Borkovec and G. J. M. Koper, *Langmuir,* **10**, 2863 (1994).
25. (a) M. Cernik, M. Borkovec, and J. C. Westall, *Environm. Sci. Technol.* **29**, 413 (1995).
26. C. Contescu, V. T. Popa, and J. A. Schwarz, *J. Colloid Interface Sci.* **180**, 149 (1996).
27. (a) T. Hiemstra, J. C. M. de Wit, and W. H. van Riemsdijk, *J. Colloid Interface Sci.* **133**, 105 (1989); (b) C. Ludwig and P. W. Schindler, *J. Colloid Interface Sci.* **169**, 284 (1995); (c) C. Contescu, V. T. Popa, J. B. Miller, E. I. Ko, and J. A. Schwarz, *J. Catal.* **157**, 244 (1995); (d) K. Marshall, G. L. Ridgewell, and C. H. Rochester, *Chem. Ind. (London)* **19**, 775 (1974); (e) W. Janusz, *J. Radioanal. Nucl. Chem.* **125**, 393 (1988); (f) C. Contescu, unpublished results (1995); (g) N. Sh. Petro, N. Z. Misak, and I. M. El-Naggar, *Colloids Surfaces* **49**, 211 (1990); (h) M. Nabavi, O. Spalla, and B. Cabane, *J. Colloid Interface Sci.* **160**, 459 (1993).
28. (a) T. Morimoto, H. Yanai, and M. Nagao, *J. Phys. Chem.* **80**, 471 (1976); (b) K. Segawa, and W. Hall, *J. Catal.* **76**, 133 (1982); (c) R. Groff, *J. Catal.* **86**, 215 (1984).
29. (a) K. Tanabe and A. Ozaki, *J. Catal.* **8**, 1 (1967); (b) K. Shibata, T. Kiyoura, J. Kitagawa, T. Sumiyoshi, and K. Tanabe, *Bull. Chem. Soc. Jpn.* **46**, 2985 (1973).
30. (a) J. Take, N. Kikuchi, and Y. Yoneda, *J. Catal.* **21**, 164 (1971); (b) T. Yamanaka and K. Tanabe, *J. Phys. Chem.* **80**, 1723 (1976).
31. G. M. S. El Shafei, *J. Colloid Interface Sci.* **182**, 249 (1996).
32. (a) G. A. Parks, *Chem. Rev.* **65**, 177 (1965); (b) R. H. Yoon, T. Salman, and G. Donnay, *J. Colloid Interface Sci.* **70**, 483 (1979).
33. E. Knözinger, K. H. Jacob, S. Singh, and P. Hofmann, *Surface Sci.* **290**, 388 (1993).
34. H. Nakabayashi, N. Kakuta, and A. Ueno, *Bull. Chem. Soc. Jpn.* **64**, 2428 (1991).
35. (a) C. L. Thomas, *Ind. Eng. Chem.* **41**, 2564 (1949); (b) C. J. Plank, *J. Colloid Sci.* **2**, 413 (1947).
36. K. Tanabe, T. Sumiyoshi, K. Shibata, T. Kiyoura, and J. Kitagawa, *Bull. Chem. Soc. Jpn.* **47**, 1064 (1974).
37. K. Tanabe, M. Misono, Y. Ono, and H. Hattori, in: *Stud. Surf. Sci. Catal.*, vol. 51, B. Delmon and J. T. Yates (Eds), Kodansha, Tokyo, 1989.
38. H. Kung, *J. Solid State Chem.* **52**, 191 (1984).
39. (a) J. B. Miller, S. E. Rankin, and E. I. Ko, *J. Catal.* **148**, 673 (1994); (b) J. B. Miller, S. T. Johnston, and E. I. Ko, *J. Catal.* **150**, 311 (1994); (c) J. B. Miller and E. I. Ko, *J. Catal.* **159**, 58 (1996).
40. (a) J. Sauer, *Chem. Rev.* **89**, 199 (1989); (b) G. J. Kramer and R. A. van Santen, *J. Am. Chem. Soc.* **115**, 2887 (1993); (c) S. P. Greatbanks, P. Sherwood, and I. H. Hillier, *J. Phys. Chem.* **98**, 8134 (1994).
41. (a) W. Grabowski, M. Misono, and Y. Yoneda, *J. Catal.* **61**, 103 (1980); (b) H. Hawakami, S. Yoshida, and T. Yonezawa, *J. Chem. Soc. Faraday Trans.* **280**, 205 (1984); (c) G. M. Zhidomirov and V. B. Kazansky, *Adv. Catal.* **34**, 131 (1986).
42. (a) W. J. Mortier and R. A. Schoonheydt, *Prog. Solid State Chem.* **16**, 1 (1985); (b) D. Barthomeouf, *Mater. Chem. Phys.* **17**, 49 (1987); (c) P. O. Fritz and J. H. Lunsford, *J. Catal.* **118**, 85 (1989); (d) P. A. Jacobs, *Catal. Rev. – Sci. Eng.* **24**, 415 (1982); (e) A. Corma, *Chem. Rev.* **95**, 559 (1995); (f) R. A. van Santen and G. J. Kramer, *Chem. Rev.* **95**, 637 (1995).
43. (a) G. Connel and J. A. Dumesic, *J. Catal.* **101**, 103 (1986); (b) G. Connel and J. A. Dumesic, *J. Catal.* **105**, 285 (1987).
44. P. O. Scokart and P. G. Rouxhet, *J. Colloid Interface Sci.* **86**, 96 (1982).
45. (a) J. M. Parera and N. S. Figoli, *J. Catal.* **14**, 303 (1969); (b) C. M. M. Sachett, P. N. Da Silva, Y. L. Lam, M. Dufaux, R. Frety, and M. Primet, *Bull. Soc. Chim. France*, 357 (1989); (c) S. Srinivasan, C. R. Narayanan, A. Biaglow, R. Gorte, and A. K. Datye, *Appl. Catal. A: General* **132**, 271 (1995).
46. A. Corma, V. Fornes, and E. Ortega, *J. Catal.* **92**, 284 (1985).
47. P. M. Boorman, R. A. Kydd, Z. Sarbak, and A. Somogyvary, *J. Catal.* **96**, 115 (1985).
48. P. Berteau, M. A. Kellens, and B. Delmon, J. *Chem. Soc. Faraday. Trans.* **87**, 1425 (1991).
49. (a) K. Arata, M. Hino, and N. Yamagata, *Bull. Chem. Soc. Jpn.* **63**, 244 (1990); (b) F. R. Chen, G. Courdurier, J. Joly, and J. C. Vedrine, *J. Catal.* **143**, 616 (1993); (c) T. Riemer, D. Splielbauer, M. Hunger, G. A. H. Mekhemer, and H. Knözinger, *J. Chem. Soc. Chem. Commun.*, 1181 (1994); (d)

V. Adeeva, J. W. de Haan, J. Janchen, G. D. Lei, V. Schunemann, L. J. M. van de Ven, W. M. H. Sachtler, and R. A. van Santen, *J. Catal.* **151,** 364 (1995).
50. (a) J. H. Lunsford, H. Sang, S. M. Campbell, C. Liang and R. G. Anthony, *Catal. Lett.* **27**, 305 (1994); (b) F. Babou, G. Courdurier, and J. C. Vedrine, *J. Catal.* **152**, 341 (1995); (c) A. Corma, V. Fornes, M. I. Juan-Rajadell, and J. M. Lopez-Nieto, *Appl. Catal.* **A116**, 151 (1994); (d) F. Lonyi, J. Valyon, J. Engelhardt, and F. Mizukami, *J. Catal.* **160**, 279 (1996).
51. L. M. Kustov, V. B. Kazansky, F. Figueras, and D. Tichit, *J. Catal.* **150**, 143 (1994).
52. (a) C. Zhang, R. Miranda, and B. H. Davis, *Catal. Lett.* **29**, 349 (1994); (b) D. A. Ward, and E. I. Ko, *J. Catal.* **150**, 18 (1994); (c) C. Morterra, G. Cerato, and V. Bolis, *Catal. Today* **17**, 505 (1993).
53. M. R. Gonzales, J. M. Kobe, K. B. Fogash, and J. A. Dumesic, *J. Catal.* **160**, 290 (1996).
54. W. Kania and K. Jurczyk, *Appl. Catal.* **61**, 27 (1990).
55. I. E. Wachs and K. Segawa, in: *Characterization of Catalytic Materials,* I. E. Wachs and L. E. Fitzpatrick (Eds.), p. 69. Butterworth-Heinemann, Boston, (1992).
56. (a) G. Deo and I. E. Wachs, *J. Phys. Chem.* **95**, 5889 (1991); (b) G. Deo and I. E. Wachs, *J. Catal.* **129**, 307 (1991); (c) D. S. Kim, K. Segawa, T. Soeya, and I. E. Wachs, *J. Catal.* **136**, 539 (1992).
57. F. J. Gil-Llambias, A. M. Escudey, J. L. G. Fierro, and A. Lopez Agudo, *J. Catal.* **95**, 520 (1985).
58. (a) M. Adachi, C. Contescu, and J. A. Schwarz, *J. Catal.* **158**, 411 (1996); (b) M. Adachi, C. Contescu, and J. A. Schwarz, *J. Catal.* **162**, 66 (1996).
59. Y. Yoneda, *J. Catal.* **9**, 51 (1967).

Part 2: Characterization of the Acid-Base Properties of Materials

Acid-Base Interactions, Vol. 2, pp. 275–298
Ed. K.L. Mittal

Characterisation of acid-base properties of polymer surfaces by XPS

MOHAMED M. CHEHIMI,[1,*] MICHEL DELAMAR,[1] JAMAL KURDI,[2] FARZANEH AREFI-KHONSARI,[2] VALERIE LAVASTE[3] and JOHN F. WATTS[3]

[1] *Institut de Topologie et de Dynamique des Systèmes, de l'Université Paris 7 Denis Diderot, associé au CNRS (UPRESA 7086), 1 rue Guy de la Brosse, 75005 Paris, France*
[2] *Laboratoire de Génie des Procédés Plasmas et Traitement de Surfaces, ENSCP, 11 rue Pierre et Marie Curie, 75005 Paris, France*
[3] *School of Mechanical and Materials Engineering, University of Surrey, Guildford, GU2 5XH, UK*

Abstract—The use of X-ray photoelectron spectroscopy (XPS) in the determination of acid-base properties of molecules, polymers, metal oxides and catalysts is reviewed. The emphasis is on the XPS study of reference acidic and basic probes adsorbed on the polymer surfaces of interest. It is shown that the "retention" of such probes by the host polymers in the ultra-high vacuum can be achieved only if the adsorbent and the adsorbate interact via acid–base forces. Acid–base interactions yield negative and positive chemical shifts for the acidic and basic probes, respectively. These chemical shifts are related to the electron density transferred from the Lewis basic polymer (or probe) to the Lewis acidic probe (or polymer). The determination of the uptake of molecular probes by the host polymers is also very useful particularly in the case of polymer surface treatments. Examples of the application of XPS in the determination of acid-base properties concern model polymers, radiation-cured coating components, and plasma-treated polypropylene.

Keywords: XPS; chemical shifts; acid–base interactions; adhesion; molecular probes; polymers; metal oxides; plasma treatment.

1. INTRODUCTION

The pioneering developments of Professor Frederick M. Fowkes of the acid-base theory in adhesion has attracted the attention of several Laboratories on this subject. A Festschrift in his Honor on the occasion of his 75th birthday was published in a monograph entitled *Acid-Base Interactions: Relevance to Adhesion Science and Technology* [1]. This monograph constitutes an important step in the history of acid–base chemistry in general and adhesion science in particular as it became clear that non-dispersive interactions at interfaces were due to Lewis acid–base forces and not polar interactions. The latter contribute only 1% in the condensed phase [2] so that they are lumped with the ubiquitous dispersive interactions.

The knowledge of the acid–base properties of polymers, fibres and fillers is an important step towards the design of high performance materials (e.g. composites, coat-

*To whom correspondence should be addressed. E-mail: chehimi@paris7.jussieu.fr ings, metallized plastics, blends). The durability of such materials depends, to a great extent, on the polymer–polymer or polymer–inorganic substrate adhesion. Although adhesion can be good without adhesive or any pretreatment, in several practical situations it was conclusively shown that the durability was poor and that pretreatments of fibres or metal oxides were necessary. In this regard, surface grafting of functional groups with appropriate acidity and/or basicity leads to good and durable polymer-substrate adhesion. Conversely, it is necessary to modify the surface of apolar polymers to achieve e.g. adequate and durable metallization.

There is now a massive number of fundamental and applied examples in the literature showing that acid-base interactions are effective in improving adhesion, adsorption, wetting, solubility and mixing of polymers and other materials [1–8]. Of particular interest was an interfacial force microscopy (IFM) study of adhesion forces at the molecular scale between various mercaptan-functionalized gold probes and various substrates [9]. The maximum work of adhesion of the probe to the substrate was obtained for the acid-base ($-NH_2$ *vs.* —COOH) combinations.

The arsenal of techniques available to assess materials acid-base properties include contact angle measurements [10, 11], microcalorimetry [1], inverse gas chromatography [7, 12, 13, 14], FTIR [15] and NMR [3]. All these methods and others have been reviewed on several occasions [16, 3, 5–8, 17, 18]. Added to these techniques, the "ESCA groups" in Paris and Guildford have established, since the early 1990s, several protocals to assess the acid–base properties of polymer surfaces by XPS [5, 19–28]. By far one of the most successful approach was the XPS analysis of reference acidic and basic molecular probes adsorbed onto polymer surfaces [20, 21, 24, 25].

Traditionally, XPS has been used in adhesion science and technology to study the locus of failure of adhesive joints and for determining molecular orientation following failure [29–31], to monitor the uptake of specific ions [32, 33] at the interface, and to identify molecular species segregating at polymer-metal oxide interface [34–37]. XPS has also been found to be effective in determining adsorption isotherms of silane coupling agents onto metal oxides [38–42], and flexible polymers onto metal oxides [43] and stiff conducting polymer particles [44, 45]. In the latter case, the XPS results were interpreted in terms of acid–base interactions of the adsorbate, the conducting polymer and the casting solvent [44–46].

The success of XPS lies in its surface specificity (analysis depth of *ca.* 5 nm), low degree of degradation of materials under test, quantitative aspects, and the detection of all elements (except hydrogen) and their chemical shifts. The so-called chemical shift is the cornerstone of XPS since it enables the surface scientist to identify the chemical environments. For example, the chemical shift of the N1s core level from aminopropylsilane coupling agent was interpreted in terms of hydrogen bond or protonation of the amino group at the silane–metal oxide interface [39, 41, 42]. Moreover, the chemical shifts were used to derive material properties such as refractive indices of thin optical layers [47], the non-dispersive component of the surface energy of polymers [48], and acid-base strengths of alcohols and amines [49]. In the case of metal oxides, Delamar [50] used an XPS data bank and a compilation of isoelectric points of inorganic solids (IEPS) to establish a linear relationship between chemical shifts of the oxygen (ΔO) and those of the 2p core shell of the metal cations (ΔM) and the IEPS of the corresponding metal oxides. Figure 1 shows a plot of IEPS versus

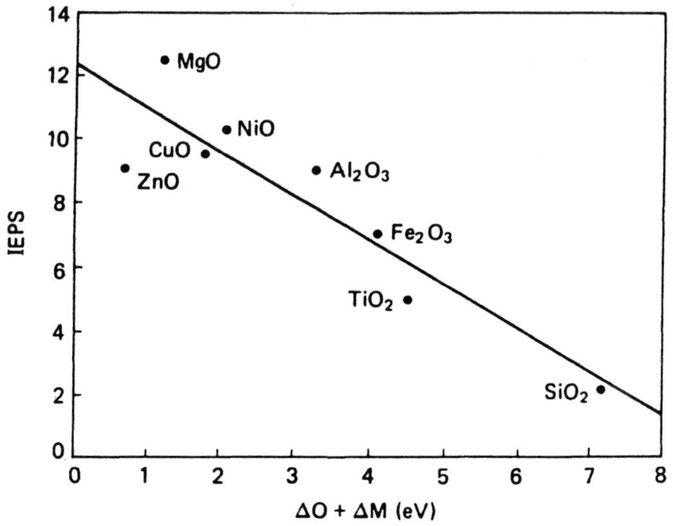

Figure 1. Plot of IEPS *vs.* (ΔO+ΔM) for a series of metal oxides. ΔO was defined as BE(O1s)-530 eV, where BE is the binding energy and 530 eV an arbitrary reference BE value. ΔM was defined as BE(M2p)-BE($M^0$2p) where M is the metal in the oxidized state, M^0 is the metal in the reference metallic state, and 2p is the core shell. (Ref. [50])

(ΔO+ΔM) for various metal oxides. This linear correlation was subsequently confirmed by Cattania *et al.* [51] in their electrophoretic and XPS characterizations of a series of metal oxides having the same history.

The aim of this paper is to review the background to the application of XPS in the assessment of the acid-base properties of model, radiation-cured and plasma-treated polymer surfaces. Some selected examples from the XPS literature related to acid-base properties of molecules and metal oxides will also be briefly presented.

2. USE OF THE MOLECULAR PROBE TECHNIQUE IN XPS

2.1. Metal Oxides

The molecular probe technique in combination with XPS has seldom been used since the early seventies and has mainly been applied to zeolites [52–55]. These studies were aimed at characterizing the Lewis acidic sites at catalyst surfaces by monitoring the binding energy (BE) shifts of N1s from adsorbed pyridine [52, 53] and ammonia [55]. In the case of ZSM-5 zeolite [53] N1s signal from the adsorbed pyridine was fitted with three components located at BEs in the 398–402 eV range. The increasing BE positions corresponded to pyridine adsorbed onto weak and strong Lewis acidic sites, and Brönsted acidic site, respectively. The N1s component around 402 eV was ascribed to the pyridinium ion formed upon adsorption of a pyridine molecule onto a protonic OH zeolite site. In-situ heat treatment of the zeolite showed that pyridine was most strongly adsorbed on Brönsted acidic sites.

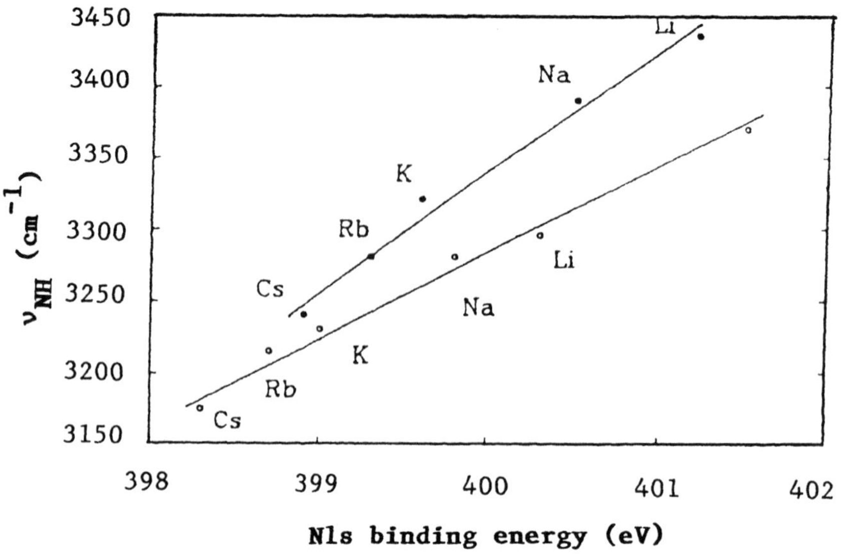

Figure 2. Linear plots of the stretching frequency of the N–H group (ν_{NH}) *vs.* N1s binding energy for pyrrole adsorbed onto a series of anionic zeolites. (Ref. [54])

For basic zeolites, pyrrole was used as a reference acidic molecular probe [54]. The authors found a linear relationship between the N1s XPS chemical shift and the stretching frequency of the N–H group (ν_{NH}) for pyrrole adsorbed onto a series of anionic zeolites (Figure 2). Since ν_{NH} is related to the heat of acid-base adduct formation in the liquid state [56], it follows that in the condensed phase N1s chemical shifts are related to the acid-base component of the heat of adsorption of pyrrole onto zeolites.

As an alternative to the molecular probe technique, ion exchange experiments were proposed to evaluate the IEPS for hydrated metal oxides by XPS [57–59]. However, some caution must be exercised in doing so and Delamar [60] suggested possible improvements to obtain data of greater reliability. Notwithstanding the shortcomings, the method has been used to assess the acid-base properties of oxidized titanium and iron surfaces under both idealized [57] and real situations [58]. In addition, Kurbatov *et al.* [59, 61] showed that surface preparation protocols induced only slight changes in the hydroxylated surface layer present on iron surfaces, thus resulting in small variation of the IEPS.

2.2. Lewis Basic Molecules

Burger and Fluck [62] established, for quickly frozen solutions of $SbCl_5$ –Lewis base complexes in 1,2-dichloroethane, a linear relationship between $Sb3d^{5/2}$ binding energy (BE) and DN, the donor number of the complexing Lewis base. On the basis of this publication, Chehimi [63] calculated ΔH^{AB} for (base–$SbCl_5$) adducts using Drago's equation:

$$-\Delta H^{AB} = E_A E_B + C_A C_B \tag{1}$$

where E and C are the susceptibilities of an acid (A) or a base (B) to undergo electrostatic and covalent bonds, respectively. Figure 3 depicts a linear correlation of $Sb3d^{5/2}$ BE versus ΔH^{AB} (base–$SbCl_5$). It was thus concluded that XPS was a potential tool for estimating Drago's parameters for polymer surfaces [63].

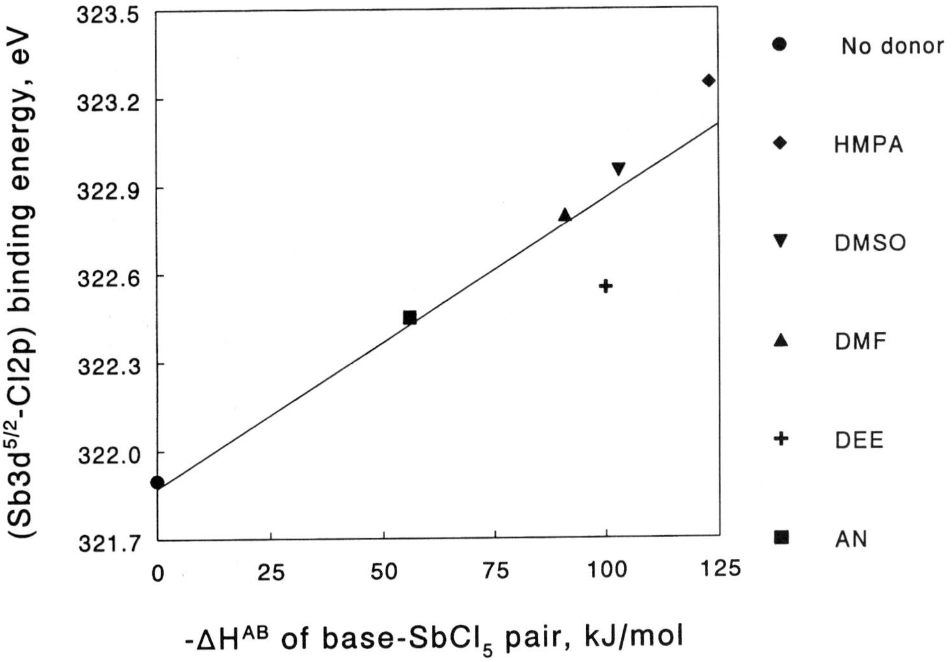

Figure 3. Plot of ($Sb3d^{5/2}$– Cl2p) BE energy difference versus ΔH^{AB} for $SbCl_5$: Lewis base adducts in quickly frozen solutions of 1,2-dichloroethane. Key to data points: AN, acetonirile; DEE, diethyether; DMF, dimethylformamide; DMSO, dimethylsulfoxide; HPMA, hexamethylphosphoramide. The data point corresponding to a zero value of ΔH^{AB} is obtained for a quickly frozen solution of $SbCl_5$ in the absence of any Lewis base. (Ref. [63])

2.3. Acid–base Properties of Model Polymer Surfaces

In the case of polymers, we have used a range of organic molecular probes to quantitatively estimate the acid-base properties of polymer surfaces [20–23, 63]. Practically, a polymer film is exposed to liquid vapours (solutes) of known acid-base properties for a few minutes. The polymer is then allowed to outgas the excess of solute and is transferred into the XPS equipment for surface characterisation. If the polymer-solute interaction is strong enough (e.g. via acid-base forces) then a residual amount of solute is detected and quantified.

Several molecular probes can be used to characterize the acid–base properties of solid surfaces by XPS. For example, chlorinated and fluorinated acidic species probe Lewis basicity, whereas pyridine and dimethylsulfoxide (DMSO) are suitable to characterize surface acidity (Table 1).

Table 1.
Molecular probes used for the XPS determination of acid-base properties of polymer surfaces

Probes	Core line	Property investigated
$CHCl_3$	Cl2p	Basicity
CH_2Cl_2	Cl2p	Basicity
HFIP	F1s	Basicity
CF_3COOH	F1s	Basicity
$SbCl_5$	$Sb3d^{5/2}$	Basicity
I_2	$I3d^{5/2}$	Basicity
pyrrole	N1s	Basicity
pyridine	N1s	Acidity
DMSO	S2p	Acidity

Figure 4 depicts a survey scan of a basic aromatic moisture-cured urethane resin (ArMCU) before and following exposure to the vapours of hexafluoroisopropanol (HFIP, $CF_3CH(OH)CF_3$), a reference Lewis acid. F1s from HFIP is easily detected indicating that it was retained by ArMCU. This retention, despite the high vacuum, is believed to be governed by acid-base interactions between the OH group from HFIP and the carbamate (HN—C=O) group from the resin. The molar ratio of solute per polymer repeat unit (%S, where S stands for solute) was evaluated and used as a measure of the uptake (or retention) of solute by the host polymer. Figure 5 shows plots of %S *vs.* AN (Gutmann's acceptor number) of the solute, for the host polymers PMMA and ArMCU. The plots are s-shaped, showing an increasing uptake of solute with AN, which denotes the basic character of both polymers. These XPS results were in full agreements with FTIR studies of the Lewis basicity of PMMA [3] and ArMCU [22]. The XPS inspection of polyethylene (PE) surface following exposure to $CHCl_3$ vapour was unsuccessful since PE interacts with $CHCl_3$ via London dispersive forces only.

In addition to the quantitative uptake of solutes, the binding energies of core electrons from the corresponding elemental markers were also investigated. In the case of chloroform, we have established a linear $Cl2p^{3/2}$ BE *vs.* ΔH^{AB} correlation (Figure 6) where ΔH^{AB} is the heat of acid–base adduct formation for the polymer-chloroform pairs in the liquid state [21]. The relationship is of the form:

$$-\Delta H^{AB}(\text{kcal}\cdot\text{mol}^{-1}) = 228.8 - 1.14\text{BE (eV)} \quad (2)$$

and was used to determine ΔH^{AB} for PPO-$CHCl_3$ adduct formation to be *ca.* 1 kcal/mol [19]. This result derived from XPS chemical shifts was combined with the IR data reported by Kwei *et al.* [64] to deduce $E_B \approx 0$ and $C_B = 9.5$ $(\text{kcal.mol}^{-1})^{1/2}$ for PPO [16, 45].

Equation (2) will be used further in the study of acid-base and adhesion properties of ammonia plasma-treated polypropylene (*cf.* 2.5.b).

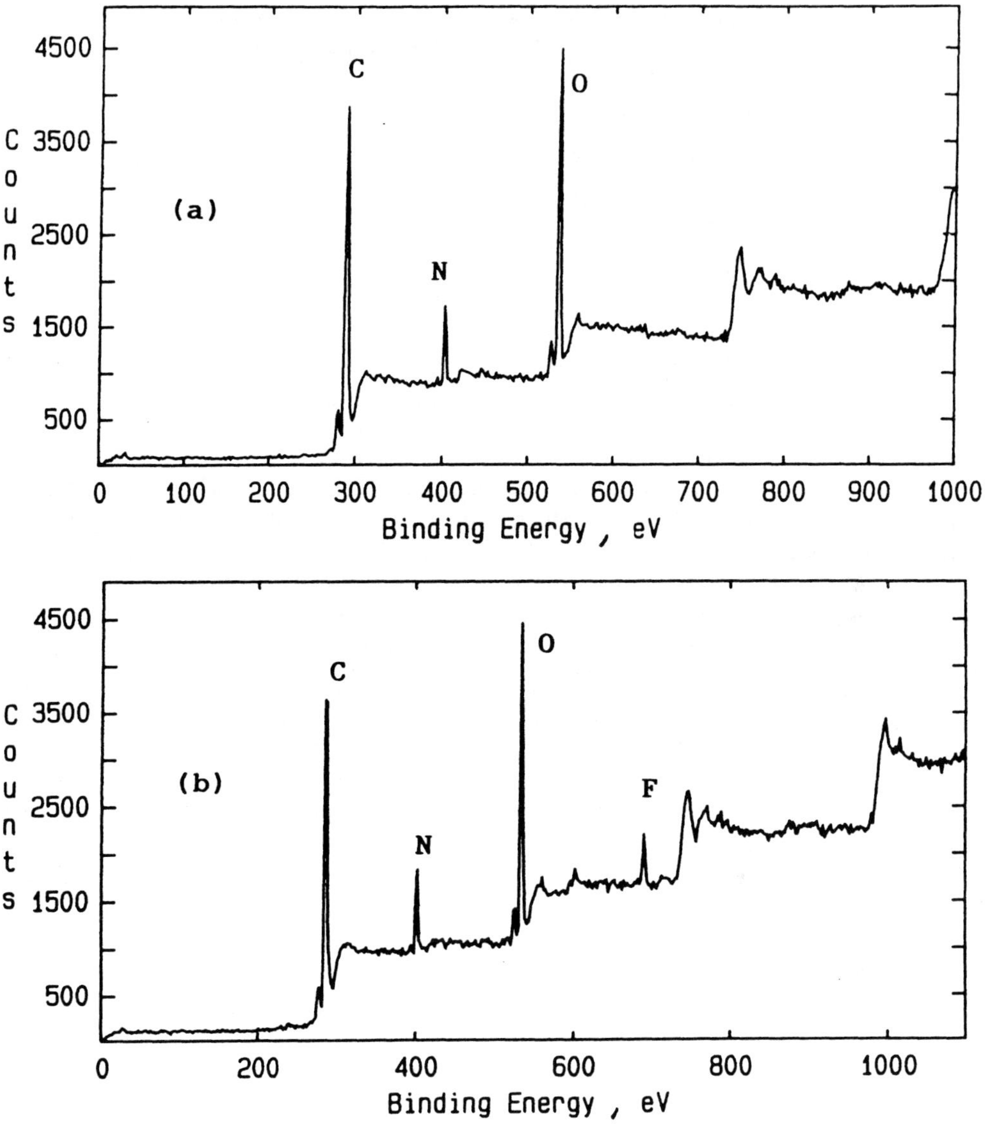

Figure 4. X-ray photoelectron survey spectra of ArMCU: (a) before, (b) after exposure to HFIP. Uptake of the Lewis acid HFIP is indicated by the presence of F1s feature.

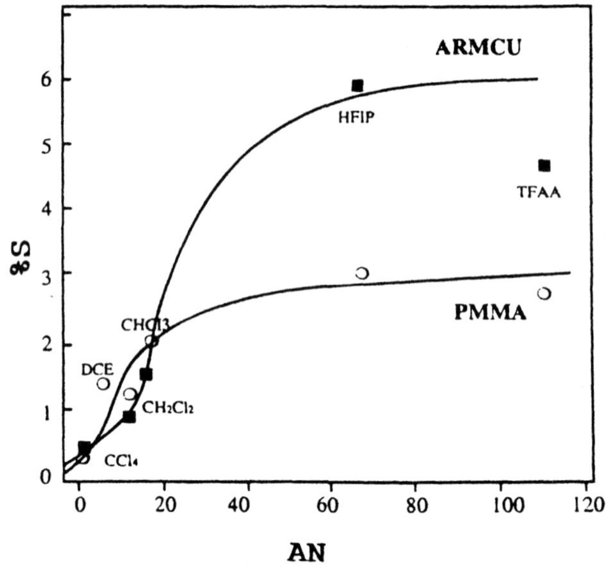

Figure 5. Uptake (%S) of Lewis acids by PMMA and ArMCU *vs.* the acidic character of the solutes. %S is the solute per repeat unit molar ratio in the case of PMMA and the solute per nitrogen atom in the case of ArMCU. The solutes were characterized by AN, the Gutmann acceptor numbers: CCl_4 (2.3); 1-2,dichloroethane, DCE (6.4); dichloromethane, DCM (13.5); chloroform or trichloromethane, TCM (18.7); hexafluoroisopropanol, HFIP (66.3); and trifluoro acetic acid, TFAA (111). (Ref. [20])

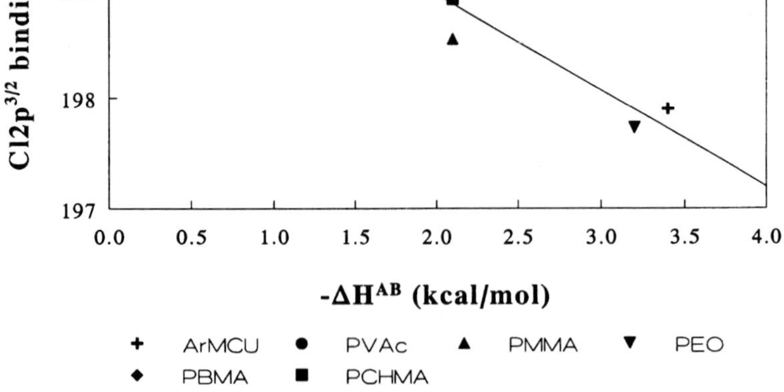

Figure 6. $Cl2p^{3/2}$ BE *vs.* ΔH^{AB} for $CHCl_3$ sorbed in ArMCU, poly(vinyl acetate) (PVAc), PMMA, poly(ethylene oxide) (PEO), poly(butyl methacrylate) (PBMA), poly(cyclohexyl methacrylate) (PCHMA). (Ref. [21])

Another Lewis acid used to probe surface acid-base properties of polymers was CF_3COOH (TFAA) for which a relationship between F1s BE and DN (donor numbers) values of the host polymers (Figure 7) was obtained [26]. DN values were computed using Gutmann's equation and the thermochemical data obtained using the infra-red OH frequency shifts observed for binary polymer blends of a poly(styrene-co-vinylphenyl hexafluoro dimethyl carbinol) [64]. The relationship between polymer DN values and the F1s BE in TFAA-polymer complexes was:

$$DN(kcal.mol^{-1}) = 3593 - 5.21BE\ (eV) \tag{3}$$

The copolymer contained 95% of styrene repeat units and its OH stretching frequency shifts were similar to those of HFIP [64]. It was thus reasonable to assign Gutmann's AN number of HFIP to this copolymer [26]. Therefore equation (3) can potentially be valuable in determining DN values for polymer surfaces of Lewis basic nature.

Attempts were also made to probe the basicity of polystyrene (PS) film with the Lewis acidic iodine vapours. PS film turned purple on exposure to iodine. However, in the high vacuum, there was a continuous desorption of iodine (the purple colour was vanishing) leading to a very weak $I3d^{5/2}$ peak intensity. Nevertheless, the $I3d^{5/2}$ peak recorded during desorption shifted from 621 to 619 eV, i.e. towards lower binding energy. This indicates that the strongly adsorbed iodine molecules were subject to

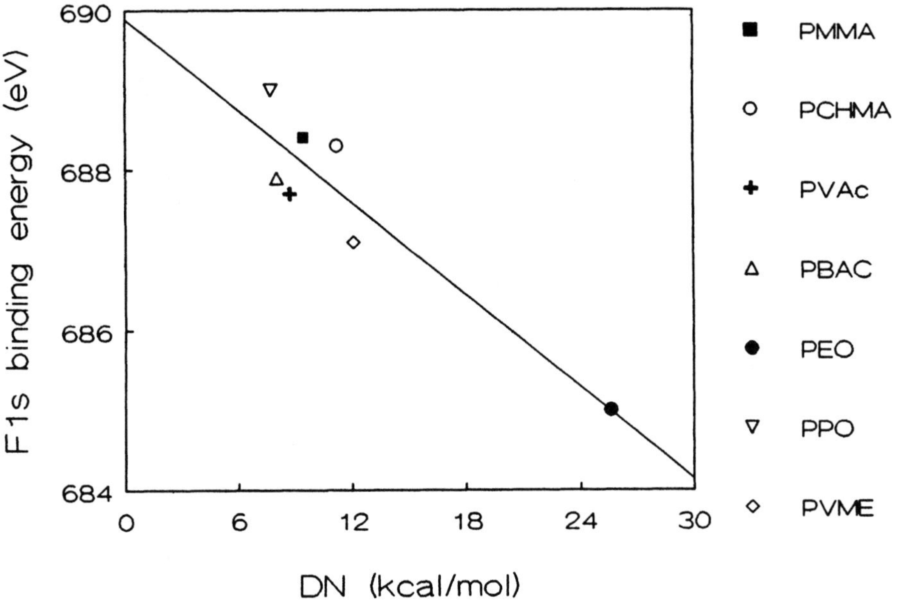

Figure 7. Plot of F1s BE *vs.* the donor numbers (DN) for trifluoroacetic acid (TFAA) sorbed in polymers characterized by their donor numbers (see text for the estimation of polymers' DN numbers). The polymers were poly(methyl methacrylate), PMMA; poly(cyclohexyl methacrylate), PCHMA; poly(vinyl acetate), PVAc; poly(bisphenol A carbonate), PBAC; poly(ethylene oxide), PEO; poly(phenylene oxide), PPO; poly(vinyl methyl ether), PVME.

electron density transfer from PS thus leading to low BE. Perhaps PS was not a strong enough Lewis base to retain iodine adsorbed in the high vacuum. Adsorption of iodine onto plasma-treated polypropylene by the group of J. J. Pireaux in Namur was apparently more successful [65].

Finally, Figure 8 depicts N1s BE for pure pyridine and that sorbed in PMMA and poly(vinyl butyral) (PVB) as a function of the acceptor number (AN) of the pure solute and the two polymers. The positive chemical shift was interpreted in terms of donation of electron density from pyridine to the carbonyl carbon of PMMA (acidic site) [20]. N1s chemical shift is even larger when pyridine is sorbed in PVB since it is predominantly acidic due to its OH pendent groups. The N1s BE positions for pyridine-polymer complexes are higher than those of the pure pyridine because pyridine-pyridine interaction occurs via dispersive forces only, because pyridine is a monofunctional species [2, 66].

To summarize, acidic ($CHCl_3$ and TFAA) and basic (pyridine) probes undergo negative and positive chemical shifts (lower and higher BEs), respectively, when they interact with host (polymer) surfaces via acid-base forces. Indeed, a Lewis acid is an electron acceptor and upon interaction with a base via acid–base forces, electron density is transfered to the former thus yielding a lower binding energy of its electrophilic site [19–21]. The opposite reasoning holds for basic probes [20, 23].

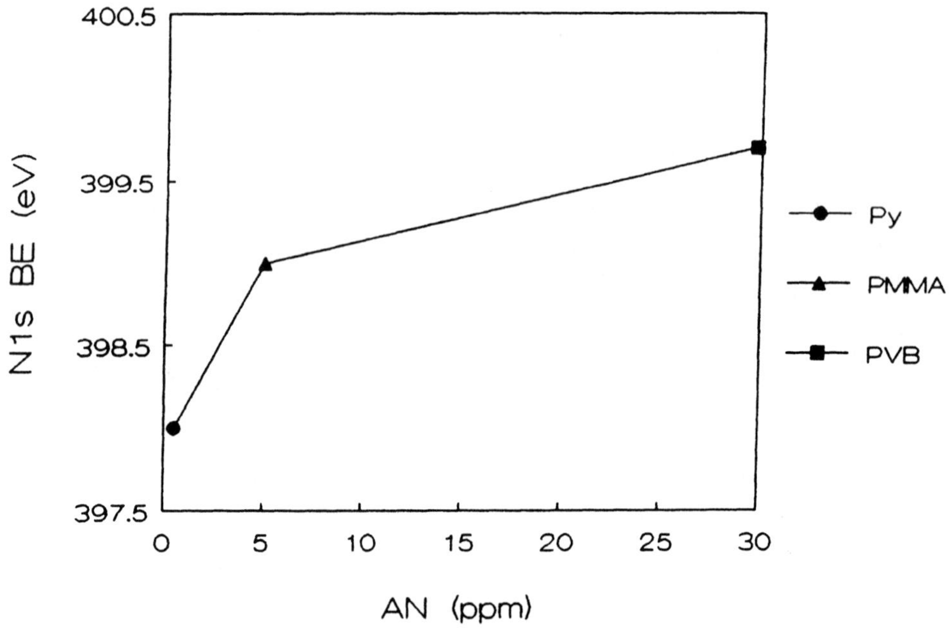

Figure 8. Plot of N1s BE *vs.* AN for pure pyridine and pyridine sorbed in PMMA and PVB. AN = 0.5 [2], 5.3 and 29.8 for pyridine, PMMA and PVB, respectively. PMMA was assigned an AN value of ethyl acetate [2], and PVB the average value of AN reported for 2-propanol [2] and the value estimated using equation (4), ΔH^{AB} for the ethyl acetate–PVB interaction [15], and DN (17.1) of ethyl acetate.

2.4. Acid–base Characteristics of Coil Coating Primer Components

A coil coating primer is a radiation-cured coating consisting of reactive (monomeric) liquid mixed with pigments and additives. Radiation cured coatings are applied to flat continuous substrates under controlled conditions determined by the nature of the metal substrate and the use of the coated coil stock. When the "wet" coating is irradiated, the reactive monomers polymerise and become a solid component of the cured coating. The advantages of this technique in terms of performances, energy costs and environmental regulations were described by Costanza *et al.* [67].

As part of an ongoing research work at the University of Surrey on the adhesion of radcure coatings to steel, it was desirable to investigate the acid-base properties of two coil coating primer components. The coil coating primer under test consisted of an epoxy resin (monomeric liquid, UVR-6110) and a photoinitiator (UVI-6990) absorbing UV radiation (Union Carbide Ltd products). The reason for investigating the photoinitiator in addition to the resin is that it has been recently shown that in some particular cases the photoinitiator segregates to the polymer-metal interface leading to a drastic loss of adhesion [68].

The epoxy resin (diepoxy cycloaliphatic resin) has the following chemical structure:

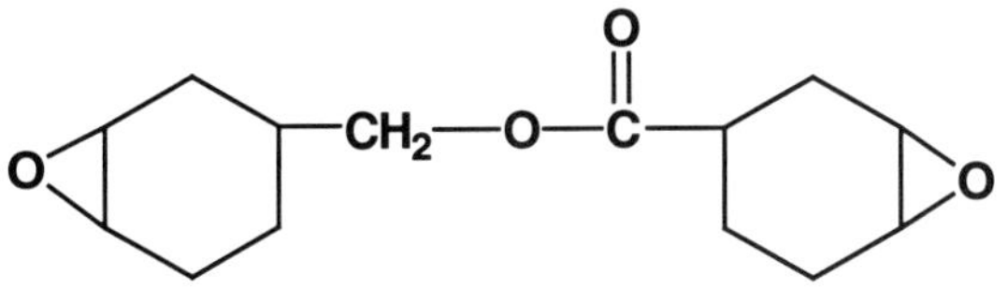

diepoxy cycloaliphatic resin (UVR-6110)

The photoinitiator was composed of a 50/50% mixture of triarylsulfonium hexafluorophosphate salt and propylene carbonate:

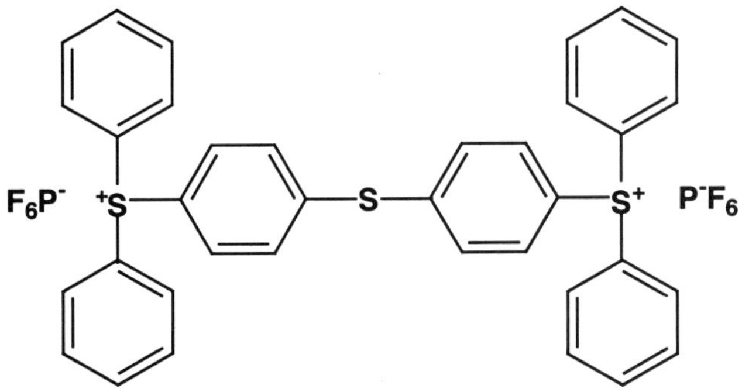

triarylsulphonium hexafluorophosphate salt

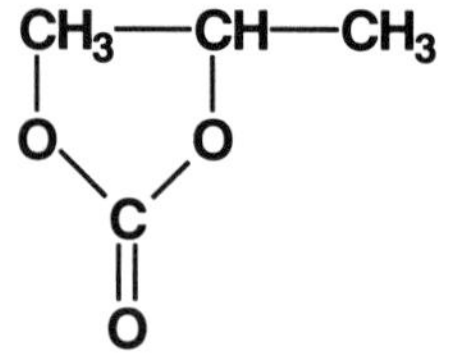

propylene carbonate

These two components were used without any dilution for spin coating onto a (pre-treated hot dip galvanised) steel substrate. The thin films were examined by XPS before and following exposure to chloroform vapours (optimum exposure time of 5 min). Table 2 reports the uptake of chloroform by the coil coating primer components, the $Cl2p^{3/2}$ binding energy for the sorbed chloroform, and the ΔH^{AB} estimated from equation (2) for the probe-coating component complexes.

Table 2.
Uptake and interaction characteristics of chloroform with the epoxy resin and the photoinitiator

	epoxy resin UVR-6110	photoinitiator UVI-6990
$CHCl_3$/ester (%)	1.1	0.27
$Cl2p^{3/2}$ BE (eV)	195.5	200.6
$-\Delta H^{AB}$ (kcal/mol)	5.93	0.17

X-ray source: Al Kα; power: 20 mA × 10 kV; pass energy: 50 eV; charge correction: C—C/C—H C1s component set at 285 eV.

The $Cl2p^{3/2}$ BE for chloroform sorbed in the resin is 5.2 eV lower than the reference value obtained for a C—Cl bond in organic species, whereas the chemical shift induced by the photoinitiator is practically nil. The epoxy resin is thus a much stronger base than the photoinitiator. It is even more basic than PMMA ($-\Delta H^{AB}$ = 2.1 kcal/mol). The Lewis basic characteristic of this resin is due to the ester functional group. However, it is reasonable to suggest that basicity is reinforced by the two strained epoxide rings which are known to react readily with acids. For example polyethylene oxide complexes with chloroform with a $-\Delta H^{AB}$ value of 3.2 kcal/mol (see Figure 6).

As far as the photoinitiator is concerned, there is a lack of basic character. On the one hand, the S^+ cationic sites confer an acidity to the aromatic triarylsulfonium cation and thus reacts with chloroform via dispersive interactions only. It is actually well known that substituted benzene rings behave as acids when they are bound to an electron-acceptor functional group. For example, toluene or benzene is a π electron donor but nitrobenzene is a π^* electron acceptor [18]. On the other hand, carbonyl group of the propylene carbonate may be orientated towards the substrate [69] and therefore cannot react with chloroform.

The results derived from the chemical shifts of the chloroform Cl2p peaks are in line with the uptake of chloroform molecules by the epoxy resin and the photoinitiator. The epoxy resin is likely to provide good adhesion to acidic (or amphoteric) oxides, meanwhile the acidic feature of the photoinitiator may affect interface bonding leading to a delamination similar to that found in [68].

2.5. Acid–base, Wettability and Adhesion Properties of Plasma-treated Polypropylene

Polypropylene is a polyolefin type of polymer, entirely built of hydrocarbon units which confer it apolar and hydrophobic characteristics. PP has excellent mechanical and dielectric properties, good chemical resistance and thermal stability. It finds applications mainly in packaging and the automobile industry. However, due to its chemical inertness, PP has very poor adhesion property. This can be improved by modifying the surface of PP using chemical or physical methods [70]. Chemical methods include the use of strong oxidants like sulfuric acid, nitric acid and sulfochromic mixtures, whereas physical methods include flame and plasma treatments, the former being preferably used for thick objects. In the case of thin sheets of isotactic PP, surface modification was rather achieved by the plasma technique [71], a dry process which can tailor polymer surfaces by grafting specific surface functional groups (acidic, basic, polar) [71–73]. For example it is well established that oxygen plasma oxidizes the surface of PP and renders it more acidic [25, 72] whilst nitrogen or ammonia plasmas introduce basic nitrogen-containing groups [24]. These "polar" groups modify the acid–base properties and thus adhesion of metal coatings to PP.

2.5.1. Oxygen-plasma Treated Polypropylene. Oxygen plasma treatment of PP surface grafts carboxylic groups as shown by capillary electrophoresis, attenuated total reflection infra-red spectroscopy (ATR-IR) and XPS [25]. These functional groups confer an acidic character to PP surface as demonstrated by the contact angle titration technique [72]. This was confirmed by XPS using DMSO as a reference Lewis base to probe the surface acidity of oxygen plasma-treated PP. S2p was used as the diagnostic core line for adsorbed DMSO. Figure 9 shows S2p signals from DMSO adsorbed onto PP for different treatment times. For the untreated PP, no S2p signal from DMSO was detected, whereas the intensity of this peak readily increases with treatment time longer than 3 s. Since oxygen plasma-treated PP bears acidic sites, it is likely that acid-base interactions are effective in retaining DMSO adsorbed onto the polymer surface in the high vacuum. This is supported by the S2p BE which was found to be in the 166.5–168 eV range. S2p BE is higher than the value of 165.8 eV reported for $(C_6H_4)_2S{=}O$ and $CH_3(C_6H_4)S{=}O$ [74], thus indicating a positive chemical shift for the S2p core line. This positive chemical shift is due to electron density transfer from the basic DMSO to the acidic surface sites of the treated PP. Therefore, XPS provides a direct evidence of acid-base adduct formation for the polymer–probe pair investigated. This XPS study was complemented by the determination of the work of adhesion of DMSO to treated PP by contact angle measurements of DMSO drops. Figures 10 and 11 depict the acid-base contribution to the DMSO-PP work of adhesion and the uptake of DMSO molecules (S/O atomic ratio) by PP, respectively: both W and S/O ratio increase with the oxygen plasma treatment time. However, the asymptotic value of W is reached very rapidly (less than 1 s) whereas that of the uptake of DMSO vapours is reached after 15–20 s. This delay in reaching the plateau value of the uptake of DMSO was related to the enrichment of surface in fumaric acid species occuring after 15 s of plasma treatment [25].

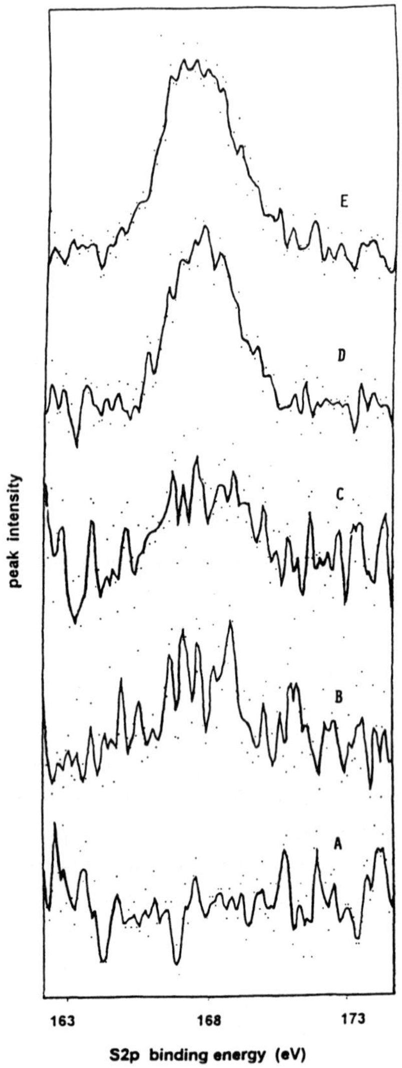

Figure 9. S2p signals from DMSO adsorbed onto oxygen plasma-treated polypropylene *vs.* the treatment. (A): untreated PP; treatment times (B) 3; (C) 8; (D) 15; (E) 30. (Ref. [25])

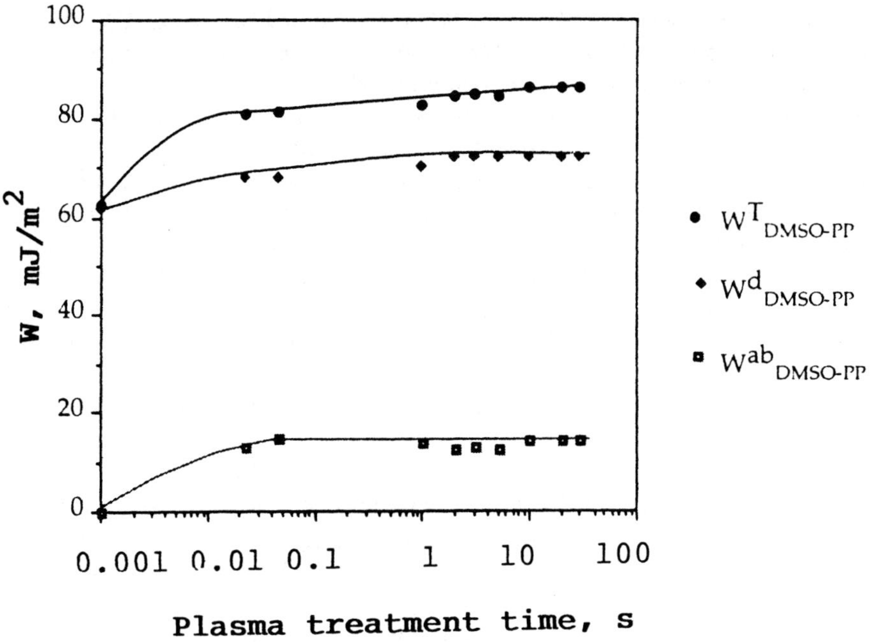

Figure 10. Variation of the total (T) work of adhesion and its dispersive (d) and acid-base (ab) contributions for the DMSO-polypropylene system *vs.* oxygen plasma treatment time of polypropylene. (Ref. [25])

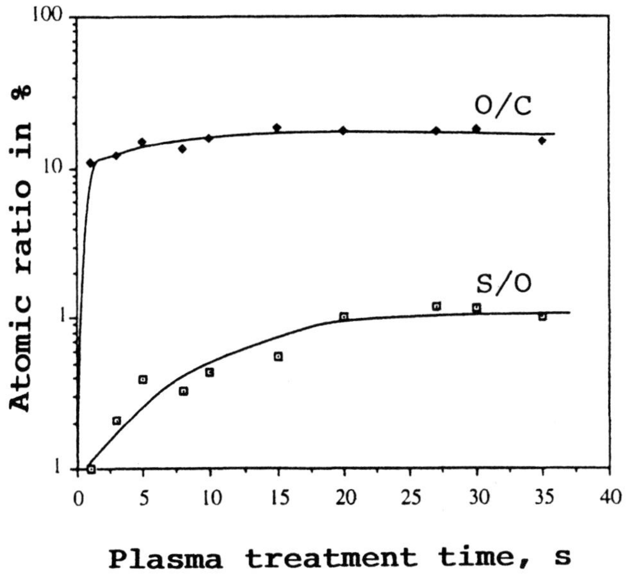

Figure 11. O/C and S/O atomic ratios (in %) *vs.* the oxygen plasma treatment time of polypropylene. (Ref. [25])

2.5.2. Ammonia-plasma Treated Polypropylene. Metallization of PP by an adherent thin aluminium layer (*ca.* 20 nm) can be achieved if PP surface is treated by an ammonia or nitrogen plasma prior to metallization [71]. For example, the ammonia plasma treatment results in grafting of nitrogen and/or oxygen (from the residual reactor atmosphere) containing groups as judged from the increase in the N/C and O/C atomic ratios [72]. Shahidzadeh-Ahmadi *et al.* [72] showed via wettability measurements that ammonia plasma-treated polypropylene (NH_3–PP) was predominantly basic, most probably due to the grafted nitrogen containing functional groups (e.g. amine and amide). This characteristic was further assessed by XPS using chloroform as an acidic molecular probe to interrogate the acid–base properties of the treated polymer surfaces [24]. Figure 12 shows that NH_3–PP surface is labelled by Cl2p core line from chloroform (Figure 12b) but not the untreated PP (Figure 12a). The retention of chloroform by the surface of NH_3–PP was interpreted in terms of strong polymer-probe acid-base interaction: $-NH_2...HCCl_3$. This was supported by the low $Cl2p^{3/2}$ BE (196.9–198 eV) due to a large electron density transfered to the probe. The $Cl2p^{3/2}$ BE-ΔH^{AB} correlation (Equation 2, Figure 6) was used to assess the ΔH^{AB} for the chloroform/NH_3–PP pairs. The ΔH^{AB} values are reported in Table 3 together with that of chloroform-ArMCU system determined via FTIR and those calculated using Drago's four parameter equation (1) for a series of chloroform-amine adducts. ΔH^{AB} value for chloroform:NH_3–PP remains roughly constant around 4.1 kcal/mol indicating that the strength of the bond does not vary significantly although the N/C atomic ratio increases with treatment time. ΔH^{AB} values for the various chloroform:NH_3–PP complexes compare fairly well with the values of 4.3–4.9 kcal/mol obtained for chloroform-amine adducts. This provides a strong supporting evidence for the specific interaction of chloroform with the nitrogen containing basic sites.

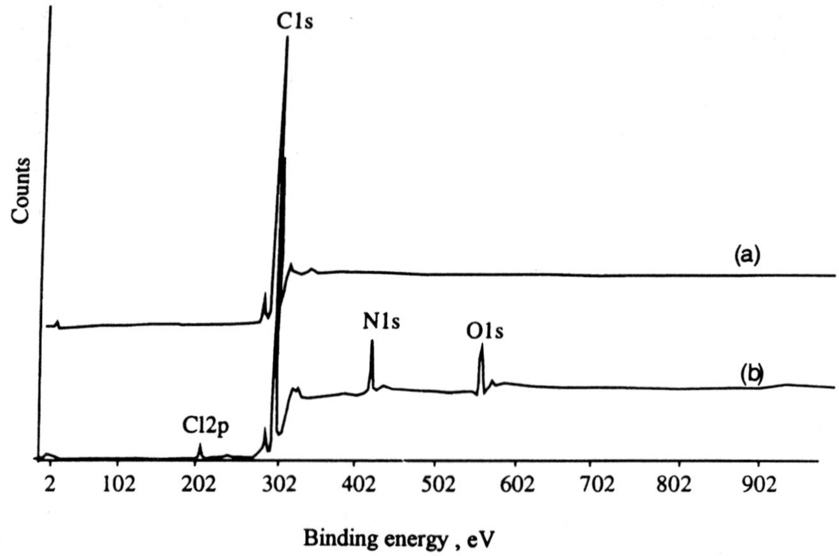

Figure 12. XPS survey scans of untreated PP (a) and 0.7 s ammonia plasma-treated polypropylene (NH_3–PP); (b) followed by 10 min exposure to $CHCl_3$ vapours. (Ref. [24])

The uptake of chloroform by the PP surface ($CHCl_3$/N molar ratio) versus the treatment time is depicted in Figure 13. For a treatment period shorter than 1 s, $CHCl_3$/N molar ratio has a plateau value equal to 5–6%, whereas for a plasma treatment longer than 1 s it decreases sharply to reach a value of 0.7% at 30 s. The unexpected variation of $CHCl_3$/N ratio *vs.* time is nevertheless in agreement with the results obtained from contact angle measurements [72] and especially the study of the adhesion of thin aluminium films to NH_3–PP [71]. NH_3–PP was metallized by thermal evaporation of aluminium [71] and subsequently subject to a U-peel test. The amount of peeled metal was assessed by an image processing system [71].

Table 3.
Drago's E and C parameters for chloroform and amines, and ΔH^{AB} values for acid–base adducts of chloroform with amines, NH_3–PP and ArMCU

	C_A	E_A	C_B	E_B	$-\Delta H^{AB}$ (kcal/mol)
Chloroform	0.159	3.02			
NH_3			3.46	1.36	4.7
CH_3NH_2			5.88	1.3	4.9
$(CH_3)_2NH$			8.73	1.09	4.7
$(CH_3)_3N$			11.54	0.81	4.3
$(CH_3CH_2)_3N$			11.09	0.99	4.8
NH_3–PP, 0.7 s					4.3
NH_3–PP, 25 s					3.1
ArMCU					3.4

ΔH^{AB} for chloroform-amine complexes were calculated using Drago's E and C parameters for chloroform and amines in equation (3) [56]. ΔH^{AB} for (chloroform–ArMCU) adduct was determined by FTIR [22].

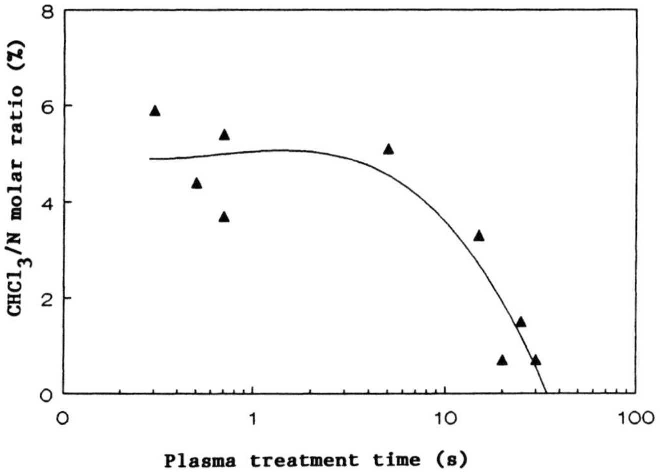

Figure 13. Chloroform to nitrogen molar ratio (in %) *vs.* the ammonia plasma treatment time of polypropylene. The plot shows that optimal treatment time is less than 1 s. (Ref. [24])

Figure 14 shows the plot of the percentage of aluminium remaining adhered *vs.* the treatment time. For the untreated PP, 95% of the aluminium was peeled off, a result that signifies a poor adhesion of aluminium to the untreated polymer. For plasma treatment as short as 0.023 s, the percentage of peeled aluminium sharply decreases to 4–5%, and remains constant with increasing plasma treatment time, an indication of a strong metal-polymer adhesion. However, the amount of peeled metal increases again to reach the value of 20% for treatment times exceeding 1 s. The similarity between the XPS study and adhesion measurements shown in Figures 13 and 14, respectively is striking and the peel test measurements were interpreted in terms of acid–base interactions. Indeed, on the one hand Arefi-Khonsari *et al.* [75] showed that aluminium was present in the alumina form at the aluminium–PP interface, on the other hand, alumina was found to be amphoteric with a predominant acidity as judged from contact angle titrations. This approach was validated using MgO as a reference basic material [76]. The amphoteric behaviour of alumina is in line with the 4.6–9 value range of its IEPS [51, 77].

There is thus a strong supporting evidence from XPS, wettability and adhesion studies that acid–base forces govern the interaction of NH_3–PP with either an acidic molecule such as chloroform or an amphoteric metal oxide such as alumina. In contrast, attempts to metallize oxygen–plasma treated PP (predominantly acidic) with aluminium resulted in a poor metal–polymer adhesion, probably due to a less favourable acid–acid interaction [78].

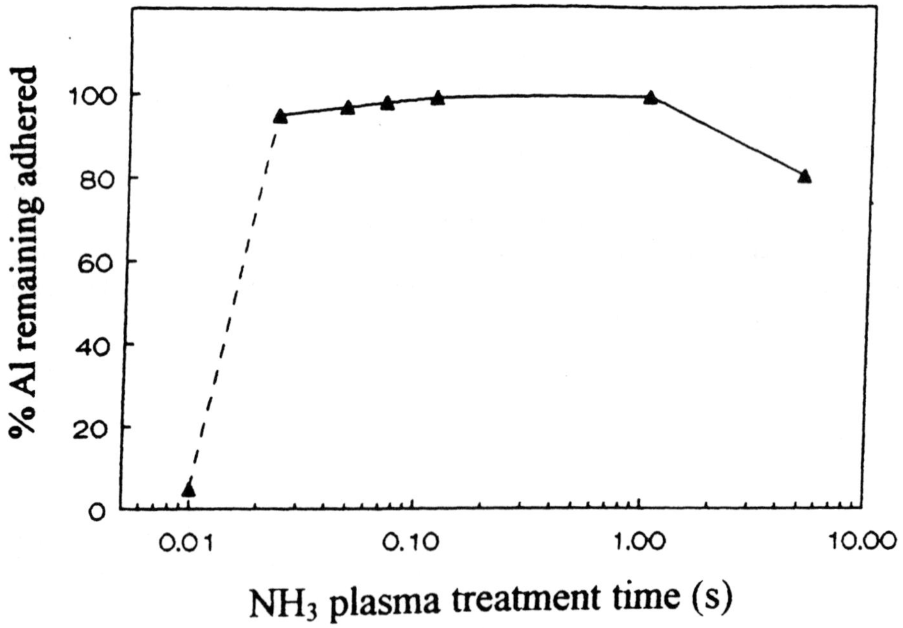

Figure 14. Adhesion of aluminium evaporated onto polypropylene *vs.* ammonia plasma treatment time of polypropylene. Adhesion was measured as a percentage of aluminium remained adhered to polypropylene following the U-peel test (adapted from ref. [71]).

2.5.3. Adhesion of Evaporated Silica to Plasma-treated Polypropylene. Following our XPS strategy, Bodino *et al.* [65] have characterized the acid–base properties of N_2 and ammonia plasma-treated PP surfaces (N_2–PP and NH_3–PP, respectively) using chloroform and iodine as reference Lewis acidic molecular probes. Their XPS derived thermochemical data are summarized in Table 4. It should be noted that whilst the ΔH^{AB} values for chloroform were obtained using Equation (2), there was no mention of the establishment of a binding energy *vs.* ΔH^{AB} plot for iodine.

Table 4.
XPS-derived ΔH^{AB} values (kcal/mol) for chloroform and iodine adsorbed onto N_2 and NH_3 plasma-treated PP surfaces [65]

	Polymer surfaces	
Molecular probes	N_2–PP	NH_3–PP
$CHCl_3$	4.5	5.1
I_2	1.8	2.6

The chloroform data agree fairly well with ΔH^{AB} values reported in Table 3 for NH_3–PP:$CHCl_3$ adducts. They also reflect a stronger interaction of chloroform with ammonia plasma-treated PP (a basic surface) than with N_2–PP (an amphoteric surface). Using Drago's equation and E and C constants for chloroform and iodine [79], Chehimi [18] estimated E_B and C_B constants (not reported in [65]!) for the treated polymer surfaces: $E_B = 1.46$, $C_B = 0.33$ and $C_B/E_B = 0.226$ for N_2–PP; and $E_B = 1.65$, $C_B = 0.91$ and $C_B/E_B = 0.552$ for NH_3–PP (E and C in $kcal^{1/2}/mol^{1/2}$). Whilst the E_B constants are in line with those reported for basic species in general [79], the C_B values and consequently the C_B/E_B ratios are too low for nitrogen-containing Lewis bases. Nevertheless, with this in hand and the $E_A = 4.39$ and $C_A = 1.14$ $kcal^{1/2}/mol^{1/2}$ for silica [80], ΔH^{AB} for silica: N_2–PP and silica: NH_3–PP pairs were predicted to be 6.8 (6.2 in ref. [65]) and 8.3 kcal/mol, respectively. These values correlate well with the larger sticking coefficients of evaporated $SiO_{1.7}$ (acidic oxide) onto NH_3–PP surface as compared to N_2–PP, O_2–PP (acidic surface, see above) and the untreated PP. Additional depth profiling of the $SiO_{1.7}$ deposits indicated a very sharp ceramic–PP interface for the untreated PP whereas an interphase with an increasing degree of diffuseness was observed for O_2–PP, N_2–PP and NH_3–PP.

2.5.4. Effect of Helium Plasma Pretreatment on the Acid–base Characteristics of Ammonia Plasma-treated Polypropylene Films. In the previous sections, it became clear that plasma treatment was a "clean" and efficient way to modify polymer surfaces. However, such surfaces are more energetic and subject to physicochemical modifications which lead to a minimization of the surface free energy, and thus to a reduced activity. This problem can be solved by a helium plasma pretreatment which is known to stabilize the surface layers via crosslinking. The so-called CASING (Crosslinking via Activated Species of Inert Gases) process consists of creating new bonds such as C=C, which improve the bond strength of the surface by forming a very cohesive and dense skin. As a result of this, the ageing effect of an ammonia-plasma treated PP has been slowed down by using a helium plasma pretreatment [81]. It was thus interesting to study the acid-base properties of plasma treated-polypropylene surface in two steps (He/NH_3–PP): a pretreatment by helium (1–30 s) followed by a treatment by ammonia (1 s). The acid-base properties of He/NH_3 treated PP surfaces were characterized using contact angle titration method and XPS in conjunction with the molecular probe technique. The former employs the sessile drop technique, but instead of pure water, unbuffered aqueous acidic and basic solutions with pH values ranging from 1 to 14 were used. Figure 15 shows plots of the work of adhesion (W) of water to the treated surfaces versus the pH of the sessile drops. For helium-treated PP, the surface reveals a dominant acidic character in contrast to ammonia plasma-treated PP, an amphoteric with a predominant basic character polymer surface. The He/NH_3–PP surfaces appear to be amphoteric with a predominant acidic character which is relatively constant with the pretreatment time. The He/NH_3 plasma treatment thus yields the combination of the acid-base properties obtained with the independent He and NH_3 plasma treatments. This "bifunctional" character was further investigated by XPS. We used chloroform and DMSO (independently) to probe the basicity and acidity of He/NH_3–PP, respectively. Both molecular probes were found to be retained at the surface of treated PP surfaces for all treatment times. The binding energies for Cl 2p from chloroform and S2p from DMSO were 197.6±0.2 eV, 168.8±0.2 eV, respectively. Clearly, chloroform (acidic) undergoes a negative chemical shift and DMSO (basic) a positive one. We have monitored the uptake of these molecular probes versus the pretreatment time of helium plasma for He/NH_3–PP surfaces. Figure 16 shows a plot of $CHCl_3$/N molar ratio versus the helium plasma pretreatment

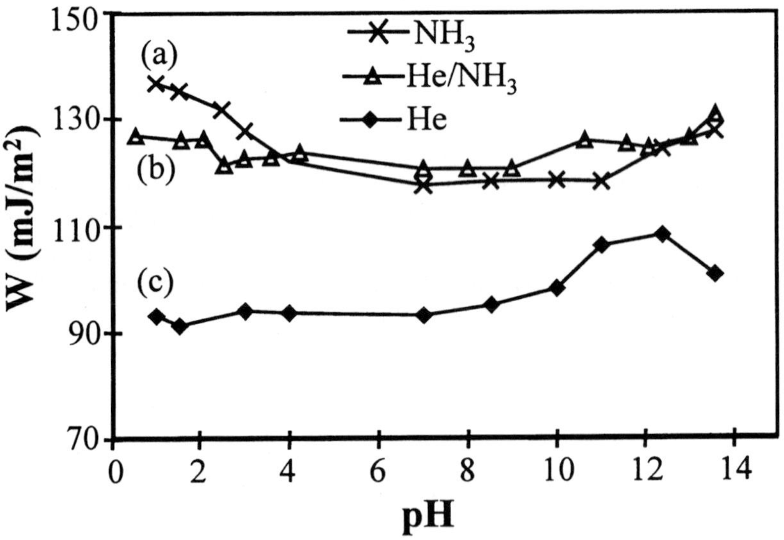

Figure 15. Variation of the total work of adhesion versus the pH of the sessile drops for (a) NH_3-treated PP, (b) He/NH_3-treated PP, and (c) He-treated PP.
Experimental conditions. NH_3 plasma: pressure = 0.75 torr, power = 9 W, flow = 100 sccm*, time = 1 s. He plasma: p = 0.75 torr, power = 1.5 W, F = 100 sccm, t = 1 s. He/NH_3 plasma: pressure = 0.75 torr, power = 1.5 W, F = 100 sccm, t = 1 s for helium and pressure = 0.75 torr, power = 9 W, flow = 100 sccm, time = 1 s for NH_3.
*sccm: standard cubic centimetre.

time. One can note that the uptake of chloroform is almost constant if the "kink" corresponding to 20 s is excluded. The stability of the amount of acidic probes retained on such amphoteric treated surfaces indicates that the basic sites introduced during the ammonia plasma treatment have a surface concentration that is constant for the whole range of time shown in Figure 16.

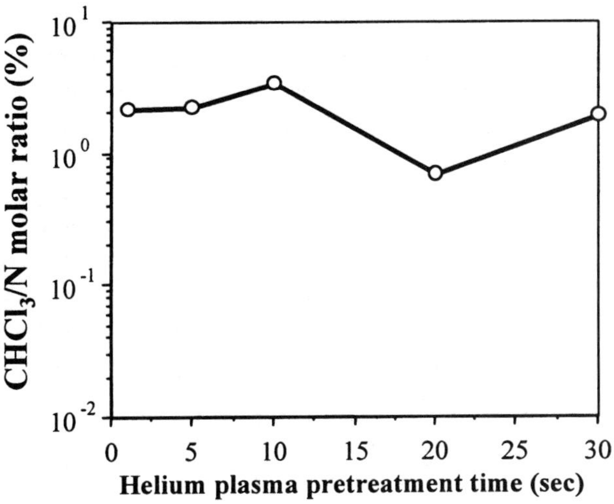

Figure 16. Plot of $CHCl_3$/N molar ratio (%) versus the He plasma pretreatment time for He/NH_3-treated PP.
Experimental conditions. He/NH_3 plasma: pressure = 0.75 torr, power = 1.5 W, flow = 100 sccm* for helium, and pressure = 0.75 torr, power = 9 W, flow = 100 sccm, time = 1 s for NH_3.
*sccm: standard cubic centimetre.

The amount of DMSO probes adsorbed onto the helium plasma-pretreated PP is of the same order of magnitude as an O_2 plasma-treated PP, showing in this way a substantial acidic character of He/NH_3–PP surfaces.

These preliminary studies of acid-base properties of He/NH_3–PP firmly demonstrate the amphoteric character of these surfaces. However, the complexity of such plasma treated surfaces renders them somehow difficult for study. Indeed, a basic plasma can introduce acidic groups on the treated surfaces which results in the adsorption of the basic probe. Work is in progress in order to fully interpret the relationship between the molecular probe technique in XPS and the acid-base character of the treated surfaces assessed by the contact angle titration technique.

3. CONCLUSION AND FUTURE PROSPECTS

The background to the use of XPS for the determination of acid–base properties of untreated and treated polymer surfaces is reviewed. Reference acidic and basic probes adsorbed on the polymer surfaces permit to assess their basic and acidic characters, respectively. For example, the XPS study of adsorbed trifluoroacetic acid can be used to determine the DN for polymers, whereas the use of chloroform as a reference probe permits to estimate the heat of acid-base bond formation (ΔH^{AB}) for the host polymer-chloroform acid–base adduct. If at least another BE–ΔH^{AB} correlation (similar to that shown in Figure 6) is established for another acidic molecular probe, Drago's E_B and C_B parameters for polymers under test will be determined by XPS. Conversely, BE–ΔH^{AB} correlations established with reference basic probes will yield E_A and C_A parameters.

The XPS study of the acid-base properties of plasma-treated PP demonstrates that this technique has come of age in this regard. The XPS is indeed shown to be a reliable and effective means in determining acid–base properties of complex organic surfaces.

With the presently available lateral resolution and imaging capabilities of modern XP spectrometers, this technique will enable the surface scientist to map acid-base properties of polymers (and other materials) at the micrometer or submicrometer level, particularly patterned polymer surfaces. However, it is of great importance to test the applicability of new molecular probes containing elements of very high sensitivity factors (to improve the detection limit and obtain an adequate chemical mapping) and which undergo large chemical shifts upon acid–base bond formations (for improving the accuracy in determining ΔH^{AB} for the polymer–probe adduct).

Acknowledgement

VL would like to thank Becker Industrial Coatings for financial support and Dr. C. Lowe for helpful discussions. JK is indebted to the French CIES (Centre International d'Etudiants et de Stagiaires) for the provision of a scholarship. MMC is indebted to the Symposium organizers and the Université Paris 7 – Denis Diderot for financial support.

REFERENCES

1. K. L. Mittal and H. R. Anderson, Jr. (Eds), *Acid-Base Interactions: Relevance to Adhesion Science and Technology.* VSP, Utrecht, The Netherlands (1991).
2. F. L. Riddle, Jr., and F. M. Fowkes, *J. Am. Chem. Soc.* **112**, 3259 (1990).
3. F. M. Fowkes, *J. Adhesion Sci. Technol.* **4**, 669 (1990).
4. C. J. van Oss and R. J. Good, *Langmuir* **8**, 2877 (1992).
5. M. M. Chehimi, in: *Handbook of Advanced Materials Testing*, N. P. Cheremisinoff and P. N. Cheremisinoff (Eds), Chap. 33. Marcel Dekker, New York (1995).
6. J. C. Berg, in: *Wettability*, J. C. Berg (Ed.), Chap. 2. Marcel Dekker, New York (1993).
7. P. Mukhopadhyay and H. P. Schreiber, *Colloids Surfaces A* **100**, 47 (1995).
8. D. M. Dwight, in: *Adhesion Science and Technology (Festschrift in Honor of K. L. Mittal),* W. J. van Ooij and H. R. Anderson, Jr. (Eds), pp. 63–80. VSP, Utrecht, The Netherlands (1998).
9. R. C. Thomas, J. E. Houston, R. M. Crooks, T. Kim and T. A. Michalske, *J. Am. Chem. Soc.* **117**, 3830 (1995).
10. M. B. Kaczinski and D. W. Dwight, *J. Adhesion. Sci. Technol.* **7**, 165 (1993).
11. K. L. Mittal (Ed.), *Contact Angle, Wettability and Adhesion*. VSP, Utrecht, The Netherlands (1993).
12. M. M. Chehimi, S. F. Lascelles and S. P. Armes, *Chromatographia* **41**, 671 (1995).
13. D. R. Lloyd, T. C. Ward and H. P. Schreiber (Eds), *Inverse Gas Chromatography: Characterization of Polymers and Other Materials.* ACS Symposium Series n° 391, American Chemical Society, Washington DC (1989).

14. M. M. Chehimi, in: *Thermodynamical Modeling and Materials Data Engineering,* J.-P. Caliste, E. Truyol and J. Westbrook (Eds), pp. 135–142. CODATA/Springer-Verlag, Berlin (1998).
15. F. M. Fowkes, D. O. Tischler, J. A. Wolfe, L. A. Lannigan, C. M. Ademu-John and M. J. Halliwell, *J. Polym. Sci., Chem. Ed.* **22**, 547 (1984).
16. F. M. Fowkes, *J. Adhesion. Sci. Technol.* **1**, 7 (1987).
17. T. B. Lloyd, *Colloids Surfaces A* **93**, 25 (1994).
18. M. M. Chehimi, in: *Adhesion Promotion Techniques: Technological Applications,* K. L. Mittal and A. Pizzi (Eds), Chap. 2. Marcel Dekker, New York (1999).
19. J. F. Watts and M. M. Chehimi, *Int. J. Adhesion Adhesives* **15**, 91 (1995).
20. M. M. Chehimi, J. F. Watts, S. N. Jenkins and J. E. Castle, *J. Mater. Chem.* **2**, 209 (1992).
21. M. M. Chehimi, J. F. Watts, W. K. Eldred, K. Fraoua and M. Simon, *J. Mater. Chem.* **4**, 305 (1994).
22. M. M.Chehimi, E. Pigois-Landureau, M. Delamar, J. F. Watts, S. N. Jenkins and E. M. Gibson, *Bull. Soc. Chim. Fr.* **129**, 137 (1992).
23. J. F. Watts, M. M. Chehimi and E. M. Gibson, *J. Adhesion* **39**, 145 (1992).
24. N. Shahidzadeh, M. M. Chehimi, F. Arefi-Khonsari, J. Amouroux and M. Delamar, *Plasmas Polym.* **1**, 27 (1996).
25. N. Shahidzadeh, M. M. Chehimi, F. Arefi-Khonsari, N. Foulon-Belkacemi, J. Amouroux and M. Delamar, *Colloids Surfaces A* **105**, 277 (1995).
26. M. M. Chehimi, M. Delamar, N. Shahidzadeh-Ahmadi, F. Arefi-Khonsari, J. Amouroux and J. F. Watts, in: *Organic Coatings*, P.-C. Lacaze (Ed.), pp. 25–31. AIP Press, Woodbury, NY (1996).
27. M. M. Chehimi and M. Delamar, *Analysis* **23**, 291 (1995).
28. S. R. Leadley and J. F. Watts, *J. Adhesion* **60**, 175 (1997).
29. J. F. Watts, *Surf. Interface Anal.* **12**, 497 (1988).
30. J. F. Watts, *Vacuum* **45**, 653 (1994).
31. J. F. Watts, in: *Handbook of Surface and Interface Analysis. Methods for Problem Solving*, J. C. Rivière and S. Myhra (Eds), Chap. 18. Marcel Dekker, New York (1998).
32. R. W. Paynter, J. E. Castle and D. K. Gilding, *Surf. Interface Anal.* **7**, 63 (1985).
33. C. A. Baillie, J. F. Watts and J. E. Castle, *J. Mater. Chem.* **2**, 939 (1992).
34. J. F. Watts and A. M. Taylor, *J. Adhesion* **55**, 99 (1995).
35. J. F. Watts and A. M. Taylor, *J. Adhesion* **46**, 161 (1994).
36. A. M. Taylor, C. H. McLean, M. Charlton and J. F. Watts, *J. Adhesion* **46**, 161 (1994).
37. S. R. Leadley, J. F. Watts, C. J. Blomfield and C. Lowe, *Surf. Interface Anal.* **26**, 444 (1998).
38. R. Bailey and J. E. Castle, *J. Mater. Sci.* **12**, 2049 (1997).
39. D. Kowalczyk, S. Slomkowski, M. M. Chehimi and M. Delamar, *Int. J. Adhesion Adhesives* **16**, 227 (1996).
40. J. Quinton, L. Thomsen and P. Dastoor, *Surf. Interface Anal.* **25**, 931 (1998).
41. M.-L. Abel, J. F. Watts and R. P. Digby, *Int. J. Adhesion Adhesives* **18**, 179 (1998).
42. C. Perruchot, M. M. Chehimi, M. Delamar and F. Fievet, *Surf. Interface Anal.* **26**, 689 (1998).
43. M. Murase and J. F. Watts, *J. Mater. Chem.* **8**, 1007 (1998).
44. M. M. Chehimi, M.-L. Abel, M. Delamar, J. F. Watts and P. A. Zhdan, in: *Organic Coatings*, P.-C. Lacaze (Ed.), pp. 351–356. AIP Press, Woodbury, NY (1996).
45. M. M. Chehimi, M.-L. Abel, B. Saoudi, M. Delamar, N. Jammul, J. F. Watts and P. A. Zhdan, *Polimery* **41**, 75 (1996).
46. M.-L. Abel and M. M. Chehimi, *Synth. Met.* **66**, 225 (1994).
47. R. H. West and J. E. Castle, *Surf. Interface Anal.* **4**, 68 (1982).
48. A. S. Kinloch, G. K. A. Kodokian and J. F. Watts, *J. Mater. Sci. Lett.* **10**, 815 (1991); idem, *Roy. Soc. Phil. Trans. Lond.* **A338**, 83 (1992).
49. R. L. Martin and D. A. Shirley, *J. Am. Chem. Soc.* **96**, 5299 (1974).
50. M. Delamar, J. Electron Spectrosc. Relat. Phenom. **53**, c11 (1990).
51. M. G. Cattania, S. Ardizzone, C. L. Bianchi and S. Carella, *Colloids Surfaces A* **76**, 233 (1993).
52. C. Defossé, P. Canesson, P. G. Rouxhet and B. Delmon, *J. Catalysis* **51**, 269 (1978).
53. R. Borade, A. Sayari, A. Adnot and S. Kaliaguine, *J. Phys. Chem.* **94**, 5989 (1990).
54. M. Huang, A. Adnot and S. Kaliaguine, *J. Catalysis* **137**, 322 (1992).
55. M. J. Remy, M. J. Genet, P. F. Lardinoix, P. P. Notté and G. Poncelet, *Surf. Interface Anal.* **21**, 643 (1994).
56. R. S. Drago, *Structure and Bonding* **15**, 73 (1973).
57. G. W. Simmons and B. C. Beard, *J. Phys. Chem.* **91**, 1143 (1987).

58. J. F. Watts and E. M. Gibson, *Int. J. Adhesion Adhesives* **11**, 105 (1991).
59. G. Kurbatov, E. Darque-Ceretti and M. Aucouturier, *Surf. Interface Anal.* **18**, 811 (1992).
60. M. Delamar, J. Electron Spectrosc. Relat. Phenom. **67**, R1 (1994).
61. G. Kurbatov, E. Darque-Ceretti and M. Aucouturier, *Surf. Interface Anal.* **20**, 402 (1993).
62. K. Burger and E. Fluck, *Inorg. Nucl. Chem. Lett.* **10**, 171 (1974).
63. M. M. Chehimi, *J. Mater. Sci. Lett.* **10**, 908 (1991).
64. T. K. Kwei, E. M. Pearce, F. Ren and J. P. Chen, *J. Appl. Polym. Sci., Polym. Phys.* **24**, 1597 (1986).
65. F. Bodino, N. Compiègne, L. Köhler, J. J. Pireaux and R. Caudano, in: *Proc. Adhesion Soc. Meeting*, pp. 41–44 (1997).
66. F. M. Fowkes, F. L. Riddle, Jr. W. E. Pastore and A. A. Weber, *Colloids Surfaces* **43**, 367 (1990).
67. J. R. Costanza, A. P. Silveri, J. A. Vona, *Radiation Cured Coatings*, Federation Series on Coatings Technology. FSCT, Blue Bell, PA (June 1986).
68. S. R. Leadley, J. F. Watts, A. Rodriguez and C. Lowe, *Int. J. Adhesion Adhesives* **18**, 193 (1998).
69. S. R. Leadley, J. F. Watts, *J. Electron Spectrosc. Relat. Phenom.* **85**, 107 (1997).
70. F. Garbassi, M. Morra and E. Ochiello, *Polymer Surfaces. From Physics to Technology*. John Wiley, Chichester (1994).
71. F. Arefi, M. Tatoulian, V. André, G. Lorang and J. Amouroux, in: *Metallized Plastics 3, Fundamental and Applied Aspects*, K. L. Mittal (Ed.), pp. 243–256. Plenum Press, New York (1992).
72. N. Shahidzadeh-Ahmadi, F. Arefi-Khonsari and J. Amouroux, *J. Mater. Chem. 5*, 2 (1995).
73. F. Arefi, V. Andre, P.Montazer-Rahmati and J. Amouroux, *Pure Appl. Chem.* **64**, 715 (1992).
74. J. F. Moulder, W. F. Stickle, P. E. Sobol and K. D. Bomben, in: *Handbook of X-Ray Photoelectron Spectroscopy*, J. Chastain (Ed.). Perkin-Elmer Corp., Eden Prairie, MN (1992).
75. F. Arefi-Khonsari, M. Tatoulian, N. Shahidzadeh, M. M. Chehimi, J. Amouroux, D. Leonard and P. Bertrand, in: *Adhesion Science and Technology (Festschrift in Honor of K. L. Mittal)*, W. J. van Ooij and H. R. Anderson, Jr. (Eds), pp. 329–353. VSP, Utrecht, The Netherlands (1998).
76. J. Kurdi, F. Arefi-Khonsari, M. Tatoulian and J. Amouroux, in: *Metallized Plastics 5 & 6, Fundamental and Applied Aspects*, K. L. Mittal (Ed.), pp. 295–319. VSP, Utrecht, The Netherlands (1998).
77. G. A. Parks, *Chem. Rev.* **65**, 177 (1965).
78. F. Arefi-Khonsari, M. Tatoulian, N. Shahidzadeh and J. Amouroux, in: *Plasma Processing of Polymers*, R. d'Agostino *et al.* (Eds), pp. 165–207. Kluwer Academic Publishers, Dordrecht, The Netherlands (1997).
79. R. S. Drago, *J. Chem. Ed.* **51**, 300 (1974).
80. F. M. Fowkes, D. W. Dwight, D. A. Cole and T. C. Huang, *J. Non-Cryst. Solids* **120**, 47 (1990).
81. M. Tatoulian, F. Arefi-Khonsari, I. Mabille-Rouger, J. Amouroux, M. Gheorgiu and D. Bouchier, *J. Adhesion Sci. Technol.* **9**, 923 (1995).

Acid-Base Interactions, Vol. 2, pp. 299–316
Ed. K.L. Mittal

Determination of the acid-base properties of solid surfaces using Inverse Gas Chromatography: Advantages and limitations

HENRI BALARD,* ERIC BRENDLE and E. PAPIRER
ICSI-CNRS, 15 rue Starcky, BP 2488, F-68057 Mulhouse, France

Abstract—Inverse Gas Chromatography, at infinite dilution conditions (IGC-ID), is a convenient method for the examination of the surface properties of a solid. A comparison of apolar (n-alkanes) and polar probe adsorption behaviors leads to an estimation of the specific (polar, acid-base) interaction capacity of the solid surface. By knowing the acid-base characteristics of the probes, one may, in principle, assess the acid-base properties of the solid surface itself. The different approaches proposed in the literature are reviewed and discussed. The necessary choice of an (empirical) acid-base scale is of importance and will also be examined. Furthermore, the probe-surface interaction depends on steric factors. Recently, we proposed a procedure to take this parameter into account. However, the limitations of the IGC-ID method for the determination of the acid-base properties of a solid surface are essentially related to the unavoidable surface heterogeneity of any solid surface. Indeed, the IGC-ID fundamental equation assumes that the solid surface is quite homogeneous. In fact, this condition is scarcely fulfilled because of the existence of surface structural defects, variation in the nature and topology of surface groups, etc... As a consequence, the sites having the highest interaction potential will principally contribute to the retention time of the probe in the chromatographic column leading to acid-base parameters that are not representative of the whole surface of the solid.

Keywords: Inverse Gas Chromatography; morphology indexes; acid-base interactions; surface heterogeneity.

1. INTRODUCTION

Inverse Gas Chromatography at Infinite Dilution Conditions (IGC-ID) is a convenient method for the examination of the surface properties of a solid. But, as in the other methods based on wettability measurements (contact angle or liquid wicking measurements), the estimation of the specific (polar, acid-base) interaction capacity of the solid surface is obtained by a comparison of the apolar (n-alkanes) and polar probes adsorption behaviors.

Then, the first problem encountered is to estimate quantitatively the respective contributions of the non-specific forces and of the acid-base interactions, because when a polar molecule interacts with a solid surface, the non-specific forces related to

* To whom correspondence should be addressed. E-mail: h.balard@univ-mulhouse.fr

the London forces and the specific ones, mainly due to acid-base interactions, take place simultaneously.

In the case of IGC-ID, for the estimation of the contribution of the specific component (ΔG_a^{sp}) to the free energy of adsorption, it was proposed to subtract from the global free energy of adsorption of the polar probe, the free energy of adsorption of a hypothetical alkane having the same molecular descriptor value. The earliest molecular descriptor used was the logarithm of the vapor pressure of the probe [1]. Later on, other molecular descriptors were suggested. Their validity will be discussed later. Finally, starting from the ΔG_a^{sp} values and calling on semi-empirical scales describing the acid-base properties of the probes, it is now possible to assess the acid-base properties of the solid surface itself.

Then, an important question is: Will the IGC-ID technique provide the same information as do wettability methods for the assessment of the acid-base characteristics of a solid surface? The latter are based on the interaction between a liquid (a condensed medium) and the solid surface, but the former, IGC-ID, is based on the measurement of the level of interaction between an isolated gaseous molecule and the solid of interest. In other words, wettability measurements will lead to an average macroscopic information whereas IGC-ID will be sensitive to local microscopic variations of the surface structure, at a molecular scale. Thus, wettability measurements will be sensitive to macroscopic geometrical factors, such as surface roughness and/or surface porosity and tortuousity of the macroporous structure. Local variations of the physico-chemical structure of the solid such as mesopores, micropores and crevices and/or the chemical composition variation will influence the IGC-ID measurements. Therefore, we may predict that wettability methods and IGC-ID techniques may provide comparable information, but only on quite homogeneous surfaces.

Coming back to IGC-ID, we shall first show how a comparison of the adsorption behaviors of linear and bulky alkanes and polar probes permits to evidence the surface roughness and acid-base properties. Thereafter, we shall discuss how surface energetic heterogeneity will affect the specific interaction of a polar probe with a solid surface.

2. DETERMINATION OF THE ACID-BASE INTERACTION USING IGC-ID

2.1. Measurement of the Specific Component of the Free Energy of Adsorption of a Polar Probe

Specific interactions include all types of interactions except the London or dispersive interactions. They may correspond to polar, ionic, electrical, magnetic, metallic, acid-base interactions, etc... Among them, acid-base interactions play a determinant role [2]. Assuming that these two types of interactions are additive, it follows that:

$$\Delta G_a = \Delta G_a^d + \Delta G_a^{sp} \quad (1)$$

According to the fundamental equation (2) of IGC-ID, the net retention volume V_n, that is the volume of carrier gas which is necessary to push the probe molecule through the chromatographic column containing the solid of interest, is directly related to the partition constant K of probe molecules, between the adsorbed state and the gas phase, and to the total adsorption area A of the solid contained in the chromatographic column:

$$V_n = K \cdot A \tag{2}$$

Then the global free energy of adsorption ΔG_a of a probe is related to V_n by the equation:

$$-\Delta G_a = RT \ln(V_n / C) \tag{3}$$

where C is a constant related to the chosen reference state for the adsorbed molecule.

As pointed out above, a comparison of the apolar (n-alkanes) and polar probe adsorption behaviors leads to the evaluation of the contribution of the specific interactions (ΔG_a^{sp}), by subtracting, from the global free energy variation ΔG_a, the contribution due to dispersive interactions. The latter is estimated from the "n-alkanes reference line" observed when plotting the global free energy of adsorption ΔG_a of n-alkanes versus a given molecular descriptor that could be, for example, the number of carbon atoms or the vapor pressure of the injected probes. The representative points of polar probes, that are more interactive with the polar surface, should be located above the so-called "alkane-line" and the contribution of specific interactions (ΔG_a^{sp}) is then given by the departure of the representative point of the polar probe from this "reference alkane line" as shown in Figure 1.

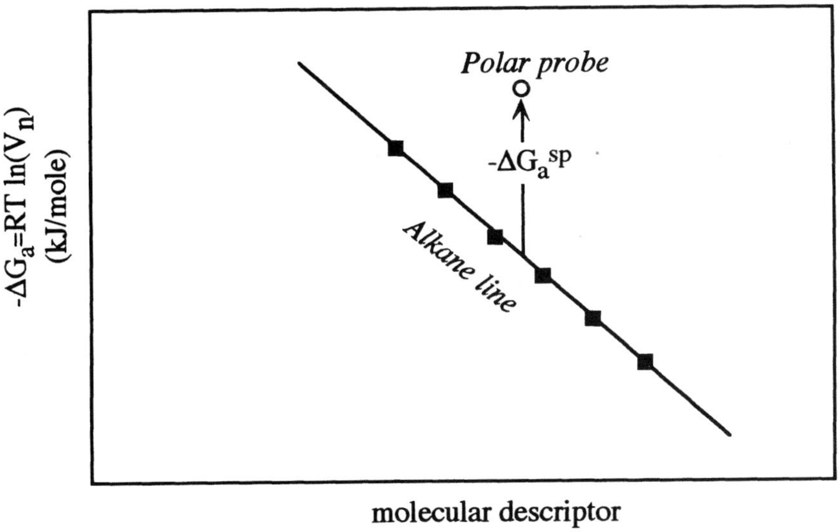

Figure 1. Principle of the determination of the specific component, ΔG_a^{sp} of the free energy of adsorption.

It is obvious that the choice of the molecular descriptor may influence notably the value of the measured ΔG_a^{sp}.

Mukhopadhyay and Schreiber [3] and more recently, Brendlé and Papirer [4] paid a special attention to the choice of the molecular descriptor, reviewing the different proposals made in the literature, namely:

- $\log(P_0)$, the logarithm of the vapor pressure [1],
- $a\ (\gamma_l^d)^{1/2}$, the product of the molecular area a by the square root of the dispersive component of the surface tension of the probe γ_l^d, taken in a liquid state [5],
- T_b, the boiling point [6],
- $\alpha_o\ (h\nu)^{1/2}$, the product of the polarisability α_o of the probe and the square root of its electronic frequency $\nu \cdot h$, where h is the Planck constant [7].
- X_t, a topological index [4] that takes into account both the geometry and the local electronic density due to the presence of heteroatoms, such as chlorine, oxygen or nitrogen atoms, in the polar probe structure.

The first descriptor was the earliest of the five and was proposed by Saint-Flour and Papirer [1] and remains up to now the most popular. The choice of this parameter was made considering that the logarithm of the vapor pressure which is closely related to the evaporation enthalpies (ΔH_v) was representative of the capacity of interactions of two identical molecules. This molecular descriptor presents several advantages since P_0 values are given in the literature or are computable even at relatively high measurement temperatures as long as these stay below the critical temperature of the probe. Yet, problems do appear with certain solids, possessing a high surface energy, since the representative points of some polar probes fall beneath the alkane-line. Chehimi and Pigois-Landureau [8] explained this observation in terms of the self-association of given polar probes and, moreover, they proposed to use the enthalpy of evaporation, ΔH_v, instead of $\log(P_0)$, correcting it by taking into account the contribution of the dispersive interactions to ΔH_v. Of course, ΔH_v and T_b are closely related physical parameters.

Later on, Schultz *et al.* [5] proposed to use the quantity $a\ (\gamma_l^d)^{1/2}$ as a molecular descriptor where a is the cross-sectional area of the adsorbed probe and γ_l^d the dispersive component of the surface tension of the probe, starting from equations established by Fowkes [9].

This procedure solved apparently some of the problems encountered using the earlier method. The limitations are due, in part, to the fact that the molecular area "a" is not exactly known and varies both with the nature of the solid [8], and the temperature and surface coverage. Furthermore, γ_l^d values are not always available from the literature at the temperature used for the IGC measurements. Both limitations are not recognized in the Schultz approach.

Finally, all preceding molecular descriptors suffer from the same weakness: they all are based on parameters that are characteristic of interactions between identical molecules in the liquid phase, whereas ΔG_a is related to the interaction between an isolated molecule and a solid surface having generally a very different chemical nature.

This observation instigated some authors to propose new molecular descriptors that are related to the molecular structure of the probe and not solely to the intermolecular interactions that take place between them in the liquid phase.

The first attempt was made by Dong *et al.* [10] and refined by Donnet *et al.* [7], who, starting from the equations describing the van der Waals interaction energy between two partners, proposed to adopt the product $\alpha_o(h\nu)^{1/2}$ as the molecular descriptor. The main difficulty encountered by these authors was due to the fact that polarizability data, measured at low frequency, are rather scarce in the literature, whereas those computed from refractive indexes according to the Debye relation are rough approximations.

Finally, Brendlé and Papirer [4] suggested that topology indexes, such as the Wiener ones [11, 12], which were extensively and successfully employed for the prediction of physicochemical molecular properties, such as boiling points, surface tensions and later for the understanding of the action of drugs [13] or crystal growth process modeling [14], could be used instead of the preceding molecular descriptors. They applied these topology indexes with success, at first, for describing the adsorption of branched alkanes used for testing, on a molecular scale, the surface roughness of the adsorbents (Size Exclusion IGC [15]). Then, calling on the method of Barysz *et al.* [16] that allows an easy extension of the Wiener indexes to polar probes, they proposed new topology indexes for the evaluation of a solid surface specific interaction potential.

For the sake of comparison, the variations of RT $\ln(V_n)$ of apolar and polar probes, measured at 110°C on a smooth fumed silica (Aerosil A130 from Degussa), with the four different molecular descriptors are shown in Figures 2 and 3.

Figures 3a and 3b look very similar. Actually, when plotting X_t versus $\alpha_o(h\nu)^{1/2}$, one obtains a quite linear correlation which is illustrated in Figure 4.

The main advantage of the X_t parameter is that it may be easily computed for each probe of interest, without any physical measurement requirement. The main molecular descriptors used for the IGC determination of the specific interactions are collected in Table 1.

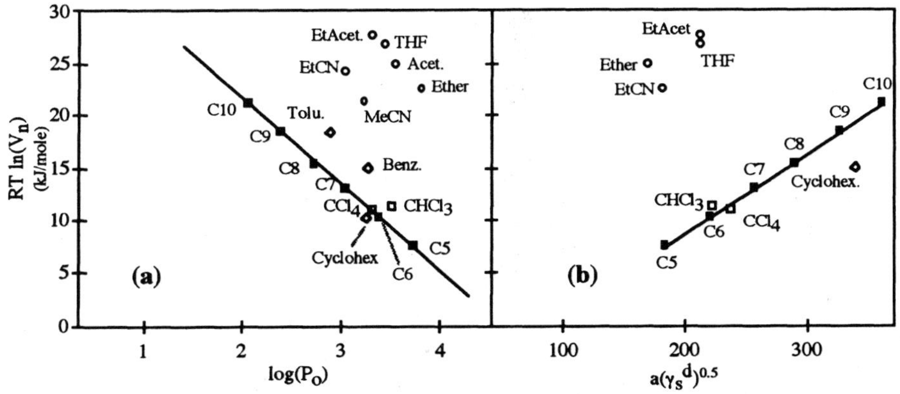

Figure 2. Variation of RT $\ln(V_n)$ of apolar and polar probes, measured at 110°C on a fumed silica (Aerosil A130 from Degussa), with $\log(P_0)$ and $a(\gamma_s^d)^{0.5}$, taken as molecular descriptors.

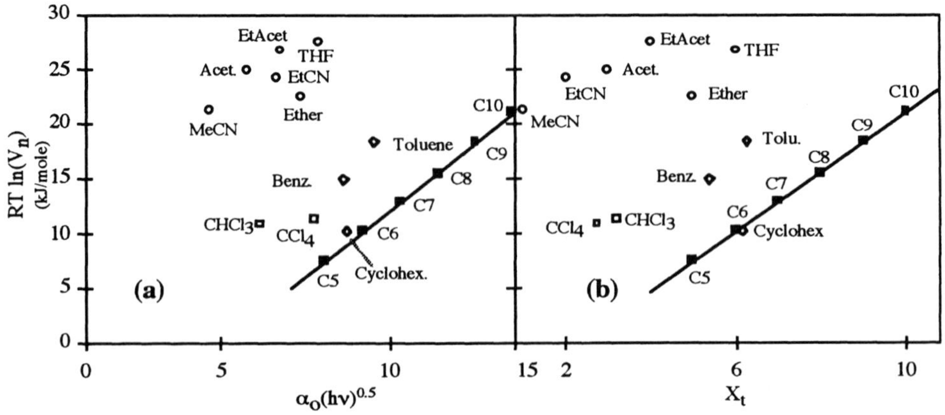

Figure 3. Variation of RT $\ln(V_n)$ of apolar and polar probes, measured at 110°C on a fumed silica (Aerosil A130 from Degussa), with $\alpha_o(h\nu)^{0.5}$ and X_t, taken as molecular descriptors.

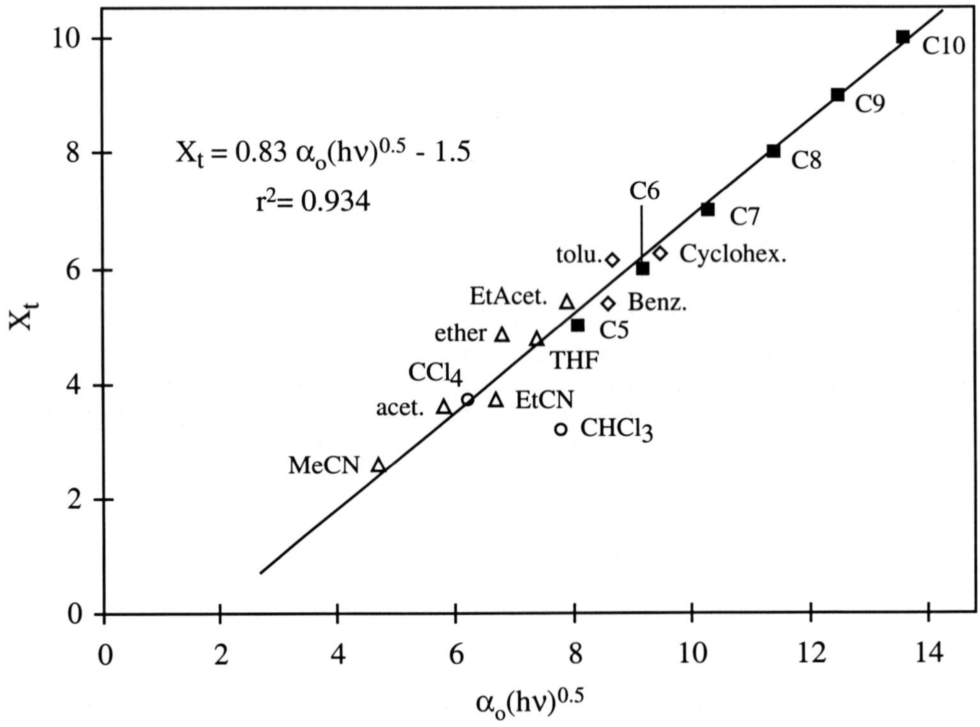

Figure 4. Variation of the topological index X_t with polarisability index $\alpha_o(h\nu)^{0.5}$.

Table 1.
Antoine's law parameters [1], indexes of surface tension [5], indexes of polarisability [7] and topology indexes [4] of some common IGC probes used for testing the acid-base properties of a solid surface

Probes	Designation*	A	B	C	$a(\gamma_s^d)^{0.5}$	$\alpha_0(h\nu)^{0.5}$ $\times 10^{49}$	X_t
Pentane	C5	6.8522	1064.6	232.00	185	8.1	5.00
Hexane	C6	6.8778	1171.5	224.34	221	9.2	6.00
Heptane	C7	6.9024	1268.1	216.90	257	10.3	7.00
Octane	C8	6.9238	1355.1	209.52	290	11.4	8.00
Nonane	C9	6.9351	1428.8	201.62	328	12.5	9.00
Decane	C10	6.9537	1501.3	194.48	363	13.6	10.00
Chloroform	$CHCl_3$	6.9033	1163.0	227.40	224	7.8	3.21
Tetra-chloro-methane	CCl_4	6.8941	1219.6	227.16	238	6.2	3.74
Cyclohexane	Cyclohex.	6.8450	1203.5	222.86	-	8.7	6.15
Benzene	Benz.	6.9057	1211.0	220.79	238	8.6	5.39
Toluene	Tolu.	6.9545	1344.8	219.48	-	9.5	6.26
Acetonitrile	MeCN	7.1226	1315.2	230.00	-	4.7	2.61
Propionitrile	EtCN	7.1522	1398.2	230.00	-	6.7	3.71
Acetone	Acet.	6.9033	1163.0	227.40	171	5.8	3.61
Ethyl Acetate	EtAcet	6.8941	1219.6	227.16	213	7.9	5.43
Ether	Ether	6.9847	1090.6	231.20	182	7.4	4.77
Tetrahydrofuran	THF	6.7538	1146.4	230.00	213	6.8	4.85

* for figure captions.
A, B, C are the parameters of Antoine's law for the calculation of the saturated vapor pressure.
$a(\gamma_s^d)^{0.5}$ is the surface tension index.
$\alpha_0(h\nu)^{0.5}\cdot 10^{49}$ is the polarisability index.
X_t is the topology index.

2.2. Assessment of the Acid-base Properties of a Solid Surface

The quantitative evaluation of acid-base interaction parameters for solid surfaces is generally based on ΔH_a^{AB}, the enthalpy of specific interaction between two molecules A and B. The different approaches in use were recently reviewed by Mukhopadhyay and Schreiber [3]. Only the Gutmann approach will be briefly recalled in this paper.

Adopting the Gutmann terminology [17], the probes are categorized as acidic, basic or amphoteric molecules by their electron-acceptor and their electron-donor numbers *AN* and *DN*, respectively (Table 2). The specific enthalpy of adsorption values ΔH_a^{sp} are computed directly from specific free enthalpies of adsorption, measured at different temperatures, according to Eq. (4).

$$\Delta H_a^{sp} = -T^2\, \partial(\Delta G_a^{sp}/T)/\partial T \tag{4}$$

Following an extension of the Gutmann equation, which allows the calculation of the interaction enthalpy between a "pure" acid and a "pure" base, the specific interaction enthalpy (ΔH_a^{sp}) between an amphoteric (having both acid and base interactions capacities) probe, characterized by both *AN* and *DN*, and an amphoteric solid surface characterized by K_A and K_D (Acceptor and Donor numbers of the solid surface) is given by equation (5).

$$\Delta H_a^{sp} = AN \cdot K_D + DN \cdot K_A \quad (5)$$

K_A and K_D are graphically determined as shown in Figure 5 for a fumed silica surface (Aerosil A130 from Degussa) and its methylated derivative, by plotting $\Delta G_a^{sp}/AN$ (instead of ΔH_a^{sp}, since ΔG_a^{sp} is proportional to ΔH_a^{sp}) versus DN/AN.

From the slopes of the lines, the K_A values are obtained whereas the ordinates at the origin give K_D's.

Table 2.
Acceptor [3] and Donor [4] numbers of some common polar probes

Probes	AN (%)	DN (kcal/mol)
Benzene	0.17	0.1
CCl_4	0.7	0.0
Chloroform	5.4	0.0
Acetonitrile	4.7	14.1
Acetone	2.5	17.0
Ethyl Acetate	1.5	17.1
Ether	1.4	19.2
THF	0.5	20.0

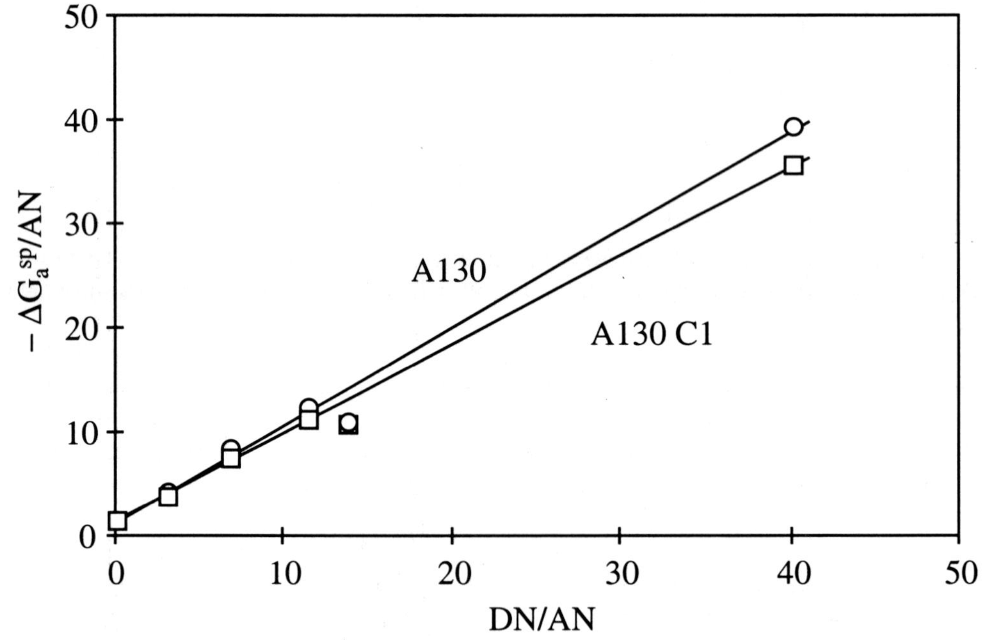

Figure 5. Determination of the K_A and K_D parameters on a fumed silica sample (Aerosil A130 - Degussa) and its methylated derivative (A130 C1).

Table 3 gathers the main characteristics of these two analyzed samples.

As expected, the methylation of the fumed silica increases its electron-donor capacity which is mainly related to the presence of siloxane bridges on the silica surface, whereas the disappearance of silanol groups, due to the esterification of the most acidic isolated silanol groups, leads to a lower electron acceptor capacity.

Table 3.
Main characteristics of a fumed silica sample (A130) and its methylated derivative (A130 C1) and their respective acceptor K_A and donor K_D parameters measured at 110°C, using IGC-ID technique

	A130	A130 C1
Specific surface area (m^2/g)	130	130
OCH_3/nm^2	0	3.4
K_A	0.94	0.85
K_D	1.21	1.29
correlation coefficient	0.989	0.994

It would be interesting to go further in this direction so as to reach a more quantitative description of the acid-base properties of solid surfaces, but this would require a knowledge of the number of surface groups, per unit surface area, which interact with the probe: a quantity that is not easy to evaluate from the sole IGC measurements.

On an energetically homogeneous surface, the number of sites visited by the molecular probe depends on both the number of these sites n_s and on their energies of interaction ε_s. The experimentally measured retention time, t_r, may be considered as the product of local residence time, τ_s, and the total number of sites visited n_s Eq. (6) [19], the two variables which cannot be separately evaluated:

$$t_r = n_s.\tau_s \qquad \text{with} \quad \tau_s = \tau_o \exp(\varepsilon_s/RT) \tag{6}$$

where τ_o is between 10^{-12} to 10^{-13} sec.

On a heterogeneous surface, the situation becomes much more complex and the influence of the surface energetic heterogeneity will be discussed more extensively later on.

The second reason that renders difficult a quantitative determination of acid-base characteristics of solid surfaces using the reported method stems from the fact that the semi-empirical acid-base scales of solutes were established considering the interactions between small organic molecules in solution. In a solvent, the two interacting molecules are free to adopt the best relative orientation leading to the highest level of interaction. This would certainly not be the case on a solid surface on which defects, steps, pores or crevices hinder the mobile probe molecule to adopt its optimal orientation, causing a decrease of the energy of interaction and this effect will also be dependent on the stereochemistry of the probe molecule itself.

Keeping in mind that the ΔG_a^{sp} determination is based on a comparison of the adsorption behaviors of n-alkanes and polar probes, it becomes evident that, if the size exclusion effects due, for instance, to given surface defects vary with the probe stereochemistry, erroneous or apparent ΔG_a^{sp} will be obtained. The influence of the geometrical surface heterogeneity on the specific interactions will be discussed now.

3. LIMITATIONS OF IGC-ID

3.1. Influence of the Surface Morphology on the Specific Interaction Capacity

According to equation (2), a comparison of the retention properties of polar and non-polar probes assumes that both types of molecules should have access to the same area of adsorption; in other words, no size exclusion effects occur during the chromatographic process.

Papirer and Balard [20, 21] showed that such assumption of surface flatness could not be accepted for solids having a lamellar structure such as crystalline silicas or talc. If, for instance, partial insertion of alkanes occurs in slot like pores, on the lateral surface of the crystal, generated by stacking defects. This insertion will lead to a very efficient adsorption process of linear alkanes. From this insertion phenomenon, non-realistic high values of the dispersive component $\gamma_S{}^d$ of the surface energy will be calculated for those materials. Then, negative I_{SP} values are also observed in that case, especially for bulky polar molecules such as chloroform or α-ω dienes that cannot access the same alkane adsorption sites [22].

Coming back to equation (2), a morphology index I_m was defined as the ratio of the true adsorption areas according to equation (7).

$$I_m = \frac{A_B}{A_L} = \frac{K_L}{K_B}\frac{V_N^B}{V_N^L} \tag{7}$$

where A_L and A_B are the areas of adsorption accessible, respectively, to linear and to the branched alkane probes and K_L and K_B their corresponding partition constants.

This takes into account the accessibility of the solid surface on the one hand and the shape or morphology of the molecular probe on the other hand. In this equation, the main problem encountered is the evaluation of the K_L/K_B ratio which obviously depends on the stereochemical structure of the branched alkane. In a first attempt, we proposed to determine this ratio for different branched alkane probes on a reference solid, a fumed silica sample (Aerosil A130). This silica may indeed be used as a model surface since independent methods such as volumetric gas adsorption, SAXS [23] and chemical approaches coupled with IGC [24] or NMR [25] had shown that its surface was smooth at the molecular level. The morphology index (I_m) we defined earlier [20] was then based on a simple comparison of the retention times of linear and branched alkane isomers on this pyrogenic silica and on the silica under investigation:

$$I_m = 100 \times \frac{V_N^B}{V_N^L}\frac{V_{N\,ref}^L}{V_{N\,ref}^B} \tag{8}$$

In this expression, V_N^L and V_N^B are, respectively, the net retention volumes of linear and branched alkanes, having the same number of C atoms, measured on the studied silica sample whereas $V_{N\,ref}^L$ and $V_{N\,ref}^B$ are the corresponding values for the reference pyrogenic silica.

A major difficulty when applying this method stems from the necessity of resorting to a reference solid surface having a different chemical structure than the solid of interest. Then, Brendlé and Papirer [15] proposed to assess this ratio K_B/K_L using new topology indexes X_t derived from Wiener's topology indexes [11]. Some X_t values [15] are reported in Table 4. Figure 6 illustrates this procedure for fumed silica (N20 from Wacker-Chemie GmbH).

Table 4.
Topology indexes X_t of some cyclic and branched alkane probes after [15]

Probes	Designation*	X_t
Cyclohexane	Cyclohex.	6.15
Cycloheptane	Cyclohept.	7.32
Cyclooctane	Cyclooct.	8.32
Methyl 2-butane	MeBut.	4.84
Methyl 2-pentane	MePent.	5.84
Methyl 2-hexane	MeHex.	6.84
Dimethyl 2,5-hexane	DMHex.	7.68
Trimethyl 2,2,4-pentane	TMPent.	7.40
Tetramethyl 2,2,3,3-butane	TeMBut.	7.09

* for figure captions

Figure 7 displays the same diagram established for H-Maggadiite, a crystalline silica having a lamellar structure.

It is seen that all cyclic and branched alkanes clearly lie below the "n-alkane line", a phenomenon due to the size exclusion effects leading to a strong decrease of the free energy of adsorption compared to the reference n-alkanes. The decrease in free energy, $-\Delta G_a^M$, equal to the departure of the representative point of the bulky alkane from the reference n-alkane straight line as depicted in Figure 7, corresponds to the contribution of the morphological factor to the loss of the interaction energy for a cyclic or a branched alkane probe.

$I_m(X_t)$, is then given by:

$$I_m(X_t) = \exp(-\Delta G_a^M/RT) \tag{9}$$

$I_m(X_t)$ was systematically measured on lamellar solids such as clay minerals [23], talc or muscovite [22] samples.

The $I_m(X_t)$ values measured on various silica samples, using different branched or cyclic alkanes, are listed in Table 5.

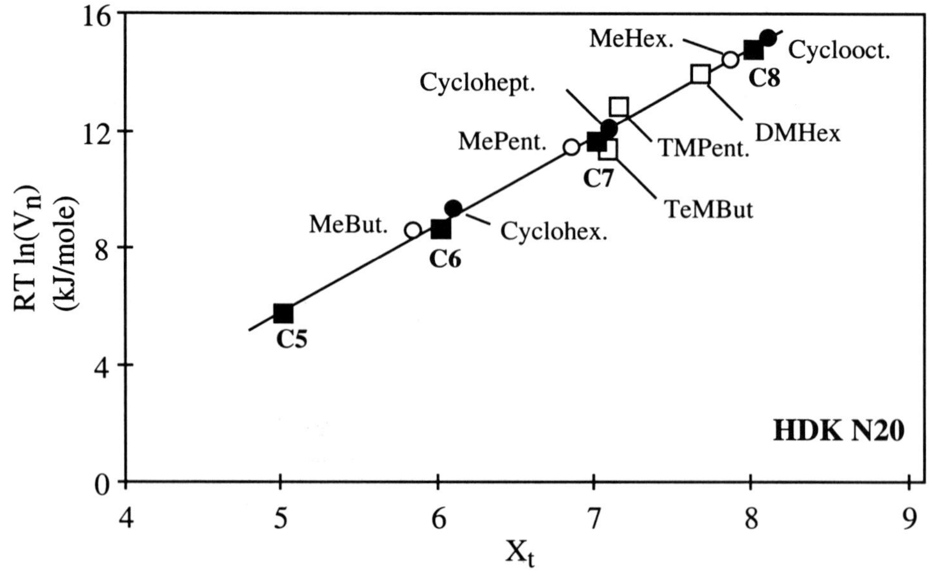

Figure 6. Variation of RT $\ln(V_n)$ with the topological index, X_t, of linear, cyclic and branched alkane probes for a fumed silica: HDK N20 from Wacker-Chemie GmbH.

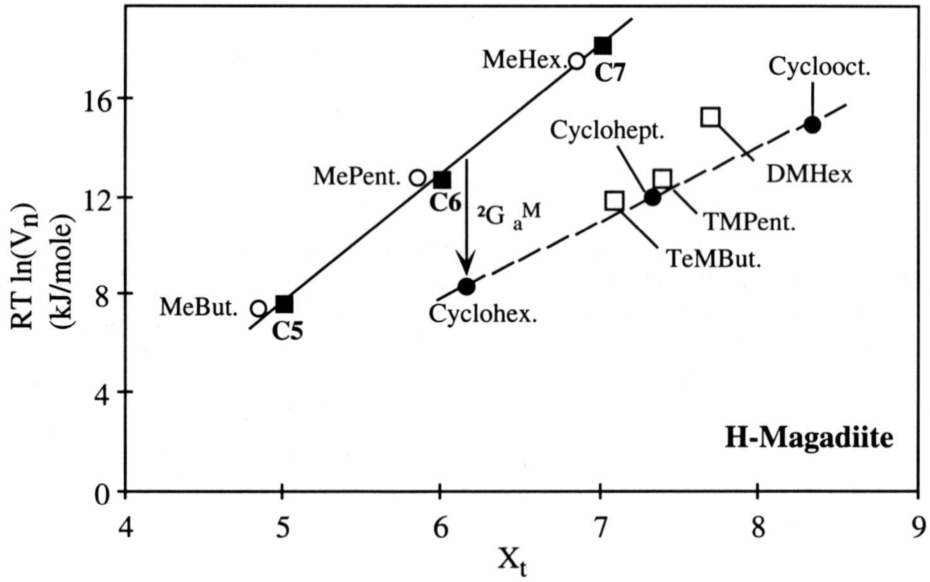

Figure 7. Variation of RT $\ln(V_n)$ with the topological index X_t of linear, cyclic and branched alkane probes for a crystalline silica: H-magadiite.

Table 5.
Decrease of the free energy of adsorption ΔG_a^M and morphology index I_m for cyclohexane and some branched alkanes measured on a fumed silica A130 (Degussa), two talc samples (Talcs de Luzenac Europe), two clays: illite and kaolinite, and a crystalline silica: H-magadiite

	CycloHex		DMHex		TMPent		TMBut	
Solids	ΔG_a^M (kJ/mole)	$I_m(X_t)$	ΔG_a^M (kJ/mole)	$I_m(X_t)$	ΔG_a^M (kJ/mole)	$I_m(X_t)$	ΔG_a^M (kJ/mole)	$I_m(X_t)$
A130	-0.2	0.95	0.3	0.91	-0.3	1.10	-0.5	1.11
Talc 00	-	-	2.0	0.52	3.5	0.32	2.8	0.41
Talc sup	-	-	2.2	0.52	3.9	0.31	3.4	0.35
Illite Vosges	-	-	41.0	0.65	3.5	0.34	3.0	0.41
H-Magadiite	-5.47	0.19	-6.8	0.13	7.8	0.09	7.1	0.11

* Talc 00 and Talc sup are industrial references of Talk de Luzenac Europe.

Bulky polar probes, such as chloroform, do not have access to the adsorption sites accessible to linear alkanes. The decrement $-\Delta G_a^M$ will subtract from the increment, related to the specific interaction $-\Delta G_a^{sp}$, leading thus to an apparent lower global parameter of interaction I_{sp}. If the $-\Delta G_a^M$ exceeds $-\Delta G_a^{sp}$, we may observe a negative I_{sp}, having no physical meaning. Therefore, each time when exclusion effects are observed, one has to keep in mind that only apparent I_{sp} values are measured.

Finally, the last factor that can affect strongly the value of I_{sp} is the surface energetic heterogeneity.

3.2. Influence of the Surface Energetic Heterogeneity on the Specific Interaction Capacity

The IGC fundamental equation (2) assumes that the value of the partition constant K does not vary whatever the adsorption site, in other words the studied solid surface is assumed to be homogeneous.

In fact, this condition is scarcely fulfilled because atomic structures present on a solid surface are rigidly connected and this rigidity is responsible for the formation of surface defects like cracks, corners, steps and porous structures. Moreover, mineral and organic impurities often contaminate the solid surface. A consequence of the existence of surface heterogeneity is that the energy of interaction will vary depending on the chemical nature or local structure of the adsorption site visited by the probe. Hence, according to equation (6), the sites having the highest energies will principally contribute to the global retention time of the probe on the chromatographic support.

Moreover, due to the gradient of interaction energy originating from the surface heterogeneity, the adsorbed molecule will migrate on the surface towards the sites having the highest energy of interaction, sites that will act as potential wells, fixing the migrating molecule.

An indirect proof of this migration of the probe in the adsorbed state was given by the study of the surface properties, using IGC-ID, of fumed silica subjected to a controlled silylation (treatment with trimethylchlorosilane) process. It was observed that using both alkane probes (or poly(dimethylsiloxane) or PDMS oligomer probes) the surface properties of the silylated silica samples changed suddenly when a coverage ratio of approximately 50% of the silica surface by the trimethylsilyl groups was reached. Such evolution is illustrated in Figure 8, in terms of the evolution of the increment of the free energy of adsorption per methylene group on fumed silicas (T30 silica from Wacker-Chemie GmbH) having an increasing trimethylsilyl coverage [26].

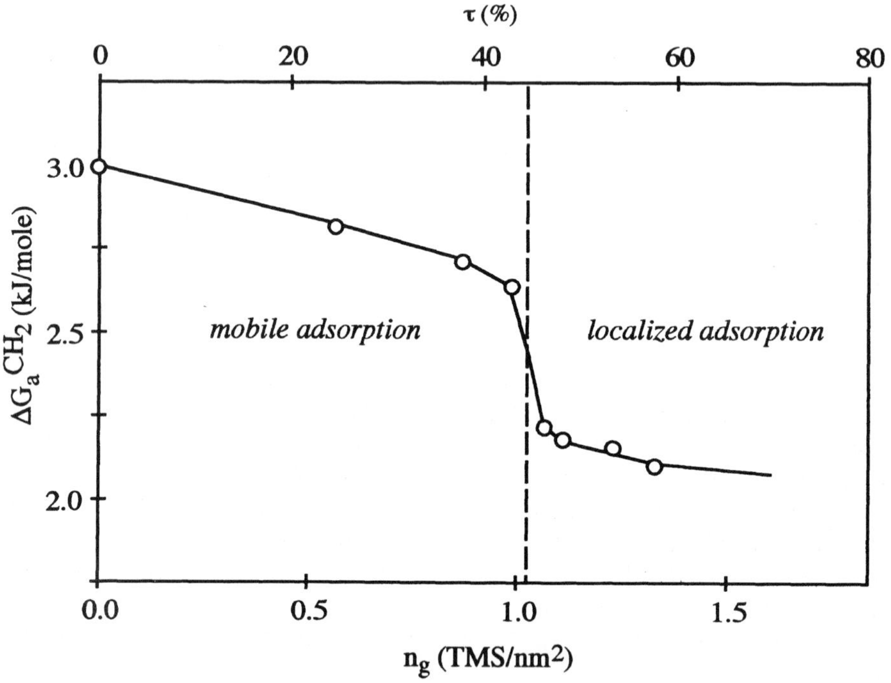

Figure 8. Variation of the increment of the free adsorption energy per methylene group $\Delta G_a^{CH_2}$ of n-alkanes with the trimethylsilyl (TMS) graft ratio n_g (in TMS/nm^2) and the surface coverage ratio (in %), calculated for a TMS molecular area of 43 Å^2, for some silylated HDK T30 fumed silicas from Wacker-Chemie GmbH (measured at 150°C).

In a chromatographic process, a given molecule is submitted to a high number of sorption and desorption steps. If the motion of a molecule along the chromatographic column is mainly due to the carrier gas (usually He) flow, the probe molecule can also migrate on the solid surface itself. If the surface is energetically homogeneous, this phenomenon will have no influence on the measured retention time. On the contrary, on a heterogeneous surface, the sites having the highest energy of interaction will strongly retain the surface migrating molecule.

After a given time, the molecule will desorb and migrate in the gas phase. Therefore, the molecule visits only a very minor part of the solid surface. In other words, the probe will be most sensitive to the surface heterogeneity on a very short distance that may be very roughly estimated to be a few molecular lengths, depending strongly, of course, on the temperature of measurement.

On a non-silylated initial silica, the adsorbed molecule can freely migrate between its site of adsorption and its site of desorption. In the case of a silylated silica sample, the trimethylsilyl grafts will act as pure geometrical barriers and will necessarily restrict the surface mobility of the probe. With increasing coverage ratios of the silica surface by the trimethylsilyl grafts, this limiting effect will increase. For coverage ratios higher than 50%, trimethylsilyl groups cover the majority of the surface and the residual accessible adsorption sites form only some "holes" in the continuous trimethylsilyl grafted layer. Consequently, the molecule can no longer migrate on the surface towards the more interactive sites and we do observe a sudden decrease of the increment in free energy of adsorption per methylene group.

Hence, the global retention time of the probe on a heterogeneous surface is a complex function of the energy of interaction of the sites present on the surface, their number and finally their capture radius because a strong interactive site can only capture a molecule that is adsorbed in its close environment. The global measured retention time t_r is then given by equation (10):

$$t_r = \Sigma\,(n_S, \varepsilon_S, r_S) \qquad (10)$$

where n_s is the number of sites having an energy of interaction ε_s and a capture radius r_S. All these parameters are presently unfortunately unknown.

Consequently, the thermodynamic values calculated from the retention time are not representative of the interaction capacity of the whole solid surface.

Moreover, the major contribution of the adsorption sites of the highest energy of interaction results in the fact that very often the most polar probes, like THF or diethylether, cannot be eluted from heterogeneous solids having, generally, a high specific surface area. Hence, only weak base probes such as aromatics or alkenes may be used in practice. Thus, it is impossible to determine meaningful acid-base properties of a solid.

A question is raised: is it possible to overcome partially these difficulties? The IGC at finite concentration conditions [27, 28] provides a possible but more time consuming way to assess the surface heterogeneity of a solid through the determination of the distribution functions of the adsorption energies of chosen solutes. Figure 9 displays such distribution functions established for octane, propanol-2 and pyridine on an industrial talc sample.

Obviously, the shapes of these distribution functions vary depending on the chemical nature of the probe: apolar like octane, acid like propanol-2 or a basic like pyridine. A challenge for the next years will be to extract from the adsorption energy distribution functions, K_A and K_D values as function of the solid surface coverage by the probes.

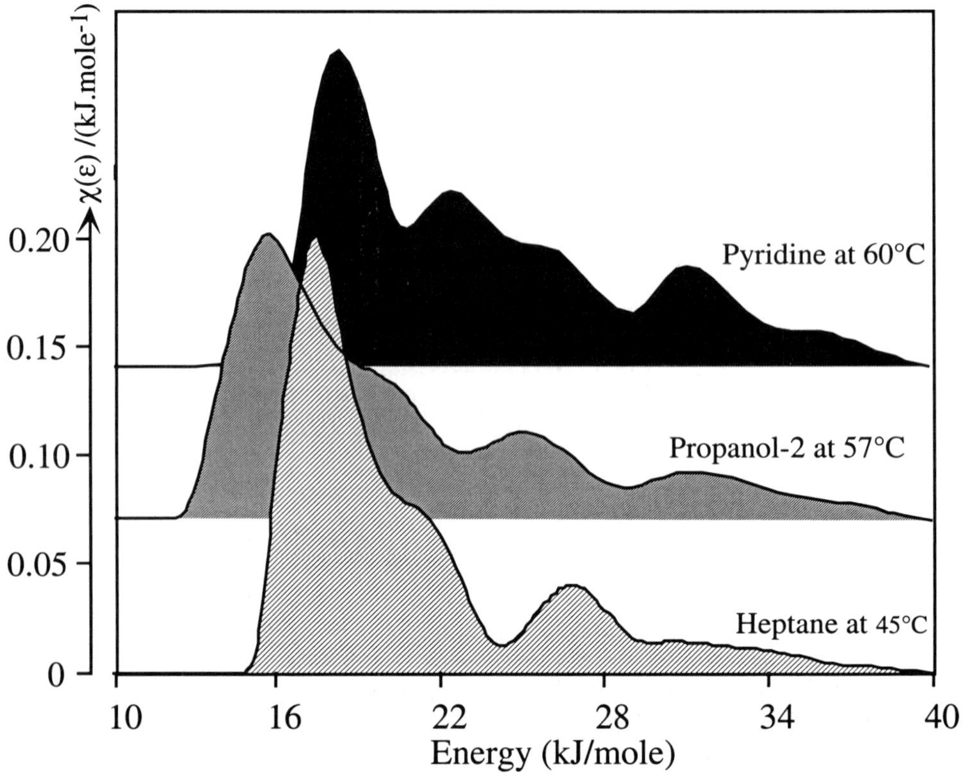

Figure 9. Distribution functions of the adsorption energies of n-heptane, propanol-2 and pyridine on a talc (Talc 00 from Luzenac Europe).

4. CONCLUSION

This paper emphasizes the importance of the choice of the molecular descriptor used for the separation of the contribution of the dispersive from the specific interactions when a polar molecule interacts with a polar solid surface. Molecular descriptors that are related to the probe/probe interactions in the liquid state have to be rejected because they will certainly be unable to take into account the interaction between an isolated molecule and a solid surface having generally a very different chemical nature.

On the contrary, polarisability index, which is only related to the probe chemical structure, will provide a good description of the dispersive interaction capacity of the IGC probe. Fortunately, although the literature data on molecular polarisability at low frequency are rather scarce, topological indexes, calculated from their molecular structures, are directly related to their polarisability indexes and can be advantageously used instead of polarisability indexes for the assessment of the specific adsorption free energies of polar probes.

By applying "classical approaches", based on a semi-empirical description of the acid-base properties of polar probes, the acid-base capacity of interaction of a solid can be estimated. However, this estimation will be valid only on quite smooth and homogeneous surfaces.

Moreover, the IGC method suffers also from some limitations. The first is related to the solid surface roughness. When this is not negligible, then the size exclusion effects will take place, which will influence the interaction capacity of the most bulky probes, in comparison to the reference linear alkanes, leading to undervalued I_{sp}.

On the contrary, surface heterogeneity may enhance I_{sp}, due to the dominant contribution, to the retention time of the probe, of the adsorption sites having the highest energy of interaction. For solids having a high specific surface area and a high surface heterogeneity such as carbon black or silicas, the main limitation of the IGC-ID method, for the assessment of the acid-base properties, is related to the fact that the interesting IGC polar probes are almost irreversibly adsorbed.

Nevertheless, the IGC-ID technique will possibly remain the most attractive and versatile method for determination of acid-base interactions on solids that are not too heterogeneous from both geometrical and energetic points of view.

REFERENCES

1. C. Saint-Flour and E. Papirer, *J. Colloid Interface Sci.* **91**, 69 (1983).
2. K. L. Mittal and H. R. Anderson Jr. (Eds.), *Acid-Base Interactions: Revelance to Adhesion Science and Technology.* VSP, Utrecht (1991).
3. P. Mukhopadhyay and H. P. Schreiber, *Colloids Surfaces A* **100**, 47 (1995).
4. E. Brendle and E. Papirer, *J. Colloid Interface Sci.* **194**, 217 (1997).
5. J. Schultz, L. Lavielle and C. Martin, *J. Adhesion* **23**, 45 (1987).
6. D. T. Sawyer and D. J. Brookman, *Anal. Chem.* **40**, 1847 (1968).
7. J. B. Donnet, S. J. Park and H. Balard, *Chromatographia* **31**, (9/10), 434 (1991).
8. M. M. Chehimi and E. Pigois-Landureau, *J. Mater. Chem.* **4**, 741 (1994).
9. F. M. Fowkes, in: *Chemistry and Physics of Interfaces*, p.1. American Chemical Society, Washington DC (1964).
10. S. Dong, M. Brendlé and J. B. Donnet, *Chromatographia* **28**, (9/10), 85 (1989).
11. H. Wiener, *J. Phys. Chem.* **52**, 425 (1948)
12. I. Motoc and A. T. Balaban, *Rev. Roum. Chim.* **26**, 593 (1981).
13. S. C. Basak, D. P. Gieschen, V. R. Magnuson and D. K. Hariss, *IRCS Medical Science* **10**, 619 (1982).
14. D. Bonchev, O. Mekenyan and J. H. Fritsche, *J. Crystal Growth* **49**, 90 (1980).
15. E. Brendlé and E. Papirer, *J. Colloid Interface Sci.* **194**, 207 (1997).
16. M. Barysz, G. Jashari, R. S. Lall, V. K. Srivastava and N. Trinajstic, in: *Studies in Physical and Theoretical Chemistry, Chemical Applications of Topology and Graph Theory*, R. B. King (Ed.) **28**, p. 222 Elsevier, Amsterdam (1983).
17. V. Gutmann, *The Donor-Acceptor Approach to Molecular Interactions,* Plenum, New York (1978).
18. F. L. Riddle, Jr. and F. M. Fowkes, *J. Am. Chem. Soc.* **112**, 3259 (1990).
19. A. W. Adamson and A. P. Gast, *Physical Chemistry of Surfaces, 6th ed.*, p 602. John Wiley and Sons, New York (1997).
20. H. Balard and E. Papirer, *Prog. Organic Coatings* **22**, 1 (1993).
21. E. Papirer and H. Balard, in: *Adsorption and Chemisorption on Inorganic Sorbents*, A. Dabrovski and T. Tertykh (Eds.), pp. 479-502. Elsevier, Amsterdam (1995).
22. G. Ligner, M. Sidqi, J. Jagiello, H. Balard and E. Papirer, *Chromatographia* **29**, 35-38 (1990).
23. H. Barthel, M. Heinemann, L. Rösch and J. Weiss, *Proc. Eurofillers Conference,* Mulhouse (France), Sept.11-14th, pp. 157-161 (1995).

24. M. Sidqi, H. Balard, E. Papirer, A. Tuel, H. Hommel and A. P. Legrand, *Chromatographia* **25**, 712 (1988).
25. A. Tuel, H. Hommel, A. P. Legrand, H. Balard, M. Sidqi and E. Papirer, *Colloids Surfaces* **58**, 17 (1991).
26. H. Balard, E. Papirer A. Khalfi, H. Barthel and J. Weis in: *Organosilicon IV: From Molecules to Materials*, N. Auner and J. Weis (Eds.), J. Wiley, Weinheim, Germany (to be published in 2000).
27. H. Balard, *Langmuir* **13**, 1260 (1997).
28. H. Balard, A. Saada, J. Hartmann, O. Aouadj and E. Papirer, *Makromol. Symp.* **108**, 63 (1996).

Acid-Base Interactions, Vol. 2, pp. 317–347
Ed. K.L. Mittal

Surface energy and acid-base properties of solids studied by immersion calorimetry

JEAN-MARC DOUILLARD and VINCENT MÉDOUT-MARÈRE
Laboratoire des Agrégats Moléculaires et des Matériaux Inorganiques, UPRESA 5072 du CNRS, CC 015, Université de Montpellier 2, Sciences et Techniques du Languedoc, Pl. E. Bataillon, 34095, Montpellier Cédex 5, France

Abstract—The calorimetric methods of determination of the enthalpies of immersion and of wetting are described in this text. The equations linking the immersion enthalpies to the surface thermodynamic parameters are described. A general model of interpretation of adsorption and wetting is then developed. This model uses three parameters which are the dispersive, acidic and basic components of the surface energy. Some experimental examples are given, in order to show that it is possible to compute the surface energy of divided solids of high surface energy. From these results, it is also possible to estimate the contact angles on low surface energy solids. Some values of the solid/vacuum and of solid/fluids interfacial tensions are given, using experimental results available in the literature. The picture presented here is in agreement with experimental contact angle results and with the classic and modern ideas from the adsorption science.

Keywords: Adsorption; contact angles; immersion calorimetry; surface energy components.

1. INTRODUCTION

From the standpoint of physical chemistry, the topics of adhesion, spreading and contact angles belong to the general problem of the intermolecular interactions studies. Hence, it is logical that the main remaining question in interface science is: How many parameters do we need to predict and explain interactions?

For instance, one can quote recent work [1] concerning prediction of Donor-Acceptor interactions where three terms are necessary: (i) the charge donation, (ii) the back donation and (iii) the charge depletion from the overlapping area, although other schemes [2] consider only two main factors, the σ-donating interactions and the π-back donation. However, in one of the most widely applied methods [3] in theoretical chemical research, the energetic contributions to the bond energy are the charge transfer, the electrostatic component, the polarisation, the exchange repulsion, along with a coupling term. The choice between several theoretical procedures is in fact a choice between a set of parameters and a search for simplicity.

This search for influencing parameters is also visible in the interpretation of enthalpies of interaction. In this case, it is clear that more than three parameters are often necessary; a recent paper [4] concerning this question shows that at least four parameters are required to describe the variation of physicochemical properties of phosphines and their complexes. Drago *et al.* [5] have shown that the Drago model [6] of two parameters, denoted E_i and C_i, is not sufficient to explain the experimental results, and that it is necessary to introduce a parameter denoted W, a constant characteristic of the acceptor. The questions posed by Pearson [7] in his book are of the same type. As Pearson writes: "If there were a single order of strengths for acids, and for bases... the Hard-Soft effect might be added..."; which means that at least four parameters are necessary to interpret acid-base interactions between two species, according to Pearson's views.

According to the best knowledge and belief of Mittal, Bolger and Michaels [8] were the first to consider that it was necessary to take into account the acid-base interactions (in terms of Brönsted acid-base) in order to interpret interfacial phenomena. Fowkes [9] has systematised this idea in terms of Lewis acid-base, and proposed to interpret contact angle results with this view. This was further developed by van Oss, Chaudhury and Good [10], with some satisfying results, and this approach has shown that three parameters are needed in contact angle analysis. But, surprisingly, it was not applied and generalised to the other interfacial aspects, e.g., adsorption and spreading. Therefore, the "state of the art" analysis of acid-base interactions across an interface is still limited by the approximations used in contact angle analysis. These approximations, which are assumed without further justifications are the following:

(i) It is possible to decompose the surface tension into components from the classical division in components used when looking at the enthalpy, because enthalpy and free enthalpy are proportional. This assumption was made by Fowkes [11].

(ii) The film pressure of the vapour onto the solid is negligible when there is a definite contact angle.

(iii) The solid-vapour surface tension does not depend on the vapour species, and is only dependent on the solid. Therefore, the solid-vapour surface tension is a constant in the fitting of results and is a characteristic of the solid surface. Moreover, the analysis does not take into account chemisorption and electrical charges phenomena. This can be considered as a fourth assumption.

All these assumptions have been shown to be oversimplified in other branches of interfacial science, namely immersion, adsorption and catalysis. Concerning point (i): recently reported results [12] on the coefficients relating the free enthalpy and enthalpy values have been shown to be varying. Regarding points (ii) and (iii), recent books [13, 14, 15] reviewing gas and vapour adsorption have reported that there is always a film pressure value at near saturation pressure, related at least to the intermolecular forces characterising the concerned vapour. Also for point (iii), the solid surface thermodynamics has shown [16] that the only surface tension characterising clearly a solid is the value in the vacuum, i.e. the surface tension of the solid in equilibrium with its own vapour. One has to emphasise here that all these results are in agreement, and are used to characterise solids of interest in the fields of materials and catalysis. They are also in agreement with the old division of solids between "high energy" and "low energy". Furthermore, all these points are used in the general theoretical analysis of adsorption. From the theoretical viewpoint [17], adsorption is due to an interaction between an attracting field, which is the solid surface energy, and the cohesion forces of the vapour, depending on the partial pressure of the vapour. This theoretical view considers that a solid surface is always an attracting field, depending on the strength of the chemical bonds in the bulk of the solid. It considers also that the intermolecular forces between the vapour molecules alone are sufficient to promote the adsorption when the vapour pressure is near saturation pressure, (that is the case when measuring contact angle, the liquid phase being present). It is the only way to interpret in all the cases the high pressure branch of the adsorption isotherm. Furthermore, recent calculations [18] of the dispersive energies of solids have shown that theoretical values of surface tension of solids are greater than the one obtained using experimental contact angles values and classical analysis. And last, but not least, starting from another point of view, de Gennes [19] has shown that the systems with contact angle are not physically different from systems showing spreading. He has systematised the problem using the spreading coefficient, S, instead of the cosine of the contact angle, and has shown that the differences in wetting behaviours and in dynamics of spreading arise from the energetic balance between the solid and the liquid.

All these points have to be introduced in a modern analysis of the contact angles. In other words, there exists in the literature many results [20] showing that there is adsorption onto "hydrophobic solids". (Moreover, there are theoretical interpretations of this adsorption [21] and computations of the film pressure on these solids.) However, it is still considered in the contact angle analysis that the film pressure is nil, or does not have to be taken into account. This contradiction is the key to the problems encountered in the analysis of contact angles, and with some controversies existing between different approaches. For instance, Neumann and co-workers [22] still consider that the solid-liquid interfacial tension is nil when wetting is perfect. It is perhaps possible that the controversies might in fact be due to the over assumptions made.

Some other experimental approaches have been used in order to determine surface energies and the influence of acid-base interactions. The most widely used techniques are the Inverse gas Chromatography [23] which uses the properties of the gas adsorption even for systems with contact angle, and the XPS (X-ray Photoelectron Spectroscopy) [24]. Flow calorimetry [25], using successive flows of solvents of different acid-base characters is also often used. An older technique, namely the immersion calorimetry, has been used successfully [26] to analyse the dispersive character of solids. Only a few experiments of adsorption isotherms have been performed from this view [27]. This is probably due to the experimental difficulties encountered when performing isotherms with polar molecules.

Owing to the easiness of immersion calorimetry measurements (if one uses a sensitive calorimeter), we have been engaged for the last five years in a program to analyse the surface energy of high energy solids of industrial interest [28]. We have proposed a new model for interpretation of immersion and adsorption using three parameters [29].

In this paper, we will show that it is possible to use our model of the interaction enthalpy to interpret immersion phenomenon concerning high energy solids. We will show that it is also possible to calculate the solid surface tension using our results and Fowkes' assumption of a simple relation between enthalpy and free enthalpy [30]. Finally, we will extend this analysis to some immersion results for Teflon [31].

2. IMMERSION AND WETTING CALORIMETRY

2.1. Surface Thermodynamic Quantities

The interfaces at a solid phase can be treated with the same intensive and extensive parameters as other interfaces as long as the solid is not submitted to a strain [16, 32]. Therefore, at equilibrium, an interface near a solid phase can be considered as an autonomous phase, characterised by the thermodynamic parameters: T, S, p, V, n_i, μ_i, γ, and A (temperature; entropy, pressure, volume, number of molecules of different species, chemical potential, surface tension, and surface area, respectively).

As a good approximation, for condensed phases [33], the surface energy E_S and the surface enthalpy H_S are not distinguished. The surface enthalpy is given as:

$$H_s = TS_s + G_s \tag{1}$$

where G_s is the surface free enthalpy.

The terms per unit area can be defined as follows:

$$H_s^{(A)} = \left(\frac{\partial H_s}{\partial A}\right)_{T,p,n_i} \quad \text{and} \quad G_s^{(A)} = \left(\frac{\partial G_s}{\partial A}\right)_{T,p,n_i} = \gamma \tag{2}$$

2.2. Wetting Enthalpy and Surface State of the Solid

The immersion experiment seems very simple. One has to immerse a solid in a liquid, at constant temperature and pressure. Generally this work is performed at atmospheric pressure (the influence of the pressure is negligible [34]).

There is an ambiguity in the above description, which is the initial surface state of the solid. If the solid is surrounded by a pure vapour phase, then, the experiment performed is one of wetting; the definition of wetting is derived from the contact of three mutually insoluble phases.

The physics of wetting is described by the Young equation [35]:

$$\gamma_{LV} \cos\theta = \gamma_{SV} - \gamma_{SL} \tag{3}$$

Where γ_{ij} is the interfacial tension between two phases i and j. The subscripts L, V and S refer, respectively, to the liquid, vapour and solid phases. The contact angle is conventionally measured in the denser fluid phase.

This equation is simply a mechanical balance between the three interfacial tensions in contact at the triple line. The practical problem in this equation is that it is not possible to measure the two solid-fluid interfacial tensions γ_{SV} and γ_{SL} [36].

The link between the mechanical and the thermal effects is very simple if one deals with pure phases. If the experiment is performed in a calorimeter, one can use the general relationship [37] between enthalpy and free enthalpy:

$$\Delta H = \Delta G - T\left(\frac{\partial G}{\partial T}\right)_p \tag{4}$$

Therefore, one can write the following two relationships defining a free enthalpy and an enthalpy of wetting, per surface area unit:

$$\Delta_W G^{(A)} = \left(\gamma_{SL} - \gamma_{SV}\right) \tag{5}$$

and

$$\Delta_W H^{(A)} = \left(\gamma_{SL} - \gamma_{SV}\right) - T\left(\frac{\partial\left(\gamma_{SL} - \gamma_{SV}\right)}{\partial T}\right)_p \tag{6}$$

These relationships imply that the heat effects involved are generally exothermic, the surface tension γ_{SV} often being greater than γ_{SL}.

Inserting the Young equation in equation (6), one obtains:

$$-\Delta_W H^{(A)} = \gamma_{LV}\cos\theta - T\left(\frac{\partial\left(\gamma_{LV}\cos\theta\right)}{\partial T}\right)_p \tag{7}$$

Hence, in the case of perfect wetting, one has:

$$-\Delta_W H^{(A)}_{\theta=0} = \gamma_{LV} - T\left(\frac{\partial\left(\gamma_{LV}\right)}{\partial T}\right)_p = H^{(A)}_{LV} \tag{8}$$

$H^{(A)}_{LV}$ is frequently denoted as H_L and is called the "surface enthalpy of the liquid". It can be determined by surface tension measurements. Rigorously speaking, it is a derivative of the actual surface enthalpy of the liquid, which does not take into account the actual number of molecules in the studied interface.

These relations define a very simple experiment which is often called "wetting calorimetry", and which has been used by Jura and Harkins [38] to determine the specific surface area of powders. But, unfortunately, the experiment is often confused with "immersion calorimetry" [39]. This confusion is linked to the fact that the solid-vapour interface (and therefore the related interfacial tension) depends on the partial pressure of the surrounding vapour [13, 40]. For instance, in a vacuum the solid-vapour surface tension vanishes, being replaced by the tension of the solid in equilibrium with its own vapour without strain, γ_S^o, well-defined theoretically [16].

In fact, it is possible to define a certain number of surface states of the solid corresponding to the different values of the vapour partial pressure. These surface states lie between two extremes: the solid under vacuum, and the solid with a vapour adsorbed in presence of the liquid phase.

2.3. Enthalpy of Immersion

Using a properly designed calorimetric device, it is possible to immerse a solid [39], maintained under a vacuum, directly into a liquid phase. The solid is maintained under vacuum in a closed glass bulb, immersed in the liquid and then the bulb is broken.

The different thermal effects have been well described in some papers [39]. The main effect is due to the disappearance of the solid-vacuum interface and to the creation of a solid-liquid interface [41]. If the liquid is in a sufficient quantity, electrical charge effects are negligible, provided that the liquid phase is pure [42].

The equations describing the exact immersion process are very close to the above ones. The solid in equilibrium with its own vapour without strain is characterised by the surface tension denoted γ_S^o. Therefore, the free enthalpy variation per surface area unit is:

$$\Delta_{imm}G^{(A)} = \gamma_{SL} - \gamma_S^o = \gamma_{SL} - \gamma_S^o + \gamma_{SV} - \gamma_{SV} = -\pi_e - \gamma_{LV} \cos\theta \quad (9)$$

$\Delta_{imm}G^{(A)}$ is the immersion free enthalpy per unit surface area. π_e is the maximum pressure of the film due to the vapour of the liquid involved (i.e.: $\pi_e = \gamma_S^o - \gamma_{SV}^{lim}$; the film pressure which varies with the surrounding pressure is denoted π_S; π_S tends to π_e [43] when p tends to p°). It is possible to compute the film pressure using a vapour adsorption experiment, with:

$$\pi_S = RT \int_{p=0}^{p} \Gamma \, d\ln p \quad (10)$$

where Γ is the surface excess.

From the above equations, the immersion enthalpy per unit area is:

$$\Delta_{imm}H^{(A)} = -\pi_e - \gamma_{LV}(\cos\theta) - T\left(\frac{\partial\left[-\pi_e - \gamma_{LV}(\cos\theta)\right]}{\partial T}\right)_P \quad (11)$$

Or, in an equivalent form:

$$-\Delta_{imm}H^{(A)} = \gamma_S{}^\circ - \gamma_{SL} - T\left(\frac{\partial\left[\gamma_S{}^\circ - \gamma_{SL}\right]}{\partial T}\right)_p \quad (12)$$

2.4. Experimental

In many experimental cases reported in the literature, a differential conduction microcalorimeter is used [44], due to the high precision and sensitivity obtained with this type of device. In our case, we used a Tian-Calvet microcalorimeter of this type [39]. A portion of the calorimeter is sketched in Figure 1 with a typical immersion setup. The powdered solid sample is located in a glass ampoule. The glass ampoule is brittle and is sealed under a controlled pressure before fixing to the end of a glass rod. The glass ampoule is placed inside the calorimetric cell. It is easy to break it by pushing

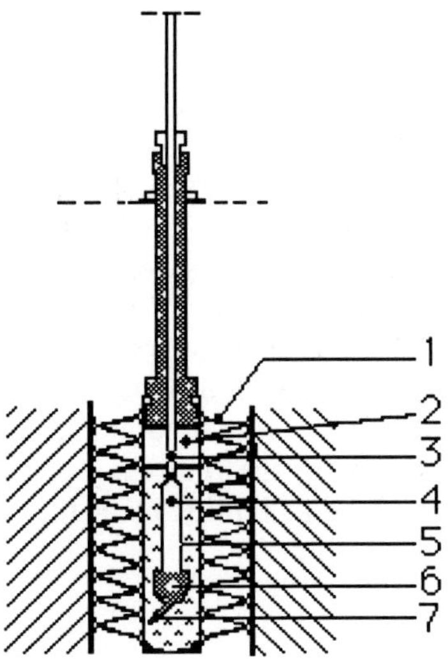

1 : thermopiles ; 2 : dead volume of the cell
3 : glass rod ; 4 : dead volume of the bulb
5 : bulb ; 6 : powder ; 7 : liquid

Figure 1. Calorimetric cell designed for immersion studies. The length of the glass cell is about 10 cm. The total length of the calorimeter is about 1.2 m.

down the glass rod. During this operation, heat due to the friction of the rod on the ring placed at the upper part of the cell is probably produced (the ring is used to maintain a seal). This heat is not recorded due to the location of the O-ring about 8 cm above the thermopile. Before the ampoule breaks, a volume v of the glass rod enters the liquid. Due to the rise of the liquid level, heat is produced which is equal to the compression of the gas phase. Another source of heat is due to the breakage of the ampoule. These two last terms are small, with an order of 1 and 5 mJ, respectively.

The main contributions which can be recorded are: the enthalpy of liquefaction of the vapour previously filling the bulb; and the enthalpy of vaporisation of the vapour which fills the volume created by the lowering of the liquid level. These effects vary with the quantities of solid and of liquid in the ampoule and the calorimetric cell. The best way to determine and remove these terms from the calculation is to use a set of blank experiments. In some cases, the experimental heat can contain some artefactual heats (due to swelling, dissolution, charge effects, ...). It is not the case in the results reported here.

The reproducibility is about 30 mJ. The values given here are the average values of more than three experiments. The calorimetric cell is kept inside the calorimeter for at least 2 hours to attain thermal equilibrium, before the actual experiment is executed. Before the experiment, the solids are kept overnight under vacuum with pressure lower than 10^{-3} torr at 150 °C. (In case of organic solids, we used lower temperatures.)

In order to obtain the enthalpy per unit area, it is useful to know the effective surface area to the specific surface area (in m^2/g). The specific surface area is measured by the classical methods, such as nitrogen adsorption [13, 15, 45] and the Harkins-Jura method [38, 39]. We use the average value [28a]. In a typical experiment, the baseline is reached after 60 minutes. The weight of solid in the ampoule is determined with a balance.

3. ENTHALPY OF IMMERSION: AN OVERVIEW OF EXPERIMENTAL RESULTS

The simple immersion experiment described in Section 2.4 gives a very complex value involving parameters which can be difficult to interpret. As a result, this type of experiment has been carried out in only a few laboratories [23, 38, 39, 46–50] with the main focus on immersion in water. In the following part, we will present the results obtained with water. The influence of liquids other than water will be examined in the second and third parts of this section.

3.1. Case of Water

Due to the presence of covalent and ionic bonds, the solids considered as "high energy" solids are always inorganic. They are composed mainly of silicates and alumino-silicates, such as: silicas, quartz, clays, zeolites, and micas. Concerning immersion, the most studied solid is quartz [38, 39, 48]. The results depend on the grinding and the washing of the sample. In Table 1 we list typical values from the literature, and in Table 2, we report some results obtained in our laboratory. All the solids reported in Table 2 are described in details elsewhere [28, 51, 52]. We have focused our present list on silicates and clays. A carbon, some sulphurs and a silicon carbide are given for reason of comparison. All these samples are of industrial interest. Our results show the great influence of preparation on the surface properties of samples which are otherwise identical. (Generally, industrial compounds are more sensitive to this type of problem than solids obtained with a rigourously defined laboratory procedure). This can be clearly seen in Table 1: for a solid of industrial interest, such as quartz, the variation is large.

The case of clays is different. Clays are compounds of the same mineralogical family, but with very different chemical composition, depending on the type of clay (illite, kaolinite, etc.) and the origin of the sample. We will see later that the quantity of aluminium present in the clay sample is a parameter of most importance to surface energy.

Table 1.
Typical enthalpy of immersion values from the literature for some solids immersed in water at ambient temperature

Solid	Enthalpy of immersion in water at 25°C (mJ/m^2)
Quartz	-450 / -850
Calcite	-480
Zeolite NaX	-500
Copper	-725
Alumina	-900

Table 2.
Specific surface area and immersion enthalpies in water at 25°C for different industrial powders

Solid	Origin	Specific Surface Area (m^2/g)	Enthalpy of immersion in water (mJ/m^2)
Dolomite $CaMg(CO_3)_2$	IFP	0.43	-1253
Chlorite	Talc de Luzenac	2.5	-868
$MgCO_3$	Talc de Luzenac	11.2	-754
α–Alumina	Union Carbide	9.2	-704
Calcite ($CaCO_3$)	Gregory, Boottley and Lloyd	2.45	-535
SiC	Microlap	1.3	-438
Quartz C800	Sifraco	5.2	-383
Illite of Hungary	IFP	127	-364
Talc from Spain	Talc de Luzenac	6.5	-356
Talc from France	Talc de Luzenac	1.9	-321
Talc from Italy	Talc de Luzenac	3.0	-311
Talc 1	Talc de Luzenac	3.0	-358
Talc 2 1 + grinding	Talc de Luzenac	6.0	-202
Talc 3 2 + grinding	Talc de Luzenac	9.2	-183
Kaolinite of Charentes	IFP	20	-308
Silica XOB 015	Rhône-Poulenc	29	-237
Silica Cabosil	Cabot	207	-170
Graphon	Cabot	82	-29
Sulphur 1	Soufres Réunis	30.4	-6
Sulphur 2	Soufres Réunis	19.7	-20
Sulphur 3	Soufres Réunis	12.4	-21

We also report the results obtained with a talc sample, that was ground twice to obtain different characteristics, and with three very pure talcs extracted from three different locations, with very different sizes of lamellae. One also has to note here that the quartz reported in Table 2 is ground and probably poorly crystalline on the surface.

In the immersion experiments, the defects present at the surface of the sample have a very strong influence on measured enthalpies. Impurities adsorbed such as surfactants and greases tend to decrease the solid surface tension and consequently the immersion enthalpy. Different methods of preparation for the same material have a pronounced effect on the immersion enthalpy.

Substitutions of surface atoms and groups by atoms or groups with a greater reactivity tend to increase the immersion enthalpy. Alumino-silicates can be viewed as intermediate products between pure silica [47] and pure alumina [53], since the obtained enthalpy of immersion lies between the average values for silica and alumina. This indicates that the experimental enthalpy can be interpreted as the sum of contributions of individual enthalpies of interaction between the different types of groups on the solid and the fluid molecules [54]. Thermal treatment also modifies the surface chemical bonds, and can also modify the surface network and therefore the surface energy. Silica has been studied extensively [23, 38, 39] and the classical interpretation of the variation of immersion enthalpy versus the treatment temperature is a result of the modification of the types of silanol (individual or geminal) and of their surface densities.

Grinding can modify the ratio between different crystallographic planes at the surface, can introduce some impurities (the metal of the grinder, mainly) and can amorphize the surface by mechanical and thermal energy transfers.

It is important to emphasize that talc is very often considered as a "hydrophobic" solid. For instance, it has been shown from wetting measurements that water gives large contact angle values on talc [52, 28d]. However, the immersion enthalpy of talc in water is of the same order of magnitude as the hydrophilic solids: quartz, hydrophilic silica, and clay minerals. This indicates great differences in behaviour between the solid-vacuum interface and the solid-air interface.

3.2. The Case of Alkanes

The results of the immersion studies of different solids in water can be considered as an estimation of the hydrophilicity of these solids. For the sake of comparison, it is interesting to report the results obtained for the immersion in n-alkanes [46d, 48]. Considerable care has to be taken to control the presence of trace water in the alkane, as this presence has an influence on the calorimetric results. Generally, samples have to be dried with molecular sieves and analysed for water content.

For comparison, when reporting our results in Table 3, we have kept the same order as in Table 2. Surprisingly, the alumino-silicates appear quite in the same order, meaning that at first sight the more hydrophilic solids are also the more hydrophobic, if one considers the affinity for apolar compounds as a scale of hydrophobicity. The trend observed shows that minerals generally give enthalpies higher than 100 mJ/m^2, and organic solids give enthalpies lower than 100 mJ/m^2. But focusing on the SiC sample, our results show that it is not possible to order the solids on such a unique parameter as an immersion enthalpy in only one liquid. As shown in the thermodynamic definition of this parameter, the solid surface enthalpy has paramount importance in the immersion enthalpy, which cannot simply be interpreted as a scale depending on a single parameter.

The results presented here are in good agreement with the literature results. Some values have been given [55] for carbons in cyclohexane (103 mJ/m^2) or in n-dodecane (100 mJ/m^2), heptane or hexadecane (120 mJ/m^2) [56]. Silica has been immersed in isooctane [46c] (55 mJ/m^2) heptane or hexadecane (50 mJ/m^2). A special case is the immersion of metals in alkanes. In this case strong interactions take place. For instance, the immersion enthalpy of copper in cyclohexane [57] is 740 mJ/m^2, the same as in water. This point is confirmed by immersion of copper in benzene or in methanol (respectively, 880 and 950 mJ/m^2).

Table 3.
Immersion enthalpies of the solid samples in decane or in n-heptane at ambient conditions

Solid	Immersion in decane (mJ/m^2)	Immersion in n-heptane (mJ/m^2)
Dolomite		-316
Chlorite		-202
$MgCO_3$		-126
α-Alumina		-162
Calcite	-206	-236
SiC		-389
Quartz C800		-104
Illite of Hungary		-103
Talc from Spain		-84
Talc from France		-103
Talc from Italy		-111
Talc 1		-119
Talc 2		-139
Talc 3		-152
Kaolinite from Charentes		-106
Silica XOB 015		-140
Silica Cabosil		-85
Graphon		-120
Sulphur 1	-73	
Sulphur 2	-106	
Sulphur 3	-44	

3.3. The Case of Polar Liquids: Acid-base Interactions

In this group, we consider the case of polar liquids, such as benzene. We observe in Table 4 the same general order as for alkanes and for water. In fact, this general order is the order of the surface energy, which decreases from "high energy" solids to "low energy" solids. This links contact angle phenomena and immersion phenomena. But as in the case of contact angles, the fact that immersion enthalpy is the result of a combination of two parameters: the pure solid surface enthalpy and the enthalpy of interaction between pure solid and fluid, explains the complexity of experimental trends. First, the immersion enthalpy increases with the solid surface enthalpy (at first approximation, the enthalpy of fusion of the solid). Thus, the seen order between the solids is explained. Secondly, the immersion enthalpy increases with the capacity of interaction between the solid and the liquid. Hence, the order between the liquids is also explained, i.e. formamide being more reactive than pyridine and benzene. However, exceptions are observed. Some of them are probably due to epitaxy phenomenon between the solid and the liquid, as for instance, in the couple talc/benzene where the bond distances between atoms of the sheets of talcs are the same as in the benzene molecule.

Table 4.
Immersion enthalpies of the solid samples in some polar liquids at ambient conditions

Solid	Immersion in benzene	Immersion in pyridine	Immersion in formamide
		(mJ/m^2)	
Dolomite	-452		
Chlorite			-1006
$MgCO_3$			-674
α–Alumina			-585
Calcite	-289	-464	
SiC			-532
Quartz C800	-157	-317	
Illite of Hungary	-151	-449	
Talc from Spain			-531
Talc from France	-281		-416
Talc from Italy			-345
Talc 1			-429
Talc 2			-284
Talc 3			-275
Kaolinite from Charentes	-171	-410	
Silica XOB 015	-183		
Silica Cabosil			-289

4. ANALYSIS OF RESULTS USING A THREE PARAMETER MODEL

4.1. Introduction

As seen above in equation (12), the enthalpy of immersion can be viewed as a difference between a value characteristic of the solid and an intermolecular enthalpy. The first term depends only on the solid, and the second term depends both on the liquid and on the solid:

$$-\Delta_{imm}H = \left[\gamma_S^0 - T\left(\frac{\partial \gamma_S^0}{\partial T}\right)_P\right] - \left[\gamma_{SL} - T\left(\frac{\partial \gamma_{SL}}{\partial T}\right)_P\right] \quad (13)$$

The left term of this difference has been precisely defined and discussed [16a]. It is clear that it is strongly correlated to the average bond energy of the solid, with some modification being due to rearrangements of the electronic density at the surface and to the differences in the average number of neighbouring atoms. Moreover, solid surfaces are not homogeneous planes, but present steps, kinks and all kinds of physical defects. A very simple way to estimate this surface tension and the corresponding enthalpy is by considering this enthalpy as a fraction (say 80%) of the enthalpy of vaporisation of the solid in question [58]. The interesting point here is that this value does not change with different immersing liquids.

However, on changing the liquid, the solid-liquid enthalpy is expected to change. Intermolecular forces theory [59] allows the interpretation of these observed variations.

It is possible to consider the solid as a heterogeneous phase composed of individual atoms, and to obtain the total interaction energy by summing the different interactions between each atom and the liquid. This summation has to be averaged by a term reflecting the different orientations of the atoms in different locations, such as on the faces and edges of the solid grain. The total immersion enthalpy can also be divided into energetic terms taking into account the actual differences between adsorption sites, i.e. due to the location of crystallographic planes, compensation cations in case of clays, chemical groups and so on. But application of this approach is very difficult, because the different locations and orientations of actual solids are generally unknown. Moreover, the precise number of surface atoms and groups, e.g. the real thermodynamics of the surface compared to the bulk, is unknown. Therefore, a more phenomenological approach to calculate intermolecular energies is preferred. In this sense, we propose a model in order to interpret immersion, adsorption and wetting.

4.2. The Model

4.2.1. General Case. For a long time, it appeared that the intermolecular energy between two species could be attributed to different additive terms, each characteristic of one specific type of interaction. Unfortunately, as each type of interaction acts over different distance, the different types of interactions do not act in the same way, it follows that the final equations describing an intermolecular energy can be complicated.

Following Kitaura and Morokuma [60], the intermolecular energy between species 1 and 2 can be written as:

$$\Delta E_{12} = \Delta E^{ES} + \Delta E^{PL} + \Delta E^{EX} + \Delta E^{CT} + \Delta E^{MIX} \tag{14}$$

where ES refers to the electrostatic component of the energy, PL to polarisation, EX to exchange repulsion, CT to charge transfer and MIX to coupling.

More generally, one can write:

$$\Delta E_{12} = \sum \Delta E^{i} \tag{15}$$

where:

$$\Delta E^{i} = E^{i}_{12} - E^{i}_{1} - E^{i}_{2}$$

and

$$E^{i}_{12} = f_i(E^{i}_{1} + E^{i}_{2})$$

f_i being a characteristic function of each interaction i.

In our particular case, where we study interactions between condensed phases, it is possible to assume the following equality between internal energy and enthalpy [33, 37]:

$$\Delta E \approx \Delta H$$

and consequently to write the intermolecular interaction enthalpy in the following form:

$$\Delta H_{12} = \Delta H^{ES} + \Delta H^{PL} + \Delta H^{EX} + \Delta H^{CT} + \Delta H^{MIX} \tag{16}$$

However, it is difficult to link the theoretical interaction terms to the actual experiments. A molecule gives rise to more than one interaction. This point led a large number of researchers to propose semiempirical equations linking the theoretical interactions to parameters which can be estimated in an experimental way. The most useful experimental term is the interaction enthalpy. Many semiempirical models have been proposed, for instance the one due to Drago and co-workers [5, 6]. Such equations are numerical derivations of exact theories. The parameters, or the "components" used do not directly correspond to the theoretical components; they are functions of them.

4.2.2. General Application to a Surface. In actual cases, numerical values in the equations given above depend on the phase density. A surface can be considered as an autonomous phase [37] if at chemical equilibrium, and we have to note that its density can be different from the density of the bulk phases [61]. However, it is possible to use the above equations to describe the interfacial interactions, and to predict the behaviour of the system.

In a liquid-solid or liquid-liquid interface (denoted σ), between two pure species, denoted 1 and 2, we can write:

$$\Delta H_{12}^{\sigma} = (\Delta H^{ES})^{\sigma} + (\Delta H^{PL})^{\sigma} + (\Delta H^{EX})^{\sigma} + (\Delta H^{CT})^{\sigma} + (\Delta H^{MIX})^{\sigma} \tag{17}$$

Consequently, the interaction enthalpy between a species A and a species B in the interface phase can be deduced using an equation of the Drago form:

$$\Delta H_{AB}^{\sigma} = k^{\sigma}(E_A E_B + C_A C_B) \tag{18}$$

k^{σ} being a factor taking into account the interface density, written here to keep in mind that it is an equation used in the surface phase. In this form it is not necessary to define the terms E_I^{σ} and C_I^{σ}, characteristic of the interface, which could be confusing.

Surprisingly, such an equation can also be used to calculate the energy of a liquid-liquid interface, assuming that the solute-solvent complex formation enthalpy ΔH_{AB}^{σ} is similar to the interfacial enthalpy ΔH_{12}^{σ}. Indeed, Fowkes [11] used it to calculate an interaction term between water and benzene, using Drago's values, and a k^{σ} factor of 1. He obtained $\Delta H_{12}^{\sigma} = 5$ kJ/mol. This choice of k^{σ} implies that the interaction at the interface is equal to the interaction in the bulk, but we will admit heuristically that this point is not so important. Starting from a molar energy [58], it is quite easy to compute the energy related to a surface area. Using the simplest calculation, established from the density and mass of the molecules, and considering that the interface is an equimolar mixture of water and benzene, we obtain:

$$k^{\sigma}N_A\,(m/d)^{2/3} = 1.16\cdot10^5\ m^2/mol \tag{19}$$

with: N_A, Avogadro's number, $k^{\sigma} = 1$, and m the average molecular mass: $(78+18)/2N_A$.

We then obtain an interfacial enthalpy of 43.1 mJ/m^2 which compares very satisfactorily to the interfacial tension (35 mJ/m^2), and justifies the heuristic approximation used above.

These terms can also be obtained by microcalorimetry as follows. As we have seen above, the interaction between a solid and a vapor is related to an adsorption enthalpy, and the interaction between a solid and a liquid is related to an immersion enthalpy or a wetting enthalpy, depending on the experimental pressure at the beginning of the experiment.

Following the recent evolutions of the type of parametrisation used by Drago and co-workers [5], and also some developements from the recent literature on colloid science [62], we will consider here that it is necessary and sufficient to characterise a molecular species by a set of three terms. Starting from calorimetric experiments, this allows intermolecular interactions to be computed. Therefore, we propose a general analysis of immersion, adsorption and wettability in terms of 3 surface components, namely dispersive, acidic and basic, generalising the Good-van Oss-Chaudhury (GvOC) method [63]. But, one has to realize that, generally speaking, the point concerning the real number of interactions taking place at the surface phase is not so important, mainly because experiments are performed with real molecules, and real molecules are representative of more than one type of interaction. Therefore, the number of interactions deduced from trends of experimental results is not the actual number of interactions defined in a theoretical approach.

One of the main points of the model we propose is the possibility to interpret experimental enthalpies related to the surface in terms of "surface enthalpy components". This approach is valid for high energy solids, if there is only physical adsorption, in the studied system; the components taking into account chemisorption are not used in the following relationships.

4.2.3. Basic Equations. The basic ideas of the general model we propose are the same as in the GvOC model. Repulsive interactions are fully additive, although attractive interactions add following some exclusion rules. In this view, polar interactions cannot interact with dispersive forces. Some points can be accepted from the point of view of theoretical chemistry. However, some points are controversial. For instance, from the viewpoint of chemical reactivity, alkanes can be considered as bases.

The fundamental equation is the definition of the surface enthalpy of interaction between a molecule 1 and a molecule 2:

$$H_{12}^{\sigma} = H_1^{\sigma} + H_2^{\sigma} - 2[(H_1^{\sigma^{LW}} \cdot H_2^{\sigma^{LW}})^{1/2} + (H_1^{\sigma^+} \cdot H_2^{\sigma^-})^{1/2} + (H_1^{\sigma^-} \cdot H_2^{\sigma^+})^{1/2}] \quad (20)$$

with

$$H_i^{\sigma} = H_i^{\sigma^{LW}} + 2(H_i^{\sigma^+} \cdot H_i^{\sigma^-})^{1/2} \quad (21)$$

The term denoted LW represents the apolar contribution (the letters L and W correspond to the initials of Lifshitz and van der Waals). The acidic and basic contributions are, respectively, indicated by plus and minus. The modern view of acidity [59] implies that these terms are not physical terms but only parameters related to the theoretical acid strengh S_A [7], without knowing in details the relationship.

The varying term in the measurement of the enthalpy of immersion is the solid-liquid enthalpy per unit surface area, which is an intermolecular term. It can be written as follow:

$$H_{SL}^{\sigma} = H_S^{\sigma} + H_L^{\sigma} - 2[(H_S^{\sigma^{LW}} \cdot H_L^{\sigma^{LW}})^{1/2} + (H_S^{\sigma^+} \cdot H_L^{\sigma^-})^{1/2} + (H_S^{\sigma^-} \cdot H_L^{\sigma^+})^{1/2}] \quad (22)$$

where we can substitute, in agreement with the idea of two pure species interacting, and with surface thermodynamics [16]:

$$H_S^{\sigma} = H_S^{o(A)} \text{ and } H_L^{\sigma} = H_{LV}^{(A)}$$

From this, we can deduce the following equation, expressed per unit area, without using the notation (A) for the sake of simplicity:

$$-\Delta_{imm}H = -H_{LV} + 2[(H_S^{o^{LW}} \cdot H_{LV}^{LW})^{1/2} + (H_S^{o^+} \cdot H_{LV}^{-})^{1/2} + (H_S^{o^-} \cdot H_{LV}^{+})^{1/2}] \quad (23)$$

If one determines the enthalpy of immersion in a liquid without polar interactions, the equation is greatly simplified, and the mesurement allows the determination of one unknown term, the surface apolar enthalpy as:

$$-\Delta_{imm}H = -H_{LV} + 2[(H_S^{o^{LW}} \cdot H_{LV}^{LW})^{1/2}] \quad (24)$$

Proceeding in this way with different probe liquids, carrying the three different types of interactions with different percentages (their reduced surface enthalpy can be determined by tensiometry measurements at different temperatures), it is possible to determine the unknown parameters.

This 3 parameter model has been chosen after a phenomenological analysis of our results. We will also see below that immersion results are in agreement with the trends of vapour adsorption isotherms. This indicates a general behavior of surface intermolecular forces, more or less in agreement with a classical analysis of solvation phenomenon [7, 59, 62, 64].

4.2.4. Numerical Application. The point now is the choice of probe liquid molecules, and the determination of values of corresponding reduced surface enthalpies, for the probe liquid molecules. There is an abundant literature on the choice of surface tension components [65, 66], even though there is no general agreement on the number of components. Many scientists divide the surface tension in polar and apolar terms, the word apolar referring in fact to the interactions "like an alkane". The users of the GvOC model confuse the value of the apolar component with the "Lifshitz-van der Waals" term, meaning that both terms describe the same interactions. Their acid-base scale assumes that water is amphoteric, i.e. the acidic and basic components for water are equal. This idea is controversial [67]; in other fields of chemistry water is considered a hard base [7]. But this choice is not so important, because such models do not describe the theoretical interactions, as studied by theoretical chemistry and physics. They are only scales, and each scale needs a starting point. It is easy to go from a well-defined scale to another [59]. On this idea, we have followed the GvOC model, and have used this scale to establish the reduced surface enthalpy of the probe liquids such as heptane, decane, benzene, water, formamide, chloroform, etc.... Pyridine is a special case, discussed elsewhere [68, 28b]. The basic and acidic components we obtain in the calculation is only a relative trend, the significance of which depends on the GvOC assumptions. Therefore, we do not focus on this point, but only on the apolar/polar ratio, and on the total surface enthalpy we obtain for each solid studied, which is a very rare and useful information in view of wettability understanding. The values used are reported in Table 5.

Table 5.
Values of the enthalpy components (in mJ/m^2) for the different liquid probes

Liquid	H_{LV}	H^{LW}	H^+	H^-
Heptane	48.9	48.9	-	-
Decane	50.8	50.8	-	-
Water	118	36	41	41
Benzene	66.7	66.7	-	4.56
Pyridine	75.5	20.5	4.1	5.3
Formamide	82.1	55.2	3.23	56

4.3. Determination of Solid Surface Tension

It is possible to deduce from surface enthalpies the solid surface tension. Indeed, it is possible to write the following relation between enthalpy and the free enthalpy at constant temperature:

$$\Delta G^\sigma = f\,\Delta H^\sigma \tag{25}$$

with the following definition of the term f:

$$f = 1 - T\,(\Delta S^\sigma/\Delta H^\sigma) \tag{26}$$

In our case, we can write:

$$f = 1 - T\,(\Delta_{imm}S/\Delta_{imm}H) \tag{27}$$

f cannot be independent of the interfacial system studied, and obviously f depends on the temperature. However, using immersion calorimetry and adsorption isotherms, f can be calculated. We have carried out some experiments on vapour adsorption to determine the adsorption isotherms. An example is given in Figure 2. This shows an interesting trend, e.g. when comparing isotherms for the same solid, an increase in the quantity adsorbed due to the polarity of the adsorbing vapour is observed. This trend is obvious in the literature results [40, 69, 70] even though it has never been highlighted. It is in clear agreement with the view of a hierarchy of interaction types, going from simple van der Waals interactions to polar effects. Starting from our experimental results, it can be remarked that f is quite constant, with a value near 0.45 (the experiments were performed with silicates and usual solvents: some alkanes, aromatics, water). The assumption that f is constant is often made, and more frequently in theory [71]. To the best of our knowledge, in surface science, Fowkes was the first to make this assumption, even if Fowkes in fact wrote this parameter in a different manner [11]. He introduced another parameter called "interfacial concentration of acid-base interactions", allowing him to obtain an f parameter, according our definition, of 0.38.

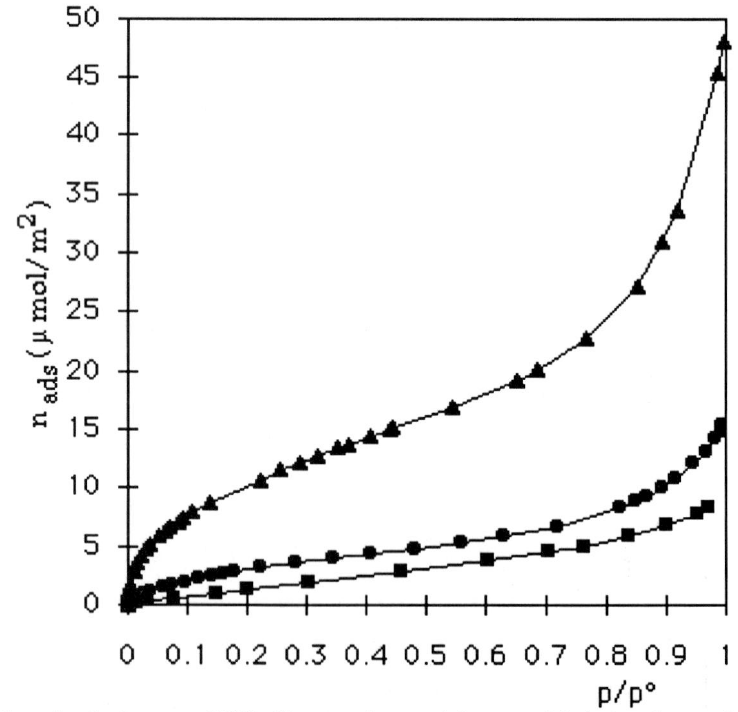

Figure 2. Adsorption isotherms at 35°C of heptane (squares), benzene (circles) and water (triangles) onto the C800 quartz sample. Experimental conditions and procedures are described in reference [29b].

Experimentally, f can also be computed starting from results obtained by Mortier and co-workers [72]. f varies between 0.3 and 0.4 for some systems, hexane/zeolites and benzene/zeolites, at ambient temperature. This small difference from our results suggests a similarity of structure between the different mineral solids studied (silicates and alumino-silicates), for the low range of temperature concerned (near ambient). On the basis of experimental trends, we have fixed a value of f=0.41 for the Lifshitz-van der Waals component, and of 0.5 for the acidic component obtained using results with benzene. Finally, the values obtained using results with polar molecules are analysed with f= 0.5.

Assuming f to be constant allows the following equation:

$$fH_{12}^{\sigma} = f(H_1^{\sigma} + H_2^{\sigma}) - 2[(H_1^{\sigma^{LW}} \cdot H_2^{\sigma^{LW}})^{1/2} + (H_1^{\sigma^+} \cdot H_2^{\sigma^-})^{1/2} + (H_1^{\sigma^-} \cdot H_2^{\sigma^+})^{1/2}] \tag{28}$$

which is equivalent to:

$$G_{12}^{\sigma} = G_1^{\sigma} + G_2^{\sigma} - 2[(G_1^{\sigma^{LW}} \cdot G_2^{\sigma^{LW}})^{1/2} + (G_1^{\sigma^+} \cdot G_2^{\sigma^-})^{1/2} + (G_1^{\sigma^-} \cdot G_2^{\sigma^+})^{1/2}] \tag{29}$$

where the G_{ij}^{σ}'s are free enthalpies per unit area, and to:

$$\gamma_{12} = \gamma_1 + \gamma_2 - 2((\gamma_1^{LW} \cdot \gamma_2^{LW})^{1/2} + (\gamma_1^{+} \cdot \gamma_2^{-})^{1/2} + (\gamma_1^{-} \cdot \gamma_2^{+})^{1/2}) \tag{30}$$

which is the basic equation of the Good-van Oss-Chaudhury model.

This indicates clearly that the Good-van Oss-Chaudhury model is simply a first-approximation of the more general model developed above. It assumes that the f coefficient, i.e. the variation of entropy during the contact of the three phases at the triple line is the same for all the molecules in contact. The small variation of f experimentally observed is in quite good agreement with this idea, even if one expects that the small variation we observe can have great consequence on the results.

4.4. Concerning the Physical Background of the Chosen Scale

Recent developments in research concerning acid-base interactions, as reviewed by Huheey *et al.* [59], have shown the difficulty to build a simple scale of acidity, in one dimension. First, even if the definition of the "perfect acid" is simple and operative (the proton H^+), the definition of the perfect base remains a difficulty: the electron is not a chemical entity easy to study in the chemical laboratory. Secondly, a one-dimensional acidity scale is not sufficient to explain the chemical reactions. Thus, it is necessary to take into account the hard-soft character of the concerned molecules for a total description of the chemical processes. This was the purpose of the scale proposed by Drago and Wayland [6], which links the ratio E_i/C_i with the hardness of the chosen species. But currently, the total thermodynamic description of the influence of hardness is not achieved. The equation proposed by Pearson [7]: $\Delta_{AB}H = S_A S_B + \sigma_A \sigma_B$, is not completely satisfactory, because it combines a first-order parameter (the strength S_i) with a second-order parameter (the hardness σ_i). Indeed, the physical description of the mechanisms of the acid-base interaction has been given, in terms of frontier orbitals description [7] and has shown that the hardness is a derivative of the chemical potential of the electrons of the species. Therefore, the physical meaning of these two parameters (strength of reactivity and hardness) is very clear, but their respective influence on the enthalpy of interaction remains to be sorted out from the thermodynamic standpoint. The main points appearing from these discussions are: (i) it is necessary to take into account the differences between dispersive and polar and electrostatic interactions; this has been emphasized recently by Ninham and Yaminsky [62]. (ii): at least two parameters are necessary to interpret acid-base interactions. These two points are satisfied in the GvOC scheme. This is the reason why we derived our model from the GvOC one. The choice of the values given to the probe liquids used is purely arbitrary, and has been shown not to be of much importance [65]. Therefore, our model provides three thermodynamic parameters that can be considered as correct: the surface enthalpy (without taking into account the part due to the chemical potential) of the solid in equilibrium with its own vapor related to the surface tension $H_S{}^{\circ}$; the dispersive part $H_S{}^{\circ LW}$ and the polar part of this energy $H_S{}^{\circ AB}$. The exact influence of the electronegativity and the hardness of the studied species in the surface phase remains unknown. The next development of the model should be the assesment of relations between the *ad hoc* acid-base parameters we use, following the GvOC model, and the physical parameters which are electronegativity and hardness.

5. SURFACE ENERGY OF HIGH ENERGY SOLIDS

5.1. Surface Enthalpy

5.1.1. Results. Using the model described above, it is easy to compute the components of the surface enthalpy per unit area for each solid sample. The values are listed in Table 6. The results for $H_S{}^{\circ}$ (per m^2) show clearly an order, mainly the same as the order of the enthalpies of fusion of the selected solids. For instance, it can be recalled that the enthalpies of fusion of quartz and alumina are about 14 kJ/mol (1470 °C) and 109 kJ/mol (2045 °C) respectively, to be compared, for instance, with an enthalpy of fusion of 146 J/g for paraffin. More precisely, this order of surface bond energy also reflects an order of surface states, influenced on the one hand by the details of the bulk solid network (bond energy, crystallographic structure), and on the other by the details of the surface bonds. There is also some influence of the repartition of chemical groups present at the surface, and of some aspects of its history (formation, mechanical strain).

Moreover, the balance between the acidic and basic enthalpies is totally dependent on the choice of assignment we have made for water, and consequently reflects a ratio due to a certain reference more than a reality. This point suggests to consider the ratio between polar and apolar components as the more representative value. Even in this type of experimental trend, the differences between the same family of solids, such as the group of illites or the group of kaolinites are clearly marked (see reference [28b]).

Table 6.
Surface enthalpy and its components (in mJ/m^2) at ambient temperature using our model and immersion results. The values of the Solid Surface Enthalpies for CaO and Sodium Chloride are taken from reference 40 (p. 315)

Solids	H_S^{LW}	H_S^+	H_S^-	$H_S°$
Dolomite	681	469	3705	3318
Chlorite	322	2832	47	1052
$MgCO_3$	156	1174	251	1241
α–Alumina	227	531	728	1470
CaO				1310
Calcite	415	29	701	700
SiC	491	62	62	615
Quartz C800	119	199	218	535
Illite of Hungary	118	89	324	458
Talc from Spain	90	553	21	305
Talc from France	118	484	4	206
Talc from Italy	131	237	54	357
Talc 1	144	444	23	346
Talc 2	180	115	2.7	215
Talc 3	206	90	0.3	216
Kaolinite from Charentes	122	180	89	376
Silica XOB 015	182	47	67	294
Sodium Chloride				276
Silica Cabosil	91	124	5.7	144

The difference between "high energy solids" and "low energy solids" is also very apparent. Silicas, talcs, kaolinites and illites are "middle energy solids" [73].

The values of the apolar enthalpy do not vary in the same proportion as the polar enthalpy. They depend more on the nature of the phase (solid or liquid, and within the solid group, lamellar, crystalline or amorphous) than on the chemical nature of the species. It is easy to see from liquids data: the dispersive surface tension approximately varies between 20 and 30 mN/m for the majority of liquids [40] at ambient temperature (except metal liquids).

5.2. Discussion

5.2.1. Influence of the Atom Type and of the Structure. As seen before, the solid surface energy is only one of two parameters governing the enthalpies of immersion. In this sense the energetic order of the immersion enthalpies is not strictly obeyed by the solid surface enthalpies. The solid surface enthalpies are, strictly speaking, characteristic of the solid as a pure phase. Therefore, it is logical that the solid surface enthalpies follow the order of the bonding energies of the different solids, a term which can be related to the electronegativity of the atoms constituting the solid [7, 59]. The difference observed in our results between carbon and carbide, silicates and aluminosilicates is very apparent. Moreover, the influence of the quantity of aluminium is of paramount importance (see reference 28b)). These results are in perfect agreement with the orders of reactivity as determined in the field of catalysis science [74].

We also observe that for the same type of solid, specifically talc and chlorite, the nature of the bond involved in the surface phase by a certain atom is of great importance. Chlorites are characterised by the presence of hydroxyls (i.e. silanol groups) on one of the external sheets of their structure. Talcs present only dehydroxylated oxygens on their external faces [75]. The result of this is very clear in the respective enthalpy values. This is due to the fact that silanol groups are reactive. However, these results show that talc is highly polar, essentially because of its strong surface acidity. This is not surprising when compared with the strong surface acidity of the clay minerals like kaolinite [76], attapulgite [76, 77] and montmorillonite [76]. It appears from other results of the adsorption of strong bases [78–80], that alumino-silicates are more acidic than magnesia-silicates. Kaolinite and illite, which are alumino-silicate clays, have a higher Lewis acid component, when determined by our method, than talc, which is a magnesium-silicate clay. Generally, the surface acidity is attributed to the proton donating ability of the surface hydroxyl groups [81, 82] of AlOH, MgOH and SiOH. Obviously, the surface acidity of these talcs increases with their surface hydroxyl group content, i.e., when the lateral surface area percentage increases.

The total picture can be described as follows. With the exception of the massive crystals found in rocks, solids usually exist as colloidal particles, with a maximum dimension less than 1 μm. Such particles have a high surface-to-volume ratio. The surface hydroxyl groups are the principal functional groups of the solids studied. If one considers adsorption and wetting as quasi-chemical reactions, they are the reactive chemical entities at the surface of the solid. They possess a double pair of electrons together with a dissociable hydrogen atom which enables them to react with both acids and bases. Due to this fact, the majority of the solids we have studied are amphoteric. Bulk structure considerations indicate that the surface hydroxyl groups may be coordinated to different atoms (Fe, Al, Si, C, Mg, etc.). Moreover, they may be coordinated to one, two or three underlying atoms. Consequently, some different types of hydroxyls may appear: simple, geminal (i.e. doubly bonded to the same atom) and doubly or triply coordinated. The overall density of these groups depends on both the crystal structure and on the extent of the development of different crystal faces. Due to the difference between the numbers of underlying atoms that are coordinated to the surface functional groups and to the distance between the chemical groups, the electronegativity and the hardness of the different types of surface groups vary. On this basis, the surface of a solid can be viewed as a patchwork composed of a certain number of chemical groups, separated by a certain distance. This distance can be quantified by the "surface density" of the groups. One could, therefore, distinguish between an attraction for the liquid molecule which is related to the influence of isolated groups and an attraction due to the interaction between groups. The first term has an influence on the surface enthalpy by the net number of surface groups and is clearly related to the composition of the solid (i.e. the stoichiometry of the different groups). The second term is more global and is related to the actual morphology of the crystal. In fact, if the crystals were perfect, the ratio between different groups located at different places such as faces or edges could be known; and the relation between the first and the second terms could be established. Generally, the actual solid grains are broken, fractured and the proportions between the different types of bonds are difficult to estimate *a priori*.

5.2.2. Influence of the Lamellarity. The immersion enthalpies in n-heptane, water and formamide of the three pure talc samples of different lamellae sizes are reported in Table 2. These three samples are well described in reference [28d]. One of them, from Italy, is characterized by sheets longer than 50 μm. The talc coming from the French deposit of Luzenac is less lamellar, with sheets of 10 μm length. The Spanish talc presents smaller lamellae of the order of the average particle diameter (1 μm). For these three samples, heat of immersion in a very branched alkane, isooctane, is also reported in Table 7.

Table 7.
Enthalpies of immersion of three different pure talcs in isooctane at 25°C

Talc Samples	$\Delta_{imm}H$ (mJ/m^2)
Spanish	- 84
Luzenac (lamellar)	- 103
Italian (very lamellar)	- 111

When the lamellarity varies from one sample to another, the ratio between the basal and lateral surface areas changes. It is an important effect, influencing the immersion enthalpy results. This modifies the partition of polar groups on the surface of the grain more than the force of the chemical bond, and explains the results obtained in the polar solvents, where the immersion enthalpy decreases with the sample lamellarity.

Our results in the case of the n-heptane, which is apolar, are difficult to explain. Generally there is a weak influence of the polarity of the solid on the immersion enthalpies in apolar solvents. Nevertheless, one observes an increase in the immersion enthalpies in n-heptane with the sample lamellarity. Such an effect comes from the orientation of the solvent molecules on the solid surface, the talc samples being characterised by a large planarity. Accordingly, one can assume that n-heptane molecules are more organized on the regular, large and plane basal surfaces of Italian talc than on the short basal surfaces of Spanish talc. Indeed, the three talc samples give similar heats of immersion values in isooctane. The difference between the immersion enthalpies in n-heptane and in isooctane is nil for Spanish talc and it increases considerably when the lamellarity of the studied samples increases. These results show a large influence of the surface geometry of the adsorbed molecule, due to the crystal morphology, on the interaction energy. For computation of enthalpies and surface tensions, we use in this case isooctane values (the liquid-vapour surface enthalpy of isooctane H_{LV} is 45.4 mJ/m^2, which is totally apolar).

5.3. Surface Tension

5.3.1. General Phenomenology. Starting from equations (28) and (29), it is possible to compute the surface tension of solids if the adsorption of probe vapours is carried out. This has been done for some samples of our set. We have observed a general trend which is in agreement with the literature results. As we have seen, high energy solids are characterised by the existence of three types of intermolecular potential energy. Therefore, they possess an attractive energy for dispersive liquids rather than for polar liquids. Obviously, the interaction energy which influences the number of molecules adsorbed per unit area is greater for polar molecules than for apolar molecules. But even for apolar molecules, adsorption is never negligible. Hence, in each case we expect to obtain adsorption isotherms per unit area, where the quantity of alkane adsorbed is always lower than that of polar (or polarizable) molecules (if molecular cross-sectional areas are in the same range). Also, molecules with low dielectric constant are expected to be less adsorbed than molecules of high dielectric constant. Experimentally, in general, water is adsorbed in the greatest quantity, due both to the strengh of interaction and the small cross-sectional area of the molecule.

The results of the computation of the free enthalpy are listed in Table 8. These values are in good agreement with the literature results as reported by Adamson [40], and make sense. It is clear that the lamellarity of the sample has a great influence on the surface tension of the solid: the surface tension of mica is the highest among all the solids studied. Crystals partially soluble in water do not give very high values of surface tension. The other trends have been discussed before. But, from this table, it is difficult to predict the behaviour with mixtures, solute molecules or impurities and

Table 8.

Surface tension components and total surface tension ($\gamma_S{}^\circ$) of the solids (in mN/m) in equilibrium with their own vapours using our model. The values of the surface tension for Mica, Copper and Paraffin are taken from reference 40 (p. 314)

Solids	$\gamma_S{}^{LW}$	$\gamma_S{}^+$	$\gamma_S{}^-$	$\gamma_S{}^\circ$
Mica				4500
Dolomite	306	234	1852	1623
Copper				1370
Chlorite	145	1416	23	506
$MgCO_3$	70	587	125	612
Alumina	102	265	364	723
Calcite	187	14	350	240
SiC	218	31	31	279
Quartz	53	99	109	261
Illite	53	44	162	222
Talc from Spain	40	276	10.5	148
Talc from France	53	242	2	97
Talc from Italy	59	118	27	172
Talc 1	65	222	11	164
Talc 2	81	57	1.3	98
Talc 3	93	45	0.1	97
Kaolinite from Charentes	55	90	44	181
Silica XOB 015	82	23	33	137
Silica Cabosil	41	62	3	55
Solid Paraffin				50

Table 9.
Surface tension components and total surface tension of some low surface energy solids (in mN/m) in equilibrium with their own vapours using our model

Solids	γ_S^{LW}	γ_S^+	γ_S^-	γ_S^o
Teflon	24	0.3	0.3	25
Graphon	66	-	-	66
Sulphur 1	34	0.3	0.3	35
Sulphur 2	54	-	-	54
Sulphur 3	20	5	5	30

contaminants in actual conditions. The solute-solvent interactions shall have an importance. Roughly speaking, the three parameters of the surface tension which have an influence are: apolarity, acidity and basicity. A weak base will adsorb preferentially on a solid of a strong acid character, although strong bases should show no difference between acidic solids. Macromolecules, even polar, would tend to adsorb preferentially on solids with a strong apolar character, but the distance between surface sites has to be known to predict the outcome.

6. CONTACT ANGLES FROM THE POINT OF VIEW OF THE PRESENT ANALYSIS OF ENTHALPIES OF IMMERSION

All this analysis which has been shown to be valid for "high energy" solids such as silicates or clays, can be extended to "low energy" solids, presenting contact angles with water. In fact, such an application is useful in order to analyse the importance of film pressure upon contact angle analysis.

Many results of heat of immersion of hydrophobic solids into water are available in the literature, and we have performed such experiments in our laboratory, mainly with silicas and sulphurs [83].

The main interesting point is the comparison between the heat of immersion and the heat of wetting. On such systems, called hydrophobic, which present large contact angles with water (around 90°), it is admitted in contact angle literature that there is no adsorption of vapour [for example ref. 40, p. 406], and therefore that the film pressure is negligible. This point is in disagreement with the analysis of the isotherms available in the literature made by Brunauer *et al.* [84], Gregg and Sing [13] or by the I.U.P.A.C. [85]. The low energy solids give type III and type V isotherms. These isotherms are characterised by a left part (low pressure) which is near zero (no adsorption) but also by a right part (high pressure range) which presents a substancial adsorption. This substancial adsorption is due to the adsorbate-adsorbate interactions in the vapor phase near the phase transition, promoting the deposit of vapor molecules on the solid surface.

Moreover, a hypothetical absence of adsorption would imply an absence of heat of adsorption. This fact is proven by the thermodynamic relation between the adsorption enthalpy and the shape of the isotherm. Anyway, in such a case, an equality between the heat of wetting and the heat of immersion should be expected. No such equality has been reported in the literature. Furthermore, large differences are seen between the reported values [26, 28, 52], and it appears that the differences observed between the heat of immersion and the heat of wetting are nearly the same for both high energy as well as low energy solids. This surprising point, from the standpoint of contact angle science, was observed by Zettlemoyer [86].

In view of the model we have used, such an observation is very simple to analyse. The values obtained for Teflons FEP and PTFE are interesting to discuss because their immersion and wetting heats were reported some forty years ago [31, 46c]. Two values of immersion enthalpies for Teflon are reported, respectively -58 mJ/m^2 in n-heptane and -6 mJ/m^2 in water. These values are very typical of low energy solids. We report some examples of solids studied in our laboratory in Tables 2 and 3. These very low enthalpies of immersion in water are classically interpreted [87] in terms of repulsion between the solids in question and the structure of water. But this point is controversial in the sense that the immersion enthalpy of graphitized carbon in water is low but still exothermic. This indicates, in fact, for hydrophobic solids, the same type of interactions as between minerals and water, but to a lower extent.

Using our approach for Teflon, a value of 58 mJ/m^2 is expected for the dispersive enthalpic component H_S^{oLW} (calculated using the result in n-heptane). This gives a computed immersion enthalpy value of -26 mJ/m^2 for water, showing some disagreement with the experimental result, which suggests a very weak polar value (with two components, H_S^{o+} and H_S^{o-} around 0.6 mJ/m^2. We cannot distinguish between acidic and basic components, with only one experimental result). This presence of polar forces on such a surface is probably due to the deposit of some impurities during the preparation of the powder. This phenomenon has been discussed by Good and co-workers [31], and has been often observed during our work with talc, a very soft material.

The interpretation of the results of adsorption and contact angle can start from the value of the surface tension of PTFE computed by Drummond *et al.* [18]. This value is 27 mN/m. It is in agreement with the value of 25 mN/m calculated by Zettlemoyer and co-workers [46c]. The comparison between this value and the value of H_S^{o} which we computed (59.2 mN/m) gives a value of the coefficient f of 0.46, in good agreement with the values reported above for high surface energy solids.

Using the thermodynamic definition, the free enthalpy of immersion is:

$$-\Delta_{imm}G = \pi_e + \gamma_{LV}\cos\theta \tag{31}$$

and using our equations:

$$-\Delta_{imm}G = -\gamma_{LV} + 2(\gamma_S^{oLW} \cdot \gamma_{LV}^{LW})^{1/2} = -23 + 2(27 \cdot 23)^{1/2} \approx 27 \text{ mJ/m}^2$$

To obtain π_e, one has to substract the term $\gamma_{LV}\cos\theta$ (a value of 39° is reported for the contact angle of n-heptane upon Teflon). One obtains a value of 9 mN/m for the film pressure of n-heptane. This value corresponds exactly to the experimental one given by Melrose [88]. With a lighter alkane, such as octane (θ=28°), one should obtain $-\Delta_{imm}G$ = 27 mJ/m^2, and this gives a π_e of 7 mN/m, closer to the experimental value given by Whalen and Hu (2.7 mN/m) [89].

This agreement between the calculated and experimental values indicates that our model makes sense, if using the more recent calculations of the dispersive surface tension.

6.1. The Case of Water

At this point for the interactions between water and Teflon, it does not seem necessary to take into account, at this level of approximation, the polar components observed in immersion. The powders are not exactly the plates used in contact angle experiments. What is expected for plates with water, using a value for $\gamma_S°$ of 27 mN/m, is a value of $\Delta_{imm}G$ of -23.5 mJ/m^2 . Different values are given in the literature for the contact angle [90]. Some very high values are reported, about 118° [63b]. This type of Teflon can be considered as very "hydrophobic". π_e values between 1 and 10 mN/m are obtained, if the two extreme contact angles are taken: 110° and 118°. This gives high values of γ_{SV}, lying between 26 and 17 mN/m. They are comparable to the values of 20 mN/m for octane and 18 mN/m for n-heptane. The values of γ_{SL} goes from 0.75 (octane) to 0.125 mN/m (n-heptane) and about 50 mN/m for water. Thus, the values of γ_{SL} can be large, which is not surprising when considering the value between water and ice (about 20 mN/m) [32]. The main points here are that γ_{SV} is not constant when changing the vapour surrounding the solid and that γ_{SL} does not necessarily decrease when the contact angle decreases. All the intermolecular interfacial terms vary when changing the species in contact; the only constant term is $\gamma_S°$.

6.2. Consequence on Classical Contact Angle Theories

One of the most important results of our work is the range of values obtained for the solid-liquid interfacial tension. Even though this value varies from "low-energy" solids to "high energy" solids, it seems clear that a solid-liquid interfacial tension is always positive and in many cases with an order of magnitude of 20 mN/m. This point has an important consequence for the classical theories of contact angle. Following Fox and Zisman [91] (in fact Zisman himself has emphasized the influence of π_e [92]) and Good *et al.* (here too, many researchers have forgotten that Good *et al.*, using crude but correct approximations, have given a value of about 56 mJ/m^2 for the solid-liquid surface enthalpy for Teflon/water [31] which implies a positive and large value of the solid-liquid interfacial tension), the majority of researchers have considered that the solid-liquid interfacial tension is nil when the liquid spreads on the solid. This assumption has allowed researchers to consider that the maximal surface tension of species which are able to spread on a solid surface, in a family of molecules (for instance alkanes), is the surface tension of the solid. Our results clearly demonstrate that this surface tension is not the solid-vapour surface tension; the film pressure cannot be neglected. More generally, all the theories considering that the solid-liquid interfacial tension can be nil cannot give a solid-vapour surface tension of the solid. They only give a value in the correct order of magnitude. This point casts a doubt on the theory of Neumann and co-workers [90, 93], named the "equation-of-state approach".

In fact, this assumption of a zero interfacial tension between a solid and a liquid was surprising because surface tension is linked to the mutual solubility of phases. For example, one can consider a zero interfacial tension in the case of the mixture of vapour phases, or for a critical interface at the critical point. But these cases are limiting cases, and one generally expects a positive value of the interfacial tension for non-mixing phases. Thus, the solid-liquid interfacial tension is probably related to the solubility product of the solid concerned in the chosen liquid. The more the solid is soluble, the lower must be the interfacial tension. But this question of the solubility of solids is a very complex one. For ionic solids, as an example [59], it is the result of a balance between the Madelung energy of the crystal, which depends on the radius of the ions, and the hydration energy of the ions, which also depends on the radius of each ion, but following a different expression. The key point here is that what is obtained using contact angles and assumptions such as used in the Zisman plot, is a complicated function of the solid-vapour surface tension.

7. CONCLUSION

The following conclusions can be deduced from the experimental results and numerical trends we have obtained.

(1) It is possible to divide solids into three classes, namely, low energy, middle energy and high energy.

(2) The pertinent parameter to describe solids is the surface energy (either enthalpy or free enthalpy) of the solid in equilibrium with its own vapour. The critical surface tension γ_c is clearly more a fitting parameter than a thermodynamic value.

(3) From a comparison between the immersion enthalpy values for powders and contact angle values for plates, it seems that our 3 parameter model makes sense, indicating that in many cases solid surface tensions are underestimated in the literature, due to the assumption that π_e is negligible. The fact that the assumption of π_e being negligible is false was actually very clear in the I.U.P.A.C. isotherms classification. Low-energy solids correspond to type III and V isotherms. For such isotherms, the adsorbed quantity is never zero. It increases drastically for high values of the partial pressure, i.e. for high activities of the vapour. In this view, our model gives phenomenological values in agreement with calculations, adsorption isotherms, contact angle experiments and immersion and wetting calorimetry. However, a precision better than 2 mJ/m^2 seems presently difficult to obtain, considering the difficulties encountered in the preparation of samples and in the different experiments.

(4) Immersion enthalpy results can be simply interpreted on the basis of a scale of surface energy of the solid powders, and by using the idea that the strength of interaction between a solid and a liquid or a vapour, is dependent on the surface energy of the solid. This point has been controversial, and is, for example, denied by Neumann and co-workers. We think that our results are important in this debate.

REFERENCES

1. S. Dapprich and G. Frenking, *J. Phys. Chem.* **99**, 9352 (1995).
2. (a) J. S. Dewar, *Bull. Soc. Chim. Fr.* **18**, c79 (1951); (b) J. Chatt and L. A. Duncanson, *J. Chem. Soc.*, 2939 (1953).
3. K. Morokuma, *J. Chem. Phys.* **55**, 1236 (1971).
4. A. Fernandez, C. Reyes, M. R. Wilson, D. C. Woska, A. Prock and W. P. Giering, *Organometallics* **16**, 342 (1997).
5. R. S. Drago, N. Wong and G. C. Vogel, *Inorg. Chem.* **26**, 9 (1987).
6. R. S. Drago and B. B. Wayland, *J. Am. Chem. Soc.* **87**, 3571 (1965).
7. R. G. Pearson, *Chemical Hardness*, Wiley-VCH, New York (1997).
8. J. C. Bolger and A. S. Michaels, in: *Interface Conversion*, P. Weiss and G. D. Cheevers (eds.). Chap. 1, Elsevier, New York (1969).
9. F. M. Fowkes, *J. Phys. Chem.* **72**, 3700 (1968).
10. C. J. van Oss, M. K. Chaudhury and R. J. Good, *Adv. Colloid Interface Sci.* **28**, 35 (1987).
11. F. M. Fowkes, *J. Adhesion Sci. Technol.* **4**, 669 (1990).
12. J. M. Douillard, *J. Colloid Interface Sci.* **188**, 511 (1997).
13. S. J. Gregg and K. S. W. Sing, *Adsorption, Surface Area and Porosity*, Academic Press, London (1982).
14. W. Rudzinski and D. H. Everett, *Adsorption of Gases on Heterogeneous Surfaces*, Academic Press, London (1992).
15. J. Fraissard and C. W. Conner (eds.), *Physical Adsorption: Experiment, Theory and Applications*, NATO ASI Series, Kluwer, Dordrecht, The Netherlands (1997).
16. (a) R. G. Linford, *Chem. Rev.* **78**, 81 (1978); (b) D. M. Mohilner and T. R. Beck, *J. Phys. Chem.* **83**, 1160 (1979).
17. W. A. Steele, *Surface Sci.* **36**, 317 (1973).
18. C. J. Drummond, G. Georgaklis and D. Y. C. Chan, *Langmuir* **12**, 2617 (1996).
19. P. G. de Gennes, *Rev. Mod. Phys.* **57**, 827 (1985).
20. E. A. Müller, L. F. Rull, L. F. Vega and K. E. Gubbins, *J. Phys. Chem.* **100**, 1189 (1996).
21. S. L. Sowers and K. E. Gubbins, *Adsorption*, in press.
22. D. Y. Kwok, D. Li, and A. W. Neumann, *Langmuir* **10**, 1323 (1994).
23. A. P. Legrand, H. Hommel, A. Tuel, A. Vidal, H. Balard, E. Papirer, P. Levitz, M. Czernichwi, R. Erre, H. Van Damme, J. P. Gallas, J. F. Hemidy, J. C. Lavaley, O. Barres, A. Burneau and Y. Grillet, *Adv. Colloid Interface Sci.* **33**, 91 (1990).
24. C. Vittoz, M. Mantel and J. C. Joud, *J. Adhesion* **67**, 347 (1998).
25. A. J. Grosjek and S. Partyka, *Langmuir* **9**, 2721 (1993).
26. A. C. Zettlemoyer, in: *Chemistry and Physics of Interfaces*, p. 137, American Chemical Soc., Washington DC (1965).
27. R. L. Every, W. H. Wade and N. Hackerman, *J. Phys. Chem.* **65**, 25 (1961).
28. (a) J. M. Douillard, M. Elwafir and S. Partyka, *J. Colloid Interface Sci.* **164**, 238 (1994); (b) T. Zoungrana, A. Berrada, J. M. Douillard. and S. Partyka, *Langmuir* **11**, 1760 (1995); (c) H. Malandrini, R. Sarraf, B. Faucompré, S. Partyka and J. M. Douillard, *Langmuir* **13**, 1337 (1997); (d) H. Malandrini, F. Clauss, S. Partyka and J. M. Douillard, *J. Colloid Interface Sci.* **194**, 183 (1997); (e) V. Médout-Marère, H. Belarbi, F. Morato, J. C. Giuntini and J. M. Douillard, *J. Colloid Interface Sci.* **202**, 139 (1998).
29. J. M. Douillard and H. Malandrini, *Compt. Rend. Acad. Sci.*, Paris, **IIc**, 1 (1999).
30. J. M. Douillard, T. Zoungrana and S. Partyka, *J. Petroleum Sci. Eng.* **14**, 51 (1995).
31. R. J. Good, L. A. Girifalco and G. Kraus, *J. Phys. Chem.* **62**, 1418 (1958).
32. R. Defay, I. Prigogine, A. Bellemans and D. H. Everett, *Surface Tension and Adsorption*, Longmans, London (1966).
33. J. A. V. Butler, *Chemical Thermodynamics*, Macmillan, London (1962).
34. J. C. Eriksonn, *Acta Chem. Scand.* **16**, 2199 (1962).
35. T. Young, *Miscellaneous Works*, Vol. 1, G. Peacock (ed.), Murray, London (1855).
36. D. H. Everett, *Basic Principles of Colloid Science*, Royal Society of Chemistry, London (1988).
37. I. Prigogine and R. Defay, *Chemical Thermodynamics*, Longmans, London (1954).
38. G. Jura and W. D. Harkins, *J. Am. Chem. Soc.* **66**, 1356 (1944).
39. S. Partyka, F. Rouquerol, and J. Rouquerol, *J. Colloid Interface Sci.* **68**, 30 (1979).

40. A. W. Adamson, *Physical Chemistry of Surfaces,* 5th edition, pp. 313–318, Wiley, New York (1976).
41. G. Jura and T. L. Hill, *J. Am. Chem. Soc.* **74**, 1598 (1952).
42. V. Médout-Marère, H. Malandrini, T. Zoungrana, J. M. Douillard and S. Partyka, *J. Petroleum Sci. Eng.* **20**, 223 (1998).
43. P. C. Hiemenz, *Principles of Colloid and Surface Chemistry,* Marcel Dekker, New York (1986).
44. E. Calvet and H. Prat, *Microcalorimétrie,* Masson, Paris (1956).
45. A. Lecloux and J. P. Pirard, *J. Colloid Interface Sci.* **70**, 265 (1979).
46. (a) A. C. Zettlemoyer, G. J. Young, J. J. Chessick, J. J. and T. H. Healey, *J. Phys. Chem.* **57**, 649 (1953); (b) G. J. Young, J. J. Chessick, F. H. Healey, A. C. Zettlemoyer and G. J. Young, *J. Phys. Chem.* **58**, 313 (1954); (c) T. H. Healey, J. J. Chessick, A. C. Zettlemoyer and G. J. Young, *J. Phys. Chem.* **58**, 887 (1954); (d) J. J. Chessick, A. C. Zettlemoyer, T. H. Healey and G. J. Young, *Can. J. Chem.* **33**, 251 (1955).
47. (a) W. H. Wade, R. L. Every and N. Hackerman, *J. Phys. Chem.* **64**, 355 (1960); (b) W. H. Wade, and N. Hackerman, *J. Phys. Chem.* **64**, 1196 (1960); (c) W. H. Wade, and N. Hackerman, *J. Phys. Chem.* **65**, 1681 (1961).
48. (a) J. W. Whalen, *Advances in Chemistry Series* **33**, American Chemical Soc., Washington DC, p. 281, (1961); (b) J. W. Whalen, *J. Phys. Chem.* **65**, 1676 (1961); (c) J. W. Whalen, *J. Phys. Chem.* **66**, 511 (1962).
49. L. Robert, *Mém. Soc. Chim. France* **412**, 147 (1967).
50. P. Roy and D. W. Fuerstenau, *J. Colloid Interface Sci.* **26**, 102 (1968).
51. S. Lagerge, P. Rousset, J. M. Douillard and S. Partyka, *Colloids Surfaces A.* **80**, 261 (1993).
52. H. Malandrini, PhD Thesis, Montpellier II, France (1995).
53. D. A. Griffiths and D. W. Fuerstenau, *J. Colloid Interface Sci.* **80**, 271 (1981).
54. E. M. Arnett, Q. Liu and M. Gumkowski, *Energy Fuels* **2**, 295 (1988).
55. M. L. Gonzalez-Martin, B. Janczuk, L. Labajos-Broncano and J. M. Bruque, *Langmuir* **13**, 5991 (1997).
56. R. Denoyel, J. Fernandez-Colinas, Y. Grillet and J. Rouquerol, *Langmuir* **9**, 515 (1993).
57. F. E. Bartell, and R. Murray Suggitt, *J. Phys. Chem.* **58**, 36 (1954).
58. J. T. Davies, and E. K. Rideal, *Interfacial Phenomena*, Academic Press, London (1961).
59. J. E. Huheey, E. A. Keiter, and R. L. Keiter, *Inorganic Chemistry: Principles of Structure and Reactivity,* Harper Collins, New York (1993).
60. K. Kitaura, and K. Morokuma, *Int. J. Quantum Chem.* **10**, 325 (1976).
61. E. Tronel-Peyroz, J. M. Douillard, L. Ténebre, R. Bennes, and M. Privat, *Langmuir* **3**, 1027 (1987).
62. B. W. Ninham, and V. Yaminsky, *Langmuir* **13**, 2097 (1997).
63. (a) C. J. van Oss, R. J. Good, and M. K. Chaudhury, *J. Colloid Interface Sci.* **111**, 378 (1986); (b) C. J. van Oss, R. J. Good, and M. K. Chaudhury, *Langmuir* **4**, 884 (1988).
64. V. Gutmann, *The Donor-Acceptor Approach to Molecular Interactions,* Plenum Press, New York (1978).
65. C. Della Volpe and S. Siboni, *J. Colloid Interface Sci.* **195**, 121 (1995).
66. A. Holländer, *J. Colloid Interface Sci.* **169**, 493 (1995).
67. L. H. Lee, in: *Contact Angle, Wettability and Adhesion*, K. L. Mittal (ed.), pp. 45–96, VSP, Utrecht (1993).
68. J. M. Douillard, H. Malandrini, T. Zoungrana, and S. Partyka, in: *Proceedings of the 3rd International Symposium on Evaluation of Reservoir Wettability and its Effect on Oil Recovery*, N. R. Morrow (ed.), pp. 21–25, U. of Wyoming, Laramie (1996).
69. C. S. Brooks, *J. Phys. Chem.* **64**, 532 (1960).
70. J. J. Jurinak, and D. H. Volman, *J. Phys. Chem.* **65**, 150 (1961).
71. M. D. Vrbanac, and J. C. Berg, in: *Acid-Base Interactions: Relevance to Adhesion Science and Technology,* K. L. Mittal and H. R. Anderson Jr. (eds.), VSP, Utrecht, p. 67 (1991).
72. J. Jänchen, H. Stach, and W. J. Mortier, *J. Phys. Chem.* **100**, 12489 (1996).
73. M. E. Schrader, in: *Modern Approaches to Wettability,* M. E. Schrader and G. Loeb (eds.), Plenum Press, New York, p. 101 (1992).
74. D. A. King and D. P. Woodruff (eds), *The Chemical Physics of Solid Surfaces and Heterogeneous Catalysis*, Elsevier, Amsterdam (1982).
75. J. H. Rayner, and G. Brown, *Clays Clay Minerals* **21**, 103 (1973).
76. H. A. Benesi, *J. Phys. Chem.* **61**, 970 (1957).
77. D. H. Salomon, J. D. Swift, and A. J. Murphy, *J. Macromol. Sci.* **A5**, 441 (1971).

78. M. Deeba, and W. K. Hall, *J. Catalysis* **60**, 417 (1979).
79. G. I. Kapustin, T. R. Brueva, A. L. Klyachko, S. Beran, and B. Wichterlova, *Appl. Catal.* **42**, 239 (1988).
80. M. Kojima, J. C. Q. Fletcher, and C. T. O'Connor, *Appl. Catal.* **28**, 169 (1986).
81. D. Nyfeler and T. Armbruster, *Am. Mineral.* **83**, 119 (1998).
82. A. M Youssef. *Surface Technol.* **9**, 187 (1979).
83. J. M. Douillard, S. Pougnet, B. Faucompré, and S. Partyka, *J. Colloid Interface Sci.* **154**, 113 (1992).
84. (a) S. Brunauer, L. S. Dewing, W. E. Dewing and E. Teller, *J. Am. Chem. Soc.* **60**, 1723 (1940); (b) S. Brunauer, *The Adsorption of Gases and Vapours*, Oxford University Press (1945).
85. J. Rouquerol, D. Avnir, C. W. Fairbridge, D. H. Everett, J. M. Haynes, N. Pernicone, J. D. Ramsay, K. S. W. Sing and K. K. Unger, *Pure Appl. Chem.* **66**, 1739 (1994).
86. A. C. Zettlemoyer, *J. Colloid Interface Sci.* **28**, 343 (1968).
87. C. Tanford, *The Hydrophobic Effect: Formation of Micelles and Biological Membranes,* Wiley, New York (1973).
88. J. C. Melrose, in: *Contact Angle, Wettability, and Adhesion*, Adv. Chem. Ser. n°43, p. 158, American Chemical Society, Washington DC (1964).
89. J. W. Whalen and P. C. Hu, *J. Colloid Interface Sci.* **65**, 460 (1978).
90. A. W. Neumann, Adv. *Colloid Interface Sci.* **4**, 1 (1974).
91. H. W. Fox and W. A. Zisman, *J. Colloid Sci.* **5**, 514 (1950).
92. W. A. Zisman, in: *Contact Angle, Wettability, and Adhesion*, Adv. Chem. Ser. n°43, p. 1, American Chemical Society, Washington DC (1964).
93. D. Li and A. W. Neumann, *Adv. Colloid Interface Sci.* **39**, 299 (1992).

Acid-Base Interactions, Vol. 2, pp. 349–362
Ed. K.L. Mittal

Comparison of two procedures for assessing the acidity of polymer surfaces graded in acid content

THOMAS B. LLOYD* and JOHN A. COLEMAN
Department of Chemistry, Lehigh University, Bethlehem, PA 18015

Abstract—This study was aimed primarily at testing surface acidity measuring procedures using contact angles. Plaques of ethylene copolymer (EAC) containing acrylic acid (AA) ranging from 0% to 20% were pressed and the three-liquid contact angle approach was used to grade the surface acidity and basicity. The data were analyzed according to the two procedures: (1) van Oss, Good and Chaudhury (vOGC) and (2) Chang and Chen (CC). In measuring the acidity, the polar pair, water (pH 6.3) / dimethylsulfoxide gave values of the acidic component $\sqrt{\gamma_S^+}$ (vOGC model) and acidic y (CC model) which tracked the % AA content better than the pair, water (pH 6.3) / ethylene glycol. Since some of the $\sqrt{\gamma_S^+}$ values are negative, squaring to get positive values of γ_S^+ destroyed the linearity with % AA. In assessing basicity, the two polar pairs were essentially equal and gave good linearity of γ_S^- with acrylic acid (C=O) content. However, values of basic z (CC model) at 15.5% and 20% AA failed to rise as expected. Presumably at these carboxyl concentrations hydrogen bonding between carboxyls is influencing the acid / base interactions with the liquid probes considerably.

Keywords: Surface acidity / basicity; contact angle (three-liquid procedure); work of adhesion models; contact angle titration.

1. INTRODUCTION

Models to quantify interfacial interactions in terms of dispersion plus acid / base forces have been proposed by van Oss, Good and Chaudhury (vOGC) [1] and by Chang and Chen (CC) [2]. These models attempt to account for the relatively non-specific dispersion attractive force as well as specific Lewis acid (electron acceptor)–base (electron donor) interactions across interfaces. For a fuller understanding, references [1] and [2] are recommended as well as other papers in this symposium proceedings.

The dispersion component of the work of adhesion is expressed for a non-polar solid by the Good-Girifalco-Fowkes equation:

$$W_a^{LW} = \gamma_L(1 + \cos\theta) = 2\sqrt{\gamma_S^{LW}\gamma_L^{LW}} \tag{1}$$

*To whom correspondence should be addressed. E-mail: tbl0@lehigh.edu

Here γ_L is the total surface tension and γ_L^{LW} its dispersion component, of a liquid having a contact angle, θ, on the non-polar solid whose surface tension is γ_S^{LW}. LW was coined by Good and coworkers for the dispersion term, Lifshitz–van der Waals [1a].

The vOGC approach proposes that the acid / base component of interfacial work of adhesion is:

$$W_a^{AB} = 2\sqrt{\gamma_S^+ \gamma_L^-} + 2\sqrt{\gamma_S^- \gamma_L^+} \tag{2}$$

where γ_S^+ and γ_S^- are the acidic and basic energetic parameters, respectively, of the solid surface. The values of γ_S^+ and γ_S^- are determined by measuring the contact angles of pairs of polar liquid probes of predetermined γ_L^+ and γ_L^- (see reference [1c]). These values are relative to the assigned values for water, viz, $\gamma_W^+ = \gamma_W^- = 25.5$ mJ / m^2; thus for water, $\gamma_W^{AB} = 2\sqrt{\gamma_W^+ \gamma_W^-} = 51.0$ mJ / m^2. γ_S^+ and γ_S^- are established using the combining rule proposed by vOGC which involves the contact angles of liquid probes, i.e.,

$$-\Delta G = W_a^{TOTAL} = \gamma_L(1 + \cos\theta) = 2\sqrt{\gamma_S^{LW} \gamma_L^{LW}} + 2\sqrt{\gamma_S^+ \gamma_L^-} + 2\sqrt{\gamma_S^- \gamma_L^+} \tag{3}$$

Here the LW term for the solid, γ_S^{LW}, is established in a straightforward way using non-polar liquids of known γ_L^{LW} and in which case $\gamma_L^+ = \gamma_L^- = 0$. Since there are two unknowns (γ_S^+ and γ_S^-), two (or more) polar liquids must be used for the contact angle measurements. Equation (3) is applied simultaneously to each pair of the polar liquid probes, one of which must be water, and as we will see later its pH is important when ionizable groups exist on the surface.

The CC model defines the LW energy parameters for liquids and solids as: $a = \sqrt{2\gamma_L^{LW}}$ and, $x = \sqrt{2\gamma_S^{LW}}$ in accord with the established practice for the LW (d) force. The energy parameters (constants) for the liquid probe are: b (basic) and c (acidic) while those for the solid are y (acidic) and z (basic). The work of adhesion for a liquid on a solid is then given by:

$$W_a = ax - by - cz \tag{4a}$$

or

$$W_a = ax - bz - cy \tag{4b}$$

whichever gives the larger value of W_a (i.e., the most negative ΔG_a). Chang and Chen used a nonlinear least squares fitting program to establish the probe liquid constants assuming that y and z for polyethylene (PE) were zero. Thus acids and bases have a common reference point. Literature contact angles were used. When values of both y and z are positive the material is said to be acidic and, it is basic when both are negative. If one is positive and the other is negative that makes the material bipolar. Chang and Chen found that, except for water, all other liquid molecules had stronger basic (b) component than acidic (c). Furthermore, polymers do not have very positive acidic (y) values. Acidic components (c and y) are increasingly positive as acidity increases. On the other hand, basic values (b and z) become more negative with increasing basicity.

The CC model attempts to take into account the repulsive forces between acid-acid and base-base interactions [2a].

As with the vOGC procedure, the W_a is determined for two polar liquid probes of known b and c and the equations solved simultaneously for y and z of the solid. We realize that precise values of the parameters are not likely because of the heterogeneous nature of our polymer surface, both chemical and roughness, influencing the contact angle. However, useful evaluations are expected since correlations of W_a^{TOTAL} with fracture energy of interfaces similar to those used here have been seen [3, 4] and our surfaces were relatively smooth.

A series of plaques of ethylene copolymer containing acrylic acid graded from zero to 20% acrylic acid (AA), were prepared for analysis of the surface free energy via the contact angle approach. The idea was to test and compare the models of van Oss, Good and Chaudhury, vOGC, and that of Chang and Chen, CC. We were especially interested in the acid / base nature of the solid surfaces with varying content of AA. Earlier we reported a brief investigation of this nature [5] and this study is an expansion of that one. In that case there was the indication that $\sqrt{\gamma_S^+}$ was a better measure of surface acidity than γ_S^+ when negative values of the square root term were involved. Equation (3) was solved for the solid's acidic and basic parameters in their square root form and sometimes $\sqrt{\gamma_S^+}$ was negative. The choice of the ethylene copolymer with the graded content of carboxylic acid was intended to shed light on this issue.

Good and van Oss discussed the meaning of negative values of $\sqrt{\gamma_S^+}$ in reference [1c] (where it appears to them as an anomaly) and suggested that they might be valid. They state that, when such are negative, these must be combined in the operational form

$$\gamma_{SL}^{AB} = 2\sqrt{\gamma_S^+}\sqrt{\gamma_L^-} \tag{5}$$

and as such Equation (2) should be rewritten as:

$$W_{SL}^{AB} = 2\sqrt{\gamma_S^+}\sqrt{\gamma_L^-} + 2\sqrt{\gamma_S^-}\sqrt{\gamma_L^+} \tag{6}$$

with the result that the first term on the right is negative. A negative value for W_{SL}^{AB} is allowable if $W_{SL}^{AB} < W_{SL}^{LW}$, i.e., the total work of adhesion is positive.

2. MATERIALS AND PROCEDURES

The ethylene acrylic acid copolymer (EAC) specimens for the contact angle measurements were hot pressed onto glass slides using a poly(ethylene terephthalate) release film which was designed to attract the carboxylic acid groups to the surface. The EAC series ranging from 20% to 3% AA were Primacors (Dow Plastics) and the EAC homolog with 1.8% AA content was Dexon-XEA-6 (Exxon). The pure polyethylene with 0% AA, was Tenite 1870A (Eastman Chemical Co.).

The probe liquids were: distilled water, pH 6.3, and ACS certified organic liquids (Fisher), 1-bromonaphthalene (1-BN), ethylene glycol (EG) and dimethylsulfoxide (DMSO). The contact angle titrations were done with Metrepak buffers (Micro Essentials Lab., Brooklyn, New York) at seven pH values from 2 through 12.

Advancing contact angles were measured with a Ramé-Hart goniometer with an environmental chamber at 23 °C. Five to ten drops of the probe liquid per surface were used to measure the contact angles on both sides and this was done on three surfaces for each EAC composition. Reproducibility ranged from ±0.4° for 1-BN to ±1.7° for water with an overall average of ±0.9°.

FTIR (Mattson Polaris) spectra were taken on samples of EAC with 0%, 3.0%, 6.5% and 20% AA content to verify bulk composition using the absorptivity of the peak at 1250 cm^{-1} (O–H bending). There was a perfect correlation with the stated level of acrylic acid. Differential Scanning Calorimetry (Perkin-Elmer DSC-7) was used to determine melting points of these four samples, again to calculate the %AA levels in the copolymer. Using the concepts of Flory [6], the DSC measurements gave an average deviation of 3% for the EAC copolymers which had 20% and 6.5% AA content but the copolymer with 3.0% AA was calculated to 4.7%.

3. RESULTS AND DISCUSSION

The contact angles, θ, of the probe liquids are plotted in Figure 1 along with the $\cos\theta$ against the % acrylic acid content of the copolymer. Note that for 1-BN, the contact angle, θ, does not vary with the % AA content of the EAC copolymer but for the other liquids the contact angle θ declines linearly with increasing % AA content of the EAC copolymer. Table 1 gives the surface free energy parameters of the probe liquids used to establish those of the solids in the EAC series.

3.1. Van Oss, Good and Chaudhury Parameters

Table 2 lists the vOGC parameters of the EAC series calculated as stated above (see Equation (2)). The polar pair, EG / water is compared to DMSO / water. It is acknowledged that water must be one of each polar liquid pair.

The basic parameter γ_S^- tracks the % AA content of the EAC linearly for both paired polar liquid probes. Thus the concentration of double bonded oxygen of the carboxyls at the surface is revealed. The acid parameter, γ_S^+, is very low and

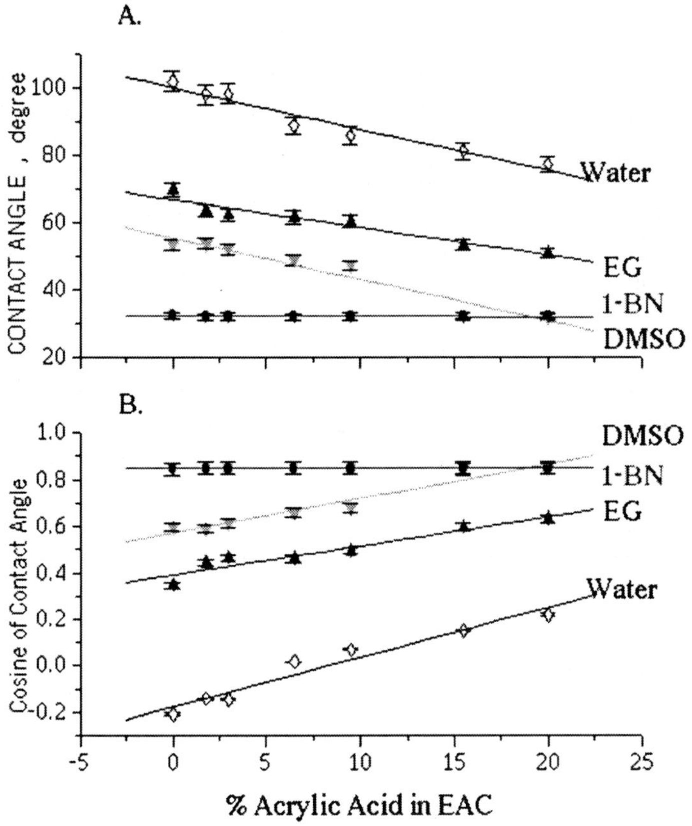

Figure 1. Contact angle, θ, (A) and cosine θ (B) for various liquids as a function of % acrylic acid in EAC.

Table 1.
Surface Energetic Parameters of the Test Liquids

	vOGC Model (mJ/m^2)					CC Model $(mJ/m^2)^{0.5}$			
		APOLAR	ACIDIC	BASIC	Ratio	APOLAR	BASIC	ACIDIC	Calc.
Liquids	γ_L^{TOTAL}	γ^{LW}	γ^+	γ^-	$\sqrt{\gamma^-/\gamma^+}$	a	b	c	γ_L^{TOTAL}
1-BN	44.4	43.5	~1260	~1260	–	9.3	−2.4	0.9	45.4
DMSO	44.0	36	0.5	32	8	9.1	−3.0	0.6	43.8
EG	47.7	29	1.92	47	5	7.7	−4.8	3.6	46.9
Water	72.8	21.8	25.5	25.5	1	6.8	−6.6	7.5	72.6

1-BN = 1 Bromonaphthalene, DMSO = dimethylsulfoxide and EG = ethylene glycol.

for both liquid pairs it is essentially constant, but the $\sqrt{\gamma_S^+}$ value obtained with the DMSO/water pair, while low, does increase regularly with the % AA content of the EAC. The negative values of $\sqrt{\gamma_S^+}$, in this case, seem to be real. However, the

Table 2.
Surface Free Energy Components of the EAC Series (vOGC model)

	1-BN	EG / WATER				DMSO / WATER			
Substrate	γ_S^{LW}	$\sqrt{\gamma_S^+}$	γ_S^+	$\sqrt{\gamma_S^-}$	γ_S^-	$\sqrt{\gamma_S^+}$	γ_S^+	$\sqrt{\gamma_S^-}$	γ_S^-
EAC 20.0%	37.1	0.3	0.1	2.8	7.9	0.4	0.16	2.7	7.5
EAC 15.5%	37.1	0.3	0.1	2.4	5.6	0.4	0.16	2.2	4.9
EAC 9.5%	37.1	0.0	0.0	2.1	4.4	−0.2	0.04	2.3	5.3
EAC 6.5%	37.1	0.0	0.0	1.7	2.9	−0.3	0.09	2.0	3.9
EAC 3.0%	37.1	0.2	0.1	0.3	0.1	−0.3	0.09	0.8	0.7
EAC 1.8%	37.1	0.2	0.0	0.4	0.1	−0.4	0.16	0.9	0.9
Corr. Coeff.		*0.46*	*0.53*	*0.93*	*0.98*	*0.95*	*−0.08*	*0.88*	*0.92*

The $\sqrt{\gamma_S^+}$ and $\sqrt{\gamma_S^-}$ energy terms have as unit $(\text{mJ}/\text{m}^2)^{0.5}$ whereas γ_S^+ and γ_S^- terms have (mJ/m^2).

Table 3.
Surface Energetic Parameters of the EAC Series (CC model)

	Ethylene glycol / Water				DMSO / Water			
Substrate	x	y	z	Polarity	x	y	z	Polarity
EAC 20.0%	8.6	−1.8	−5.6	Basic	8.6	0.2	−3.8	Bipolar
EAC 15.5%	8.6	−1.4	−4.6	Basic	8.6	0.3	−3.1	Bipolar
EAC 9.5%	8.6	−2.7	−4.9	Basic	8.6	−2.5	−4.8	Basic
EAC 6.5%	8.6	−2.1	−3.9	Basic	8.6	−2.7	−4.5	Basic
EAC 3.0%	8.6	1.1	0.5	Acidic	8.6	−3.1	−3.2	Basic
EAC 1.8%	8.6	0.4	−0.2	Bipolar	8.6	−3.5	−3.6	Basic
Corr. Coeff.		*−0.60*	*−0.84*			*0.96*	*0.03*	

The energetic parameters of the CC model have as units $(\text{mJ}/\text{m}^2)^{0.5}$.

EG / water pair is not appropriate for assessing the acid parameter. We have here an indication of the importance of the choice of the second polar liquid (to complement water). Figure 2 shows these results graphically.

3.2. Chang and Chen Parameters

The y and z parameters of the EAC series are given in Table 3. As was true for the vOGC acidic parameter, $\sqrt{\gamma_S^+}$, the DMSO / water pair yields y values which become more positive (more acidic) linearly as % AA increases. On the other hand, the y values from the EG / water pair are not well-behaved.

The basic z parameter is expected to become more negative as % AA increases, i.e., show increasing basicity. For the DMSO / water pair this is true up to 9.5% AA content in the copolymer but the value of the z parameter falls off in the case of the EAC homologs with higher % AA contents. We suspect that the increased hydrogen bonding at these higher carboxyl levels is competing with the liquid probes. This is

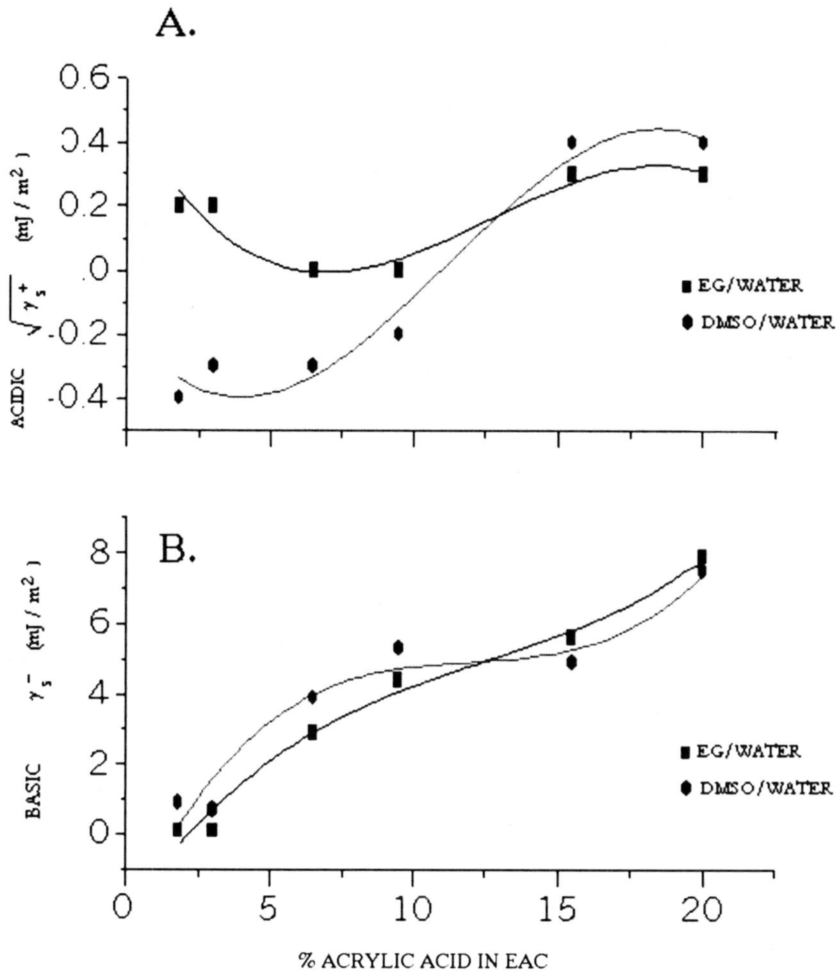

Figure 2. Acidic $\sqrt{\gamma_S^+}$ (A) and basic γ_S^- (B) components for the polar pairs EG / water and DMSO / water as a function of % acrylic acid in EAC.

more noticeable with the DMSO which is basic, than with the EG with its greater acidity. These data are plotted in Figure 3.

3.3. Acid / Base Work of Adhesion of Probe Liquids

In Table 4 the acid / base work of adhesion (W_a^{AB}) values of the polar probe liquids on the EAC surfaces is given for the vOGC and CC models. For each model, the two polar pairs are compared. That is, the acidic and basic parameters of the solid were calculated using each polar pair separately; then the W_a^{AB} for each of the three polar liquids was calculated. The vOGC model is written as $W_a^{AB} = 2\sqrt{\gamma_S^+}\sqrt{\gamma_L^-} + 2\sqrt{\gamma_S^-}\sqrt{\gamma_L^+}$ where the $\sqrt{\gamma_S^+}$ was negative in Table 2, the negative value was used. In such cases this improves the linear correlation slightly.

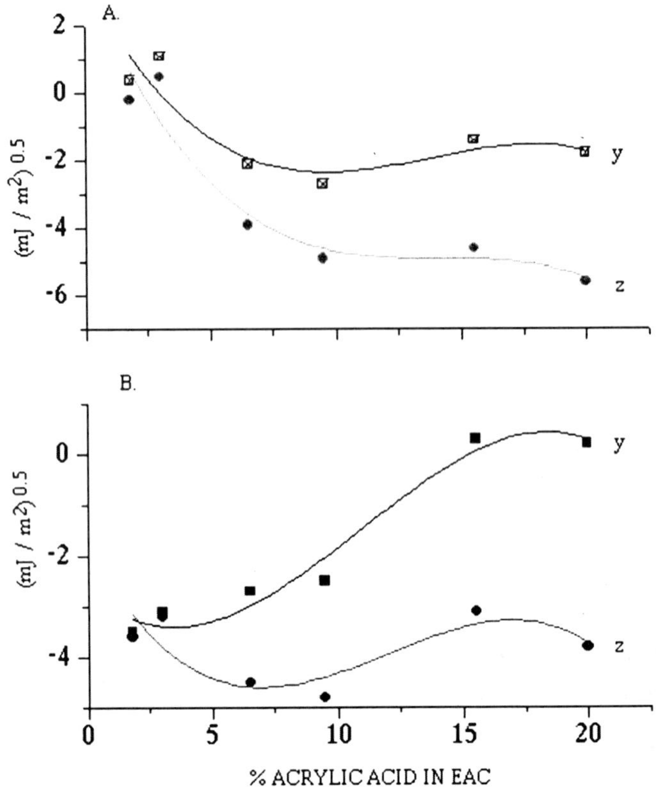

Figure 3. Acidic y and basic z versus acid content for (A): EG / water and (B): DMSO / water.

Table 4.
Work of Adhesion of Paired Polar Liquids on EAC Series

	vOGC Model $W_{SL}^{AB} = 2\sqrt{\gamma_S^+}\sqrt{\gamma_L^-} + 2\sqrt{\gamma_S^-}\sqrt{\gamma_L^+}$						CC Model W_a^{AB} = - by - cz					
	EG / Water			DMSO / Water			EG / Water			DMSO / Water		
Substrate	Water	EG	DMSO	Water	EG	DMSO	Water	EG	DMSO	Water	EG	DMSO
EAC 20.0%	31.4	11.0	7.4	31.7	13.0	8.4	30.1	11.5	−2.0	29.8	14.6	2.9
EAC 15.5%	27.3	10.8	6.8	26.3	11.6	7.6	25.3	9.8	−1.4	25.2	12.6	2.8
EAC 9.5%	21.2	5.8	3.0	21.2	3.6	1.0	18.9	4.7	−5.2	19.5	5.3	−5.2
EAC 6.5%	17.2	4.7	2.4	17.2	1.4	−0.56	15.4	4.0	−4.0	15.9	3.2	−5.4
EAC 3.0%	5.1	3.6	2.7	5.1	−1.9	−2.3	3.5	3.5	3.0	3.5	3.4	−7.4
EAC 1.8%	6.1	3.9	2.8	4.0	−3.0	−3.3	4.1	2.6	1.3	3.9	−3.8	−8.3
Corr. coeff.	0.966	0.976	0.919	0.964	0.983	0.964	0.972	0.976	−0.44	0.965	0.952	0.966

The W_a^{AB} term has the unit (mJ / m^2).

Good and van Oss recognize that negative values may be valid and recommend this means of calculating the acid-base interaction [1c]. Note: whereas negative values of W_a^{AB} are seen in Table 4, $W_a^{TOTAL} = W_a^{LW} + W_a^{AB}$ is always positive. In all cases, except column 10, the increase in W_a^{AB} is quite linear with increase in % AA.

Table 5.
Contact Angles of Aqueous Buffers on EAC series

Substrate	pH 2	pH 4	pH 5	pH 6.3	pH 7	pH 9	pH 10	pH 12
EAC 20.0%	76.9	77.9	77.0	77.5	76.9	70.7	70.1	69.0
EAC 15.5%	82.1	80.5	79.1	81.4	83.4	76.7	72.8	73.3
EAC 9.5%	87.3	87.0	87.6	86.0	87.3	77.5	73.7	73.2
EAC 6.5%	88.3	87.0	85.8	89.0	88.9	86.6	82.2	82.0
EAC 3.0%	93.5	95.1	93.2	98.4	94.9	92.5	86.7	88.1
EAC 1.8%	98.0	97.8	98.7	98.2	98.2	95.7	87.1	88.5

The contact angles are measured in degrees.

The W_a^{AB} values for water are close in all cases, i.e., comparing the two models and for the two liquid pairs. For the W_a^{AB} of EG, the same is true but there is somewhat more scatter in the values especially at the two lowest % AA levels. However, for the EG there is a good agreement between the vOGC and CC models within each of the polar pairs.

The picture in the case of W_a^{AB} for DMSO is less satisfactory. The CC model gives lower (or more negative) values than the vOGC model and, indeed for the EG / water pair, the CC model does not track the % AA content. The data obtained with the DMSO / water pair do track the W_a^{AB} for DMSO in both models. Here we note a repulsion in both models although the values are different. The repulsion increases as the % AA decreases. Since DMSO is quite basic ($\sqrt{\frac{\gamma^-}{\gamma^+}} = 8$, $\frac{b}{c} = -5.0$ compared to $\sqrt{\frac{\gamma^-}{\gamma^+}} = 5$ and $\frac{b}{c} = -1.3$ for EG), one would expect that this means that the surfaces lower in carboxyl group concentrations are showing more basic sites to the DMSO. It is possible that, as hydrogen bonding decreases, there are more free carboxyl oxygens. And, if the hydroxyl groups are oriented away from the surface, one might see increasing surface basicity with decreasing % AA content.

These data indicate that caution should be exercised if one is to determine acid / base parameters for a polymer surface relative to its interaction with a second phase, such as an adhesive. The second polar probe liquid (with water) should model the opposing phase as closely as possible, i.e., be as similar chemically as possible to the second phase.

3.4. Contact Angle Titrations

Aqueous buffers from pH 2 to 12 were used in contact angle measurements on the EAC series. The results are given in Table 5. Cosine of the contact angles, θ, are plotted versus pH in Figure 4.

These are typical pH titration curves and from their shapes we can deduce that the plaques of EAC with 1.8% and 3.0% AA have surfaces which are readily ionizable, i.e., they are "stronger" acids. As the % AA level increases in the EAC the slope at the end point decreases in steepness, indicating decreasing degree of ionizability of

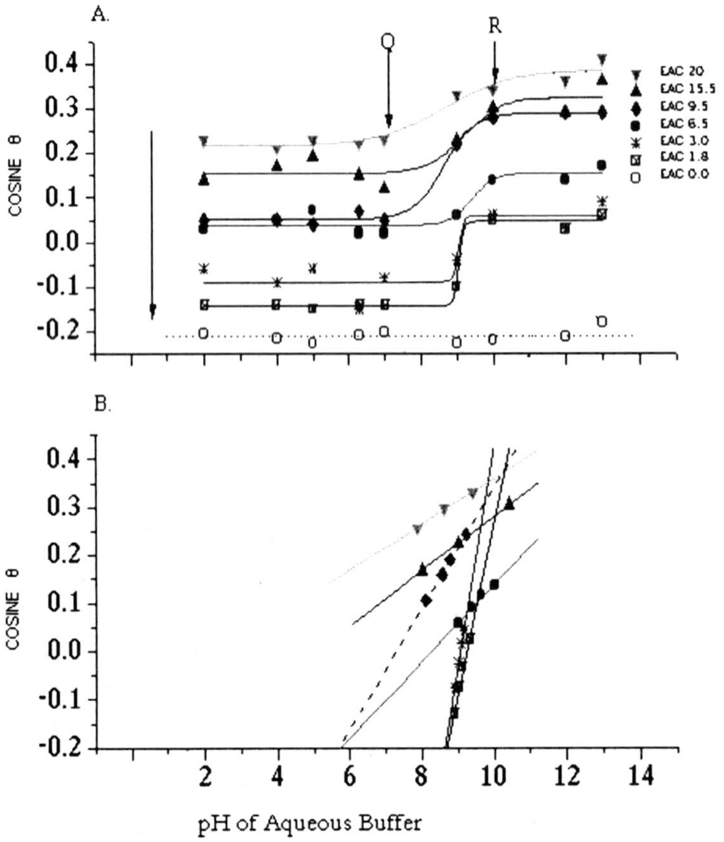

Figure 4. (A): Cosine θ pH versus buffer on the EAC series. (B): Enlargement of the Q-R region of the upper plots.

the carboxyl protons. This is due to increasing hydrogen bonding in the surface as carboxyl concentration increases.

Whitesides and his coworkers [7, 8], postulated that the cosine θ of an aqueous solution on an ionizable solid surface is related to the pH of the solution and the pK_a of the surface in a complicated manner for the surface carboxyl groups, i.e.,

$$\cos\theta = \cos\theta_{(pH1)} + \frac{2.3nkT}{\gamma_{LV}}(pH - pK_a) \tag{7}$$

where $\theta_{(pH1)}$ is the contact angle of the buffer probe with the lowest pH, n is the number of carboxylic acid groups per unit area of the surface, k is the Boltzmann constant and T is the temperature in degree Kelvin. Equation (7) is rearranged into the convenient linear form as:

$$\Delta\cos\theta = \cos\theta - \cos\theta_{(pH1)} = \frac{2.3nkT}{\gamma_{LV}}pH - c \tag{8}$$

Table 6.
The Relationship Between pKa and The Population Density of Carboxylic Acid / Carboxylate Ions on EAC Surface

Substrate	Δ cosine θ	Slope	pK_a	n (10^{13})
EAC 20.0%	0.178	0.05	8.8	4.0
EAD 15.5%	0.226	0.06	9.0	4.4
EAC 9.5%	0.245	0.13	9.0	9.7
EAC 6.5%	0.141	0.08	9.0	6.1
EAC 3.0%	0.146	0.46	9.0	0.4
EAC 1.8%	0.195	0.36	9.0	0.3
Acrylic acid in free solution			4.25	
Poly (acrylic acid), PAA			6.8	

$\Delta \cos\theta = \cos\theta_{(\text{given buffer})} - \cos\theta_{(\text{pH1})}$.
Slope $= \frac{2.3nkT}{\gamma_{LV}}$ (see Equation (8)).
n = number of carboxylic acid groups per cm^2 (see Equation (7)).

Table 7.
The W_a^{AB} of adhesion of Aqueous Buffer on the EAC Series

Substrate	pH 2	pH 4	pH 5	pH 6.3	pH 7	pH 9	pH 10	pH 12	pH 13
EAC 20.0%	32.4	31.2	32.4	31.7	32.5	40.1	40.7	42.0	45.5
EAC 15.5%	25.9	28.0	29.7	26.8	24.3	32.6	37.4	36.8	42.1
EAC 9.5%	19.3	19.7	19.0	21.0	19.4	31.6	36.3	37.0	37.1
EAC 6.5%	18.1	19.7	21.3	17.2	17.3	20.2	25.8	26.0	28.3
EAC 3.0%	11.4	9.4	11.8	5.3	9.7	12.7	20.2	18.3	22.1
EAC 1.8%	5.8	6.0	4.9	5.4	5.5	8.6	19.6	18.0	19.9

The W_a^{AB} term has the unit (mJ / m^2).

where the slope is $\frac{2.3nkT}{\gamma_{LV}}$ and the intercept, c, is $\frac{2.3nkT}{\gamma_{LV}} pK_a$. The value of n was evaluated from the slope and the pK_a for each EAC homolog was determined by differentiation. The results are listed in Table 6 and and the first derivative of $\cos\theta$ in Figure 5. Note that n increases up from EAC with 0% AA through 9.5% AA and then drops off indicating a decrease in available or ionizable carboxylic protons. The pK_a values for the EAC surfaces were different from that of acrylic acid in free solution which is 4.25 [9, 10], and it is also different from that of poly(acrylic acid) with pK_a 6.8. Our pK_a values were much higher ranging between 8.8 and 9.0, which is suggestive of basic surfaces for all of the EAC homologs.

The acid / base work of adhesion (W_a^{AB}) at the various pH levels was determined by establishing W_a^{TOTAL}, Equation (2), using the contact angles from Table 5, and assuming that the surface tension of the buffers was constant [10]. W_a^{LW} was assumed constant (see Table 2) and it was subtracted from the W_a^{TOTAL} to yield W_a^{AB}. W_a^{AB} is given in Table 7 for the EAC series and plotted in Figure 6.

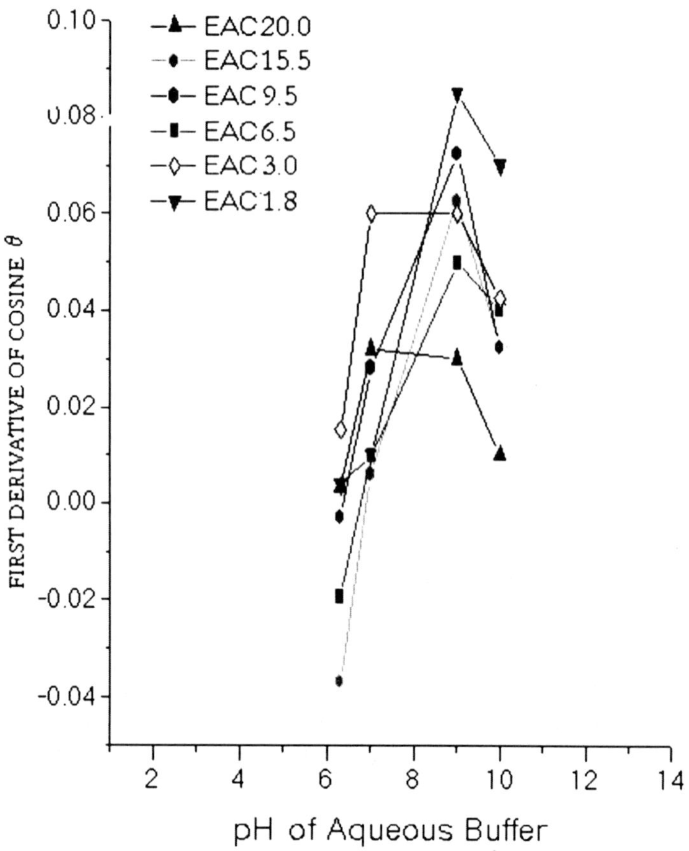

Figure 5. The first dervative of cosine θ versus pH of aqueous buffer.

As expected W_a^{AB} increases sigmoidally as pH increases and the basic buffers increase the extent of ionization of the carboxylic acid groups. Also, W_a^{AB} increases at all pH levels as the % AA increases. Note too that the spread in W_a^{AB} between 1.8% and 20% AA remains fairly constant as the pH increases.

4. CONCLUSIONS

4.1. Acid and Base Parameters

Probing the EAC series with EG/water gave γ_S^- (vOGC model) values that increased linearly with increase in % AA. The DMSO/water pair did likewise, giving γ_S^- values essentially equal to the EG/water pair results. The meaningful acidic parameter was $\sqrt{\gamma_S^+}$ and with the DMSO/water pair it went from negative to positive, tracking % AA increase. With EG/water neither $\sqrt{\gamma_S^+}$ nor γ_S^+ showed the expected rise with increase in % AA.

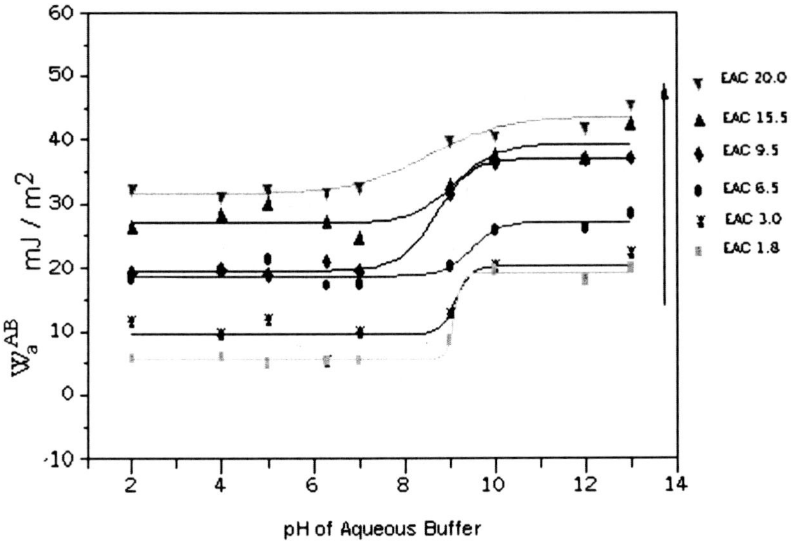

Figure 6. W_A^{AB} versus the pH of the aqueous buffers on the EAC series.

The acidic *y* parameter (CC model) tracked % AA very well when DMSO / water was used but not for EG / water pair. Basic *z* parameter was not tracked by DMSO / water but it was with EG / water, albeit with some scatter.

The choice of the second polar pair liquid is important and preliminarily tests with several liquids should be made.

4.2. W_{L-EAC}^{AB}

The evaluation of the acid / base work of adhesion of water on the surfaces of the EAC plaques, $W_{Water-EAC}^{AB}$, gave essentially the same values by both vOGC and CC models. Also, acid / base parameters resulting from use of the polar probe pairs, EG / water and DMSO / water, resulted in comparable $W_{Water-EAC}^{AB}$ values. Further, these values increased linearly with carboxylic acid content. The same was true for EG but here the correspondence was not quite as good.

For $W_{DMSO-EAC}^{AB}$ the values tracked the acid content of the EAC best when data from the probe pair, DMSO / water were used. However, the vOGC model gave higher values. Both models gave values which became increasingly negative with decreasing carboxylic acid content in the solid polymer. This indicated an increasing basicity available for repulsion of the DMSO as the carboxyl content of the polymer decreased. The probe pair EG / water was not appropriate for assessing the DMSO-EAC interaction.

4.3. pH Titration

As expected, the contact angles of water, buffered from pH 2 to 12, decreased abruptly as the pH went up, ionizing the carboxylic acid groups on the surface.

The lower the carboxyl concentration, the sharper the decrease because of the lesser relative hydrogen bonding. The pK_a was essentially constant at 9.0 for all the EAC homologs irrespective of the % AA content.

Acknowledgement

We thank Professor Ned Heindel of Lehigh University for his financial support of JAC.

REFERENCES

1. (a) C. J. van Oss, M. K. Chaudhury and R. J. Good, *Chem. Rev.* **88**, 927 (1988); (b) R. J. Good, *J. Adhesion Sci. Technol.* **6**, 1269 (1992); (c) R. J. Good and C. J. van Oss, in: *Modern Approaches to Wettability – Theory and Applications*, M. E. Schrader and G. I. Loeb (Eds) Ch. 1. Plenum, New York (1992).
2. (a) W. V. Chang and F. Chen, *Polym. Mater. Sci. Eng.* **61**, 607 (1989); (b) F. Chen and W. V. Chang, *Langmuir* **7**, 2401 (1991).
3. A. Gent and J. Schultz, *J. Adhesion* **3**, 281 (1972).
4. R. A. Pearson and T. B. Lloyd, *Mater. Res. Soc. Symp. Proc.* **445**, 247 (1997).
5. T. B. Lloyd, K. E. Ferretti and J. LaGow, *J. Appl. Polym. Sci.* **58**, 291 (1995).
6. P. J. Flory, *J. Chem. Phys.* **17**, 223 (1949).
7. S. R. Holmes-Farley, R. H. Reamey, T. J. McCarthy, J. Deutch and G. M. Whitesides, *Langmuir* **1**, 725 (1985).
8. C. D. Bain and G. M. Whitesides, *Langmuir* **5**, 1370 (1989).
9. J. McMurry, *Organic Chemistry*, Brooks / Cole Publishing, New York, 2nd edition, p. 721 (1988).
10. K. J. Hüttinger, S. Höhmann-Wien and G. Krekel, *J. Adhesion Sci. Technol.* **6**, 317 (1992).

Acid-Base Interactions, Vol. 2, pp. 363–383
Ed. K.L. Mittal

Comparison of acid-base characterization techniques on lignocellulosic surfaces

D. J. GARDNER,* S. Q. SHI and W. T. TZE
Advanced Engineered Wood Composites Center, University of Maine, Orono, ME 04469

Abstract—This study compares the acid-base character of lignocellulosic surfaces using the Good-van Oss-Chaudhury (GvOC), and Chang models for contact angle analysis (CAA) data, and the Papirer, and Schultz methods for inverse gas chromatography (IGC) data. For the CAA data, it is shown that using all probe liquids to first determine the dispersive surface energy component is the best way for calculating the surface energy components from both the GvOC and Chang models. The Chang model is more robust than the GvOC model because it is able to measure the attractive-repulsive character of acid-base interactions which is important for processes involving lignocellulosic surfaces. The GvOC model can also reflect both the repulsive and attractive effects on the surface if the square root terms of the acid and base components are directly used to calculate the polar portion of the surface energy (GvOC*). The Chang model and GVOC* calculations provide nearly identical results on the surface energy components determined for lignocellulosic fibers. For the IGC data, a good correlation exists between the Schultz, and Papirer methods to determine the acid-base parameters.

Keywords: Acid-base; contact angle analysis; inverse gas chromatography; Chang model; Good-van Oss-Chaudhury model; Schultz method; Papirer method.

1. INTRODUCTION

An understanding of lignocellulosic surface chemistry is important since wood and paper are adhesive bonded or coated in a majority of applications. Improvements in adhesive bonding technology and lignocellulosic composite material development require a thorough knowledge of surface chemical characteristics. The successful adhesive bonding of wood and the manufacture of paper are dependent on strong secondary valence forces, i.e., acid-base interactions. The importance of acid-base character in paper manufacture is easily demonstrated because of the dependence of fiber-fiber interactions on hydrogen bonding, which is a subset of acid-base interactions. Poor adhesive bond formation in wood can often be traced to a surface that exhibits both low surface energy and acid-base character.

The lignocellulosic or wood surface can be categorized as a heterogeneous polymer composite composed of cellulose, hemicellulose, lignin, and extractives. Cellulose makes up 50% of wood weight. Hemicellulose makes up 25 to 35% of the weight of

* To whom correspondence should be addressed. E-mail: doug_gardner@umenfa.maine.edu
Present address: 5793 AEWC Building, University of Maine, Orono, ME 04469.

wood and lignin makes up 16 to 33% of the weight of wood [1]. The amounts of hemicellulose and lignin in wood differ depending on whether it comes from an angiosperm or gymnosperm source. Extractives in wood also vary depending on the source, and can comprise from 1 to 10% of wood weight [2]. Cellulose and hemicellulose are polymers comprised of monomeric carbohydrates including glucose, galactose, mannose, xylose, *etc.* Lignin is a polymeric material based on phenylpropanoid (C-9) units. The major difference in the chemistry of lignin is whether the source of lignin is based entirely on guaiacyl monomers (gymnosperm or softwood lignin) or a 50:50 mixture of guaiacyl and syringyl monomers (angiosperm or hardwood lignin). The extractives in lignocellulosics are low molecular weight byproducts of physiological processes and include a vast array of chemical compounds such as fatty acids, waxes, and polynuclear aromatic compounds. The extractives tend to dominate the surface chemistry of wood [3], but are typically removed during pulping processes. Thermomechanical pulp fibers have a lignin-rich surface while chemical pulps like bleached kraft pulp fiber surfaces are dominated by cellulose and hemicellulose. At ambient conditions most lignocellulosic surfaces also have a monomolecular layer of water associated with them.

The primary techniques used to characterize the acid-base character of surfaces include contact angle analysis and inverse gas chromatography (IGC). There are an array of elemental shapes and sizes encountered in lignocellulosic materials including timber, veneer, flakes, fibers, and particles. It is now possible to make contact angle measurements on many types of lignocellulosic elements by choosing the proper contact angle technique. Contact angle measurements can be made using image analysis of static or dynamic liquid droplets on plates (lumber), Wilhelmy (force) measurements on thin plates (veneers) and individual fibers, as well as capillary rise (wicking) measurements using particles or fibers packed in capillary columns. For contact angle analysis, two models have been established to determine the acid-base character of a surface: Good-van Oss-Chaudhury (GvOC) theory [4], and the Chang model [5, 6]. However, differences have been shown in the acid-base properties of the wood surfaces between the two models [7]. In addition, different calculation methods may also give different results for the surface energy components [8].

In IGC analysis, the retention time of acidic, basic and amphoteric probes relative to the retention time of nonpolar (alkane) probes in the gaseous state are used to calculate the enthalpy of adsorption. By knowing the enthalpy of adsorption, and the acceptor (AN) and donor (DN) members of the probes, the acidic K_A and basic K_B character of the surface can be determined [9]. IGC measurements on lignocellulosic materials require the use of either particles or fibers packed into columns. In the case of fibers and particles, it is possible to make direct comparisons between contact angle and IGC analyses. It should be noted that a primary difference between contact angle and IGC analyses is the environmental conditions used to conduct the experiments. Contact angle measurements are usually made under ambient conditions, i.e., room temperature and humidity, while IGC measurements are made under dry conditions and temperatures ranging from 30 to 80°C. Moisture contents of lignocellulosic particles used in contact angle measurements are typically 6 to 8% while particles used in IGC have a moisture content of 0 to 2%. It has been demonstrated that moisture content differences in lignocellulosic particles will inherently result in different surface energies.

It is the intent of the authors to provide a clear understanding of the methodologies used in performing acid-base calculations for lignocellulosic materials. Reasons for this are as follows: 1) There appears to be a lack of concensus among researchers in the surface science field as to which model or equation is most appropriate for determining the acid-base components of surface energy. 2) To discuss any problems associated with the applications of the acid-base calculations for lignocellulosic materials along with recommendations for future study. 3) Lastly, clear instructions for performing acid-base calculations are desirable because it improves the potential for international standards development, and routine use of these analyses in lignocellulosic materials science research and development.

In this paper, we will discuss the results of research on determining the acid-base character of a variety of lignocellulosic surfaces using the Good-van Oss-Chaudhury (GvOC) and Chang methods for contact angle data, and the Schultz, and Papirer methods for IGC data. A detailed analysis of the calculation methods used to determine acid-base character when applying the GvOC and Chang methods will be presented. This paper also compares the results and correlates the acid-base data between contact angle analysis and IGC.

2. MATERIALS AND METHODS

2.1. Materials

Six wood species, thermomechanical pulp (TMP) fibers, bleached kraft pulp fibers, and papers treated with different sizing agents were used in the experiments. The wood species used in the experiment were heartwood of ash (*Fraxinus americana* L.), cherry (*Prunus serotina* Ehrh.), hard maple (*Acer saccharum* Marsh.), red oak (*Quercus rubra* L.), white oak (*Quercus* spp.), and walnut (*Juglans nigra* L.). Both wood veneer and particles were used. The veneers were commercially dried on a screen dryer with gas-fired jets, after which they were conditioned to eight percent moisture content. The heartwood portion of the veneers was ground to 60 mesh using a Wiley mill. The particles had a moisture content of approximately seven to nine percent, based on the oven-dry weight of the wood. The TMP fibers, composed of 75% aspen and 25% mixed hardwood species, were obtained from Georgia Pacific Company in Phillips, Wisconsin. Bleached kraft pulp made of hardwood was obtained from Mead Paper Co. in Escanaba, Michigan. The sized papers used in this study were softwood pulp; rosin size (1% fortified rosin soap based on dry pulp was added); AKD size (0.3% AKD based on dry pulp was added as a 0.1% solution to the paper slurry); AKD-alum size (after AKD sizing, 0.5% alum was then added after 5 min.). The detailed information about the sized papers is described in Reference [10].

2.2. Contact Angle Analysis

2.2.1. Contact Angle Data. The contact angles on the six wood veneers were obtained in reference [11]. The contact angles on the wood particles used in this study were determined by column wicking measurements [7, 12]. The Washburn equation is usually used to determine the contact angles by wicking measurements. Lignocellulosics are swelling materials. When polar liquids are used in the wicking measurements, they will interact with the swelling materials and heat of adsorption will occur. The energy loss due to the heat of adsorption will delay the capillary rise. To eliminate the energy loss effect due to the heat of adsorption, the capillary rise model established by Shi and Gardner [13] was used to determine the contact angles on the wood particles as:

$$h^2 = \frac{r_s^2 \gamma_L \cos\theta}{2\eta r} t - \frac{C\delta_V}{4\eta\pi} \left(\frac{r_s}{r}\right)^2 t \tag{1}$$

h is the height to which liquid has risen in time t,
r is the effective interstitial pore radius between the packed particles,
γ_L is the surface tension of the liquid,
θ is the contact angle,
η is the viscosity of the liquid,
δ_V is the volumetric swelling of particles in a certain liquid. The swelling values used in this calculation were obtained by extrapolation based on the published data in [14],
C is the energy loss coefficient (J/m),
r_s is the swelled average capillary radius which can be obtained from the following equation,

$$r_s = \sqrt{\frac{\pi R^2 \rho_m - (1+\delta_v) G_m}{\pi \rho_m R^2 - G_m}} \cdot r \tag{2}$$

in which R is the radius of the column, ρ_m is the material density, and G_m is the unit mass in the column. In Equation (1), the volumetric swelling (δ_V) is a known parameter or can be measured. The original average capillary radius (r) can be determined using hexane for which $\delta_V = 0$, $r_s = r$, and $\cos\theta = 1$. The energy loss coefficient (C) can be determined using methanol or other polar liquids with a low surface tension [13]. The capillary rise model shown in Equation (1) was applied to the contact angle determination of the other polar liquids: ethylene glycol, formamide, and water. Contact angle of a non-polar liquid, α-bromonaphthalene, was also determined using the Washburn equation. All these liquids were of either ACS reagent or HPLC grade. The values of the surface tension components and parameters of all the probe liquids are shown in Table 1.

The contact angles on TMP fibers using α-bromonaphthalene, ethylene glycol, formamide, glycerol, and water were obtained from column wicking measurements. The details of the determination procedure are given by Shi *et al.* [12]. The contact angles on Kraft fibers in methylene iodide, ethylene glycol, formamide, and water were obtained from dynamic contact angle (DCA) analysis. Single fibers were immersed into the probe liquid contained in a beaker at a speed of 12 μm/sec, with a probe depth of one mm. The force exerted on the fiber was automatically recorded from a balance. The details of the determination procedure are given by Liu *et al.* [15]. The contact angle data for the kraft fibers were obtained by Tze [16].

The contact angles on the sized papers in methylene iodide, ethylene glycol, formamide, and water were also obtained from DCA analysis. Detailed information on the sized papers is given by Huang *et al.* [10].

Table 2 summarizes the contact angle data on all the lignocellulosic materials used in this paper.

Table 1.
The property values of the probe liquids used in the experiments [6, 7]

Liquids	γ_L	γ^{LW}	γ^{AB}	$\gamma+$	$\gamma-$	η	P^d	P^a	P^b
		(mJ/m^2)			(mN·s/m^2)			((mJ)^{½}/m)	
Diiodomethane	50.8	50.8	0	0	0	2.76	11.6	-4.11	-4.12
α-bromonaphthalene	44.4	44.4	0	0	0	4.89	10.5	-2.67	-3.82
Ethylene glycol	48.0	29.0	19	1.92	47.0	19.9	7.5	3.69	-5.44
Formamide	58.0	39.0	19	2.28	39.6	4.55	7.3	6.92	-4.64
Glycerol	64.0	34.0	30	3.92	57.4	1490	8.0	3.40	-9.30
Water	72.8	21.8	51	25.5	25.5	1.00	6.6	6.88	-7.40

Table 2.
Contact angles on the lignocellulosic materials used in this study [7, 10–12, 15]

Materials	Probe liquids (Average contact angles and standard deviation)					
	BN	EG	FM	GL	WT	MI
Ash (veneer)	9.8 (2.4)	17.5(7.7)	0(0)	59.8(8.9)	23.7(10.0)	
Cherry (veneer)	0(0)	14.0(5.2)	0(0)	50.3(10.7)	44.4(5.2)	
Maple (veneer)	0(0)	14.3(4.5)	0(0)	48.4(2.1)	34.9(7.1)	
Red oak (veneer)	0(0)	16.2(4.9)	28.5(4.7)	56.0(5.0)	40.1(2.7)	
White oak (veneer)	20.4(6.1)	55.5(5.9)	33.1(9.1)	58.5(3.0)	59.6(9.4)	
Walnut (veneer)	14.3(5.7)	36.1(5.9)	32.9(10.5)	43.4(11.0)	26.3(12.9)	
Ash (particles)	47.7(6.8)	47.0(6.1)	62.7(0.7)		73.3(0.5)	
Cherry (particles)	0(0)	23.1(19.6)	54.4(1.6)		65.4(1.8)	
Maple (particles)	6.6(14.8)	18.5(14.3)	58.7(1.6)		71.1(0.6)	
Red oak (particles)	6.6(14.8)	38.4(4.0)	52.4(1.0)		69.0(0.9)	
White oak (particles)	10.0(22.4)	16.2(20.3)	54.4(0.9)		68.1(1.0)	
Walnut (particles)	0(0)	0.9(1.8)	50.5(0.9)		68.3(1.4)	
Kraft fiber		43.9(5.6)	33.8(5.2)		30.6(3.4)	54.9(2.3)
TMP fiber	3.8(5.5)	33.5(2.9)	56.5(2.9)	66.5(3.6)	81.6(1.0)	
Softwood		24.7	20.4		31.1	49.5
AKD-sized paper		56.8	70.6		136	64.5
AKD-alum sized paper		76.9	94.6		137	75.3
Rosin sized paper		95.3	121		141	51.6

BN: α-bromonaphthalene; EG: ethylene glycol; FM: formamide; WT: water; GL: glycerol; MI: methylene iodide. The standard deviation of the contact angles for the softwood and sized papers was not available.

2.2.2. Models Used to Determine the Acid-base Components. Both the Good–van Oss–Chaudhury (GvOC) model (Lifshitz–van der Waals approach) and Chang model were evaluated using the contact angle data from the different lignocellulosic materials. Both models are based on the Geometric mean equation which can be expressed as:

$$W_a = (1 + \cos\theta)\gamma_L = 2(\gamma_L{}^d\gamma_S{}^d)^{1/2} + 2(\gamma_L{}^p\gamma_S{}^p)^{1/2} \quad (3)$$

in which γ^d is the non-polar portion of the surface energy, γ^p is the polar portion of the surface energy, γ_L and γ_S are the surface tensions of liquid and solid respectively, and W_a is the work of adhesion.

van Oss *et al.* [4] proposed a generalization of the Fowkes surface tension component approach and claimed that for any liquid or solid, the total surface tension could be uniquely characterized by two different surface tension components: the Lifshitz–van der Waals (LW) component γ^{LW} and the acid-base component γ^{AB}:

$$\gamma = \gamma^{LW} + \gamma^{AB} \quad (4)$$

in which

$$\gamma^d \approx \gamma^{LW} \quad \text{and} \quad \gamma^p \approx \gamma^{AB} \quad (5)$$

The acid-base component can be separated into Lewis acid (+) and Lewis base (-) terms, i.e.

$$\gamma^{AB} = 2(\gamma^+)^{1/2} \cdot (\gamma^-)^{1/2} \quad (6)$$

Therefore, the following relationship can be established:

$$(1 + \cos\theta)\gamma_L = 2[(\gamma_L{}^{LW}\gamma_S{}^{LW})^{1/2} + (\gamma_L{}^+\gamma_S{}^-)^{1/2} + (\gamma_L{}^-\gamma_S{}^+)^{1/2}] \quad (7)$$

After the non-polar portion (dispersive component) was determined by the geometric mean method, the two unknowns $\gamma_S{}^+$ and $\gamma_S{}^-$ can be determined by multiple regression.

The Chang model is an empirical acid-base model which takes into account both the repulsive and attractive interactions between acidic and basic components [6]. The following equation was proposed:

$$W = (1 + \cos\theta)\gamma_L = P_L^d P_S^d - P_L^a P_S^b - P_L^b P_S^a \quad (8)$$

in which $P_L{}^d$ and $P_S{}^d$ are the dispersive parameters of the liquid and the solid respectively, which can be obtained by the following equation:

$$P_L^d = \sqrt{2\gamma_L^d} \quad \text{and} \quad P_S^d = \sqrt{2\gamma_S^d} \quad (9)$$

$P_L{}^a$ and $P_L{}^b$ are the principal acid-base values of the liquid which were determined by the above empirical model using the non-linear least squares method based on the published data [5, 17]. The values of $P_L{}^d$, $P_L{}^a$ and $P_L{}^b$ for different liquids are listed in Table 1.

P_S^a and P_S^b are the principal acid-base parameters of the solid. The surface is called acidic if both the parameters are positive, basic if both are negative, and amphoteric if they are opposite. The acid-base portion of the total surface energy can be obtained by the following equation:

$$\gamma_S^{AB} = -P_S^a P_S^b \tag{10}$$

2.2.3. Determination of the Acid-base Parameters. According to the GvOC and Chang models described in Equations (7) and (8) respectively, the following mathematical approaches can be used to obtained the surface energy components.

1) Multiple regression:

The three unknowns in both GvOC and Chang models are determined using multiple regression directly.

2) One-liquid method:

The dispersive component is first determined by one non-polar liquid, such as α-bromonaphthalene or diiodomethane, using the Geometric mean equation which is:

$$W_a = (1+\cos\theta)\gamma_L = (2\gamma_L^d \gamma_S^d)^{1/2} \tag{11}$$

As γ_S^d is known, there are only two unknowns, acid-base components (parameters), in GvOC and Chang models which are determined using the multiple regression method.

3) All-liquid method:

This is the same as the one-liquid method except that the dispersive energy is determined by all the probe liquids instead of only one non-polar liquid. Equation (3) is used to determine the γ_S^d and γ_S^p using multiple regression. Once the dispersive component is known, the acid-base components (parameters) in both GvOC and Chang models can be determined using multiple regression.

2.3. Inverse Gas Chromatography

Inverse gas chromatography (IGC) is an extension of conventional gas chromatography (GC), in which a non-volatile material to be investigated is immobilized within a GC column. The stationary phase is then characterized by monitoring the passage of volatile probe molecules of known properties as they are carried through the column via an inert gas.

In this experiment, chromatography measurements at infinite dilution were carried out at three different temperatures, 30°C, 45°C, and 60°C. Helium was used as the inert carrier gas. The probes, along with their properties used in the IGC experiment, are shown in Table 3.

Table 3.
Properties of the probes used in the inverse gas chromatography experiment [12]

Probes	a (Å^2)	γ_L^d (mJ/m^2)	DN (kcal/mol)	AN (arbitrary unit)	AN* (kcal/mol)	Vapor Pressure (kPa)			Specific Characteristics
						30°C	45°C	60°C	
n-hexane	51.5	19.4	-	-	-	23.4	42.2	72.1	
n-heptane	57	20.3	-	-	-	7.5	14.8	27.3	Neutral
n-octane	63	21.3	-	-	-	2.3	4.8	9.6	
n-nonane	69	22.7	-	-	-	0.8	1.8	3.8	
Acetone	42.5	16.5	17.0	12.5	2.5	36.2	64.9	110.3	Amphoteric
EA	48	16.5	17.1	9.3	1.5	15.5	29.9	54.3	
DEE	47	15.0	19.2	3.9	1.4	85.7	147.3	240.9	Basic
THF	45	22.5	20.1	8.0	0.5	26.9	48.2	82.1	
Benzene	46	26.7	0.1	8.2	0.17	13.5	25.8	46.8	
CH_2Cl_2	31.5	27.6	-	20.4	3.9	70.2	123.9	207.8	Acidic
CH_3NO_2	37.8	26.2	2.7	20.5	4.3	26.7	51.4	93.1	
$CHCl_3$	44.0	25.9	-	25.1	5.4	30.7	55.6	95.5	

EA: ethyl acetate; DEE: diethyl ether; THF: tetrahydrofuran.

From the retention time measured by IGC, the net retention volume (the volume of carrier gas required to elute a zone of solute vapor), per gram of adsorbent, can be determined by the following equation [9].

$$V_g = \frac{273.15}{T_C} \frac{1}{W} Q(t_r - t_i) \tag{12}$$

T_c is the column temperature,
W is the amount in grams of adsorbent packed into the column,
Q is the corrected flow rate of the helium gas,
t_r and t_i are the retention times of probe and inert gas (propane), respectively.

2.3.1. Determination of the Acid-base Parameters. The specific interaction I_{sp} represent the contribution of forces other than dispersive, which includes hydrogen, polar, and acid-base interactions. The specific interaction can be calculated by the following Equation [9]:

$$I_{sp} = \Delta G_A^{sp} = RT \ln \frac{V_g}{V_g^*} \quad (13)$$

V_g* and V_g are the net specific retention volumes of the alkane and the polar probe, respectively. ΔG_A^{sp} is the free energy of adsorption.

Two methods were used to determine the specific interaction, Papirer's method [18] and Schultz's method [19]. Both methods used similar $RT \ln V_g$ values for the ordinate but different abscissa parameters. By comparing the ordinate difference between the representative point of a given probe and the alkane reference line, one obtains an estimation of the level of the specific interactions with the chromatography support, and hence the acid-base character can be evaluated. The higher the difference between a basic polar probe representative point and the corresponding point on the alkane line, the higher the acidic character of the solid surface and vice-versa. In Schultz's method, I_{sp} is obtained through a plot of $RT \ln V_g$ versus $[a(\gamma_L^d)^{1/2}]$ in which γ_L^d is the surface tension and a is the surface area of the probe molecule. The acid-base surface properties can be characterized by examining the respective position of a representive point above the alkane reference line. In Papirer's method, I_{sp} is obtained by the plot of $RT \ln V_g$ versus the logarithm of the saturation vapor pressure (log P_o). The free energy of adsorption (ΔG_A) can be determined according to the following thermodynamic function:

$$\frac{\Delta G_A}{T} = \frac{\Delta H_A}{T} - \Delta S_A \quad (14)$$

ΔH_A is the enthalpy of adsorption, and ΔS_A is the entropy of adsorption. By plotting $\Delta G_A^{sp}/T$ versus $1/T$ should yield a straight line with a slope of ΔH_a^{sp} [9].

According to Gutmann's approach [20], the acid-base probes are characterized by their donor-acceptor numbers (see Table 3). The donor number (DN) defines the basicity or electron donor ability and is estimated by the molar enthalpy of the reaction of the donor with the acidic reference $SbCl_5$. It is expressed in kcal/mole. The acceptor number (AN) gives the acidity or electron acceptor ability and corresponds to NMR chemical shift of ^{31}P after reaction of triethyphosphine $(C_2H_5)_3PO$ with the acceptor [21] and has an arbitrary unit. An acid is able to attract electrons while a base releases electrons. Because DN and AN are expressed in different units, Riddle and Fowkes [22] redefined a new AN* on the same scale and with the same units as DN, taking into consideration the amphoteric character of molecules.

By using this acid-base concept, the following relation was proposed [23]:

$$-\Delta H_A^{sp} = K_A DN + K_B AN^* \quad (15)$$

in which K_A and K_B, respectively, correspond to *AN* and *DN** of molecules; ΔH_A^{SP} is the enthalpy of adsorption corresponding to the specific interactions which can be obtained from the difference of ordinates between the n-alkane line and the polar probes.

Equation (15) can be rewritten as:

$$\frac{-\Delta H_A^{sp}}{AN^*} = K_A \frac{DN}{AN^*} + K_B \tag{16}$$

K_A and K_B can be determined from the slope and the intercept at the origin of the straight lines.

3. RESULTS AND DISCUSSION

3.1. Evaluation of the Calculation Methods for both GvOC and Chang Models

Figures 1 and 2 show the effect of the three different calculation methods: multiple regression, one-liquid method, and all-liquid method, on the surface energy values of the red oak wood particles for both the GvOC and Chang models. For the GvOC model shown in Figure 1, the results from the multiple regression method are similar to that from the one-liquid method. However, for the Chang model shown in Figure 2, the total surface energy and dispersive energy components determined by the multiple regression method are lower than that from other two methods. Some researchers have recommended determination of dispersive component prior to the determination of acid-base parameters for both the GvOC model [24] and the Chang model [6]. Table 4 shows the dispersive component values of relatively higher energy wood surfaces (contact angle is zero), determined by both one-liquid and all-liquid methods. Theo retically, the one-liquid method and all-liquid method should give the same results. However, it is seen from Table 4 that the one-liquid method gives the same dispersive component as that of a non-polar liquid for all these surfaces. Therefore, the one-liquid method may not be reliable for the dispersive component determination, especially for the high energy wood surface.

From the above analysis for the different calculation methods, it appears best to determine the dispersive component first using all the probe liquids followed by determination of the acid-base parameters.

Table 4.
Comparison of dispersive component (mJ/m^2) determined by one-liquid and all-liquid methods

Material	GvOC model		Chang model	
	All-liquid	One-liquid	All-liquid	One-liquid
Maple (veneer)	38.3	44.4	34.4	44.4
Red oak (veneer)	37.6	44.4	33.3	44.4
Cherry (particles)	40.2	44.4	40.2	44.4
Walnut (particles)	42.2	44.4	42.2	44.4

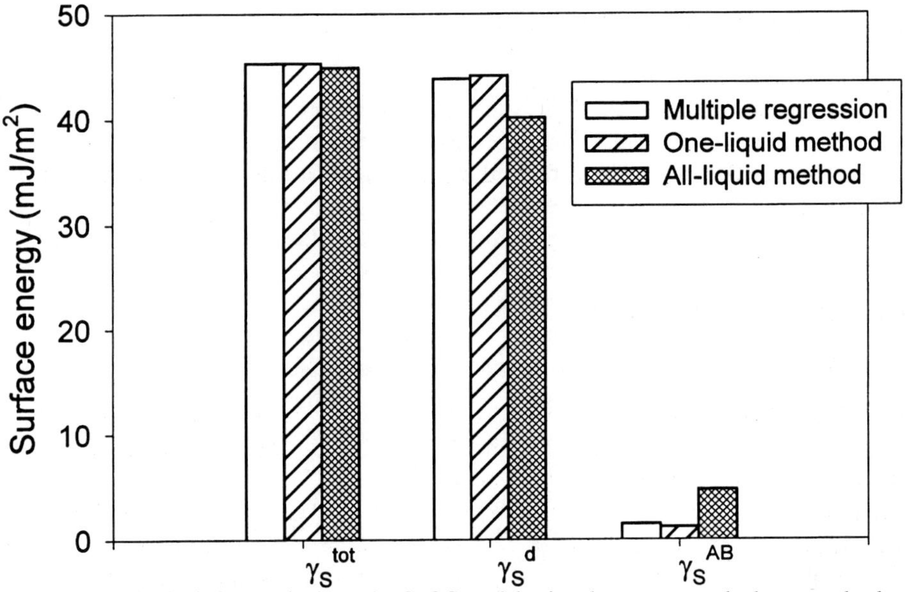

Figure 1. Effect of calculation methods on the GvOC model using the contact angle data on red oak particles.

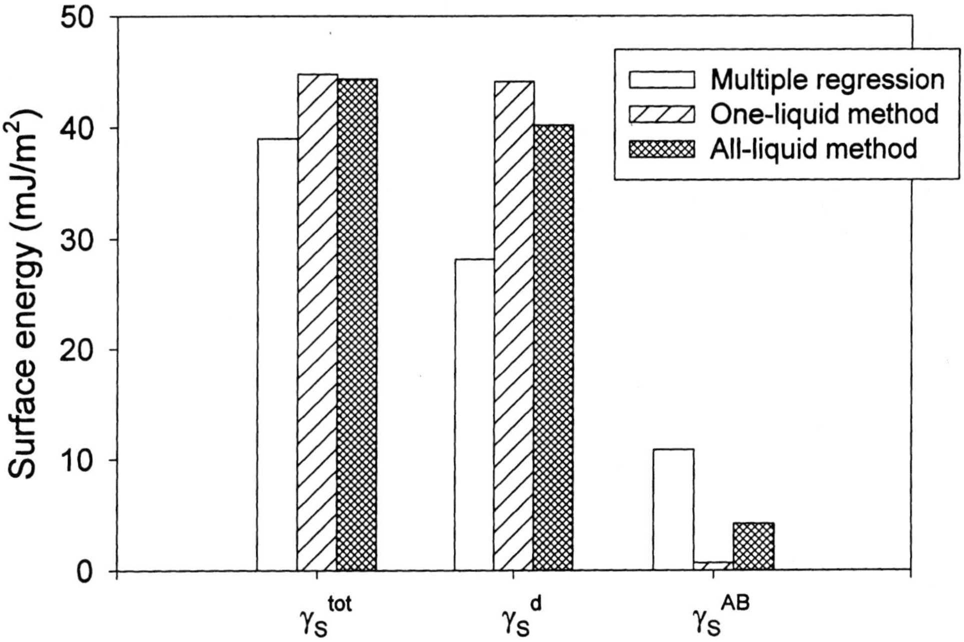

Figure 2. Effect of calculation methods on the Chang model using the contact angle data on red oak particles.

3.2. Evaluation of the Surface Energy Models using Veneer and Particle Surfaces

The surface energies of wood particles were compared to those of wood veneer based on the values calculated from either the GvOC model or the Chang model (Table 5). From the comparison, the total surface energies of wood particles are lower than those of veneers regardless of which surface energy model was used in the calculation. This observation is true for all the species examined except for white oak which exhibits similar total surface energy values for both veneer and particles. While the outlying nature of white oak is unclear, the discrepancy in surface energy between the veneer and particles can be generally attributed to the sample preparation method. Particle preparation, such as grinding, is likely to induce thermal changes into the sample. When the temperature of the particle surface rises to 60°C, the lignin component of the wood undergoes a glass transition [25]. The glass transition of lignin increases the free volume; and therefore, increases the diffusion in the wood. The increased diffusion possibly allows the migration and deposition of the low-energy extractive molecules to the wood surface [25]. Thus the total surface energy of wood particles is decreased. The overall surface changes in the wood can be successfully expressed using either the GvOC or the Chang model which provide total surface energies within 1 mJ/m^2 of each other.

Both the GvOC and Chang models are potentially useful for indicating the physical (apolar) and chemical (polar) aspects of the wood polymer surface. A typical separation of the surface energy components of the veneer and particles is represented by red oak in Figure 3 for the GvOC model, and Figure 4 for the Chang model. For both models, a low acid-base component of the surface energy for the particles was responsible for the low value of total surface energy. Although the surface energy of extractives as pure compounds is not well known, it can be observed from this study that the extractives dominated the particle surface, and possessed a low magnitude of acid-base character. The acid-base component of surface energy for both veneer and particles resulted from the electron donating sites which dominated the lignocellulose surface ($\gamma^- > \gamma^+$ as in Figure 3). While the GvOC model shows that basicity dominates the lignocellulose surfaces, the Chang model reveals that lignocellulose surfaces are more amphoteric (with one principal value positive, the other negative as in Figure 4).

Table 5.
Total surface energy (mJ/m^2) of different wood species determined by the GvOC and Chang models

Material geometry	Model	Maple	Red oak	White oak	Ash	Walnut	Cherry
Veneer	GvOC	53.73	50.12	43.68	50.86	50.38	53.01
Veneer	Chang	53.41	50.48	43.19	51.85	51.89	52.02
Particles	GvOC	45.54	44.87	46.55	35.26	47.98	46.28
Particles	Chang	45.24	44.33	46.09	34.74	47.26	46.06

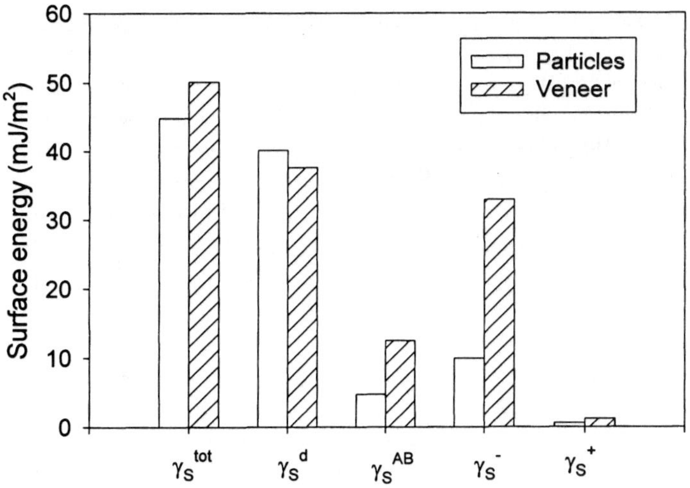

Figure 3. Comparison of wood surface energy components determined by the GvOC model.

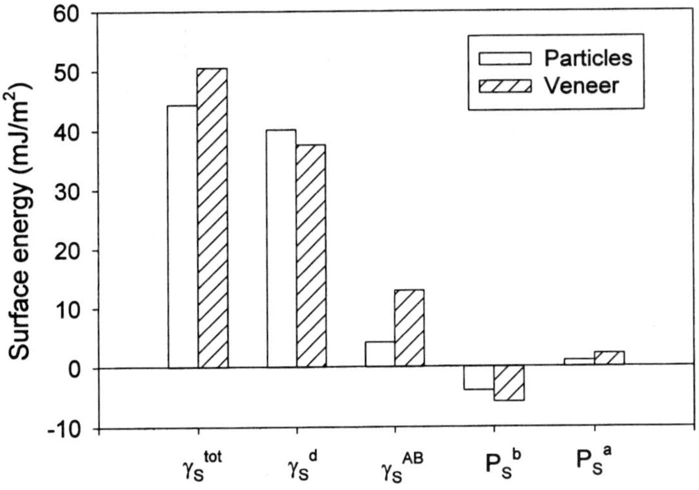

Figure 4. Comparison of wood surface energy components determined by the Chang model.

3.3. Evaluation of Surface Energy Models using TMP and Bleached Kraft Pulp Fiber Surfaces

The surface properties of TMP are compared to that of bleached kraft pulp in Table 6. The TMP had a higher dispersive component, a lower acid-base component, and a slightly lower resultant total surface energy compared to the bleached kraft pulp. The difference in the surface energy can be related to the lignin which is present in the TMP, a high-yield pulp. This postulation is supported by the surface energy values of hardwood lignin (γ^{tot}= 45.9 mJ/m^2, γ^d= 43.5 mJ/m^2, and γ^{AB} = 2.4 mJ/m^2 [26]) which are very close to those of TMP used in this study. Although the previous work used glycerol and α-bromonaphthalene as probe liquids and the geometric-mean approach as the calculation model, the close proximity of values with the present study does not seem to be a coincidence. While the surface lignin is postulated to influence the surface energy of TMP fibers, the cellulose must have certain effects on the surface properties of the bleached kraft pulp, a pulp that is high in cellulose content. This postulation is again strengthened by Lee and Luner [26] who generally found cellulose to have a lower dispersive component, a much higher polar component, and a higher resultant total surface energy than lignin, all of which agree well with the observations from this study.

Both the GvOC and Chang models make physical sense when applied to the pulp fiber surface. From the GvOC model, the bleached kraft fibers had both electron donating and accepting parameters that are higher than those of TMP (Table 6). This characteristic agrees with the molecular structure of cellulose, a major compound in bleached kraft fibers. In the cellulose molecule, the hydrogen atoms of the hydroxyl groups act as electron acceptors (Lewis acid) and the oxygen atoms act as electron donors (Lewis base) [27]. Such a bipolar nature verifies the Chang model which reflects an amphoteric surface in the pulp fibers. As the hydrogen bonding sites in TMP fibers are blocked by lignin, these fibers possess less electron donating and accepting ability than that of the bleached kraft fibers [28].

Table 6.
Comparison of the surface properties of TMP and kraft fibers determined by different methods

Surface energy	GvOC model		Chang model		IGC	
	TMP	Kraft	TMP	Kraft	TMP	Kraft
γ_s^{tot}	45.2	48.4	44.4	50.3		
γ_s^{d}	43.4	25.5	43.4	25.5	43.3	43.7
γ_s^{ab}	1.81	22.9	1.07	24.8		
γ_s^{+}, K_A	0.36	2.58			-0.001	0.09
γ_s^{-}, K_B	2.30	50.7			-0.13	0.35
P_s^a			0.33	4.14		
P_s^b			-3.18	-5.99		

Note: K_A and K_B have no unit and they were obtained from Schultz's method [18]; P_s^a and P_s^b are in (mJ/m^2)$^{1/2}$; all other units are in mJ/m^2.

3.4. Evaluation of Surface Energy Models using Different Sized Papers

The values of the surface energy components for sized and un-sized papers determined by different models are shown in Table 7. Figure 5 shows the comparison of total surface energy of the different sized papers. Both the GvOC and Chang models show that the total surface energy of paper decreases after sizing. Also, as shown in Figures 6 and 7, the decreased total surface energy is mainly caused by the decrease of the polar portion (γ_S^{ab})of the surface energy. However, differences in acid-base character for the two models are shown in Figures 6 and 7. The GvOC model shows the basic character to be dominant on the surfaces of both sized and unsized papers, while the Chang model shows an amphoteric character present on the unsized paper surface (one principal value is positive and the other is negative). Also, according to the Chang model, the amphoteric character of the paper surface was changed to basic after AKD and AKD-alum sizings. As shown in Table 7, negative values for both the P^a and P^b were found on the AKD and AKD-alum sized papers. The Chang model makes allowances for both repulsive and attractive interactions. The repulsion arises from the interaction between acid-acid or base-base, while the attraction results from the interaction between the acid and base.

The conventional calculation for the GvOC model does not allow the negative value of the acid-base components because these components are obtained by squaring the coefficients in the GvOC equation (Equation 7). Therefore, the acid-base component (γ_S^{ab}) is always positive for the GvOC model. From Table 7, it is seen that the coefficients, $(\gamma^+)^{1/2}$ and $(\gamma^-)^{1/2}$, in the GvOC model are negative for the surfaces of both AKD and AKD-alum sized papers. The negative square root for the GvOC model was also reported by Schneider [8]. Squaring these negative coefficients to obtain the acid-base components is misleading. It indicates that the GvOC theory is not theoretically robust for evaluating the entire range of material surfaces encountered in nature. This suggests the need for a better acid-base theory for contact angle data.

However, if the γ_S^{ab} in the GvOC model is calculated directly by the acid-base coefficients calculated by the multiple regression (Equation 7), there is a possibility that the γ_S^{ab} is negative if the two acid-base coefficients have the opposite signs. We call this calculation method the GvOC*. It is interesting to see in Table 7 that the surface energy components from the GvOC* method are very comparable to those determined by the Chang model.

Table 7.
Surface energy of sized and unsized paper fibers

Fiber	Model	γ_s^{tot}	γ_s^{d}	γ_s^{ab}	γ_s^{+}	γ_s^{-}	$(\gamma_s^+)^{1/2}$ or P_s^a	$(\gamma_s^-)^{1/2}$ or P_s^b
Unsized	GvOC	53.5	30.0	23.5	3.27	42.2		
	GvOC*	53.5	30.0	23.5			1.81	6.50
	Chang	54.4	30.0	24.4			3.93	-6.21
AKD-sized	GvOC	41.0	35.8	5.13	0.39	16.8		
	GvOC*	30.7	35.8	-5.13			0.63	-4.10
	Chang	30.6	35.8	-5.25			-2.85	-1.84
AKD-alum sized	GvOC	25.1	24.3	0.75	0.02	7.69		
	GvOC*	23.6	24.3	-0.75			0.14	-2.77
	Chang	22.5	24.3	-1.83			-2.18	-0.84
Rosin-sized	GvOC	36.9	30.6	6.33	2.09	4.79		
	GvOC*	36.9	30.6	6.33			-1.45	-2.19
	Chang	35.4	30.6	4.86			-3.08	1.58

Note: All units are in mJ/m^2 except for the two principal values of the Chang model: P_s^a and P_s^b, which are in $(mJ/m^2)^{1/2}$.

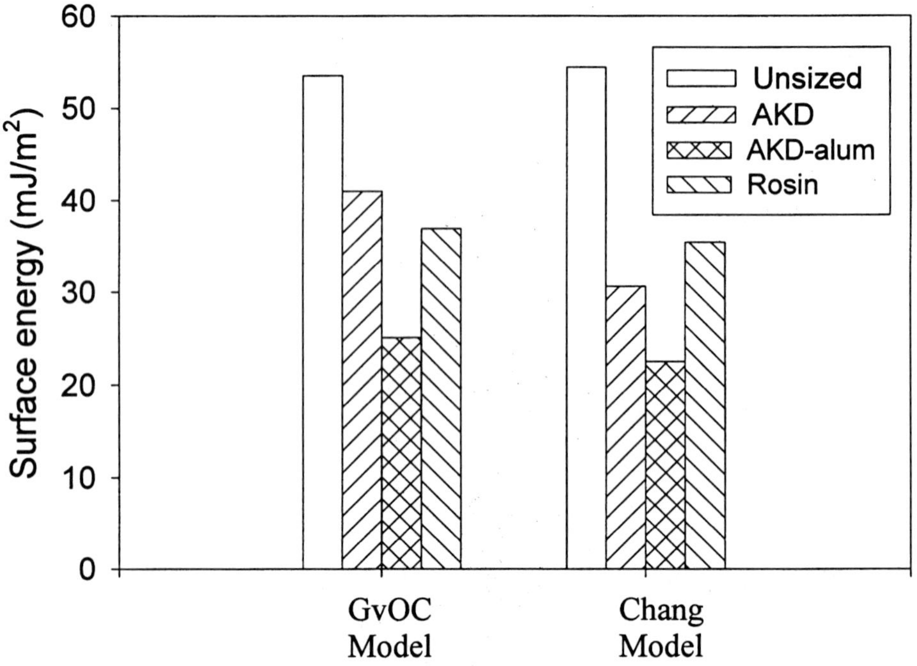

Figure 5. Total surface energy of sized and unsized papers.

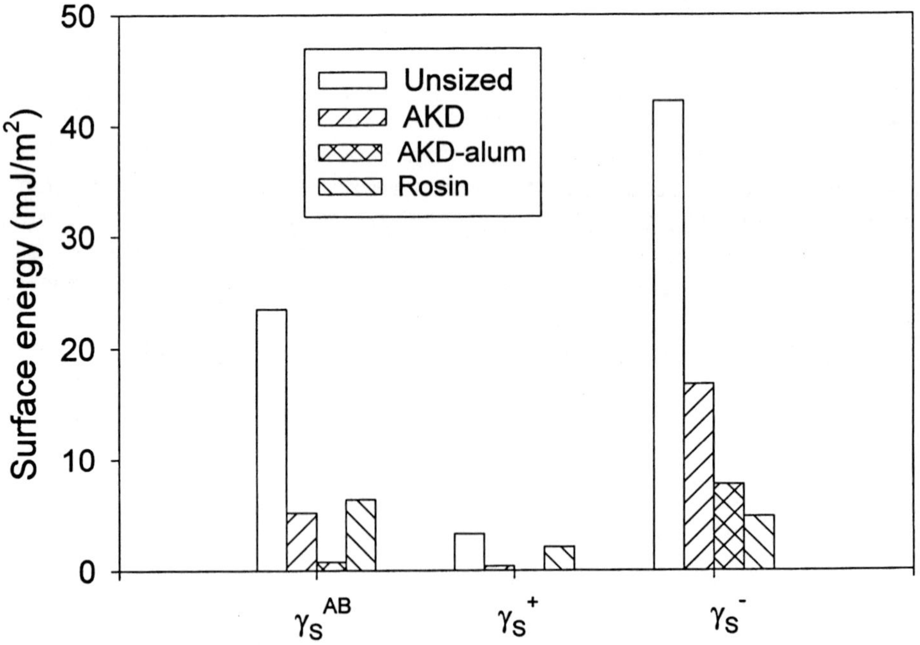

Figure 6. Surface energy components of sized and unsized papers determined by the GvOC model.

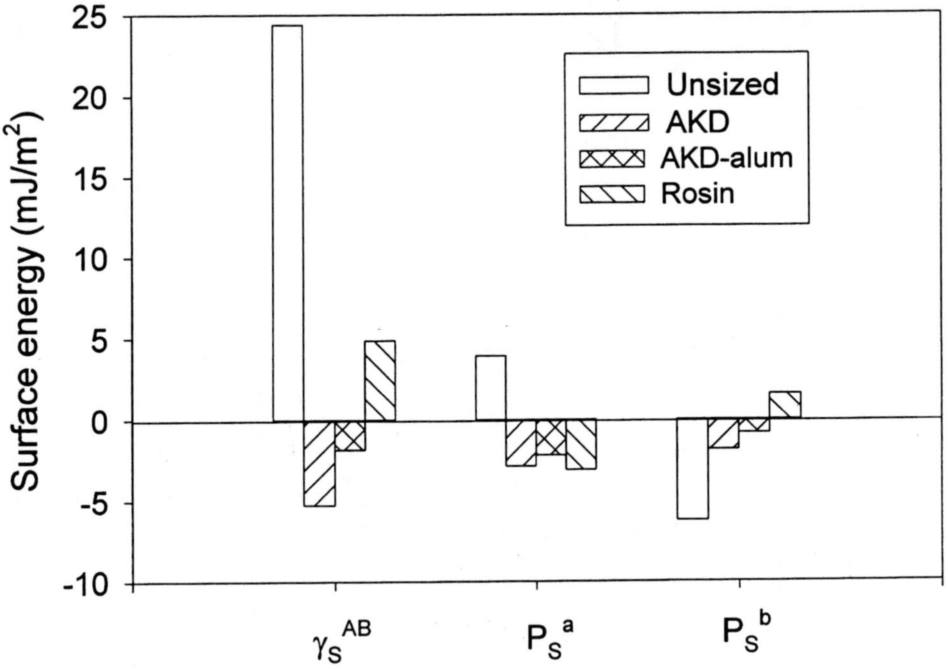

Figure 7. Surface energy components of sized and unsized papers determined by the Chang model.

Figure 8 shows the correlations of total surface energy of the sized papers and the sizing performance. The total surface energy of the sized papers determined by the Chang model correlates well with the HST test with 1% formic acid (FA) ink. This result is not unexpected because sizing performance should be directly correlated to surface energy. For aqueous-based inks, a low energy surface should inhibit ink wetting and, consequently, absorption in sized paper as measured by the sizing test. However, a poor correlation exists between the surface energy determined by the GvOC model and the sizing performance. This poor correlation is mainly due to the incorrect surface energy values determined by the GvOC model for the surfaces of AKD and AKD-alum sized papers on which the repulsive phenomenon was found by the Chang model. However, a significant improvement in the correlation of γ^{tot} and the HST test with 1% FA ink is shown for the GvOC* calculation method as compared to the conventional GvOC calculations.

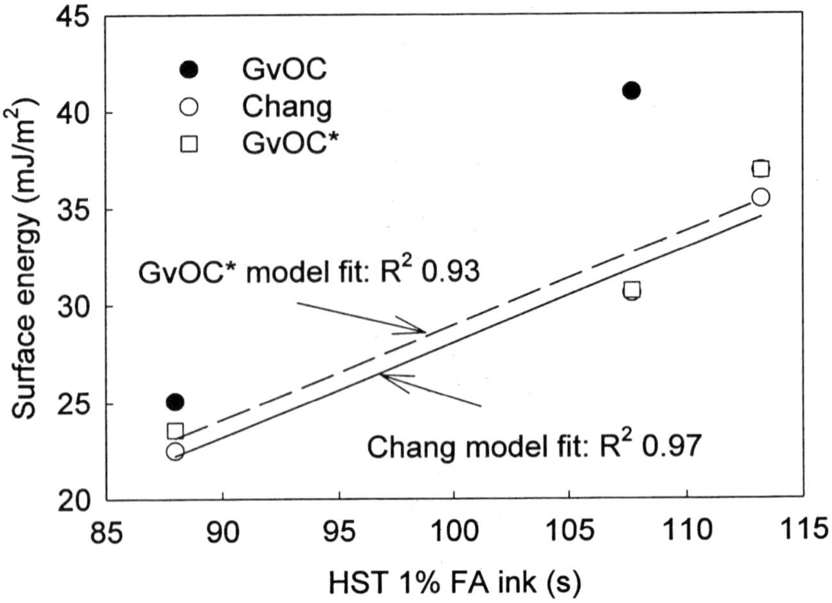

Figure 8. Correlation of total surface energy determined by the different models and sizing performance.

3.5. Evaluation of Vapor Adsorption Calculation Methods using Lignocellulose Surfaces

Enthalpies determined from both the Papirer [18] and Schultz [19] methods exhibit negative values for wood particles in the specific adsorption of several probe liquids. A set of enthalpies of specific adsorption on wood particles are presented in Table 8. The negative enthalpy indicates that phenomenon other than surface adsorption must have occurred. The phenomenon was probably due to the solubility of extractives on the particle surface by certain probe liquids. The sample was observed to start leaching as early as the column conditioning stage; the base line of the signals experienced several rises and falls before finally stabilizing. Because volatile material as well as the vapors of the probe liquids left the column, multiple peaks appeared in the chromatogram which complicates analysis.

Table 8.
Enthalpy of specific adsorption (kJ/mol) determined by Schultz [18], and Papirer [19] methods for maple particles with various probes

Probe liquid	$-\Delta H_A^{sp}$ (Papirer)	$-\Delta H_A^{sp}$ (Schultz)
DEE	4.18	5.53
$MeCl_2$	-12.76	-7.42
Benzene	-4.69	-3.69
Acetone	-18.24	-13.01
Chloroform	0.13	-0.82
THF	4.48	4.35
Ethyl acetate	1.79	8.35
Nitromethane	-6.85	-0.29

The negative enthalpy of the specific adsorption, when taken for calculation, may result in negative acid-base constants (K_A or K_B). Problems with negative enthalpy did not exist when IGC analysis was carried out on extracted birch wood meal [9]. However, there was no attempt in this study to remove extractives from wood particles prior to chromatography. Such a consideration may reduce the practicality of our analysis as wood particles are more likely to be processed in non-extracted forms in commercial operations. Therefore, IGC is limited in measuring the acid-base properties of wood materials.

A good correlation exists between the acid-base parameters determined using the Schultz method [18] and that determined using the Papirer method [19]. The Schultz method employs probe molecule areas to convert the molar average Gibbs free energy into an area average quantity so that the free energies of the acid-base probe adsorptions can be directly compared to that of the Lifshitz-van der Waals probe adsorptions [29]. On the other hand, the Papirer method [18] skips the need of knowing the probe molecule areas, and uses the probe vapor pressure to compare between adsorptions of n-alkanes and acid-base probes. Acid-base parameters calculated from both Schultz and Papirer methods are plotted in Figure 9. A good correlation exists and it is independent of the types of lignocellulose, i.e. particles of different wood species, TMP, or kraft fibers. Although the Papirer method has been commented as not having a theoretical basis [29], the method is still useful in the determination of acid-base properties of lignocellulose surfaces.

There is a poor correlation for the acidic (or basic) parameters between the wetting measurements and the IGC results. Examples of such correlations are presented in Figure 10 which show the relationship between K_A (Schultz) and γ^+ (GvOC), and that between K_B (Schultz) and γ^- (GvOC). The R^2 values are only 0.19 for γ^+ vs. K_A, and 0.12 for γ^- vs. K_B. The poor correlation in both cases is probably related to the nature of the technique whereby IGC predominantly detects high-energy sites which are normally small in population but not representative of the average values as obtained from the wetting measurements. It should also be pointed out that in the wetting measurements, the equilibrium spreading pressures of the liquids on the solid surfaces are ignored in the surface energy determination which may cause errors in the contact angle analysis.

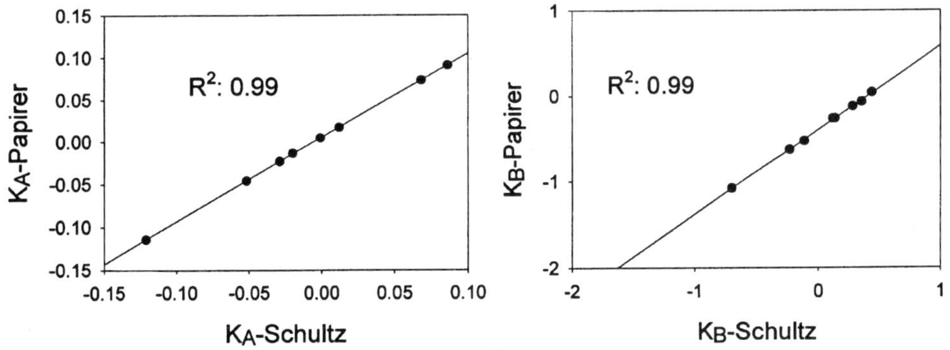

Figure 9. Correlation of the acid-base parameters determined by Schultz [18], and Papirer [19] methods.

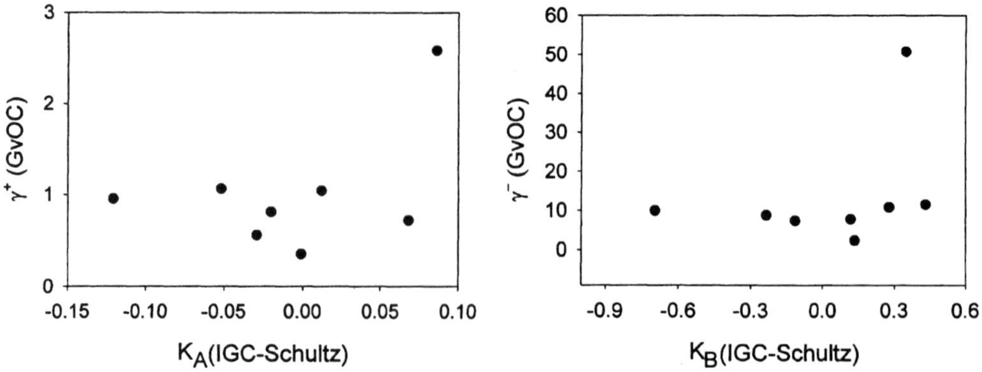

Figure 10. Correlation of acid-base parameters determined by CAA-GvOC and IGC-Schultz.

4. CONCLUSIONS

Different lignocellolusic materials were used in the evaluation of the GvOC and Chang models to determine the acid-base characteristics of the surfaces. Different calculation methods: multiple regression, one-liquid method, and all-liquid method, were evaluated for both models. Inverse gas chromatography was also used for the surface energy determination and compared to the contact angle analysis.

It is shown from the study that using all liquids to first determine the dispersive component is the most optimum way for calculating the total surface energy for both the GvOC and Chang models. The surface energy analysis of sized paper shows the importance of being able to measure attractive-repulsive character in acid-base measurements. Based on the results, it is apparent that the Chang model is more robust than the GvOC model, and is recommended that the Chang model be used to study lignocellulosic surfaces.

The GvOC model can also reflect both the repulsive and attractive effects on the surface if the square root terms of the acid and base components are directly used to calculate the polar portion of the surface energy (GvOC*). The Chang model and GvOC* calculation provide nearly identical results for the surface energy components of the lignocellolusic fibers.

In IGC measurement, a good correlation was found between the Schultz and Papirer methods in determining acid-base parameters.

REFERENCES

1. United States Department of Agriculture (USDA). *Wood Handbook: Wood as an Engineering Material.* U.S. Government Printing Office, Washington, DC (1987).
2. A. J. Stamm. *Wood and Cellulose Science*. The Ronald Press Company, New York (1964).
3. D. J. Gardner, in: *Wood Adhesives 1995*, A. W. Christiansen and A. H. Conner (Eds). pp. 29–36. Forest Products Society, Madison, WI (1996).
4. C. J. van Oss, M. K. Chaudhury, and R. J. Good, *Chem. Rev.* **88**, 927 (1988).
5. W.V. Chang and F. Chen, *Polym. Mater. Sci. Eng.* **61**, 607 (1989).
6. X. Qin and W.V. Chang, *J. Adhesion Sci. Technol.* **9**, 823 (1996).

7. D. J. Gardner, W. T. Tze, and S. Q. Shi. *Progress in Analytical Methodologies Applied to Lignocellulosic Materials*, pp. 277–307. TAPPI Press, Atlanta (1999).
8. R. P. Schneider. *J. Adhesion Sci. Technol.* **11**, 65–93 (1997).
9. D. P. Kamdem, S. K. Rose, and P. Luner. *Langmuir* **9**, 3039–3044 (1993).
10. Y. Huang, D. J. Gardner, M. Chen, and C. J. Biermann. *J. Adhesion Sci. Technol.* **9**, 1403–1411 (1995).
11. D. J. Gardner. *Wood Fiber Sci.* **28**, 422–428 (1996).
12. Q. Shi, D. J. Gardner, and J. Wang. *Proc. of The Fourth International Conference on Wood Fiber-Plastic Composites*, p. 245. Forest Products Society, Madison (1997).
13. S. Q. Shi, and D. J. Gardner. *J. Adhesion Sci. Technol.* (1999). In press.
14. G. I. Mantanis, R. A. Toung, and R. M. Rowell. *Holzforschung*. **48**, 480–490 (1994).
15. F. P. Liu, M. P. Wolcott, D. J. Gardner, and T. G. Rials. *Composite Interfaces* **2**, 419–432 (1994).
16. W. T. Tze, *The Effects of Swelling Treatments on the Surface Chemistry of Recycled Wood Pulp Fibers.* MS Thesis. Michigan Technological University, Houghton, MI (1998).
17. F. Chen and W.V. Chang, *Langmuir* **7**, 2401 (1991).
18. C. Saint Flour and E. Papirer. *Ind. Eng. Chem. Prod. Res. Dev.* **21**, 666–669 (1982).
19. L. Lavielle, J. Schultz, and K. Nakajima. *J. Appl. Polym. Sci.* **42**, 2825–2831 (1991).
20. V. Gutmann, *The Donor-Acceptor Approach to Molecular Interactions.* Ch. 2, Plenum Press, New York (1978).
21. L. Lavielle, and J. Schultz, *Langmuir* **7**, 978 (1990).
22. F. L. Riddle and F. W. Fowkes, *J. Am. Chem. Soc.* **112**, 3259 (1990).
23. P. N. Jacob, and J. C. Berg, *J. Adhesion* **54**, 115 (1995).
24. W. Wu, R. F. Giese, Jr., and C. J. van Oss, *Langmuir* **11**, 379 (1995).
25. D. W. Gunnells, D. J. Gardner, and M. P. Wolcott. *Wood Fiber Sci.* **26**, 447–455 (1994).
26. S. B. Lee and P. Luner. *Tappi* **55**, 116 (1972).
27. M. A. Tshabalala. *J. Appl. Polym. Sci.* **65,** 1013–1020 (1997).
28. C. J. Biermann. *Essentials of Pulping and Papermaking*. Ch. 3. Academic Press, New York (1993).
29. P. N. Jacob and J. C. Berg. *Langmuir* **10**, 3086–3093 (1994).

Acid-Base Interactions, Vol. 2, pp. 385–397
Ed. K.L. Mittal

Assessment of acid–base character of polymer surfaces from contact angle and other surface chemical data

F. M. ETZLER,* J. SIMMONS, N. LADYZHYNSKY, V. THOMAS and S. MARU
Boehringer-Ingelheim Pharmaceuticals, Inc., 900 Ridgebury Road, Ridgefield CT 06877, USA

Abstract—The performance of many products depends on the adhesion or abhesion of component materials. In the pharmaceutical industry, for instance, the adhesion between drugs, excipients, delivery devices and packaging materials may affect product performance. Fowkes, in particular, has stressed the importance of acid-base chemistry at interfaces to adhesion. Over the years, several investigators have proposed models to describe the acid–base nature of interfaces. Good and van Oss have championed the notion of calculating surface free energy components from contact angle data. Their components, γ^+ and γ^-, reflect respectively electrophilicity and nucleophilicity of surfaces. Chang and Chen have suggested another model for determining the acid–base character of surfaces. Chang and Chen's model, in contrast to that of Good and van Oss, is able to account for acid–acid and base–base repulsions. Fowkes has offered an approach dissimilar to both that of Good and van Oss and that of Chang and Chen which incorporates the ideas of Drago or Gutmann regarding measurement of electrophilicity and nucleophilicity. Those employing inverse gas chromatography (IGC) to assess surface acid-base character have often used Fowkes' approach. In each of the above models different scales for electrophilicity and nucleophilicity are used. Because certain models are more easily associated with particular experimental approaches and because certain materials lend themselves to investigation by particular experimental methods, it would be useful to know if the various possible approaches give the same general assessment of surface acid-base character. In this study we have measured the contact angles of various liquids on poly (methyl metyhacrylate) [PMMA], poly (vinyl chloride) [PVC] and polyethylene [PE]. The heats of adsorption of both phenol and pyridine have been measured on PMMA and PVC. The results of this work suggest that a given model may produce results which suggest different qualitative assessments of surface acid–base character than that obtained using another model. It is clear that knowledge of the precise relations between the various theoretical quantities calculated using the above models would be particularly helpful when comparing the conclusions based on different models. At present it does not appear that an unambiguous model independent assessment of surface acid-base character can be made.

Keywords: Contact angle; adhesion; acid–base; surface chemistry.

1. INTRODUCTION

The work of adhesion is related to the surface free energy of each material. [see Ref. 1 for instance] Surface free energy is the extra free energy caused by the creation of a surface. Adhesion results from a systems minimization of this extra free energy when surfaces are brought into contact. Because adhesion is an interfacial phenomenon, it depends upon the precise nature, mobility and orientation of surface molecules. The

* To whom correspondence should be addressed. E-mail: fetzler@rdg.boehringer-ingelheim.com

ability for molecules to come into intimate contact also influences adhesion. Factors such as surface texture and the mobility of interfacial atoms and molecules may affect the degree of intimate contact that is allowed to occur. In other words, practical adhesion depends on both the quality and quantity of interaction. In this work, chemical factors affecting adhesion are addressed [1–6].

Young's equation describes the relation between contact angle and the surface free energies for a liquid drop placed on a solid surface. Young's equation is expressed as

$$\gamma_{sv} = \gamma_{sl} + \gamma_{lv}\cos\theta, \tag{1}$$

where γ is the surface free energy, θ is the contact angle, and the subscripts sv, sl and lv refer to the solid-vapor, solid-liquid and liquid-vapor interfaces, respectively. The work of adhesion, W_a, between two surfaces is defined below:

$$W_a = \gamma_{lv} + \gamma_{sv} - \gamma_{sl}. \tag{2}$$

Combining Young's equation with Equation (2) results in the following expression relating contact angle and the work of adhesion:

$$W_a = \gamma_l(1+\cos\theta). \tag{3}$$

Equations (1)–(3) describe the basic thermodynamics of adhesion [1].

An early approach for estimating the surface free energy of solids was offered by Zisman [see Ref. 7, for instance]. Zisman found that a nearly linear dependence existed between the cosine of the contact angle and the surface tension of the wetting liquid. The intercept of the linear plot with $\cos\theta = 1$ is called the Zisman critical surface tension, γ_c. Although not rigorously correct, γ_c is often considered to be an estimate of the surface free energy of a solid. Zisman's approach can sometimes be useful for solving adhesion problems but gives no direct information regarding the molecular events contributing to adhesion.

Fowkes [3] suggested that total surface tension or surface free energy (γ^T) of a system could be divided into contributions from different components resulting from different classes of intermolecular interactions: thus, $\gamma^T = \Sigma\,\gamma_i$. Seven possible surface free energy components were identified. Schultz *et al.* [4] suggested dividing the total surface free energy into two components—polar (dipole–dipole interactions) and non-polar (dispersive) components:

$$\gamma^T = \gamma^d + \gamma^p. \tag{4}$$

The non-polar component results from London forces which result primarily from the polarizability of atoms and molecules. The polar component was considered to result from dipole–dipole (Keesom) interactions and other non-dispersion interactions. More recently, van Oss and Good [6] following the notions of Fowkes regarding the importance of Lewis acid–base interactions suggested an alternate to Equation (4):

$$\gamma^T = \gamma^{LW} + \gamma^{AB}. \tag{4a}$$

In Equation (4a) the superscript LW contains all the electrodynamic interactions (London–van der Waals) including dipole–dipole interactions, and the remaining interactions presumed to result from Lewis acid-base interactions. Kinloch [5] gives brief historical and technical discussion concerning the determination of surface free energy components. Van Oss and Good in Equation (5) model the Lewis acid and base contributions:

$$\gamma^{AB} = 2(\gamma^{+}\gamma^{-})^{1/2}. \tag{5}$$

On this basis the Young–Good–Girifalco [2, 6] equation defines the work of adhesion for a polar system as

$$W_a = \gamma_l(1+\cos\theta) = 2[(\gamma_l^{LW}\gamma_s^{LW})^{1/2} + (\gamma_l^{+}\gamma_s^{-})^{1/2} + (\gamma_l^{-}\gamma_s^{+})^{1/2}]. \tag{6}$$

Here, θ is the contact angle, γ^{LW} is the surface free energy due to London–van der Waals forces, γ^{+} is the Lewis acid surface free energy component, and γ^{-} the Lewis base surface free energy component.

For a nonpolar system, the work of adhesion reduces to

$$W_a = \gamma_l(1+\cos\theta) = 2(\gamma_l^{LW}\gamma_s^{LW})^{1/2}. \tag{7}$$

Equations (6) and (7) may be used to find the dispersive (γ_s^{LW}) and Lewis acid–base components. Equation (7) describes the work of adhesion resulting from only dispersion forces and is common to all the models discussed here. γ^{LW}, γ_s^{+} and γ_s^{-} may be determined using contact angles of various probe liquids. Good and van Oss consider Equation (4) now to be obsolete.

Chen and Chang [8] have more recently developed another method to evaluate acid and base character of surfaces. Their equation is

$$W_a = ax - by - cz = \gamma_l(1+\cos\theta), \tag{8}$$

where a is $(2\gamma_l^{LW})^{1/2}$, x is $(2\gamma_s^{LW})^{1/2}$, b, c are the principal values of liquid and y, z are the principal values of solid.

The principal values are related to the acid-base character of the materials in question. If a, b, and c are known for several liquids and x can be found using γ_s^{LW} from the nonpolar liquid contact angles, next y and z may be determined from contact angles of polar probes. A surface is considered acidic if y and z are both positive, basic if they are both negative and bipolar if they are of opposite signs. Unlike the van Oss–Good model, Chang and Chen's model allows for acid–acid and base–base repulsions. Fowkes [9, 10, see also Ref. 11] has suggested a different approach to evaluating the acid–base character of surfaces. As stated previously,

$$W_a = W_a^{LW} + W_a^{AB}. \tag{9}$$

W_a^{AB} is then, according to Fowkes, expressed by the following relation,

$$W_a^{AB} = -f \cdot N \cdot \Delta H^{AB}, \tag{10}$$

where N is the number of interacting sites per unit area and

$$f = \left[1 - \frac{\partial \ln W_a^{AB}}{\partial \ln T} \right]^{-1} \tag{11}$$

frequently

$$f \approx 0.2 \ldots 1.0.$$

When using the Fowkes approach some authors have taken *f* as unity although this does not seem to be a good approximation. [see Ref. 11 for a discussion] Because *f* is not generally known, direct calculations of the work of adhesion are not often made. Determination of ΔH^{AB} for multiple probe liquids on a given solid together with models by Drago [12] or Gutmann [13] can be used, however, to assess the acid-base nature of a surface.

Lewis acid–base interactions encompass hydrogen bonding, electron donor-acceptor and organic nucleophile-electrophile interactions. Mulliken theory [14] is based on a valence bond model for the ground state energy of the donor-acceptor complex. This model considers the complex in terms of two resonance forms—non-bonded and ionic (electrons fully transferred). This notion of covalent and Coulombic contributions forms the basis for Drago parameters.

According to Drago [12]

$$-\Delta H^{AB} = E_A \cdot E_B + C_A \cdot C_B \tag{12}$$

where E_i are the electrostatic susceptibility parameters and C_i are the covalent susceptibility parameters. Drago parameters allow one to easily consider the hardness of the acid–base interaction. Drago's model does not, however, allow one to consider easily the amphoteric nature of many materials [For a discussion of Hard and Soft Acids see Refs. 15 and 16].

Many adsorbates are amphoteric; therefore, it becomes necessary to recognize the dual polarity of many compounds. Gutmann introduced the notion of electron donor numbers (DN) and electron acceptor numbers (AN, AN^*). These parameters are similar to the van Oss surface tension parameters as they both describe the same molecular parameters but from different points of view. In 1966 Gutmann *et al.* [17] introduced the donor number based on the interaction with $SbCl_5$. DN has units corresponding to enthalpy (e.g. kJ/mole). In 1975, Mayer *et al.* [18] introduced the acceptor number based on the relative ^{31}P shift induced by triethylphosphine oxide. AN has arbitrary units. In 1990 Riddle and Fowkes [19] removed the dispersion component from AN. The corrected AN^* values have the usual units of enthalpy. According to Gutmann water is somewhat more nucleophilic. This conclusion contrasts with van Oss and Good's assumption that water is equally electrophilic and nucleophilic. (The question of the electrophile/nucleophile balance for water remains an open question as does the exact relation between each of the various acidity scales proposed in the literature.)

Using Gutmann's notions,

$$\Delta H^{AB} = -(K_A DN + K_B AN^*). \tag{14}$$

Here, K_B and K_A are additional parameters, which describe, respectively, electron donor and acceptor characteristics of a solid adsorbent. K_A and K_B are often determined using either inverse gas chromatography (IGC) or flow microcalorimetry (FMC) [20–23, also see Ref. 24 for general discussion].

Because the models described above are frequently associated with particular experimental methods and these methods may be more suitable to samples in a given physical form (e.g. powders versus sheets), it was decided to determine whether or not the models discussed above give the same general assessment of surface acidity. Morra [25] has previously indicated that the assessment may not be the same for all models.

2. EXPERIMENTAL

The contact angles of various liquids on polyethylene, poly(vinyl chloride) [PVC] and poly(methyl methacrylate) [PMMA] were measured at 20°C using the Wilhelmy plate method on a Cahn-315 Dynamic Contact Angle Analyzer with a stage speed of 150 μm/s. The polymers were obtained from Modern Plastics, Bridgeport, CT. The contact angle was not noticeably dependent on stage speed over a wide range of speeds surrounding the chosen stage speed. Measured liquid surface tensions were consistent with literature values. Advancing contact angles were tabulated. The 0.075 mm thick polyethylene, PVC and PMMA sheets were cut into 2.54 cm squares for analysis. The surface free energies were calculated from advancing contact angles using several probe liquids whose properties are listed in Table 1. The liquids were obtained from Aldrich Chemical Co. Aqueous buffer solutions with pH 1.5, 3.8, 7.1, 8.8, and 9.9 were also used. Three pieces of polyethylene, PVC and PMMA were tested for each probe liquid and pH solution [See Ref. 26, for instance regarding effects of pH on contact angle].

Heats of adsorption of 20 mM pyridine and phenol solutions in iso-octane were measured on a Microscal Model 3Vi Flow Microcalorimeter, equipped with two syringe pumps and a Thermo Separation Products refractoMonitor® IV Refractive Index Detector. The signals from both the calorimeter and RI detector were interfaced to a computer using Caldos v2.4 software. The Caldos software was not used for data analysis as the algorithms used for data analysis were not adequately documented. Data were analyzed as described earlier in the literature [21, 22]. Heats of adsorption of these materials can be used with Drago's model to assess surface acid-base character. The use of flow microcalorimetry to study surface chemistry has been described in the literature [21, 22]. In order for the polymers to be used within the calorimeter it was necessary to grind the sheets of polymer into a powder. Approximately 50 mg of powder were used for each sample. Duplicate measurements for each probe-polymer pair were made. 50 mg samples filled the calorimeter cell.

Table 1.
Probe liquids for contact angle measurements. Surface free energy values at 20°C (mJ/m^2) [6, 11]

Probe liquid	γ	γ^{LW}	γ^+	γ^-
1-Bromonaphthalene	44.4	43.5	0	0
Chlorobenzene	33.6	32.1	0.9	0.61
Diiodomethane	50.8	50.8	0	0
Dimethylsulfoxide	44	36	0.5	32
Ethylene glycol	48	29	1.92	47
Formamide	58	39	2.28	39.6
Glycerol	64	34	3.92	57.4
Nitrobenzene	43.9	41.3	0.26	6.6
Pentadecane	27.07	27.07	0	0
Tritiolyl phosphate	40.9	35.3	-	-
Pyrrole	37.6	32.6	-	-
Water	72.8	21.8	25.5	25.5

Table 2.
Principal values of probe liquids for Chen and Chang theory [8]

Probe liquid	a	b	c
1-Bromonaphthalene	9.3	-2.4	0.9
Diiodomethane	10.1	-2.8	-0.2
Dimethylsulfoxide	9.1	-3.0	0.6
Ethylene glycol	7.7	-4.8	3.6
Formamide	8.4	-6.2	3.8
Glycerol	8.7	-6.8	3.8
Tritiolyl phosphate	8.4	-3.2	7.5
Water	6.8	-6.6	1.8

3. RESULTS AND DISCUSSION

The results of the measurements described above are given in the Tables 1 to 7 and Figures 1 to 3.

The assessment of acid–base properties may be a valuable tool for the assessment of the compatibility of component materials used in the formulation of various products. Because component materials may exist in different physical forms (e.g. powders and sheets) more than one experimental method may be used for the assessment of acid–base properties of various components used in a given formulation.

Earlier, Morra [25] suggested that the use of different experimental techniques and theoretical models might lead to different conclusions. Morra's findings are summarized in Table 3. From this table it may be reasonable to conclude that PVC is acidic and PMMA is basic from the determination of W_a^{AB} for Pyrrole (acid) and DMSO (base) and indeed this is the expected result. Looking at the Good and van Oss parameters, γ^- is larger than γ^+ for both materials leading one to perhaps conclude that both PVC and PMMA are basic. It may be worthwhile to note that comparing only the values of γ^- to each other leads to the expected conclusion with regard to the relative acidity of PMMA and PVC.

Table 3.
Morra's [24] surface free energy data

	W^{AB} DMSO	W^{AB} Pyrrole	γ^{LW}	γ^+	γ^-
PVC	8.6	0.1	43	0.04	3.5
PMMA	-1.0	3.8	39–43	0	9.5–22.4

Like Vrbanac and Berg [11], we find neutral polyethylene to have sometimes significant values for W_a^{AB} that are calculated from contact angles. It is not entirely clear why W_a^{AB} significantly deviates from zero for some polar liquids although surface oxidation is likely to be at least, in part, responsible. According to the Good–van Oss model polyethylene appears to have a small basic component. Using the Chang–Chen model both principal values are negative suggesting a basic surface. Note, W_a^{AB} has rather large values for formamide, glycerol and water (see Table 4). The form of the Chang–Chen model accommodates the negative values of W_a^{AB} found for formamide and glycerol, for example.

Contact angle measurements have also been made on PVC and PMMA. Using these measurements, the acid-base properties of these polymers have been determined using both the Good–van Oss model and Chang and Chen model. W_a^{AB} were also calculated for the various standard probe liquids and pyrrole as well. The Good–van Oss parameters calculated from the data collected here closely match those reported by Morra [25]. Both polymers have a basic component but that reported for PVC is very small and similar to that found for polyethylene. W_a^{AB} values (Table 4) are generally similar to those found for PE except for diiodomethane and 1-bromonaphthalene which have high values considering their presumed purely dispersive character. Using the Chang and Chen Model, the principal values are negative for PVC and positive for PMMA. This result suggests that PVC is a base and PMMA is an acid—this is, indeed, the opposite of the result expected. Our values for W_a^{AB} for pyrrole and DMSO are also consistent with this conclusion. Also, W_a^{AB} values for glycerol and formamide are more positive for PMMA.

Table 4.
Contact angle values (degrees) and W^{AB} for PVC, PMMA and PE with various probe liquids

Probe Liquid	Material	Mean Angle	Standard Deviation	$W^{AB\,*}$ (mJ/m^2)
1-Bromonaphthalene	PVC	22.98	4.10	10.08
	PMMA	---	---	---
	PE	50.90	1.32	3.27
Chlorobenzene	PVC	15.53	3.50	1.37
	PMMA	---	---	---
	PE	---	---	---
Diiodomethane	PVC	40.72	3.51	8.04
	PMMA	33.23	1.68	2.58
	PE	61.43	5.64	1.10
Dimethylsulfoxide	PVC	58.09	3.16	-1.15
	PMMA	15.29	5.62	10.07
	PE	69.24	2.53	-2.78
Ethylene glycol	PVC	72.76	2.61	0.82
	PMMA	49.17	3.01	10.84
	PE	81.94	1.71	-1.14
Formamide	PVC	83.61	3.94	-6.75
	PMMA	54.58	2.36	12.13
	PE	92.92	1.82	-9.74
Glycerol	PVC	96.32	2.42	-9.52
	PMMA	82.25	0.39	-1.59
	PE	102.28	2.44	-10.10
Nitrobenzene	PVC	41.30	5.02	3.61
	PMMA	39.57	5.22	-4.06
	PE	52.27	2.50	4.09
Water	PVC	94.73	4.83	13.56
	PMMA	77.91	1.72	28.63
	PE	97.96	5.27	14.31
Pentadecane	PVC	21.11	8.72	-7.00
	PMMA	---	---	---
	PE	17.12	6.12	-1.00
Hexadecane	PVC	---	---	---
	PMMA	---	---	---
	PE	17.64	6.02	-0.73
Tetradecane	PVC	---	---	---
	PMMA	---	---	---
	PE	17.79	4.57	-1.76
Pyrrole	PVC	16.25	8.46	8.62
	PMMA	63.39	7.42	-18.21
	PE	---	---	---
Tritolyl phosphate	PVC	49.20	1.94	-0.12
	PMMA	---	---	---
	PE	57.77	1.04	1.68

$^*W^{AB} = \gamma_l(1+\cos(\theta)) - 2(\gamma^{LW}_l \gamma^{LW}_s)^{1/2}$.

Table 5.
Contact angles (degrees) of aqueous buffer solutions with various pH values on PVC and PMMA

pH value	Material	Mean angle	Standard deviation
1.5	PVC	94.37	2.30
	PMMA	78.82	2.11
3.8	PVC	95.84	1.40
	PMMA	76.82	1.15
7.1	PVC	98.74	0.21
	PMMA	77.01	0.84
8.8	PVC	99.34	2.31
	PMMA	79.64	4.69
9.9	PVC	100.19	1.06
	PMMA	79.33	0.84

Table 6.
Good–van Oss surface free energy components (in mJ/m^2) for PVC, PMMA and PE

	PVC	PMMA	PE
γ^{LW}	32.53	40.52	26.91
γ^{-}	0.35	6.83	0.33
γ^{+}	0	0	0

Table 7.
Chang and Chen principal values for PVC, PMMA and PE

	PVC	PMMA	PE
y	-2.25	2.50	-2.57
z	-2.53	2.22	-3.30

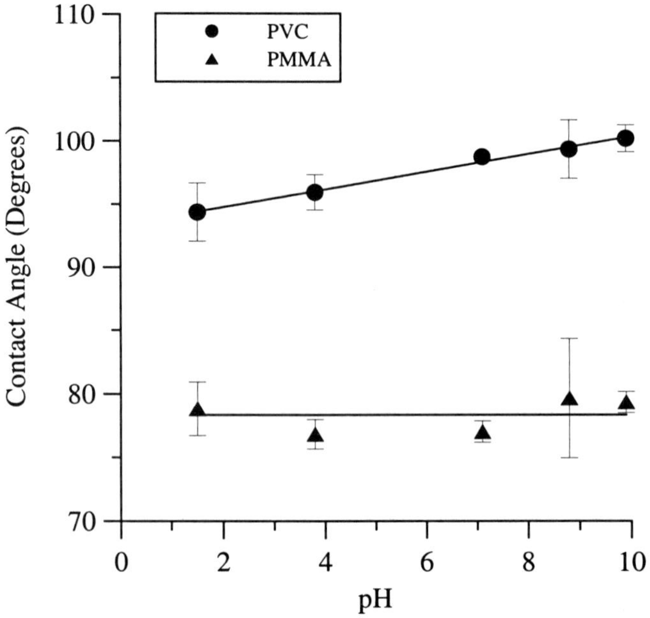

Figure 1. Contact angle (degrees) *versus* pH for aqueous buffers. Note PVC contact angle rises with increasing pH; such behavior would suggest a basic surface chemistry. Aqueous buffers probe the Brønsted acid–base character of the surface.

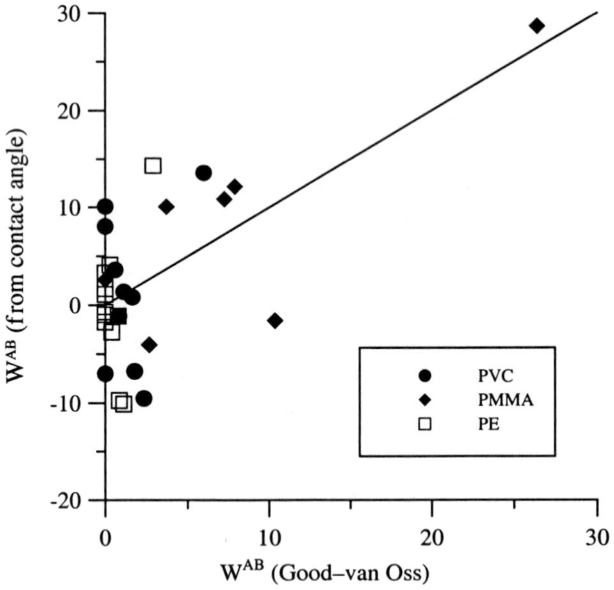

Figure 2. W^{AB} calculated from contact angle *versus* that calculated using the Good–van Oss model. Line represents Y=X. Note that points are not well distributed along the line.

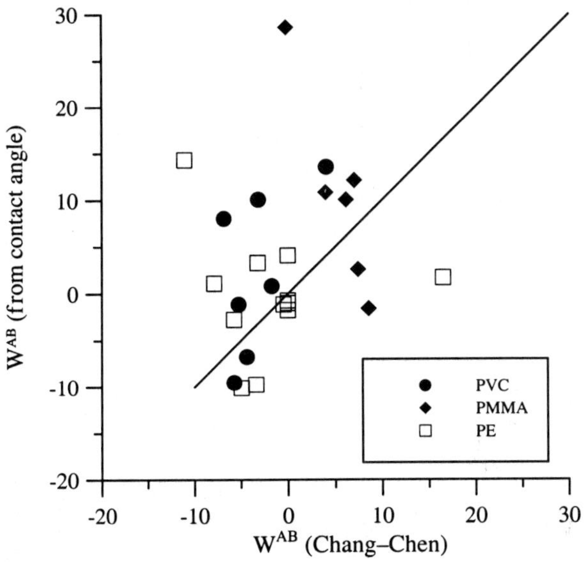

Figure 3. W^{AB} calculated from contact angle *versus* that calculated using the Chang–Chen model. Line represents Y=X. Note points more evenly distributed along the line. The Chang–Chen model represents the data somewhat better than does the Good–van Oss model. The water contact angle appears to receive less weighting in the fit than it does using the Good–van Oss approach.

The contact angles of aqueous buffer solutions of various pH values were also measured. Buffer solutions monitor the Brønsted acid–base character of surfaces. For the case of PVC a small dependence of contact angle on pH was noted. Contact angle was found to rise with increasing pH. (see Table 5 and Fig. 1) This observation might be expected for a basic surface yet PVC is considered to be acid.

W_a^{AB} calculated for the three polymers investigated is plotted versus that predicted by both the Chang–Chen and Good–van Oss models (see Figs. 2 and 3). Although neither model appears to be an excellent predictor of the experimentally observed W_a^{AB} it appears that the Chang–Chen model is the better predictor as it can account for negative or repulsive interactions.

Flow microcalorimetry was also used to characterize PMMA and PVC using phenol (acid) and pyridine (base) as probes (see Table 8). The results here indicate that indeed PMMA is a base and PVC slightly acidic as predicted on the basis of their respective molecular formulae. The different assessment of acid-base character found by flow microcalorimetry may, in part, reflect differences between the surface chemistry of the polymer sheets and that of the ground powder made from the same sheets.

Table 8.
ΔH^{AB} for pyridine and phenol adsorption measured on PVC and PMMA by FMC (kcal/mole)

	Pyridine	Phenol
PVC	0.3	0.5
PMMA	0.08	9.4

4. SUMMARY

In this work the contact angles of various probe liquids on polyethylene, PVC and PMMA were measured. Flow microcalorimetry (FMC) was also used to characterize the surface chemistry of PVC and PMMA. From the contact angle data, W_a^{AB} of the probe–polymer interfaces has been calculated. It was noted that several negative values were observed suggesting the possibility of acid–acid and base–base repulsions. Some significant values of W^{AB} were noted for "neutral" polyethylene. The contact angle data were further analyzed by the Chang–Chen and Good–van Oss models. The results of such analysis differ from that found by FMC and that expected from the molecular structure. The Good–van Oss model suggests that both materials are basic but PMMA is more so. The Chang–Chen model predicts that PVC is basic and PMMA acidic. It appears that the Chang-Chen model is a better predictor of the observed W_a^{AB} as it can accommodate $W_a^{AB} < 0$ but does not give the expected characterization of acid-base chemistry. W_a^{AB} data, furthermore, suggest that the various probe parameters (see Tables 1 and 2) may not form a self-consistent set.

The various models investigated here do not give a consistent assessment of the acid–base character of surfaces. Comparison of assessments of surface acid–base character obtained using different methods must be made with considerable caution. It is clear that further critical work regarding the limitations of the Chang–Chen and Good–van Oss models is required as is a detailed understanding of each group's choice of acidity scales. An explanation for differences seen when comparing data obtained from contact angles to that obtained by IGC/FMC is also needed. Such differences may either reflect limitations of one or more models or reflect genuine surface chemical effects resulting from the necessary sample preparation methods. At present, it does not appear that an unambiguous model independent assessment of surface acid–base character can be made.

REFERENCES

1. A. W. Adamson, *Physical Chemistry of Surfaces*, 5th edn. John Wiley, New York (1990).
2. L. A. Girifalco and R. J. Good, *J. Phys. Chem.* **61**, 904–909 (1957).
3. F. M. Fowkes, *J. Phys. Chem.* **66**, 382 (1962); *J. Colloid Interface Sci.* **28**, 493 (1968).
4. J. Schultz, K. Tsutsumi, and J. B. Donnet, *J. Colloid Interface Sci.* **59**, 277 (1977).
5. A. J. Kinloch, *Adhesion and Adhesives: Science and Technology*. Chapman and Hall, New York (1987), see Chapter 2.
6. R. J. Good, in: *Contact Angle, Wettability and Adhesion,* K. L. Mittal (Ed.), p. 3. VSP, Utrecht, The Netherlands (1993); C. J. van Oss, *Interfacial Forces in Aqueous Media*. Marcel Dekker, New York (1994).
7. W. A. Zisman, in: *Contact Angle, Wettability and Adhesion*, ACS Adv. Chem. Ser., No. 43, p. 1. American Chemical Society, Washington, DC (1964).
8. F. Chen and W. V. Chang, *Langmuir* **7**, 2401 (1991); F. Chen and W. V. Chang, unpublished.
9. F. M. Fowkes and M. A. Mostafa, *Ind. Eng. Chem. Prod. Res. Dev.* **17**, 3 (1978).
10. F. M. Fowkes, *J. Adhesion Sci. Technol.* **1**, 7 (1987).
11. M. D. Vrbanac and J. C. Berg, in: *Acid-Base Interactions: Relevance to Adhesion Science and Technology*, K. L. Mittal and H. R. Anderson, Jr. (Eds), p. 67. VSP, Utrecht, The Netherlands (1991); J. C. Berg, in: *Wettability*, J. C. Berg (Ed.), p. 75. Marcel Dekker, New York (1993).
12. R. S. Drago, G. C. Vogel and T. E. Needham, *J. Amer. Chem. Soc.* **93**, 6014 (1971).

13. V. Gutmann, The Donor–Acceptor Approach to Molecular Interactions. Plenum Press, New York (1978).
14. R. S. Mulliken, *J. Chem. Phys.* **19**, 514 (1951).
15. R. G. Pearson, *J. Amer. Chem. Soc.* **85**, 3533 (1963).
16. L. H. Lee, *J. Adhesion Sci. Technol.* **5**, 71 (1991).
17. V. Gutmann, A. Steininger and E. Wychera, *Montash. Chem.* **97**, 460 (1966).
18. U. Mayer, V. Gutmann and E. Wychera, *Montash. Chem.* **106**, 1235 (1975).
19. F. L. Riddle, Jr. and F. M. Fowkes, *J. Amer. Chem. Soc.* **112**, 3259 (1990).
20. D. P. Kamdem, S. K. Bose and P. Luner, *Langmuir* **9**, 3039–3044 (1993).
21. F. M. Fowkes, K. L. Jones, G. Li and T. B. Lloyd, *Energy Fuels* **3**, 97 (1989).
22. R. S. Farinato, S. S. Kaminski and J. L. Courter, *J. Adhesion Sci. Technol.* **4**, 633 (1990).
23. J. Schultz and L. Lavielle, in: *Inverse Gas Chromatography: Characterization of Polymers and Other Materials,* ACS Symp. Ser., Vol. 391, D. R. Lloyd, T. C. Ward and H. P. Schreiber (Eds), pp. 185–202. American Chemical Society, Washington, DC (1989).
24. L.-H. Lee, in: *Contact Angle, Wettability and Adhesion*, K. L. Mittal (Ed.), pp. 45–90. VSP, Utrecht, The Netherlands (1993).
25. M. Morra, *J. Colloid Interface Sci.* **182,** 312 (1996).
26. G. M. Whitesides, H. A. Biebuyck, J. P. Folkers and K. L. Prime, in: *Acid-Base Interactions: Relevance to Adhesion Science and Technology*, K. L. Mittal and H. R. Anderson, Jr. (Eds), p. 229. VSP, Utrecht, The Netherlands (1991).

Acid-Base Interactions, Vol. 2, pp. 399–418
Ed. K.L. Mittal

Characterization of acid–base properties of surfaces by contact angle titration: Application to the adhesive bonding of silicon carbide

SOPHIE HELT, JÉRÔME EVIEUX, YVES BAZIARD,* VALERIE NASSIET and JACQUES-ALAIN PETIT

Laboratoire Génie de Production—Génie des Matériaux, Ecole Nationale d'Ingénieurs, B.P. 1629, 65016 Tarbes Cedex, France

Abstract—The acid–base properties of the surface of a sintered silicon carbide (SSiC) subjected to two different surface treatments, chemical etching and thermal air oxidation, have been evaluated using surface titrations (Hüttinger's method) based on contact angle measurements. On the other hand, the chemical composition of ceramic substrates were investigated by X-ray photoelectron spectroscopy and microscopies (SEM, AFM), along with three-dimensional profilometry. After the thermal treatment the SSiC surface was converted into silica, but not after the etching treatment. However, both treatments yielded clean surfaces with a relatively low carbon contaminant concentration. Irrespective of the surface treatment, typical stepwise titration curves were found for a contaminant concentration lower than 35% thus showing Brönsted acid–base interactions, with no modification of the surface roughness at the micrometer scale. For this same value of 35%, a transition between interfacial and cohesive ruptures in the adhesive was observed for both thermal air and etched SSiC substrates bonded with a structural epoxy adhesive.

Keywords: Titration of surfaces; wettability; Brönsted acid–base interactions; carbon contaminants; adhesive bonding; epoxy adhesive.

1. INTRODUCTION

A knowledge of physicochemical properties of substrate surfaces—and especially acid–base properties—is, among many parameters, of great importance to obtain a good adhesive bond [1]. Acid–base properties of surfaces can be determined by means of several techniques such as Electron Spectroscopy for Chemical Analysis or X-ray Photoelectron Spectroscopy (XPS) [2], Atomic Force Microscopy [3, 4], zeta potential [5–7], Inverse Gas Chromatography [8] and also wettability or contact angle technique. The last method is today a classical technique widely used to determine the work of adhesion (WSL) at a Solid–Liquid interface. Fowkes [9, 10] was the first to propose that WSL could be separated into two components:

$$W_{SL} = W_{SL}{}^{D} + W_{SL}{}^{AB} \tag{1}$$

* To whom correspondence should be addressed. E-mail: baziard@enit.fr

due to dispersion forces (W_{SL}^D) and non-dispersion or acid–base forces (W_{SL}^{AB}), with a geometric mean relationship giving $W_{SL}^D = 2(\gamma_S^D.\gamma_L^D)^{1/2}$ [11], the acid–base component being given by:

$$W_{SL}^{AB} = \Sigma_i\, f_i n_i^{AB}(-\Delta H_i^{AB}) \tag{2}$$

where $-\Delta H_i^{AB}$ is the enthalpy of reaction to form an acid–base complex (i), n_i^{AB} is the concentration (number) of the acid–base complexes formed at the solid–liquid interface, f_i is a correction factor between enthalpy and free energy (W_{SL}^{AB}).

In the Fowkes approach, the acid–base interactions term is obtained by subtracting the dispersion term from the total work of adhesion which, taking into account the Young–Dupré equation:

$$W_{SL} = \gamma_L(1 + \cos\theta) + \pi \tag{3}$$

can be written as:

$$W_{SL}^{AB} = \gamma_L(1 + \cos\theta) + \pi - 2(\gamma_S^D.\gamma_L^D)^{1/2} \tag{4}$$

in which θ the contact angle at thermodynamic equilibrium, γ_L^D the dispersive component of the total surface free energy of the liquid (γ_L), γ_S^D the dispersive component of the surface free energy of the solid surface, π the spreading pressure of the liquid on the solid surface that influences the spreading of the liquid drop; π is due to the formation of a film of this liquid in front of the solid–liquid–vapour triple line.

In the one-liquid sessile drop model proposed by Young [12], the equation yielding W_{SL}^{AB} can be simplified by neglecting the spreading pressure π, for low energy surfaces only, or by measuring the advancing contact angle which is equal to the static contact angle on a film-free surface. Recently, Hüttinger *et al.* [13, 14] after the first approach of Whitesides and Laibinis [15], developed a new method for determining the acid–base work of adhesion, based on the ideas of Fowkes together with the classic wetting technique, in which pure water is replaced by aqueous solutions with pH values ranging from 1 to 14. Actually, this method involves titration of surfaces where θ = f(pH) and hence W_{SL} = f(pH) curves obtained are titration curves, showing stepwise increases or decreases of W_{SL} with increasing pH values corresponding, respectively, to acidic or basic sites present at the solid surface.

In this paper, Hüttinger's method was used to determine the acid–base properties of surfaces of Sintered Silicon Carbide (SSiC) subjected to different surface treatments. This work was a part of the thesis work of Helt [16] about the structural adhesive bonding of SSiC, following the first study of Levallois on the same subject [17, 18]. Besides the contact angle technique, XPS measurements were performed for the study of the chemical composition of SSiC surfaces, along with Scanning Electron Spectroscopy (SEM), Atomic Force Microscopy (AFM) and three-Dimensional Profilometry (3DP) for the surface roughness and morphology. The experimental results were compared and analysed, showing the influence of acid–base properties of surfaces investigated and also that of carbon contaminants on the adhesive bonding of SSiC.

2. EXPERIMENTAL

2.1. Silicon Carbide Substrates

The study of the physicochemical properties of surfaces was carried out on a Sintered Silicon Carbide (trade name SiC100) supplied by the French Society Céramiques & Composites. This SSiC is obtained by sintering, in argon atmosphere at 2070°C, of an α-SiC powder (average grain diameter: 5 μm), boron and carbon being the main sintering agents. Its chemical composition before sintering along with the surface roughness parameters of solvent (1,1,1-trichloroethane) degreased as-received SSiC samples are given elsewhere [17].

As-received samples showed relatively smooth surfaces, with the average and total surface roughness values (R_a = 0.5 μm and R_t = 7.0 μm, respectively). Furthermore, the weak negative value of the skewness (S_k = –0.78) of the distribution of heights indicates a surface with slightly more hollows than peaks and this surface is uniform because of the spread of the distribution of heights or kurtosis (E_k = 5.70) being higher than 3 (E_k = 3 for a Gaussian distribution corresponding to a random surface). After solvent degreasing (with 1,1,1-trichloroethane) for 5 minutes under ultrasonic motion, two "cleaning" treatments were performed independently on SiC100 samples:

- a chemical treatment with the etching aqueous solution of Murakami [$K_3Fe(CN)_6$ (225 g/l) + NaOH (200 g/l)]; this solution was used at ambient or at boiling temperature. This treatment was originally used to treat surfaces of chromium or tungsten carbide [19, 20], with the view to increase the adhesion of diamond layers on these carbides by Chemical Vapor Deposition.
- a thermal treatment in air atmosphere, for 1 h at temperatures ranging from 20°C (as-received samples) to 1600°C.

Detailed procedures are given elsewhere for Murakami's treatment [16] and thermal treatment [16–18]. It must be mentioned that the roughness parameters of SiC100 surfaces were not altered after these two treatments, at the micrometer scale however.

2.2. Surface Analysis

XPS analysis of SiC100 substrates was performed using Surface Science Instruments spectrometer with a Leybold optical source; the spot size was 600 μm^2 in diameter and the residual pressure was 5.10^{-8} Pa. Surface roughness parameters were measured using a 3D profilometer (S6P instrument from the French society Perthen), with a 5 μm radius of curvature diamond tip sensor and a vertical sensitivity equal to ± 0.01 μm. Details about the surface parameters investigated are given in [18].

2.3. Wettability Measurements

Contact angles were measured with the Digidrop goniometer from the French Society GBX-Instruments. This apparatus is equipped with a video monitor and a pattern recognition software for the liquid drop. All the experiments were carried out using the one-liquid sessile drop method (Fowkes procedure for the calculation of $\gamma_S{}^D$ and $W_{SL}{}^{AB}$, neglecting the spreading pressure π of the liquid) and using also the method of titration of surfaces. More details about this method and particularly explanations on the titration curves are given in [14]. Note that this method is based on Equations (1) and (2) due to Fowkes. The dispersive term of the surface tension of the liquid (γ_L^D) being independent of pH [14], the work of adhesion WSL can be derived in relation to pH and, taking into account Equations (1) to (3):

$$(dW_{SL}/dpH) = (dW_{SL}^{AB}/dpH) = \gamma_L(d\cos\theta/dpH) = \Sigma_i\, f_i(-\Delta H_i^{AB})(dn_i^{AB}/dpH) \qquad (5)$$

The contact angle variation as a function of pH is thus directly linked to the variation of the number of acid–base bonds formed at the solid–liquid interface.

In view of the large number of measurements performed in this study, contact angle measurements were made less than 15 s after the drop was placed on the surface, at constant temperature T = 23 ± 1°C and Relative Humidity (RH) ranging from 60 to 80%. Under these conditions, the contact angle value was very close to that of the advancing contact angle (θ_a) for low energy surfaces (because there was almost no variation of q *vs* time in this case) and lower by around 2 to 4 degrees than θ_a for higher energy surfaces. In the latter case, the liquid drop "spreads" and θ decreases with time, θ_a being the value extrapolated to t = 0. Actually, this time dependence of θ was observed for water drops, as was reported elsewhere [1, 14].

For the titration of surfaces, aqueous alkaline or acidic solutions were prepared using distilled water and analytical grade NaOH or HCl to obtain solutions with $1 \leq pH \leq 12$. The surface tension (γ_L) of these aqueous solutions was measured using the pendant drop method. Table 1 shows that γ_L remains constant and stays close to that of pure water despite some scatter in γ_L values.

Table 1.
Values of the surface tension (γ_L) of aqueous solutions of various pH

pH	1.2	3.1	5.7	8.0	10.0	12.0
γ_L (mJ/m^2)	70.8	71.4	72.0	71.2	72.4	70.9

2.4. Adhesive Bonding and Mechanical Characterization

A compression-shear rupture test [21–23] was used on adhesively-bonded SSiC substrates subjected to Murakami or thermal treatment, to determine the rupture load (F_R) and the critical strain energy release rate in the shear mode (G_{IIC}). The method of Williams [24] was used for the determination of G_{IIC}, as we had done in a previous work on the same bonded assemblies [20]. Adhesive bonding was performed using a classical structural epoxy (DGEBA) – dicyandiamide (DDA) system (AV119 from Ciba). The procedure of bonding was described in [18].

3. RESULTS AND ANALYSIS

3.1. Etching Treatment of SiC100

The etching effect of Murakami's solution was difficult to observe on rough "as-received" SiC100 samples. So, some samples were polished and then subjected to the etching treatment. AFM views of a polished surface before and after Murakami's treatment are shown in Figure 1. One can see that the polished surface subjected to the solution of Murakami presents holes (in black) and plateaus (in light grey) corresponding to SiC grains at different heights. During the treatment, the solution of Murakami well etches the SiC grains but at different rates and this leads to the surface morphology observed. However, the modification of the surface morphology after treatment was weak and concerned only the external layer of polished substrates to a depth less than 1 μm. Therefore, when Murakami's treatment was applied on non-polished substrates, such a depth of etching was not able to modify the random morphology of these substrates, and hence the roughness parameters (especially the average roughness) at the micrometer scale.

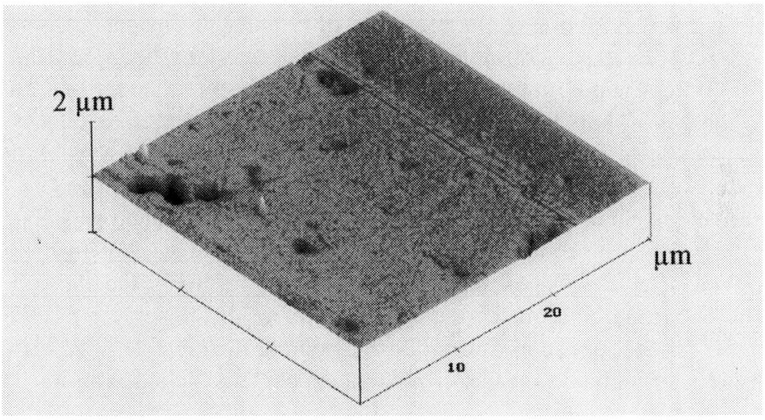

(a)

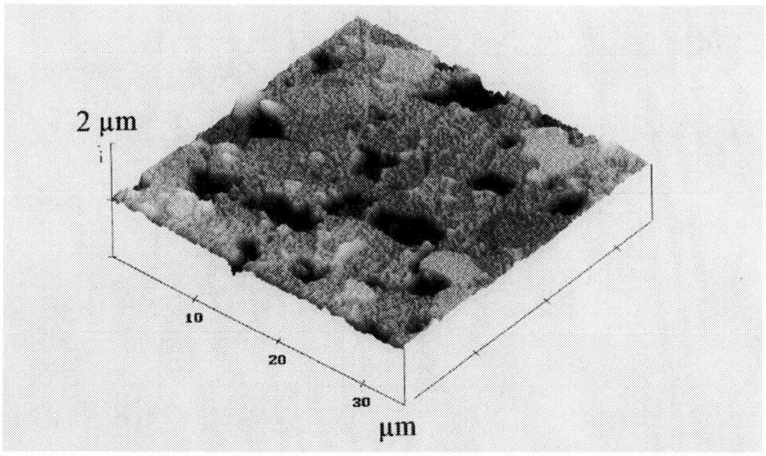

(b)

Figure 1. AFM views of a polished SiC100 surface before (a) and after (b) treatment with the etching solution of Murakami.

Figure 2 shows an increase in the work of adhesion (W_{SL}) for alkaline and acidic aqueous solutions vs treatment time, for SiC100 substrates subjected to the solution of Murakami at ambient temperature. This increase was sensitive only after 3 h of treatment and WSL tended to a constant value after 140 h, for both alkaline as well as acidic water drops. The increase in WSL was observed in the same way with the boiling Murakami solution, but the time to obtain the maximal value of 145 mJ/m^2 for W_{SL} was considerably reduced to around 10 min [16].
The chemical composition of the SiC100 surface varied widely during this treatment. The carbon contaminant concentration (C_{contam}) was high for as-received samples (58%) and after 3 hours of treatment (74.2%) but afterwards it decreased strongly, showing therefore the cleaning effect of Murakami's treatment (Table 2). The initial slight increase of C_{contam} was not observed when the boiling solution was used, but this was not surprising owing to the high rate of etching in this case. Contrary to carbon contaminants, as expected, included-in-SiC carbon and silicon concentrations increased with time, along with the quantity of oxygen despite the fact that no silica was detected at the SiC100 surface. The variation of oxygen cannot be clearly related to the variation of the other elements, but the XPS spectrum shows a weak peak at 288 eV for oxygen bonded to carbon contaminants (O-C=O), besides the C-C/C-H peak and the peak due to SiC (Fig. 3). A very small quantity of nitrogen was also present at the surface, originating probably from the Murakami solution, but this nitrogen was easily removed after 2 min of ion (Ar^+) sputtering (Table 2). On the other hand, it must also be mentioned that the dispersive component of the surface free energy (γ_S^D) of SiC100 was almost constant during all the treatment time: γ_S^D = 42 mJ/m^2 for solvent degreased as-received samples to γ_S^D = 44 mJ/m^2 for etched samples.

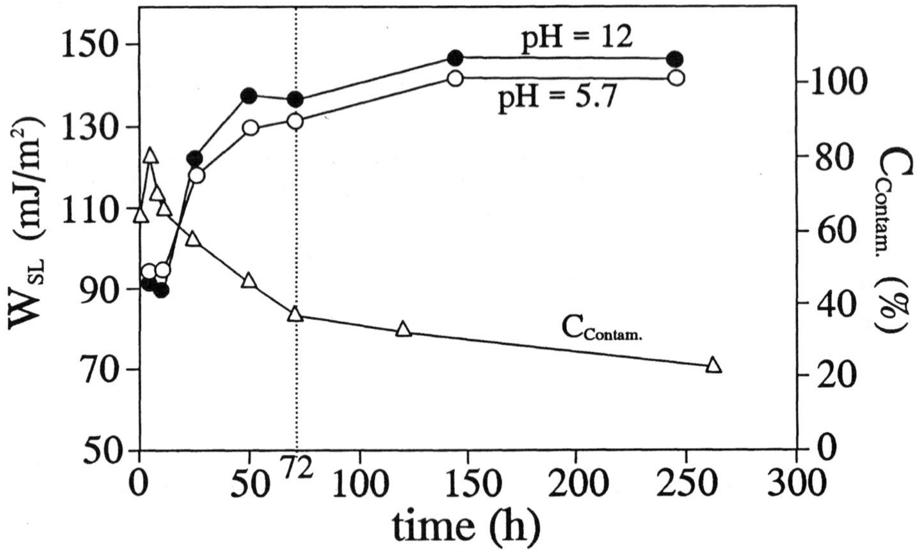

Figure 2. Time dependent variation of the work of adhesion (W_{SL}) of distilled and basic water (pH=5.7 and 12, respectively) and of the percentage of carbon contaminants (C_{contam}), for SiC100 substrates subjected to the solution of Murakami at ambient temperature.

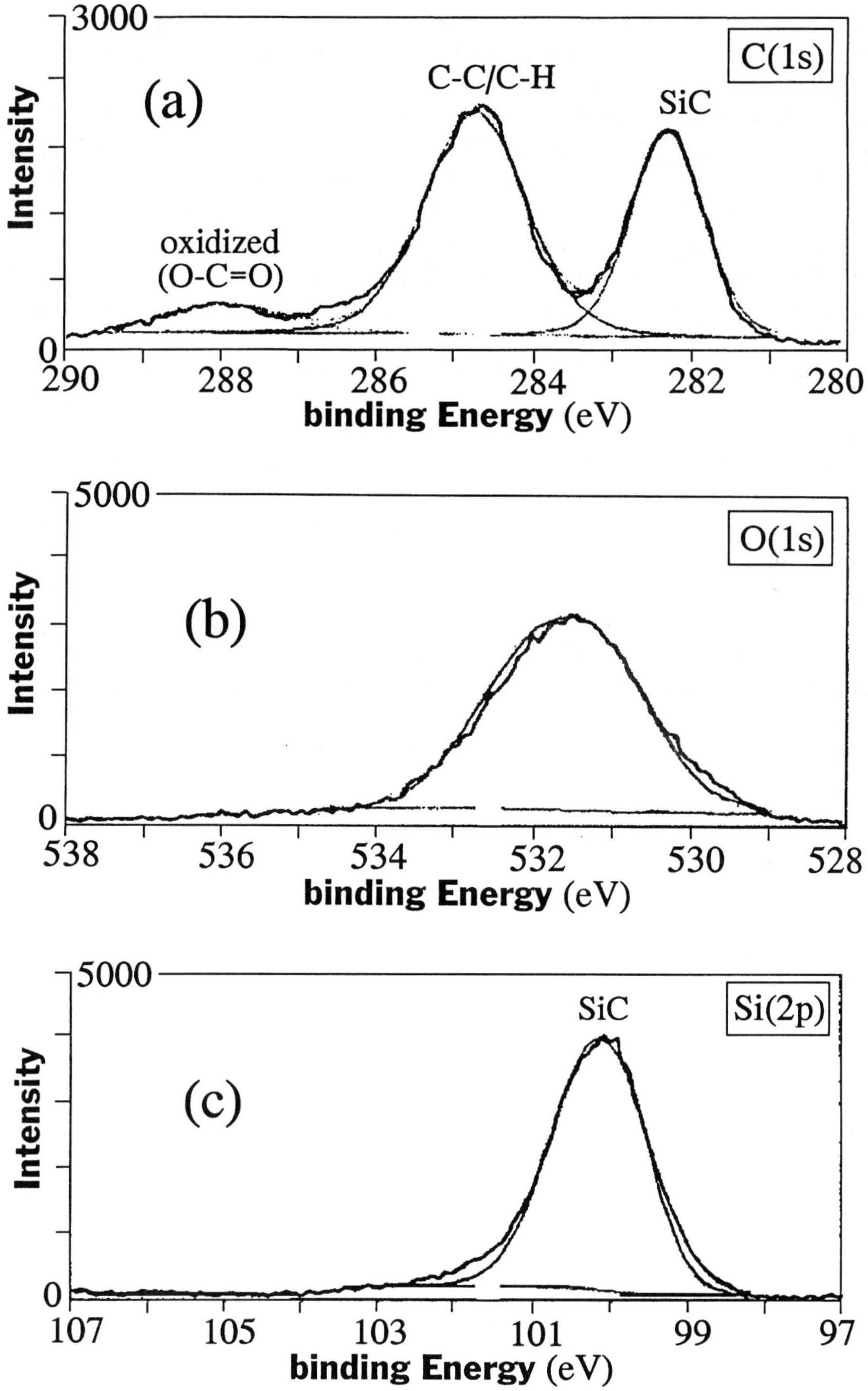

Figure 3. C1s (a), O1s (b) and Si2p (c) photopeaks from the surface of SiC100 subjected to Murakami's treatment.

Table 2.
Atomic percentages of chemical elements analyzed by XPS, for as-received and for treated SiC100 substrates with the solution of Murakami at ambient temperature

Sample	C (atomic%)		Si (atomic%)		O	N
	Included in SiC	Not in SiC	Included in SiC	Not in SiC	(atomic%)	(atomic%)
As-received	15.8	58	20.7	0	5.5	0
Murakami 3h	8.2	74.2	10.6	0	6.4	0.6
Murakami 5 h	11.8	64.5	15.6	0	7.4	0.7
Murakami 10 h 20 min	12.9	59.7	17.1	0	9.5	0.8
Murakami 25 h	15.4	53	19.8	0	11.2	0.7
Murakami 47 h 30 min	17.8	44.9	24.7	0	11.2	1.4
Murakami 72 h 20 min	19.1	38.1	29.1	0	12.9	0.8
Murakami 120 h	21.5	35.2	30	0	12.4	0.8
Murakami 265 h	25.8	22.5	35	0	15	1.1

If one looks now at θ = f(pH) curves performed after different times (t) of treatment with the solution of Murakami at ambient temperature, several remarks can be made (Fig. 4):

- for $t \leq 3$ h, θ was constant as a function of pH and stayed the same maximal value of around 70°;
- for $3h < t \leq 72$ h, θ was still constant vs pH but decreased with time, showing thus an increase of the work of adhesion;
- for $t > 72$ h, θ was constant for pH < 8 and still decreased with time, but the large difference from the previous results came here from the decay of θ for pH > 8: this means that the work of adhesion due to acidic sites was higher than that of basic sites at the SiC100 surface.

These results show that the surface polarity increases after 3 h of treatment and it can be assumed that the increase in the work of adhesion is due to Lewis acid–base interactions. On the other hand, for t > 72 h and a surface concentration of contaminants of around 35% (Fig. 2), typical stepwise titration curves are observed showing Brönsted acid–base interactions, with an increase of the surface acidity of SiC100 treated with the solution of Murakami.

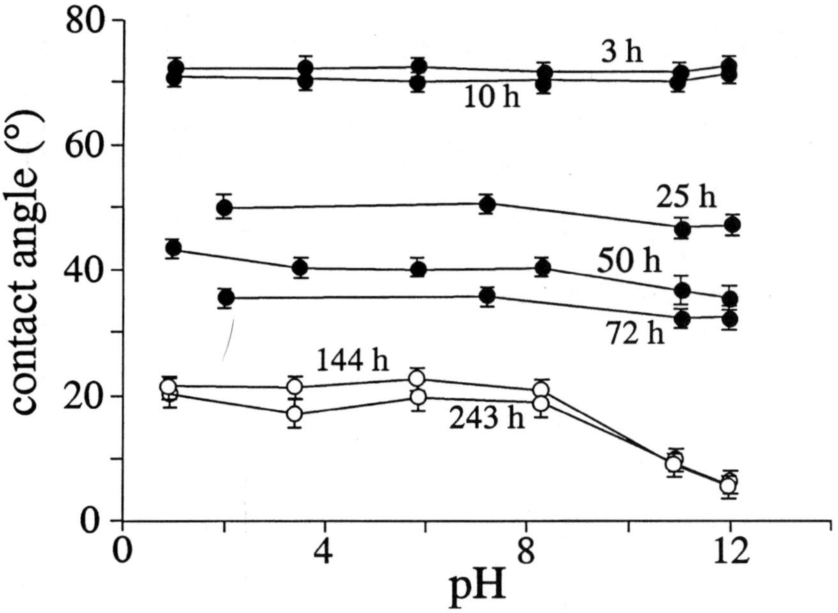

Figure 4. Contact angle titration curves performed on SiC100 after different times of treatment with the solution of Murakami at ambient temperature.

3.2. Thermal Treatment of SSiC

Like Murakami's treatment, the thermal treatment in air modified strongly the surface chemistry of SiC100. In a previous work, Levallois *et al.* [18] who studied this treatment for temperatures ranging from 20 to 800°C showed that the carbon contaminant decay and the silica formation were the two main modifications of the surface chemistry of SiC100. Figure 5 confirmed these results for treatment temperatures used up to 1600°C: silica formed considerably above 400°C, its concentration increased markedly between 400 and 800°C and stabilized for treatment temperatures higher than 800°C. In the last case, XPS measurements indicated that whole surface of SiC100 was probably coated by silica. At 1200°C, as also expected after the work of Frish *et al.* [25], crystalline silica (in white) was observed at the surface as shown by the SEM photomicrograph in Fig. 6. As for the carbon contaminant concentration, it decreased strongly between 400 and 600°C and stabilized from 600 to 1600°C. Because of the large quantity of contamination at the surface of as-received SiC100 samples, a reliable value of the contamination layer thickness seemed difficult to determine using XPS measurements, but it was possible to observe the decrease of this thickness for high treatment temperatures. Indeed, a comparison between XPS measurements at the external layer of SiC100 (C_{contam}) and after 2 min of Ar^+ ion sputtering ($C_{contam} + Ar^+$) has shown that 2 min were enough to remove all the contamination for temperatures higher than 1200°C (Fig. 5). This means that the contamination layer thickness was considerably reduced at these treatment temperatures and that the thermal treatment in air atmosphere acted well as a surface cleaning treatment.

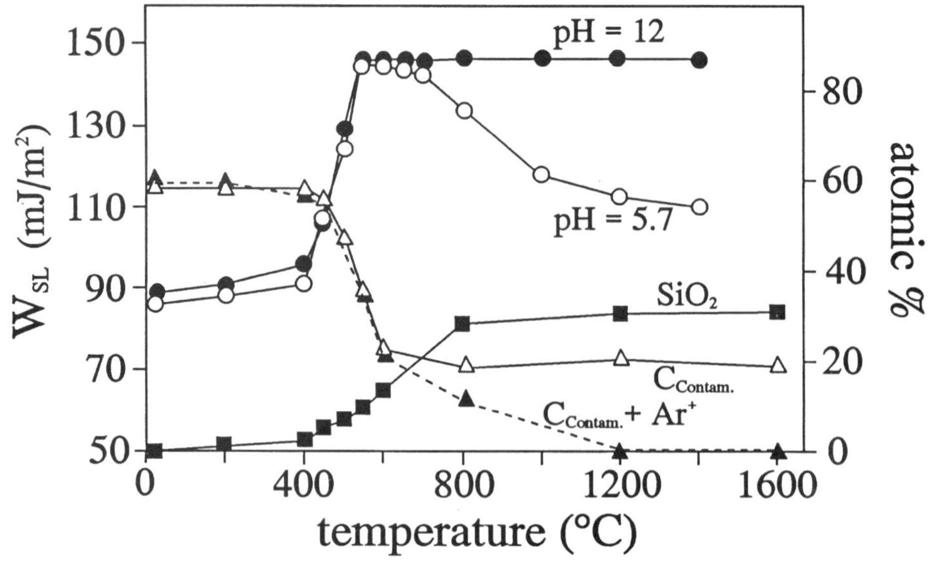

Figure 5. Variations of the work of adhesion (W_{SL}) of aqueous solutions (pH=5.7 and pH=12) and of atomic percentages of carbon contaminants and silica at the SiC100 surface versus temperature of thermal treatment in air atmosphere (Ar^+: Argon ion sputtering of the surface).

Figure 6. SEM photomicrograph of the SiC100 surface after the thermal air treatment for 1 h at 1200°C.

As was previously shown by Levallois *et al.* [18], the work of adhesion at the solid-liquid interface (W_{SL}) was linked to the carbon contaminant concentration at the SiC100 surface. The most important effect of this dependence is the sharp increase of WSL between 400 and 550°C, inversely to the decay of contaminant concentration (Fig. 5). This increase of W_{SL} is mainly due to the acid–base component ($W_{SL}{}^{AB}$) of the work of adhesion, because the variation of the dispersive component of the surface free energy ($\gamma_S{}^D$) – and hence that of WSLD – is low: $\gamma_S{}^D$ decreases from 44 to 40 mJ/m^2 for as-received to oxidized samples (Fig. 7). On the other hand, the liquid probes used here showed the same variations of $W_{SL}{}^{AB}$ from 20 to 1400°C, except for the (acidic) distilled water for which $W_{SL}{}^{AB}$ decreases from 600°C because of the stronger acidity of the oxidized SSiC surface, as will be demonstrated in the following.

Let us now look at θ = f(pH) curves performed after different temperatures (T) in the thermal air treatment:

- for 20°C < T ≤ 400°C, θ was constant *vs.* pH with the maximal value around 80° (Figure 8.a);
- for 400°C < T ≤ 550°C, θ was still constant *vs.* pH but decreased with temperature (Figure 8a), showing thus an increase in the work of adhesion;
- for 550°C < T ≤ 1400°C, the contact angle was constant for pH < 8 and still decreased with temperature, but only to 700°C because there was a new increase of θ from 700 to 1400°C (Figure 8b). On the other hand, θ decreased for pH > 8, more noticeably as treatment temperature increased: the work of adhesion due to acidic sites was thus higher than that due to basic sites at the SiC100 surface.

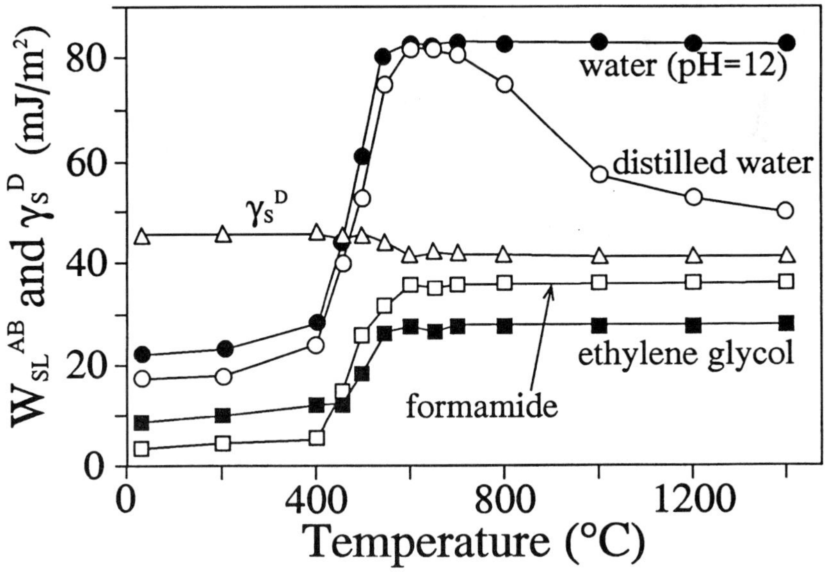

Figure 7. Variations of the acid–base component of the work of adhesion of various liquids ($W_{SL}{}^{AB}$) and of the dispersive component of the surface free energy ($\gamma_S{}^D$) of SiC100 as a function of the temperature of thermal treatment in air atmosphere.

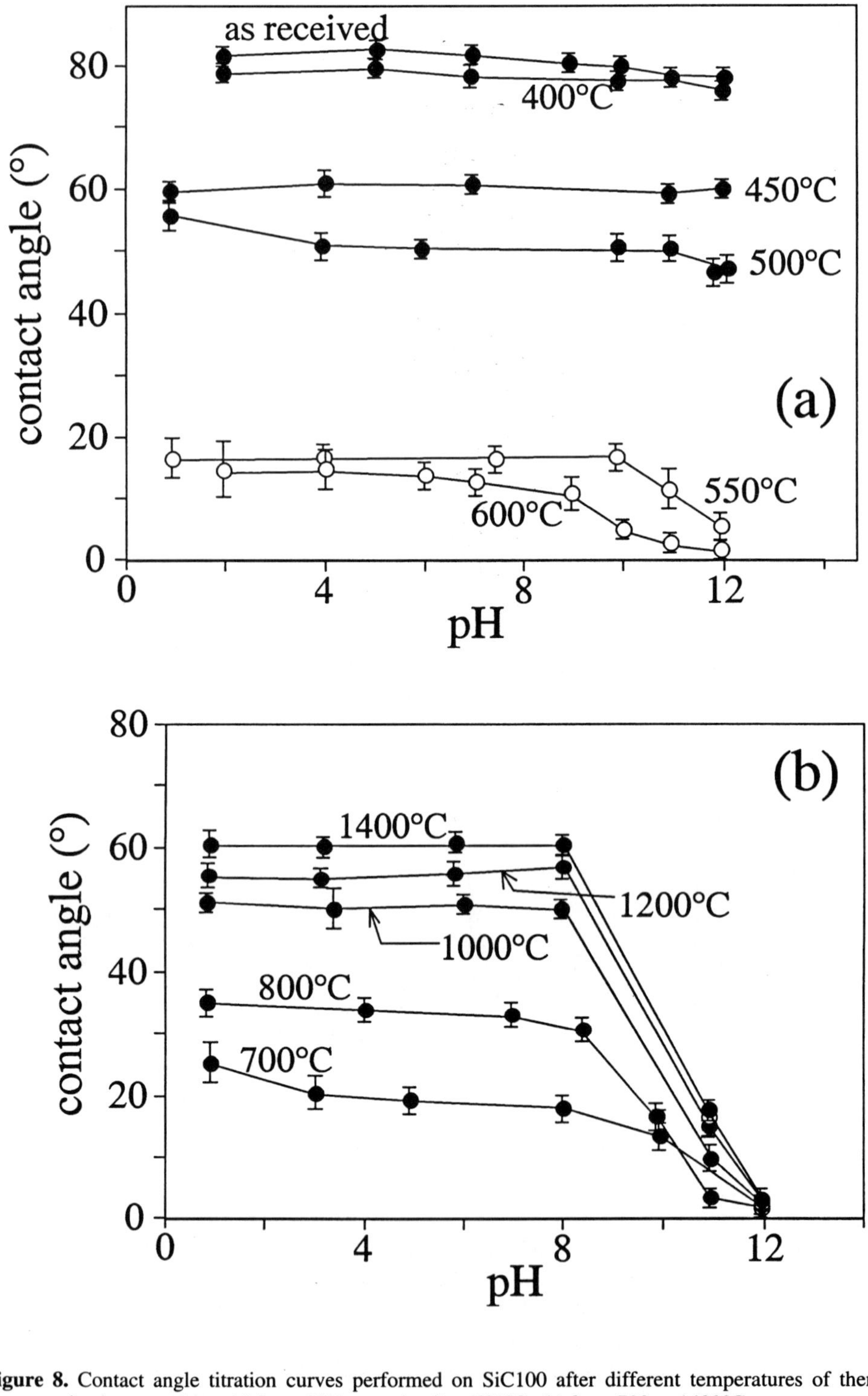

Figure 8. Contact angle titration curves performed on SiC100 after different temperatures of thermal treatment in air atmosphere, (a) from 20 (as-received) to 600°C, (b) from 700 to 1400°C.

The conclusions were similar to those made after Murakami's treatment but here the driving parameter was the treatment temperature: the surface polarity increased and Lewis acid–base interactions occurred if the temperature was lower than 550°C, but beyond this temperature stepwise titration curves were observed showing Brönsted acid–base interactions. It is interesting to note that the Brönsted acid–base interactions occurred for a carbon contaminant concentration of around 35%, as in the case of Murakami's treatment.

3.3. Adhesive Bonding

Figure 9 shows the variation of the rupture force (F_R) of adhesively-bonded SSiC as a function of the acid–base component of the work of adhesion at the solid-distilled water interface (W_{SW}^{AB}). The rupture test used was the compression-shear test [21], the structural adhesive was AV119 from Ciba and the SiC100 substrates were subjected to the following surface treatments marked with a number in Figure 9:

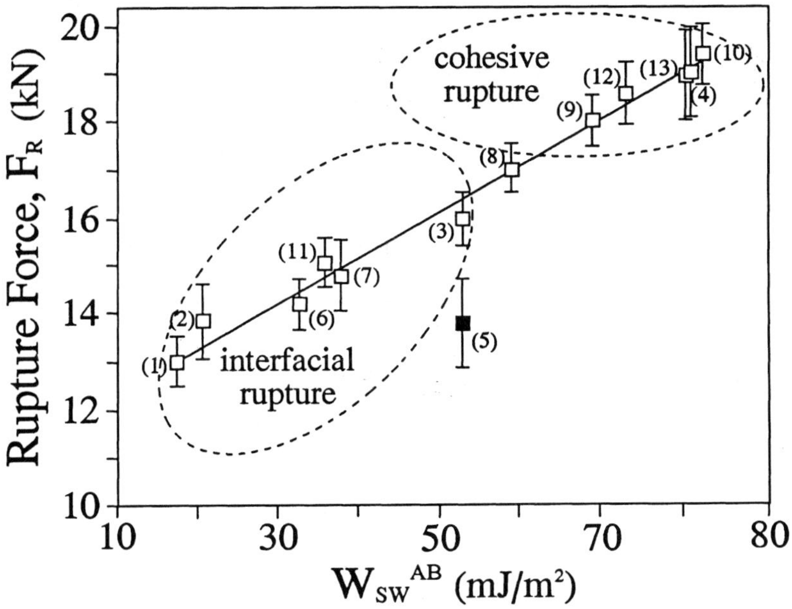

Figure 9. Variation of the rupture force (F_R) of single-lap adhesively-bonded SiC100 joints as a function of the acid–base component of the work of adhesion of distilled water (W_{SW}^{AB}) with SiC100 substrates subjected to various treatments (Numbers 1–13 correspond to each surface treatment of substrates).

(1): as-received SiC100 only solvent degreased with 1,1,1-trichloroethane;
(2) to (5): thermal treatment for 1 h at 300, 500, 700 and 1200°C, respectively;
(6) to (10): Murakami's treatment at ambient temperature for 1h, 5h 30 min, 24 h, 60 h and 144 h respectively;
(11) to (13): "hot" Murakami's treatment: the SiC100 substrates were dipped in the etching solution at ambient temperature, the solution was then heated to a given tem-

perature and substrates were then taken out and rinsed; the temperatures were 72°C (11), boiling (12) and substrates maintained in the boiling solution for 15 min (13).

Irrespective of the surface treatment, i.e. solvent degreasing, Murakami or thermal, the $F_R = f(W_{SL}^{AB})$ plot in Fig. 9 shows a linear dependence between F_R and W_{SL}^{AB} for distilled water but not for ethylene glycol and formamide (Figure 10), with interfacial ruptures for low values of W_{SL}^{AB} and cohesive ruptures in the adhesive for the higher values of W_{SL}^{AB}. In the latter case, the ruptures were cohesive close to the adhesive/adherend interface, with a thin adhesive layer (thickness of around 5 μm) remaining at the SiC100 surface. Such a result was found by Levallois *et al.* [18], with thermal air treated SiC100 substrates bonded with a structural epoxy adhesive, different from the one used in this study. A particular result has been reported in Figs 9 and 10: it is a matter of bonded assemblies made with substrates heated at 1200°C (point 5) for which the rupture loads (F_R = 13.5 kN) were as weak as those obtained with interfacial ruptures corresponding to lower treatment temperatures (T < 500°C) (the same value around 13.5 kN was also found with substrates oxidized at 1400°C; not reported here). However, the ruptures corresponding to 1200 and 1400°C appeared different from interfacial ruptures. Indeed, the bonded assemblies made with substrates oxidized at these high temperatures showed most of the time total cohesive ruptures in the substrates, and thus it was impossible to locate precisely the original failure. However, one can think that the interfacial zone becomes fragile because of the presence of crystalline silica formed at the SSiC surface at 1200°C (Fig. 6) and 1400°C, and moreover because of the creation of a fragile amorphous/crystalline silica interface, which can lead to substrate cohesive ruptures. This last point will be discussed later in Section 4.3.

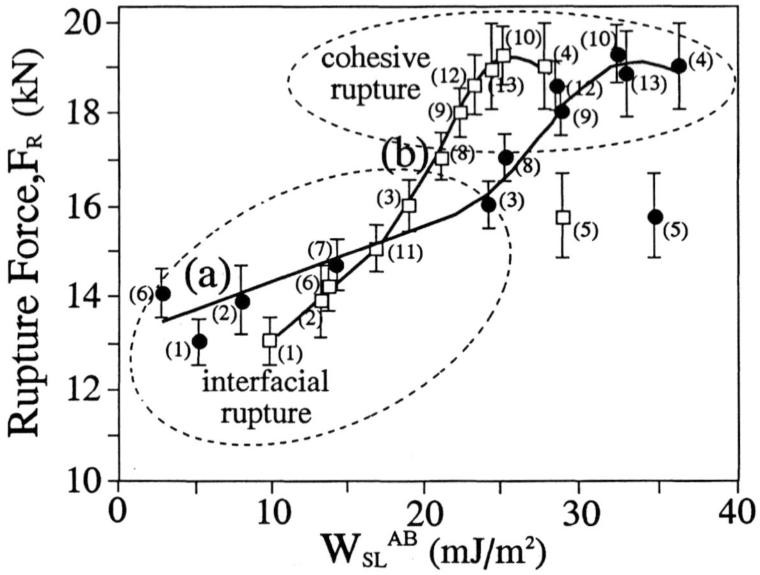

Figure 10. Same variation as in Figure 9, for formamide (a) and ethylene glycol (b) (Numbers 1–13 correspond to each surface treatment of substrates).

The linear increase of F_R with W_{SL}^{AB} is a priori surprising because there is a transition between interfacial and cohesive ruptures in the adhesive joints and no transition is observed in Fig. 9. However, from a mechanistic viewpoint the energy (U_R) involved at rupture (U_R = area subtended by the force (F)-displacement (d) curve in the (F,d) plane) and the critical strain energy release rate in shear mode (G_{IIC}) yield more precise information than F_R on the mechanical behaviour of bonded assemblies at rupture. Thus, using the Williams method with known defects of different lengths (0 to 5 mm) placed at the adhesive/substrate interface [20] yielded different UR values for bonded assemblies made with thermal treated SiC100 substrates (Fig. 11). Low energy interfacial and high energy cohesive ruptures in the adhesive are well separated in Fig. 11, because a clear sharp transition in rupture energy appears at 550°C. Levallois *et al.* [18] found this transition for similar thermal treated SiC100 assemblies bonded with two other epoxy adhesives, but the value of the transition temperature here is more precise and the mixed interfacial-cohesive ruptures (if they occur) can be observed only in the interval 500 to 550°C. Moreover, the Williams method allowed to calculate G_{IIC}, showing therefore quite higher values for cohesive ruptures (G_{IIC} ~ 3000 J/m^2) than for interfacial ruptures (G_{IIC} ~ 800 J/m^2), values in agreement with the previous results [21, 26, 27].

Figure 11 shows also that the transition at 550°C corresponds well to the carbon contaminant decay, as it was previously demonstrated [18], and for this temperature the work of adhesion W_{SL} of the surface with water (acidic or basic) is equal to 145 mJ/m^2, this value corresponding also to the emergence of Brönsted acid–base interactions at the solid–liquid interface.

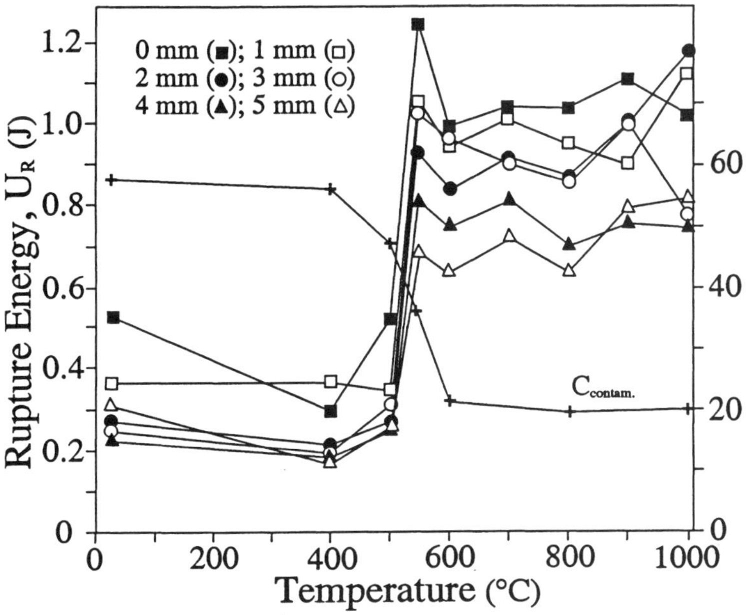

Figure 11. Variations of the rupture energy (U_R) and of the carbon contaminant concentration as a function of the temperature of thermal treatment of SiC100 substrates in air atmosphere, for adhesively-bonded SiC100 joints containing defects of various lengths (0 to 5 mm) at the adhesive/substrate interface.

4. DISCUSSION

Our attention is focused on the presence of carbon contaminants and on the formation (or not) of silica at the SiC100 surface, its acid–base properties being highly dependent on these two parameters. Comments will be also made on the mechanical results in relation to the surface properties.

4.1. Layers at the SiC100 Surface

First, a silica layer is formed at the thermally air treated SSiC surface, but not after the etching treatment of Murakami. After 2 min of Ar^+ sputtering of the thermally treated surface (800°C, 1 h), i.e. the ion beam penetrated this silica layer: this means that under these conditions the silica layer was around 10 nm thick, and it was 200 nm for a treatment for 1h at 1600°C. This silica is amorphous for treatment temperatures lower than 870°C and a part of it is converted into crystalline silica above this temperature [25]. SEM observations of the SiC100 surface showed crystalline silica for temperatures higher than 1000°C, but we have not verified its polymorphism. For 867°C < T < 1470°C however, one can expect that this crystalline silica was tridymite rather than crystobalite [28].

On the other hand, a significant quantity of carbon contaminants was measured by XPS at the surface of as-received SiC100 samples. Both the Murakami and the thermal treatments cleaned the SiC100 surface. This cleaning effect resulted in a considerable decrease of carbon contaminants. If the ceramic substrate was treated at ambient temperature with the solution of Murakami for a long time (11 days), its surface contamination decreased from 58% (as-received) to 22.5%. The same phenomenon was observed with the thermal treatment and the contaminant decay was strong between 400 and 600°C, to result in relatively low contamination (around 20%) for treatment temperatures higher than 800°C. Contrary to the silica layer, too much uncertainty was present in the calculation of the contamination layer thickness under our XPS experimental conditions and thus we did not try to evaluate this thickness. However, XPS measurements showed clearly that the thickness of the contamination layer decreased with time (Murakami) or temperature (thermal treatment).

4.2. Acid–base Properties of the SiC100 Surface

Titration of surfaces using the method of Hüttinger allowed to highlight the acid–base properties of the SiC100 surface. The θ = f(pH) curves showed no variation of the contact angle with increasing pH, for treatment times less than 72h with Murakami's solution at ambient temperature. According to equation (4), this means that the acid–base component of the work of adhesion (W_{SL}^{AB}) was constant with pH, assuming the presence of an equal quantity of Lewis acidic and basic sites at the ceramic surface. Beyond 72h, however, the ceramic surface became more acidic in the Brönsted sense, owing to the stepwise decrease of titration curves, but this surface acidity seemed weak because the contact angle decay *vs.* pH was not significant (around 10°).

The case of thermally treated SiC100 surfaces was more clear. For temperatures lower than 550°C, the total work of adhesion of aqueous solutions with the SiC100 surface was quasi-constant *vs.* pH and increased from approximately 80 (as-received samples) to 145 mJ/m^2 (T = 550C) (Fig. 12a). Note that this value of 145 mJ/m^2 was

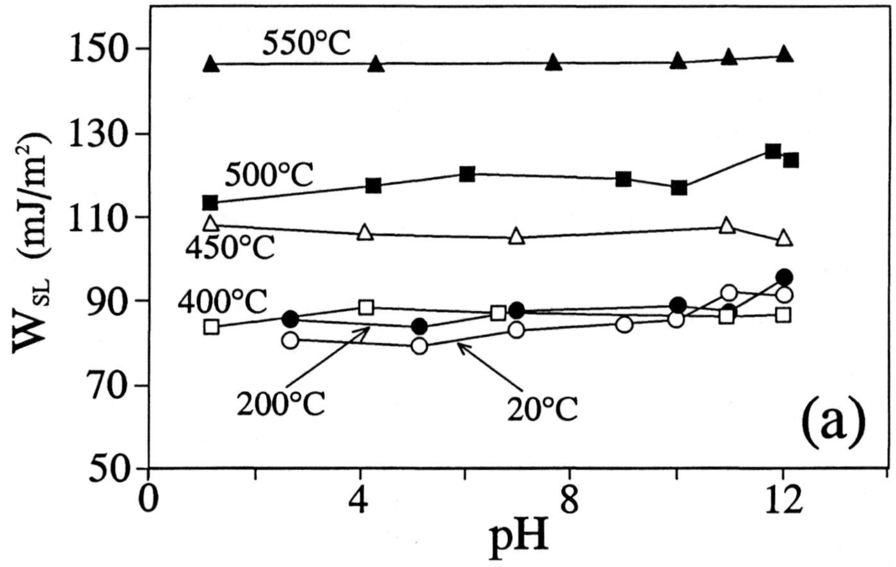

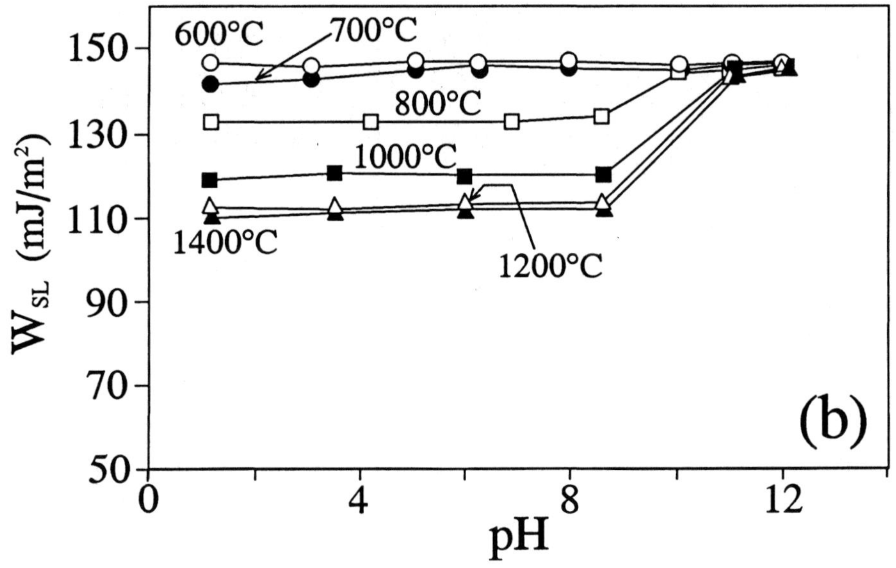

Figure 12. Variation of the work of adhesion of aqueous solutions as a function of pH, for SiC100 substrates subjected to thermal treatment in air atmosphere at: (a) 20°C < T < 550°C and (b) 600°C < T < 1400°C.

equal to that found by Hüttinger *et al.* on hydrolysed crystalline silica fibers [14]. As in the case of the ambient Murakami treatment for short time of treatment, either the surface was chemically inert or the work of adhesion of the basic sites (W_{SL}^{AB}) was equal to that of acidic sites (W_{SL}^{AB}) at the ceramic surface. For temperatures going from 550°C to 1400°C, the maximal work of adhesion remained constant (as did W_{SL}^{AB}) and equal to 145 mJ/m^2 for basic pH, whilst W_{SL}^{AB} and hence W_{SL} both decreased for acidic pH (Fig. 12b). So, in agreement with the Hüttinger criterion, W_{SL}^{AB} (basic pH) > W_{SL}^{AB} (acidic pH) and the ceramic surface became more acidic in the Brönsted sense as the treatment temperature increased.

For both treated SiC100 surfaces (Murakami and thermal), Brönsted acid–base interactions were observed for the same carbon contaminant concentration at the ceramic surface, i.e. around 35%. Actually, these interactions were observed only if the thickness of the contaminant layer was low enough. In other words, according to this result, contaminated low polarity surfaces of as-received only degreased SiC100 samples resulted in long-range Lifshitz–van der Waals interactions with water, whilst clean high polarity Murakami or thermally treated surfaces resulted in short-range acid–base interactions, with a transition between Lewis and Brönsted interactions for a particular thickness of the contamination layer. Most certainly, the presence of silica at the ceramic surface was important because silica made the ceramic surface more acidic when the SiC100 surface was oxidized, but Brönsted interactions were also observed without silica thus showing that silica was not the main parameter that governed the acid–base properties of the surface.

4.3. Surface Properties and Mechanical Results

It is interesting to note that the Lewis-Brönsted transition corresponds also to the mechanical transition between interfacial and cohesive ruptures in the adhesive, for treated (Murakami or thermal) SiC100 samples bonded with an epoxy adhesive. In the case of thermal treatment, the transition was precisely at 550°C and almost no mixed (interfacial + cohesive) rupture was observed. It must be mentioned that the correspondence between the mechanical transition and the Lewis-Brönsted acid–base transition is rather surprising, because this would mean that the epoxy adhesive is sensitive to the type of acid–base interactions on the one hand, and that Brönsted interactions are needed to obtain cohesive ruptures on the other hand.

To make this last point clear, some of the results in Section 3.3 can be commented on and connected to the type of interactions previously discussed (Section 4.2). These results concern the type of ruptures obtained for bonded assemblies made with SiC100 substrates oxidized at high temperatures (T=1200°C and 1400°C). In these cases, the brittle cohesive rupture of ceramic substrates could be attributed to a fragile amorphous/crystalline silica interface. Such a hypothesis seems appropriate, but does not provide any explanation about the two important points which are the level of rupture forces obtained and the actual initial failure location, i.e. interfacial or cohesive. Indeed, in Section 3.3, it was mentioned that the rupture forces F_R were equal to 13.5 kN for bonded assemblies made with substrates oxidized at 1200 or 1400°C, the same F_R values as those obtained for lower treatment temperatures (T < 500°C) which corresponded to clear interfacial ruptures. Afterwards, F_R follows the same variation as W_{SL}^{AB} for acidic distilled water as a function of the treatment temperature of substrates (Fig. 7). Such a result is not contradicted by the point (5) in Fig. 9, because this point corresponds to a special rupture behaviour of bonded assemblies; the rupture force value at point (5) (T = 1200°C) would probably be the same as the one at point (3) (T = 500°C) with interfacial ruptures (see UR at 500°C in Fig. 11), if there was no fragile amorphous/crystalline silica interface. Thus, one can assume that the ruptures at 1200 and 1400°C were mixed interfacial and cohesive ruptures, with a probable initiation located at the adhesive/adherend interface and a propagation failure towards zones of less mechanical strength, i.e. the amorphous/crystalline silica interface. Such a hypothesis is supported by the fact that in single-lap ceramic assemblies subjected to compression, the maximal stresses are located in the ceramic substrates, close to the free edges of the joint and at the ceramic/adhesive interface [22].

Another point can also be made from this study: the non-linear dependence of F_R *vs.* W_{SL}^{AB} for ethylene glycol and formamide (Fig. 10). Actually, according to what has just previously been explained, the straight line $F_R = f(W_{SL}^{AB})$ was observed only for acidic distilled water, and not for ethylene glycol, formamide or even basic water. Ethylene glycol and formamide showed the same variations as basic water (pH=12) *vs.* treatment temperature (Fig. 7), and hence the same acid–base behaviour toward SiC100 substrates. Figure 10 thus confirms, as have been shown in Figures 7 and 9, that the mechanical properties of epoxy-dicyandiamide (AV 119) bonded SiC100 assemblies are linked to the surface acidity of substrates on the one hand and that the adhesion of AV119 epoxy-dicyandiamide adhesive is not favourated when it is applied to an acidic surface on the other hand.

5. CONCLUSION

This work concerned the study of acid–base properties of the surface of a sintered SiC ceramic subjected to an etching treatment or a thermal treatment in air, in relation to the adhesion of a structural epoxy adhesive at this surface. Techniques for characterization of surfaces (three-dimensional profilometry, SEM, AFM, XPS and wettability) were used, as well as mechanical characterization (single-lap compression-shear rupture test) of bonded ceramic assemblies was performed. The study has shown the following important results:

- the Hüttinger method used with the one-liquid wettability technique is a sensitive and powerful means for the characterization of acid–base properties of surfaces;
- the acid–base properties of SSiC surfaces are dependent on their hydrocarbon contamination and of the creation of silica (if it occurs), but acid–base properties can be found without the presence of silica at SSiC surface;
- a Lewis-Brönsted acid–base transition appears for carbon contaminant concentrations lower than 35% (XPS atomic percentage) and this transition corresponds also to the mechanical transition between interfacial and cohesive ruptures in the adhesive, irrespective of the surface treatment used (etching Murakami's in aqueous solution or thermal treatment in air);
- for long times of etching or high temperatures of oxidation the contaminant concentration stays around 20%, and this corresponds to a lower thickness of the contamination layer; however, it is not sure that the contamination thickness is lower than 0.5 nm (short-range AB interactions). So, contamination thickness measurements (by XPS) must be made on these surfaces in the future to confirm this important point;

- the treated SiC100 surface becomes more acidic (without the formation of silica) for long times of etching treatment, but the phenomenon is more significant for high temperatures of oxidation in air when a considerable quantity of silica is formed on the SiC surface (crystalline silica for T > 870°C theoretically on SiC, but visible for T > 1000°C on SiC100 in our experimental conditions);
- and last the mechanical properties of epoxy-dicyandiamide (AV 119) bonded SiC100 assemblies depend on the surface acidity of substrates, and the adhesion of AV119 is not favored when it is applied to an acidic surface.

REFERENCES

1. K. L. Mittal and H. R. Anderson, Jr. (Eds), *Acid–Base Interactions: Relevance to Adhesion Science and Technology*. VSP, Utrecht (1991).
2. M. Mantel and J. P. Wightman, *Surface Interface Anal.* **21**, 595–605 (1994).
3. X.-Y. Lin, F. Creuzet and H. Arribart, J. Phys. Chem. **97**, 7272–7276 (1993).
4. K. Hue and A. J. Bard, *Langmuir* **13**, 5114–5119 (1997).
5. F. M. Fowkes, *J. Adhesion Sci. Technol.* **1**, 7 (1987).
6. C. Vittoz, *Thesis*, Institut National Polytechnique, Grenoble, France (1997).
7. M. G. Barthès and J. C. Joud, in these proceedings.
8. H. Balard and E. Papirer, in these proceedings.
9. F. M. Fowkes, *J. Phys. Chem.* **68**, 1863 (1962).
10. F. M. Fowkes, *Ind. Eng. Chem.* **56**(12), 40 (1964).
11. L. A. Girifalco and R. J. Good, *J. Phys. Chem.* **61**, 904–909 (1957).
12. T. Young, *Phil. Trans. Roy. Soc.* **95**, 65 (1805).
13. K. J. Hüttinger, S. Höhmann and M. Seiferling, *Carbon* **3**, 449–455 (1991).
14. K. J. Hüttinger, S. Höhmann-Wien and G. Krekel, J. Adhesion Sci. Technol. **6**, 317–331 (1992).
15. G. M. Whitesides and P. E. Laibinis, *Langmuir* **6**, 87–96 (1990).
16. S. Helt, Thesis no. 1410, Institut National Polytechnique, Toulouse, France (1998).
17. F. Levallois, *Thesis* no. 1026, Institut National Polytechnique, Toulouse, France (1995).
18. F. Levallois, S. Helt, Y. Baziard and J. A. Petit, *J. Adhesion Sci. Technol.* **13**, 273–287 (1999).
19. F. Deuerler, H. Van den Berg, R. Tabersky, A. Freundlieb, M. Pies and V. Buck, *Diamond and Related Materials* **5**, 1478 (1996).
20. R. Haubner, S. Kubelka, B. Lux, M. Griesser and M. Grasserbauer, *J. Physique IV, Colloque C5, suppl. to the Journal de Physique* **II**, 5 (June 1995).
21. Y. Baziard, R. El Abdi, D. Amara, J. A. Petit and F. Levallois, *Intl. J. Adhesion Adhesives* **15**, 155–160 (1995).
22. D. Amara, F. Levallois, Y. Baziard and J.-A. Petit, *J. Adhesion Sci. Technol.* **10**, 1153–1164 (1996).
23. F. Levallois, R. El Abdi, Y. Baziard and J. A. Petit, *J. Adhesion* **60**, 1–14 (1997).
24. J. G. Williams, *Fracture Mechanics of Polymers*. Ellis Horwood, Chichester (1984).
25. B. Frish, W. R. Thiele, R. Drumm and B. Münnich, *Ceramic Forum Intl.* **65**(8-9), 177 (1988).
26. F. Schmit and P. Fraysse, *Matériaux et Techniques* **4-5**, 55 (1992).
27. M. Papini, G. Fernlund and J. K. Spelt, *Intl. J. Adhesion Adhesives* **14**(1), 5 (1994).
28. W. D. Kingery, H. K. Bowen and D. R. Uhlmann, *Introduction to Ceramics,* 2nd Edition. John Wiley (1976).

Acid-Base Interactions, Vol. 2, pp. 419–437
Ed. K.L. Mittal

Thin layer wicking — Methods for the determination of acid-base free energies of interaction

EMIL CHIBOWSKI*

Department of Physical Chemistry, Faculty of Chemistry, Maria Curie-Skłodowska University, 20-031 Lublin, Poland

Abstract—Theoretical and practical aspects of a modified "thin layer wicking" method for solid surface free energy components determination are discussed. This method is based on measurements of the imbibition rate of a probe liquid in a thin porous layer of the studied solid deposited on a substrate (usually a glass slide). The theory of the process is based on Washburn's equation which at constant temperature predicts a linear relationship between wicked squared distance and the time of wicking. The slope of the relationship is determined by the liquid viscosity, apparent radius of the interparticle capillaries, and most importantly by the free energy changes occurring at the liquid front. A proper recognition and description of the energy changes allows calculation of the solid surface free energy components; the apolar and Lewis acid-base: electron donor and electron acceptor. In the case when the contact angle method cannot be applied (powdered solids) this method can be satisfactorily used. Some examples of experimental results are presented and discussed.

Keywords: Surface free energy determination; thin layer wicking theory and experiment.

1. INTRODUCTION

About a decade ago van Oss, Good and Chaudhurry [1] proposed a new approach to the surface and interface free energy determination in which Lewis acid-base interactions were formulated in a quantitative way as follows for the first time.

$$\gamma_i^{TOT} = \gamma_i^{LW} + \gamma_i^{AB} = \gamma_i^{LW} + 2(\gamma_i^{+}\gamma_i^{-})^{1/2} \qquad (1)$$

where γ_i^{TOT} is the phase i total surface free energy, γ_i^{LW} is the apolar Lifshitz-van der Waals component, γ_i^{AB} is the Lewis acid-base component, and γ_i^{+}, γ^{-} are the electron acceptor (proton donor, acidic) and electron donor (proton acceptor, basic) parameters, respectively. For most solids and liquids this formulation gives low values of experimentally determined acidic parameter and high values of the basic parameter. This may be partially attributed to relative values of the acid-base parameters of the probe liquids used for the contact angle measurements, such as formamide, glycerol, ethylene glycol and others. The parameters are based on the assumption that for water $\gamma^{+} = \gamma^{-} = 25.5$ mJ/m^2 [2–4]. Although the problem of experimental determina-

* To whom correspondence should be addressed. E-mail: emil@hermes.umcs.lublin.pl

tion of the acid-base interactions is not solved yet, even the relative changes in the experimentally determined parameters can give insight into the processes taking place at the interfaces in which the free energy interactions play an essential role. It should be stressed that the assumption about equality of γ^+ and γ^- for water does not influence the resulting values of the work of adhesion and interfacial free energy, because these parameters cancel out in the calculations [4].

In case when a flat surface of the studied solid can be obtained, the contact angle measurements for three probe liquids (at least two of which are polar) allow determination of the solid surface free energy components mentioned above. This can be achieved by solving simultaneously three equations relating the contact angle and the work of adhesion, W_a, for the three probe liquids used [2–4]. It should be mentioned that no film pressure behind the drop is taken into account in the equations, which may significantly affect the calculated results.

$$\gamma_l (\cos\theta) = W_a = 2\,[(\gamma_s^{LW}\gamma_l^{LW})^{1/2} + (\gamma_s^{+}\gamma_l^{-})^{1/2} + (\gamma_s^{-}\gamma_l^{+})^{1/2}] \tag{2}$$

where subscripts “s” and “l” mean solid and liquid, respectively. Recently van Oss *et al.* [5] calculated the extent to which the surface of poly(methyl methacrylate) (PMMA) could be covered with adsorbed vapors of several probe liquids. Similar calculations were also done for water vapor on poly(ethylene oxide) (PEO). They found that at 20°C the PMMA surface percentage coverage was: 1.53 for diiodomethane, 2.14 for 1-bromonaphthalene, 3.55 for water, and only 0.075 for formamide, while for water on PEO it was 3.90. These coverages may affect the contact angles to an extent of 0.3–1.5 deg. The calculations were based on the liquids vapor pressure at 20°C and an adsorption equation of the Langmuir type, as well as the determined components of the surface free energy of the solids. The authors [5] consider the calculated coverages as maximum ones, but they did not take into account condensation of the vapor on the adsorbed molecules, as well as any possible migration of the liquid molecules from the drop over the solid surface. In case of a liquid with definite contact angle, these processes will increase the solid surface free energy, and this will appear in contact angle hysteresis, i.e. the receding contact angle will be less than the advancing one. In the process of the vapor condensation nano-size droplets will grow up to micro-size and even to macro-size ones, depending on the process duration and the temperature. The final effect of such process, for example, can be seen for water on a glass surface (dew).

However, in case of precipitated solids, clays, soils, etc., the contact angle method cannot be applied. The compressed discs always show some porosity, which affects the measured contact angles. Another method for contact angle measurement on powdered solids relies on deposition of the powder suspension onto a porous membrane by suction and then drying it. But, thus obtained surface of the deposit is rough, again affecting the measured contact angles. This method was, however, successfully applied for mammalian cells and bacteria [4]. However, in the case of powdered solids the method based on the measurement of the rate of probe liquid penetration into a porous layer of the solid can be applied. For the calculation of the surface free energy components Washburn’s equation [6] is applied. Recently van Oss and coworkers [7] proposed a modification of the method. They deposited a thin porous layer of the investigated solid on a glass substrate and called the method “thin-layer wicking”.

2. THEORY

The theory of the thin layer wicking method is based on the well-known Washburn equation [6, 8, 9], which in the case of a liquid entering into an empty capillary reads:

$$v = \frac{r\gamma_l \cos\theta}{2\eta x} \tag{3}$$

where v is the liquid entry rate into the capillary, r is the capillary radius, γ_l is the liquid surface tension, η is the viscosity of the liquid, x is the penetrated distance, and θ is the angle between the tangent to the meniscus surface and the capillary wall. The magnitude of this angle is often interpreted to be equal to the contact angle in the Young equation.

A more useful form of Eq. (3) can be obtained when the entry rate v is expressed as $v = x/t$ (t is the time needed to reach the distance x), then:

$$x^2 = \frac{rt}{2\eta}\gamma_l \cos\theta \tag{4}$$

As mentioned above, for practical purposes the penetration rate of a liquid into the porous thin layer of a powdered solid, for example deposited on a glass slide substrate, can be measured. However, to use Eq.(4) for the contact angle calculation one has to know the effective radius (an average apparent radius) R of the interparticle capillaries, which are formed by the solid particles (the porous layer) deposited on the glass substrate.

But, there is more fundamental problem with the use of this equation for contact angle determination. As it was already shown in the literature [10–15] thus calculated contact angle is not equal to that of Young's equation and therefore cannot give true values of the solid surface free energy components. Nevertheless Eq.(4) is still used for this purpose [16–19]. Lately, Constanzo *et al.* [20] have shown that the contact angles calculated from Washburn's equation are practically the same as those measured directly on the thin layer of monosized cuboid hematite particles deposited on a glass plate. In the same paper [20] they presented that contact angles of probe liquids measured on a flat surface of calcite, dolomite and glass were very different from those calculated from the wicking experiments conducted on the same solids. This might be partially due to changes in the surface properties caused by the grinding process [21]. Nevertheless, there are also some methodology reasons, which are outlined below.

Recently, we have proposed a general form of Washburn's equation [14, 15, 22–24] for the determination of the solid surface free energy components; i.e,

$$x^2 = \frac{Rt}{2\eta}\Delta G \tag{5}$$

in which ΔG means specific (per unit area) change in the free energy accompanying the process of the liquid entering into the porous layer. In this notation a positive value of the energy change means that the process is spontaneous. For particular liquid wicking the porous layer of the solid at constant temperature, the relationship $x^2 = f(t)$ should be linear with the ΔG slope. From theoretical considerations and experimental results [14, 15, 22–24] it was deduced that ΔG depended on the solid/liquid interactions resulting from the liquid and the solid surface free energy components, i.e. apolar Lifshitz-van der Waals, γ_s^{LW}, and polar acid-base: the electron donor, γ_s^-, and electron acceptor, γ_s^+. Thus, considering a model system of a liquid moving into a single capillary having radius r, four cases can be distinguished in which the reduction in free energy *dG*/*dx* is different.

I. For completely wetting liquid and bare surface of the capillary wall (the porous layer) it is:

$$dG/dx = 2\pi r(\gamma_s - \gamma_{sl} - \gamma_l) = 2\pi r(W_a - W_c) = 2\pi r W_s \tag{6}$$

Where $W_c = 2\gamma_l$ is the work of cohesion, W_a is the work of adhesion, and W_s is the work of spreading. This case is illustrated in Fig. 1a.

II. The solid surface is covered with the duplex film of the same liquid [16], the film pressure of which equals: $\pi_e = W_s = W_a - W_c$, and $\gamma_{sv} = \gamma_s - \pi_e$, then:

$$dG/dx = 2\pi r\ \{[(\gamma_s - W_s) - \gamma_{sl} - \gamma_l = 0] + \gamma_l\} \tag{7}$$

The presence of the duplex film means that there is a layer of the liquid parallel to the capillary surface and as a consequence a spherical meniscus must form and thus cause the pressure difference across the interface proportional to γ_l (see below) [25]. For a circular capillary and spherical meniscus, the capillary radius equals that of meniscus $r = r_m$ [25]. This situation is illustrated in Fig. 1b.

III. For partially wetting liquid (the liquid droplet forms contact angle on the flat solid surface) and bare surface of the capillary, the liquid moves at a definite dynamic advancing angle θ_a:

$$dG/dx = 2\pi r \gamma_l \cos\theta_a \tag{8}$$

For such liquid and spherical meniscus, a relationship between the radius of the capillary and that of the meniscus also exists. Now the two radii are not equal but interrelated: $r_m = r/\cos\theta_a$ [25]. This system is depicted in Fig. 1c.

IV. For the same partially wetting liquid as in III when the liquid duplex film is present on the surface (definite dynamic receding wetting angle).

$$dG/dx = 2\pi r \gamma_l \cos\theta_r \tag{9}$$

In this system a similar relationship as in III between the two radii holds: $r_m = r/\cos\theta_r$, and it is shown in Fig. 1d.

Then, the Laplace equation relates the energy reduction with the pressure across a curved surface ΔP [9, 25] as:

$$\Delta P = \frac{1}{\pi r^2}\frac{\partial G}{\partial x} \tag{10}$$

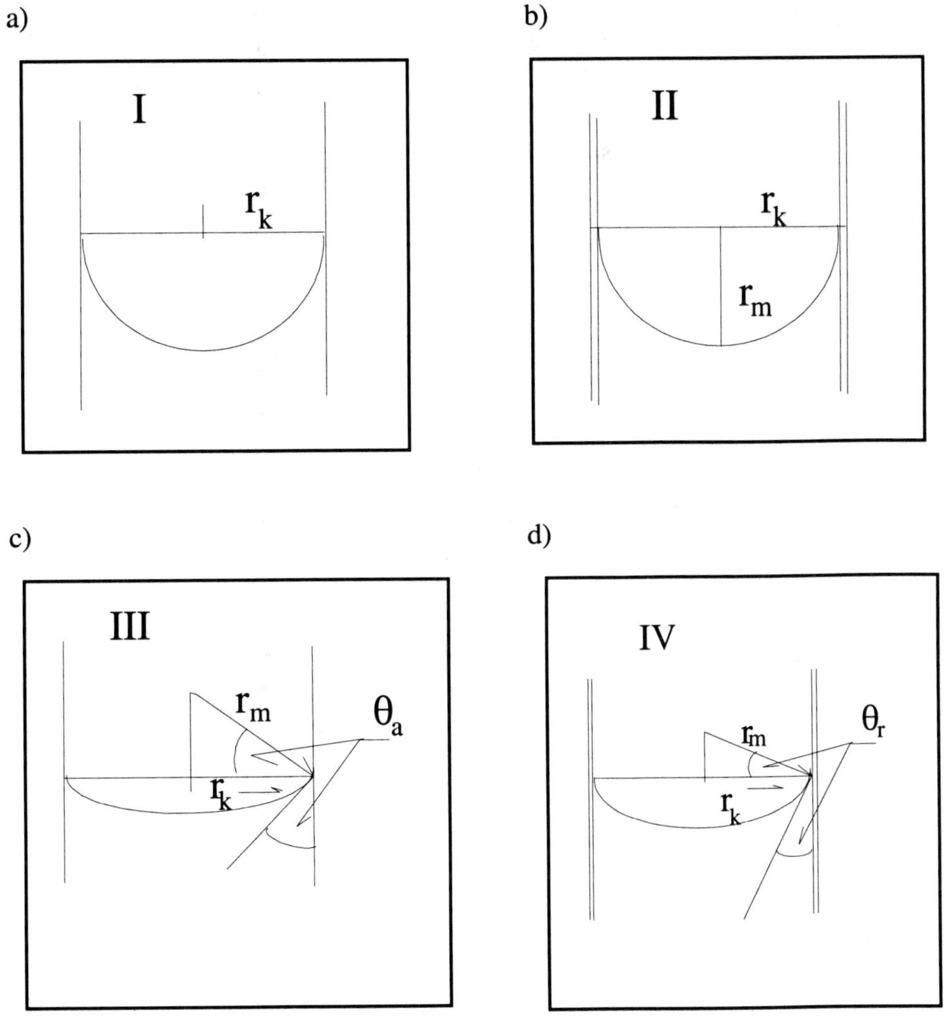

Figure 1. Schematic representation of the discussed solid/liquid systems I–IV (see text).

Hence for these particular cases one obtains:

(I)	(II)	(III)	(IV)
$\Delta P = \frac{2W_s}{r}$	$\Delta P = \frac{2\gamma_l}{r}$	$\Delta P = \frac{2\gamma_l \cos\theta_a}{r}$	$\Delta P = \frac{2\gamma_l \cos\theta_r}{r}$

The relationship between θ_a and θ_r of cases III and IV will be discussed further.

In the case of a thin porous layer an average apparent radius R of the interparticle capillaries (pores) can be used instead of the capillary radius r: Combining relationships (6–9) with Poiseuille's equation for the liquid motion into a capillary:

$$\frac{dx}{dt}=\frac{\Delta P r^2}{8\eta x} \quad \Rightarrow \quad \Delta P=\frac{8\eta x}{r^2}\frac{dx}{dt}=\frac{4\eta}{r^2}\frac{dx^2}{dt} \tag{11}$$

appropriate equations are obtained for the cases discussed:

I. $$\frac{4\eta}{R^2}\frac{dx^2}{dt}=\frac{2W_s}{R} \quad \Rightarrow \quad x^2=\frac{Rt}{2\eta}(W_a-W_c) \tag{12}$$

Hence in Eq.(5): $\Delta G = \Delta G_b = W_a - W_c$

II. $$\frac{4\eta}{R^2}\frac{dx^2}{dt}=\frac{2\gamma_l}{R} \quad \Rightarrow \quad x^2=\frac{Rt}{2\eta}\gamma_l \tag{13}$$

In this case, from Eq. (5) it results: $\Delta G = \Delta G_p = \gamma_l$.

III. $$\frac{4\eta}{R^2}\frac{dx^2}{dt}=\frac{\gamma_l\cos\theta_a}{R} \quad \Rightarrow \quad x^2=\frac{Rt}{2\eta}\gamma_l\cos\theta_a \tag{14}$$

Here, for partially wetting liquid and the bare surface, in Eq.(5): $\Delta G = \Delta G_b = \gamma_l\cos\theta_a$

IV. $$\frac{4\eta}{R^2}\frac{dx^2}{dt}=\frac{\gamma_l\cos\theta_r}{R} \quad \Rightarrow \quad x^2=\frac{Rt}{2\eta}\gamma_l\cos\theta_r \tag{15}$$

Finally, for the same liquid but the prewetted surface: $\Delta G = \Delta G_p = \gamma_l\cos\theta_r$.

As can be seen Eqs. (14) and (15) are similar to Eq.(4), except that the contact angle has been specified. These contact angles are dynamic wetting angles and cannot be directly used for the solid surface free energy calculations. Here, a question arises whether thus calculated contact angles are useful for the determination of the surface free energy components at all. To answer this question many experimental results on static advancing and receding contact angles were analyzed, obtained both from the literature [26–29] as well by us [30, 31]. It was found that there was a functional relationship between the values of advancing and receding contact angles of the liquid on the same solid surface, which is called hysteresis of contact angle [32]. Assuming that in a dynamic condition the same relationship holds as in a static one, Young equation can be applied to find the relationship between the angles in Eqs. (14) and (15).

Thus for a bare surface:

$$\gamma_l\cos\theta_a = \gamma_s - \gamma_{sl} \tag{16}$$

and for the precontacted surface (equilibrium film present on the surface):

$$\gamma_l \cos\theta_r = \gamma_{sf} - \gamma_{sl} \quad (17)$$

Next, if:

$$\gamma_{sf} = \gamma_s - \pi_e = \gamma_s - W_s = \gamma_s - (W_a - W_c) \quad (18)$$

$$\gamma_l \cos\theta_r = \gamma_s - W_a - W_c - \gamma_{sl} \quad (19)$$

Subtracting Eq.(19) from Eq.(16) results in:

$$\gamma_l (\cos\theta_a - \cos\theta_r) = W_a - W_c \quad (20)$$

or considering Eq.(15) and (16) one obtains:

$$\Delta G_b - \Delta G_p = W_a - W_c \quad (21)$$

So, to determine the solid surface free energy components, at least six wicking experiments have to be conducted by using three probe liquids, and at least two of them should be polar in nature, like water and formamide [1–4].

First the wicking experiment, using a low surface energy liquid, should be carried out for system II, to determine the effective radius R, which in this system is the only unknown in Eq. (13), if the surface has been equilibrated with the liquid vapor. The best liquids for this purpose are short chain-length n-alkanes (heptane to decane) [16, 21], which completely spread on all solid surfaces, and thus:

$$R = \frac{2x^2\eta}{t\gamma_l} \quad (22)$$

Then, the same n-alkane may be used for the apolar component determination, γ_s^{LW}, from system I (Eq. (12)), and considering the expression for the work of adhesion W_a and the work of cohesion W_c it reads.

$$\Delta G_b = W_a - W_c = 2\,[(\gamma_s^{LW}\gamma_l^{LW})^{1/2} + (\gamma_s^{+}\gamma_l^{-})^{1/2} + (\gamma_s^{-}\gamma_l^{+})^{1/2}] - 2\gamma_l \quad (23)$$

Here the polar terms in the brackets equal zero for apolar liquid.

To determine the polar components, two wicking experiments, for example with water and formamide, for bare and precontacted plates (systems III and IV) should be conducted and (Eqs.(14) and (15)) and two Eq.(21) solved simultaneously.

$$\{\Delta G_b - \Delta G_p\}_{water} = W_a - W_c = 2\,[(\gamma_s^{LW}\gamma_w^{LW})^{1/2} + (\gamma_s^{+}\gamma_w^{-})^{1/2} + (\gamma_s^{-}\gamma_w^{+})^{1/2}] - 2\gamma_w \quad (24)$$

$$\{\Delta G_b - \Delta G_p\}_{form.} = W_a - W_c = 2\,[(\gamma_s^{LW}\gamma_f^{LW})^{1/2} + (\gamma_s^{+}\gamma_f^{-})^{1/2} + (\gamma_s^{-}\gamma_f^{+})^{1/2}] - 2\gamma_f \quad (25)$$

Obviously, to learn about the reproducibility of the experiments at least 2-3 sets of the wicking experiments (cases I–IV) should be carried out.

3. EXPERIMENTAL RESULTS

To be sure that required conditions in the casess I–IV are fulfilled, first of all wicking experiments should be carried out in a closed chamber. The schematic of the sandwich chamber used in planar chromatography (Chromdes, Lublin, Poland) is shown in Fig. 2. The probe liquid from the container to the edge of the thin porous layer is transported using a glass slide. It ensures proper amount of the transported liquid without any excess of it. The excess amount of the liquid would cause a hydrostatic pressure of the liquid appearing at the edge, thus affecting the penetration rate. It would happen if the liquid was transported by a flannel wick from the vessel. Using the sandwich chamber, the glass plate with the deposited porous layer can be placed up side down. This considerably reduces contact of the solid surface with the probe liquid vapor. This condition is important in case of the bare surface of the solid wicked with the probe liquid. To show how the results may be affected if the wicking experiments are conducted in an open area instead in a closed sandwich chamber, in Fig. 3a and b wicking results are presented for silica with n-octane and water, respectively, both for the plates kept horizontally and vertically during imbibition [15]. These results clearly show that the more volatile the liquid the more significant is the usage of the closed space for the wicking. As can be seen (Fig. 3 a and b) at a wicked distance the rate of penetration is equal to the rate of evaporation, thus no wicking occurs. It is especially seen for imbibition of n-octane.

Next, in Figs. 4–6 are shown the penetration rates for n-alkane series [22], from pentane to hexadecane. From hexane to decane (Figs. 4 and 5) wicking experiments were conducted both on bare and precontacted SiO_2 plates. For longer chain n-alkanes, the experiments were carried out only on bare plates because the volatility of the alkanes is too low. To verify the above described approach, wicking experiments were also conducted for diiodomethane and α-bromonaphthalene both on bare and precontacted plates (Cases III and IV, definite contact angle). These results are shown in Fig. 7. As can be seen, practically in all systems linear relationship $x^2 = f(t)$ was observed. The standard deviations are marked on the experimental points, which are average of 3–4 experiments. For practical reason, in the figures reversed relationship is presented ($t = f(x^2)$), because actually the time was recorded for particular distances wicked. These results allowed calculation of Lifshitz-van der Waals component, γ_s^{LW}, for the silica surface by applying the procedure described above. For the calculations, surface tensions and viscosities of the probe liquids were necessary. They are listed in Table 1 [22, 33–35], together with the boiling point of the liquids. Thus calculated values of the effective radius and γ_s^{LW} component are collected in Table 2 [22]. On analyzing these results, first of all it can be stated that R value cannot be determined from n-pentane wicking because of its extremely high volatility (the boiling point 36°C, Table 1). Also, using the R value determined from higher n-alkanes, for the same reason, the γ_s^{LW} calculated from pentane was too low (Table 2). The R values determined from hexane and heptane are somewhat lower than those determined from octane to decane (Table 2). However, using these R values to calculate γ_s^{LW} from wicking of the bare plates by these alkanes, the calculated values fit the range of the γ_s^{LW} values determined with higher n-alkanes (Table 2). What is important is that the values of this component determined from n-alkanes (System II) and diiodomethane and α-bromonaphthalene (Cases III and IV) agree very well, and the standard deviation in the average value 41.1 is only 2.4 mJ/m^2. This strongly supports the described procedure. Moreover, having determined ΔG from Eq.(5) for n-alkanes (as well for diiodomethane and α-bromonaphthalene) on the bare plates, one can ormally calculate

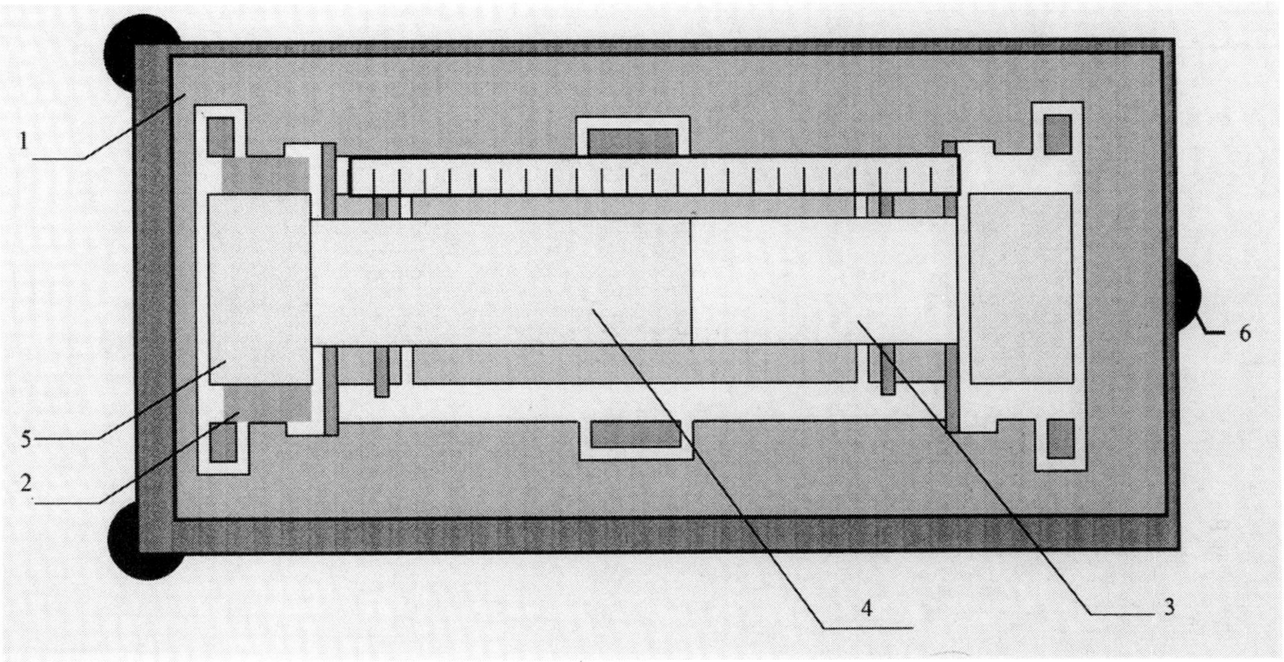

Figure 2. Scheme of the sandwich chamber type Chromdes.
1 – Teflon body, 2 – probe liquid container, 3 – thin layer of the solid, 4 – wicked part of the thin layer, 5 – glass slide cover, 6 – leveling screw.

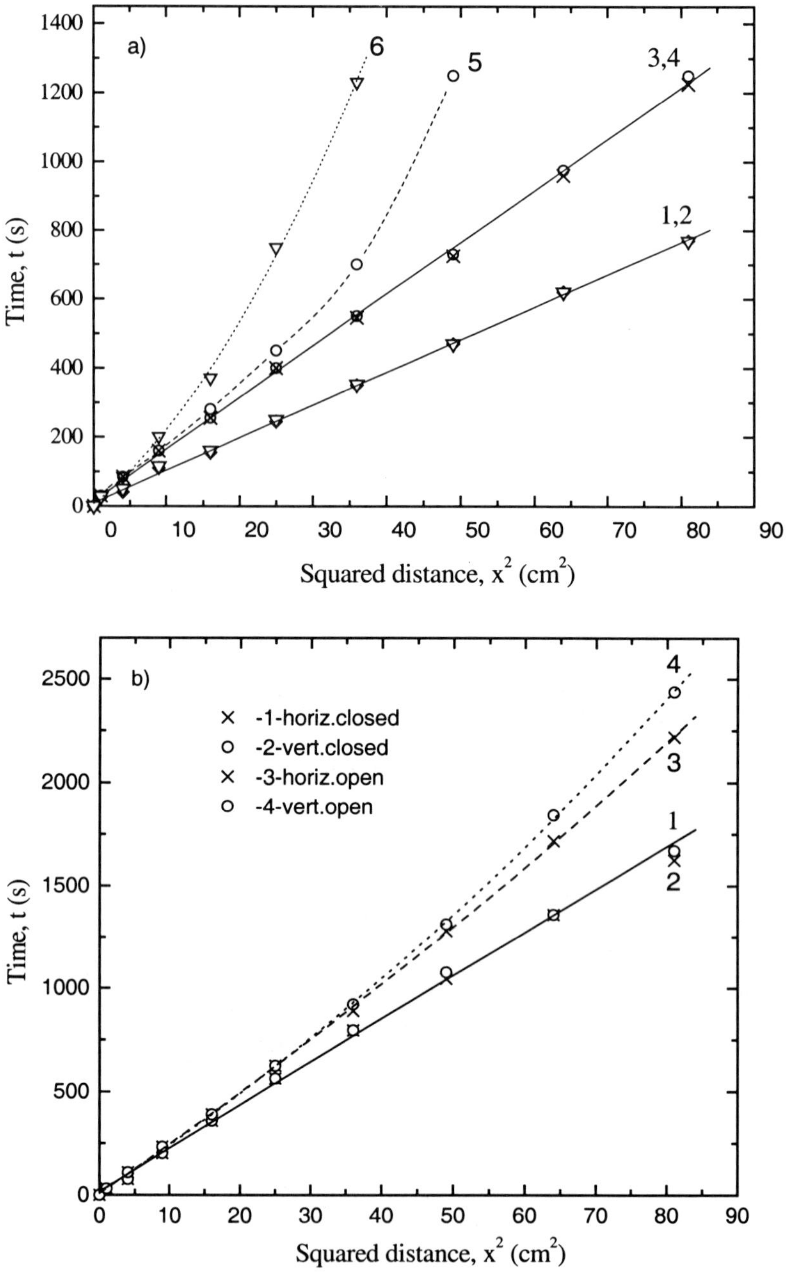

Figure 3. Penetration time vs. squared distance of n-octane (a) and water (b) in silica plates.
Key: a) 1– precontacted horizontal plate in sandwich chamber, 2 – precontacted vertical plate in sandwich chamber, 3 – bare horizontal plate in sandwich chamber, 4 – bare vertical plate in sandwich chamber, 5 – bare horizontal plate in an open area, 6 – bare vertical plate in open area, b) bare plates as indicated in the figure.

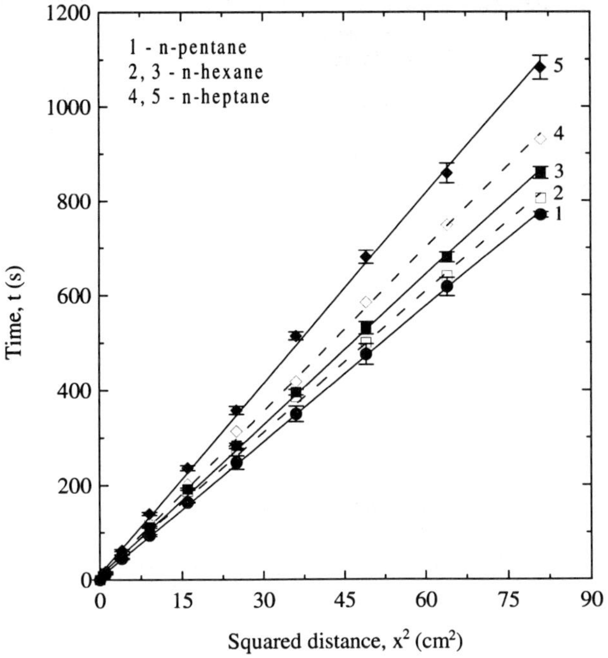

Figure 4. Penetration time vs. squared distance for n-pentane to n-heptane in SiO_2 porous layer. The solid lines are for bare and the dashed lines are for precontatcted plates.

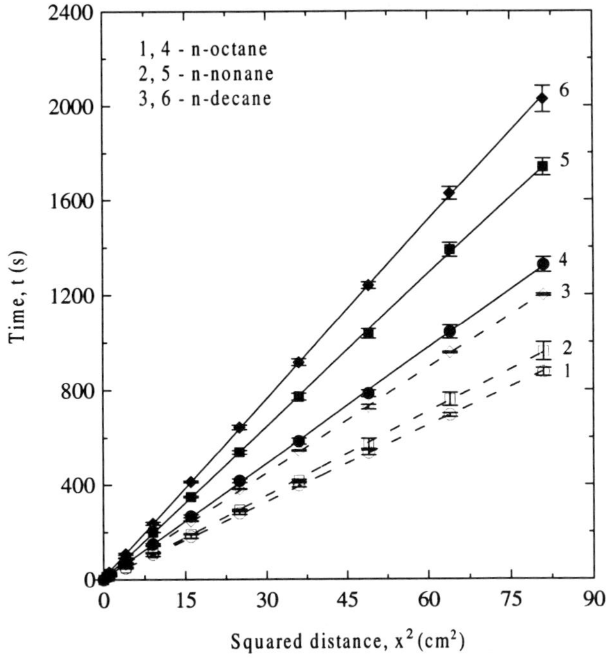

Figure 5. Penetration time vs. squared distance for n-octane to n-decane in SiO_2 porous layer. The solid lines are for bare and the dashed lines are for precontatcted plates.

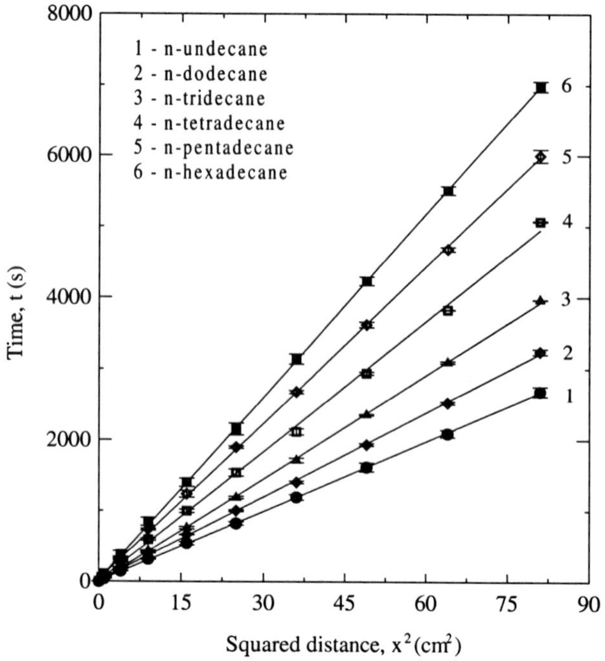

Figure 6. Penetration time vs. squared distance for n-undecane to n-hexadecane in SiO_2 porous layer for bare plates.

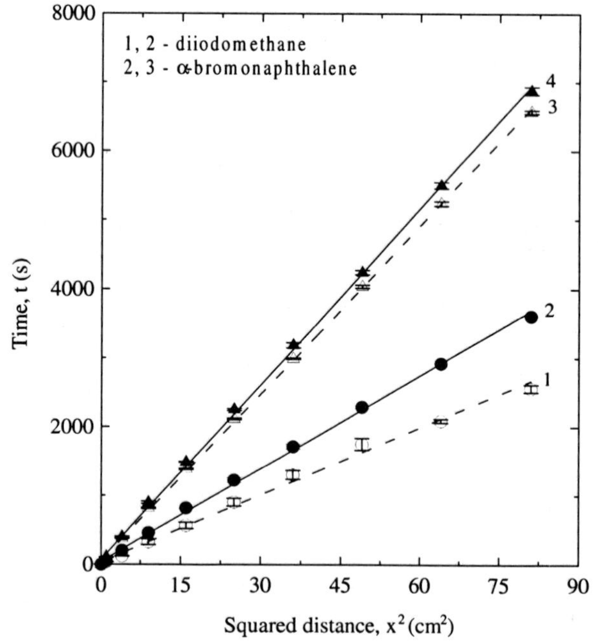

Figure 7. Penetration time vs. squared distance for diiodomethane and α-bromonaphtalene in bare (solid lines) and precontacted (dashed lines) SiO_2 plates.

Table 1.
Surface tension, viscosity (at 20°C) and boiling point of the probe liquids

Liquid	Surface tension, γ_l[a] (mN/m)	Viscosity, η[b] (cP)	Boiling point (°C)
Pentane	16.00	0.240	36
Hexane	18.49	0.307	69
Heptane	20.30	0.417	98.5
Octane	21.80	0.546	125
Nonane	22.91	0.714	150.5
Decane	23.90	0.907	174
Undecane	24.70	1.182	195
Dodecane	25.08	1.492	216.5
Tridecane	25.38	1.878	234
Tetradecane	25.60	2.322	252
Pentadecane	25.80	2.841	270.5
Hexadecane	26.35	3.451	288
α-Bromonaphthalene	44.40[c]	4.890[c]	281
Diiodomethane	50.80[c]	2.821[c]	188

[a] ref. [33], [b] ref. [34], [c] ref. [35].

the dynamic advancing contact angles by equating the free energy change to $\gamma_l \cos\theta_a = \Delta G$ (Case III, Eq.(14)). It appeared that for all n-alkanes $\cos\theta_a$ was less than 1, which means that at the moving front of the alkane a dynamic advancing contact angle could appear, even though the driving force at the front line (probably invisible) was the work of spreading (Case II). The contact angles thus calculated are listed in Table 2. The appearance of the contact angles can be understood because in a capillary there must be concave meniscus (Fig. 1b) behind the spreading front of a completely wetting liquid (no contact angle on the flat surface of this solid), despite the fact that at the liquid front line the energy changes are determined by the work of spreading (Eq. (12)). In fact, these calculations also validate the approach presented above, because no doubt, the energy changes occur at the liquid front and in the case of bare SiO_2 surface ΔG in Eq. (5) is less than γ_l, thus giving contact angles higher than zero. If this condition is fulfilled one can calculate the maximum γ_s^{LW} for the silica, taking $\Delta G = \Delta G_b = W_a - W_c$ (System I, Eqs. (12) and (23)) and, for example for hexane, $\gamma_l = \gamma_l^{LW} = 18.49$ mN/m , and 26.35 mN/m for hexadecane. From the calculations it results that the γ_s^{LW} of SiO_2 should be less than 43.5 and 59.3 mJ/m^2, respectively, which is the case indeed (Table 2). It is reasonable that the contact angle increases with increasing the chain length. It varies from 20–24° for pentane and hexane, 27–31° for heptane, 49–52° for octane, and then it levels off at 53–57°, for rest of the n-alkanes, depending on which average value of R was taken for ΔG calculation (Table 2). Some other relations can be found in ref. [22].

Table 2.
Effective radius R, γ_S^{LW} component of SiO surface and calculated dynamic advancing contact angle θ_a of the test liquids

Liquid	Effective radius ($R \times 10^{-5}$ cm)	γ_s^{LW} (mJ/m^2)	Dynamic advancing contact angle (deg)
Pentane	3.341[a]	~~34.6~~	20
	3.454[b]	~~33.9~~	24
Hexane	3.341[a]	40.0	20
	3.454[b]	39.2	24
Heptane	3.567[c]	41.5	31
	3.454[b]	42.3	27
Octane	4.624[d]	38.6	49
	5.015[e]	37.1	52
Nonane	5.278[g]	36.5	57
	5.015[e]	38.1	55
	5.210[f]	37.5	56
Decane	5.142[h]	40.0	54
	5.015[e]	40.5	53
	5.210[f]	39.8	54
Undecane	5.015[e]	41.1	55
	5.210[f]	40.5	56
Dodecane	5.015[e]	42.4	53
	5.210[f]	41.7	55
Tridecane	5.015[e]	43.4	53
	5.210[f]	42.9	54
Tetradecane	5.015[e]	43.3	54
	5.210[f]	42.3	55
Pentadecane	5.015[e]	43.6	53
	5.210[f]	42.8	55
Hexadecane	5.015[e]	44.8	53
	5.210[f]	44.1	54
α-Bromonaphthalene	5.015[e]	44.2	59
	5.210[f]	43.4	
Diiodomethane	5.015[e]	41.7	60
	5.210[f]	40.9	
Average		41.1± 2.4	

Effective radius R calculated from: [a] hexane, [b] average from hexane and heptane, [c] heptane, [d] octane, [e] average from octane, nonane and decane, [f] average from nonane and decane, [g] nonane, [h] decane.

Another confirmation for this method are the results presented in Table 3, where for SiO_2-originated materials the values of apolar γ_s^{LW} component are collected as determined from contact angles of diiodomethane or α-bromonaphthalene [22, 31, 36–39]. The average value is 41.4 ±1.8 mJ/m^2, which is practically the same as that in Table 2 (41.1±2.4 mJ/m^2), determined from the wicking method. Moreover, if one would apply the original Washburn equation (Eq. (4)) and calculate the contact angle for these two liquids, and then calculate γ_s^{LW} from the Young equation, then the values thus obtained (also shown in Table 3) are much lower than those determined from contact angles measured on a flat surface of these materials, or determined by the above described method (Table 2).

To show the usefulness of this method for acid-base parameters determination, some wicking results for silica gel 100 (Merck) used for column gas chromatography are discussed next [40, 41]. The samples were heated at 200, 400, 600, 800 and 1000°C for 2 h and then the following probe liquids: n-octane, n-nonane, n-decane, diiodomethane, water and formamide were employed for wicking experiments. In Fig. 8 are shown the wicking results for water, formamide and diiodomethane on the sample heated at 600°C. The straight-line dependencies show that the Washburn equation is obeyed, but it is worth mentioning that here instead of a thin porous layer a thin column packing in glass tubes (9 cm long and 2 mm in diameter) was used [41]. The theoretical approach was obviously the same. Taking the literature data for the number of hydroxyl groups per nm^2 for this material heated at the above mentioned temperatures [43], the changes in the surface free energy components, γ_s^{LW}, γ_s^- and γ_s^+ were plotted against –OH density. These are shown in Fig. 9. It is striking that practically linear dependence of γ_s^- was found, except for the sample heated at 1000°C (not plotted), where structural changes would take place [41]. From these studies [40, 41], it results that during an unavoidable contact of the sample with the atmosphere water molecules probably adsorb physically on the silica surface and therefore the surface is more basic (high γ_s^-) than acidic. It may be a result of the interaction of the hydrogen from the water molecule with the oxygen of the surface –OH group [42]. Such mechanism was suggested by Grivtsov *et al.* [43]. Moreover, for example for the sample heated at 200°C the component $\gamma_s^- = 58.4$ mJ/m^2 = 5.84×10^{-20} J/nm^2. The energy of hydrogen bonding lies in the range of 10–40 kJ/mole [44]. Taking the lowest value it results that this energy amounts to 1.66×10^{-20} J/–OH group. Hence, dividing 5.84×10^{-20} (J/nm^2) by 1.66×10^{-20} (J/–OH) one obtains 3.52 –OH/nm^2, which is very close to the value 4.6, given by Zhuravlev [42]. Also good agreement was obtained for other samples [41].

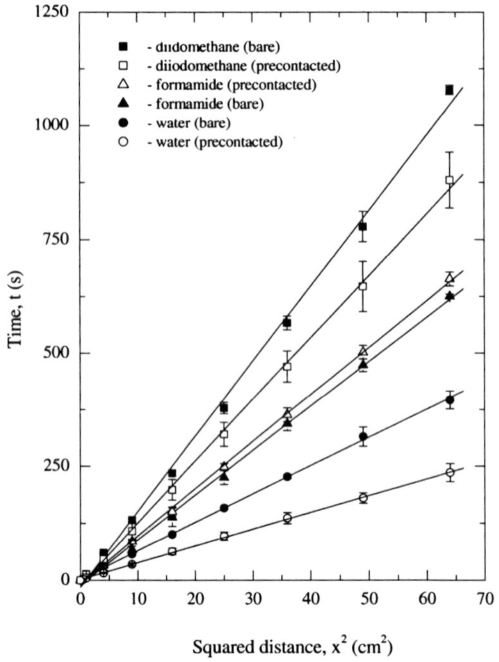

Figure 8. Penetration time vs. squared distance of the probe liquids for bare and precontacted SiO_2 samples heated at 600°C for 2 h.

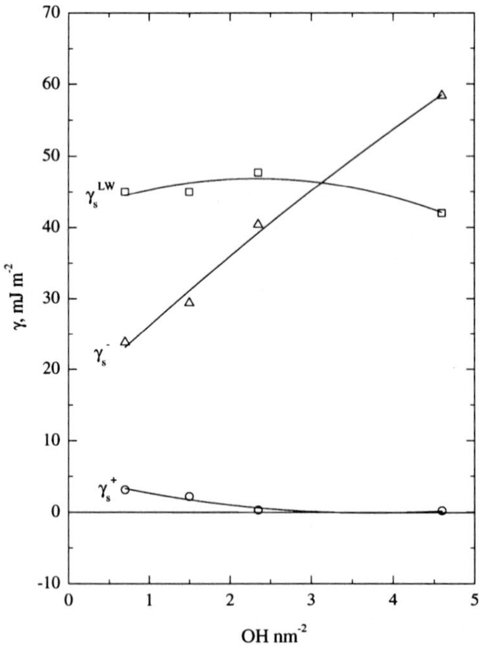

Figure 9. Surface free energy components, apolar Lifshitz-van der Waals γ^{LW}, and polar electron donor γ_s^- and electron acceptor γ_s^+ as a function of the –OH groups density on the SiO_2 surface.

Table 3.
Experimental contact angles θ of diiodomethane and α-bromonaphthalene measured on plates of "SiO_2-originated" materials

Material	Liquid	Contact angle θ (deg)	γ_s^{LW} (mJ/m^2)	Ref.
Fused silica plates, 200°C	CH_2I_2	31	43.8	36
Quartz plate, 105°C, 1 h	CH_2I_2	36.5	41.3	37
Thermisil, 200°C, 2 h	CH_2I_2	40 ± 2	39.4	
	α-BrN*	25 ± 2	40.3	38
Microscope glass plate,	CH_2I_2	31 ± 1	43.8	
110°C, 2 h	α-BrN	14 ± 2	43.1	31
Microscope glass plate,	CH_2I_2	36 ± 2	41.6	
200°C, 2 h	α-BrN	21 ± 2	41.5	31
Glass plate from TLC SiO_2 plate (Merck),	CH_2I_2	43 ± 3	38.1	
130°C, 2 h	α-BrN	26 ± 1	40.8	39
Average			**41.4±1.8**	
SiO_2 from TLC plates (Merck), pellets compressed at 100 kg/cm^2	CH_2I_2	17.5	48.5	22
SiO_2 TLC plates (Merck)	CH_2I_2 **	60 - b	28.5	
		47 - p	36.1	22
	α-BrN **	59 - b	25.7	
		57 - p	26.4	

*– α-BrN – α-bromonaphthalene,
**– contact angle values calculated from the wicking experiment and Eq. (4),
b – for bare surface, p – for precontacted surface.

4. SUMMARY

To summarize briefly, being aware of all the problems with the determination of the absolute values of the Lewis' acid-base components, which were extensively discussed during this Symposium, it can be stated that still the relative values can help understand the processes occurring at interfaces. As was mentioned above, these relative values allow calculation of absolute values for interfacial free energy or the work of adhesion. In the case of powdered solids, for which contact angle technique cannot be successfully applied, the thin layer or thin column wicking techniques can be employed satisfactorily if the experimental procedure and the theoretical background described above are used. More experimental details about the method can be found in Hołysz's paper, where the porous layer thickness [45] or the nature of the substrate (glass, aluminum, PVC) [39], were investigated to find out whether they would affect the determined surface free energy components of the deposited powdered solid. This method has also been successfully used in the studies of radiofrequency electric field (RF) effect on $CaCO_3$ properties [46, 47] and its deposition on glass [48], adhesion of colloidal ZnS on silicon [49], effect of adsorbed ions on Al_2O_3 surface free energy components (also after RF irradiation) [50], adsorption properties of natural and polymeric fibers [51–53], surface energetic properties of cellulose [54], as well bacteria adhesion to a mineral support [55] and others.

REFERENCES

1. C. J. van Oss, M. K. Chaudhury and R. J. Good, *Chem. Rev.* **88**, 927 (1988).
2. C. J. van Oss, M. K. Chaudhurry and R. J. Good, *Sep. Sci. Technol.* **24**, 15 (1989).
3. C. J. van Oss, R. J. Good and M. K. Chaudhury, *Langmuir* **4**, 884 (1988).
4. C. J. van Oss, *Interfacial Forces in Aqueous Media*, Marcel Dekker, New York, 1994.
5. C. J. van Oss, R. F. Giese and W. Wu, *J. Dispersion Sci. Technol.* **19**, 1221 (1998).
6. E. W. Washburn, *Phys. Rev. Ser.* **17**, 273 (1921).
7. R. F. Giese, P. M. Costanzo and C. J. van Oss, *J. Phys. Chem. Minerals* **17**, 611 (1991).
8. A. W. Adamson, *Physical Chemistry of Surfaces, Fifth Edition,* p. 495, John Wiley, New York, (1990).
9. R. J. Good, *J. Colloid Interface Sci.* **42**, 473 (1973).
10. L. R. Fisher and P. D. Lark, *J. Colloid Interface Sci.* **69**, 486 (1979).
11. Y-W. Yang and G. Zografi, *J. Pharm. Sci.* **75**, 719 (1986).
12. Y-W. Yang, G. Zografi and E. E. Miller, *J. Colloid Interface Sci.* **122**, 35 (1988).
13. G. E. Parsons, G. Buckton and S. M. Chatham, *J. Adhesion Sci. Technol.* **7**, 95 (1993).
14. E. Chibowski and L. Hołysz, *Langmuir* **8**, 710 (1992).
15. E. Chibowski and F. González-Caballero, *Langmuir* **9**, 330 (1993).
16. C. J. van Oss, R. F. Giese, Z. Li, K. Murphy, J. Norris, M. K. Chaudhury and R. J. Good, *J. Adhesion Sci. Technol.* **6**, 413 (1992).
17. Z. Li, R. F. Giese, C. J. van Oss, J. Yvon and J. Cases, *J. Colloid Interface Sci.* **156**, 279 (1993).
18. Z. Li, R. F. Giese and C. J. van Oss, *Langmuir* **10**, 330 (1994).
19. C. J. van Oss, W. Wu, R. F. Giese and J. O. Naim, *Colloids Surfaces:B* **4**, 185 (1995).
20. P. M. Constanzo, W. Wu, R. F. Giese, Jr. and C. J. van Oss, *Langmuir* **11**, 1827 (1995).
21. W. Wu, R. F. Giese and C. J. van Oss, *Powder Technol.* **89**, 129 (1996).
22. E. Chibowski and L. Hołysz, *J. Adhesion Sci. Technol.***11**, 1289 (1997).
23. E. Chibowski, *J. Adhesion Sci. Technol.* **6**, 1069 (1992).
24. L. Hołysz and E. Chibowski, *Langmuir* **8**, 717 (1992).
25. A. W. Adamson, *Physical Chemistry of Surfaces, Fifth Edition,* p. 10, John Wiley, New York (1990).
26. R. J. Good and M. N. Koo, *J. Colloid Interface Sci.* **71**, 283 (1979).
27. R. J. Good, N. R. Srivatsa, M. Islam, T. L. Huang and C. J. van Oss, *J. Adhesion Sci. Technol.* **4**, 607 (1990).
28. H. J. Busscher, Doctoral thesis, p.48. Groningen University, The Netherlands, 1985.
29. E. J. McNally, C.-s. Gau and G. Zografi, *J. Adhesion Sci. Technol.* **6**, 445 (1992).
30. E. Chibowski and F. González-Caballero, *J. Adhesion Sci. Technol.* **7**, 1195 (1993).
31. E. Chibowski, L. Hołysz, A. Zdziennicka and F. González-Caballero, in: *Surfactants in Solution*, A. K. Chattopadhyay and K. L. Mittal, (Eds), pp. 31–53, Marcel Dekker, New York (1996).
32. R. J. Good, in: *Surfaces and Colloid Science*, R. J. Good and R. R. Stromberg (Eds), Vol. 11, Plenum Press, New York (1979).
33. B. Jańczuk E. and Chibowski, *J. Colloid Interface Sci.* **95**, 268 (1983).
34. Textbook of Chemist, Izd. Khim. Lit., Moscow, 1962 (in Russian).
35. C. J. van Oss, M. K Chaudhury and R. J. Good, *Sep. Sci. Technol.* **24**, 15 (1989).
36. J. Rayss, A. Gorgol, W. Podkościelny, J. Widomski and M. Chołyk, *J. Adhesion Sci. Technol.* **12**, 293 (1998).
37. B. Jańczuk and A. Zdziennicka, *J. Mater. Sci.* **29**, 3559 (1994).
38. L. Hołysz, *Adsorption Sci. Technol.* **14**, 89 (1996).
39. L. Hołysz, Colloids Surfaces A., submitted.
40. L. Hołysz and E. Chibowski, paper presented at the *9th International Conference on Surface and Colloid Science*, Sofia (1997).
41. L. Hołysz, *Colloids Surfaces: A* **134**, 321 (1998).
42. L. T. Zhuravlev, *Colloids Surfaces: A* **74**, 71 (1993).

43. A. G. Grivtsov, L. T. Zhuravlev, G. A. Gerasimova and L. G. Khazin, *J. Colloid Interface Sci.* **126**, 397 (1988).
44. J. Israelachvili, *Intermolecular and Surface Forces*, Academic Press, New York, 1992.
45. L. Hołysz, *J. Mater. Sci.* **33**, 445 (1998).
46. L. Hołysz and E. Chibowski, *J. Colloid Interface Sci.* **164**, 245 (1994).
47. E. Chibowski, L. Hołysz and W. Wójcik, *Colloids Surfaces: A* **92**, 79 (1994).
48. J. D. G. Durán, A. Ontiveros, A. V. Delgado, F. González-Caballero and E. Chibowski, J. Adhesion Sci. Technol. **10**, 847 (1996).
49. A. Ontiveros, J. D. G. Durán, A. V. Delgado, F. González-Caballero and E. Chibowski, *J. Adhesion Sci. Technol.* **10**, 999 (1996).
50. E. Chibowski and A. Sołtys, in: *Mittal Festschrift on Adhesion Science and Technology*, W. J van Ooij and H. R. Anderson, Jr. (Eds) pp. 837–855, VSP, Utrecht (1998).
51. A. Ontiveros-Ortega, M. Espinosa-Jiménez, E. Chibowski and F. González-Caballero, *J. Colloid Interface Sci.* **202**, 189 (1998).
52. A. Ontiveros-Ortega, M. Espinosa-Jiménez, E. Chibowski, and F. González-Caballero, *J. Colloid Interface Sci.* **199**, 99 (1998).
53. E. Chibowski, M. Espinosa-Jiménez, A. Ontiveros-Ortega and E. Giménez-Martin, *Langmuir* **14,** 5237 (1998).
54. D. Fernando, F. M. Gama, E. Chibowski and M. Mota *J. Adhesion Sci. Technol.* **12**, 1081 (1998).
55. P. Teixeira, J. Azeredo, R. Oliveira and E. Chibowski, *Colloids Surfaces: B* **12**, 69 (1998).

Acid-Base Interactions, Vol. 2, pp. 439–452
Ed. K.L. Mittal

Determination of the acid–base properties of the human skin surface: Influence of skin lipids

ALAIN MAVON*

Service de Pharmacocinétique Cutanée, Institut de Recherche Pierre Fabre (IRPF), BP 74, Vigoulet-Auzil, 31322 Castanet Tolosan Cedex, France

Abstract—The purpose of this work was to determine the surface free energy of human skin, on the volar forearm, (a sebum-poor area), and on the forehead, (a sebum-rich area) and to assess the influence of skin lipids on acid–base properties using advancing contact angle measurements of water, formamide and diiodomethane. The acid–base components of skin surface free energy were calculated according to the Good–van Oss–Chaudhury (GvOC) theory. Advancing contact angles were measured on both skin surfaces of 10 healthy volunteers before and after ether-treatment, and on the forehead before and during casual sebum level restoration following skin washing by soap, of 4 oligo- (OS), 4 normo- (NS), and 4 hyperseborrheic (HS) volunteers. From the surface free energy calculations, the forearm skin surface was found to be an almost apolar surface (γ_s^{LW} = 35 mJ·m^{-2}; γ_s^- = 4 mJ·m^{-2}) and the forehead skin surface a monopolar basic surface (γ_s^{LW} = 35 mJ·m^{-2}; γ_s^- = 28 mJ·m^{-2}). Ether-treatment caused both surfaces to become more apolar, mostly by reducing the basic component. On the forehead a low sebum level (OS skin) does not deprive the skin surface of a normal γ_s^{LW} component. On the other hand, the basic component is related to the sebum level restoration, with a distinction between OS skin (γ_s^- = 11 mJ·m^{-2}) and NS or HS skins (γ_s^- = 21 mJ·m^{-2}). Skin surface lipids, mainly sebum, give the skin surface a basic hydrophilic and not an acidic character. This could be explained by the presence of the free fatty acids, which are abundant in the sebum, and according to Good's proposal: an outward orientation of their carboxylic groups and a masking of acidic moieties by the adsorbed water molecules. But is there a relationship with the well-known acidic pH of the skin surface (between 4 to 6.3)? Our results confirm doubts about the reliability of the GvOC theory. However, and in spite of the controversy about this theory, we have shown that the skin lipids mainly change the basic component and thus the skin water wettability. Thus these data give a new insight into the so-called “skin hydration”.

Keywords: Wettability; surface free energy; acid–base interactions; advancing contact angle; human skin surface; sebum level; skin surface lipids.

1. INTRODUCTION

An important function of human skin is to minimize water loss and entry of foreign matter. This barrier property is exerted by its outermost layer, the stratum corneum, a dead layer made up of stacked, flattened horny cells stuck together by intercellular lipid bilayers. It is generally agreed that these lipid bilayers make up the main path of skin permeation; hence an extensive amount of research (literature review in [1]) is

* To whom correspondence should be addressed. E-mail: alain.mavon@pierre-fabre.com

devoted to its understanding. Percutaneous absorption is a diffusion phenomenon [2] and, accordingly, is dependent on the contact area between the permeant and the skin surface, which, in turn, is related to the skin wettability. In addition, interfacial phenomena play a major role in the constitution and maintenance of the skin surface ecosystem, adhesion of microorganisms [3], and desquamation. Moreover, they are involved in the interaction between sebum, an oily mixture [4], and sweat, a watery secretion [5], both produced in the same place.

Despite such an extensive role in skin physiology, medicine and cosmetic science, the wettability of human skin has been poorly investigated (literature review in [6]). A few of these studies are related to skin physiology. Furthermore, they are mainly descriptive and do not provide any understanding of the wetting phenomena. This prompted us to undertake experiments aimed at gaining insight into the physical-chemical properties of the skin surface, using the Good–van Oss–Chaudhury (GvOC) theory [7–9], by calculating the surface free energy components, *in vivo*, on two sites of major interest for skin physiology and cosmetology:

1. The volar forearm, selected as a "dry skin" because of the low density of sebaceous orifices, less than 50/cm^2. The lipids present in this area, the epidermal lipids (EL), only originate from horny layer desquamation [10].
2. The forehead, chosen as a "greasy skin" because the sebaceous orifice density is more than 300/cm^2 and has a much higher sebaceous gland activity. Accordingly, the higher level of lipid on the forehead surface was accounted for by the presence of sebum [11]. Moreover, this sebum level is remarkably constant on the forehead in a given individual and is also called the "casual sebum level". It is restored within 2 hours [11, 12] following washing or degreasing, through a sudden and sustained flow of sebum from the sebaceous glands [13]. The overwhelming prevalence of sebum over EL on the forehead leads to the usual expression of oligo-, normo-, or hyperseborrheic skins to characterize the sebum level [14].

2. THEORY

Lifshitz [15] showed that in a condensed phase, the three electrodynamic forces (namely London "dispersion", Keesom dipole–dipole and Debye dipole-induced dipole forces) decayed with distance at the same rate and obeyed the same combining rules. Good, van Oss and Chaudhury [7–9] pointed out that the surface tension arising from these three electrodynamic forces, when combined together, gave rise to a surface tension component that is called the Lifshitz–van der Waals (LW) component. If suitable chemical groups are present, significant contributions to interfacial energy can arise from hydrogen bonding, i.e. Lewis acid–base or electron-acceptor/electron-donor interactions.

Fowkes [16] suggested that the surface tension γ could be considered as the sum of the independent terms, each representing a particular intermolecular force. Accordingly, γ can be expressed as the sum of Lifshitz–van der Waals, γ^{LW}, and acid-base, γ^{AB} components:

$$\gamma = \gamma^{LW} + \gamma^{AB} \tag{1}$$

The surface tension of a solid surface can be readily calculated by measuring the contact angles of two different liquids of known surface tension components. A more complex approach was developed by van Oss *et al.* [7–9] by incorporating the electron-donor and the electron-acceptor parameters in the acid-base component which is expressed as:

$$\gamma^{AB} = 2\,(\gamma^{+}\,\gamma^{-})^{1/2} \tag{2}$$

The γ^{+} and γ^{-} denote the electron-acceptor and the electron-donor components respectively. Finally, the three components of the interfacial free energy between materials 1 and 2 can be defined by the following equation [8–9]:

$$\gamma_{12}=[\gamma_1^{LW}+\gamma_2^{LW}-2(\gamma_1^{LW}\gamma_2^{LW})^{1/2}]+2[(\gamma_1^{+}\gamma_1^{-})^{1/2}+(\gamma_2^{+}\gamma_2^{-})^{1/2}-(\gamma_1^{+}\gamma_2^{-})^{1/2}-(\gamma_1^{-}\gamma_2^{+})^{1/2}] \tag{3}$$

Solid surface free energy can be determined from contact angle measurements, in appropriately chosen systems using Young's equation:

$$\gamma_l \cos\theta = \gamma_s - \gamma_{sl} - \pi_e \tag{4}$$

where γ_s is the solid surface free energy, γ_{sl} the solid–liquid interfacial free energy, γ_l the liquid surface tension, θ the contact angle and π_e the film pressure of the liquid. After combining Equation (3) with Equation (4) and neglecting π_e [16], the following equation is obtained:

$$(1 + \cos\theta)\,\gamma_l = 2(\gamma_s^{LW}\gamma_l^{LW})^{½} + 2(\gamma_s^{+}\gamma_l^{-})^{½} + 2(\gamma_s^{-}\gamma_l^{+})^{½} \tag{5}$$

The three parameters characterizing the solid surface free energy can be determined by measuring the contact angles of at least three standard liquids, provided the relevant values of the surface tension components of these liquids are known [9, 17].

3. EXPERIMENTAL

3.1. Liquids

Formamide and diiodomethane (analytical grade, minimum 98% purity) were supplied by Fluka (St Quentin Fallavier, France). Water was distilled twice and deionized (Milli-Q reagent water system, Millipore) yielding a resistivity of 18 MΩ cm^{-1} at 25°C (Table 1).

Table 1.
Literature values ($mJ \cdot m^{-2}$) for Lifshitz–van der Waals (γ_l^{LW}), acidic (γ_l^{+}) and basic (γ_l^{-}) components of standard liquids (γ_l) used for the determination of skin surface free energy components

Liquid	γ_l	γ_l^{LW}	γ_l^{+}	γ_l^{-}	Ref.
Water (W)	72.8	21.8	25.5	25.5	[9, 17]
Formamide (F)	58	39	2.28	39.6	[9, 17]
Diiodomethane (D)	50.8	50.8	0.72	0	[9, 17]

3.2. Volunteers (Skin Surfaces) and Skin Surface Treatment

Volunteers were previously informed of the harmful and corrosive potential of formamide and diiodomethane on contact with the skin, and they gave their informed consent. In our experimental conditions, during contact angle measurements, the liquid drop stayed on the skin surface for a maximum of 20 s. Then, the drop was immediately removed using absorbent paper. At the end of the experimentation, the skin surfaces were washed using soap and rinsed with water.

In these conditions and according to the skin barrier properties [2] we can consider that the body exposure to these liquids is negligible, and does not involve danger for the subjects.

3.2.1. Volar Forearm/Forehead. Contact angle measurements were carried out on two skin areas, the volar forearm and the forehead of 10 healthy volunteers (5 males, 5 females, age range 20–30 years). They were measured on the skin kept unwashed for at least 8 hours and after 2 gentle wipings (10 s each) with a tissue soaked in ether [18, 19]. The ether was purchased from Coopération Pharmaceutique Française (Melun, France).

3.2.2. Forehead of Different Skin Types. 12 volunteers (age range 23–31 years) were selected in each of the previously defined skin types [14], namely oligoseborrheic (OS), normoseborrheic (NS) and hyperseborrheic (HS) whose sebum levels were < 80 AU, 120–160 AU and > 200 AU, respectively. These sebum levels were assessed by the photometric method [20] using the Sebumeter® SM 801 instrument (Courage & Khazaka GmbH, Cologne, Germany) and given in arbitrary units (AU).

Contact angle and sebum level measurements were carried out on the forehead before washing the skin kept unwashed for at least 8 h (Baseline); just after washing (0 min) and after recovery of the sebum level for: 20, 40, 60, 90 and 120 min. The forehead was washed by scrubbing twice for 10 s with a gauze pad impregnated with "savon doux" (amphoteric soap purchased from Laboratoires Anios, Lille, France) followed by rinsing three times with a gauze pad soaked in water and drying by three applications of an absorbent paper without rubbing.

3.3. Contact Angle Measurements

Advancing contact angles, θ_A, (i.e. the maximum value of the contact angle when the drop increases in volume) were measured according to Zisman's method [21, 22]. Whereas Zisman used a fine platinum wire, our procedure used a Hamilton microsyringe with a fine needle (0.18 mm, internal diameter) held about 1mm above the skin. The volume of the drop placed on the skin surface was increased by addition of droplets: the pending droplet detached from the needle and was incorporated into the drop on the skin. The final volume was 5 μl. Contact angles were assessed using a surgical microscope (Wild Heerbrugg® M650), with a magnification of 25×, fitted with a slanted mirror and a video camera (CCD-IRIS, Sony®) connected to a personal computer. The computer was equipped with video capture hardware Videospigot® and digital video software (MS®). This system insured the observation, acquisition, storage and computation of θ_A. From the video sequence, θ_A's were measured from the tangent on both sides of the drop (7 measurements for each liquid). The measurements were carried out at room temperature (21±1.5°C).

3.4. Skin Roughness and Skin Temperature

Although the rationale developed in the GvOC theory [7–9] was based on a smooth, homogeneous, plane surface, human skin is known to have a rough and random surface [23]. In spite of many works carried out on the effects of roughness on contact angle measurements and surface wetting [24–28], no reliable methods for correcting for these parameters are available for a rough surface. Therefore, our study did not take into account the skin roughness parameters although they were quantitatively assessed.

It is well known [25] that contact angle measurements depend on temperature. The temperature of the skin surface was measured with a thermocouple (Thermodig N 800) and found to be 34.6±1.2°C on the forehead and 33.2±1.1°C on the forearm. Since the θ_A video sequence was captured for about 20 s, the drop (21°C) had time to warm in contact with the skin.

However, whatever the errors in our θ_A values due to skin topography and changes in drop temperature, these were similar in magnitude in all experiments and accordingly could not strongly alter the results. These difficulties stem from the fact that we worked with a living surface.

4. RESULTS

4.1. Advancing Contact Angles on the Skin Surfaces

θ_A values on the forearm and forehead are shown in Table 2. Those on the forehead for the three skin types are shown for water in Table 3, for formamide in Table 4 and for diiodomethane in Table 5. A full discussion of these values can be found in ref. [6, 29].

Table 2.
Advancing contact angles (in degrees) of standard liquids on the volar forearm and on the forehead of 10 subjects (mean and sd), before and after ether treatment. Bef. ET: before ether treatment; After ET: after ether treatment

	Volar forearm		Forehead	
	Bef. ET	After ET	Bef. ET	After ET
Water	87.4±12.1	100.9±6.5	55.0±11.6	83.3±5.9
Formamide	64.4±13.9	74.5±7.5	45.4±7.9	64.0±6.3
Diiodomethane	41.4±12.5	55.3±7.4	33.7±7.2	48.0±7.4

Table 3.
Advancing contact angles (in degrees) of water on the forehead, of 4 normoseborrheics, NS, 4 hyperseborrheics, HS and 4 oligoseborrheics, OS, (mean and sd), before (at baseline), after washing (t = 0) and during the recovery of sebum level (t = 20 to 120 min)

	NS	HS	OS
Baseline	56.3±3.3	58.5±4.0	73.3±6.7
t = 0 min	80.6±7.0	83.5±7.0	101.6±4.1
t = 20 min	76.3±5.0	80.1±8.8	93.2±4.9
t = 40 min	75.5±8.6	78.6±10.6	90.9±2.9
t = 60 min	66.2±7.7	71.3±7.1	84.6±5.0
t = 90 min	63.0±7.2	66.0±6.9	77.6±2.4
t = 120 min	57.9±4.9	60.7±5.6	75.8±1.7

Table 4.
Advancing contact angles (in degrees) of formamide on the forehead 4 normoseborrheics, NS, 4 hyperseborrheics, HS and 4 oligoseborrheics, OS, (mean and sd), before (at baseline), after washing (t = 0) and during the recovery of sebum level (t = 20 to 120 min)

	NS	HS	OS
Baseline	39.6±2.5	43.6±4.1	57.6±8.7
t = 0 min	63.9±11	63.4±9.9	77.5±6.2
t = 20 min	54.0±8.6	58.2±12.4	75.1±6.3
t = 40 min	58.5±8.4	62.5±12.1	75.1±6.3
t = 60 min	51.7±7.2	56.4±7.8	68.1±6.9
t = 90 min	47.9±7.3	49.5±10.2	62.3±4.1
t = 120 min	42.9±6.2	47.2±9.0	60.7±1.0

Table 5.
Advancing contact angles (in degrees) of diiodomethane on the forehead, of 4 normoseborrheics, NS, 4 hyperseborrheics, HS and 4 oligoseborrheics, OS, (mean and sd), before (at baseline), after washing (t = 0) and during the recovery of sebum level (t = 20 to 120 min)

	NS	HS	OS
Baseline	29.1±3.5	31.7±4.5	31.4±1.2
t = 0 min	43.4±1.7	44.3±2.1	49.6±5.6
t = 20 min	36.0±4.0	34.8±2.3	47.4±7.0
t = 40 min	32.5±4.0	34.76±2.3	43.0±5.1
t = 60 min	28.4±3.0	32.0±2.0	36.8±2.0
t = 90 min	28.5±2.1	30.9±3.1	32.5±1.0
t = 120 min	28.7±2.6	29.7±3.6	30.5±1.6

4.1.1. Before Ether Treatment. On the forearm (Table 2), the contact angles were namely 87°, 64° and 41° for water, formamide and diiodomethane, respectively. On the forehead (Table 2), the contact angles were 55°, 45° and 34° for the same liquids, respectively. Between the forearm and the forehead, the differences in contact angles for the three polar liquids were significant whereas this was not the case for the apolar liquid. These results suggest that although the surfaces of both areas are lipophilic to the same extent, only the forehead is hydrophilic.

4.1.2. After Ether Treatment. On both skin areas, ether treatment produced a significant increase in contact angles whatever the standard liquid used (Table 2). Contact angles for diiodomethane increased from 34° to 48° and from 41° to 55° on the forehead and on the forearm, respectively. Accordingly, both surfaces became lipophobic. The forehead surface became hydrophobic and the forearm surface more hydrophobic than before ether treatment. The two sites differed mostly in their wettability for water.

4.1.3. On the Forehead of the 3 Skin Types. At baseline (before washing), the water contact angles (Table 3) were the same (57°) in NS and HS skins but much higher in OS skin (73°). After degreasing, they all had increased by about 25° whatever the skin type. Baseline values were recovered in 2 h. The difference of 15°, observed between OS skin and the two other types before degreasing, was still present at this time. As observed for water, the formamide contact angle (Table 4) differed, at baseline, in the NS and HS skins (41°) from the OS skin (58°). Washing increased the values by around 20° in all skin types. For 2 h, the values decreased continuously to baseline, keeping a final difference of about 15° between OS skin and the other two skin types. At baseline, the contact angles of the diiodomethane (Table 5) were the same for the 3 skin types (30°). Just after degreasing, it was high and similar (42°) in NS and HS skins, but still higher in OS skin (50°). By 20 min, NS and HS contact angles had strongly decreased. They had fallen to baseline values by 60 min. Conversely, on the OS skin, the contact angles slowly decreased over 2 h, and only by the end of this time, was the baseline value recovered.

4.2. Surface Free Energy Components of the Skin Surfaces

The skin surface free energy components (volar forearm, forehead, and forehead of the three skin types) were calculated by solving Equation (5) using the contact angles (Tables 2–5) obtained for the three standard liquids shown in Table 1.

4.2.1. Forearm Surface. On the forearm (Table 6), a sebum-poor surface, the bulk of γ_s originated from the Lifshitz–van der Waals component, γ_s^{LW}. Both acidic and basic components were low. Ether treatment induced a moderate decrease in both overall γ_s and its γ_s^{LW} component whereas the basic component became one-third. Consequently, this area should be referred to as a weakly monopolar basic surface ($\gamma_s^- = 4.2\ mJ{\cdot}m^{-2}$).

4.2.2. Forehead Surface. On the forehead (Table 6), a sebum-rich area, the surface free energy was higher than on the forearm. The value of the γ_s^{LW} component was the same as on the forearm and the main difference was in the γ_s^- component which was considerably higher ($\gamma_s^- = 28$ mJ·m^{-2}). Consequently this area can be better referred to as a monopolar basic surface. After ether treatment, γ_s decreased from 42.5 to 34.5 mJ·m^{-2} and the γ_s^{LW} component decreased only slightly. The γ_s^+ component remained very low and the decrease in γ_s was mostly related to the dramatic drop in its γ_s^- component from 28 to 6.0 mJ·m^{-2}.

Table 6.
Lifshitz–van der Waals (γ_s^{LW}), acidic (γ_s^+), basic (γ_s^-), acid–base (γ_s^{AB}) components and total surface free energy (γ_s), in mJ·m^{-2}, of skin surfaces of 10 subjects (mean and sd)

	γ_s	γ_s^{LW}	γ_s^+	γ_s^-	γ_s^{AB}
Volar forearm					
Before ether treatment	38.7±6.4	35.6±5.8	0.8±0.7	4.2±3.0	3.4
After ether treatment	32.4±4.8	30.6±4.6	0.4±0.4	1.2±1.2	1.4
Forehead					
Before ether treatment	42.5±3.9	35.0±3.1	0.5±0.2	28.3±11.5	7.5
After ether treatment	34.5±4.1	32.0±3.4	0.3±0.3	5.9±2.6	2.7

4.3. Surface Free Energy Components and Sebum Level on the Forehead of 3 Skin Types

4.3.1. Sebum Level. The volunteers were classified into three groups according to their forehead casual sebum level (at the baseline): 70 AU, 140 AU and 210 AU in OS, NS and HS skins, respectively. Thus, the 3 skin types were clearly defined (Fig. 1). At time 0, just after washing, the sebum level was about 10 AU. This level differed from zero for two reasons. First, using soap, degreasing was gentle; secondly, the sebaceous excretion was continuous and had already resumed during the 2 min between degreasing and the sebum-level measurement. Afterwards the sebum level increased and the baseline value was reached in 90 min, irrespective of the skin type. By 120 min it was at its maximum and slightly higher than at baseline [30].

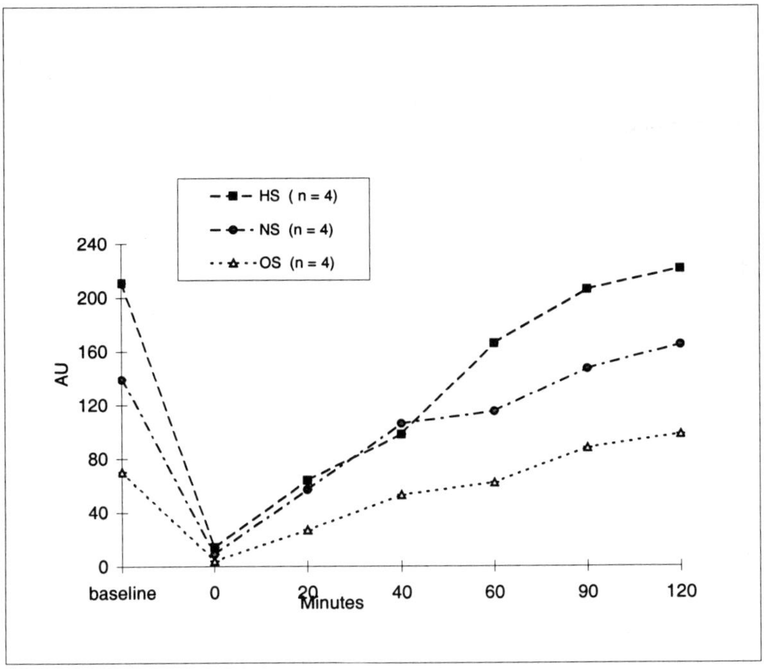

Figure 1. Baseline, washing (t = 0) and sebum level restoration on forehead in OS (oligoseborrheic), NS (normoseborrheic) and HS (hyperseborrheic) skins. Baseline: causal sebum level, before washing. The sebum level measured every 20 and 30 minutes (from 0 to 120 min). AU: Arbitrary Units.

4.3.2. Surface Free Energy Components.

Lifshitz–van der Waals component, γ_s^{LW}. At the baseline, γ_s^{LW} was 37.5 mJ·m^{-2} for all three skin types (Tables 7–9). The degreasing induced a slight decrease (3 mJ·m^{-2}) in NS and HS skins, and a greater decrease in the OS skin (5 mJ·m^{-2}). On HS and NS skins, the baseline value was reached in 20 min although the sebum level was restored in 90 min. This was not the case with OS skin, where γ_s^{LW} increased in parallel with sebum excretion and by 90 min the baseline value was reached for both the sebum level and the γ_s^{LW} component.

Electron acceptor or acidic component, γ_s^{+}. The acidic component ranged from 0 to 1 mJ·m^{-2} (Tables 7–9) and was considered negligible, at all times, for each skin type and whether the skin had just been washed or the casual sebum level had recovered.

Electron donor or basic component, γ_s^{-}. At the baseline (Tables 7–9), all skin types exhibited an electron donor character, of similar value in NS and HS skins (21 mJ·m^{-2}) but much lower in OS skin (12 mJ·m^{-2}). Also in all skin types, it dropped sharply immediately after degreasing by the same amount (11 mJ·m^{-2}). However, only OS skin became almost apolar, because its basic component dropped to about 0 mJ·m^{-2}. For the OS skin, γ_s^{-} component restoration started immediately and was continuous over 120 min but only increased after 20 min in other skin types. For all skin types, restoration of γ_s^{-} was achieved after sebum level restoration.

Table 7.
Total surface free energy (γ_s) and Lifshitz–van der Waals (γ_s^{LW}), acid–base (γ_s^{AB}), acidic (γ_s^+), and basic (γ_s^-) components in mJ·m^{-2}, on forehead of 4 normoseborrheics, NS, (mean and sd), before (at baseline), after washing (t = 0) and during the recovery of sebum level (t = 20 to 120 min)

	γ_s	γ_s^{LW}	γ_s^{AB}	γ_s^+	γ_s^-
Baseline	46.3±1.1	37.5±1.1	8.8±1.8	1.0±0.5	21.0±4.3
t = 0 min	38.2±5.6	33.4±1.8	4.4±3.3	0.8±0.9	8.3±0.5
t = 20 min	41.3±2.7	37.2±1.9	4.1±1.8	0.7±0.4	7.2±2.1
t = 40 min	40.6±3.0	38.3±1.4	2.8±2.3	0.2±0.2	10.5±5.4
t = 60 min	41.4±2.7	38.4±1.0	3.2±2.8	0.2±0.2	18.1±6.9
t = 90 min	43.4±3.2	38.5±0.4	5.2±2.5	0.4±0.3	18.3±5.4
t = 120 min	44.8±3.7	38.0±0.9	7.6±1.4	0.7±0.5	23.8±6.1

Table 8.
Total surface free energy (γ_s) and Lifshitz–van der Waals (γ_s^{LW}), acid–base (γ_s^{AB}), acidic (γ_s^+), and basic (γ_s^-) components in mJ·m^{-2}, on forehead in 4 hyperseborrheics, HS, (mean and sd), before (at baseline), after washing (t = 0) and during the recovery of sebum level (t = 20 to 120 min)

	γ_s	γ_s^{LW}	γ_s^{AB}	γ_s^+	γ_s^-
Baseline	44.4±2.3	36.6±2.4	8.0±0.4	0.8±0.4	20.6±3.3
t = 0 min	37.4±3.7	34.0±2.2	3.2±2.0	0.4±0.4	8.0±4.6
t = 20 min	42.2±2.1	38.1±0.9	3.9±1.9	0.7±0.5	6.3±2.8
t = 40 min	41.6±3.3	37.3±1.1	4.4±2.2	0.7±0.7	10.7±4.0
t = 60 min	40.5±1.9	37.4±1.1	3.2±1.9	0.2±0.2	13.6±4.4
t = 90 min	42.6±4.9	37.4±1.2	6.5±3.4	0.6±0.7	17.0±2.7
t = 120 min	42.9±4.9	37.0±1.7	7.5±2.7	0.6±0.6	20.4±1.8

Table 9.
Total surface free energy (γ_s) and Lifshitz–van der Waals (γ_s^{LW}), acid–base (γ_s^{AB}), acidic (γ_s^+), and basic, (γ_s^-) components in mJ·m^{-2}, on forehead of 4 oligoseborrheics, OS, (mean and sd), before (at baseline), after washing (t = 0) and during the recovery of sebum level (t = 20 to 120 min)

	γ_s	γ_s^{LW}	γ_s^{AB}	γ_s^+	γ_s^-
Baseline	41.8±2.4	38.6±0.7	2.8±2.4	0.3±0.4	11.6±2.4
t = 0 min	34.6±4.0	33.5±4.2	1.2±0.9	0.4±0.2	1.3±1.1
t = 20 min	34.8±4.9	33.1±4.8	1.7±1.3	0.3±0.3	3.7±2.1
t = 40 min	38.2±3.1	34.6±3.1	3.8±1.0	0.6±0.3	6.5±1.8
t = 60 min	39.5±1.9	37.3±1.1	2.7±1.4	0.3±0.5	6.8±1.8
t = 90 min	40.1±1.5	38.3±0.4	2.1±0.5	0.1±0.1	10.2±0.7
t = 120 min	40.2±0.7	38.8±0.4	1.8±0.5	0.1±0.1	11.8±1.5

5. DISCUSSION

5.1. Surface Free Energy Components

The free energy of the skin surface was related to its lipid coating. On both forehead and volar forearm, degreasing (with ether-treatment or soap washing) reduced the γ_s, demonstrating that both epidermal lipids, originating from horny layer desquamation [10], but especially sebum increased the skin surface free energy and its wettability for both apolar or polar liquids.

At the baseline, the γ_s^{LW} component was identical for both skin locations, although the former is a sebum-rich and the latter a sebum-poor surface, only coated with epidermal lipids. Thus, both sebum and epidermal lipids have the same effect on the skin surface γ_s^{LW}. Following the removal of the skin surface lipids, γ_s^{LW} decreased only slightly, (3–5 mJ·m^{-2}). This is in accordance with previous data [18, 31] showing that washing the skin with acetone or soap reduces its critical surface tension γ_c (which can be considered as equivalent to γ_s^{LW} [9] by only 4 mN·m^{-1}. Therefore, the bare (i.e. without added lipids) skin surface γ_s^{LW} component was about 32 mJ·m^{-2}, and lipids caused only a minor change. On the forehead, the γ_s^{LW} baseline level (Tables 7 and 8) was re-established 60 min before casual sebum levels (Fig. 1) in NS and HS skins, whereas for the OS skin (Table 9), the baseline γ_s^{LW} and casual sebum level (Fig. 1) were reached at the same time. This suggests that the lower sebum flow in the OS skin is responsible for the delay in γ_s^{LW} baseline level re-establishment as compared to NS and HS skins, but a low sebum level is enough to maintain a normal skin surface lipophilicity.

The largest difference found in this study is for the basic component of suface free energy: 4 mJ·m^{-2} and 28 mJ·m^{-2}, respectively, between the volar forearm and the forehead; and on the forehead: 21 mJ·m^{-2} and 11 mJ·m^{-2}, respectively, between the HS or NS skins on the one hand and OS skin on the other hand. Due to the difference of normal sebum level between the NS and HS skin; 140 AU and 210 AU, respectively; the γ_s^- level at the skin surface might result not only from the absolute amount of lipid, but probably also from its amphiphilic lipid content. Indeed, the most amphiphilic components of skin surface lipids are free fatty acids (FFAs) [11] which arise from the hydrolysis of triglycerides during the time they reside on the skin surface or within the follicular canal [32]. The FFA fraction amounts to 30% of skin surface lipids of sebum origin [11]. From our calculations, it can be assumed that a low basic component is the result of a low lipid level and low FFA concentration. Above the NS sebum level, the FFA concentration is sufficient, and has no influence on the γ_s^- component. After washing, γ_s^- was still about 6–8 mJ·m^{-2}, in HS and NS skins despite a very low sebum level (about 10 AU). This suggests that some FFA's still remain tightly bound to the skin surface.

Conversely, degreasing seems more efficient for the forearm and OS skin, probably because the hydrophilic molecules are less numerous on the skin surface. Finally, the γ_s^- level was restored later than the sebum level for all skin types, possibly as a result of the time needed for the hydrolysis of native sebum triglycerides into FFAs [4]. We can conclude that a high FFA fraction in a high sebum level is needed to produce a normal γ_s^- level, i.e. hydrophilicity on the forehead skin surface.

Concerning the acidic component, our data show that the forearm and forehead skin surfaces, regardless of their lipid levels, exhibited a very low acidic character (γ_s^+). Good and van Oss [9] observed that some carbohydrates were γ_s^- monopoles although on the basis of their structures, they are expected to be bipolar because the protons of the hydroxyl groups should be electron acceptors. One possible explanation [9] for this behaviour is surface hydration. The presence of water in the form of a vapor coat over the whole body surface is a common fact: water molecules originate either from diffusion through the epidermis, named "transepidermal water loss" [33] or, especially on the forehead, from inconspicuous sweating [5]. According to this hypothesis [9] and owing to the presence of adsorbed water molecules tightly bound to the surface, the Lewis acid character of the "hydrated carboxyl groups", may not be expressed and the skin surface may be found to have a very low acidic character.

But, if we consider the current debate in the recent literature about the reliability of the acid–base properties of solids predicted by the GvOC approach [34, 35], this hypothesis that the acidic character is masked by adsorbed water molecules is not valid. Actually the use of the GvOC approach to the evaluation of the acidic and basic components of the solid surface free energy shows that most solid surfaces, such as polymers [9, 36–38], psoralen [39] or wood surfaces [40], are overwhelmingly basic with a small or negligible acidic component. Our data are in line with these findings especially for oily skin surfaces. Morerover, these findings are against conventional knowledge because the skin surface exhibits an acidic pH, from 4.5 to 6.5 [41]. Recently, it has also been demonstrated that a relationship exists between casual sebum level and skin pH [42] and shows a trend for an oily skin surface to exhibit a lower pH than non-oily skin surfaces.

Thus, further calculations and studies are needed to analyse the donor-acceptor properties of the skin surface. Della Volpe and Siboni [35] proposed that a complete and coherent definition of the calculation procedure is required, with the use of a wider set of liquids including monopolar acids. Tate *et al.* [38] suggested a revised donor-acceptor analysis, assuming that the entire process of adhesion resulted from hydrogen bonding interactions [43]. In the specific context of the skin surface, with a well-known acidic pH, the contact angle titration [44, 45] seems to be a useful technique to examine the ionization of functional groups at the skin-water interface.

6. CONCLUSION

The wettability of human skin was analysed, *in vivo*, by calculating its surface free energy and its components γ_s^{LW}, γ_s^+, γ_s^-, γ_s^{AB} using advancing contact angle measurements and the GvOC theoretical approach. The forearm showed a hydrophobic or a weak monopolar basic surface and the forehead showed a hydrophilic or monopolar basic surface. The major characteristic of epidermal lipids and especially sebum was found to be their salient basic component which increased skin wettability by polar liquids (water). However these lipids were responsible only for a minor change in the γ_s^{LW} component. The very low acidic component found is inconsistent with the well-known acidic pH of the skin surface. Thus this result includes our work in the debate about the reliability of the GvOC approach. Further studies, such as contact angle titration should be done to evaluate the donor-acceptor properties of the skin. However, using this theoretical approach, the results demonstrate that the skin lipids give the surface a hydrophilic character, which can be assumed to be a parameter describing the hydrophilic properties of the skin surface. This gives a new insight into the so-called "skin hydration" and shoud be useful in adhesion studies of dressings and transdermal drug-delivery systems.

REFERENCES

1. S. M. Jackson and P. M. Elias, in: *Dermatology in General Medicine*, T. B. Fitzpatrick and A. Z. Eisen (Eds), 4th edn. International Edition (1993).
2. R. L. Bronaugh and H. I. Maibach (Eds), *Percutaneous Absorption*. Marcel Dekker, New York (1989).
3. M. H. Schmid and H. C. Korting, *Dermatology* **191**, 276–280 (1995).
4. D. T. Downing, J. S. Strauss and P. E. Pochi, *J. Invest. Dermatol.* **53**, 322–327 (1969).
5. K. Sato, in: *Dermatology in General Medicine*, T. B. Fitzpatrick and A. Z. Eisen (Eds), 4th edn. International Edition (1993).
6. A. Mavon, H. Zahouani, D. Redoules, P. Agache, Y. Gall and P. Humbert, *Colloids Surfaces B: Biointerfaces* **8**, 147–155 (1997).
7. C. J. van Oss, R. J. Good and M. K. Chaudhury, *Separation Sci. Technol.* **21**(1), 1–24 (1987).
8. C. J. van Oss, R. J. Good and M. K. Chaudhury, *Langmuir* **4**, 884–891 (1988).
9. R. J. Good and C. J. van Oss, in: *Modern Approaches to Wettability*, M. E. Schrader and G. I. Loeb, (Eds). Plenum Press, New York (1992).
10. D. T. Downing, M. E. Stewart, P. W. Wertz and J. S. Strauss, in: *Dermatology in General Medicine*, T. B. Fitzpatrick and A. Z. Eisen (Eds), 4th edn. International Edition (1993).
11. G. E. Piérard, C. Piérard-Franchimont and D. Saint-Léger, *Ed. Tech. E.M.C. Dermatol.* **12223** A10 (1991).
12. H. Eberhardt, *Arch. Dermatol. Forsch.* **251**, 155–164 (1974).
13. D. Blanc and P. Agache, *J. Cosmet. Sci.* **2**, 243–250 (1980).
14. Y. Gall and J. P. Chappuis, in: *Cosmetic Dermatology*, R. Baran and H. I. Maibach (Eds). Dunitz, UK (1994).
15. I. E. Dzyoloshinksii, E. M. Lifshitz and L. P. Piteavskii, *Advan. Phys.* **10**, 165 (1961).
16. F. M. Fowkes, *Ind. Eng. Chem.* **56**(12), 40–52 (1964).
17. B. Janczuk, E. Chibowski, J. M. Bruque, M. L. Kerkeb and F. Gonzalez-Caballero, *J. Colloid Interface Sci.* **159**, 421–428 (1993).
18. A. El Khyat, A. Mavon, P. Agache and P. Humbert, *Skin Res. Technol.* **2**, 91–96 (1996).
19. F. H. Herrmann, P. H. Prose and M. B. Sulzberger, *J. Invest. Dermatol.* **21**, 397–419 (1951).
20. D. St-Léger, C. Berrebi, C. Duboz and P. Agache, *Arch. Dermatol. Res.* **265**, 79–89 (1979).
21. C. O. Timmons and W. A. Zisman, *J. Colloid Interface Sci.* **22**, 165 (1966).
22. R. J. Good, *J. Adhesion Sci. Technol.* **6**, 1269–1302 (1992).
23. P. G. Agache, J. Mignot and S. Makki, in: *Cutaneous Aging*, A. M. Kligman and Y. Takase (Eds). University of Tokyo Press (1988).
24. R. N. Wenzel, *Ind. Eng. Chem.* **28**, 988–994 (1936).
25. A. W. Neumann, *Adv. Colloid Interface Sci.* **4**, 105 (1974).
26. R. H. Dettre and R. E. Johnson, Jr., in: *Contact Angle, Wettability, and Adhesion*, Adv. Chem. Ser. No 43. American Chemical Society, Washington, D.C. (1964).
27. J. F. Oliver, C. Huh and S. G. Mason, *Colloids Surfaces* **1**, 79–104 (1976).
28. A. M. Cazabat and M. A. Cohen-Stuart, *J. Phys. Chem.* **90**, 5845–5849 (1986).
29. A. Mavon, D. Redoules, P. Humbert, P. Agache and Y. Gall. *Colloids Surfaces B: Biointerfaces* **10**, 243–250 (1998).
30. M. B. Jani, *Ph. D. thesis, Besançon*, no.92-014 (1992).
31. M. E. Ginn, C. M. Noyes, and E. Jungermann, *J. Colloid Interface Sci.* **26**, 146–151 (1968).
32. M. E. Stewart, *Seminars in Dermatology* **11**(2), 100–105 (1992).

33. D. R. Wilson and H. I. Maibach, in: *Cutaneous Investigation in Health and Disease Aging*, J. L. Levêque (Ed.). Marcel Dekker, New York (1989).
34. M. Morra, *J. Colloid Interface Sci.* **182**, 312–314 (1996).
35. C. Della Volpe and S. Siboni, *J. Colloid Interface Sci.* **195**, 121–136 (1997).
36. F. Boury and J. E. Proust, *J. Adhesion Sci Technol.* **6**, 1359–1369 (1992).
37. M. D. Vrbanac and J. C. Berg, *J. Adhesion Sci. Technol.* **4**, 255–266 (1992).
38. M. L. Tate, Y. K. Kamath, S. P. Wesson and S. B. Ruetsch, *J. Colloid Interface Sci.* **177**, 579–588 (1996).
39. A Saïd, A. Mavon, S. Makki, P. Humbert and J. Millet, *Colloids Surfaces B: Biointerfaces* **8**, 227–237 (1997).
40. D. J. Gardner, *Wood Fiber Sci. Technology* **28**(4), 422–428 (1996).
41. G. Yosipovitch and H. Maibach, *Cosmetics & Toiletries* **111**, 101–102 (1996).
42. S. Edgerly, G. Nole, A. Johnson and A Znaiden, *Drug & Cosmetic Industry* **162**(5), 46–50 (1998).
43. W. Gutowski, in: *Fundamentals of Adhesion*, L. H. Lee (Ed.). Plenum, New York (1991).
44. G. M. Whitesides, H. A. Bieduyck, J. P. Folkers and K. L. Prime, in: *Acid–Base Interactions: Relevance to Adhesion Science and Technology*, K. L. Mittal and H. R. Anderson, Jr. (Eds), pp. 229–241. VSP, Utrecht, The Netherlands (1991).
45. J. C. Joud, J. P. Petit and M. Mantel, *La revue de Métallurgie,* 643–654 (May 1997).

Acid-Base Interactions, Vol. 2, pp. 453–463
Ed. K.L. Mittal

Acid–base characterization of metallic materials and the use of model molecules in the study of adhesion mechanisms

M.-G. BARTHES-LABROUSSE[1,*] and J.-C. JOUD[2]

[1]*CNRS – Centre d'Etudes de Chimie Métallurgique, 15 rue Georges Urbain, 94407 Vitry Cedex, France*
[2]*Institut National Polytechnique de Grenoble – Laboratoire de Thermodynamique et Physico-Chimie Métallurgique, 38402 Saint Martin d'Hères, France*

Abstract—This paper illustrates through a few examples how the parameters such as the point-of-zero charge (PZC), the maximum surface charge and/or the acidic (basic) nature of the reactive sites (Brønsted or Lewis) on a flat inorganic surface can be derived from wettability measurements and/or XPS investigations of the adsorption of probe molecules. Such a precise acid–base characterisation of the surface should lead to a better understanding of interfacial chemistry.

Keywords: Acid–base interactions; reactive sites (Brønsted or Lewis); work of adhesion; XPS; adsorption of probe molecule; mimic molecules.

1. INTRODUCTION

As acid–base chemistry plays an important role in determining interfacial interactions, so it is of paramount importance in various fields such as adhesion, corrosion, catalysis...

Several experimental and theoretical methods have been proposed to characterise the acidic/basic character of flat inorganic solids [1] and its influence on the formation of interfacial bonds. In this paper, we will show through a few examples that (i) wettability experiments can be used for a quantitative determination of the surface charge (point of zero charge PZC, maximum charge σ_m) on metallic (metals or alloys) materials covered with an oxide layer; (ii) characterisation of the nature (Brønsted or Lewis) of the acidic (basic) sites can be deduced from X-ray Photoelectron Spectroscopy (XPS) measurements by following the adsorption of basic (acidic) probe molecules on the substrates; (iii) XPS investigations of the adsorption of model molecules can provide a useful insight in the chemistry of polymer-metal interfaces; and (iv) model studies can help in understanding "real-world" interfaces.

* To whom correspondence should be addressed. E-mail: Marie-Genevieve.Barthes@glvt.cnrs.fr

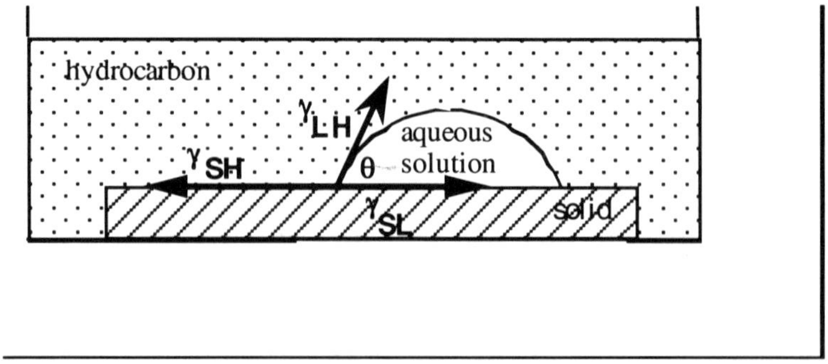

Figure 1. Contact angle in a two-liquid configuration.

2. ACID–BASE CHARACTERISATION USING WETTABILITY EXPERIMENTS

Wettability measurements have been extensively used to determine the apolar (dispersive) and polar (acid–base) components of the work of adhesion at a solid-liquid interface. By comparison with a thermodynamical evaluation of the interfacial tension, a method to deduce the point-of-zero charge (PZC) and the maximal charge (σ_m) has recently been proposed [2].

2.1. Components of the Work of Adhesion

Following the pioneer work by Fowkes [3], it is now generally accepted that the work of adhesion W_{SL} at a solid (S)–liquid (L) interface can be written as a sum of two contributions due to long range (Lifshitz–van der Waals or dispersive) and short range (acid–base) interactions:

$$W_{SL} = W_{SL}^{D} + W_{SL}^{AB} \quad (1)$$

The dispersive component of the work of adhesion is determined from the geometric mean of the dispersive components of the surface energies γ_S^D and γ_L^D of the solid and the liquid respectively:

$$W_{SL}^{D} = 2(\gamma_S^D \gamma_L^D)^{1/2} \quad (2)$$

In a two-liquid configuration (see Fig. 1), the work of adhesion at the solid-apolar hydrocarbon (alkane) interface (SH) is generally assumed to be only dispersive ($\gamma_H = \gamma_H^D$), thus leading to the following Dupré expressions at the solid–alkane (SH) and solid–aqueous solution (SL) interfaces:

$$\gamma_{SH} = \gamma_S + \gamma_H^D - 2(\gamma_S^D \gamma_H^D)^{1/2} \quad (3)$$

$$\gamma_{SL} = \gamma_S + \gamma_L - 2(\gamma_S^D \gamma_L^D)^{1/2} - W_{SL}^{AB} \quad (4)$$

where γ_{SH} and γ_{SL}, respectively, stand for the solid-alkane and solid-liquid interfacial energies and γ_H^D is the dispersive component of the surface energy of the alkane.

When the equilibrium is reached, the Young equation can be written in the form:

$$\gamma_{SH} = \gamma_{SL} + \gamma_{LH} \cos\theta \quad (5)$$

where γ_{LH} is the aqueous solution-alkane interfacial energy.

Replacing γ_{SH} and γ_{SL} in Equation (5) by their expressions in Equations (3) and (4) leads to:

$$\gamma_L - \gamma_H + \gamma_{LH}\cos\theta = 2\,(\gamma_S{}^D)^{1/2}\,[(\gamma_L{}^D)^{1/2} - (\gamma_H{}^D)^{1/2}]\ W_{SL}{}^{AB} \tag{6}$$

As initially proposed by Schultz *et al.* [3], plotting the left hand side of Equation (6) *vs.* $[(\gamma_L{}^D)^{1/2} - (\gamma_H{}^D)^{1/2}]$ for a series of alkanes gives a straight line. Both components of the work of adhesion can then be derived from the slope and the y-intercept.

This method has been used in a study of the effect of lithium ions on ageing of zirconium alloys [5]. Zirconium alloys are used as cladding tubes in Pressurized Water Reactors (PWR). In operating conditions they are immersed in a coolant composed of water (at 10–15 MPa and 280–340°C) with added LiOH. However, lithium ions are known to have a detrimental effect on the corrosion resistance of zirconium alloys [6, 7]. The evolution of the surface properties has been studied for two different alloys: Zircaloy-4 (98.21% Zr in weight, containing 1.34% Sn) and Zr-Nb (98.92% in weight, containing 0.96% Nb). Samples taken from sheets supplied by Cezus-Framatome were electrochemically polished to a mirror finish and subsequently oxidised in water stream (28 days at 415°C, 10.3 MPa). This treatment lead to the formation of a 2.5–3.5 μm thick oxide. Carbon contamination was minimised using standard cleaning procedure with water and acetone, followed by an argon plasma treatment (50W, 5 min). Samples were then immersed for 20 days in an aqueous solution containing 1000 ppm of boron (in H_3BO_3 form) and 2ppm of lithium (in LiOH form) at pH ~ 7. Contact angle measurements were performed using a Krüss goniometer in a two-liquid configuration. Distilled water (pH ~ 5.8) and a series of alkanes of various chain lengths (C_7H_{16}, C_8H_{18}, $C_{10}H_{22}$ and $C_{16}H_{34}$) were used, respectively, for the sessile drop and the hydrocarbon phase. The values of contact angles are summarised in Table 1. Figure 2 shows the results obtained using the Schultz representation.

The values obtained for the dispersive and acid–base components of the work of adhesion are given in Table 2. It can be seen that the acid–base component of the work of adhesion is higher for the Zr-Nb alloy than for Zy-4 and that, for both alloys, its value is higher when ageing is performed in the "PWR" solution (i.e. the solution containing boron and lithium) rather than in air. The high uncertainty indicated for the values of W^D, compared to those of W^{AB}, originates from the large error bars in the Schultz representation.

Table 1.
Average contact angles (over 20 runs) of distilled water droplets on Zy-4 and ZrNb alloys aged 20 days in PWR solution

	θ ZrNb (deg)	Δθ ZrNb (deg.)	θ Zy4 (deg.)	Δθ Zy4 (deg.)
heptane	33.7	3	41.8	4
decane	41.2	4	49.2	3
hexadecane	50.7	5	60.6	3

Table 2.
Dispersive and acid–base components of the work of adhesion as deduced from wettability measurements in a two-liquid configuration

Alloy	Ageing in	W^D (mJ.m^{-2})	W^{AB} (mJ.m^{-2})
Zy-4	Air	110±10	52.5±1
	"PWR" solution	130±15	86.5±1
ZrNb	Air	110±10	67.0±1
	"PWR" solution	115±15	92.0±1

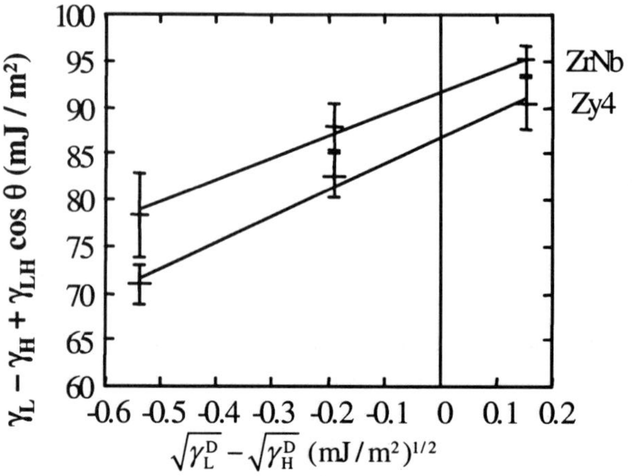

Figure 2. Schultz representation of wettability results obtained on zirconium alloys immersed 20 days in PWR solutions.

2.2. Surface Charge

Whitesides and Laibinis [8] and Hüttinger *et al.* [9] have shown that the work of adhesion of a solid surface can be determined using aqueous solutions with systematically varied pH values instead of pure (distilled) water in the wetting measurements. By combining the Young–Dupré equation in a two-liquid configuration and the Dupré expression for the work of adhesion and taking into account that W_S^{LD}, γ_L and γ_{LH} are not influenced by the addition of small amounts of mineral acids or bases [9, 10], the variation of the work of adhesion can be expressed as a function of the pH of the aqueous solution:

$$\frac{dW_{SL}}{dpH} = -\frac{d\gamma_{SL}}{dpH} = -\gamma_{LH}\sin\theta\frac{d\theta}{dpH} \tag{7}$$

As developed elsewhere [2], the variation of the interfacial tension γ_{SL} between an oxide and an aqueous solution can be calculated using the Gibbs adsorption. This leads to a relationship in which $d\gamma_{SL}$ is expressed as a sum of an electrical and a chemical contributions:

$$d\gamma_{SL} = -\sigma d\Phi = -2\sigma \frac{RT}{F} d\ln\left[H^+\right] \quad (8)$$

where σ is the surface charge, Φ the surface potential, F the Faraday constant and $[H^+]$ the proton concentration of the solution. Using the Grahame relationship to express the surface charge in a simple Gouy-Chapman interface model [11], integration of expression (8) gives a general relationship between the evolution of the interfacial tension γ_{SL} and the pH of the aqueous solution.

Two limiting situations are of special interest (see Fig. 3):
i) when the pH of the aqueous solution is close to the PZC of the oxide, a parabolic evolution of γ_{SL} as a function of pH is observed:

$$\Delta\gamma_{SL} = -8C\frac{RT}{K}(PZC - pH)^2 \quad (9)$$

where C is the concentration of the aqueous solution and 1/K the Debye screening length.

This approach has been applied to the study of ageing of zirconium alloys in air or in "PWR" solutions. Investigations in the pH range of 3–12 have been carried out (droplets immersed in heptane). NaOH and HNO_3 were used to modify the pH of aqueous solutions. The results are shown in Fig. 4 (Zircaloy-4) and Fig. 5 (ZrNb). The most striking feature in these figures is the difference in shapes observed for the two alloys. For Zircaloy-4, the shape is characteristic of an amphoteric behaviour. The PZC can be estimated to 8.5±1 for the "aged-in-air" samples (Fig. 4a) and around 7 for the "aged-in-PWR" samples (Fig. 4b). They are in good agreement with values found in the literature for zirconium oxide [12, 13]. For ZrNb, two humps can be observed both for the "aged-in-air" (Fig. 5a) and the "aged-in-PWR" (Fig. 5b) samples, which can be, respectively, ascribed to zirconium oxide (PZC around 9) and niobium oxide (PZC around 2). The presence of niobium oxide in the oxide layer has been further evidenced using XPS [5].

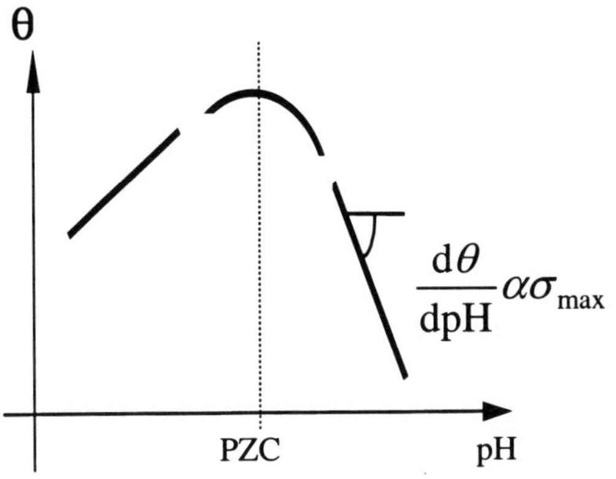

Figure 3. Theoretical variation of θ as a function of pH of the aqueous solution (amphoteric oxide). The maximum value of the surface charge can be deduced from the slope of the variation of θ as a function of pH (see text).

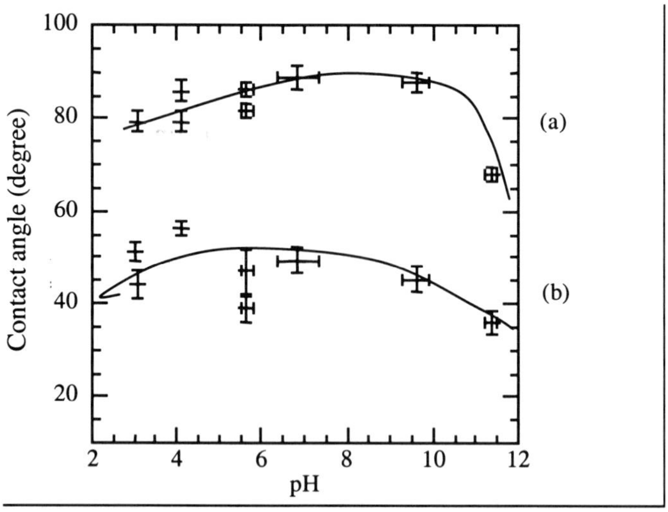

Figure 4. Contact angle variations with pH for the oxidized Zy4 alloy: (a) aged in laboratory air, (b) immersed in a PWR solution.

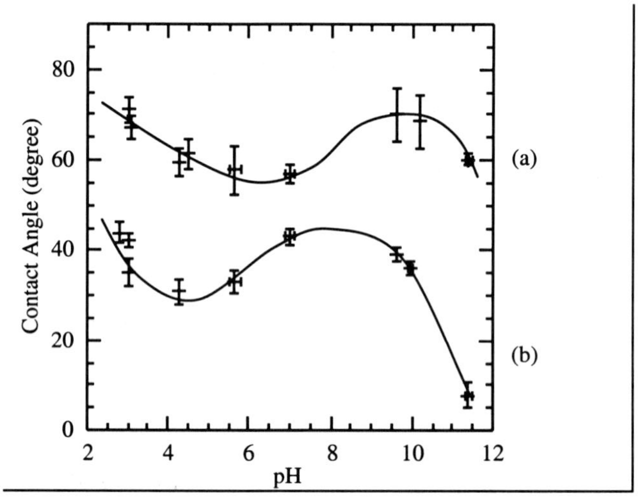

Figure 5. Contact angle variations with pH for the oxidized ZrNb alloy: (a) aged in laboratory air, (b) immersed in a PWR solution.

ii) when the surface charge is close to its maximum value σ_{max}, the interfacial tension varies as a linear function of the pH as:

$$\frac{\Delta\gamma_{SL}}{\Delta pH}=4.6\frac{RT}{F}\sigma_{max} \tag{10}$$

$$\sigma_{max}=\frac{F}{4.6\,RT}\gamma_{LH}\sin\theta=\frac{d\theta}{dpH} \tag{11}$$

Figures 6 and 7 show results obtained from the wettability experiments performed on iron (Fig. 6) and chromium (Fig. 7) oxide/hydroxide surfaces [14]. In both cases, the shape of the plot is characteristic of an amphoteric behaviour. The PZC values are close to 7-9 (iron oxide/hydroxide) and 8-9 (chromium oxide/hydroxide). The maximum negative surface charges are –67 $\mu C.cm^{-2}$ (iron oxide/hydroxide) and –2.5°$\mu C.cm^{-2}$ (chromium oxide/hydroxide).

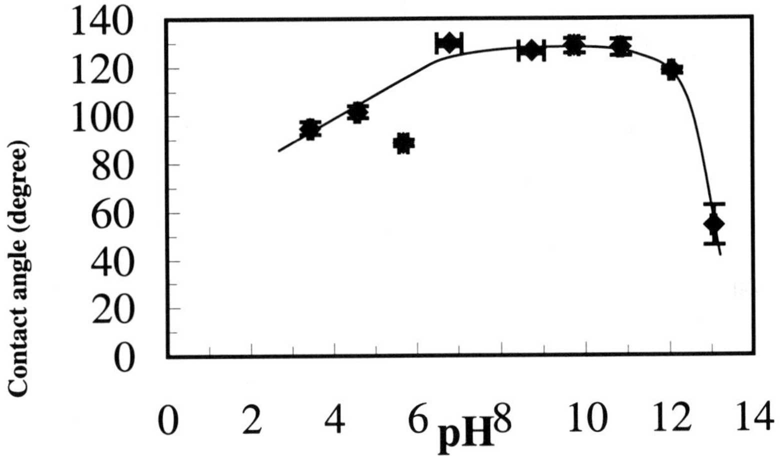

Figure 6. Contact angle variations with pH for a Fe oxide/hydroxide.

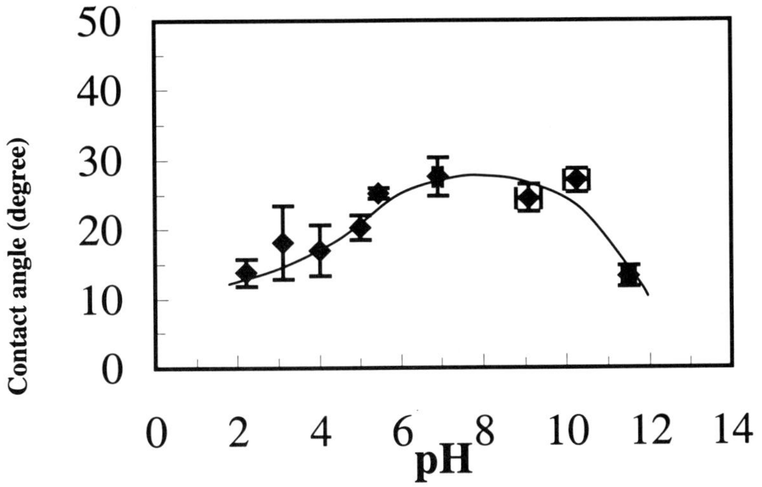

Figure 7. Contact angle variations with pH for a Cr oxide/hydroxide.

3. ACID–BASE CHARACTERISATION BY FOLLOWING ADSORPTION OF PROBE MOLECULES WITH XPS

The binding energies of inner core electrons of atoms such as carbon, nitrogen, or oxygen can vary over a range of several eV and increase from basic (electron-donating) to acidic (electron-accepting) character of the atoms. Following the adsorption of probe molecules of known acidic (basic) properties using XPS can thus provide a useful insight in the nature of basicity (acidity) of the surface.

Fig. 8 shows the results of an XPS investigation [15] where 1,2-diaminoethane (ethylene diamine) has been used as a basic probe molecule to study the influence of ageing on the nature of acidity of anodised titanium. Titanium specimens were prepared from a titanium sheet (99.6%, Goodfellow-Sarl) polished to a high mirror finish using successively increasing grade silicon carbide disks, diamond polishing and polishing with finely divided silica in potassium hydroxide solution (PO-S, Struers). The specimens were then anodised at 10 mA.cm^{-2} with a 10 V potential maximum in 0.5 M H_2SO_4 followed by rinsing with deionised water and drying with warm air. These specimens will be further referred as "freshly prepared" samples (FP). Ageing was done by annealing the "fresh specimens" at 230°C for 40 h under vacuum (10^{-8} torr). These last specimens will be referred as "thermally treated samples" (TT). Adsorption of 1,2-diaminoethane on samples was obtained by ultrasonic treatment for 5 min in a 10% solution of 1,2-diaminoethane (99,5%+) in carbon tetrachloride (purity > 99.8%), followed by ultrasonic rinsing 3 x 5 min with carbon tetrachloride. It can be seen in Fig. 8 that, following molecular adsorption, the nitrogen 1s peak can be decomposed into two contributions corresponding, respectively, to Lewis bonded nitrogen (N(L)) at 400.4± 0.2 eV binding energy and protonated Brønsted nitrogen (N(B)) at 402.1± 0.2 eV binding energy [16–18].

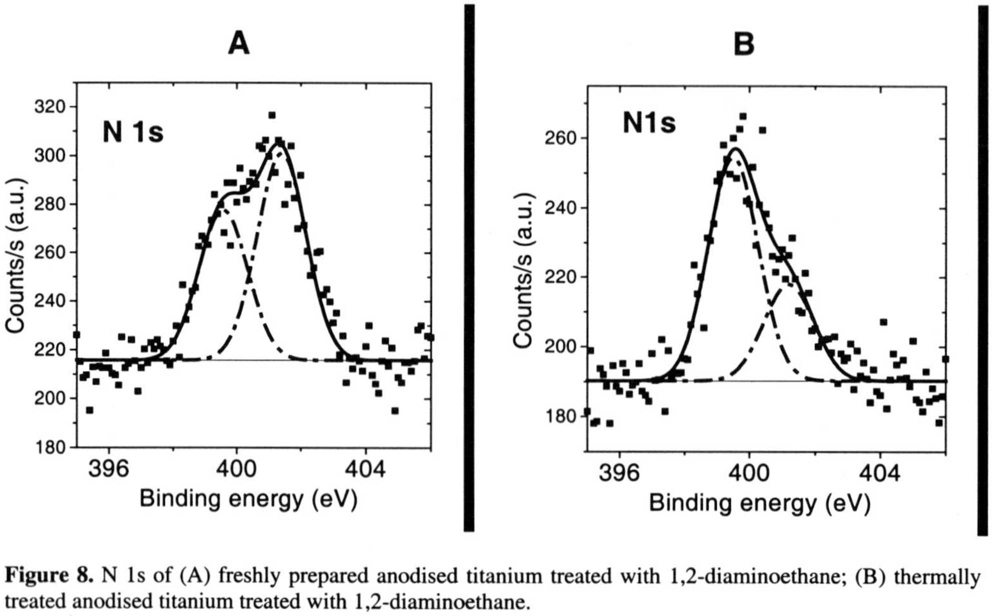

Figure 8. N 1s of (A) freshly prepared anodised titanium treated with 1,2-diaminoethane; (B) thermally treated anodised titanium treated with 1,2-diaminoethane.

Table 3 shows the results of curve-fitting analysis carried out on respectively freshly prepared Ti and thermally treated anodised Ti before (FP Ti and TT Ti) and after (FP Ti + EDA and TT Ti + EDA) adsorption of 1,2-diaminoethane assuming linear background and pure Gaussian lineshapes.

Table 3.
Peak fitting for O 1s and N 1s on freshly prepared Ti and thermally treated Ti (FP Ti and TT Ti) and after adsorption of 1,2-diaminoethane on freshly prepared Ti and thermally treated Ti (FP Ti + EDA and TT Ti + EDA). N(L) and N(B) correspond, respectively, to Lewis bonded nitrogen and protoned Brønsted bonded nitrogen

	FP Ti		FP Ti + EDA		TT Ti		TT Ti + EDA	
XPS Feature	Energy (eV) ±0.2 eV	Area (%) ±5%	Energy (eV) ±0.2 eV	Area (%) ±5%	Energy (eV) ±0.2 eV	Area (%) ±5%	Energy (eV) ±0.2 eV	Area (%) ±5%
O^{2-}	530.8	70	530.8	70	530.9	90	530.9	90
OH^-	532.7	30	532.8	30	532.9	10	533.0	10
N(L)	-		400.4	30	-		400.4	70
N(B)	-		402.1	70	-		402.1	30

From the values (in percentage) of the areas of the different contributions shown in Table 3, it can be seen that the change in dominance of Brønsted to Lewis interactions observed from the N 1s response corresponds to a substantial decrease in hydroxylation between freshly prepared and thermally annealed anodised titanium. This clearly indicates that at least some of the hydroxyl groups are Brønsted acidic sites.

4. ADHESION MECHANISMS

Adsorption of simple model molecules as analogs to more complex systems can lead to a better understanding of the chemistry of an interface and has often been used, in particular to investigate polymer-metal interfaces. The chemical specificity of XPS again provides a powerful means to access the interface bonding mechanism. For example, adsorption of simple molecules which can mimic the various components of an epoxy/polyamine polymer onto anodised titanium has recently been investigated using XPS [19]. The molecules chosen to represent the various components of an uncured, fully cured epoxy/polyamine system and the amine hardener were phenyl-2,3-epoxypropyl ether, N-methyldiethanolamine and 1,2-diaminoethane. Phenyl-2,3-epoxypropyl ether is representative of bisphenol A, the basic building block of almost all epoxy resin systems. N-methyldiethanolamine can mimic the structures formed as an epoxy/amine polymer approaches its final state of cure. 1,2-diaminoethane is a multifunctional amine and can be used to represent an amine hardener. Titanium specimens were prepared from a titanium sheet (99.6%, Goodfellow-Sarl) as indicated in Section 3 for "freshly prepared" titanium. Adsorption of the different molecules was performed by ultrasonically treating the specimens in a 10% solution of the model molecule in carbon tetrachloride for 5 min, followed by ultrasonic rinsing (3 x 5 min) with carbon tetrachloride. From the examination of the Ti 2p, C 1s, O 1s and N 1s peaks, it has been shown that the interaction with thermally treated specimens is rather poor with all molecules except 1,2-diaminoethane. On freshly prepared specimens, as previously discussed (see Section 3), 1,2-diaminoethane interacts with hydroxylated anodised titanium primarily with protonation of the amine functionality by the acid surface hydroxyl groups. Lewis bonding to surface cationic sites occurs as a minor secondary bonding mechanism. The Brønsted-like bonding mechanism begins to break down as we move to N-methyldiethanolamine, for which a weakening of the protonic interaction of the amino functionality is observed. Phenyl-2,3-epoxypropyl ether is not adsorbed on freshly prepared titanium specimens. Such differences in the mechanims of interaction of the surface with molecules containing aminous nitrogen was not observed for copper and aluminium substrates and seems specific of titanium [19].

5. RELEVANCE TO "REAL-WORLD" INTERFACES

The question of the relevance of applying the conclusions drawn from fundamental acid–base properties as deduced from adsorption studies to "real-world" interfaces is still open. However, it can be shown that results obtained on model systems can provide a useful guideline to understanding of complex behaviours.

The shape of the plots of contact angles versus pH observed for zirconium alloys (see Figs 4 and 5) will have strong implications in terms of corrosion of these alloys when immersed in the coolant solution of a PWR. In the case of Zircaloy-4, the pH of the coolant solution is close to the PZC value of the surface and the surface charge is close to zero, thus minimising electrostatic interactions and adsorption of charged species such as lithium ions. On the other hand, the wettability of Zr-Nb alloy surfaces is maximum in this pH-region. Moreover, the positive charge observed in this case following immersion in the aqueous solution containing boron and lithium suggests adsorption of lithium ions. This would mean that lithium adsorption is more likely to happen on Zr-Nb alloys than on Zircaloy-4. Further evidence of such a preferential adsorption has been given using Auger Electron Spectroscopy [5].

On titanium substrates, the pronounced strength of the interaction between the substrate and molecules containing aminous nitrogen, compared to epoxy molecule, could indicate that when an epoxy/polyamine coating or adhesive is applied, preferential segregation of the amine or polyamine agent in the vicinity of the interface could occur, with the resulting formation of an interphasial zone between the bulk polymer and the substrate with a structure substantially different from the bulk polymer. The formation of such interphasial regions means that this region could be the locus of failure, as opposed to the polymer/metal interface. This observation is further supported by a recent study of the variation of the glass transition of an epoxy resin (DGEBA+IPD) as a function of the thickness of the polymer deposited on a TA6V substrate, where a sharp transition was observed for a critical thickness of 600μm and ascribed to the existence of an interphasial region [20].

6. CONCLUSION

We have illustrated through a few examples that characterisation of acid–base character of flat inorganic solid surfaces can be achieved through wettability measurements and adsorption of probe molecules. Information at a macroscopic (PZC) or molecular scale can be obtained and this can provide a useful insight in the understanding of the reactivity of solid surfaces, in particular in relation to corrosion of inorganic solids or adhesion of polymers. Model experiments, carried out with simple molecules, can help in understanding the behaviour of more complex "real-world" interfaces.

REFERENCES

1. See for example *Acid–base Interactions: Relevance to Adhesion Science and Technology*, K. L. Mittal and H. R. Anderson, Jr. (Eds). VSP, Utrecht, The Netherlands (1991).
2. C. Vittoz, M. Mantel and J. C. Joud, *J. Adhesion* **67**, 347 (1998).
3. F. M. Fowkes, *Ind. Eng. Chem.* **56** (12), 40 (1964).
4. J. Schultz, K. Tsutsumi and J. B. Donnet, *J. Colloid Interface Sci.* **59**, 272 (1977).
5. C. Vittoz, P. Bossis, F. Lefebvre and J. C. Joud, *J. Adhesion Sci. Technol.* **13**, 1045 (1999).
6. R. A. Perkins and R. A. Busch, in: *Zirconium in the Nuclear Industry*, ASTM STP-1132, Philadelphia (1991).
7. B. Cox, *J. Nuclear Mater.* **249**, 87 (1997).
8. G. M. Whitesides and P. E. Laibinis, *Langmuir* **6**, 87 (1990).
9. K. J. Hüttinger, S. Höhmann-Wien and G. Krekel, *J. Adhesion Sci. Technol.* **6**, 317 (1992).
10. R. Defay and I. Prigogine, *Surface Tension and Adsorption*. Longmans, London (1966).
11. See J. O'M. Bockris and A. K. N. Reddy, *Modern Electrochemistry – 2*. Plenum Press, New York (1977).
12. G. A. Parks, *Chem. Rev.* **65**, 177 (1965).
13. J. P. Jolivet, M. Henry and J. Livage, *De la solution à l'oxyde*. InterEditions, Editions du CNRS, Paris (1994).
14. P. E. Dubois, C. Vittoz, M. Mantel and J. C. Joud, paper presented at EURADH'98, Garmish-Partenkirschen, Germany (1998).
15. J. Marsh, L. Minel, M. G. Barthés-Labrousse and D. Gorse, *Appl. Surf. Sci.* **99**, 335 (1996).
16. M. R. Horner, F. J. Boerio and H. M. Clearfield, *J. Adhesion Sci. Technol.* **6**, 1 (1992).
17. J. A. Kelber and R. K. Brow, *Appl. Surf. Sci.* **59**, 273 (1992).
18. R. H. Bradley, G. Beamson, X. Ling and I. Sutherland, *Appl. Surf. Sci.* **72**, 273 (1993).
19. J. Marsh, L. Minel, M. G. Barthés-Labrousse and D. Gorse, *Appl. Surf. Sci.* **133**, 270 (1998).
20. M. Benabdi and A. A. Roche, *Le Vide: Science, Technique et Applications*, Supplément au n°**284**, 11 (1997).

Acid-Base Interactions, Vol. 2, pp. 465–480
Ed. K.L. Mittal

Acid-base characterization of porous silica particles

JAN BÄCKMAN, TOM EKLUND and JARL B. ROSENHOLM*

Department of Physical Chemistry, Åbo Akademi University, Porthansgatan 3-5, FIN-20500 Turku, Finland

Abstract—The number, type and strength of the surface sites of porous silica particles have been characterized. The surface acidity/basicity of the silica particles have been investigated by adsorbing organic probe molecules from cyclohexane. The probe molecules have been classified in terms of Lewis acidity and basicity using Gutmann's acceptor and donor numbers. A thermodynamically more sound approach is, however, to determine the number and the strength of the adsorption sites from adsorption isotherms. The fit of the experimental data to the Langmuir adsorption isotherm is compared with the fits to the Freundlich and Henry-Langmuir adsorption isotherms. It was found that the adsorption of benzylamine and phenylpropionic acid with predominantly Brönsted sites was described by the simple Langmuir isotherm. However, in order to describe the adsorption of phenol the Freundlich isotherm or a combined Henry-Langmuir isotherm had to be applied. It is concluded that both Brönsted and Lewis interactions have to be considered when evaluating the adsorption behavior of phenol.

Keywords: Acidity/basicity; silica; adsorption isotherms: Langmuir, Freundlich, Henry isotherms.

1. INTRODUCTION

Porous silica particles are used in a variety of applications [1] such as, e.g. fillers in polymers [2], as chromatographic beads [3] and as support material for the metallocene catalysts [4]. Surface modifiers or organic catalyst molecules are usually bound to or reacted with surface hydroxyl groups. Since the silanol groups successively desorb upon heating, the amount of, or the distance between, the mole-cules adsorbed may be conveniently controlled by heat treatment [5–8].

The hydroxyl groups represent the Brönsted acidic/basic sites on the surface. It is, however, of interest also to determine whether other sites exist on the surface showing an acid/base activity in the Lewis sense [8–10]. The number and strength of the surface sites, as well as the (re)activity of the solid surface may thus be enhanced by thermal pretreatment or variations in the composition of the inorganic support material. We have chosen to characterize the surface acidity/basicity of silica used for metallocene catalyst support with a method based on the Lewis acid/base concept previously used for characterization of cellulose [12], road stone aggregates [13] and silicon nitride powder [14]. In this method the adsorption of probe molecules of different basicity or acidity on the surface sites is studied in non-aqueous solutions. Cy-

* To whom correspondence should be addressed. E-mail: Jarl.Rosenholm@abo.fi

clohexane is used as solvent in order to eliminate the effect of hydrolysis and of dispersive (apolar) interactions between the probes and the solid surface. This option is not available, e.g. in gas adsorption analysis although similar overall effects have been recorded [9, 10].

2. MATERIALS AND METHODS

2.1. Silica

Silica (Si S8MY) from W.R. Grace was used in the adsorption experiments. Before use it was calcined at 420°C for 20h. The manufacturer gives the value of 303 m^2/g for the specific surface area and 70 μm as the particle size. These values will be used later in the calculations reported in the Results Section. In the calibration of the experimental techniques other silica samples were also used.

2.2. Adsorption of Gaseous Nitrogen

The gas sorption isotherms were run on a Carlo Erba Sorptomatic 1900-instrument using standard procedures and nitrogen gas as adsorbate.

2.3. Thermal Gravimetric Analysis

The desorption of water and the number of hydroxyl groups as a function of temperature were followed by thermal gravimetric analysis (Netzsch TG 209) between 80 and 800°C. An initial isothermal period at 80°C for 80 minutes was followed by a temperature increase of 3°C/min. All samples were mounted into a platinum sample holder. Dry nitrogen was used to flush the sample during the TG run.

2.4. Adsorption of Organic Probe Molecules from Liquid Cyclohexane

Four different monosubstitued benzene probe molecules (Table 1) were supplied by Fluka Chemie AG. Molecular sieves (0.3 nm) were used to dry the liquids before use. Table 1 gives the estimated donor numbers (DN) and acceptor numbers (AN) for the probes used in this study [13, 14].

Cyclohexane was used as the solvent in all the experiments, because it can be assumed that it will interact with silica and the probes only through dispersion forces [9, 10]. Hence, cyclohexane will not affect the adsorption strength of the probes. Furthermore, cyclohexane is geometrically similar to the chosen probe molecules.

The adsorption was carried out on 0.2g silica in glass test-tubes with approximately 20 ml of solvent and probe concentrations ranging from 5 to 10 mM. The reaction time for the probe molecules was determined to be 60–90 minutes by kinetic studies. A traymixer was used for stirring. The solution was separated from the powder by centrifugation and the supernatant was thereafter carefully removed with a pipette. The sediment was dried for 16 hours at 80°C in ambient atmosphere.

The concentration of probe molecules was determined with a Schimazdu UV-240 UV-VIS spectrometer. From the decrease of probe concentration in the solution, the adsorbed amount per square meter of dried silica was calculated.

Table 1.
Some characteristics of organic probe molecules used in the experiments

Sorbate	Molecular mass ($gmol^{-1}$)	DN	AN	λ_{max} (nm)	Purity
Benzylamine	107.15	~55	~15	259	≥99%(GC)
Toluene	92.15	0.1	0.5	262	≥99.5%(GC)
Phenol	94.11	~20	~32	270	≥99.5%(GC)
3-Phenylpropionic acid	150.19	~20	~50	259	≥98%(HPLC)
Cyclohexane	84.16	0	0	-	≥99.5%(GC)

DN and AN are estimated Gutmann´s donor and acceptor values [13, 14]

Samples for Diffuse Reflectance Infrared Fourier Transform (DRIFT) spectroscopy (Bruker IFS66 instrument) were made by mixing approximately 15 wt% sample with KBr in a WIG-L-Bug capsule without the steel ball for two minutes. These samples were compacted in the DRIFT sample cup by gentle tapping.

3. RESULTS

3.1. Heat Treatment of Silica Particles

The influence of a heat treatment on silica has been thoroughly documented in a vast number of publications [e.g. 1, 5–8]. A continuous decrease in the IR band intensity was, as expected, found when the silica powder was heated to 1000°C. The predominant contribution of heavily associated water and hydroxyl groups was found in the range of 3260–3640 cm^{-1}. Additionally, a small shoulder representing the inaccessible intrinsically hydrogen bonded silanol groups was observed in the range 3640–3680 cm^{-1} and the isolated geminal hydroxyl groups in the range 3730–3740 cm^{-1}. In particular the very sharp peak representing isolated hydroxyl groups was found in the range 3740–3750 cm^{-1}. The total amount of water can be investigated separately by recording the scissors vibration band at 1640–1650 cm^{-1}.

The weakly and strongly physisorbed water is considered to be removed in 110–130°C and 225–300°C ranges, respectively. A one step adsorption at about 190°C [6, 16] and a final desorption (or condensation of bonded hydroxyl groups) at 400°C [3, 5] have also been reported. However, in the present study the physisorbed water cannot be resolved from the themogravimetry printout. Instead, the heating seems to remove water and/or surface hydroxyls more or less continuously. This is to be expected from the broad distribution of hydroxyl groups left on the silica surface as judged from the DRIFT spectra.

3.2. Determination of the Pore Structure of the Silica Particles

The porous structure of the solid support is of crucial importance for the efficiency of the chromatographic beads and the activity of the catalyst. Typically the pores are subdivided into micropores (type I or V) the width (w) of the pores being < 2nm, mesopores (type IV or V) 2nm<w<50 nm and macropores (type III) w> 50 nm. In some cases, the micropores are further subdivded into micropores w< 0.6–0.7 nm and super micropores 0.6–0.7<w<1.5–1.6 [15].

For meso- and macropores the specific pore volume can be determined, e.g., by applying the BET isotherm analysis, while the average radius of the pores can be determined using the Kelvin equation [17]. For the Grace silica treated at 150°C, a BET surface area of 373 m^2/g was found. According to the sorptogram the porosity is due to interparticle voids with a "pore volume" of 1.70 cm^3/g and an average "pore diameter" of 18.0 nm, but with a large distribution towards smaller values.

3.3. Spectroscopic Measurements of the Adsorption of Organic Probe Molecules

The characteristic bands for aromatic C–H stretching are observed between 3090 cm^{-1} and 3035 cm^{-1}. and for the aliphatic C–H stretching in the 2860 cm^{-1} to 2930 cm^{-1} range. At about 750 cm^{-1} we find an out-of-plane aromatic C–H band, and at 697 cm^{-1} we observe a ring C=C bend, as seen in Figure 1c [18]. The asymmetric and the symmetric vibration peaks of the amine group of benzylamine are clearly discerned at 3360 cm^{-1} and 3300 cm^{-1}, respectively.

The lack of aromatic absorbance peaks for toluene, and the data obtained from adsorption isotherms indicate that no toluene is adsorbed. For benzylamine, 3-phenylpropionic acid (3-PPA) and phenol absorbance peaks are found, showing that these compounds were adsorbed. The spectrum of calcined silica with adsorbed toluene has still a sharp peak at 3747 cm^{-1} due to isolated non-interacting silanol groups. This supports the view that this probe is not interacting through its π-electrons of the benzene ring (weak Lewis basicity) [9, 10, 19]. When the other probes are adsorbed onto silica the peak at 3747 cm^{-1} disappears due to the reaction with the surface SiOH groups.

3.4. Microanalysis of Surface Hydroxyl Groups

Trimethyl aluminium reacts quantitatively with the available surface hydroxyl groups on silica dispersed in toluene. According to Collins *et al.* [20] the number of hydroxyl groups can be calculated from the volume of methane gas liberated. In our measurements the silica was dispersed in freshly distilled, sodium wire dried toluene and kept under nitrogen atmosphere. Unfortunately, although the experimental procedure was calibrated with known amounts of water added to the toluene and checked with Karl Fischer titrations the method was found to be of insufficient accuracy for our purposes.

However, the FTIR analysis seems to be sensitive enough to provide a quantitation of the amount of hydroxyl groups remaining as evidenced by the spectra recorded before and after the complete removal of the hydroxyl groups. The spectrum of silica had no peak at 3747 cm^{-1} after reacting it with an excess of trimethyl aluminium in toluene showing that all the hydroxyl groups were used up during the titration. Metal chlorides have also been used for quantitative reactions with the silanol groups [5, 6, 8].

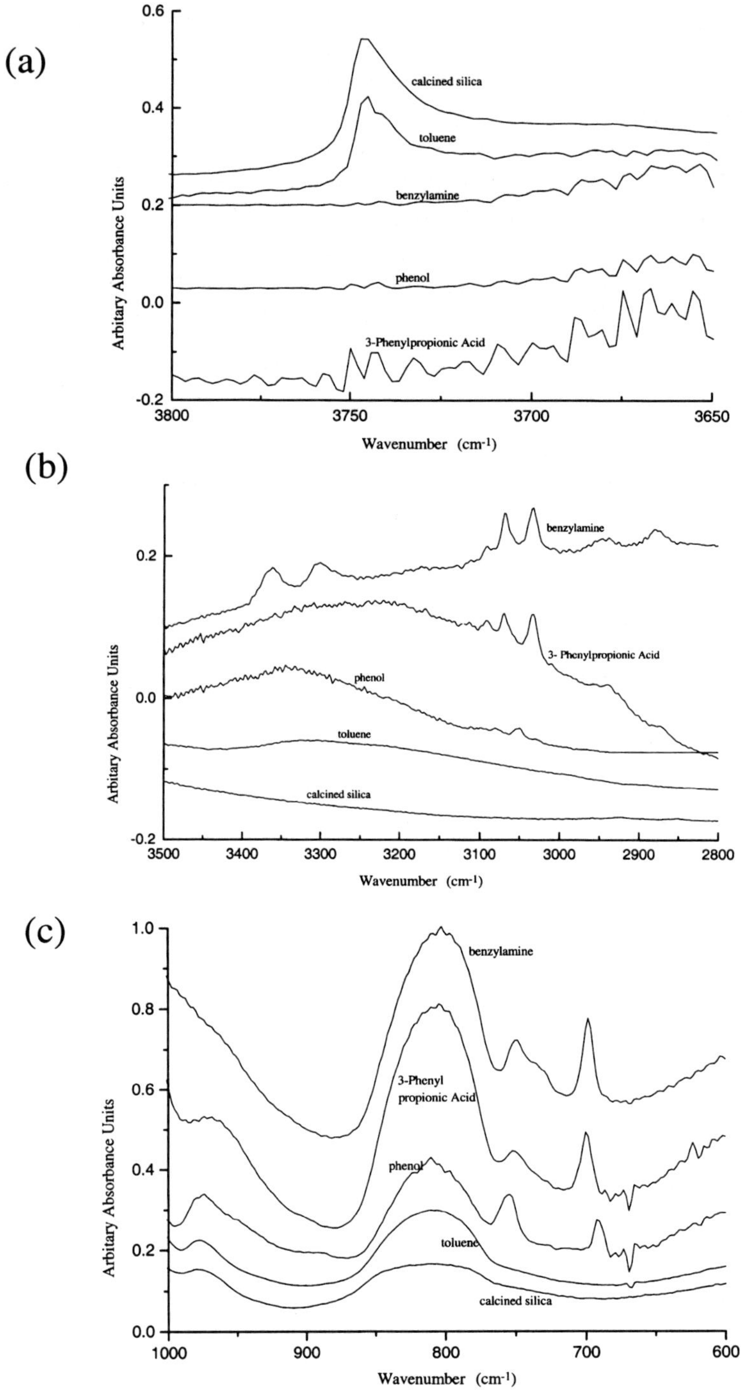

Figure 1. The DRIFT spectra (a) 3800–3650 cm^{-1} range, (b) 3500–2800 cm^{-1} range, c) 1000–600 cm^{-1} range of untreated silica and silica after treatment with probe molecules in cyclohexane.

3.5. Analysis of Adsorption Data

The experimental surface excess of probe molecules (Γ, given as the amount or number of moles per unit surface area) adsorbed onto silica is plotted versus the equilibrium concentration in solution in Figure 2. These results show that the probes with different functional groups have distinctly different adsorption affinities. benzylamine adsorbs to the largest extent onto silica. Initially the adsorption of 3-phenylpropionic acid is more extensive than phenol, but at higher equilibrium concentrations the reverse relation is found. At very high concentrations the adsorption of phenol even exceeds that of benzylamine. Toluene has no acidic or basic functional groups and therefore only weak adsorption is expected due to its π-electron interaction with acidic surface sites. This is also found as shown in Figure 2.

Assuming that the adsorbate (A) molecules successively displace the solvent (liquid, L) molecules from the surface, we may write the equilibrium constant;

$$K_L = \frac{a_A^S \, a_L}{a_L^S \, a_A} \tag{1}$$

where a denotes the activity and the superscript (S) the surface layer. If we further assume that the surface molecules remain ideal we obtain;

$$K_L = \frac{x_A^S \, a_L}{x_L^S \, a_A} = \frac{x_A^S a_L}{(1 - x_A^S) a_A} \tag{2}$$

where x denotes mole fraction. The adsorption is furthermore assumed to be determined by the number of sites and not by the size of the molecules.

Provided that the equilibrium is obtained for dilute solutions, the activity of the solvent may be assumed to be $a_L \approx 1$ and the activity of the adsorbate (solute) to be approximated by its concentration (arbitary scale). The rearrangement of equation (2) in terms of equilibrium molar concentration then gives;

$$x_A^S \cong \frac{K_L c}{1 + K_L c} \tag{3}$$

where x_A^S equals the surface coverage, i.e. neglecting the subscript (A) we define $x^S = (n^S/n^S{}_m) = (\Gamma / \Gamma_m) = \Theta$, where Γ_m represents the maximum surface excess. For a nonspecific adsorption the surface coverage, $\Theta = x^S$ A, where A = total area of the particle. Similar relationships have been derived for the chromatographic analysis of probe molecule retention [3].

The adsorption can also be described in terms of the "specific amount adsorbed" n^s/w [17]:

$$\frac{c}{n^s / w} = mc + b \tag{4}$$

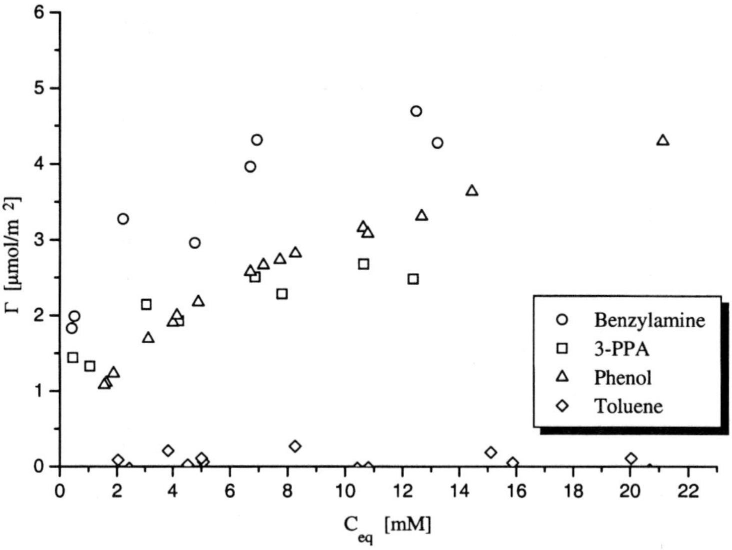

Figure 2. The experimental surface excess of the probe molecules adsorbed onto silica as a function of their equilibrium concentration in cyclohexane at 25°C.

where (n^s/w) is the amount of solute adsorbed per unit weight of adsorbent, c is the equilibrium molar concentration of the solute. The constants m and b are; $m=1/(n^s_m/w)=1/(A_{sp}\Gamma_m)$ and $b= m/K_L$. By plotting $c/(n^s/w)$ versus c straight lines were obtained for benzylamine and 3-phenylpropionic acid (Figure 3) which indicates a Langmuir type of adsorption. For phenol the linearity applies only in the low concentration region where the adsorption to free silanol groups predominates [3, 9, 10]. The intercepts and slopes obtained by linear regression of the linearized Langmuir adsorption isotherm are reported in Table 2.

Table 2.
The intercept, slope and regression constant (R) obtained from a fit of the experimental data to the linearized Langmuir adsorption isotherm (eq. 4)

Probe	Intercept (g/dm^3)	Slope (10^3g/mole)	R
Benzylamine	0.80	0.69	0.986
3-Phenylpropionic acid	1.14	1.21	0.992
Phenol	3.48	0.82	0.997

Assuming that all the sites are available for the probe molecules the site density can be calculated using equation 5;

$$\sigma_{site} = \frac{N_A}{mA_{sp}} = \Gamma_m N_A \quad (5)$$

where m is the slope, N_A Avogadro's number and A_{sp} the specific area of the adsorbent (303 m^2/g). The equilibrium constant, for the Langmuir isotherm, K_L can be calculated according to equation 6 where b is the intercept:

$$K_L = \frac{m}{b} \tag{6}$$

Taking K_L as a dimensionless value we can calculate the standard free energy of adsorption.

$$\Delta G^{\theta} = -RT \ln K_L \tag{7}$$

The results of the calculations are given in Table 3. It should be emphasized that the absolute value is dependent on the concentration scale chosen! In order to convert the equilibrium constant to an absolute scale it should be divided by the concentration of the liquid at infinite dilution ($c_L \approx \rho_L/M_L$, ρ = density, M = molar mass).

Table 3.
Values of pK_a [22], σ_{site}, equilibrium constants and standard free energy of adsorption for the probe molecules

Probe	pK_a	σ_{site} (N_{site}/nm^2)	$K_L(10^3 l/mole)$	$-\Delta G^{\theta}_{ads}$(kJ/mole)
Benzylamine	9.33	2.9	0.87	16.8
3-phenylpropionic acid	4.37	1.6	1.06	17.3
Phenol	9.89	2.4	0.24	13.6

As shown in Figure 3 phenol did not follow initially the Langmuir type of adsorption at higher equilibrium concentrations and therefore the Freundlich type of isotherm was fitted to the adsorption data. The Freundlich isotherm is given by equation 8:

$$\left(\frac{n^s}{w}\right) = K_F c^{1/d} \tag{8}$$

By plotting ln (n^s/w) versus ln c a straight line for phenol should be obtained (Figure 4). A rather good fit is obtained over the entire concentration range. The slope gives the constant 1/d and the intercept the constant ln K_F. The exponent (d) of the Freundlich isotherm gives an indication of the amount adsorbed, being equal to unity in the case of nonspecific adsorption.

The Langmuir isotherm (eq. 3) may be rewritten as:

$$\left(\frac{n^S}{w}\right) = K_L c \left[\frac{(n_m^S / w)}{1 + K_L c}\right] \tag{9}$$

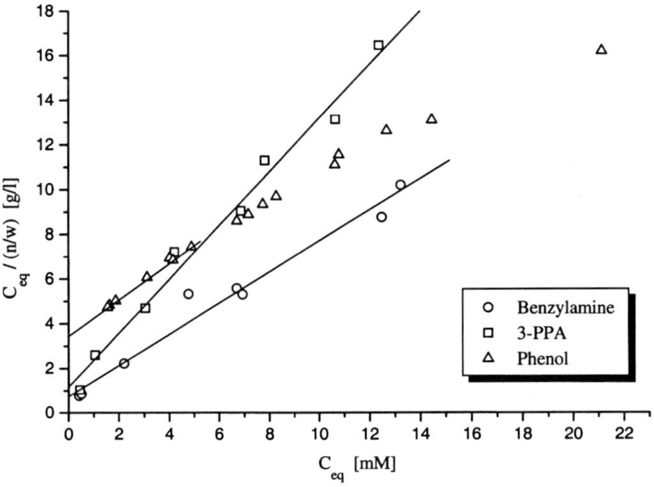

Figure 3. The linearized Langmuir plot of the probe molecules adsorbed onto silica from cyclohexane at 25°C.

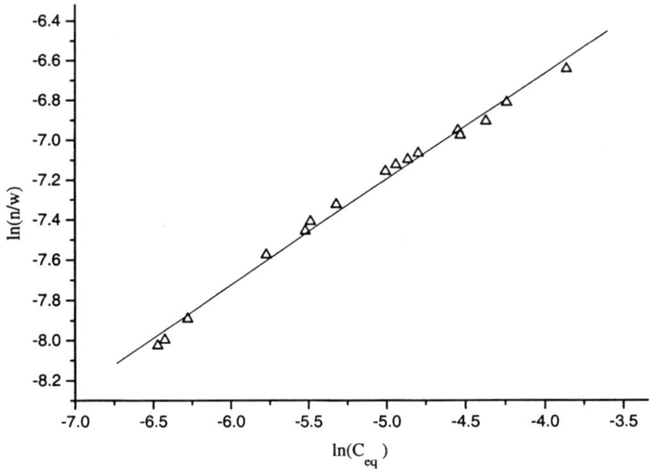

Figure 4. The linearized Freundlich isotherm for phenol adsorbed onto silica from cyclohexane at 25°C.

For dilute solutions and/or small K_L equation (9) is simplified to:

$$\left(\frac{n^S}{w}\right) = K_L c\left(\frac{n_m^S}{w}\right) \tag{10}$$

For a nonspecific adsorption (d=1) it is obvious that the Freundlich and Langmuir isotherms are related through:

$$K_F = K_L \left(\frac{n_m^S}{w} \right) \tag{11}$$

The intercept, slope and the correlation constant R are given in Table 4 as are the equilibrium constant and the Gibbs free energy.

Since a slight systematic deviation from the linearized Freundlich isotherm was observed (Figure 4) it was further evaluated whether the Freundlich adsorption isotherm was composed of a Langmuir type adsorption and additionally a nonspecific adsorption. The latter is described by the Henry isotherm as;

$$\left(\frac{n^S}{w} \right) = K_H c \tag{12}$$

which corresponds to the Freundlich isotherm with the exponent, d=1. The plot of (n^s/w) against c through the experimental points in the high concentration range (Figure 5a) gives the intercept and slope given in Table 4.

Disregarding the intercept, the adsorption given by the Henry isotherm is subtracted from the experimental (total) adsorption before applying the Langmuir isotherm.

$$\left(\frac{n^S}{w} \right)_{corr} = \left(\frac{n^S}{w} \right) - K_H c = \left(\frac{n_m^S}{w} \right)_{corr} \frac{K_{L,corr} c}{1 + K_{L,corr} c} \tag{13}$$

In the inverted form, where $m_{corr} = 1/(n^S_m/w)_{corr}$ and $b_{corr} = m_{corr}/K_{L,corr}$, it resembles equations (4) and (9). The Henry isotherm and the linearized corrected Langmuir isotherm are presented in Figure 5 (a and b). The intercept, slope, σ_{site}, equilibrium constant ($K_{L,corr}$) and standard free energy of adsorption extracted from the plots are given in Table 4.

Table 4.
The intercept, slope, regression constant (R) and the derived quantities of the Freundlich and the combined Henry and Langmuir isotherms for phenol

	Freundlich	Henry	Langmuir-corr
Intercept	-4.549	0.552 (mmol/g)	2.18 (g/dm^3)
Slope	0.529	0.036 (l/g)	1.66 (10^3g/mole)
R	0.996	0.996	0.994
K	10.60×10^{-3}	0.036 (dm^3/g)	0.76 (10^3 dm^3/mole)
$-\Delta G^\theta$ (kJ/mole)	11.28		16.4
σ_{site} (N_{site}/nm^2)			1.2

(a)

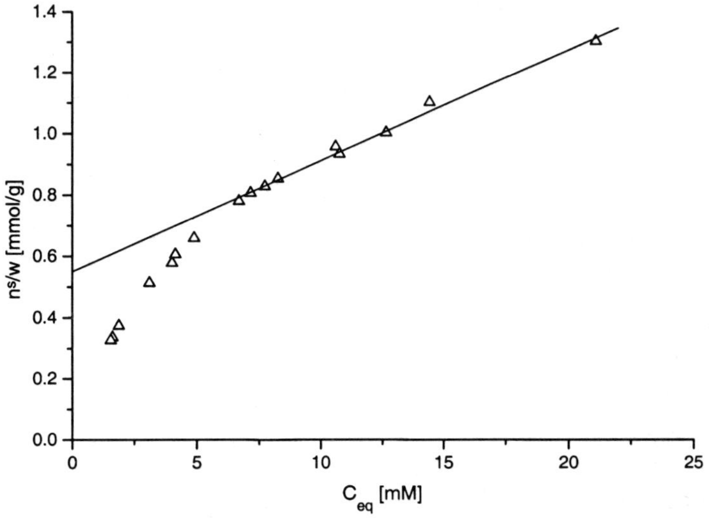

(b)

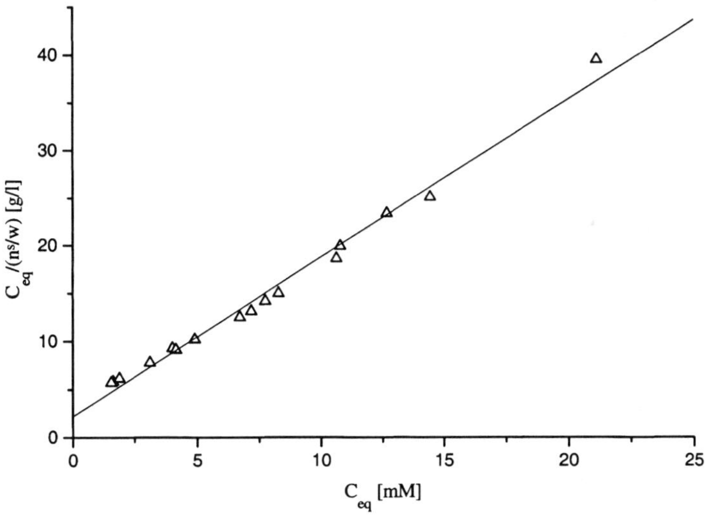

Figure 5. The Henry plot (a) and the linearized (and corrected) Langmuir plot (b) for phenol adsorbed onto silica from cyclohexane at 25°C.

4. DISCUSSION

The amount of hydroxyl groups on the silica surface can be controlled by thermal treatment which provides a convenient way to adjust the amount of probe molecules adsorbed by proton exchange or hydrogen binding. The σ_{site} value of benzylamine is thought to represent the surface density of hydroxyl groups or the Brönsted activity of porous silica and agrees with the value reproted, e.g. by Zhuravlev [6, 16]. The small difference of *ca.* 1 OH-group/nm^2 from the total average value reported by others [5, 7, 8] may be attributed to inaccessible hydroxyl groups of the silica samples [6–8]. Alternatively the cross-sectional area of the probe molecules (0.30–0.35 nm^2) does not allow for adsorption onto all the surface sites. The molecular size is, however, by more than an order of magnitude smaller than the interparticle pore size of the silica particles.

The aromatic character of the benzene ring (weak Lewis basicity [9, 10]) does not contribute to any appreciable extent to the adsorption of probe molecules. As evidenced by the lack of aromatic peaks in the DRIFT spectra of toluene and the thermodynamic parameters extracted from the adsorption isotherm of toluene the π-electron interaction of the aromatic ring is not capable to induce adsorption in the dispersive (cyclohexane) medium.

The extensive adsorption of benzylamine is expected to be due to the matching between the acidic sites on the silica surface with the highly basic amine groups. This is shown both by the high pK_a value (Brönsted proton accepting capability) and the estimated donor number of 55 (Lewis activity). The 3-phenylpropionic acid is a Brönsted acid having a low pK_a value (Brönsted donating capability) and an estimated donor number of only 20 (Table 1) and thus adsorbs less extensively. There is, however, a clear correlation between the Brönsted acidity (pK_a-value) and the σ_{site} value. Benzylamine being a Brönsted base accepts a large number of protons from the acidic silanol groups. 3-phenylpropionic acid being a modestly strong Brönsted acid is capable of donating protons to less than half of those sites. Therefore, the surface excess of benzylamine exceeds the adsorption of 3-phenylpropionic acid. Assuming that the equilibrium constant (K_L) is indicative of the strength of the Brönsted sites of the solid, only small differences can be found between the probes. The strength of adsorption is slightly lower for benzylamine as compared with 3-phenylpropionic acid. In the latter case silica may also be acting as hydrogen bond donor to the carbonyl group. The Gibbs free energies found are of the same magnitude as found for silica [3] and siliceous minerals, such as quartz [13].

Phenol can be considered to be both a Lewis acid and a Lewis base and is expected to interact weakly primarily by hydrogen (acidic) and less probably by oxygen (basic) bonding. The energy of such bonds should be smaller than that of the bonds formed after proton exchange. The σ_{site} value is found to be only slightly lower than that of benzylamine despite the fact that its lesser tendency to donate protons (pK_a) is reduced further from that of benzylamine. Since the aromatic ring does not contribute to the bond strength it seems that it is primarily the Lewis (hydrogen) bonding that contributes to the adsorption. This seems to be typical for alcohol bonding [11]. The site density indicates that the binding is extensive but the equilibrium constant shows, as expected, that the bond strength is weak.

Using the values in Table 2 lines have been drawn through the experimental data in order to provide a visual estimate of the quality of the fit. For benzylamine and 3-phenylpropionic acid we obtain a good fit with the Langmuir isotherm (Figure 6a, b). For phenol the Langmuir fit agrees only at low equilibrium concentrations (Figure 7a) where the adsorption occurs predominantly to the free silanol groups [5, 9, 10].

(a)

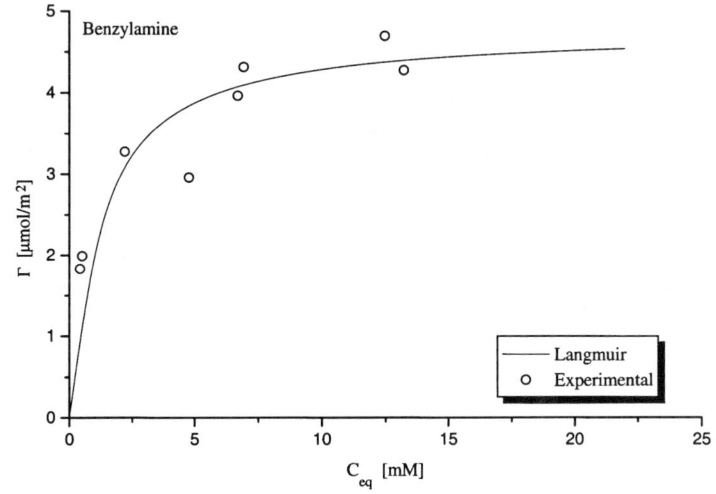

(b)

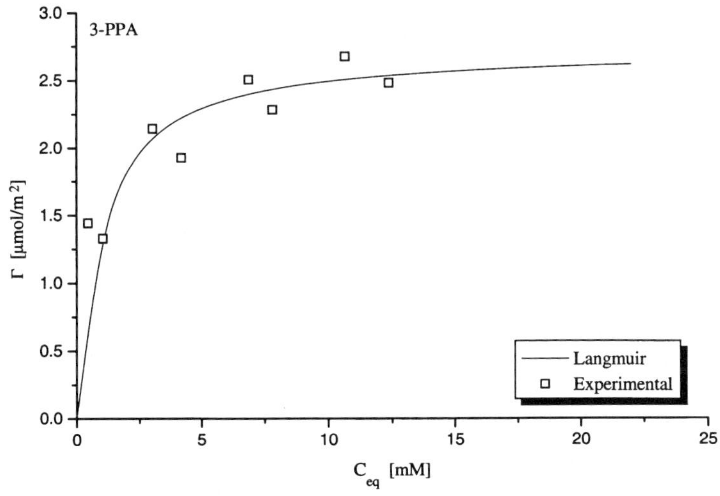

Figure 6. The experimental surface excess of (a) benzylamine and (b) 3-phenyl-propionic acid adsorbed onto silica as a function of their equilibrium concentration in cyclohexane at 25°C. The full drawn line represents the fit provided by the Langmuir isotherm.

The adsorption of phenol seems not to be restricted to a pure monolayer plateau. If all experimental points are equally considered the Freundlich isotherm (equation 8, Figure 4) provides a relatively good fit. From a double logaritmic representation the equilib-rium constant of $10.6 \cdot 10^{-3}$ (assumed dimensionless) is obtained from the in-tercept of -4.55 ±0.06 and the slope gives the inverse of the constant d as 0.53±0.01. The constant (d≈ 2) obtained gives the relation between the proportion of the specific amount ad-sorbed and the molecular weight of the adsorbate, which indicates that the maximum adsorption corresponds to a bilayer formation. As shown in Figure 7a the fit to all the experimental points is considerably improved when the Freundlich iso-therm is used to represent the adsorption of phenol.

(a)

(b)

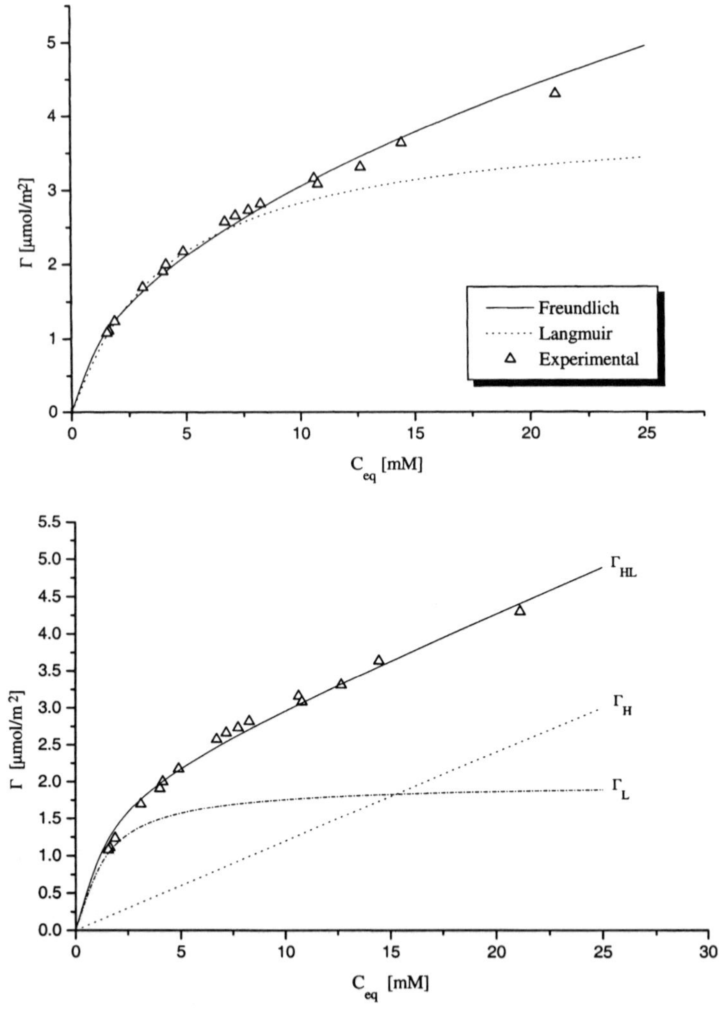

Figure 7. The Freundlich isotherm (full drawn line) and the Langmuir isotherm (dotted line, a); and the Henry-Langmuir or corrected Langmuir isotherm (full drawn line) and the component isotherms (dotted lines, b). The experimental adsorption data of phenol adsorbed onto silica at 25°C are shown as triangles.

An alternative explanation can be sought in the enhanced dimerization tendency of monosubstituted benzenes for which the functional group is directly conjugated with the benzene ring [14]. This seems to suggest that phenol may, at high concentrations, adsorb onto the surface sites as dimers [21]. In order to accommodate all the molecules, a corresponding packing arrangement must prevail over the entire surface. The amine and the acid groups of benzylamine and 3-phenylpropionic acid are not directly conjugated to the benzene ring and thus should show a low tendency for dimerization. Since benzylamine has a relatively high σ_{site} value and 3-phenylpropanoic acid a relatively low σ_{site} value the dimerization does not seem to influence the degree of adsorption, but rather the strength of adsorption as shown by the shape of the adsorption isotherm and the equilibrium constant. The estimated overall (Freundlich) adsorption affinity (equilibrium constant) for the phenol aggregates (dimers or (hemi)micelles) is given by the Gibbs free energy being -11.28 kJ/mole which is slightly lower than the value found for the primary Langmuir adsorption of monomers (Table 3).

No improvement is achieved (Figure 7b) when the Freundlich isotherm is subdivided into Henry- and Langmuir-component isotherms (Figure 5a, b) which has been used to distinguish the two adsorption components relating to surface silanol groups and adsorption into pores [12]. It is thus not possible to give a further interpretation of the mechanism giving rise to the non-Langmuirian adsorption. It is equally probable that the monolayer adsorption onto the surface sites is followed by a multilayer adsorption or adsorption into the pores [3].

5. CONCLUSIONS

- The adsorption of "Brönsted active" and "Lewis active" probes provided a realistic picture of the number and strength of the surface sites of the silica particles dispersed in cyclohexane.
- Only a fair correlation was found between the estimated (Lewis) AN-DN values and the (Brönsted) pK_a values on the one hand, and the experimentally determined equilibrium constants for the probe molecules on the other hand. Clearly, only an experimentally determined number (obtained from the maximum adsorption capacity) and strength (expressed through the equilibrium constant) of surface sites are thermodynamically sound.
- The relative behaviors of benzylamine and 3-phenylpropionic acid suggest that the Brönsted (proton exchange) activity is important in the non-aqueous cyclohexane medium. However, the equally large σ_{site} values and the considerably smaller equlibrium constants of the benzylamine and phenol pair indicate an extensive Lewis (hydrogen bond) interaction at low concentrations. When the concentration is increased the phenol molecules adsorb onto multiple surface sites.

Acknowledgements

This research has been funded as a part of the Materials and Structure Research Program (MATRA) by the Academy of Finland.

REFERENCES

1. R. K. Iler, *The Chemistry of Silica.* John Wiley, New York, 1979.
2. F. M. Fowkes, *J. Adhesion Sci. Technol.* **1**, 7 (1987).
3. C. Souteyrand, M. Thibert, M. Caude and R. Rosset, *J. Chromatography* **262**, 1 (1983).
4. H. Rahiala, I. Berroies, T. Eklund, K. Hakala, R. Gougeon, P. Trens and J.B. Rosenholm, *J. Catalysis* (1999), in press.
5. J. B. Peri and A. L Hensley Jr., *J. Phys. Chem.* **72**, 2926 (1968).
6. L. T. Zhuravlev, *Colloids Surfaces* **A74**, 71 (1993).
7. S. Haukka, E.-L. Lakomaa and A. Root, *J. Phys. Chem.* **97**, 5085 (1993).

8. S. Haukka and A. Root, *J. Phys. Chem.* **98**, 1695 (1994).
9. A. V. Kiselev, *Discuss. Faraday Soc.* **40**, 205 (1965).
10. G. Curthoys, V.Ya. Davydov, A.V. Kiselev, S.A. Kiselev and B.V. Kuznetsov, *J. Colloid Interface Soc.* **48**, 58 (1974).
11. H. Knözinger and W. Stählin, *Progr. Colloid Polym. Sci.* **67**, 33 (1980).
12. A. Larsson and P. Stenius, *Nordic Pulp Paper Res. J.* **3,** 87 (1987).
13. H. Ardebrant and R.J. Pugh, *Colloids Surfaces* **53,** 101 (1991).
14. L. Bergström, *Colloids Surfaces* **69,** 53 (1992).
15. S. J. Gregg and K. S. W. Sing, *Adsorption, Surface Area and Porosity*, 2nd ed. Academic press, NewYork, 1982.
16. L. T. Zhuravlev, *Langmuir* **3**, 316 (1987).
17. P. C. Hiemenz, *Principles of Colloid and Surface Chemistry.* Marcel Dekker, New York, 1986.
18. R. M. Silverstein, *Spectrometric Identification of Organic Compounds.* John Wiley, New York, 1981.
19. R. S. S. Murthy and D.E. Leyden, *Anal. Chem.* **58,** 1228 (1986).
20. S. Collins, W.M. Kelly and D.A. Holden, *Macromolecules* **25,** 1780 (1992).
21. L.-N. Lin, S.D. Christian and E.E. Tucker, *J. Phys.Chem.* **82,** 186 (1964).
22. D. R. Lide (Ed.), *Handbook of Chemistry and Physics*, 77th ed. CRC Boca Raton, FL, 1996.

Acid-Base Interactions, Vol. 2, pp. 481–493
Ed. K.L. Mittal

Experimental characterization of hydroxyl groups and acid–base properties on flat oxide surfaces

HERVÉ ARRIBART, MARIE-JOSÉ AZZOPARDI, JEAN-MARC BERQUIER, VALÉRIE COUSTET, FRANÇOIS CREUZET, XUE-YUN LIN and JACQUES JUPILLE*

Laboratoire CNRS/Saint-Gobain "Surface du Verre et Interfaces", 39, Quai Lucien Lefranc. 93303 Aubervilliers, France

Abstract—The knowledge about oxide surfaces mostly refers to finely divided samples, instead of flat surfaces which, because of their technological relevance, deserve a specific research effort. In the present work, three experimental approaches on such surfaces are presented. For well-defined MgO and Al_2O_3 crystals surfaces, the ability of oxide substrates to become hydroxylated is discussed by comparing experimental results to quantum calculations. A connection between the OH stretching frequency and the basicity of the surface oxygen atoms is established in the case of the magnesium oxide, in agreement with theory. A direct determination of the point of zero charge of oxide surfaces is then achieved by probing the surfaces by either adhesion of colloidal particles, or force measurements which are performed by atomic force microscopy. These methods open a way for a local characterization of surfaces. Finally, the role of OH groups in grafting and adhesion to oxide substrates is examined by *in situ* infrared spectroscopy study. On silica surfaces, the carbonyl groups of poly(methyl methacrylate) are shown to be hydrogen-bonded to the surface silanol groups, while organosilanes are shown to graft via siloxane Si-O-Si covalent bonds.

Keywords: Surfaces; hydroxylation; point of zero charge; interface bond; magnesium oxide; alumina; silica; silane; poly(methyl methacrylate); electron energy loss spectroscopy; photoemission; infrared spectroscopy; atomic force microscopy.

1. INTRODUCTION

Many structural and functional materials are oxides in nature. Alumina and zirconia are amongst the most common technologically important ceramics. Industrial glasses, used for the production of glazings, fibers, bottles, and tubes for lamps and TV, are all made of a mixture of several oxides among which silica is the most abundant. Functional layers on glazings or optical elements generally comprise one or several oxide thin films. As combination of different materials has been a general trend in the materials industry for a few decades, it has become crucial to develop an in-depth knowledge of the surface properties of these oxides, in particular in the field of adhesion of organic molecules. Bonding to oxide surfaces is assumed to depend strongly on the presence of hydroxyl groups as exemplified by the grafting of organosilanes onto sil-

* To whom correspondence should be addressed. E-mail: jacques.jupille@sgr.saint-gobain.com

ica [1]. The electron donnor-acceptor properties of these groups [2], which govern partly the acid–base character of the surfaces, are expected to play a direct role in the interfacial bonding of the oxide substrates with other materials.

The knowledge accumulated in the past about oxide surfaces mostly refers to finely divided samples [3]. Despite the technological interest, little is known about flat surfaces. In particular, experimental methods well adapted to their study are lacking, which justifies the development of a specific research effort. In the present work, the part of the activity of our laboratory dealing with acid–base properties of flat surfaces is reviewed. First, evidences for the presence of hydroxyl groups on the well-defined surfaces of oxide crystals are produced by means of ultra-high vacuum experiments, and trends in the acid–base character of surfaces are derived from the examination of their adsorption behaviour. A more general determination of this acid–base character is then presented, which consists in a direct measurement of the point of zero charge of flat surfaces by probing the surfaces, in an electrolyte as a function of pH, by either adhesion of colloidal particles, or atomic force microscopy. Finally, the direct role of OH groups in grafting and adhesion is illustrated by *in situ* infrared spectroscopy studies of the bonding of organic molecules to silica surfaces.

2. EXPERIMENTAL

2.1. Surface Studies

Hydroxylation of well-defined crystal surfaces has been performed in a ultra-high vacuum (UHV) system equipped with a high-resolution electron-energy-loss spectrometer (HREELS) LK2000 provided by Larry Kesmodel Technologies and a X-ray photoelectron spectrometer (XPS) Clam 2 from Vacuum Generator. Samples could be heated by a backside electron bombardment and cooled by circulating liquid nitrogen. Three substrates where studied, the MgO(100) surface, the α-Al_2O_3(0001) surface [4–6] and the surface of vitreous silica [7]. After introduction in the UHV chamber, samples were cleaned by heating, at 1000 K for α-Al_2O_3 and silica and at 1200 K for MgO, under a partial pressure of oxygen of 10^{-3} to 10^{-2} Pa which was obtained by placing the sample surface near a gas dosing pipe. The cleanliness of the surface was checked by XPS.

2.2. Zero Point Charge Measurements

As oxide surfaces in an aqueous solution are generally electrically charged due to the amphoteric dissociation of surface hydroxyl groups M–OH, the Brönsted acidity or basicity of an oxide surface can be characterized by the point of zero charge (pzc). This value corresponds to the pH value required to achieve zero net surface charge. Electrophoretic measurements and potentiometric titration are generally used to measure the pzc of oxide particles in suspension. Only a few studies related to the pzc of planar oxide surfaces have been carried out by streaming potential measurements [8, 9] and zeta-potential measurements in the plane interface technique [10]. We have developed two methods to measure the pzc of flat surfaces. They consist in probing the electrochemical double-layer resulting from the redistribution of electrical charges in the neighbourhood of the substrate placed in an aqueous medium. The first one consists in probing the surface with latex particles. The pzc of the substrate can be obtained from the kinetics of adhesion of the latex particles onto the substrate surface [11]. In the second approach, the probe is the Si_3N_4 tip of an Atomic Force Microscope (AFM). The interaction force between the tip and the substrate is measured as a function of the tip-substrate distance and the pzc of the substrate is determined [12].

2.3. Infrared Attenuated Total Reflection

To identify the nature of the bond formed at the surface of the oxide during either the adsorption or the grafting of foreign molecules, a specific infrared spectroscopy method has been developed [13]. The oxide to be studied is deposited as a thin layer on the surface of a prism designed as an internal reflection element in an infrared (IR) attenuated total reflection (ATR) experiment. The surface is then placed in contact with the solution under study. Under these conditions, the evanescent wave of the internal reflection directly probes the interface between the oxide layer and the organic medium. Infrared spectra can be recorded in real time and it is then possible to follow *in situ* the progress of the chemical reaction taking place at this interface. This technique can be extremely sensitive, because the infrared element can be designed such that the number of internal reflections is very high, yielding a high signal-to-noise ratio. The detection of less than one monolayer of bonded molecules is easily achieved. The details of the experimental procedure are given elsewhere [13].

3. HYDROXYL GROUPS AND TRENDS IN ACID–BASE CHARACTER

The adsorption of protons and hydroxyl groups can characterize the acid–base character of surfaces. The proton adsorption energy increases from SiO_2 to TiO_2, MgO, CaO, SrO, BaO while the OH adsorption energy decreases along the same series [14]. Our objectives are to follow the process of dissociative adsorption of water, starting from well-defined and clean oxide surfaces, and to characterize the nature, chemical activity and geometry of the obtained hydroxyl groups.

3.1. Dissociative Adsorption of Water

Silica, alumina and magnesium oxide surfaces can be hydroxylated in ultra-high vacuum conditions by dissociative adsorption of water. However, high coverages can only be obtained after exposure to water vapour at pressures above threshold pressures (of the order of 10^{-1} Pa for MgO [15] and 10^2 Pa for α-Al_2O_3 [16]) or, alternatively by adsorbing an ice multilayer below 150 K and desorbing it at room temperature [6]. After exposure of silica to water vapour at 300 K, a ν_{OH} stretching mode in the region 3720 cm^{-1} to 3750 cm^{-1} was observed by HREELS [7], in agreement with what was found for silanol groups on free silica surfaces [3]. On α-Al_2O_3(0001) surfaces exposed to H_2O, the loss due to the OH stretching vibration was seen at 3720 cm^{-1} in the HREELS spectrum [4]. The relative intensity of this loss compared to the intensities of the surrounding phonon losses can be strongly enhanced with the help of a resonance scattering (Fig. 1) [4].[1] The above value of the ν_{OH} frequency is close to those found on hydroxylated alumina films [17, 18] and it is within the frequency range where the stretching modes associated with hydroxyl groups bonded to octahedral aluminium cations of alumina are expected to be [19]. Further insight is gained

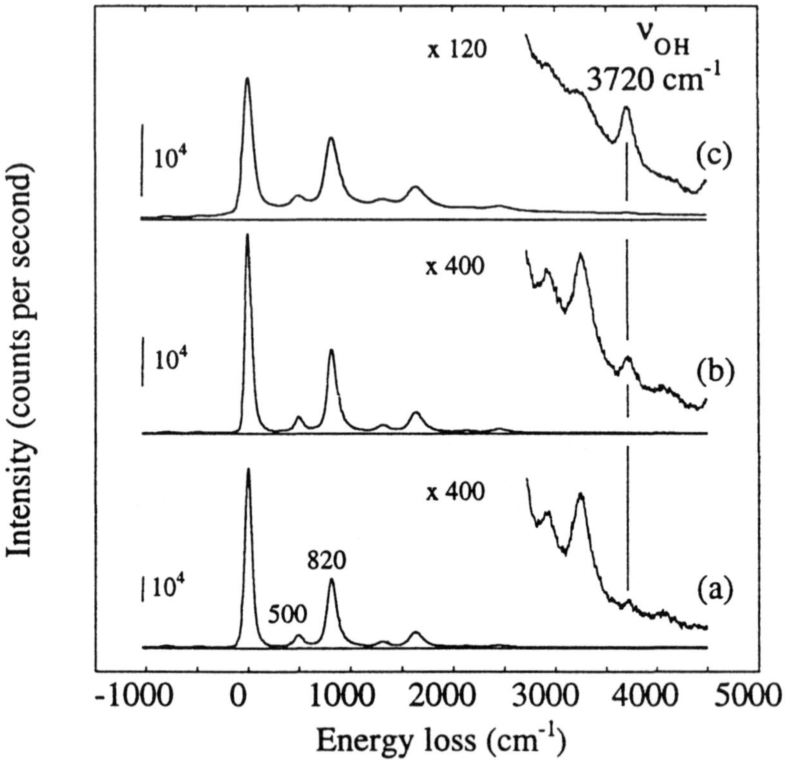

Figure 1. High-resolution electron-energy-loss spectrum (HREELS) of hydroxylated α-Al_2O_3(0001) at 300 K under vacuum. The hydroxylated surface has been prepared by adsorption of a multilayer of water molecules at 110 K, followed by a desorption at 300 K. The stretching mode ν_{OH} at 3720 cm^{-1} is distinguished from the phonon modes via a low energy resonance scattering, whose maximum is observed at an electron impact energy E_p of 3 eV [4]. The intensity of the ν_{OH} loss peak depends on the different experimental conditions: (a) far from the resonance energy, at E_p = 12 eV, the analysis being performed in the specular direction with incident angle θ_i = scattering angle θ_s = 60°; (b) close to the resonance energy, E_p = 6 eV, analysis in the specular direction with $\theta_i = \theta_s$ = 60°; (c) close to the resonance energy, E_p = 6 eV, analysis away from the specular direction with θ_i = 60° and θ_s = 45°: the intensity of the loss peaks due to dipolar scatterers (such as phonons) is much reduced with respect to that in the specular direction, while the resonant mode keeps similar intensity.

[1] In HREELS, electron energy losses occur via either dipole scattering (which follows the same selection rules as infrared spectroscopy), or impact or resonance scattering. The latter occurs when the incident electron is trapped in an empty orbital of the surface molecule prior to being scattered in a direction which depends on the nature of this orbital and may strongly differ from the specular direction in which dipole scattering takes place. Figure 1 gives the example of a resonance scattering of a ν_{OH} stretching loss observed at low electron impact energy and in an off-specular direction.

through the XPS analysis of the adsorption of water at low temperature [6]. The hydroxyl groups give rise to an extra O 1s component that is shifted by 1.6 eV with respect to the O 1s level corresponding to the bulk of the alumina substrate, while the shift associated with the molecular adsorption of water on the same substrate amounts to 2.9 eV [6]. Therefore, this chemical shift and the OH stretching mode unambiguously characterize the hydroxyl groups adsorbed on the α-Al_2O_3(0001) surface. The finding of a dissociative adsorption of water on this surface agrees with recent cluster calculations [20]. However, evidences for the hydroxyl adlayer observed by XPS and HREELS to be ordered and to involve the site suggested by the calculation are lacking.

Similar observations were made in the case of the exposure of MgO surfaces to water vapour. In the present context, magnesium oxide surfaces are of great interest since they offer the possibility to compare theoretical predictions to experimental observations. Following *ab initio* calculations, the dissociative adsorption of water is not expected to occur on five-coordinated surface sites – i.e. on (100) faces – but only on four- and three-coordinated sites [14, 21]. The experimental observation is somewhat more complex. When a freshly cleaved MgO surface at room temperature is exposed to water vapour at pressures lower than 10^{-4} Pa, the water coverage of the surface does not exceed a few percent of a monolayer. This is consistent with a dissociative adsorption on steps and kinks only, as predicted by theory. However, at higher pressures of water vapour, the amount of dissociated water increases strongly and reaches a maximum close to 1 monolayer at a pressure of 10^{-1} Pa [6, 15] (1 monolayer corresponds to 1 adatom per surface atom). Similar coverage can be achieved by desorbing a thick ice layer [6] which has previously been adsorbed below 150 K. In Fig. 2, the

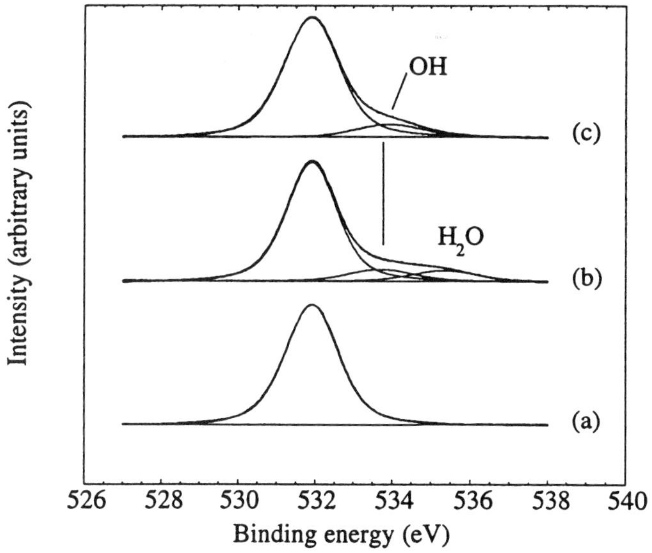

Figure 2. Hydroxylation of a cleaved MgO(100) surface. O1s XPS spectra: (a) clean surface; (b) after the adsorption of a water multilayer at 110 K and a partial desorption between 150 K and 300 K: the shifted peak is due to both molecular water and hydroxyl groups; (c) the sample is now at room temperature: molecular water is fully desorbed and the shifted peak is only due to hydroxyl groups.

XPS O 1s core level shift associated with dissociative and molecular adsorptions are recorded at 1.9 and 3.6 eV, respectively [6]. Hydroxyl groups on MgO surfaces are also characterized by a ν_{OH} HREELS loss near 3700 cm^{-1} [5, 6]. The hydroxylation of the (100) surface also involves a surface restructuring accompanied by a decrease in reactivity toward water dissociation, which suggests that the low energy (100) face of MgO transforms into the low energy hydroxylated (111) face at the contact with water [22].

3.2. Correlation between OH Stretching Frequency and Oxygen Basicity

On MgO surfaces, hydroxyl groups can arise from hydroxyl groups bonded to surface cations and from protons bonded to surface anions, along the following schematic reaction:

$$H_2O + MgO_{surface} \text{->} HO\text{-}Mg_{surface} + H\text{-}O_{surface}$$

Within the hydroxylated layer, the higher the charge on the proton, the higher the ν_{OH} stretching frequency and the more basic the oxygen atom. Indeed, interactions between protons and hydroxyl groups are expected to result in an increase in oxygen basicity (OH stretching frequency) when (i) surfaces become more open and (ii) when the coverage decreases on a given surface [23]. This prediction meets the occurence of a blue shift of the ν_{OH} stretching mode associated with adsorbed OH on MgO surfaces – from 3685 to 3740 cm^{-1} – when the OH coverage decreases [6]. This shift has already been recorded on hydroxylated MgO powder by infrared spectroscopy [24]. Its observation on a well-defined MgO(100) surface on which the presence of molecular water can be unambiguously discarded [6] demonstrates that it is a true effect within the hydroxylated layer.

Due to the decrease in the maximum hydroxyl coverage which occurs along cycles of adsorption-desorption of water on a freshly cleaved MgO surface [22], the acid–base character of such a surface cannot be unambiguously described. This points to the fact that, depending on the environment, the properties of oxide surfaces may significantly change on ageing. It is worth noting that the restructuring that occurs on the hydroxylated MgO(100) surface [22] indicates that water dissociation on that surface is not a checkerboard adsorption, although the order within the adlayer has not been found yet. In fact, to date, no evidence has been produced for the existence of any ordered hydroxyl adlayer at the surface of any crystalline oxide. A challenge to surface science is to find out such surfaces which could then be used as references system for an in-depth study of the properties related to the hydroxylation.

4. ACID–BASE CHARACTER OF OXIDE SURFACES

According to the site complexation model [25], when an oxide is immersed in an aqueous solution, its surface becomes charged because M–OH groups can exchange protons with the solution to give $M\text{–}O^-$ if the solution is basic or $M\text{–}OH_2^+$ if the solution is acidic. For a given intermediate pH, i.e. the pzc value, the amounts of $M\text{–}O^-$ and MOH_2^+ become equal and the surface becomes neutral.

Within the framework of the Derjaguin–Landau–Vervey–Overbeek (DLVO) theory [26], the electrostatic force between two surfaces of dissimilar materials in an aqueous electrolyte is related to the corresponding surface electrostatic charges. The force is repulsive when the charges are of the same sign and attractive if they are of opposite signs. One then expects the electrostatic force to switch from repulsive to attractive, or vice versa, if the pH value crosses the pzc of one of the two surfaces. If the pH is gradually varied from very acidic to very basic, one expects the force to be repulsive, then switch to attractive for a pH equal to the first pzc, and then switch again to repulsive for a pH equal to the second pzc. These switches can be probed by the two above-mentioned methods.

4.1. Point of Zero Charge Measurements with the Particles Adhesion Method

The deposition rate of amidine-grafted polystyrene latex particles (diameter 0.60 μm) onto various oxide surfaces was measured in well-controlled hydrodynamical conditions in solutions of HNO_3 and KOH of different pH's. In independent experiments, the pzc of the latex particles was measured to be 6.0±0.2. The detailed experimental procedure for particle adhesion method is described in ref. [11]. As an example, Fig. 3 shows the variation of the deposition rate of latex particles onto an indium tin oxide (ITO) substrate as a function of pH. As expected, three successive regimes can be

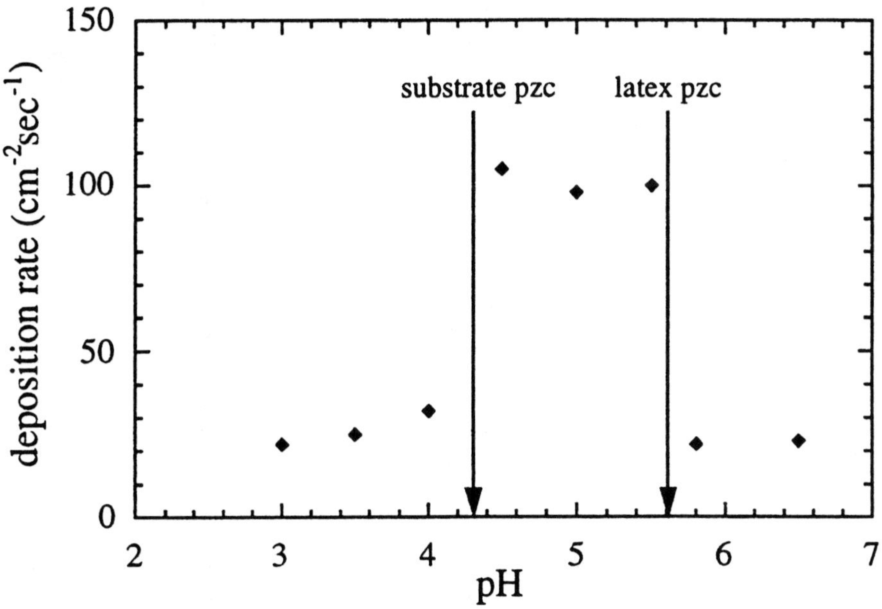

Figure 3. Deposition rate of amidine latex particles onto an indium(90%)-tin(10%) oxide thin film as a function of the pH of the HNO_3 + KOH solution. At pH lower than 4.3 (higher than 5.6), the deposition rate of latex particules is slow since both the surface and the particules are positively (negatively) charged. For pH ranging between 4.3 and 5.6, above the pzc value of the substrate and below that of the particules, an attractive interaction takes place and the deposition rate of particules is fast.

distinguished, for which slow, fast, and slow deposition rates are observed. (The deposition rate is not zero in the two extreme regimes because, at short range, the repulsive electrostatic interaction is dominated by the van der Waals interaction which is attractive [26]. It results in an energy barrier for particle deposition and hence a non-zero rate due to activated jumps). The transitions between the three regimes occur for pH values of 4.3±0.3 and 5.7±0.1. According to the principle described above, these two values can be assigned to the two pzc's. The latter is consistent with the one measured for the amidine latex by electrophoresis. Therefore, we assign the former to the pzc of the ITO thin film.

The same general behavior holds for the other tested oxides, allowing for an evaluation of their pzc. The pzc values measured for seven oxide substrates are given in Table 1 (all oxides were in thin film form, except silica and glass which were in bulk form). The error bar depends on errors in determining the pH values and on the inaccuracy inherent to the method [11].

Table 1.
Values of pzc of various oxide substrates determined by the particle adhesion method

Substrate	Measured pzc (unit of pH)
Iridium oxide	<1.5
Tungsten oxide	1.5±0.5
Silica	2.0±0.3
Soda-lime silica glass	2.8±0.2
Tin oxide	4.0±0.5
Indium(90%)-tin(10%) oxide	4.7±0.2
Aluminium oxide	8.0±0.5

4.2. Point of Zero Charge Measurements using the Atomic Force Microscopy

The experiment consists in the measurement, using an AFM, of the force between the Si_3N_4 tip of the apparatus and the oxide substrate, both immersed in an aqueous solution, as a function of the tip-substrate distance, and for different pH's. The principle of the method and the experimental details are given in ref. [12].

Depending on the pH, repulsive or attractive curves are obtained. As in the previous experiment, the interaction is found to be repulsive at extreme pH's and attractive in an intermediate pH range. This is illustrated in Figure 4 in the case of a monocrystalline α-Al_2O_3 substrate. The value of the force measured at the arbitrarily fixed distance of 15 nm is given as a function of pH. The electrostatic interaction is found to be attractive for pH between 5.8±0.2 and 8.2±0.4 and repulsive for pH values outside this range. In independent experiments, we have measured the pzc of the AFM tip to be 6.2±0.2 [12]. The change in the sign of the interaction around pH 6 can, therefore, be ascribed to the change in the sign of the surface charge of the Si_3N_4 tip. Accordingly, the second transition is attributed to the change in sign of the surface charge of

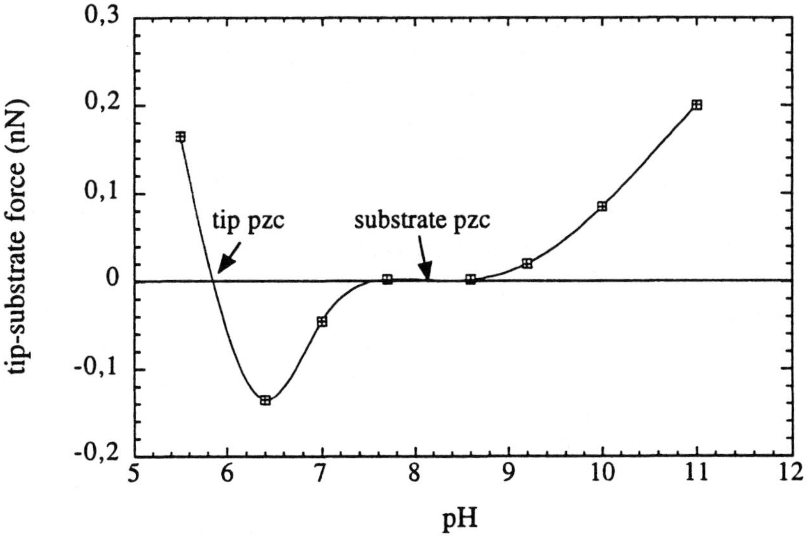

Figure 4. Electrostatic force measured in an aqueous solution between α-Al_2O_3 substrate and the Si_3N_4 tip of an AFM apparatus separated by 15 nm as a function of the pH of the solution.

the α-Al_2O_3 substrate. Therefore, the pzc of the α-Al_2O_3 surface is estimated to be 8.2±0.4. This value is in agreement with the value determined from other measurements [9].

Compared to other methods, the AFM-based pzc measurement has the advantage of being very local. We evaluate the lateral resolution of the method to a few tens of nanometers [12]. It could therefore be used to map the acid–base properties of heterogeneous oxides which are often found in industrial practice.

5. ADHESION OF POLYMERS TO OXIDE SURFACES

On the oxide surfaces, the hydroxyl groups are the active chemical functional groups allowing for adhesion of polymers. Some polymers show strong natural adhesion to oxides because they possess polar groups that form covalent bonds or hydrogen bonds with the hydroxyl groups of the surface. But many polymers (e.g. polyethylene, polystyrene) are non-polar and thus are unable to form such bonds. In that case, adhesion promoters (or coupling agents) are frequently used. These small molecules have the property of developing chemical interactions with the organic polymer at one end and with the hydroxyl groups of the oxide at the other end. Among them, the organosilane family is the most widely used [1, 27]. These molecules are organized around a silicon atom and possess both organic and inorganic functionalities. Their general chemical formula is $R_{4-x}SiX_x$, where x equals 1, 2 or 3, R is the organic function, chosen to be compatible with the polymer, and X is either OH or a hydrolyzable group such as Cl, OMe_3 or OEt_3. When hydrolyzed, the molecule can react with an hydroxyl group of the oxide, forming either a covalent bond or a hydrogen bond, as written below for the case where x = 1.

$$R_3Si\text{–}OH + HO\text{–}M(oxide) \rightarrow R_3SiOM(oxide) + H_2O \qquad (1)$$

or,

$$R_3Si\text{–}OH + HO\text{–}M(oxide) \rightarrow R_3SiOH......HOM(oxide) \qquad (2)$$

In the two cases which are described in the following, infrared attenuated total reflection experiment [13] allows to determine the nature of the bonding of organic molecules to a silica substrate.

The infrared experiment described here is rather versatile. It can work with any oxide provided one can deposit it in the form of a thin layer (about 10 nm) on the surface of an ATR prism. Recent results have shown that it is possible to characterize the acidity of the surface sites of various oxides with this same infrared experiment by choosing pyridine as a probe molecule in solution [28].

5.1. Hydrogen Bonds at the PMMA/Silica Interface

Figure 5 gives the IR spectrum of the interface between the silica layer and a solution of poly(methyl methacrylate) (PMMA) in CCl_4. Due to the substraction of the spectrum before reaction, IR vibration bands are shown downwards (upwards) for species which disappear (appear) [13]. In the 1600–1800 cm^{-1} region, which is the region of the C=O stretching mode, the band of the carbonyl groups from the polymer presents a shoulder at 1716 cm^{-1} on the low wavenumber side of the main band at 1731 cm^{-1}, which is due to free carbonyl groups. This shoulder can be assigned [29] to those

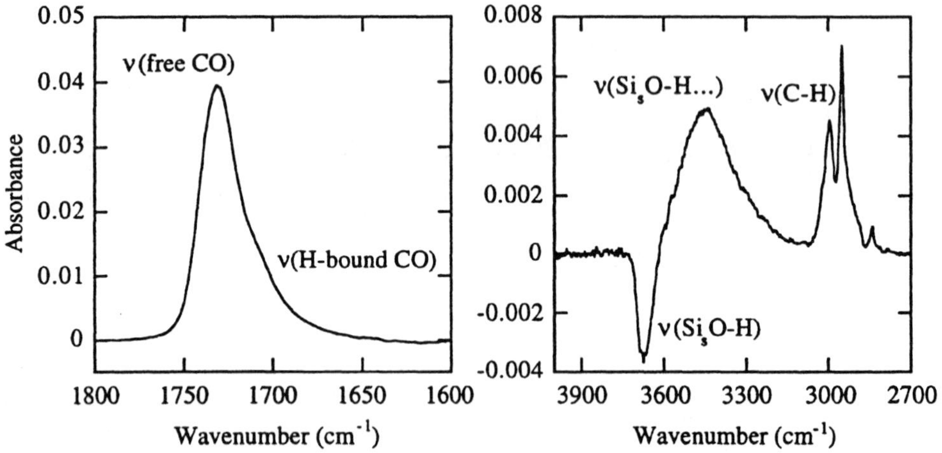

Figure 5. Infrared attenuated total reflection difference spectrum (PMMA-covered surface minus clean surface) collected through a germanium wafer covered with a thin layer of silica (8 nm) in contact with a PMMA solution. The shoulder on the low frequency side of the C=O stretching peak (left) corresponds to carbonyl groups bonded to the silanol groups of the surface via hydrogen bonds. In the SiOH stretching region (right), the negative peak at 3665 cm^{-1} is assigned to originally free silanol groups which are now involved in hydrogen bonds and give rise to the broad band centered around 3420 cm^{-1}.

carbonyl groups of the PMMA chains which are in interaction with the silica surface. The shift of this component by 15 cm^{-1} is typical of the formation of a hydrogen bond [30]. Hydrogen bonding is also responsible for the shift of the stretching mode of the hydroxyl groups (silanol groups: Si–OH) of the silica surface. This is observed in the 2700–4000 cm^{-1} region of the spectrum in Fig. 5. The downward band at 3665 cm^{-1} is assigned to the surface silanol groups which were free before the PMMA solution was present and which have subsequently interacted with the carbonyl groups of the PMMA. After this interaction, they give the upward broad band around 3420 cm^{-1}, which is typical of silanol groups involved in hydrogen bonding.

Therefore, we can conclude that the PMMA chains are linked to the oxide surface by hydrogen bonding between some of the carbonyl groups of the chain and some silanol groups of the surface.

5.2. Grafting of Organosilanes onto a Silica Surface

The surface of the silica layer was placed in contact with a solution of a triethylsilanol (TES), a model organosilane of formula $(C_2H_5)_3SiOH$, in CCl_4. Figure 6 displays the IR spectrum of the silica/solution interface after reaction, substracted from the spectrum before reaction [13]. The upward bands between 2800 and 3000 cm^{-1} are assigned to the ethyl groups of TES. They indicate that some TES molecules have reacted with the surface. As in the case of PMMA, the downward band at 3665 cm^{-1} indicates that the reaction has occurred with some silanol groups of the silica surface which were free before the reaction began. In contrast with the case of PMMA, no broad band appears in the bonded OH stretching mode region. Therefore, the formation of hydrogen bonds can be precluded. Accordingly, the appearance of a band at 1060 cm^{-1} [31], i.e. at a frequency characteristic of the antisymetric Si–O–Si mode,

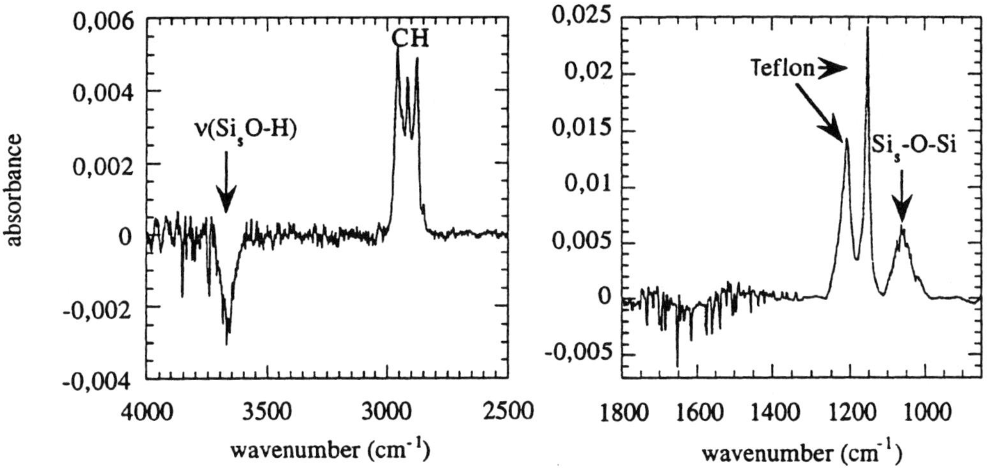

Figure 6. Infrared attenuated total reflection difference spectrum collected through a germanium wafer covered with a thin layer of silica (8 nm) in contact with a solution of triethylsilanol. The OH stretching region only shows the disappearance of the band assigned to free silanols (left), while a peak corresponding to a Si-O-Si stretching mode appears around 1100 cm^{-1} (right).

demonstrates that the grafting occurs by formation of a siloxane covalent bond as given by reaction (1) if M = Si.

Beyond these two cases, one may wonder (i) whether PMMA and organosilane can be representative of molecules with the same functional groups, carbonyl group and hydroxyl group, respectively and (ii) whether the occurrence of bonds of either type, as well as their density and strength, depend on the acid–base characteristics of the substrate and of the molecule.

6. CONCLUSION

We have presented various methods for the characterization of the hydroxyl groups at the surface of oxides, in the case where samples are flat.

- Conditions in which well-defined MgO(100) and α-Al_2O_3(0001) surfaces can be hydroxylated under controlled conditions have been found. In the former case, the present finding is directly conflicting with theory which predicts that the surface should not dissociate water molecules. The restructuring that occurs on the hydroxylated MgO(100) surface reveals that ageing may affect the properties of a surface, including the acid–base character. The challenge is now to obtain model surfaces covered with ordered layers of hydroxyl groups which, as in the case of many adsorbate-covered metal or semiconductor surfaces, could serve as references for the study of surface reactions.
- The point of zero charge of flat surfaces has been measured in a manner analogous to electrophoresis techniques used for powders, either via the deposition of charged particles on a substrate or by Atomic Force Microscopy. Using the latter method in the case of heterogeneous materials, it might be even possible to obtain an acid–base imaging of the surfaces by probing the substrate with spatial resolution.
- Finally, the nature of the chemical bonds formed by some organic molecules at the interface with silica has been identified. The carbonyl groups of the poly(methyl methacrylate) form hydrogen bonds with the silanol groups of the silica surface. The OH groups of triethylsilanol molecules react with those of the silica surface to give rise to covalent bonds.

The experiments reported here point toward two directions for future work. On the one hand, to understand acid–base properties at an atomic level, there is a clear need for an in-depth knowledge of well-defined oxide surfaces. On the other hand, further work remains to be done to take into account the chemical and structural complexities of most of the real oxides, i.e. those used or produced in the materials industry, like glass.

REFERENCES

1. E. P. Pluddemann, *Silane Coupling Agents*. Plenum Press, New York (1982).
2. K. L. Mittal and H. R. Anderson, Jr. (Eds), *Acid–base Interactions: Relevance to Adhesion Science and Technology*. VSP, Utrecht, The Netherlands (1991).
3. B. A. Morrow, *Stud. Surf. Sci. Catal.* **57**, A161 (1990).
4. V. Coustet and J. Jupille, *Surf. Sci.* **307-309**, 1161 (1994).
5. V. Coustet and J. Jupille, *Surf. Interface Anal.* **22**, 280 (1994).
6. V. Coustet and J. Jupille, *Il Nuovo Cimento* **19 D**, 1657 (1997).
7. C. Mazzara and J. Jupille, unpublished results.
8. J. M. Horn, Jr. and G. Y. Onoda, Jr., *J. Amer. Ceramic Soc.* **61**, 523 (1978).
9. L. Bousse, S. Mostarshed, B. van der Shoot, N. F. de Rooij, P. Gimmel and W. Göpel, *J. Colloid Interface Sci.* **147**, 22 (1991).
10. A. Doren, J. Lemaitre and P. G. Rouxhet, *J. Colloid Interface Sci.* **130**, 146 (1989).
11. X. Y. Lin, E. Fahri and H. Arribart, *J. Adhesion* **51**, 181 (1995).
12. X. Y. Lin, F. Creuzet and H. Arribart, *J. Phys. Chem.* **97**, 7272 (1993).
13. M. J. Azzopardi and H. Arribart, *J. Adhesion* **46**, 103 (1994).
14. J. Goniakowski, S. Russo and C. Noguera, *Surf. Sci.* **284**, 315 (1993).
15. P. Liu, T. Kendelewicz, G. E. Brown, Jr. and G. A. Parks, *Surf. Sci.* **412/413**, 287 (1998).
16. C. E. Nelson, J. W. Elam, M. A. Cameron, M. A. Tolbert and S. M. George, *Surf. Sci.* **416**, 341 (1998).
17. J. G. Chen, J. E. Crowell and J. T. Yates, *J. Chem. Phys.* **84**, 5906 (1986).
18. B. G. Frederick, G. Apai and T. N. Rhodin, *Surf. Sci.* **244**, 67 (1991).
19. H. Knözinger and P. Ratnasamy, *Catal. Rev. Sci. Eng.* **17**, 31 (1978).
20. K. C. Hass *et al.*, *Science* **282**, 265 (1998).
21. W. Langel and M. Parrinello, *J. Chem. Phys.* **103**, 3240 (1995).
22. D. Abriou, P. Nael and J. Jupille, *Surface Sci. Lett.* **430**, L527 (1999).
23. C. Noguera, *J. Adhesion* **57**, 91 (1996).
24. A. J. Coluccia, R. L. Tench, J. Segall, *Chem. Soc., Faraday Trans. I* **75**, 1769 (1979); S. Coluccia, L. Marchese, S. Lavagnino, M. Anpo, *Spectrochemica Acta* **43A**, 1573 (1987).
25. D. E. Yates, S. Levine and T. W. Healy, *J. Chem. Soc., Faraday Trans. I* **70**, 1807 (1974).
26. J. N. Israelachvili, *Intermolecular and Surface Forces: with Applications to Colloidal and Biological Systems.* Academic Press (1989).
27. K. L. Mittal (Ed.), *Silanes and Other Coupling Agents.* VSP, Utrecht, The Netherlands (1992).
28. L. Guéneau, *Thesis*, Université Paris XI (1998).
29. J. M. Berquier, in: *Chemically Modified Surfaces*, J. J. Pesek and I. E. Leigh (Eds), p. 210. Royal Society of Chemistry, Cambridge (1994).
30. W. Hertl and M. L. Hair, *J. Phys. Chem.* **75**, 2181 (1971).
31. M. J. Azzopardi, *Thesis*, Université Paris VI (1994).

Part 3: Applications of Acid-Base Interactions

Acid-Base Interactions, Vol. 2, pp. 497–512
Ed. K.L. Mittal

Acid–base characteristics of polymer surfaces: Relevance to bioadhesion

MARCO MORRA* and CLARA CASSINELLI
Nobil Bio Ricerche, Str. S. Rocco 32, 14018 Villafranca d"Asti, Italy

Abstract—This paper comments on the relationship between the results of non-specific bacterial adhesion experiments, using both a Gram positive and a Gram negative bacterial strains, and electron-donor/electron-acceptor properties of a series of hydrophilic surfaces. Tested surfaces involved glass with different degrees of surface cleaning and wettability and a polysaccharide (alginic acid) coated glass. Electron-donor/electron-acceptor properties of the tested surfaces are calculated from wettability measurements and the use of different scales. It is shown that the improved scales such as those suggested by Lee or Della Volpe and Siboni give a more satisfactory description of the surface properties of hydrophilic surfaces than the original approach suggested by van Oss and coworkers. The fundamental role played by the reference values used to calculate liquid and solid surface free energy components is emphasized.

Keywords: Surface acid-base properties; electron-donor electron acceptor parameters; surface free energy components; bacterial adhesion; interfacial interactions; interactions in water.

1. INTRODUCTION

Bioadhesion phenomena, i.e. adhesional interactions between biological species and surfaces of synthetic materials play an important role in many technological fields. Bacterial adhesion to polymer surfaces, for instance, is still the major barrier to the extended use of many medical devices [1–3] and the development of surfaces that can resist bacterial adhesion is actively pursued in many laboratories worldwide.

The description of interfacial interactions between living species and synthetic materials in the frame of the physico-chemical knowledge of adhesion science has been the subject of many studies [4–7]. Practically all theories developed in the continued effort of several decades of investigations on the wettability–surface energetics relationship have been applied to the description of bacterial adhesion, but no approach has been completely satisfactory. In fact, the general inconsistency of results has contributed to a certain skepticism about the usefulness of wettability-based approaches as valuable descriptors of bioadhesion phenomena [7].

The recognition of the pivotal role played by surface acid–base (in the Lewis sense) properties in interfacial interactions of surfaces bearing various functional groups (e.g. OH, NH_2, COOH, etc.) [8, 9] has brought new life to the study of bioad-

* To whom correspondence should be addressed. E-mail: mmorra@tin.it

hesional interactions in aqueous media by interfacial energetics. It is well accepted that water owes its unique properties, as a solvent and as the medium in which the evolution of biomolecules and biomolecular interactions has taken place, to its molecular architecture coupled to the asymmetry of charge distribution of the O-H bond [10]. This molecular imbalance and the ensuing orientation-dependent properties of water molecules offer a convenient way to propagate in the aqueous environment the perturbations induced by the presence of a surface [11]. Hydrogen bonding, as a specific sub-set of Lewis acid–base interactions, promoted by the electron-donor/electron-acceptor properties of the relevant surfaces, is the language different phases use to communicate in and with the aqueous environment. Framed in this scenario, the capability to measure and calculate the electron-donor/electron-acceptor properties of surfaces could be an invaluable tool in the surface engineering of medical devices and in the control of bioadhesional interactions.

The approach to adhesion-resistant surfaces we are carrying out in our laboratory is based on the working hypothesis that strong interaction with interfacial water molecules is the way nature prevents non-specific interactions. According to this view, the resistance of synthetic surfaces to, for instance, bacterial adhesion could be improved by a surface structure able to create a "water shield" over the underlying substrate. Coatings based on highly hydrophilic polysaccharides, such as hyaluronic acid or alginic acid, have given excellent results in several *in vitro* and *in vivo* testings [12, 13].

The "hydrophilic way" to resistance to bioadhesion has often been described in the literature, and can be qualitatively rationalized by thinking that, for a given adherend, it will be more difficult to displace water molecules and interact with a wettable, as opposed to a non-wettable, surface [14]. However, a further sub-division among hydrophilic surfaces seems to exist: clean glass, for instance, is very hydrophilic, the water contact angle can be as low as zero, yet it is a good substrate for cells and bacterial adhesion. The same is true for plasma treated or Tissue Culture Polystyrene (TCPS), which is normally moderately hydrophilic, with an advancing water contact angle of about 40–50 degree (because of well-known aging effects of plasma-treated PS [15]). Yet, its adhesion behavior towards cells is absolutely identical to that displayed by a freshly plasma treated PS, with an advancing water contact angle of 10–13 degrees (this observation applies, by the way, to most plasma treated polymers).

In the excellent theoretical description of hydration forces by a bcc lattice model for water [11] at interfaces it is shown that the relative abundance of electron donor and electron acceptor sites on hydrophilic surfaces has marked effects on short-range hydrogen-bond forces. Based on the previous considerations, it could be anticipated that a quantitative characterization of acid–base properties of hydrophilic surfaces showing different degrees of resistance to bioadhesion could shed some light on the reason for the different behaviors and thus improve the present status of wettability-based models for bioadhesional interactions. This appealing approach is still hampered by existing shortcomings of the theory for the calculation of the Lifshitz–van der Waals (γ^{LW} the electron acceptor (γ^{+}) and the electron donor (γ^{-}) character of surfaces, as originally proposed by Good, van Oss and Chaudhury (GvOC) [16, 17], and thoroughly described by van Oss in the specific case of interactions in aqueous media [18]. As already described in a number of papers, the main problems arise from some evident quantitative problem that can be found in the otherwise excellent qualitative frame of the theory [19–22]. Briefly, the calculated surface free energy components, and, thus, the electron-donor and electron-acceptor properties of surfaces are definitely dependent on the probe liquids used for calculation, without any clear explanation of the why and how this happens [18]. Moreover, most solid surfaces appear "monopolar" basic, even in cases where chemistry suggests that some acidic contribution should be expected [16, 17]. And even if it seems quite clear that at least part of the reason for these findings is due to the mathematical treatment of the data rather than true chemico-physical characteristics [19–22], the laws of interactions between hydrophilic phases in aqueous media are modeled in the frame of ubiquitous monopolar basic surfaces [18].

The recognition of these shortcomings has prompted several authors to suggest improved approaches to the calculation of Lewis acid–base components of surfaces [20, 22]. The aim of the present work was to test these methods and to try to comment on the perspective they yield regarding the relationship between bioadhesion phenomena and surface acid-base characteristics. In particular, we have used the original GvOC scale [18], the improved scale suggested by Lee [20], and the more recent development proposed by Della Volpe and Siboni (DVS) [23] to calculate the Lifshitz–van der Waals, the electron donor and the electron acceptor properties of some hydrophilic surfaces, namely: a glass slide carrying a chemically coupled alginic acid (AA) layer, as a model, low-adhesion, hydrophilic surface; and uncoated glass slides, subjected to different degrees of surface cleaning, as highly adherent hydrophilic surfaces. It must be noted that the choice of a polysaccharide surface is of direct relevance for the ongoing debate on the correctness of the GvOC approach, because from the molecular structure of polysaccharides a certain acidic contribution should be expected [17, 19, 21], contrary to published findings [18]. Bacterial adhesion experiments involving clinically relevant Gram negative (*Escherichia coli*) and Gram positive (*Staphylococcus epidermidis*) bacterial strains suspended in PBS have been performed to evaluate the effect of the different surface structures and of the calculated acid-base parameters on bacterial adhesion.

2. EXPERIMENTAL

2.1. Samples Preparation and Materials

Microscope glass slides were used in three different forms. (i) without cleaning, i.e. as supplied and equilibrated with the normal laboratory atmosphere. The typical advancing water contact angle on these slides is around 60 deg, showing a certain degree of contamination from ubiquitous hydrocarbons; this sample will be coded G; (ii) same as (i) but with 1 minute air plasma cleaning, coded GC; and (iii) same as (i) but with 5 minute air plasma cleaning, coded GCC. As is well known, air plasma can remove and/or oxidize the overlayer of contaminants. After plasma cleaning, samples were subjected to the relevant experiments (bacterial adhesion or contact angle measurement) within a maximum of 2 hours.

As to the alginic acid coated sample (AA), a glass slide was subjected to 5 min air plasma cleaning followed by 3 min deposition from propylene plasma. The hydrocarbon overlayer was then oxidized for 15 seconds in an air plasma. After this treatment, the coated glass slide was placed in 5 ml of a 0.5% (w/v) aqueous solution of polyethyleneimine (PEI) and adsorption was performed for two hours. The irreversible adsorption of PEI to the oxidised substrate allows to introduce a high density of amino groups on the sample surface [24, 25]. After prolonged rinsing in doubly distilled water, samples were placed in 5 ml of a 0.5% (w/v) solution of AA in water. Carboxyl groups of AA were covalently coupled to amino groups of PEI by carbodiimide condensation at R.T. overnight. Samples were then extensively rinsed, stored overnight in doubly distilled water and dried under a laminar flow hood.

Propylene, High molecular weight PEI (average $M_w \approx 25000$), and "medium viscosity", water soluble Na salt of AA (obtained from Macrocystis pyrifera) were from Sigma-Aldrich.

2.2. XPS Analysis

XPS analysis was performed with a Perkin Elmer PHI 5500 ESCA system. The instrument is equipped with a monochromatic X-ray source (Al K_α anode) operating at 14 kV and 250 W. The diameter of the analyzed spot is approximately 400 μm, and the base pressure 10^{-8} Pa. The angle between the electron analyzer and the sample surface was maintained at 68°. The curve fitting and quantification of elements were accomplished using the software and sensitivity factors supplied by the manufacturer. In the high-resolution spectra, all binding energies were referenced by setting the CH_x peak maximum in the resolved C1s spectra at 285 eV.

2.3. Contact Angle Measurements

Contact angle measurements were performed either by the sessile drop technique using a Kruss, G 23 contact angle goniometer (Kruss, Hamburg, Germany) or by the Wilhelmy plate method (Cahn DCA 312). In the former case, advancing and receding contact angles were measured by increasing or decreasing the drop volume until the movement of the three phase boundary, keeping the capillary pipette of the microsyringe immersed in the probe liquid for the duration of the measurement. Five measurements were carried out on each sample. The stage velocity in DCA measurement was kept at 40 μm/s.

Four probe liquids were used, namely: water (w) (HPLC grade), Methylene Iodide (MI) (> 99% purity), α-bromonaphthalene (Br) (> 99%), and Ethylene Glycol (EG) (>99%). All liquids were purchased from Sigma-Aldrich.

2.4. Calculation of Surface Free Energy Components

The calculation of the surface free energy componets was performed as described in ref. [23]. Briefly, an overdeterminated set of equation was written using measurements from the four probe liquids and results were obtained by a best fitting routine. The surface free energy components (in mJ/m^2) of the probe liquids used for the calculations are the following: GvOC: $\gamma_W^{LW} = 21.8$, $\gamma_W^+ = 25.5$, $\gamma_W^- = 25.5$; $\gamma_{MI}^{LW} = 50.8$, $\gamma_{MI}^+ = 0$, $\gamma_{MI}^- = 0$; $\gamma_{Br}^{LW} = 43.5$, $\gamma_{Br}^+ = 0$, $\gamma_{Br}^- = 0$; and $\gamma_{EG}^{LW} = 29$, $\gamma_{EG}^+ = 1.92$, $\gamma_{EG}^- = 47$. Lee: $\gamma_W^{LW} = 21.8$, $\gamma_W^+ = 34.2$, $\gamma_W^- = 19$; $\gamma_{MI}^{LW} = 50.8$, $\gamma_{MI}^+ = 0$, $\gamma_{MI}^- = 0$; $\gamma_{Br}^{LW} = 43.5$,

$\gamma^+_{Br} = 0$, $\gamma^-_{Br} = 0$; and $\gamma^{LW}_{EG} = 29$, $\gamma^+_{EG} = 2.6$, $\gamma^-_{EG} = 34.8$. DVS: $\gamma^{LW}_{W} = 26.2$, $\gamma^+_{W} = 48.5$, $\gamma^-_{W} = 11.2$; $\gamma^{LW}_{MI} = 50.8$, $\gamma^+_{MI} = 0$, $\gamma^-_{MI} = 0$; $\gamma^{LW}_{Br} = 44.4$, $\gamma^+_{Br} = 0$, $\gamma^-_{Br} = 0$; and $\gamma^{LW}_{EG} = 33.9$, $\gamma^+_{EG} = 0.966$, $\gamma^-_{EG} = 51.6$.

2.5. Force vs. Distance Curves

Force-distance curves were obtained using a Nanoscope III (Digital) Atomic Force Microscope, equipped with a liquid cell, using an A scanner. Conventional silicon nitride tips were used. Measurements were performed using the long (200 μm), "narrow" leg tip, whose force constant is approximately 0.06 N/m. The same tip was used for all measurements. Since no attempt was made to measure the actual force constant of the cantilever, force data in the relevant figures should not be taken as representative of the actual force involved. Measurements were performed under two different NaCl solutions: 0.001 M and 1 M. Data were acquired and processed using the software supplied by the instrument producer.

2.6. Bacterial Adhesion Experiments

Two different bacterial strains were used for testing. The first one was *Staphylococcus epidermidis* RP62A (ATCC 35984). It is a slime and capsular polysaccharide producer. It was routinely maintained both on tryptic soy agar (TSA, SIGMA) plates and transferred to new plates monthly, and as frozen suspension in liquid culture medium supplemented with 10% glycerol. Bacterial suspension was obtained by inoculating 100 ml tryptic soy broth (TSB, SIGMA) and incubating overnight at 37°C. After incubating overnight, the suspension was washed three times with phosphate buffered saline (PBS, SIGMA), and finally resuspended in PBS. PBS used for this experiment had the following composition: 0.2 g/l KCl, 0.2 g/l KH_2PO_4, 8.0 g/l NaCl, 1.15 g/l Na_2HPO_4, pH at RT= 7.3 ±0.3, Osmolality (mOsm/kg H_2O) = 290±5%. The bacterial suspension, based on previous experiments, was spectrophotometrically adjusted to the optical density required to obtain 5×10^8 colony forming units (cfu)/ml. Finally, four replicates for each sample were introduced into 100-mm-diameter polystyrene Petri dishes containing the bacterial suspension and incubated for 2 hours at 37°C. After incubation time, samples were carefully rinsed with PBS in order to remove non-adherent bacterial cells, then fixed in a 5% glutaraldheyde–PBS solution and dehydrated using increasing concentrations of ethanol in water-ethanol solutions up to 100% ethanol. Final dehydration step was performed with hexamethyldisilazane (HMDS, Aldrich). Dehydrated samples were gold sputter-coated (using an Auto Sputter Coater produced by Agar Scientific) and observed with an LEO 420 scanning electron microscope (SEM). Adherent bacteria on each replicate were counted in ten different fields (each field = 3174 μm^2) and normalized to number of bacteria per square centimeter.

The strain *Escherichia coli* used in this study was isolated from a freshly explanted biliary stent. It was routinely maintained both on TSA plates and as frozen suspension in TSB supplemented with 10% glycerol. Bacterial suspension was obtained by inoculating 100 ml TSB and incubating overnight at 37°C. After incubating overnight, the suspension was washed three times with PBS and finally resuspended in PBS with the above-described composition. Bacterial suspension, based on previous experiments, was spectrophotometrically adjusted to the optical density required to obtain $5x10^8$ colony forming units (cfu)/ml. Also in this case four replicates for each sample were introduced into 100-mm-diameter polystyrene Petri dishes containing the bacterial suspension and incubated for 2 hours at 37°C. After incubation time, samples were processed and observed as described above.

3. RESULTS

3.1. XPS Analysis

The XPS analysis of the AA coated glass slide was performed to confirm the linking of theE polysaccharide overlayer to the glass substrate. The analysis of the AA sample yielded the following surface composition (at.%): 60.3 C, 34.8 O, and 4.9 N. The quantitative results stemming from the XPS analysis of polysaccharides surface-immobilized by a PEI intermediate layer have been thoroughly discussed elsewhere [26]. Briefly, this procedure results in a surface-modified layer thinner than the XPS sampling depth. The XPS signal contains convoluted information from the underlying oxidized hydrocarbon layer, from the intermediate PEI layer and from the topmost AA overlayer (whose chemical composition, since XPS does not detect hydrogen, is 50% C and 50% O). By the use of a conventional overlayer model the theoretical, XPS-detected surface composition was calculated as a function of the fractional coverage of AA. It was shown that the O/C ratio increased and the N/C ratio decreased with increasing fractional coverage and that values such as those obtained in this case corresponded to a practically complete AA overlayer (fractional coverage = 1). More details on the model used can be found in ref. [26].

3.2. Contact Angle Measurements

The results of contact angle measurements are reported in Table 1. As a general observation it is interesting to note that the advancing water contact angle is always lower than that of the apolar liquids Methylene Iodide and α-Bromonaphthalene, contrary to what could be expected on the basis of a simple liquid surface tension-wettability relationship. This feature is typical of hydrogen-bonded liquids on hydrophilic surfaces. Obviously in this case hydrogen bonding supplies an extra energy of interaction that is not available when the wetting liquid “communicates” with the surface only by mean of Lifshitz–van der Waals interactions.

The effect of cleaning of the glass slide is clearly shown by the increase of water wettability. A difference of 40 degrees is observed between the advancing water contact angle on glass exposed to the laboratory atmosphere and to plasma cleaning.

Table 1.
Measured advancing contact angles (°). Typical error is ± 1° in the case of Wilhelmy plate, and ± 3° in the case of sessile drop data

Sample	w	MI	Br	EG
G[a]	59	66	49	61
GC[b]	24	41	32	22
GCC[b]	19	38	30	10
AA[c]	18 (20)	55 (52)	37 (35)	15 (13)

Note:
w = water; MI = methylene iodide; Br = α-Bromonaphthalene; EG = ethylene glycol, G = glass slide without pre-cleaning; GC = glass slide 1 min. air plasma cleaning; GCC = glass slide 5 min air plasma cleaning; AA = alginic acid coated glass slide.
a = measured by the Wilhelmy plate method; b = measured by the sessile drop method; c = data in parentheses were obtained with the sessile drop method, other data by the Wilhelmy plate.

3.3. Calculation of Surface Free Energy Components

The surface free energy components calculated from the contact angle data of Table 1 are reported in Table 2. Three different scales were used: (1) the original GvOC scale, where the reference value for probe liquids was taken by assuming equal electron-donor electron-acceptor properties for water [18]. (2) the improved Lee scale, where the acidic component of water is larger than the basic one [20]. (3) the DVS scale, based on a further increase of the ratio between the acidic and basic components of water and a higher value of the Lifshitz–van der Waals component [23]. Details on the reasons and the physical meaning of these choices can be found in the relevant references.

The values obtained using the three scales are in general qualitative agreement, in the sense that both the acidic and the basic components increase with increasing cleanliness of glass. A further increase is also observed on the AA coated surface, except in the case of the basic component calculated using the Lee scale.

Table 2.
Surface free energy components calculated using three different scales (see the Experimental section for the components of the liquids. All values are expressed in mJ/m^2)

Sample	Scale	γ^{LW}	γ^{ab}	γ^{+}	γ^{-}
G	GvOC	27.50±0.07	2.33±0.23	0.03±0.15	39.17±0.03
GC	GvOC	38.54±0.07	4.70±0.16	0.09±0.13	60.05±0.18
GCC	GvOC	39.68±0.14	6.67±0.24	0.18±0.15	60.37±0.32
AA	GvOC	33.57±0.06	11.55±0.10	0.52±0.07	63.73±0.12
G	Lee	23.45±0.07	0.59±0.27	<0.01	29.45±0.30
GC	Lee	32.55±0.14	7.60±0.39	0.31±0.35	46.55±0.42
GCC	Lee	33.60±0.13	10.01±0.17	0.53±0.21	46.59±0.11
AA	Lee	28.05±0.06	18.00±0.12	1.78±0.11	45.43±0.12
G	DVS	27.50±0.07	1.35±0.15	0.03±0.13	15.81±0.18
GC	DVS	38.54±0.14	7.28±0.22	0.51±0.18	25.89±0.25
GCC	DVS	39.68±0.14	8.50±0.17	0.68±0.13	26.36±0.20
AA	DVS	33.57±0.06	12.34±0.06	1.31±0.05	28.97±0.07

From a quantitative point of view, considering the absolute values of the data, the GvOC scale yields very high value of the basic component and very low values of the acidic one, as compared to the other two scales. All three scales indicate that AA is the most acidic among the samples tested.

3.4. Force vs. Distance Curves

The force-distance curves measured in 0.001 M and 1 M NaCl on the AA sample are shown in Fig. 1. It is noted that in both cases the curves show a "hydrophilic behavior", i.e. no significant jump to contact is observed [27]. It is interesting to note that, despite the significant difference between the Debye lengths of the two solutions (10 nm for the 0.001 M, 0.3 nm for the 1 M) no significant effect on the range of the repulsive force is detected, contrary to what happens in the case of charged surfaces, such as mica, alumina or glass [27]. These data confirm what is expected from a chemical point of view, i.e. carboxyl groups of alginic acid are involved in chemical bonding with the underlying amino groups, and cannot build a significant negative charge on this surface.

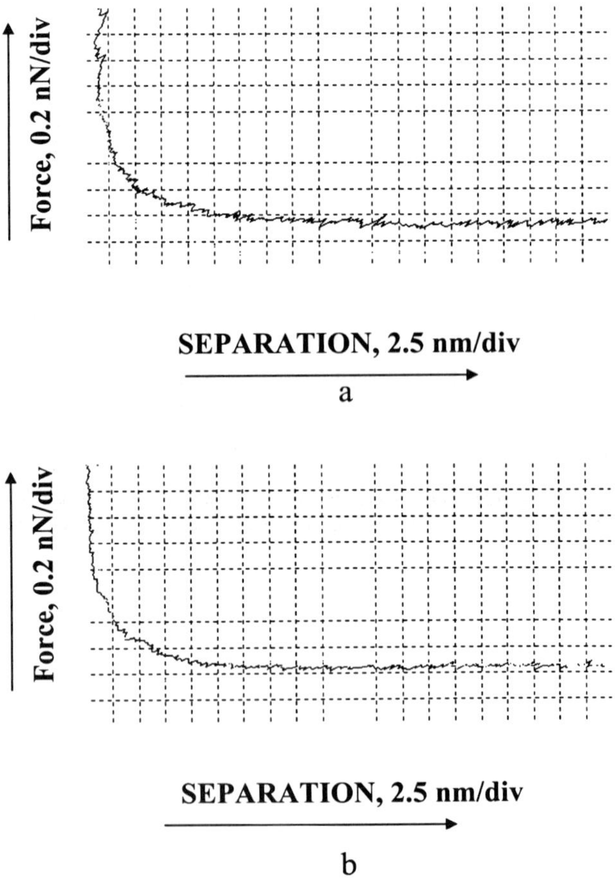

Figure 1. Force *vs* distance curve for AA in (a) 0.001 M NaCl and (b) 1M NaCl.

3.5. Results of Bacterial Adhesion Experiments

The results of bacterial adhesion experiments are shown in Fig. 2. Despite the different nature of the bacterial wall of Gram positive as opposed to Gram negative bacteria, the results clearly follow a common pattern. There is no statistically significant difference as far as adhesion to the different glass surfaces is concerned, while a significant reduction of bacterial adhesion is observed in the case of the AA coating. SEM images shown in Fig. 3 give a visual indication of the different adhesion behaviors observed on GCC (as representative of glass samples) and AA.

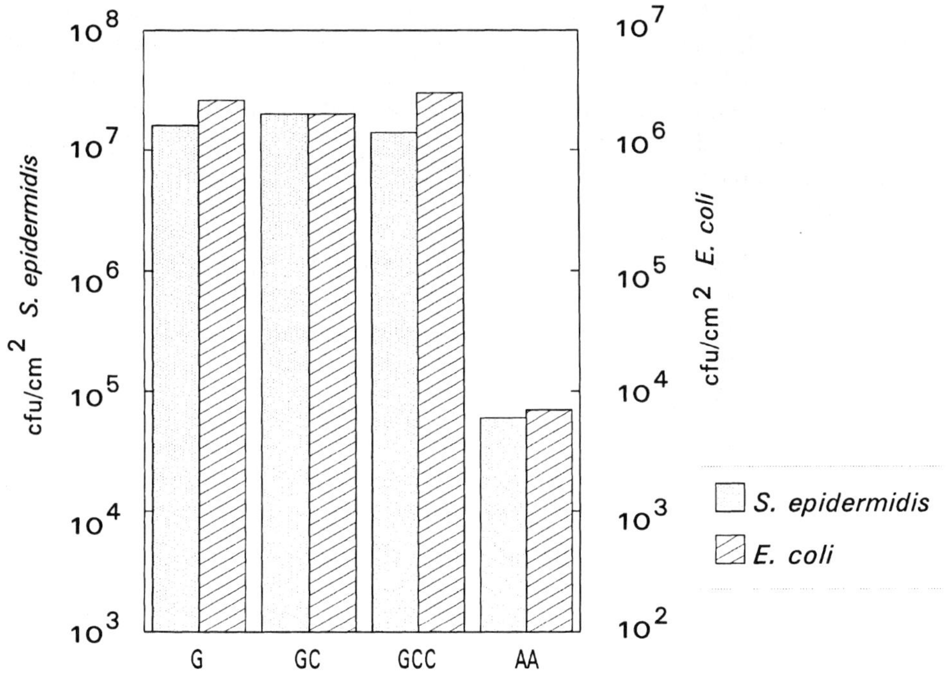

Figure 2. Results of bacterial adhesion testing. Differences between the results obtained on G, GC and GCC substrates are not statistically significant. Differences between the results obtained on AA and G, GC, GCC substrates are statistically significant.

(a)

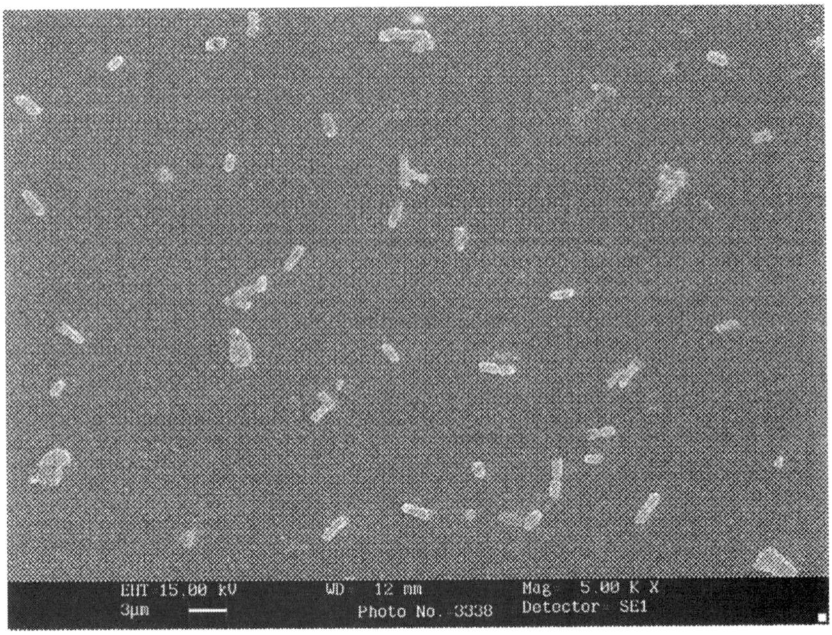

(b)

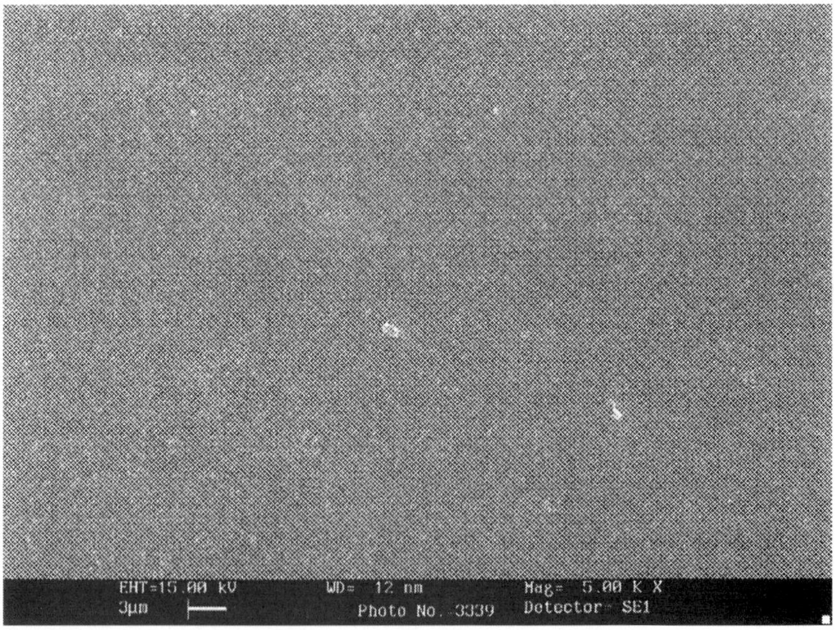

Figure 3. SEM images from bacterial adhesion experiments: (a) *E. coli* on GCC, (b) *E. coli* on AA, (c) *S. epidermidis* on GCC, and (d) *S. epidermidis* on AA.

(c)

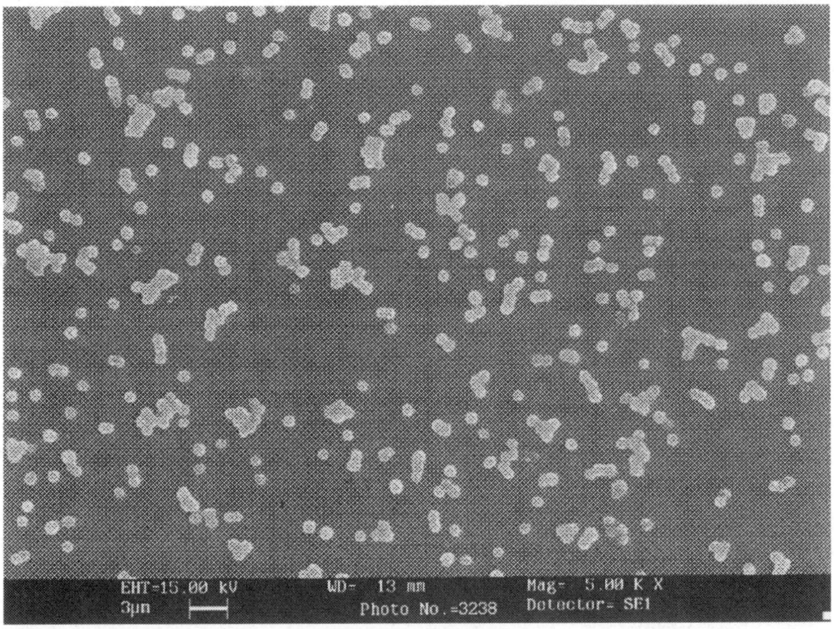

(d)

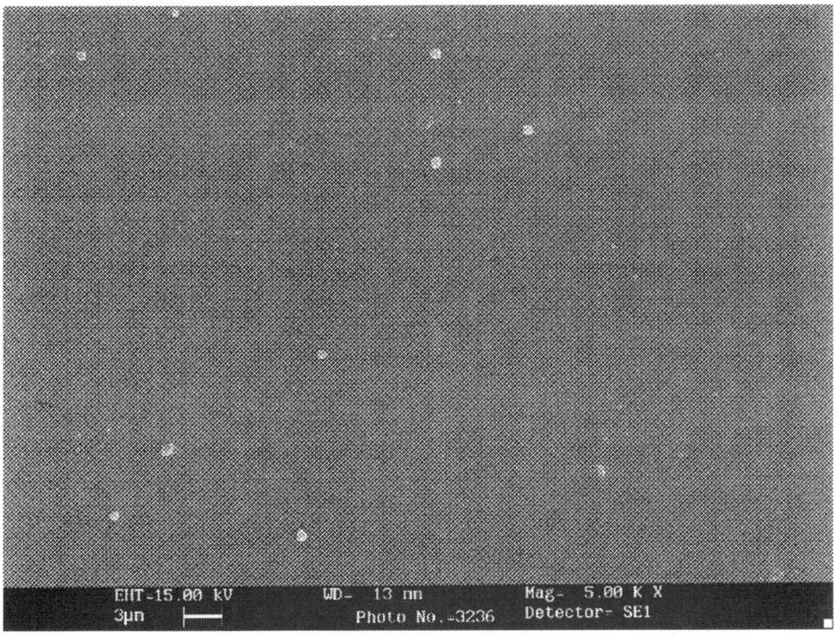

Figure 3. (Continued).

4. DISCUSSION

Experiments were performed to compare adhesion of two completely different bacterial strains from PBS suspensions to several "hydrophilic" surfaces. This kind of experiment is concerned with non-specific bacterial adhesion mechanisms (i.e. mechanisms not mediated by specific interaction of the ligand-receptor type) and has been the subject of many studies in the literature [4–7].

Data show that the bacterial adhesion is unaffected by the differences in wettability of the "rigid", hydrophilic glass surfaces, at least within the hydrophilic range of the wettability spectrum probed using the present samples (actually, "very dirty" glass slides, with higher water contact angles than the present ones, yield different results, typical of hydrophobic surfaces). The AA coated surface, even if its water contact angle is similar to that of GCC, shows definitely less adhesion towards both bacterial strains. Force-distance curves data (Fig. 1) show that these results cannot be accounted for by electrostatic repulsion. Carboxyl groups of the polyanionic AA yield amide bonds with the amino groups of the underlying PEI in the coupling reaction (even in the absence of chemical reaction they will probably be charge-neutralized by ionic coupling with PEI. Actually, ionic interaction alone is enough to immobilize polyanionic polysaccharides such as AA to PEI [26]). Short range, repulsive acid-base "hydrophilic interactions", as described by van Oss [18], are the most likely reason for the improved resistance of the AA surface to bacterial adhesion.

The specific aim of this work was to evaluate the effectiveness of the present theories in the description of electron-donor electron-acceptor properties of hydrophilic surfaces and their ability to predict non-specific bioadhesional interactions. In this respect, the polysaccharide coated AA surface is of particular interest: from a chemical point of view, carbohydrates are expected to bear a significant electron acceptor or Lewis acid contribution, because of hydrogen atoms on hydroxyl groups. The complex pattern of water-carbohydrate interactions (in aqueous solution) has been recently described by molecular dynamics simulation in the case of maltose [28].

Considering data of Table 2, the AA surface will be classified "monopolar basic" according to the rigorous van Oss interpretation because: (i) the acid-base component of the surface free energy is, from a quantitative point of view, overwhelmingly dominated by the electron-donor component, and (ii) the electron-acceptor component is lower than one [18]. This interpretation, however, does not take into account: (i) that quantitative values of electron-acceptor/electron-donor are definitely dependent on the choice made for the reference liquid [22], and (ii) because of the lack of any absolute value as a reference, it does not make sense to compare directly the acidic and the basic components of the same compound. The only meaningful comparison is between acidic vs. acidic and basic vs. basic components of different compounds. These fundamental aspects have been discussed thoroughly in two recent papers [22, 23].

The improved approach to surface free energy components of liquids introduced by the Lee and DVS scales [20, 23] is nicely reflected in the data obtained. The AA clearly stands up as the most acidic compound with a significant electron-acceptor contribution. Here the water-surface AA layer interaction appears in a chemically more satisfactory view, with electron-acceptor sites on the large number of available hydroxyl groups ready to interact with the electron-rich portions of the water molecules [28]. By the way, this is true also when the GvOC scale is used but, because of the very nature of this scale, basic components are so magnified and thus acidic ones are so compressed to appear (if one forgets that the direct comparison of components has actually no meaning, as discussed before) almost negligible.

Concerning the use of calculated components in the identification for the physical reason for the different behaviors towards bacteria of the glasses vs. the AA surface, some consideration can be made, namely: the reduction of non-specific interactions by hydration can be considered from the point of view of the water-surface interaction, as described in the Introduction. The free energy of interfacial interaction of a given surface in water can be taken as a measure of its hydrophilicity, i.e. of its propensity to interact with water. Table 3 reports the values of the free energy of interfacial interaction ΔG_{iwi}, calculated from the following equation [18]:

$$\Delta G_{iwi} = -2((\gamma_i^{LW})^{1/2}-(\gamma_W^{LW})^{1/2})^2-4((\gamma_i^+\gamma_i^-)^{1/2}+(\gamma_W^+\gamma_W^-)^{1/2}-(\gamma_i^+\gamma_W^-)^{1/2}-(\gamma_i^-\gamma_W^+)^{1/2}) \quad (1)$$

Table 3.
Calculated free energy of interaction (mJ/m^2) of sample surface i in water

Sample	Scale	ΔG_{iwi}
G	GvOC	22.97
GC	GvOC	46.57
GCC	GvOC	45.01
AA	GvOC	48.25
G	Lee	24.56
GC	Lee	50.00
GCC	Lee	47.98
AA	Lee	42.20
G	DVS	17.03
GC	DVS	41.20
GCC	DVS	41.13
AA	DVS	46.49

From the analysis of the data, it appears that both the GvOC and the DVS scales show a correct trend, i.e. the hydrophilicity of the surface, and hence its ability to interact with the medium and keep away incoming bacteria, while in the Lee scale this trend is not observed. Also, in the DVS scale the difference between the "adherent surfaces" and the AA one is greater than in the case of the GvOC scale, i.e. the unique nature of this surface among the set of analyzed samples is more clearly underlined. This observation, coupled with the improved description of the electron-acceptor character of the AA surface, leads to the conclusion that, at least within the set of data stemming from the present experiments, the DVS scale represents a significant improvement over the existing approaches.

From a more general point of view, it is interesting to note that the comparison between the raw contact angle data of water and of apolar liquids on the test surfaces gives further clues about the importance of the role of the medium in the overall interaction. The MI or α-Bromonaphthalene contact angles on the AA surface are much higher than those measured on GC and GCC samples, which have a similar water contact angle. This indicates a strong driving force for the interaction between water and AA, driven by mechanisms of interaction (hydrogen bonding) that are more effective on the AA than on the GC and GCC surfaces, in agreement with ΔG_{iwi} data obtained with the GvOC and DVS scales. It is possible to calculate the acid-base work of adhesion in water from [8, 19]:

$$W_{ab} = \gamma_W (1+\cos\theta_W) - 2(\gamma_i^{LW}\gamma_W^{LW})^{1/2} \tag{2}$$

and:

$$\gamma_i^{LW} = \gamma_{MI}(1+\cos\theta_{MI})^{2/4} \tag{3}$$

The ratio between the acid-base work of adhesion in water (Equation 2) and the "dispersive" work of adhesion (the second term of Equation 2) for the four tested surfaces is shown in Fig. 4. Again, despite the similar wettability by water, as displayed by the advancing water contact angle, the unique nature of AA-water interaction is clearly evident, and underlines again the importance of specific water-related effects in the outcome of non-specific adhesion of bacterial cells to hydrophilic surfaces.

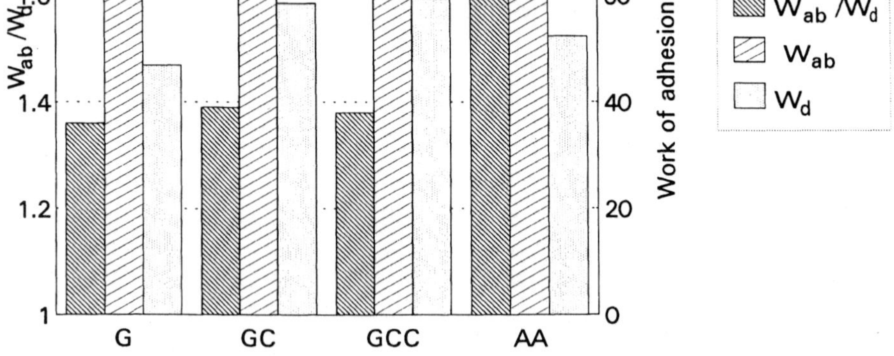

Figure 4. Ratio of the acid–base to the "dispersive" work of adhesion and total work of adhesion, calculated according to the data of Table 1 (θ_W, θ_{MI}) and Equations 2 and 3 in the text.

5. CONCLUSIONS

In conclusion, results stemming from the calculation of electron-donor/electron-acceptor properties of several hydrophilic good and poor substrates for non-specific bacterial adhesion, using different approaches, suggest that:

(i) The "monopolar basic" nature of hydrophilic surfaces is mostly the consequence of a scale (GvOC) that magnifies the basic component and compresses the acidic one. When this mathematical consequence of the scale used is coupled with the assumption that components below 1 mJ/m² should be considered negligible, all surfaces become indeed "monopolar basic".

(ii) The polysaccharide coated AA surface bears an acidic component, as expected from its chemical nature. As described in (i) this component is very low on the GvOC scale, but its value increases when improved scales (Lee and DVS) are used. Actually, the significant acidic character is the main distinction between the AA and other water-wettable samples (GC and GCC).

(iii) In agreement with the hypothesis of a significant role of short-range acid-base interactions between the sample surface and the surrounding water molecule in the reduction of non-specific bacterial adhesion, the low-adhesion AA surface shows a significantly higher ΔG_{iwi}, using the DVS scale (Table 3), and a significantly greater contribution of acid–base work of adhesion with water (Fig. 4).

REFERENCES

1. A. G. Gristina, *Science* **237**, 1583–1595 (1987).
2. A. F. Von Recum and E. Barth, *J. Invest. Surg.* **2**, 351–352 (1989).
3. B. Sugarman and E. J. Young, *Surg. Clin. North Amer.* **68**, 167 (1988).
4. M. Paulsson, I. Gouda, O. Larm and A. Ljungh, *J. Biomed. Mater. Res.* **28**, 311–317 (1994).
5. I. W. Wang, M. J. Danilich, J. M. Anderson and R. E. Marchant, *Trans. Soc. Biomaterials* **XVII**, 368 (1993).
6. G. Reid, L. A. Hawthorn, A. Eisen and H. S. Beg, *Colloids Surfaces* **42**, 299–311 (1989).
7. M. Morra and C. Cassinelli, *J. Biomater. Sci. Polym. Ed.* **9**, 55–74 (1997).
8. F. M. Fowkes, in: *Surface and Interfacial Aspects of Biomedical Polymers*, J. D. Andrade (Ed.), vol. 1, pp. 337–372. Plenum Press, New York (1985).
9. K. L. Mittal and H. R. Anderson, Jr. (Eds), *Acid Base Interactions: Relevance to Adhesion Science and Technology*. VSP, Utrecht, The Netherlands (1991).
10. E. A. Vogler, *Adv. Colloid Interface Sci.* **74**, 69–117 (1998).
11. N. A. M. Besseling, *Langmuir* **13**, 2113–2122 (1997).
12. M. Morra, C. Cassinelli, L. Benedetti, L. Callegaro, L. Ambrosio and L. Nicolais, *Trans. 5th World Biom. Cong.*, vol. 2, p. 171. University of Toronto Press, Toronto, Canada (1996).
13. A. Pavesio, D. Renier, C. Cassinelli and M. Morra, *Med. Dev. Technol.* **8**, 20–27 (1997).
14. Y. Ikada, *Adv. Polym. Sci.* **57**, 103–140 (1984).
15. E. Occhiello, M. Morra, P. Cinquina and F. Garbassi, *Polymer* **33**, 3007–3015 (1992).
16. C. J. van Oss, M. K. Chaudhury and R. J. Good, *Adv. Colloid Interface Sci.* **28**, 3564 (1987).
17. R. J. Good and C. J. van Oss, in: *Modern Approach to Wettability: Theory and Application*, M. E. Schrader and G. Loeb (Eds), pp. 1–27. Plenum Press, New York (1991).
18. C. J. van Oss, *Interfacial Forces in Aqueous Media*. Marcel Dekker, New York (1994).
19. J. C. Berg, in: *Wettability*, J. C. Berg (Ed.), pp. 75–148. Marcel Dekker, New York (1993).
20. L. H. Lee, *Langmuir* **12**, 1681–1687 (1996).
21. M. Morra, *J. Colloid Interface Sci.* **182**, 312–314 (1996).
22. C. Della Volpe and S. Siboni, *J. Colloid Interface Sci.* **195**, 121–136 (1997).

23. C. Della Volpe and S. Siboni, these proceedings.
24. C. Brink, E. Osterberg, K. Holmberg and F. Tiberg, *Colloids Surfaces* **66**, 149–156 (1992).
25. J. J. Chance and W. C. Purdy, *Langmuir* **13**, 4487–4489 (1997).
26. M. Morra and C. Cassinelli, *Surface Interface Anal.* **26**, 742–746 (1998).
27. H. J. Butt, *Biophys. J.* **60**, 1438–1444 (1991).
28. J. W. Brady and R. K. Schmidt, *J. Phys. Chem.* **97**, 958–966 (1993).8.5

Acid-Base Interactions, Vol. 2, pp. 513–524
Ed. K.L. Mittal

Acid–base interactions in microbial adhesion to hexadecane and chloroform

R. BOS,* H. J. BUSSCHER, G. I. GEERTSEMA-DOORNBUSCH and H. C. VAN DER MEI

Department of Biomedical Engineering, Division of Bioadhesion, University of Groningen, Bloemsingel 10, 9712 KZ Groningen, The Netherlands

Abstract—Acid-base interactions play an important role in adhesion, including microbial adhesion to surfaces. Qualitatively acid-base interactions in microbial adhesion can be demonstrated by comparing adhesion to hexadecane (a negatively charged interface in aqueous solutions, unable to exert acid-base interactions) and chloroform (able to exert acid-base interactions) from aqueous suspensions of microorganisms. For 31 different microbial strains (4 *Escherichia coli*, 4 *Enterococcus faecalis*, 4 lactobacilli, 10 staphylococci, 9 streptococci) in low and high ionic strength potassium phosphate buffers, adhesion to chloroform was initially faster and in a stationary end-point more extensive than to hexadecane, due to acid-base interactions additionally operative from chloroform. However, a quantitative surface thermodynamical analysis of acid-base interactions based on water, formamide, methylene iodide and α-bromonaphthalene contact angles failed to explain the adhesion data obtained at low ionic strength, likely because of interference of electrostatic interactions. A weak relationship between initial removal rates and the acid-base interfacial free energy of adhesion was observed for the high ionic strength data, provided the analysis was confined to strains with an absolute zeta potential less than 15 mV. Similarly, a weak relationship was obtained between the Gibbs free energy of partitioning, calculated from the stationary end-point adhesion of the strains, and the acid-base interfacial free energy of adhesion. It is concluded that acid-base interactions play a definitive role in microbial adhesion, but the extent to which thermodynamically determined acid-base interactions eventually become operative in microbial adhesion may vary from strain to strain due to the presence of structural appendages and chemical heterogeneities on microbial cell surfaces.

Keywords: Microbial adhesion; hexadecane; chloroform; acid-base interactions; surface thermodynamics.

1. INTRODUCTION

The classical DLVO-theory [1, 2] describes adhesion between two surfaces interacting in an aqueous fluid as a result of Lifshitz-van der Waals and electrostatic interactions. It has been employed to explain qualitative aspects of microbial adhesion to surfaces [3], but both structural and chemical heterogeneities on microbial cell surfaces have impeded a more quantitative analysis of microbial adhesion data on the basis of the DLVO-theory [4–6].

The extended DLVO-theory [7] not only accounts for Lifshitz-van der Waals and electrostatic interactions, but also acknowledges a role for acid-base interactions in

* To whom correspondence should be addressed. E-mail: r.bos@med.rug.nl

microbial adhesion. In fact, the extended DLVO-theory, more rigorous than the classical DLVO-theory, combines thermodynamic analyses, using contact angle data [8, 9], with electrokinetic data, i.e. the zeta potentials of the interacting surfaces [10]. Unfortunately, however, verification of the extended DLVO-theory in microbial adhesion is hampered by a lack of knowledge of the interaction parameters on a microscopic level [11, 12].

Pending this situation, Bellon-Fontaine *et al.* [13] experimentally confirmed a definitive role for acid-base interactions in microbial adhesion to surfaces by comparing microbial adhesion to hexadecane, hexane and decane (no acid-base interactions) with adhesion to chloroform (electron-accepting), diethyl ether and ethyl acetate (electron-donating) in a two phase partitioning assay [14], named MATH (microbial adhesion to hydrocarbons) or MATS (microbial adhesion to solvents), respectively. In a study comprising only two dairy strains, and neglecting electrostatic interactions, it was demonstrated that microbial strains possessing electron-donating properties adhered better to the slightly acidic chloroform and less to the basic solvents than to the hydrocarbons, although no general conclusions may be drawn from a study comprising only two strains. Furthermore, the neglect of electrostatic interactions may not be allowed *a priori*, since it is known that hydrocarbons in aqueous solutions can be highly negatively charged [15–17], as are most microbial cell surfaces [18]. As a consequence, highly negatively charged strains experience electrostatic repulsion by hydrocarbons and are then erroneously classified as being hydrophilic due to their lack of adhesion and most strains studied to date demonstrate a clear maximum in their adhesion to hydrocarbons near their isoelectric point [19].

The aims of this study were:

1. To establish qualitatively the role of acid-base interactions in microbial adhesion by comparing adhesion to hexadecane and chloroform of a wide variety of 31 different microbial strains.
2. To determine a possible relationship between adhesion data and the acid-base component of the interfacial free energy of adhesion ΔG^{AB}, as derived from contact angle data, while accounting for electrostatic charge interactions.

2. MATERIALS AND METHODS

2.1. Strains and Growth Conditions

Table 1 gives an overview of the culture media, and growth conditions of the bacteria used in this study. All bacteria were streaked, from frozen stock, onto an agar plate and incubated at 37°C. To prepare subcultures, culture medium was inoculated with a colony from plate and incubated overnight under the conditions indicated in Table 1.

Table 1.
Culture media and growth conditions of the strains used in this study

Strain	Medium type	Growth conditions
Escherichia coli	Brain Heart Infusion broth	aerobic
Enterococci	Brain Heart Infusion broth	aerobic
Lactobacilli	De Man, Rogosa and Sharp broth	atmosphere containing 5% CO_2
Staphylococci	Tryptone Soy Broth	aerobic
Streptococci	Todd Hewitt Broth	aerobic

2.2. Microbial Adhesion to Hexadecane and Chloroform

Since the originally described MATH test [14] has been criticized for not being sufficiently quantitative due to the neglect of kinetic effects, the so-called kinetic MATH (or MATS) test as proposed by Lichtenberg *et al.* [20] has been employed at room temperature. Briefly, bacteria were suspended to an optical density, A_0 (at 600 nm), of between 0.4 and 0.6 in a 10 or 100 mM potassium phosphate buffer, with pH adjusted to 7. Next, 150 μl of hexadecane or chloroform was added to 3 ml of bacterial suspension, and the two-phase system was vortexed for 10 s, allowed to settle for 10 min and the optical density A_t of the suspension measured. The latter procedure was repeated until the total vortexing time amounted to 60 s and log (A_t/A_0 x 100) was plotted against the vortexing time. Linear least squares fitting subsequently yielded the initial removal rate R_0 (min^{-1}) as a measure of the adhesion of the cells to the hexadecane or chloroform.

The absorbance values after the last vortexing were taken as a measure of adhesion in a stationary end-point. The equilibrium constant K which expresses the partitioning of organisms between the aqueous and hexadecane or chloroform phases can be calculated from

$$K = \frac{(A_0 - A_{eq}) x (3000/150)}{A_{eq}} \tag{1}$$

in which A_0 and A_{eq} are the absorbances (at 600 nm) of the aqueous phase before and after 60 s vortexing, respectively and

$$\Delta G_{par} = -kT \ln K \tag{2}$$

in which ΔG_{par} the Gibbs free energy of partitioning expressed in kT units per organism. Note that a factor of 20 appears in Equation (1) to correct for the different amounts of the aqueous phase and hexadecane or chloroform [21].

2.3. Zeta Potential and Contact Angle Measurements on the Bacteria and Surface Thermodynamic Analysis

Zeta potentials of the bacteria ζ_b in both high and low ionic strength buffers were measured at room temperature with a Lazer Zee Meter 501 (PenKem, Bedford Hills, NY, USA) which uses scattering of incident laser light to enable detection of the bacteria at relatively low magnifications. The absolute electrophoretic mobilities can be derived directly from the velocity of the bacteria in the applied electric field, the applied voltage, and the dimensions of the electrophoresis cell [22]. Although the conditions required to convert the measured electrophoretic mobilities into zeta potentials may not have been completely met in the present study, it is considered preferable here to use more appealing zeta potential values rather than electrophoretic mobilities. Zeta potentials were measured for the bacteria suspended to a concentration of 10^7 cells·ml^{-1}. The experimental reproducibility (i.e. the standard deviaton over six independent measurements on one suspension) of all zeta potentials reported was ± 2 mV.

The method for measuring 'plateau contact angles' on bacterial layers deposited on membrane filters has been described in detail elsewhere [9, 23]. Briefly, for contact angle measurements a suspension of washed cells in water was sucked through a cellulose-acetate filter by negative pressure until a multilayer of bacterial cells was deposited. Subsequently, filters were air-dried for approximately 1h until the so-called 'plateau contact angles' could be measured. 'Plateau contact angles' represent the microbial cell surfaces in its hydrated state, although the thermodynamic status of the contact angle thus obtained is admittedly not clear. Water, formamide, methylene iodide and α-bromonaphthalene contact angles on the filters were determined by the sessile drop technique, always on two filters prepared out of one culture.

Surface free energies were subsequently estimated by least-squares fitting of the data to the geometric-mean equation, using the concept of Lifshitz-van der Waals (LW) and, acid-base (AB) components, while neglecting spreading pressures [24]. The LW components reported are always an average of the results, calculated from the methylene iodide and α-bromonaphthalene contact angles, while in case of negative square roots of a surface free energy parameter in the calculations this parameter was taken to be zero [25].

The bacterial surface energy components and parameters γ^{LW}, γ^{+}, and γ^{-} were subsequently employed to calculate the interfacial free energies of adhesion of the strains from an aqueous suspension to hexadecane and chloroform, ΔG_{adh}. Negative values of ΔG_{adh} indicate that adhesion is thermodynamically favourable, while positive ΔG_{adh} values indicate unfavourable conditions for adhesion. ΔG_{adh} can be separated into an LW and an AB part according to

$$\Delta G_{adh} = \Delta G_{adh}^{LW} + \Delta G_{adh}^{AB} \tag{3}$$

where

$$\Delta G_{adh}^{LW} = -2\left(\sqrt{\gamma_{bv}^{LW}} - \sqrt{\gamma_{lv}^{LW}}\right)\left(\sqrt{\gamma_{sv}^{LW}} - \sqrt{\gamma_{lv}^{LW}}\right) \tag{4}$$

and

$$\Delta G^{AB}_{adh} = +2\Big[\Big(\sqrt{\gamma^{\oplus}_{bv}} - \sqrt{\gamma^{\oplus}_{sv}}\Big)\Big(\sqrt{\gamma^{-}_{bv}} - \sqrt{\gamma^{-}_{sv}}\Big) - \Big(\sqrt{\gamma^{\oplus}_{bv}} - \sqrt{\gamma^{\oplus}_{lv}}\Big)\Big(\sqrt{\gamma^{-}_{bv}} - \sqrt{\gamma^{-}_{lv}}\Big) - \Big(\sqrt{\gamma^{\oplus}_{sv}} - \sqrt{\gamma^{\oplus}_{lv}}\Big)\Big(\sqrt{\gamma^{-}_{sv}} - \sqrt{\gamma^{-}_{lv}}\Big)\Big] \quad (5)$$

in which the subscripts *b, l, s,* and *v* denote the bacteria, water, solvent (hexadecane or chloroform) and vapour, respectively. The γ^{LW}, γ^{+}, and γ^{-} values for water were taken 21.8, 25.5 and 25.5 $mJ \cdot m^{-2}$, for hexadecane 27.7, 0 and 0; and for chloroform 27.7, 3.8 and 0 $mJ \cdot m^{-2}$ [26].

3. RESULTS

The cell surface characteristics of the different microbial species and strains employed in this study are listed in Table 2 and [27]. Zeta potentials in the 10 mM buffer (κ^{-1} = 2.3 nm) are highly negative for the enterococci and the staphylococci, but for the lactobacilli zeta potentials are almost zero while major variations for the zeta potentials in the 10 mM buffer exist for *E. coli* and streptococci. Generally, zeta potentials are less negative in the 100 mM buffer (κ^{-1} = 0.8 nm), except for the *E. coli* it becomes even more negative, possibly due to preferential phosphate adsorption. LW surface free energy components usually vary between 30 and 40 $mJ \cdot m^{-2}$, but for some strains like *L. plantarum* ATCC 8014 can become as low as 21.8 $mJ \cdot m^{-2}$. In almost all cases the electron-donating surface free energy parameter exceeds the electron-accepting parameter, although for *S. epidermidis* 3081 and 3294 both parameters are almost equally small.

Microbial interactions with hexadecane and chloroform are summarized in Table 3 and demonstrate that strains generally adhere better or equally well (higher or equal initial removal rates) in 100 mM buffer than in 10 mM buffer, while also adhesion to chloroform is higher than to hexadecane, as judged from the initial removal rates. For adhering strains the Gibbs free energies of partitioning vary between 0 and -6 kT with an average around -3 kT, indicative of a relatively weak interaction. Non-adhering strains demonstrate positive Gibbs free energies of partitioning, that can become infinitely positive.

The LW interfacial free energy of adhesion γG^{LW} of the strains to both hexadecane and chloroform was always negative and ranged up to -2 $mJ \cdot m^{-2}$.

In Figs. 1 and 2 the initial removal rates and Gibbs free energies of partitioning by hexadecane and chloroform are plotted as a function of the acid-base interfacial free energies of adhesion for the combined 10 mM and 100 mM buffers data. At first glance, no straightforward relationships can be discerned, but when considering only those strains for which intervening electrostatic interactions can be neglected (absolute values of bacterial zeta potentials less than 15 mV), reasonable relationships are observed (see the closed symbols in Figs. 1 and 2).

Table 2.
Zeta potentials in 10 (ζ_{10}) and 100 (ζ_{100}) mM potassium phosphate buffers pH 7.0, together with the LW, electron-accepting and electron-donating surface free energy components of the microbial strains used in this study[a]

Strain	ζ_{100} (mV)	ζ_{100} (mV)	γ^{LW} ($mJ \cdot m^{-2}$)	$\gamma^{\oplus}$ ($mJ \cdot m^{-2}$)	γ^{-} ($mJ \cdot m^{-2}$)
E. coli 02K7	-48	-51	25.8	4.0	48.5
E. coli 0111K58	-11	-33	27.5	3.6	54.5
E. coli 083K?	-41	-55	32.8	2.5	26.0
E. coli 0161K-	-34	-50	33.8	0.2	59.1
E. faecalis 1131	-37	-17	39.1	0.3	57.4
E. faecalis C1030	-27	-13	35.8	0	6.2
E. faecalis IC14	-38	-37	43.4	0	104.9
E. faecalis 29212	-30	-20	39.0	0	99.4
L. casei ssp. rhamnosus NCFB 2964	-3	-3	39.5	0.7	30.2
L. casei spp. casei ATCC 393	-4	+8	36.8	1.8	36.6
L. fermentum ATCC 23271	0	+6	29.2	4.4	43.8
L. plantarum ATCC 8014	-37	-38	21.8	0	91.8
S. epidermidis HBH2	-28	-10	38.2	0	80.6
S. epidermidis 169	-29	-17	37.8	0.4	52.8
S. epidermidis HBH154	-21	-15	36.8	1.8	41.0
S. epidermides ATCC35984	-45	-14	30.0	2.9	32.8
S. epidermides 3081	-29	-15	37.9	1.4	1.3
S. epidermides 3294	-49	-14	33.5	3.7	2.6
S. epidermides 3059	-45	-8	34.9	0	64.9
S. epidermides 3399	-38	-29	34.9	0	80.6
S. epidermides 26585	-36	-10	34.5	0	69.1
S. epidermides 26512	-36	-13	34.2	1.1	4.8
S. mutans HG1003	-11	-6.5	36.5	1.2	47.9
S. sobrinus HG1025	-21	-8	35.7	0.6	49.5
S. thermophilus B	-7	0	38.7	0.5	56.5
S. mitis BA	-25	-13	38.3	0	0
S. sanguis 10556	-16	0	35.9	1.7	50.4
S. sanguis CR311	-25	-10	34.3	0.1	17.1
S. sanguis PSH	-33	-12	37.6	1.2	8.5
S. salivarius HB	-25	-11	36.2	0	55.9
S. salivarius HB-C12	-21	-13	39.4	0.6	51.3

[a] Contact angles used in the calculations were taken from [27].

Table 3.
Microbial removal rates by hexadecane and chloroform in 10 ($R_{0,10}$) and 100 ($R_{0,100}$) mM potassium phosphate buffers and the free energies of partitioning ΔG_{par} in 10 and 100 mM phosphate buffers

Strain	hexadecane				chloroform			
	$R_{0,10}$(min^{-1})	$\Delta G_{par,10}$(kT)	$R_{0,100}$(min^{-1})	$\Delta G_{par,100}$(kT)	$R_{0,10}$(min^{-1})	$\Delta G_{par,10}$(kT)	$R_{0,100}$(min^{-1})	$\Delta G_{par,100}$(kT)
E. coli 02K7	0	∞	0	∞	0	∞	0.4	-1.4
E. coli 0111K58	0	∞	0	∞	0	∞	0	+0.9
E. coli 083K?	0	∞	0	+0.8	0	+0.1	0.1	-2.2
E. coli 0161K-	0	∞	0	-0.3	0	+0.2	0.2	-2.7
E. faecalis 1131	0	∞	0	∞	0	∞	0.1	-1.4
E. faecalis C1030	0	-0.7	0	+0.9	0	-0.4	0.2	-2.2
E. faecalis IC14	0	∞	0	-0.3	0	∞	0	+1.3
E. faecalis 29212	0	∞	0	+1.4	0	∞	0	∞
L. casei ssp. Rhamnosus NCFB 2964	0	∞	0	∞	0	∞	0	∞
L. casei spp. Casei ATCC 393	0	∞	0	∞	0	∞	0	∞
L. fermentum ATCC 23271	0	0.03	0	0.6	0	+3.4	0.2	-2.8
L. plantarum ATCC 8014	0	∞	0	0.3	0	∞	0	∞
S. epidermidis HBH2	0.2	-2.3	0.4	-2.5	1.0	-4.0	0.9	-4.4
S. epidermidis 169	0.4	-2.7	0.4	-1.4	0.9	-3.8	1.3	-3.9
S. epidermidis HBH154	1.3	-4.8	1.8	-6.0	2.2	-2.0	1.9	-4.4
S. epidermidis ATCC35984	0.2	-2.1	0.5	-2.9	0.5	-3.3	0.9	-3.0
S. epidermidis 3081	0.9	-4.3	0.8	-3.5	0.9	-4.2	1.2	-4.4
S. epidermidis 3294	0.6	-2.9	0.5	-2.6	1.1	-3.7	0.8	-3.0
S. epidermidis 3059	0.2	-2.1	0.5	-3.4	0.6	-3.4	1.0	-4.2
S. epidermidis 3399	0	∞	0	-0.4	0	∞	0	∞
S. epidermidis 26585	0.3	-2.8	0.4	-2.8	0.5	-3.3	0.7	-4.5
S. epidermidis 26512	0.3	-1.8	0.3	-2.3	0.7	-4.4	1.0	-3.0
S. mutans HG1003	0.2	-2.2	0.3	-2.9	0.5	-3.5	0.6	-3.7
S. sobrinus HG1025	0	∞	0.1	-1.2	0.3	-2.4	1.1	-2.3
S. thermophilus B	0	∞	0	∞	0.6	-1.8	0.8	-2.5
S. mitis BA	0.8	-3.6	0.7	-4.1	0.7	-4.6	0.9	-4.9
S. sanguis 10556	0.3	-2.7	0.3	-2.7	0.3	-4.1	1.3	-3.5
S. sanguis CR311	0.2	-2.7	0.4	-3.5	0.4	-4.4	0.8	-4.8
S. sanguis PSH	0	∞	0.1	-2.1	0.1	-1.8	0.5	-3.6
S. salivarius HB	0.3	-3.1	0.5	-3.5	0.5	-4.6	2.2	-5.9
S. salivarius HB-C12	0	∞	0	∞	0	∞	0	-2.7

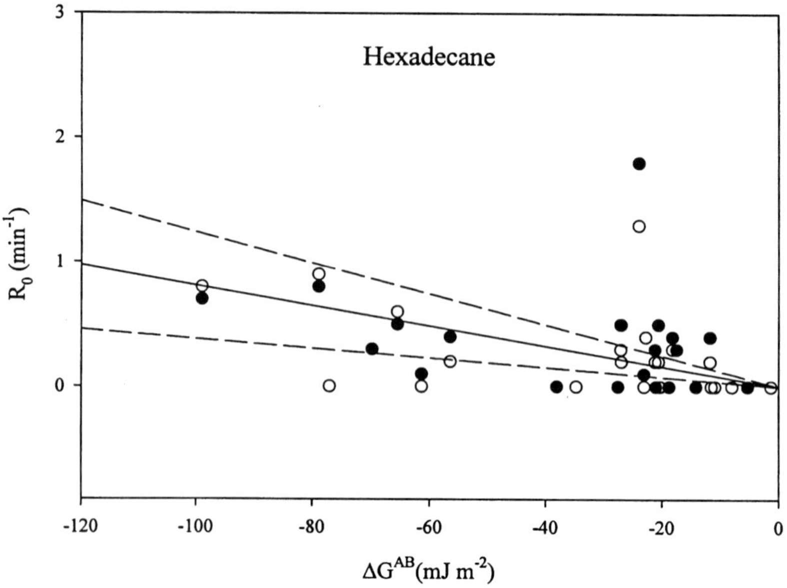

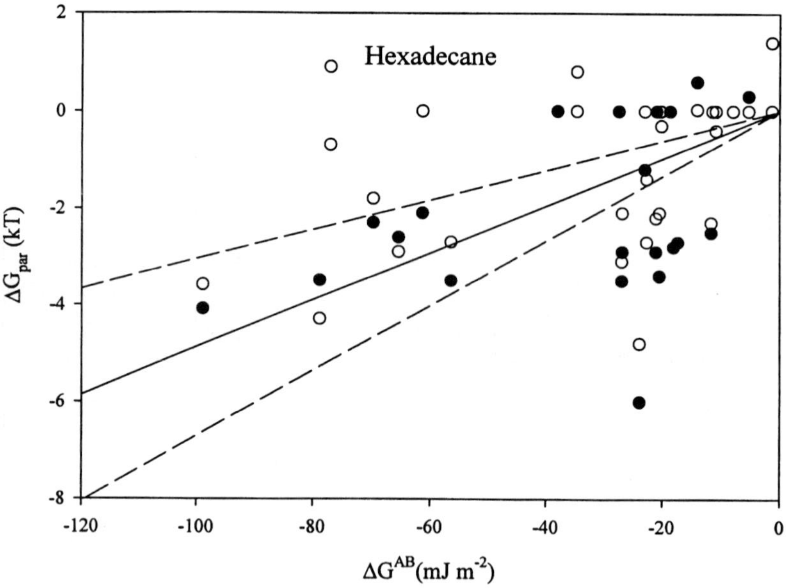

Figure 1. The removal rate (R_0) and the free energy of partitioning (ΔG_{par}) by hexadecane as a function of the acid-base interfacial free energy component (ΔG^{AB}) of adhesion derived from contact angle data (data for experiments in 10 and 100 mM phosphate buffers combined). Closed symbols represent cases for which electrostatic interactions can be neglected (absolute value of bacterial zeta potentials $|\zeta| \leq 15$ mV), while the open symbols represent cases for which the bacterial zeta potential is below -15 mV. Drawn lines indicate linear relationships through the closed symbols with the broken lines indicating the 95 % confidence intervals.

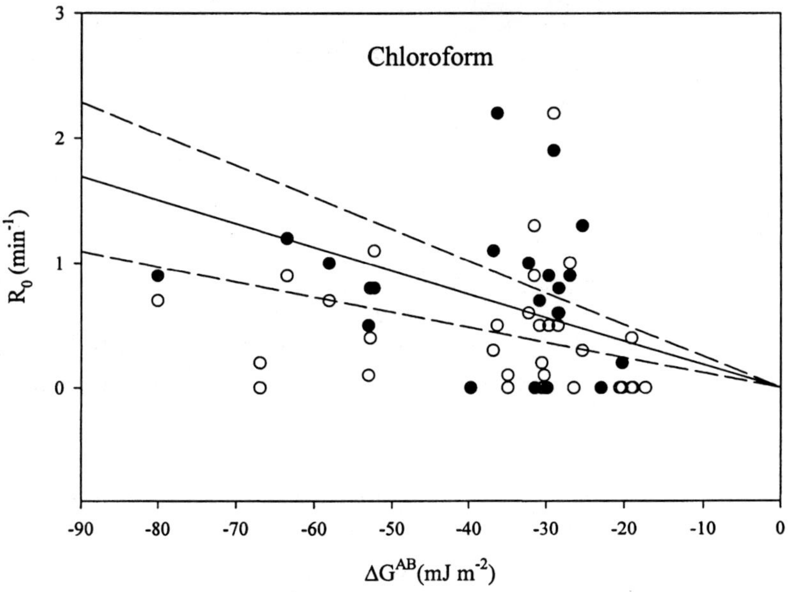

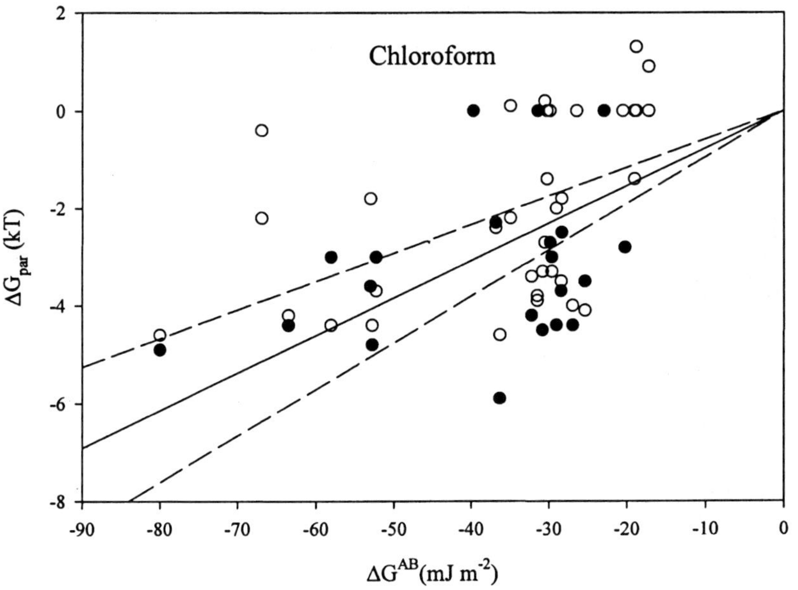

Figure 2. The removal rate (R_0) and the free energy of partitioning (ΔG_{par}) by chloroform is a function of the acid-base interfacial free energy component (ΔG^{AB}) of adhesion derived from contact angle data (data for experiments in 10 and 100 mM phosphate buffers combined). Closed symbols represent cases for which electrostatic interactions can be neglected (absolute value of bacterial zeta potentials $|\zeta| \leq 15$ mV), while the open symbols represent cases for which the bacterial zeta potential is below -15 mV. Drawn lines indicate linear relationships through the closed symbols with the broken lines indicating the 95 % confidence intervals.

Similarly, also for the Gibbs free energy of partitioning in Figs. 1 and 2 the data appear completely scattered, while an expected trend with the acid-base interfacial free energy of adhesion is observed for those strains with an absolute zeta potential less than 15 mV (see the closed symbols in Figs. 1 and 2).

4. DISCUSSION

In this study, we have investigated the adhesion of a wide variety of bacterial strains to hexadecane and chloroform with the aid of a two phase partitioning assay, the so-called MATH or MATS assays. As an advantage of this assay, adhesion of a large number of strains can be studied in a relative short period of time. More importantly, the interfaces to which the bacteria adhere are probably much smoother, and more homogeneous than any solid substratum. In addition, the comparison between hexadecane and chloroform is interesting because they have the same Lifshitz-van der Waals surface free energy components, while hexadecane is non-polar and chloroform has only an electron-accepting surface energy parameter. However, as drawbacks of these assays, both the size and number of micro-droplets created during vortexing depend, amongst others, on the duration of vortexing, the power input, dimensions of the test tube and the ratio between the aqueous and organic phases, which are all difficult to control. Furthermore, and partly as a consequence of the above, mass transport conditions are poorly defined. Nevertheless, when carried out under standardized conditions, i.e. use of the same type of tubes, vortex apparatus etc., experimental results coinciding within 20 % are feasible.

Recently, Pelletier *et al.* [28] also investigated adhesion of a collection of eight lactobacillus strains by both MATH and MATS and observed that all strains displayed a higher affinity for chloroform than for hexadecane or the basic solvent ethyl acetate. Although it was concluded qualitatively here that the cell surfaces of these lactobacilli had a basic character, no quantitative surface thermodynamical analysis based on contact angle measurements was done. Our data demonstrate that a basic character of a bacterial cell surface, as derived from contact angles, not necessarily coincides with a higher affinity for chloroform as compared with hexadecane, not even when a subset of data of one particular strain is considered. For instance, *S. epidermidis* 3081 and 3059 have comparable removal rates and $\Delta G_{par,100}$ values for hexadecane as well as for chloroform but possess markedly different acid-base surface free energy parameters.

The merits of this paper are in our opinion not so much in terms of creative ideas forwarded, since the science is a repetition of previous works by different authors, but its value must be appreciated from the number of strains included. To our knowledge, this is the first report on an evaluation of MATH and MATS in relation to surface thermodynamic analyses, comprising more than 10 strains. In this respect, Figs. 1 and 2 not only demonstrate that generalizations are difficult to make in microbial adhesion, but furthermore it indicates the risk of relying too heavily on relationships. Random selection of data for four strains out of all data presented and calculation of a linear correlation coefficient for these data reveal a 30% chance of an acceptable (linear correlation coefficient better than 0.5) relationship. For a random selection of six strains, this chance reduces to 15%, which must still be considered as appreciable. Non-random selection of data on the basis of negligible electrostatic interactions reveals weak relationships between the acid-base interfacial free energy of adhesion ΔG^{AB}, calculated from contact angle data, and both the removal rate R_0 and ΔG_{par} (see Figs. 1 and 2). This confirms the role of electrostatic interactions in microbial adhesion to hexadecane and chloroform [16]. In addition, interpretation of adhesion data by calculations of acid-base parameters from contact angles is complicated and partly dictated by liquids used and the mathematics involved [25].

Still, even for this non-random selection comprising 21 data points, correlations are weak. Interpretation of data is complicated by the fact that microorganisms are not perfectly smooth or chemically homogeneous. Streptococci, for instance, display a great variety of structural surface appendages [4] while staphylococci often possess a glycoproteinaceous capsule [29]. Depending on their relative prevalence, these microscopic chemical and structural heterogeneities are not always reflected in macroscopic contact angle data to the same extent to which they influence adhesion. Furthermore, due to the presence of surface appendages, the concepts of distance and contact area between interacting surfaces lose their meanings upon close approach, hampering physico-chemical interpretations of bacterial adhesion data.

5. CONCLUSION

Summarizing, in this study we have qualitatively demonstrated the role of acid-base interactions in bacterial adhesion to hexadecane and chloroform, but no quantitative relationships between surface thermodynamic, acid-base interactions and adhesion to hexadecane and chloroform were found, except when a non-random selection of data was taken for which electrostatic interactions could demonstrably be neglected. However, due to structural surface appendages and chemical heterogeneities on the cell surfaces, these relationships remained relatively weak.

REFERENCES

1. B. V. Deryagin and L. Landau, *Acta Phys. Chim. URSS* **14**, 633 (1941).
2. E. J. W. Verwey and J. Th. G. Overbeek, *Theory of the Stability of Lyophobic Colloids*. Elsevier, Amsterdam (1948).
3. P. R. Rutter and B. Vincent, in: *Microbial Adhesion to Surfaces,* R. C. W. Berkeley, J. M. Lynch, J. Melling, P. R. Rutter and B. Vincent (Eds), pp. 79–91. Ellis Horwood Limited, London (1980).
4. P. S. Handley, *Biofouling* **2**, 239–264 (1990).
5. H. J. Busscher, M. M. Cowan and H. C. van der Mei, *FEMS Microbiol. Rev.* **88**, 199–210 (1992).
6. H. J. Busscher and A. H. Weerkamp, *FEMS Microbiol. Rev.* **46**, 165–173 (1987).
7. C. J. van Oss, M. K. Chaudhury and R. J. Good, *Chem. Rev.* **88,** 927–941 (1988).
8. D. R. Absolom, F. V. Lamberti, Z. Policova, W. Zingg, C. J. van Oss and A. W. Neumann, *Appl. Environ. Microbiol.* **46,** 90–97 (1983).
9. H. J. Busscher, A. H. Weerkamp, H. C. van der Mei, A. W. J. van Pelt, H. P. de Jong and J. Arends, *Appl. Environ. Microbiol.* **48,** 980–983 (1984).
10. N. Mozes, F. Marchal, M. P. Hermesse, J. L. van Haecht, L. Reuliaux, A. J. Leonard and P. G. Rouxhet, *Biotechnol. Bioeng.* **39**, 1161–1170 (1987).
11. C. J. van Oss, *J. Molecular Recognition* **3**, 128–136 (1990).

12. C. J. van Oss, *Colloids Surfaces B: Biointerfaces* **5**, 91–110 (1995).
13. M.-N. Bellon-Fontaine, J. Rault and C. J. van Oss, *Colloids Surfaces B: Biointerfaces* **7**, 47–53 (1996).
14. M. Rosenberg, E. Rosenberg, H. Judes and E. Weis, *FEMS Microbiol. Lett.* **20**, 1–5 (1983).
15. K. B. Medrczycka, *Colloid Polym. Sci.* **269**, 85–90 (1991).
16. H. C. van der Mei, J. de Vries and H. J. Busscher, *Appl. Environ. Microbiol.* **59**, 4305–4312 (1993).
17. D. J. Shaw, *Colloid and Surface Chemistry.* Butterworth Heinemann, Oxford (1992).
18. B. A. Jucker, H. Harms and A. J. B. Zehnder, *J. Bacteriol.* **178**, 5472–5479 (1996).
19. H. C. van der Mei, B. van de Belt-Gritter and H. J. Busscher, *Colloids Surfaces B: Biointerfaces* **5**, 116–126 (1995).
20. D. Lichtenberg, M. Rosenberg, N. Sharfman and I. Ofek, *J. Microbiol. Methods* **4,** 141–146 (1985).
21. R. Bos and H. J. Busscher, *Colloids Surfaces B: Biointerfaces,* in press.
22. P. C. Hiemenz, *Principles of Colloid Surface Chemistry.* Marcel Dekker, New York (1977).
23. C. J. van Oss, C. F. Gillman and A. W. Neumann, *Phagocytic Engulfment and Cell Adhesiveness.* Marcel Dekker, New York (1975).
24. H. J. Busscher, A. W. J. van Pelt, H. P. de Jong and J. Arends, *J. Colloid Interface Sci.* **95**, 23–27 (1983).
25. C. Della Volpe and S. Siboni, *J. Colloid Interface Sci.* **195**, 121–136 (1997).
26. C. J. van Oss, *Interfacial Forces in Aqueous Media.* Marcel Dekker, New York (1994).
27. H. C. van der Mei, R. Bos and H. J. Busscher, *Colloids Surfaces B: Biointerfaces* **11**, 213–221 (1998).
28. C. Pelletier, C. Bouley, C. Cayuela, S. Bouttier, P. Bourlioux and M.-N. Bellon-Fontaine, *Appl. Environ. Microbiol.* **63**, 1725–1731 (1997).
29. H. C. van der Mei, B. van de Belt-Gritter, G. Reid, H. Bialkowska-Hobrzanska and H. J. Busscher, *Microbiology* **143**, 3861–3866 (1997).

Acid-Base Interactions, Vol. 2, pp. 525–537
Ed. K.L. Mittal

Laser surface modification of ultra-high-strength polyethylene fibers: Correlation between acid–base interactions and adhesion to epoxies

A. N. NETRAVALI* and Q. SONG

Fiber Science Program, Cornell University, Ithaca, NY 14853-4401, USA

Abstract—Fibers based on ultra-high-strength polyethylene (UHSPE) such as Spectra® have excellent mechanical properties for use in advanced composites. They have the highest specific strength among all high strength fibers. However, their poor adhesion to any common resins limits their use in composites. This paper describes the use of pulsed XeCl UV excimer laser (308 nm wavelength) to increase the acid–base characteristics and the surface roughness of the UHSPE fibers to increase the interfacial shear strength (IFSS) with epoxy resins. Spectra® 1000 fibers were treated with the excimer laser radiation in different environments including air, ammonia, argon, and helium. The changes in surface energies and wettability parameters including the dispersive and acid–base components and work of adhesion were measured by the Wilhelmy technique. Surface topographies of control and treated fibers were characterized using scanning electron microscopy. The interfacial shear strength (IFSS) with epoxy was measured using single fiber pull-out tests. The results indicate that laser treatments when carried out in air incorporate oxygen on the surface through oxidation and the radicals generated which later combine with the atmospheric oxygen when exposed to air. Fibers treated in ammonia environment incorporate both nitrogen and oxygen on the surface. This results in an increase in the acid–base interaction component of the fiber surface energy and makes it more polar. The laser treatment is also responsible for the surface ablation and increased roughness. Laser treatments carried out in inert environments do not alter the acid–base characteristics but do increase the surface roughness resulting in a 100% increase in the IFSS. The combined effect of the increased acid–base interaction and the surface roughness is to improve the fiber/epoxy IFSS by about 300%. Since the increased roughness alone is shown to improve the IFSS by about 100%, it may be concluded that a significant contribution to the IFSS (about 200%) is made by the acid–base component of the surface energy.

Keywords: Surface modification; ultra-high-strength polyethylene fibers; pulsed excimer laser; adhesion; fiber/matrix interface; acid–base interactions; surface energy; photodissociation; ablation.

1. INTRODUCTION

High performance fibers such as graphite, glass, aramid, and silicon carbide are commonly used in aerospace and other structural reinforcement applications because of their excellent mechanical properties. Advanced composites made from these fibers which possess high strength and light weight are being routinely used in place of

* To whom correspondence should be addressed.

more conventional metals and metallic alloys in such applications as sports gear and automobile parts.

Ultra-high-strength polyethylene (UHSPE) fibers, which have been developed in the last decade, possess the highest specific strength and fracture toughness compared to any other high strength fibers including graphite, glass, and aramid. This is because of their excellent mechanical properties combined with the lowest density among any fibers of 0.97 g/cm^3. They also have excellent chemical resistance to acids and alkalis because of their inherently inert nature [1]. As a result, UHSPE fibers are currently being used in a variety of applications including anti-ballistic materials, sail fabrics, surgical gloves, and as reinforcing element in synthetic replacement bones and teeth. However, their inert nature precludes any covalent and hydrogen bonding with resins of interest and their non-polar nature and low surface energy keeps any resin from wetting them. Besides, these fibers have smooth surfaces, which do not permit any mechanical bonding, and their relatively large diameters reduce the specific surface area. As a result, they have the poorest adhesion amongst all reinforcing fibers to any resin and have not been used extensively in composite materials.

It is well accepted that the fiber/matrix interface plays an important role in achieving superior tensile properties of a composite [2, 3]. The tensile strength of the composite is dependent on its ability to transfer the tensile load from the broken fibers to the surviving ones through matrix shear and the interface. Improvement in the interfacial bond strength may increase a composite's tensile strength, but this is generally accompanied by a decrease in the impact strength and toughness [2, 3]. On the other hand, weak interfacial bonding may encourage energy absorbing modes of crack propagation at the interface and thereby increase fracture toughness of the composite [2].

It is generally acknowledged that good adhesion is achieved through covalent chemical bonding and acid–base interactions (hydrogen bonding), favorable wettability, large specific surface area, and mechanical bonding [4]. As mentioned earlier, lack of any of these factors in the UHSPE fibers results in their poor interfacial strength. Most of the surface modification techniques available for other high strength fibers may not be available for UHSPE fibers because of their inert nature. As a result, few chemical modifications have been attempted. Chemical oxidation [5] and sulphonation with sulphonic acid [6] have been useful to some extent. Plasma surface treatments have also been investigated by several researchers [7–13]. Nguyen and coworkers have treated UHSPE fibers with various plasmas [7]. Their results show a significant increase in the interlaminar shear strength of UHSPE fiber/epoxy composites which is indicative of improved interfacial bonding. Holmes and Schwartz [8] treated UHSPE fabrics using an ammonia plasma. Their results showed improved wetting of the fabrics by the epoxy resin indicating an increase in the fiber surface energy. They also obtained over 40% increase in adhesion with resin through T-peel test, after the plasma treatment. Li and Netravali [9–11] used polymer-forming allylamine plasma deposition on single UHSPE fibers. Their results showed that the plasma polymer coatings improved the fiber/epoxy interfacial shear strength (IFSS) by a factor of three or more without significantly affecting the fiber strength. They believed that the amine groups on the fiber surface, created by the plasma deposition, were responsible for covalent bonding with the epoxy resin. Li *et al.* [12], who studied effects of ammonia plasma treatment observed a four-fold increase in IFSS with epoxy. However, the maximum IFSS values after the treatment were only in the range of 2.0 MPa. Miller and Schwartz [13] characterized the effects of oven aging in air and in water on the long-term effectiveness of the plasma treatments. Their results indicated that even though there was no change in the wettability of the ammonia plasma treated UHSPE fibers, the oven aging caused a significant decrease in the IFSS. However, the oxygen and argon plasma treated fibers did not show any change in their IFSS. This indicates that the surface chemical changes brought about by the ammonia plasma treatment were lost during aging whereas the effects of the mechanical bonding promoted by the oxygen and argon plasma etching still remained after the aging.

Song and Netravali [14, 15] treated UHSPE fibers with XeCl pulsed UV excimer lasers (308 nm wavelength) in air and ammonia environments to improve the surface wettability and polarity as well as to increase the surface roughness, to enhance their bonding to epoxy. Their results showed an increase in bond strength by over 300% as a result of increased acid–base component of the total surface energy and increased surface roughening. Song and Netravali [16] also treated Spectra® 2000 fibers with and without kink bands. In single fiber pull-out tests both control and treated fibers with kinks showed higher IFSS than fibers without kinks indicating mechanical interlocking due to kinks. The control fibers with kinks had 37% higher IFSS compared to fibers without kinks. However, after the laser treatment the difference between the two values narrowed to only 14%. This was probably due to the kinks being ablated and becoming less pronounced and thus providing less mechanical interlocking after the laser treatment. The present study reports the efforts to separate the contributions of the acid–base component and surface roughness to the UHSPE fiber/epoxy resin IFSS of XeCl pulsed laser treated fibers.

2. EXPERIMENTAL

2.1. Materials

Ultra-high-strength polyethylene (UHSPE), Spectra® 1000, fibers were obtained from AlliedSignal Corporation. The diglycidyl ether of bisphenol-A (DGEBA)-based epoxy resin (DER 331) and tetraethylenepentamine (TEPA), DER 26, curing agent were obtained from Dow Chemical Company.

2.2. Specimen Preparation and Laser Treatment

UHSPE fibers were placed in a parallel array between two pieces of cardboard having 5 mm × 4 mm windows. When exposed to laser beam, only the part of the fiber within the window received the energy. After exposing the first side with 250 pulses at the required energy density, the cardboard holder was flipped 180° and the other side was treated by an equal number of pulses for a total of 500 pulses. The pulse rate was maintained at 9 pulses per second (9 Hz) which was the highest available. The schematic of the laser treatment setup is shown in Figure 1. In order to expose fibers to ammonia, argon, or helium environment, individually, during the laser treatment,

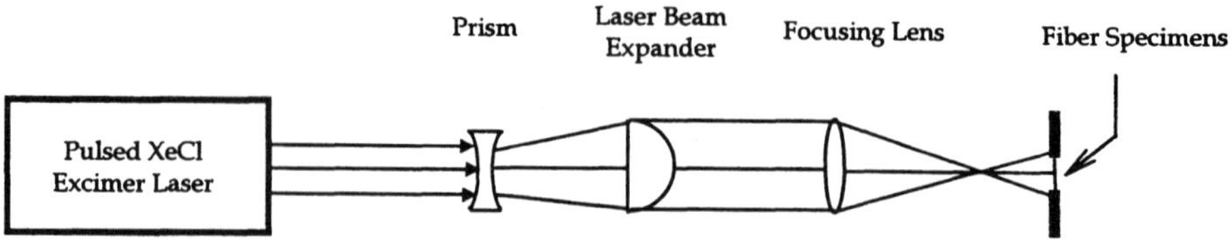

Figure 1. Experimental setup for laser irradiation study with pulsed excimer laser [14].

a Plexiglas™ chamber with quartz windows on both front and back sides was built [14]. The quartz windows had excellent UV transmission characteristics. The cardboard holder with the fibers was placed in the chamber. To remove all traces of air the chamber was continuously flushed with the desired gas for at least 5 minutes before the laser treatment. The gas flow was then maintained at a rate of 20 ml/min during the entire laser treatment. Further details of the chamber design may be found elsewhere [14].

The pulsed XeCl excimer laser (308 nm) treatment was carried out using a Lambda Physik LXP 200 laser apparatus. The size of the beam was adjusted to be slightly larger than the window. The laser operating parameters (energy density per pulse, repetition rate, and number of pulses) and the environments are listed in Table 1.

Table 1.
Laser treatment conditions used for surface modification of UHSPE fibers

Environment	Energy density (mJ/cm^2 per pulse)	Pulse repetition rate (Hz)	No. of pulses
Air*	280	9	500
Air*	450	9	500
Air	770	9	160
Ammonia*	450	9	500
Argon	280	9	500
Argon	450	9	500
Argon	770	9	500
Helium	280	9	500
Helium	450	9	500
Helium	770	9	500

*Data from References 13 and 14

2.3. Fiber Surface Characterization

The surface topographies of the control and laser treated fibers were observed using a Leica scanning electron (SEM) microscope, model 440X. Surface Science Instruments, Model SSX 100, X-ray photoelectron spectrometer (XPS) was used to obtain the fiber surface elemental composition. Aluminum Kα X-rays were used to scan the fiber specimens from 0 to 1000 eV binding energy. All measurements were made at 45° take-off angle.

2.4. Characterization of the Wettability Properties and Acid–Base Interaction

The Wilhelmy technique was used to characterize the wettability of the control and treated fibers. The experiments were carried out using a TRI/Princeton wettability apparatus. A more complete description of the technique has been provided by Kamath *et al.* [17]. A 10 mm long piece of the specimen was attached to a steel wire hook using Duco® cement. The fiber specimen to be characterized was suspended from the electrobalance hangdown wire and its weight was tared (zeroed). A beaker containing the probe liquid, placed on the moving platform, was raised until first contact with the fiber was made. To collect the wetting force data in the advancing mode the probe liquid was raised at the rate of 0.0015 cm/s until 3 mm depth of the fiber was immersed. In a similar fashion the receding mode data were collected while lowering the probe liquid at the same rate. To obtain the acid–base and dispersive components of the surface energy, the three probe liquid method, described by Good *et al.* [18], was used. Other wettability parameters such as contact angle, work of adhesion, and hysteresis were computed for all three probe liquids. The three well-characterized probe liquids included high purity distilled water and formamide, both polar liquids, and methylene iodide, a nonpolar liquid. Table 2 lists the surface energy components of these liquids [18].

Table 2.
Surface energy* components (in mJ/m^2) of the probe liquids used in this study [18]

Probe liquid	γ	γ^d	γ^{AB}	γ^+	γ^-
Distilled water	72.8	21.8	51.0	25.5	25.5
Formamide	58.2	39.5	18.7	2.28	39.6
Methylene iodide	50.2	50.2	0	0	0

γ: surface energy; γ^d: dispersive component; γ^{AB}: hydrogen bonding (acid–base interaction) component; γ^+: acidic component; γ^-: basic component.

2.5. Fiber/Matrix Interface Characterization

The single-fiber pull-out test was used to characterize the UHSPE fiber/epoxy resin interfacial shear strength (IFSS). Single-fiber pull-out specimens were based on the mold technique developed by Li and Netravali [10]. To make a pull-out specimen a single fiber was embedded and fixed in a slit cut in the silicone rubber mold to half the depth of the mold cavity. The molds were prepared to maintain the embedded length of 1.54 mm. DER 331 epoxy resin and DEH 26 curing agent were mixed thoroughly, in a stoichiometric ratio. Prior to filling the mold cavity the resin/hardener mixture was degassed in a vacuum oven to remove any trapped air. Care was taken not to disturb the fiber while pouring the resin. The mold along with the resin and fiber was kept in an air circulating oven at 80°C for 3 h to cure the resin and allowed to cool slowly overnight to avoid residual stresses at the interface. The specimen was then removed carefully from the silicone mold. The lower part of the fiber, below the resin specimen, was cut with a razor and the upper part of the fiber was wound (5.5

loops) on the capstan jaw. Fiber pull-out test was performed on an Instron tensile testing machine. A detailed description of the pull-out test can be found elsewhere [12].

The average interfacial shear strength, τ, was calculated using the following equation:

$$\tau = F_p/\pi dl \quad (1)$$

where F_p is the fiber pull-out load, l is the embedded length and d is the fiber diameter. Equation (1) assumes that the shear stress is uniformly distributed along the entire embedded length.

3. RESULTS AND DISCUSSION

3.1. Surface Chemical Analysis

Figure 2 presents the FTIR-ATR spectra of control and laser treated UHSPE fibers. The spectrum of control Spectra 1000® fiber shows major peaks around 2915 cm^{-1}, 2848 cm^{-1}, 1470 cm^{-1}, and 716 cm^{-1}. All these peaks relate to the carbon/hydrogen stretching, bending and/or rocking modes and are consistent with the pure hydrocarbon nature of the fiber [16]. No significant changes are evident for the fibers treated in air (Figure 2B). This is possibly because of the insensitivity of the FTIR-ATR to detect the low concentration of functional groups present. The XPS spectra, however, had indicated a significant increase in oxygen concentration [14]. The fiber treated in ammonia (Figure 2C) shows lower peaks at 1470 cm^{-1} and 716 cm^{-1}. However it also shows a broad peak indicating the presence of -NH, around 3380 cm^{-1}. Earlier, the XPS data had indicated a strong presence of nitrogen in the ammonia treated fibers [14, 15]. The FTIR-ATR spectra of fibers treated in argon and helium shown in Figures 2D and 2E, respectively, show no changes in any major peaks. However, the spectrum for the fiber treated in argon does show several additional peaks in the 800 cm^{-1} to 1200 cm^{-1} range which may not be easily attributable to specific chemical groups. In summary, FTIR-ATR spectroscopy is able to detect some changes but is not sensitive enough to detect low concentrations of functional groups on the laser treated UHSPE fiber surfaces.

3.2. Fiber Surface Topography

The surface topographies of the control and pulsed excimer laser treated UHSPE fibers were characterized using scanning electron microscopy (SEM) technique. Typical SEM photomicrographs of control and excimer laser treated fiber surfaces at a fluence of 450 mJ/cm^2 per pulse are presented in Figure 3. The control (untreated) UHSPE fiber, shown in Figure 3A, exhibits very smooth surface with some small striations parallel to the fiber axis. In comparison, fiber treated in air shows significant surface roughness as seen in Figure 3B [14]. However, the depth of the surface roughness seems to be only a fraction of a micrometer. Fiber treated in ammonia environment, Figure 3C, shows several striations that are deeper than the control fiber. Fibers treated in argon and helium, seen in Figures 3D and 3E, respectively, although smoother than the fiber treated in air, show deeper striations. Song and Netravali [14,

15] who studied the effects of different laser parameters on the same Spectra® 1000 fibers had concluded that energy density level and number of pulses were also significant factors, along with the environment, in determining the fiber surface roughness. At low energy density levels and low number of pulses not enough energy (photons) is put into the fiber. As a result, no ablation or surface roughness occurs. In the present case, the combination of 450 mJ/cm^2 of energy density and 500 pulses are sufficient to create the roughness. With increased energy and/or pulses there should be even higher surface roughness. In air, with oxygen present, the surface also undergoes significant oxidation. In summary, it can be concluded that XeCl laser treatment induces surface roughness and the extent of roughness depends on the environment and the total amount of photons (energy) deposited.

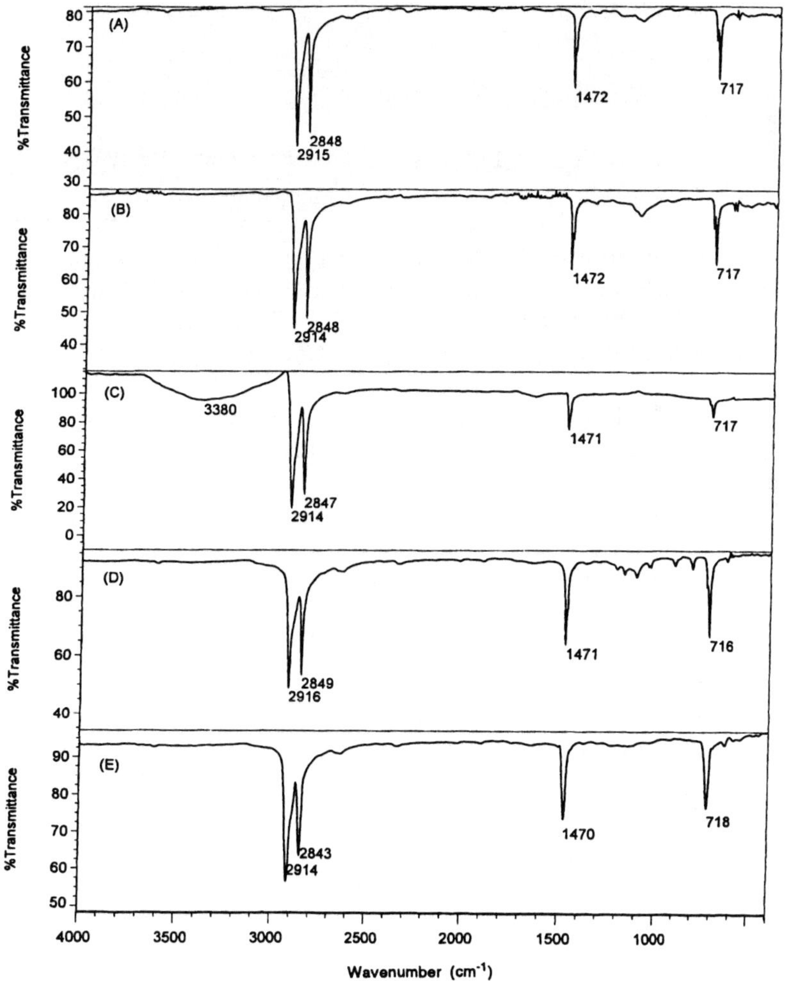

Figure 2. Typical FTIR-ATR spectra of laser-treated UHSPE fibers: (A) control, (B) in air, (C) in ammonia, (D) in argon, and (E) in helium.

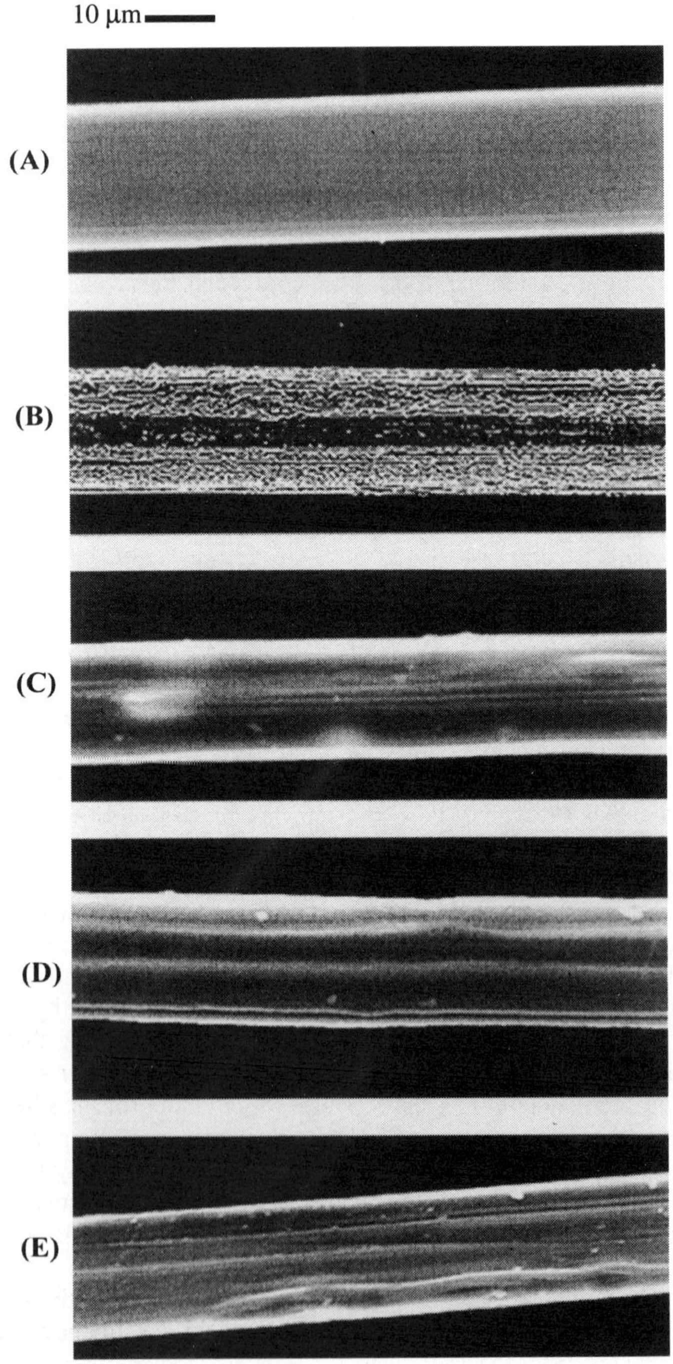

Figure 3. Typical SEM photomicrographs of control and laser-treated UHSPE fibers: (A) control, (B) in air [14], (C) in ammonia [15], (D) in argon, and (E) in helium.

3.3. Wettability Properties and Acid–Base Interaction Measurements

The wettability parameters for treated and control fibers were determined by the Wilhelmy technique. The work of adhesion, W, advancing contact angle, θ_{adv}, and hysteresis data of UHSPE fibers, in water, are presented in Table 3. Similar data for the receding mode are presented in Table 4. Typical wetting charts for all treatment conditions are shown in Figure 4. It can be seen from Tables 3 and 4 that the excimer laser treatments in air and ammonia environments reduce the contact angles significantly and increase the work of adhesion in the advancing mode making the surface much more wettable. The changes in the receding mode are even more significant. These data indicate that the fiber surface becomes significantly more polar after the laser treatments. The XPS results had indicated the presence of significantly higher amounts of surface oxygen in fibers treated in air, primarily due to the oxidation and to some extent due to the radicals generated by the UV laser that combine with the oxygen when exposed to the air [14, 15]. Fibers treated in ammonia showed increased oxygen as well as nitrogen on the surface [15].

In the case of argon and helium the changes in the contact angle and work of adhesion were not as significant as those observed in the case of the fibers treated in air and ammonia. This was expected mainly due to the inert environment presented by argon and helium. However, helium environment seems to be less inert compared to argon. The small increase in the wettability is again attributed to combining of the radicals, generated during the treatment, with oxygen in the air after the fibers are taken out of the chamber [14]. These results are also similar to those observed by Waldman *et al.* [19] for plasma treated UHSPE fibers.

Typical wetting charts for the control and laser treated fibers are presented in Figure 4. It can be seen from the advancing curve for the control fiber (bottom part of the curves), Figure 4A, that the amplitude of fluctuation for fibers, before treatment, is very small. Also the hysteresis is very small. In comparison, the fibers treated in air and ammonia seen in Figures 4B and 4C, respectively, show significantly higher amplitude fluctuations. For these fibers, two types of chatter can be observed from the advancing curves. The first type of chatter had lower frequency but higher amplitude of fluctuations originating from surface roughness, as confirmed by the SEM observations. The second type with high frequency and low amplitude that can be attributed to the heterogeneous nature of the chemical sites on the fiber surface created by oxygen and nitrogen containing groups. The wetting charts for the fibers treated in argon and helium shown in Figure 4D and 4E, respectively, show particularly more of the lower frequency and higher amplitude fluctuations confirming the increased surface roughness but less of the high frequency chatter that relates to the chemical groups. The receding mode curves further confirm these observations. This is also indicated in the data presented in Tables 3 and 4.

Table 5 summarizes the surface energy and its components, in the advancing mode, for the UHSPE fibers treated under different conditions. As mentioned earlier, these data were calculated from the wettability data obtained from three probe liquids: water, formamide, and methylene iodide [17, 18]. Similar data for the receding mode are presented in Table 6. It can be seen from Table 5 that the total surface energy increased only slightly for the fibers treated in air and ammonia. However, the dispersive component decreased from 30.5 mJ/m^2 to about 25.00 mJ/m^2 and that decrease was more than compensated by an increase of over 8 mJ/m^2 in the acid–base component from near zero for control fibers. Many-fold increase is also seen in the electron acceptor, γ^+, (acid) and donor, γ^-, (base) sites after the laser treatment in air and ammonia. The data presented in Table 6 for the receding mode confirm these observations. It is widely recognized that the acid–base interaction, or hydrogen bonding, across an interface makes a major contribution to the interfacial bond strength [20–22]. In the case of the fibers treated in argon and helium, it can be seen from Tables 5 and 6 that the surface energy remains unchanged compared to that of the untreated fibers. The decrease in the dispersive components as well as the increase in the acid–base components is small, approximately 2.0 mJ/m^2, compared to the fibers treated in air and ammonia. Changes in the surface energy and acid–base component in the receding mode are also not significant. This was expected because of the inert environment presented by these gases. From these data it may be concluded that the acid–base characteristic (polar nature) of the fiber surface increases significantly as a result of the laser treatment in air and ammonia. This increase is the result of oxygen and nitrogen incorporation at the UHSPE fiber surface, resulting in polar functional groups [14, 15]. However, no changes are seen when UHSPE fibers are treated in inert argon and helium environments. The surface roughness, however, increases with all treatments.

Table 3.
Wettability parameters, in advancing mode, of UHSPE fibers treated under different conditions

Experimental conditions			W_{adv}	θ_{adv}	Hysteresis
Environment	ED*	No. of pulses	(mJ/m^2)	(deg)	(mJ/m^2)
Control**	450	500	64.5	93.4	1.8
Air**	450	500	91.8	71.0	18.4
Ammonia**	450	500	90.0	75.1	17.1
Argon	450	500	68.1	92.6	8.0
Helium	450	500	74.3	87.6	11.0

* Energy density (ED) in mJ/cm^2
** Data from [14, 15]

Table 4.
Wettability parameters, in receding mode, of UHSPE fibers treated under different conditions

Experimental conditions			W_{rec}	Θ_{rec}	Hysteresis
Environment	ED*	No. of pulses	(mJ/m^2)	(deg)	(mJ/m^2)
Control**	450	500	66.2	91.9	1.8
Air**	450	500	109.9	57.7	18.4
Ammonia**	450	500	107.1	60.3	17.1
Argon	450	500	76.1	86.1	8.0
Helium	450	500	85.3	78.7	17.1

* Energy density (ED) in mJ/cm^2
** Data from [14, 15]

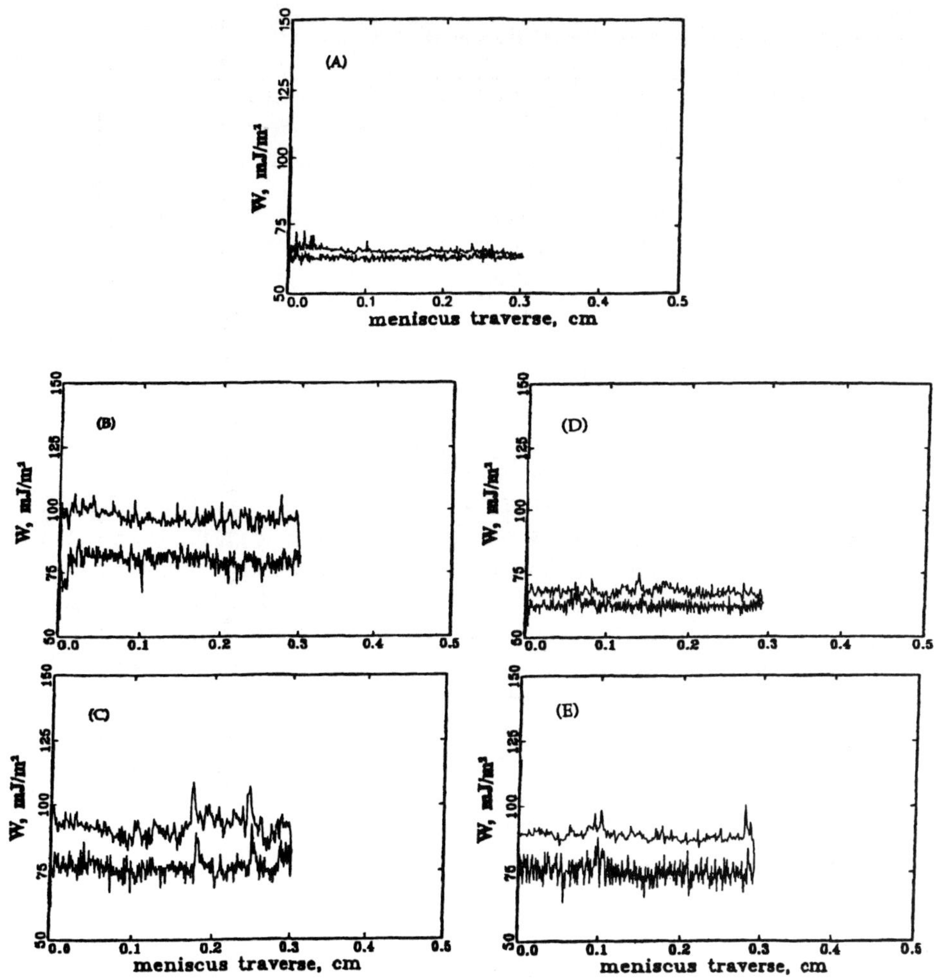

Figure 4. Typical wetting charts for the control and laster-treated UHSPE fibers: (A) control [14], (B) in air [14], (C) in ammonia [15], (D) in argon, and (E) in helium.

Table 5.
Surface energy and its components for UHSPE fibers in advancing mode

Experimental conditions			Dispersion γ^d	Acid–base γ^{ab}	Acid γ^+	Base γ^-	Surface energy γ^s
Environment	Energy density	No. of pulses					
Control*	450	500	30.50	0.20	0.01	1.32	30.80
Air*	450	500	24.40	8.10	1.53	10.59	32.46
Ammonia*	450	500	25.80	8.60	6.04	3.08	34.45
Argon	450	500	26.61	1.30	0.16	2.48	27.86
Helium	450	500	27.60	2.80	0.74	2.69	30.40

* Data from [14, 15]

Table 6.
Surface energy and its components for UHSPE fibers in receding mode

Experimental conditions			Dispersion	Acid–base	Acid	Base	Surface energy
Environment	Energy density	No. of pulses	γ^d	γ^{ab}	γ^+	γ^-	γ^s
Control*	450	500	35.73	0.40	0.59	0.07	36.13
Air*	450	500	34.64	11.69	2.17	15.74	46.33
Ammonia*	450	500	30.38	15.04	6.24	9.07	45.42
Argon	450	500	33.40	2.14	0.54	2.12	35.55
Helium	450	500	34.81	1.06	0.04	7.85	35.87

* Data from [14, 15]

3.4. Interfacial Shear Strength

Table 7 summarizes the UHSPE fiber/epoxy resin IFSS obtained for untreated and excimer laser treated fibers using a single fiber pull-out test. The control fiber has an IFSS value of only 0.96 MPa. This is very low and results from its non-polar nature which does not allow any hydrogen bonding and the smooth surface that prevents any mechanical bonding (interlocking). The laser treatments, particularly in air and ammonia environments, increase the IFSS by approximately 325% (3.13 MPa) and 400% (3.76 MPa), respectively. These values are also much higher than 1.82 MPa obtained by Li and Netravali by after depositing allylamine plasma polymer on Spectra® 900 fibers [10]. As discussed earlier, these treatments increase the surface energy and particularly the acid–base component, significantly boosting the hydrogen bonding capability with epoxy resin. The increased surface roughness is also a contributing factor. In the case of UHSPE fibers treated in argon and helium, there is only a slight increase in the surface energy or the acid–base component. As a result, the increase in the IFSS, of about 200%, can be concluded as primarily the contribution of the increased surface roughness with only a small contribution of the acid–base interaction. Thus it may be concluded that the acid–base interaction alone may be responsible for approximately 200% increase in the IFSS. It must be made clear that although the effects of the surface roughness and the acid–base interaction are cumulative, the two effects may not be considered as simple additive.

Table 7.
Summary of the UHSPE fiber/epoxy resin interfacial shear strength treated under different conditions

Experimental conditions			IFSS	CV	Increase in
Environment	ED*	No. of pulses	(MPa)	(%)	IFSS (%)
Control**	450	500	0.96	10.6	-
Air**	450	500	3.13	4.3	326
Ammonia**	450	500	3.76	31.9	392
Argon	450	500	2.38	9.6	248
Helium	450	500	2.09	5.6	218

* Energy density (ED) in mJ/cm^2
** Data from [14, 15]

4. CONCLUSIONS

The UHSPE fiber surface was modified with a pulsed XeCl excimer laser (308 nm) in different environments to modify the surface acid–base characteristics and the roughness. The results suggest that both roughness and acid–base components increase after the laser treatment. However, the environment has a significant effect on the extent of these changes. Both these factors contribute to the UHSPE fiber/epoxy IFSS. Although their individual contributions cannot be precisely calculated from the results of this study, it can be concluded that the acid–base component is responsible for over 200% improvement in the IFSS.

Acknowledgements

The authors would like to thank Prof. Michael Thompson of Cornell University for his valuable assistance in conducting the laser experiments. Thanks are also due to Dr. Yash Kamath of TRI/Princeton and Dr. Sheldon Wesson of Ynir Instrumentation Software for their valuable discussions.

REFERENCES

1. P. J. Lemstra, R. Kirschbaum, T. Ohta and H. Yasuda, in: *Developments of Oriented Polymers-2*, I. M. Ward (Ed.). Elsevier Applied Science, London (1987).
2. C. C. Chamis, in: *Interfaces in Polymer Matrix Composites*, E. P. Plueddemann (Ed.), pp. 31–77. Academic Press, New York (1974).
3. M. R. Piggott, in: *Progress in Science and Engineering of Composites*, Proceedings of the 4th International Conference on Composite Materials, ICCM-IV, T. Hayashi, K. Kawata and S. Umekawa (Eds), p. 193. Japan Society for Composite Materials, Tokyo (1982).
4. D. Scola, in: *Interfaces in Polymer Matrix Composites*, E. P. Plueddemann (Ed.), pp. 217–284. Academic Press, New York (1974).
5. N. H. Ladizesky and I. M. Ward, *J. Mater. Sci.* **18**, 533 (1983).
6. A. R. Postema and A. J. Pennings, in: *High Modulus Polymers-Approaches to Design and Development*, A. E. Zachariades and R. S. Porter (Eds), p. 431. Marcel Dekker, New York (1988).
7. N. X. Nguyen, G. Riahi, G. Wood and A Poursartip, *Proc. 33rd International SAMPE Symposium*, p. 1721 (1988).
8. S. Holmes and P. Schwartz, *Composites Sci. Technol.* **38**, 1 (1990).
9. Z.-F. Li and A. N. Netravali, *J. Appl. Polym. Sci.* **44**, 319 (1992).
10. Z.-F. Li and A. N. Netravali, *J. Appl. Polym. Sci.* **44**, 333 (1992).
11. A. N. Netravali and Z.-F. Li, in: *Polymer and Fiber Science: Recent Advances*, R. E. Fornes, R. D. Gilbert and H. F. Mark (Eds), Chapter 6, p. 67. VCH Publishers, New York (1992).
12. Z.-F. Li, A. N. Netravali, and W. Sachse, *J. Mater. Sci.* **27**, 4625 (1992).
13. A. Miller and P. Schwartz, *Plasma and Polymers* **2**, 115 (1997).
14. Q. Song and A. N. Netravali, *J. Adhesion Sci. Technol.* **12**, 957 (1998).
15. Q. Song and A. N. Netravali, *J. Adhesion Sci. Technol.* **12**, 983 (1998).
16. Q. Song and A. N. Netravali, *J. Adhesion Sci. Technol.* **13**, 501 (1999).
17. Y. K. Kamath, C. J. Dansizer, S. Hornby and H. D. Weigman, *Textile Res. J.* **57**, 205 (1987).
18. R. J. Good, M. K. Chaudhury and C. J. van Oss, in: *Fundamentals of Adhesion*, L.-H. Lee (Ed.), Chapter 4, p. 153. Plenum Press, New York (1991).
19. D. A. Waldman, Y. L. Zou, and A. N. Netravali, *J. Adhesion Sci. Technol.* **9**, 1475 (1995).
20. K. L. Mittal, *Polym. Eng. Sci.* **17**, 467 (1977).
21. F. M. Fowkes, *J. Adhesion Sci. Technol.* **1**, 7 (1987).
22. K. L. Mittal and H. R. Anderson, Jr. (Eds), *Acid–Base Interactions: Relevance to Adhesion Science and Technology*. VSP, Utrecht, The Netherlands (1991).

Acid-Base Interactions, Vol. 2, pp. 539–550
Ed. K.L. Mittal

Acid–base interaction in adhesion of polymer particles to silanized silica in water/n-alcohol mixtures

KEIKO GOTOH,[1,*] JUNKO TAO[1] and MIEKO TAGAWA[2]

[1]*Kyoto University of Education, 1 Fukakusa-fujinomori-cho, Fushimi-ku, Kyoto 612-0863, Japan*
[2]*Kanazawa Institute of Technology, 7-1 Ohgigaoka, Nonoichi, Ishikawa 921-8501, Japan*

Abstract—The adhesion of polyethylene and nylon particles to silica plates was investigated in water/n-alcohol mixtures. Silica plates were treated with γ-aminopropyltriethoxysilane, methyltriethoxysilane and perfluoroethyltrimethoxysilane. The number of particles adhering to the plate increased by silanization of silica and decreased with increasing volume ratio of n-alcohol in water/n-alcohol mixtures. The acid and base parameters of the surface free energy were reduced by silanization of silica and by an addition of ethanol to water. The particle adhesion was discussed in terms of the total potential energy of interaction as the sum of the electrical double layer, Lifshitz–van der Waals, and acid–base interactions. Moreover, the relationship between the adhesion phenomenon and the free energy change in adhesion was investigated. The particle adhesion in the present systems was found to be dominated by the acid-base interaction.

Keywords: Particle adhesion; silanized silica; water/n-alcohol mixture; ζ-potential; contact angle; surface free energy; acid–base interaction.

1. INTRODUCTION

When two surfaces approach closer than a few nanometers, the classical DLVO theory often fails to describe their interaction. This theory takes into consideration only the electrostatic (EL) and Lifshitz–van der Waals (LW) forces and does not take into account the polar forces that are often predominating forces between the surfaces in polar media. The polar interactions are largely based on electron acceptor-electron donor (Lewis acid–base) interactions [1, 2]. The addition of acid–base (AB) forces to the electrostatic and Lifshitz–van der Waals forces yields a new and powerful theory capable of predicting coagulation and adhesion behaviors. In fact, it has been reported that the stability of the hectorite clay [3] and ethylcellulose latex [4] suspensions, and the adhesion of calcium carbonate to glass [5] experimentally observed agree with predictions by the extended DLVO theory (EL+LW+AB). However, we need further experimental evidences of the extended DLVO theory in various systems.

In order to clarify the role of acid–base interaction in adhesion, the adhesion of polyethylene and nylon particles to silica plates untreated and silanized with three silane coupling agents was investigated *in situ* in water/n-alcohol mixtures. The addi-

* To whom correspondence should be addressed. E-mail: keiko@wsml.kyokyo-u.ac.jp

tion of n-alcohols to water enables to change the interactive forces between the particle and the substrate. The results are discussed in terms of the total potential energy of interaction which was calculated as the sum of EL, LW, and AB interactions using experimentally determined ζ-potentials and surface free energy components.

2. EXPERIMENTAL

2.1. Materials

Silica plates ($5 \times 1 \times 25$ mm^3) were used as substrates in particle adhesion experiments. Spherical particles of polyethylene and nylon 12 used had mean diameters of 4 μm [6] and 5 μm [7], respectively. For the contact angle measurements employing the Wilhelmy technique, a single fiber is suitable because of the negligible buoyancy effect. Whereas, the fibrous assembly is appropriate for preparing a porous plug in the ζ-potential measurements employing the streaming potential method. Therefore, for experimental convenience, fused silica, polyethylene and nylon 6 fibers were used for the contact angle measurements and silica fibrous assemblies (silica filter QR–100, Advantec, Japan) were used for the measurements of the ζ-potentials. The silica fibrous assemblies, polyethylene particles, and nylon particles were successively purified by extraction with water, ethanol, and ethyl ether. The silica plates and fibers were soaked in a dichromate sulfuric acid solution, rinsed with running water, and boiled in doubly distilled water. The silica samples were silanized with γ-aminopropyltriethoxysilane (APTES), methyltriethoxysilane (MTES), and perfluoroethyltrimethoxysilane (FETMS) provided by Shinnetsu Kagaku Kougyou Co., Ltd., Japan. Silica samples were immersed in the silane coupling agent solutions for 1 h and were heated in an oven at 100±1°C for 1 h. Silanized silica was then rinsed with water, ethanol, and acetone. This procedure was repeated four times.

Water/n-alcohol (methanol, ethanol, 1-propanol, and 1-butanol) mixtures were used as suspending media. Water, diiodomethane, and ethylene glycol were chosen as probe liquids for the determination of the surface free energies by contact angle measurements. Water was deionized and doubly distilled using a borosilicate glass apparatus.

2.2. Particle Adhesion

Polyethylene and nylon particles were dispersed ultrasonically in water/n-alcohol mixtures containing $1 \cdot 10^{-3}$ mol·dm^{-3} of potassium chloride. Particle concentrations of polyethylene and nylon were $3.2 \cdot 10^9$ dm^{-3} and $3.0 \cdot 10^9$ dm^{-3}, respectively. A quartz cell ($10 \times 10 \times 45$ mm^3 in volume) was filled with the dispersion and the silica plate previously wetted with the medium was perpendicularly immersed in the dispersion. The dispersion was stirred at 100 rpm using a magnetic stirrer to prevent flotation or sedimentation of the particles. After 60 min, the microscopic images of the particles adhering to the plate were recorded on a video tape. The number of particles adhering to 0.118 mm^2 of the plate was counted over 90 different areas. Most of particles on the plate adhered as primary particles, but some aggregates consisting of several particles were observed. Hence, we obtained the total number of individual particles. The particle adhesion was expressed as the average number of particles per mm^2, n_a.

2.3. Contact Angle Measurements

Advancing contact angles were measured by the Wilhelmy technique [8, 9] using an electrobalance (Model C-2000, Cahn Instruments Inc., USA) and an inchworm motor (Model 6000, Burleigh Instruments Inc., USA). A fiber was suspended from the arm of the electrobalance and a beaker containing the liquid was placed on a platform connected to the inchworm motor. Continuous weight recording was obtained during a fiber immersion–emersion cycle at precisely controlled rate of 0.3 mm/min [8, 9]. The cycle was repeated twice, and the reproducibility of the weight recording was checked. The advancing contact angle was calculated from the weight change at the moment when the fiber just touched the liquid surface by the Wilhelmy equation [8], using the effective perimeter of the fiber which was determined from the wetting force obtained in the fiber/pentane/air system [8].

2.4. Determination of the Surface Free Energies

Considering the acid–base interactions, the Young–Good–Girifalco–Fowkes equation can be written as [10]

$$\gamma_L(1+\cos\theta) = 2\left(\sqrt{\gamma_S^{LW}\gamma_S^{LW}} + \sqrt{\gamma^+\gamma_L^-} + \sqrt{\gamma_S^-\gamma_L^+}\right), \tag{1}$$

where θ is the contact angle of the liquid, γ is the surface free energy, the subscripts S and L refer to the solid and liquid, respectively, and the superscripts LW, +, and - refer to the Lifshitz–van der Waals component and Lewis acid and base parameters of the surface free energy, respectively. In order to solve γ_S^{LW}, γ_S^+, and γ_S^-, one needs to determine the contact angles with three probe liquids [11]. Therefore, the contact angles of water, diiodomethane, and ethylene glycol were measured by the Wilhelmy technique, and γ_S^{LW}, γ_S^+, and γ_S^- were obtained from θ and the published values of γ_L^{LW}, γ_L^+ and γ_L^- [11] using Equation (1).

The surface free energy components of the water/ethanol mixture were determined experimentally as follows. The total surface free energy of the liquid, γ_L, was measured by the pendant drop technique using a video contact angle system (VCA 2500, AST Products Inc., USA). Then the contact angle of the liquid on the silica fiber coated with paraffin as an apolar solid was determined by the Wilhelmy technique and the value of γ_L^{LW} was calculated from θ, γ_L, and γ_S^{LW} for paraffin [12] using Equation (1). And then the contact angle of the liquid on the APTES treated silica fiber as a bipolar solid [11] was measured by the Wilhelmy technique. Finally, γ_L^+ and γ_L^- were calculated from θ, γ_L, γ_L^{LW}, γ_S^{LW}, γ_S^+, and γ_S^- using Equation (1) and the relation [10]

$$\gamma_L = \gamma_L^{LW} + 2\sqrt{\gamma_L^+\gamma_L^-}\,. \tag{2}$$

2.5. ζ- potential Measurements

The ζ-potentials of the substrates and the particles in water/ethanol mixtures containing $1\cdot10^{-3}$ mol dm^{-3} of potassium chloride were determined by the streaming potential method and the microelectrophoresis, respectively. The ζ-potentials of the substrates were measured using a streaming potential apparatus (ZP-10B, Shimazu, Japan) [13]. The silica fibrous assemblies were plugged in a cylindrical poly(tetrafluoroethylene) cell and porous platinum electrodes were attached to the top and bottom faces of the plug. The ζ-potential was calculated from the slope of the linear relation between the pressure difference and the streaming potential by the Helmholtz–Smoluchowski equation [14].

The ζ-potentials of the polyethylene and nylon particles were determined using a microelectrophoresis apparatus (Zeecom, Microtec Co., Ltd, Japan). The ζ-potentials were calculated from the electrophoretic mobilities at the stationary levels in the cell by the Smoluchowski equation [15].

All experiments were performed at room temperature (25±2°C).

3. RESULTS AND DISCUSSION

3.1. Effect of Volume Ratio of n-alcohol on Particle Adhesion

The number of the particles adhering to the silica plate increased with the time and reached almost constant at 50–60 min. The surface coverage of the plate by the particles at 60 min was much smaller than unity, and most of particles on the plate adhered as primary particles. Hence the adhesion is considered to be dominated by the interaction between the particle and the plate. Therefore, the number of particles adhering at 60 min, n_a, was regarded as an apparent equilibrium adhesion.

The numbers of polyethylene and nylon particles adhering to the silica plates, n_a, are shown in Figs. 1 and 2, respectively, as a function of ethanol concentration in water/ethanol mixtures. For both particles, the particle adhesion increased by the silanization of silica. In water, the n_a values for polyethylene particles on silanized silica plates were larger compared with those for nylon particles. The particle adhesion to silanized silica decreased with increasing ethanol concentration.

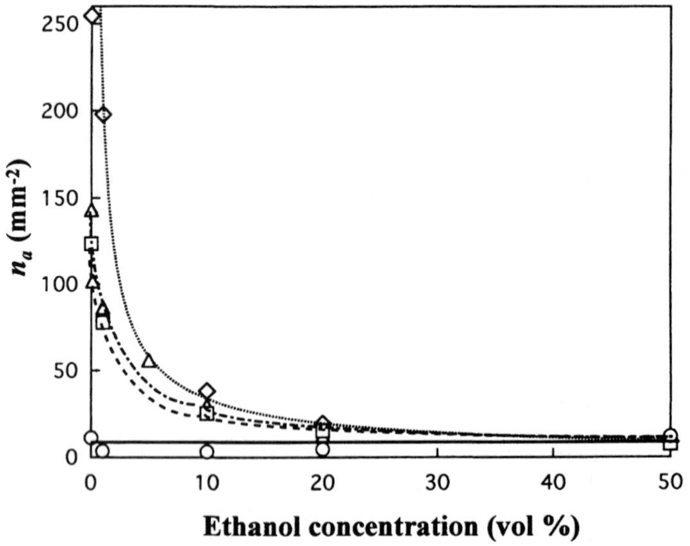

Figure 1. Effect of ethanol concentration in water/ethanol mixtures on the number of polyethylene particles adhering to the untreated (○), APTES treated (□), MTES treated (△), and FETMS treated (◇) silica plates at t = 60 min, n_a.

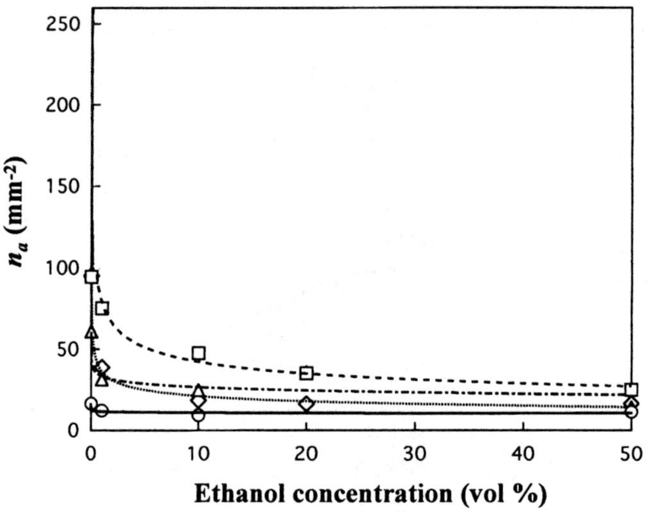

Figure 2. Effect of ethanol concentration in water/ethanol mixtures on the number of nylon particles adhering to the untreated (○), APTES treated (□), MTES treated (△), and FETMS treated (◇) silica plates at $t = 60$ min, n_a.

Figure 3 gives the effect of n-alcohol concentration on the number of polyethylene particles adhering to the MTES treated silica plate. For all alcohols, n_a decreased with increasing n-alcohol concentration. This tendency became noticeable with increasing number of carbon atoms of n-alcohol.

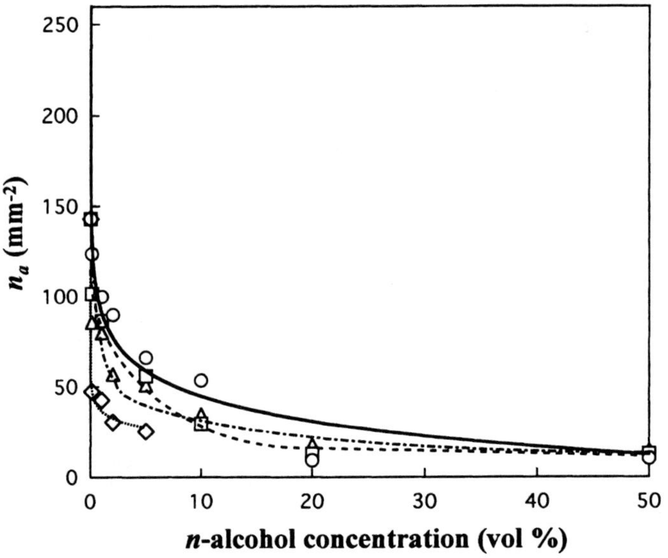

Figure 3. Effect of concentrations of methanol (○), ethanol (□), 1-propanol (△), and 1-butanol (◇) in water/n-alcohol mixtures on the number of polyethylene particles adhering to the MTES treated silica plate at $t = 60$ min, n_a.

3.2. Contact Angles and Surface Free Energies

The advancing contact angles of water on untreated silica, silanized silica, polyethylene, and nylon are shown in Table 1. After silanization of silica, the contact angles increased and hence the wettability decreased. The reproducibility of the weight recording indicates that the change in the structure of the silanized silica surface due to adsorption of the liquid into the porous silica layers did not take place.

Table 1 also lists the experimentally determined values of the surface free energy components for the substrates, the particles, and the water/ethanol mixture. The Lifshitz–van der Waals component and acid and base parameters were reduced by silanization of silica and by the addition of ethanol to water. The surface free energy components of polyethylene were smaller than those of nylon.

In the present study, γ_S^{LW} was calculated with a single LW-probe liquid, diiodomethane. Hence, the dispersive surface free energy, γ_S^d, was also evaluated by the wetting force measurements in the MTES treated silica/water/n-alkane systems [8]. The γ_S^d was 24.6 mJ m^{-2}, which was in fairly agreement with γ_S^{LW} (25.1 mJ m^{-2} in Table 1).

Table 1.
Advancing contact angle of water, θ_a, and Lifshitz–van der Waals component, γ^{LW}, and Lewis acid and base parameters, γ^+ and γ^-, of surface free energies, for substrates, particles and media

Material	θ_a (degree)	γ^{LW} (m·Jm^{-2})	γ^+ (mJ·m^{-2})	γ^- (mJ·m^{-2})
Untreated silica	30.6	33.3	0.88	50.8
APTES silica	75.5	30.0	0.81	9.30
MTES silica	85.3	25.1	1.08	4.49
FETMS silica	116.9	6.4	0.48	0.85
Polyethylene	110.7	29.7	0.2	0.2
Nylon	73.7	37.0	0.6	7.6
Water	—	21.8*	25.5*	25.5*
Water/ethanol (90/10)	—	18.4	14.8	21.7

* Taken from [11]

3.3. ζ-potentials

The effects of ethanol concentration on the ζ-potentials of the substrates and the particles are presented in Fig. 4. APTES treated silica was found to be positively charged, and the others negatively charged. In all cases, the absolute values of the ζ-potentials decreased with increasing ethanol concentration. Szczypa *et al.* [16, 17] reported that the absolute value of the surface charge of rutile and the pore glass decreased with concentration of n-alcohol in water/n-alcohol mixtures. They concluded that some fraction of surface hydroxyls underwent esterification by the adsorbed n-alcohol and these hydroxyls did not take part in the formation of the surface charge. The decreases in the ζ-potentials with respect to n-alcohol concentration in Fig. 4 may also be associated with the adsorption of n-alcohols as well as the change in ionization of the dissociated groups.

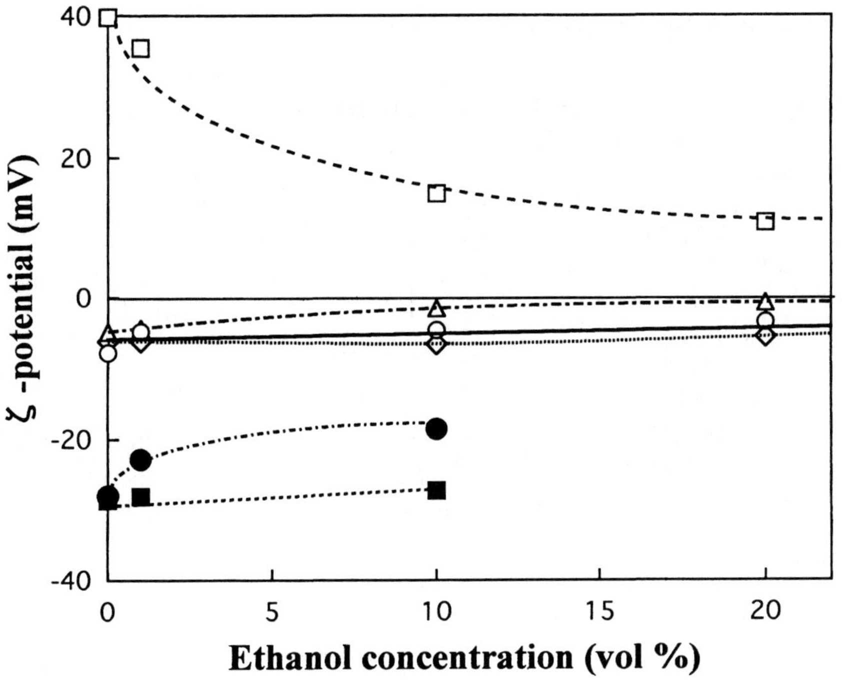

Figure 4. ζ-potentials of the untreated (○), APTES treated (□), MTES treated (△), and FETMS treated (◇) silica plates, and polyethylene (●) and nylon (■) particles as a function of ethanol concentration in water/ethanol mixtures.

3.4. Total Potential Energies of Interactions

According to the extended DLVO theory [3], the total potential energy of interaction between two macroscopic bodies, V_T, can be obtained as the sum of three contributions:

$$V_T = V_{EL} + V_{LW} + V_{AB}, \tag{3}$$

where V_{EL}, V_{LW}, and V_{AB} are the potential energies due to the electrical double layer, Lifshitz–van der Waals, and acid–base interactions, respectively. The terms V_{EL} and V_{LW} for a sphere-plate model are given by the heterocoagulation theory [5, 18]. The overall Hamaker constant, A_{132}, necessary for calculation of V_{LW}, is related to the Lifshitz–van der Waals component of the surface free energy, γ_{LW}, by the following Equation [5]:

$$A_{132} = -12\pi d_0^2 \left(2\sqrt{\gamma_1^{LW}\gamma_3^{LW}} + 2\sqrt{\gamma_2^{LW}\gamma_3^{LW}} - 2\sqrt{\gamma_1^{LW}\gamma_2^{LW}} - 2\gamma_3^{LW}\right), \tag{4}$$

where the subscripts 1, 2 and 3 refer to the plate, sphere and medium, respectively. d_0 is the minimum contact distance between the surfaces.

Concerning the Lewis acid–base interaction, the following expression has been proposed [5].

$$V_{AB} = 4\left[\sqrt{\gamma_3^+}\left(\sqrt{\gamma_1^-}+\sqrt{\gamma_2^-}-\sqrt{\gamma_3^-}\right)+\sqrt{\gamma_3^-}\left(\sqrt{\gamma_1^+}+\sqrt{\gamma_2^+}-\sqrt{\gamma_3^+}\right)\right.$$
$$\left.-\sqrt{\gamma_1^+\gamma_2^-}-\sqrt{\gamma_1^-\gamma_2^+}\right]\pi a\lambda \exp\left(\frac{d_0-d}{\lambda}\right), \qquad (5)$$

where γ^+ and γ^- the Lewis-acid and base parameters of the surface free energy, respectively, λ is the decay length of the liquid molecules [3, 5]. In water, the decay lengths of the hydration repulsion and the hydrophobic interaction were estimated to be 0.6–1.1 nm and 1–2 nm, respectively [19]. Using the experimentally determined surface free energy components and the ζ-potentials as approximated values for the outer Helmholtz plane potentials, the potential energies of interactions, V_{EL}, V_{LW}, V_{AB}, and V_T, were calculated as a function of the separation distance using Equations (3)–(5).

The results for the MTES treated silica plate/polyethylene particle/water system are given in Fig. 5. In Fig. 5, V_{AB} is found to be the major attractive energy at small distances of a few nanometers. Similar results were obtained for other systems.

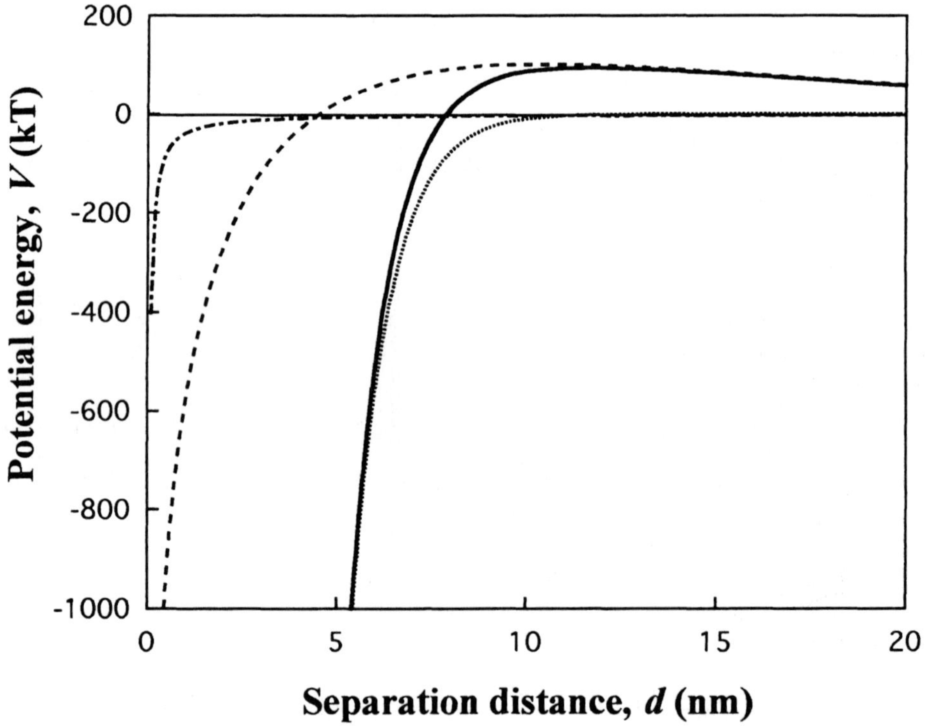

Figure 5. Calculated potential energies of the electrostatic (- - - - - -), Lifshitz–van der Waals (· - · - · - ·), and Lewis acid-base (··················), and total (———) interactions, as a function of the separation distance, *d*, in the MTES treated silica plate/polyethylene particle/water system.

Since n_a can be regarded as the apparent equilibrium particle adhesion, n_a was plotted against the potential energy at the closest approach, i.e. the depth of the primary minimum. Here, the closest approach, d_c, is assumed to be 1 nm [20].

Figure 6 shows the relationship between n_a and the depth of the primary minimum calculated using $d_0 = 0.158$ nm [2–4] and $\lambda = 1$ nm [3, 5]. A good correlation can be found: n_a increased with increasing depth of the primary minimum. The same tendency as the results in Fig. 6 was obtained using any value of 0.158–0.3 nm as d_0, 1–2 nm as λ, and 1–2 nm as d_c for the calculation of V_T. The maximum and the secondary minima of the potential energy did not correlate with n_a. As shown in Fig. 5, the depth of the primary minimum was largely dependent on the acid–base interaction energy. This indicates that the particle adhesion in the present systems is dominated by the acid–base interaction. In fact, the depth of the primary minimum calculated as the sum of only the electrostatic and Lifshitz–van der Waals interactions without the acid–base interaction did not correlate with n_a.

In the calculation of V_T, the changes in the Hamaker constant [13, 21, 22] and the surface free energy due to the adsorption of n-alcohol [23–28] were not taken into account although the change in the ζ-potential was considered. Nevertheless, a good correlation is found in Fig. 6, which indicates that the changes in the Hamaker constant and the surface free energy due to the adsorption of ethanol do not significantly influence the particle adhesion in the present study.

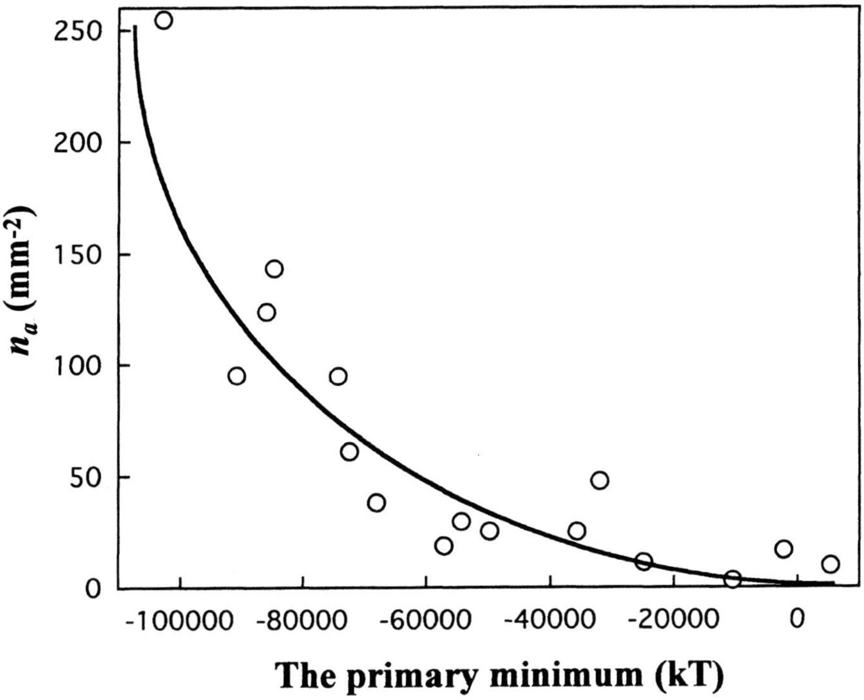

Figure. 6 Relationship between the number of particles adhering to the plate at t = 60 min, n_a, and the depth of the primary minimum assuming the closest approach as 1 nm.

3.5. Free Energy Change in Adhesion

From the above experimental findings, the particle adhesion in the present systems is expected to be related to the free energy change in adhesion.

The free energy change in adhesion, ΔG_a, is expressed as [29]

$$\Delta G_a = \gamma_{12} - \gamma_{13} - \gamma_{23}. \tag{6}$$

In this Equation, γ_{12}, γ_{13}, and γ_{23} are the interfacial free energies between the plate and the sphere, the plate and the medium, and the sphere and the medium, respectively. Substituting the surface free energy components into the interfacial free energies [3], Equation (6) can be rewritten as

$$\Delta G_a = -2\Big(\gamma_3^{LW} + \sqrt{\gamma_1^{LW}\gamma_2^{LW}} - \sqrt{\gamma_2^{LW}\gamma_3^{LW}} - \sqrt{\gamma_1^{LW}\gamma_3^{LW}} + 2\sqrt{\gamma_3^{+}\gamma_3^{-}} + \sqrt{\gamma_1^{+}\gamma_2^{-}} + \sqrt{\gamma_2^{+}\gamma_1^{-}} - \sqrt{\gamma_2^{+}\gamma_3^{-}} - \sqrt{\gamma_3^{+}\gamma_2^{-}} - \sqrt{\gamma_1^{+}\gamma_3^{-}} - \sqrt{\gamma_3^{+}\gamma_1^{-}}\Big). \tag{7}$$

The relationship between the particle adhesion, n_a, and the calculated ΔG_a is shown in Fig. 7. The particle adhesion was found to increase with increasing negative value of ΔG_a.

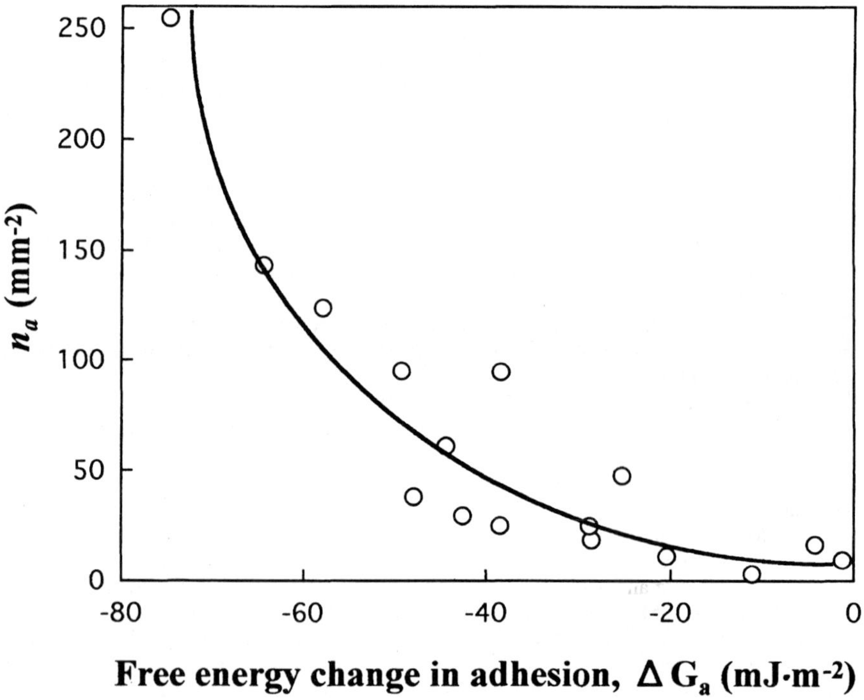

Figure 7. Relationship between the number of particles adhering to the plate at t = 60 min, n_a, and the free energy change in adhesion, ΔG_a.

4. CONCLUSIONS

The particle adhesion to silica plates in water/n-alcohol mixtures was found to increase as a result of the silanization of silica. The particle adhesion to silanized silica decreased with increasing volume ratio of n-alcohol. The Lifshitz–van der Waals component and acid and base parameters of the surface free energy of silica and water reduced by silanization of silica and by the addition of ethanol to water, respectively. The total potential energy curves which were calculated as the sum of the electrical double layer, Lifshitz–van der Waals, and acid-base interactions explained well the particle adhesion in the present systems. The particle adhesion was also correlated with the free energy change in adhesion. Thus, the acid-base interaction was found to play an important role in particle adhesion.

Acknowledgements

We wish to express our gratitude to the Ministry of Education of Japan for a Grant-in-Aid for Scientific Research and to the Space Utilization Research Promotion from the Japan Forum to carry out this work. We also would like to thank Miss Sima Kinosita, Miss Eriko Fukui and Miss Yuki Suzuki for their experimental assistance.

REFERENCES

1. F. M. Fowkes, *J. Adhesion Sci. Technol.* **1**, 7–27 (1987).
2. C. J. van Oss, M. K. Chaudhury and R. J. Good, *Chem. Rev.* **88**, 927–941 (1988).
3. C. J. van Oss, R. F. Giese and P. M. Costanzo, *Clays Clay Minerals.* **38**, 151–159 (1990).
4. P. Vera, V. Gallardo, J. Salcedo and A. V. Delgado, *J. Colloid Interface Sci.* **177**, 553–560 (1996).
5. J. D. G. Duran, A. Ontiveros, A. V. Delgado, F. Gonzalez-Caballero and E. Chibowski, *J. Adhesion Sci. Technol.* **10**, 847–868 (1996).
6. M. Tagawa, Y. Nakata and K. Gotoh, *J. Oil Chem. Soc., Japan* **38**, 88–93 (1989).
7. K. Gotoh, M. Iriya, A. Mitui and M. Tagawa, *Colloid Polym. Sci.* **261**, 805–811 (1983).
8. M. Tagawa, K. Gotoh, A. Yasukawa and M. Ikuta, *Colloid Polym. Sci.* **268**, 589–594 (1990).
9. M. Tagawa, A. Yasukawa, K. Gotoh, M. Tagawa, N. Ohmae and M. Umeno, *J. Adhesion Sci. Technol.* **6**, 763–776 (1992).
10. C. J. van Oss, R. J. Good, and M. K. Chaudhury, *Langmuir* **4**, 884–891 (1988).
11. R. J. Good, in: *Contact Angle, Wettability and Adhesion*, K. L. Mittal (Ed.), pp. 3–36. VSP, Utrecht, The Netherlands (1993).
12. J. R. Dann, *J. Colloid Interface Sci.* **32**, 302–320 (1970).
13. K. Gotoh, E. Takahashi, M. Maekawa and M. Tagawa, *J. Adhesion Sci. Technol.* **8**, 211–222 (1994).
14. J. Th. G. Overbeek, in: *Colloid Science* Vol. 1, H. R. Kruyt (Ed.), Ch. V, p. 195, Elsevier, New York (1952).
15. H. Ohshima, in: *Electrical Phenomena at Interfaces*, 2nd edn, H. Ohshima and K. Furusawa (Eds.), Ch 2, p. 21. Marcel Dekker, New York (1998).
16. J. Szczypa, L. Wasowska and M. Kosmulski, *J. Colloid Interface Sci.* **126**, 592 (1988).
17. J. Szczypa, I. Kajdewicz and M. Kosmulski, *J. Colloid Interface Sci.* **137**, 157–162 (1990).
18. K. Gotoh, T. Inoue and M. Tagawa, *Colloid Polym. Sci.* **262**, 982–989 (1984).
19. J. N. Israelachivili, *Intermolecular and Surface Forces*, 2nd edn, Ch. 13, p. 277. Academic Press, New York (1992).
20. G. Frens and J. Th. G. Overbeek, *J. Colloid Interface Sci.* **38**, 376–387 (1972).
21. K. Gotoh, T. Inoue, J. Inada and M. Tagawa, *J. Dispersion Sci. Technol.* **19**, 475–491 (1998).
22. M. J. Vold, *J. Colloid Sci.* **16**, 1–12 (1961).
23. W.-D. Einicke, U. Messow and R. Schollner, *J. Colloid Interface Sci.* **122**, 280–282 (1988).
24. M. L. Gee, T. W. Healy and L. R. White, *J. Colloid Interface Sci.* **133**, 514–516 (1989).

25. I. Regdon, Z. Kiraly, I. Dekany and G. Lagaly, *Colloid Polym. Sci.* **272**, 1129–1135 (1994).
26. T. Permien and G. Lagaly, *Colloid Polym. Sci.* **272**, 1306–1312 (1994).
27. I. Dekany, A. Farkas, I. Regdon, E. Klumpp, H. D. Narres, and M. J. Schwuger, *Colloid Polym. Sci.* **274**, 981–988 (1996).
28. G. M. Forland, T. Rahman, H. Hoiland and K. J. Borve, *J. Colloid Interface Sci.* **182**, 348–355 (1996).
29. K. Gotoh, C. Kako, T. Yoshimura and M. Tagawa, *J. Adhesion Sci. Technol,* **7,** 467–479 (1993).

Acid-Base Interactions, Vol. 2, pp. 551–580
Ed. K.L. Mittal

Effects of acid-base interactions on carbon/polycarbonate composite interfacial adhesion

RONALD E. ALLRED[1,*] and SHELDON P. WESSON[2]
[1]*Adherent Technologies, Inc., 11208 Cochiti SE, Albuquerque, NM 87123*
[2]*Ynir Instrumentation Software, 203 Mandon Court, Princeton, NJ 08540*

Abstract—Interfacial bond strength is often a performance limiting factor of carbon fiber reinforced thermoplastic composites. This limitation is due, in part, to the unreactive nature of engineering thermoplastics and to the inadequacy of commercially applied carbon fiber surface treatments. Radio frequency glow discharge plasmas are an effective means of modifying carbon fiber surface acid/base characteristics. CO_2 plasmas are especially effective compared to electro-oxidative treatments for tailoring carbon fiber surfaces as shown by titrations, Electron Spectroscopy for Chemical Analysis (ESCA), wetting, and Inverse Gas Chromatography (IGC) measurements. Plasma modified fiber, with both basic and acidic groups, showed increased adhesion with polycarbonate. We hypothesize that increased adhesion is caused by ester interchange reactions between the plasma formed surface groups and the polycarbonate chains. Use of an esterification catalyst with CO_2 plasma treated fibers resulted in further increases in interfacial adhesion. Increased interfacial adhesion also resulted in improved moisture resistance of carbon/polycarbonate composites. It appears that the primary achievement of surface treatments was to improve adhesion by increasing formal covalent bond density rather than acid/base adduct formation at the interface. The principle of promoting electron donor/acceptor interaction across the interface guided development of the surface treatments, however, and was very useful for developing diagnostic procedures such as wetting and IGC.

Keywords: Carbon; polycarbonate; interface; adhesion; plasma treatment; surface characterization; acid/base interactions.

1. INTRODUCTION

Engineering thermoplastics offer many attractive features as matrices for high performance composites. Among these features are high impact strength, good solvent resistance, the capability of being re-formed, and the potential to be produced very rapidly and economically. Among the problems that limit the use of thermoplastic matrix composites, poor fiber/matrix adhesion is one of the most difficult [1–4]. Polycarbonate, in particular, is very difficult to bond to carbon fibers [1, 2, 5, 6]. The primary reason for poor fiber/matrix adhesion in these systems is the lack of reactivity of thermoplastic polymers. Reactive curing polymers such as epoxies offer straightforward mechanisms for chemical bond formation to fiber surfaces if the appropriate sur-

* To whom correspondence should be addressed. E-mail: atiadmin@flash.net

face chemistry can be attained on the fiber [7]. Thermoplastics offer no such bonding mechanisms.

Approaches to improving interfacial bonding with thermoplastic polymers can be based on three fundamental adhesion mechanisms or combinations of these: (1) mechanical interlocking; (2) acid/base interactions; or (3) chemical bond formation through reactive intermediates.

Mechanical interlocking cannot be used to improve adhesion with carbon fibers without extensively degrading desirable bulk fiber properties because of their surface sensitivity. Acid/base bonding requires that one bulk material be acidic and the other basic [8–10]. To optimize bond strength, the strengths of the acid and base must be matched as well [8–10]. Engineering thermoplastic polymers, in general, are weak bases. As such, weakly acidic fiber surfaces are required to optimize this bonding mechanism. Chemical bond formation is more difficult to achieve and, when possible, is based upon a particular polymer chemistry.

Surface treatments for carbon and graphite fibers are generally based upon an electrooxidative process that oxidizes the fiber surface and removes weakly bonded material formed during the carbonization cycle [11, 12]. These treatments have been optimized for bonding to epoxy resins, and display adequate adhesion with most epoxy based systems. Electrooxidative treatments form a wide range of acidic and basic oxygen containing moieties on carbon surfaces including ethers, hydroxyls, lactones, ketones, carboxylic acids, and carbonates [13]. The electrooxidative process cannot be controlled to emphasize only a particular surface species. For this reason, it cannot be tailored for forming acid/base or chemical bonds with thermoplastics.

Glow discharge plasmas offer an effective means to control modification of surface chemistry [14–19]. Several patents have been filed on plasma treatment for surface chemical modification of carbon fibers [20–24]. These patents entail the use of oxidizing gases (NO_2, O_2, SO_2, CO_2, NO, H_2O, and air) or ammonia. Evans and Kuwana have studied plasma amination of bulk graphite [25]. More recently, Donnet and coworkers studied plasma modification of carbon fibers [26] as did Loh [27]. Allred and coworkers studied plasma interactions with organic and carbon fibers [28–35] and its effects on adhesion. These studies show that plasmas can effectively oxidize or aminate carbon surfaces.

In the plasma process, far ultraviolet radiation and ion impact volatilize surface species from the solid substrate creating surface active sites. Weak boundary layers and contaminants are also removed by these mechanisms. The gas phase contains numerous reactive species including ions, electrons, free radicals and molecules with a variety of electronically excited states. Surfaces may be functionalized with a narrow distribution of moieties centered around a desired chemical group by properly choosing gases and plasma operating parameters.

Plasma processes offer a technique to tailor carbon fiber surface chemistry to interact with thermoplastic resins without sacrificing fiber bulk properties [28–35]. The objectives of this study were to demonstrate the potential for improving adhesion in carbon/polycarbonate composites using plasma surface modification techniques. Plasmas for forming acidic, basic, and neutral surface groups were examined for their effects on intermediate modulus carbon fibers. Surface chemistry of the plasma treated fibers was extensively characterized. Unidirectional composite laminates were fabricated and tested in transverse tension and transverse flexure.

Results show that plasmas are a versatile technique for controlled modification of carbon fiber surfaces. Fibers with strongly acidic surfaces degrade polycarbonate chains in the interphase, presumably by a hydrolysis reaction. This reaction degrades interfacial adhesion by forming a weak boundary layer. Weaker acids and bases showed adhesion improvements over electrooxidized fiber surfaces. By the use of an appropriate esterification catalyst, these reactions produced further adhesion gains to values twice those of the controls.

2. EXPERIMENTAL

2.1. Materials

Two types of intermediate modulus carbon fibers were examined: unsized IM6 roving (12K, Lot X663-3W) and unsized IM7 roving (12K, Lot 725-3C), both from Hexcel, Inc. Both IM6 and IM7 are continuous, high-performance, intermediate-modulus, PAN-based round fibers. Polycarbonate pellets were obtained from General Electric's Pittsfield, Massachusetts Technical Service Center. This is GE's standard viscosity grade of neat, unfilled (with the exception of a ppm-level heat stabilizer) polycarbonate designated 101 (high viscosity, Lot LN4091). The exact viscosity of the polycarbonate is unknown. Laboratory chemicals and gases were standard research grade.

2.2. Plasma Treatments

Three gases were selected for initial plasma treatments. Ammonia (NH_3) was chosen for its ability to attach amine groups to reactive sites on the edges of the carbon crystallites thereby increasing the basic character of the fiber surface [29]. A CO_2 plasma was chosen to attach carboxyl and hydroxyl groups to the fiber surface and increase the acidic character of the fiber surface [35]. Finally, hydrogen was selected to act as a reducing agent on the electrochemically oxidized surface of the as-produced IM fibers.

All batch plasma treatments were performed in a Tegal Plasmod reactor. Gas pressures were set with the RF power off. The power settings were tuned using a Heathkit antenna tuner, Model 2060-A. The reaction chamber was equilibrated for five minutes with the plasma gas after sample introduction into the chamber, the power was turned on to produce the plasma for the desired treatment time, and then the chamber was equilibrated for five minutes with the plasma gas before sample removal.

Ammonia plasma was produced with 45–50 W of RF power using 800 mTorr gas pressure in the batch reactor chamber. A 60 second treatment time was standard. Some exposure times were varied to study the effects on the surface chemistry, however. The CO_2 plasma was run at 15-20 W RF power. Exposure time was 120 seconds. The IM fibers were treated for 15 seconds and 60 seconds in a 1000 mTorr hydrogen plasma at 45–50 W RF power.

Continuous fiber treatments were conducted in a custom plasma reactor with a 3 m reaction zone termed the continuous fiber treatment reactor (CFTR). Plasma operating parameters were optimized for each gas based upon surface chemical characterization results in conjunction with the plasma emission spectrum. Plasma parameters (power, pressure, flow rate) of subsequent treatments were adjusted so that the emission spectrum duplicated that of the optimized condition to provide reproducibility between multiple runs of the same chemistry. Nominal operating parameters were 500 mTorr pressure, 60 sec and 100 W power. Prepreg was fabricated on a drum winder in-line with the CFTR.

Several methods were used to characterize the IM6 and IM7 carbon fibers. Single fiber tensile tests were conducted on untreated, unsized fibers to compare the bulk properties of the two fibers. Single fiber tests were also conducted on plasma treated fibers to ensure that no degradation of the fibers occurred during the plasma treatment process. ESCA was used to determine fiber surface chemical content and the local environments of various chemical groups. Surface energetics was characterized by single filament wetting and IGC. These data were supplemented by nonaqueous titrations and ion exchange procedures.

2.3. Fiber Bulk Properties

Tests were conducted on single fibers of IM6 and IM7 to evaluate tensile strength and tensile modulus. The test method was similar to the ASTM Test D 3379-75. Twenty-five fibers of each type were tested to ensure statistically significant results.

A United FM-20 electromechanical tester with a 500 g (1.1 pound) load cell was used to conduct the tests. Single fibers were glued at the ends onto paper cards. Fibers were mounted in the grips of the testing machine, then the paper was cut and folded back to free the fiber for the test. The machine was configured to use a 10.25 cm (4.035 in.) fiber gage length. Rubber faced grips were used to eliminate stress concentrations at the grips. The fiber was carefully aligned to be vertical in the grips. The plotter was zeroed at the start of each test. After initial experiments showed that elongation rate did not affect results over the range examined, a crosshead speed of 0.5 cm (0.2 in.) per minute was chosen for the tests. Elongation data were measured with the crosshead position indicator of the testing machine.

The tensile strength data were subjected to a Weibull analysis, which describes a defect-dependent distribution. The Weibull equation is shown below as Equation 1 [36]. The Weibull scale parameter corresponds to the mean of a Gaussian distribution. The shape parameter describes the slope of the distribution. A higher shape parameter indicates a smaller defect size distribution.

$$W(X) = 1 - \exp[-L(X/X_0)^{\rho}] \tag{1}$$

where

L = fiber length
X = fiber tensile strength
X_0 = Weibull scale parameter
ρ = Weibull shape parameter.

2.4. ESCA Analysis

Unsized and treated IM6 and IM7 fibers were analyzed by ESCA. Survey spectra were taken to examine the as-received surface chemical stoichiometry. High resolution spectra were taken of significant peaks seen in the survey scans to determine possible surface bonding states. The analyses were run at Rocky Mountain Laboratories, Inc. (Golden, Colorado) using a Surface Sciences SSX-100 spectrometer with an Al Kα source. C_{1s} high resolution peaks were averaged over 10 to 15 scans using a spot size of 300 μm^2 with no flood gun. O_{1s} peaks were averaged over 30 scans. A Gaussian curve fitting routine was used to resolve high resolution photopeaks into components based on binding energy references from model compounds [37]. The spectra were not charge referenced.

2.5. Wetting Experiments

Fiber surface energy characterization was performed using the Wilhelmy technique [38]. Carbon monofilaments were glued to metal hooks and suspended vertically from the working arm of a Cahn 2000 microbalance, while a precision elevator raised and lowered the liquid surface along 0.5 mm of fiber. A computer periodically recorded the change in apparent mass caused by wetting forces at the three-phase boundary using the probe liquids listed in Table 1. Advancing work of adhesion, W_{adv}, the dispersive component of solid surface energy, γ^d, and the acid/base component of fiber wettability, $W^{a/b}$, were computed by methods previously reported [30–33].

Many of the substrate-liquid combinations spread in the receding mode and were close to spreading in the advancing mode. Since non-spreading liquids with known acid/base character, (ethylene glycol and formamide) were carefully selected for having high surface tensions, the number of additional liquid probes that could be employed is severely limited. This indicates that, while wetting studies are a valuable method for surface characterization, further characterization of these surfaces requires another technique, such as inverse gas chromatography or microcalorimetry.

Table 1.
Surface Energy (mJ/m^2) Characteristics of Probe Liquids [33]

	γ^d	$\gamma^{a/b}$	γ^T
Methylene Iodide (CH_2I_2)	48.5	1.7	50.2
Tricresyl Phosphate (TCP)	39.2	1.7	40.9
Formamide ($HCONH_2$)	39.5	18.7	58.2
Ethylene Glycol ($HOCH_2CH_2OH$)	29.3	19.0	48.3

γ^d = dispersive component of surface energy
$\gamma^{a/b}$ = acid/base component of surface energy
γ^T = total surface energy

2.6. Inverse Gas Chromatography

A complementary method for determining surface energetics is IGC. The theory of IGC characterization of fiber surfaces is presented in Ref. 30.

Columns were constructed from 0.25 in. stainless steel tubing (5.2 mm inner diameter) with passivated inner walls. Fifty-five centimeter long sections were fitted with Swagelok nuts and ferrules, and then weighed. Fiber yarns were wound into 60 cm loops that were drawn into the tubes and cut flush at the ends; the columns were then weighed again. Sample weights varied from 2 to 3 g.

A Hewlett-Packard 5880A gas chromatograph fitted with a flame ionization detector and a data acquisition output board was connected to a computer containing a 16 bit Data Translation Series 2801 A/D conversion card. Routines written in ASYST programming language collected detector voltages at frequencies between 1 and 10 Hz.

Both injector and detector temperatures were maintained at 150°C. Nitrogen carrier flow rates were measured with a Gasmet flow meter and were maintained between 22 and 24 ml/min. Measured flow rates **F** were corrected for column temperature using Eqn. 2.

$$F = \mathbf{F}(T_{col}/T_{amb}) \tag{2}$$

There was no pressure drop across these columns, obviating the need for the James/Martin correction [39]. Gas holdup times were measured with 10 μl injections of methane. Hamilton 7101NCH syringes were used to inject 2 μl volumes of pentane (neutral probe) and t-butylamine (Lewis base); injections of t-butanol (Lewis acid) were restricted to 0.5 μl in order to obtain sharp peak fronts. Three chromatograms were obtained on each column with each probe at 30°C after conditioning the column overnight at 50°C.

Chromatograms were obtained at the minimum attenuation *a* required to collect the detector signal at peak maximum. Zero attenuation was implemented after the signal fell below 1/*a* of detector saturation, permitting desorption at low coverage to be measured with maximum resolution [30]. Chromatograms using pentane, t-butylamine, and t-butanol were collected over periods of 1000, 6000, and 2000 seconds, respectively. An additional two hours elapsed between measurements with t-butylamine to allow chemisorbed probe to clear from the column.

2.7. Titration of Acid and Base Surface Sites

A procedure was developed to determine acid and base groups on the surfaces of treated carbon fibers indirectly by nonaqueous titration.

The solution composition for the blank and fiber samples was as follows: 89 parts benzene or toluene, 10 parts propylene carbonate, and 1 part of a dilute solution (approximately 10^{-2} molar in benzene or toluene) of either diphenyl phosphate or 1,3-diphenylguanidine, depending on whether the back titration was for base or acid groups on the fiber.

The solutions used in titrating were 1:20 dilutions of the above stock solutions of diphenyl phosphate or 1,3-diphenylguanidine. Two drops of the indicator, a 0.1%

w/w solution of Bromophthalein Magenta E [40] in benzene, was used per titration. One gram of fibers was used for each set of three sample titrations.

The blanks were made up of 89 ml benzene or toluene, 10 ml propylene carbonate, and 1.0 ml acid or base stock solution. Two 5-ml aliquots were titrated to the indicator endpoint. One gram of fibers was then suspended in the remaining 90 ml and allowed to react with the solution for approximately half an hour. Three 5-ml aliquots were then titrated to the indicator endpoint. The difference between the blank average and sample average was the amount required to react with surface groups on the fiber.

The difference between this procedure and others for determining surface groups [41, 42] is that, rather than reacting the indicator directly with the surface groups and then titrating a slurry of the solid, the surface groups are reacted with acid or base and aliquots of the solution are then back titrated using an indicator.

2.8. Determination of Surface Amine Sites

The effectiveness of the ammonia plasma to attach amine sites to the fiber surfaces was measured using the Ponceau 2R Dye ion exchange procedure, which is described in Appendix A of Reference 28. This procedure is a quantitative technique for determining the number of amine sites per square nanometer of fiber surface.

2.9. Prepreg and Laminate Fabrication

Prepreg with IM7 and polycarbonate was fabricated by solution impregnation of 12k tows using the CFTR. Polycarbonate prepreg was made from a 20% w/w solution in methylene chloride. Solvent volatility required extensive modifications to the impregnation end of the CFTR to produce good quality prepreg. The impregnation chamber was moved closer to the winding drum so that the impregnated tows were still wet and could be consolidated into uniform prepreg. A controlled atmosphere chamber was constructed between the plasma exit and the impregnation chamber to protect plasma treated surface before impregnation.

Impregnated fibers were wrapped on a 15.2 cm diameter aluminum mandrel with a spacing of 4.33 wraps per cm. Upon removal from the mandrel, the prepreg sheets were laid flat between two aluminum plates for drying. Each sheet was separated by fiber glass cloth and perforated Teflon release fabric. The prepreg was then dried at 204°C (400°F) for 60 minutes to drive off any residual solvents that could adversely affect the composite.

Unidirectional composite laminates were fabricated by vacuum bag hot press molding in a U-shaped mold with interior dimensions of 10.2 ×15.2 cm. The mold was designed with bolt-on sides and ends to restrain the composite during pressing and for easy removal of the consolidated part. A thermocouple mounted in the mold wall inside the vacuum bag controlled the thermal cure cycle.

Vacuum was applied at room temperature under contact pressure to remove air from the laminate stack. A pressure of 3.8 MPa (550 psi) was applied at room temperature and the mold heated to 271°C at a rate of 1°C/minute. After a 15 minute hold at 271°C, the temperature was reduced to 246°C and held 60 minutes before cooling to room temperature in 30 minutes. This process produced uniform, void free laminates with the polycarbonate system. Several other investigators have shown that molding temperatures in excess of 265°C are required to achieve adequate fiber wetting [5, 6, 43].

Samples cut from each laminate were cross-sectioned normal to the fiber direction, polished, and examined by optical microscopy before testing. Void content and fiber distribution were the physical characteristics of primary concern. Only laminates displaying low void content (< 1%) and having a uniform fiber distribution were mechanically tested.

2.10. Mechanical Testing

Two distinct tests were performed on unidirectional plates to evaluate their relative mechanical strength. Both tests were conducted on a United FM-20 electromechanical tension and compression machine with a 2,220 N (500 pound) capacity tension/compression load cell.

The first test was the transverse tensile test. Specimens were nominally 1.27 cm wide and 5 cm in length between the bonded end tabs. Strain was monitored with an extensometer mounted directly to the specimen. Each specimen was tested at a constant rate of 0.02 cm/min until failure occurred.

The second test was the transverse (90°) three-point flexural test. An aspect ratio (L/d) of 10 was used which was large enough to induce tensile failure rather than shear failure. Specimens were nominally 1.27 cm wide and spanned 1.5 cm between supports. A compression load was applied at a constant rate of 0.01 cm/min until failure occurred.

Sample fracture surfaces were examined with scanning electron microscopy using an Hitachi 850 field emission microscope.

3. RESULTS AND DISCUSSION

Because little is known about plasma-carbon fiber interactions, the objective of this study was to explore these interactions and how plasma modified surfaces affected interfacial adhesion with polycarbonate. The following sections discuss surface chemical and bulk effects, resin characterization, and composite mechanical properties.

The difficulties in analyzing the surface chemistry of fibers dictate that numerous complimentary techniques be used to determine plasma-surface interactions. Plasmas for forming acidic, basic, and neutral surface groups were examined for their effects on intermediate modulus carbon fibers. Titrations, wetting, IGC, ion exchange, and ESCA were used to characterize fiber surface chemistry and surface energy.

3.1. Surface Analysis of Amine Site Concentration

3.1.1. Batch Reactor. Amine sites on fiber surfaces can be determined quantitatively using an ion exchange procedure with an azo dye, Ponceau 2R [28]. This procedure has been used extensively on Kevlar fibers. These data provide a useful comparison for determining the effect of ammonia plasmas on carbon fibers. IM6 and IM7 carbon fibers were subjected to ammonia plasmas for varying lengths of time and the resulting amine site density was determined using the procedure detailed in Reference 28 modified to account for carbon surface area and density.

Kinetics curves for surface amination were produced by analyzing ammonia plasma treated surfaces exposed in the batch reactor for various time intervals. The maximum concentration (0.45 NH_2/nm^2) on the IM6 surface appears to occur at 60–90 seconds. The IM7 fibers show a maximum of 0.72 NH_2/nm^2 at 90 seconds. A 60-second plasma treatment was chosen for in-depth analyses. With a 60 second ammonia plasma treatment, the IM7 was still more reactive than the IM6 displaying an amine site density of 0.62 NH_2/nm^2.

Based upon ESCA data from unsized IM6 and IM7 that will be presented later, the amination results can be interpreted as amine radicals in the plasma replacing surface oxygen groups at edge sites and defects rather than reacting with the fiber surface in general. It is known that the electrooxidative treatments used by most carbon fiber manufacturers attach oxygen groups only at the edges of crystallites on the carbon surface [13]. Amine concentration follows the initial surface oxygen content and is much lower than that seen on organic surfaces [28], indicating an oxygen replacement mechanism.

Emission spectra were taken of ammonia plasmas produced in the Tegal Plasmod batch reactor. They showed that nitrogen was present in the plasma at all power settings and all pressures. Furthermore, nitrogen seemed to be the primary residual gas present in the reactor chamber at the lowest obtainable vacuum, about 5–10 mTorr. This result can be explained by quenching of the ammonia emission by the nitrogen.

3.1.2. Continuous Reactor. The emission spectra obtained for the CFTR plasmas are quite different from those obtained from the Tegal batch reactor. The CFTR employs a plasma chamber that is 3 m long compared to 0.15 m for the batch reactor. The aspect ratios (length/diameter) for the two reactors are 120 for the continuous and 1.5 for the batch. The long plasma tube of the continuous reactor is split into three separate chambers, each with an antenna and a bisecting baffle. The ammonia plasma emission spectrum for the CFTR is given in Figure 1.

The behavior of the ammonia present in the CFTR was also very different from that obtained in the small batch reactor. At low mass flow rates, a dramatic depletion of ammonia available in the plasma was detected, showing a negative gradient from the first antenna to the third antenna. The ammonia was being consumed to produce nitrogen and hydrogen, producing a positive gradient in nitrogen concentration from the first to the last antenna.

Initial kinetics studies on both Kevlar and IM7 showed a marked decrease in the amine density found on the fiber surfaces (see Table 2). We postulate that the nitrogen 337 nm line was so strong along the fiber axis that ultraviolet induced desorption of the active surface sites led to a marked decrease in the amine site density.

A series of experiments were performed to determine the conditions that enhanced the nitrogen 337 nm line, and, conversely, reduced the population of this state. A matrix of available plasma conditions was investigated. Power, pressure, mass flow rate, and the residual gas species were varied. When argon was introduced to the drying chamber at the beginning of the reactor, it reduced the overall amount of nitrogen present in the plasma chamber. All fibers were subsequently exposed to an argon drying before treatment.

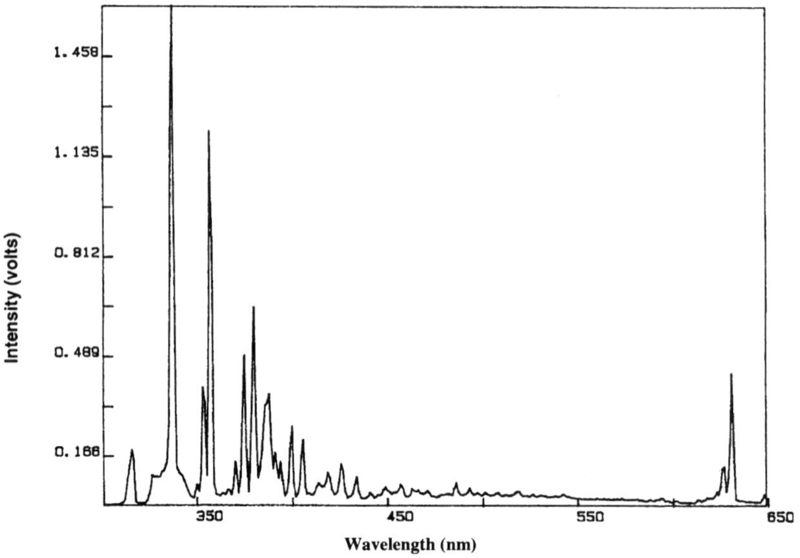

Figure 1. Emission spectrum of ammonia plasma at 4000 mTorr pressure.

Table 2.
Amination Kinetics from CFTR Reactor[a]

Fiber Type	Treatment Times, sec	Power, W	CFTR, NH_2/nm^2	Batch Reactor, NH_2/nm^2
Kevlar 49	0	0	0.236±0.052	0.19 [28]
Kevlar 49	30	50	0.505	NCA[b]
Kevlar 49	60	50	0.635	NCA
Kevlar 49	30	100	0.176	0.84[c] [28]
Kevlar 49	60	100	0.672	0.75[c] [28]
IM7	0	0	0.157±0.026	0.17
IM7	30	100	0.181	0.56[c]
IM7	60	100	0.170	0.62[c]
IM7	30	150	0.137	0.39[d]
IM7	60	150	0.208	0.41[d]

[a] All samples were run in CFTR at MKS setting 4.0 and 800 mTorr ammonia pressure
[b] No comparison available
[c] 25 W RF power
[d] 50 W RF power

An increase in RF wattage while maintaining all other variables at a constant value led to an increase in the nitrogen lasing line and a decrease in the ammonia continuum. Increasing the pressure while maintaining a consistent mass flow rate led to an increase in the ammonia band intensity and a decrease in the nitrogen 337 nm line.

Increasing the mass flow of ammonia while maintaining the same total pressure in the plasma chamber led to a decrease in the nitrogen 337 nm line as compared to the remaining nitrogen lines and also to an increase in the band intensity in the visible region of the emission spectra that arises from varying electronic excited states in the ammonia plasma.

The gas flow inside the CFTR chamber was then modified so that the gas was split into three equal mass flow rates with one third of the flow inletting into the chamber at the beginning of each of the three antennas. This modification further reduced the nitrogen 337 nm fluorescence and increased the band intensity of the visible ammonia fluorescence. Table 3 gives plasma amination results for IM7 fibers treated at higher pressures using the modified gas flow system in the CFTR.

These data indicate that the higher gas flow rate and higher gas pressures provide higher amine site density on the IM7 fiber surface. The maximum amine site density produced (0.31 sites/nm^2) represents a significant increase over the results presented in Table 2. It also appears that higher power levels slightly increase the fiber surface amine concentration. Nevertheless, these results are only 50% of the amine site concentrations attained on IM7 fibers in the batch reactor.

Table 3.
Amination Kinetics for IM7 Carbon Fiber in Modified CFTR

Gas Pressure, mTorr	Treatment Time, sec	RF Power, W	Amine Site Conc., NH_2/nm^2
Control	0	Control	0.10
750-1000	30	100	0.22
750-1000	60	100	0.27
1500-2000	60	150	0.23
750-1000	60	150	0.23
1500-2000	60	150	0.31

All samples run in CFTR at MKS setting 7.

3.2. Surface Energy Effects

Changes in surface energy often accompany plasma treatments, especially on low energy surfaces. Surface energy changes can often be used to characterize plasma effects when other techniques are unavailable. Single filament wetting measurements and IGC were used to monitor plasma-induced surface energy changes.

3.2.1. Single Filament Wetting. Wetting measurements were used to determine the acid/base component of wetting work of adhesion for the various fiber treatments.

Work of adhesion measurements with methylene iodide and tricresyl phosphate (TCP) were made on the control fibers as an independent check on the determination of the dispersive component of surface energy. A comparison of work of adhesion measurements in methylene iodide and TCP is shown in Table 4 along with the calculated dispersive surface energy components for both IM6 and IM7 fibers. Also shown for comparison are the dispersion force components calculated from a Fowkes plot [8, 9].

With the exception of the dried IM7 control fiber, the data are consistent for dispersive energy components. This agreement allows the non-dispersive or acid/base component of work of adhesion to be calculated from wetting measurements with polar liquids. It is also noted in Table 5 that the dispersive component of fiber surface energy is not substantially changed by the various plasma treatments. This result is expected, since the dispersive energy component is a function of atomic number and density only, and should not change with these types of surface modifications. Other authors have reported large changes in the dispersive energy component after similar surface treatments [26].

Methylene iodide was used as a dispersive wetting liquid because of its high surface energy, 50.2 mJ/m^2, which mitigates spreading. Using the methylene iodide-determined dispersive surface energy component, the work of adhesion with the other probe liquids can be divided into dispersive, W^d, and non-dispersive (polar), $W^{a/b}$, components [30–33]. This information for formamide (basic) and ethylene glycol (acidic) wetting fluids is presented in Table 5.

Ammonia plasma treatment changed the IM6 and IM7 surfaces only minimally. It is likely that existing basic sites on the fiber surface were replaced by amine groups,

Table 4.
Wetting Results on Carbon Fibers Using Dispersive Liquids

Fiber	Treatment	CH_2I_2		TCP		Fowkes Plot [8, 9]
		W_{adv} (mJ/m^2)	γ^d (mJ/m^2)	W_{adv} (mJ/m^2)	γ^d (mJ/m^2)	γ^d (mJ/m^2)
IM7	CONTROL (dried, 50°C)	78.2	30.5	76.6	35.9	33.5
IM7	CONTROL	90.5	40.8	77.3	36.5	38.5
IM6	CONTROL	84.4	35.5	74.2	33.7	34.5
IM7	NH_3 plasma (60 sec)	87.7	38.3			
IM6	NH_3 plasma (60 sec)	81.9	33.4			
IM7	CO_2 plasma (120 sec)	81.0	32.7			
IM6	CO_2 plasma (120 sec)	86.7	37.4			
IM7	H_2 plasma (15 sec)	81.6	33.2			
IM7	H_2 plasma (60 sec)	87.6	38.2			

W_{adv} – advancing work of adhesion

γ^d – dispersive component of surface energy

Table 5.
Surface Energy Components of Carbon Fibers

Fiber Type	Plasma Treatment	Formamide ($HCONH_2$)			Ethylene Glycol ($HOCH_2CH_2OH$)		
		W_{adv} (mJ/m^2)	W^d (mJ/m^2)	$W^{a/b}$ (mJ/m^2)	W_{adv} (mJ/m^2)	W^d (mJ/m^2)	$W^{a/b}$ (mJ/m^2)
IM7	CONTROL (dried, 50°C)	101.4	69.4	32.0	79.1	59.7	19.4
IM7	CONTROL	103.3	80.3	23.0	90.8	69.2	21.6
IM6	CONTROL	90.6	74.9	15.7	84.3	64.5	19.8
IM7	NH_3 plasma (60 sec)	108.4	77.8	30.6	90.9	67.0	23.9
IM6	NH_3 plasma (60 sec)	86.8	72.6	14.2	75.8	62.6	13.2
IM7	CO_2 plasma (120 sec)	117.7	71.9	45.8	90.9	61.9	29.0
IM6	CO_2 plasma (120 sec)	113.8	76.9	36.9	85.9	66.2	19.7
IM7	H_2 plasma (15 sec)	101.4	72.4	29.0	83.2	62.4	20.8
IM7	H_2 plasma (60 sec)	98.7	77.7	21.0	79.7	66.9	12.8

but the ammonia plasma is not strong enough to change the basicity of the fiber surface significantly. Ammonia plasma did lower the acidity of the IM6 surface, but it also lowered the basic component of work of adhesion. Despite the weak plasma-carbon fiber interactions seen with ammonia, it can be an effective treatment for replacing nonreactive sites with amine groups.

The increase in the IM7 surface acidity from exposure to ammonia plasma probably resulted from reaction of residual activated surface sites (free radicals) when the sample was removed from the plasma chamber and exposed to ambient oxygen and water. Hydroxyl or carboxyl groups could have formed, and these would increase the surface acidic group concentration. The same reactions did not occur on the IM6 treated surface, indicating its lower population of reactive sites at crystallite edges.

Hydrogen plasma was used to reduce the amount of oxidation on the surface of IM7. Treatment times were varied to determine the most effective schedule. As seen in Table 5, hydrogen plasma does not reduce the oxidized fiber surface, but rather causes a small increase in surface acidity. Again, reaction of residual active sites with oxygen or water is the mechanism likely to increase surface acidity. Only by treating a surface in a high vacuum and observing it *in situ* can the true plasma modified surface be characterized. The CFTR is designed to coat the fiber surfaces with resin before exposure to the atmosphere, which allows additional control over the interface chemistry attained in composite laminates.

CO_2 plasma had the most significant effect on both fibers. These effects are shown by the dramatic increase in the work of adhesion and the acid/base components resulting from wetting measurements with formamide. Both fibers show twofold increase in acid/base component over the manufacturer's treatments. It should be noted that the work of adhesion values measured for the CO_2 plasma treated fibers are very close to the work of cohesion of formamide, where the liquid spreads on the fiber surface. The measured value for W_{adv} is thus a minimum value.

The dramatic increase in acidity seen for the CO_2 plasma treated IM6 and IM7 fibers cannot be attributed solely to substitution onto existing oxidized sites at the crystallite edges or defects of the fiber surfaces. A more reasonable explanation is that the plasma activated CO_2 is creating reactive sites on the basal plane of the carbon fiber [35].

3.3. Inverse Gas Chromatography

Diffuse profiles of chromatograms were transformed into isotherms at every point on the chromatogram between peak maximum and the end of the tail using algorithms described in Reference 30. Points near peak maximum were discarded as being unduly influenced by axial and radial diffusion in the column, and other extraneous effects. Isotherms are displayed in log/log format to emphasize differences at low adsorbate pressure and coverage. The geometric specific surface $\Sigma_{geo} = 4/\rho d$ was computed for each carbon fiber sample using the density ρ supplied by the manufacturer, and the diameter, d, found from wetting measurements. For IM6, the specific surface area is 0.44 m^2/g. IGC isotherms of untreated IM6 and IM6 treated with the CO_2 plasma are presented in Figure 2.

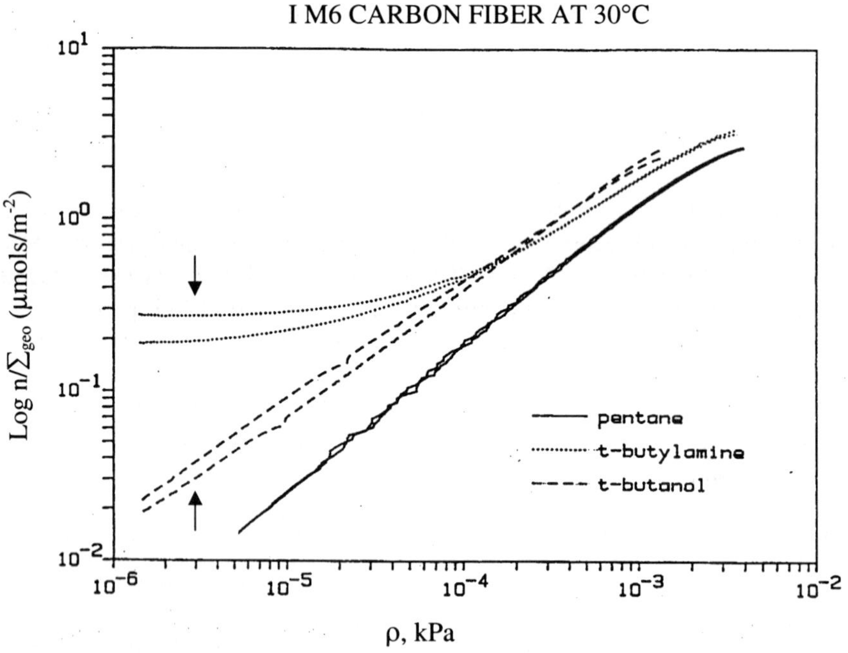

Figure 2. Pentane (lines), t-butanol (dashes), and t-butylamine (dots) on unsized IM6 fiber at 30°C. Arrows denote the CO_2 plasma treated surface.

Probe vapors used to examine the fiber surface were pentane (neutral probe), t-butanol (Lewis acid), and t-butylamine (Lewis base). These gases are analogous in purpose to methylene iodide, formamide, and ethylene glycol for the solid/liquid wetting measurements. Adsorption volumes are expressed as n/Σ_{geo} (μmoles/m^2) to normalize the isotherms for all experimental conditions except differences in adsorption energies. 10 μmoles/m^2 is approximately one monolayer for these adsorbates. Equilibrium pressure is given in kPa.

The shapes of these isotherms are controlled by surface heterogeneity. This provides a means of deducing the various modes of interfacial attraction presented by the substrates. Isotherms of pentane on unsized IM6 before and after plasma treatment are congruent. Pentane interacts only by dispersion force attraction, as does methylene iodide. The registry of these isotherms is predicted by the similarity in γ^d obtained from wetting measurements. Pentane isotherms serve as a reference to compare the effect of additional modes of solid/vapor interactions with acidic and basic probes on the shape of those isotherms.

As seen in Figure 2, isotherms for adsorption of t-butanol (the Lewis acid) on unsized and plasma treated IM6 substrates are in registry within experimental precision. The adsorption capacity of carbon fiber for the Lewis acid is greater than for pentane across the entire pressure range, indicating that dispersion force interaction is augmented with weak attraction to basic sites on the substrate. These observations corroborate trends evinced by wetting with ethylene glycol.

Isotherms for t-butylamine adsorption are in registry above 10^{-4} kPa; the plasma treated surface has a greater adsorptive capacity for the basic probe at lower pressures (Figure 2). The adsorption capacity of both specimens for the Lewis base is more than an order of magnitude greater than for acidic or neutral probes at pressures below 10^{-4} kPa. The main inferences are that both IM6 surfaces are acidic and plasma treatment increases the acid site concentration imparted during manufacture, which is in accord with the analysis of XPS spectra, titrations, and wetting measurements.

The surface area fraction of high energy sites can be quantified by applying the CAEDMON algorithm to adsorption data [30]. Isotherms for t-butylamine on unsized IM6 and unsized IM6 subjected to CO_2 plasma treatment were analyzed to obtain adsorption potential histograms. Details of the CAEDMON analysis are presented in Reference 30. Resultant adsorption energy histograms are given in Figure 3.

Oxidation during manufacture of carbon fibers creates heterogeneity, manifested as a broader range of low to medium energy sites on the unsized IM6 surface. These weak acidic sites increase the adsorptive capacity of IM6 for the basic probe at pressures above 10^{-4} kPa, compared with the dispersion force interaction shown by pentane isotherms (see Figure 2). The most important phenomenon is the appearance of a surface component with a very high adsorptive energy, constituting less than 5% of the specific surface area. Plasma treatment does not change the distribution of weak acidic sites appreciably (t-butylamine isotherms are congruent above 10^{-4} kPa), but increases the surface concentration of high energy sites to about 7% of the specific surface area. These high energy patches are similar in magnitude and proportion to C_{1s} peak components at 288.7 eV.

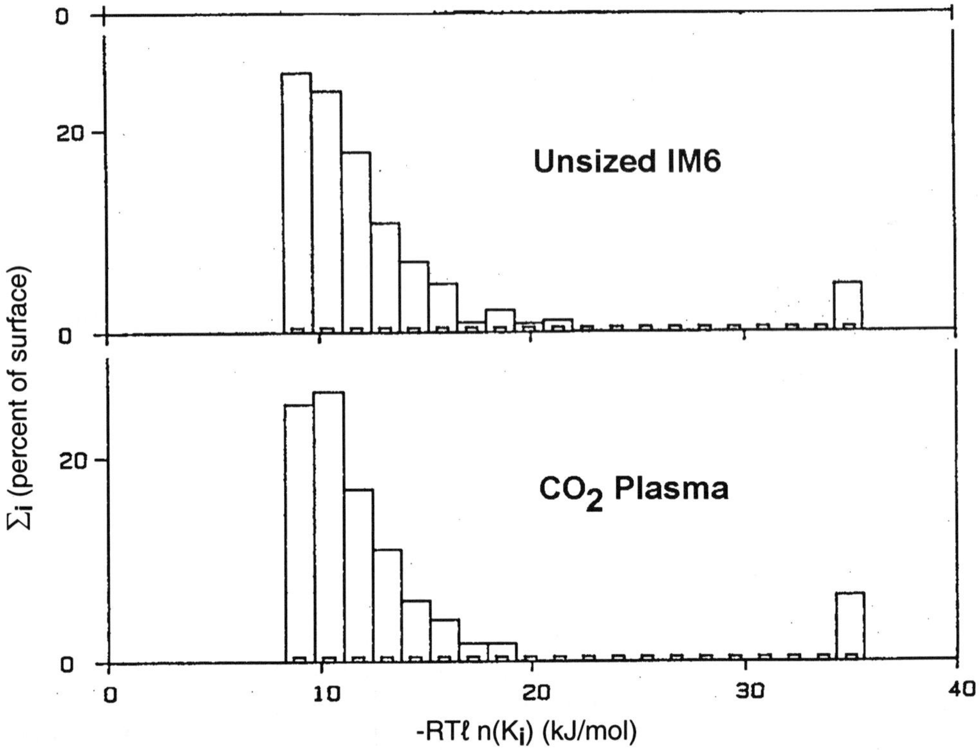

Figure 3. Adsorptive energy histograms for t-butylamine isotherms at 30°C.

The range of adsorptive energies supplied to the algorithm affects the shape of the histogram and the computed value of specific surface area. Σ is sensitive to the position of the lowest energy patch; the low energy bound is adjusted to bring Σ to within 10% of Σ_{geo} in the present analysis. The exact location of the highest energy site is unimportant when the low pressure sector of the isotherm corresponds to saturation coverage on the highest energy patch [30].

The last increment of t-butylamine takes several hours to traverse the column when IM6 surfaces are eluted. A long elution time is evidence of very strong solid/vapor interaction. Adsorption energy of the highest energy site is indeterminate in principle under these circumstances, because this patch saturates at pressures much lower than the operating range of the detector. The position of the highest energy patch is arbitrarily assigned as the lowest energy that brings the computed fit into registry with the low pressure sector of the isotherm. The patch area thus obtained is an upper limit, and a good approximation because it will not vary much if its energy assignment is modified slightly (the fit will fall out of registry with the data if the energy assignment is too high). Isotherms for the Lewis base on IM6 surfaces show zero slope at pressures below 10^{-5} kPa, which indicates that the adsorbate complexes with chemisorptive sites on the substrate.

3.4. Surface Chemical Characterization

The effects of plasma treatment on fiber surface chemistry were examined using non-aqueous titrations and Electron Spectroscopy for Chemical Analysis (ESCA). Based upon the wetting results, hydrogen plasmas were omitted from the surface chemical characterization test matrix.

3.4.1. Acid/Base Titrations. Table 6 lists the titration values for total acidic sites on IM6 and IM7 fibers. Total basic sites are given in Table 7. In addition to control and plasma treated surfaces, three drying temperatures (50°C, 100°C, and 150°C) were investigated in order to determine the thermal stability of the surface species. Thermal exposures occur during some experimental techniques, e.g. IGC, and during composite fabrication procedures. The lability of the surface functional groups can, thus, affect chemical bonding potential between the fiber and the resin.

Table 6.
Acidic Site Titration Data (sites/nm^2)

		Plasma Treatment Conditions		
Fiber Type	Drying Temperature (1 hr at 1 Torr)	Control	NH_3 (60 sec)	CO_2 plasma (120 sec)
IM7	Undried	0.87	nd*	nd
IM6	50°C	1.26	0.23	2.09
IM7	50°C	0.71	0.27	1.95
IM6	100°C	0.99	0.090	1.01
IM7	100°C	0.64	0.57	1.29
IM6	150°C	0.29	0.46	0.53
IM7	150°C	0.36	0.29	0.86

* not determined.

Table 7.
Basic Site Titration Data (sites/nm^2)

		Plasma Treatment Conditions		
Fiber Type	Drying Temperature (1 hr at 1 Torr)	Control	NH_3 (60 sec)	CO_2 plasma (120 sec)
IM7	Undried	0.94	nd*	nd
IM6	50°C	0.52	1.16	none detected
IM7	50°C	0.67	1.12	0.40
IM6	100°C	0.74	0.76	nd
IM7	100°C	0.54	0.74	0.24
IM6	150°C	0.60	0.15	nd
IM7	150°C	0.75	0.36	0.085

* not determined.

Figure 4 graphically depicts both the acid and base titration data for fibers dried at 50°C. In general, the trends of the data are very much as expected. Basic sites increase and acidic sites decrease following the ammonia (basic) plasma treatment, and acidic sites increase and basic sites decrease following the CO_2 plasma treatment. In addition, the two fibers give very similar values except for two conditions: the acidic sites on the dried control fibers and the basic sites on the CO_2 plasma treated fibers. The higher concentration of acidic sites on the IM6 control is unexpected considering the wetting data.

Acidic sites increased dramatically when the fiber surface was treated with the CO_2 plasma. These data are consistent with the previously presented argument that sites are being created on the graphite basal plane.

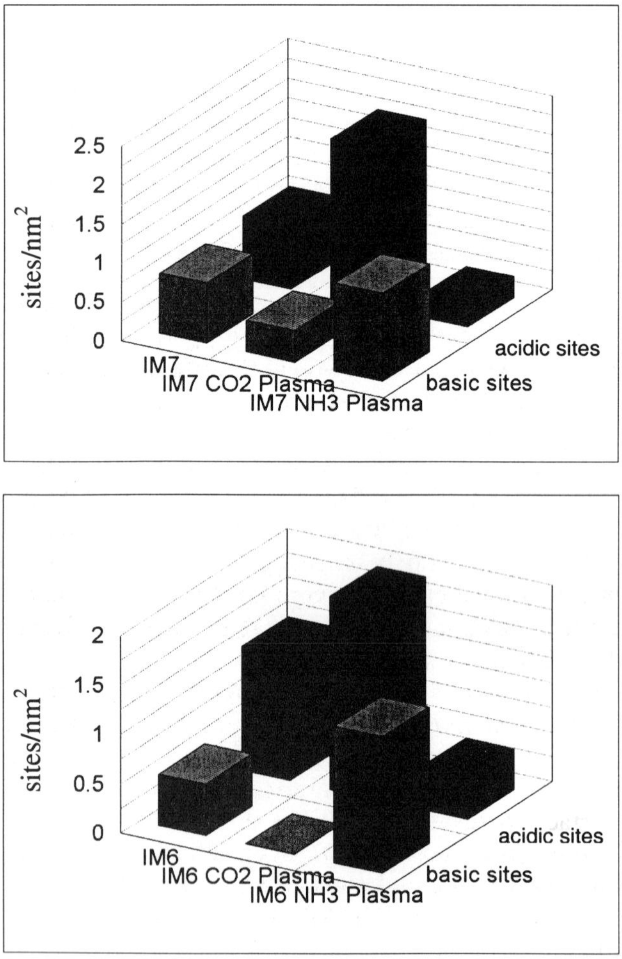

Figure 4. Acidic and basic site concentrations per nm^2 determined by nonaqueous titrations of carbon fiber surfaces dried at 50°C.

Effects of different drying temperatures shown in Table 6 suggest that the acidic sites on both fibers are thermally labile. The concentration of acidic sites on the control graphite fibers decreased consistently with the increase in drying temperature, as did the sites on the CO_2 plasma treated samples. The inconsistent data for acidic sites on the ammonia treated fiber surfaces is not easily understandable. One explanation for the increase in acidic site concentration on the ammonia plasma treated IM6 fibers dried at 150°C and the IM7 fibers dried at 100°C could be that, as the amine sites are lost with increasing temperature, available oxygen combines with the resultant free radicals to form hydroperoxides that rearrange to various acidic species.

As shown in Table 7, the basic sites on control fibers were not significantly affected by heating, except for the initial drop from undried IM7. These consistent values are particularly valuable in that they offer a baseline to evaluate the modifications of the fiber surfaces induced by the plasma treatments. The increased drying temperatures result in obvious decreases in the basic sites on all the treated fibers. The decreases in basic sites on the ammonia plasma treated fibers support the idea that basic sites on the control fiber surfaces have been replaced with amines that desorb from the surface upon heating, but that do not necessarily change the overall basicity of the fiber surface, as determined by wetting experiments.

The main conclusion from the titration results is that the fiber surfaces were altered by the plasma treatments. The data also indicate that both acidic and basic functional groups were attached, as expected, to the fiber surfaces.

3.4.2. Electron Spectroscopy for Chemical Analysis. Survey or wide scan spectra for unsized IM6 and IM7 fibers showed carbon, oxygen, and nitrogen as the primary surface components. A trace of Al is seen on both surfaces and traces of Ca and Si are seen on the IM7. These are likely contaminants from handling or processing.

Since the elements observed other than carbon, nitrogen, and oxygen do not constitute a significant component of the fiber, it is useful to ignore them for the purpose of examining plasma surface effects. The surface stoichiometries normalized to 100 percent C plus O plus N are given in Table 8 for unsized IM6 and IM7 and for CO_2 plasma treated IM6. The effects of ammonia plasma on IM6 fibers are discussed in Reference 29. Only a minimal change in core level ESCA spectra could be observed after plasma treatment. This minimal change is due to the significant nitrogen content of as-produced IM6 fibers.

The IM6 results agree well with an earlier analysis [29]. IM7 shows a much higher oxygen and nitrogen content than the IM6. Nitrogen is expected as a residue from the polyacrylonitrile starting material. Oxygen results from the oxidation treatment used by the fiber manufacturer to remove weak surface material on the as-carbonized fibers and to increase surface energy to aid wetting by potential matrix resins. The higher nitrogen content would imply that a lower carbonization temperature was used for the IM7. A higher surface oxygen content would be an indication that a more extensive oxidation treatment was applied to the IM7 than was applied to the IM6 or that the IM7 surface had a larger percentage of edge sites.

High resolution C_{1s} and O_{1s} photopeaks for unsized IM6, unsized IM7, and CO_2 plasma treated IM6 are given in Figures 5–7. Possible binding states of carbon and oxygen on the fiber surfaces were determined by resolving the C_{1s} and O_{1s} high

Table 8.
ESCA Elemental Composition (%) of Unsized Carbon Fibers

Fiber	C	O	N
IM6 Unsized	89.9	7.9	2.2
IM7 Unsized	83.2	11.5	3.9
IM6 CO_2 plasma (120 sec)	83.4	13.8	2.8

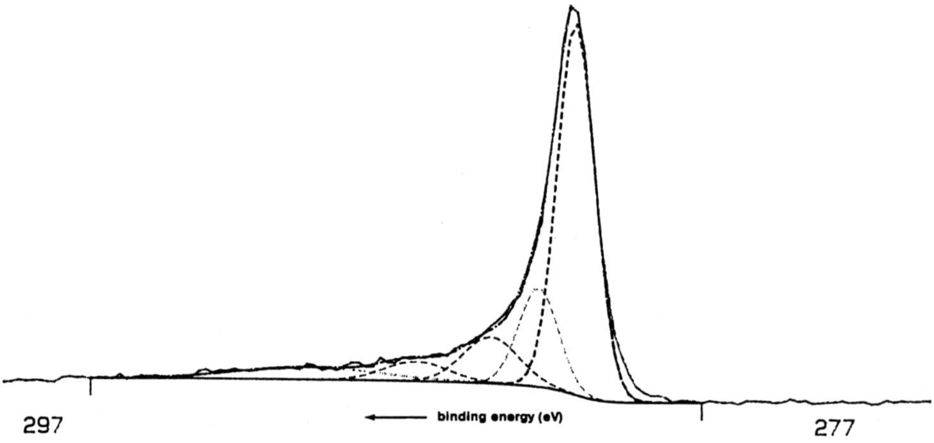

Figure 5. High resolution C_{1s} spectrum for unsized IM6.

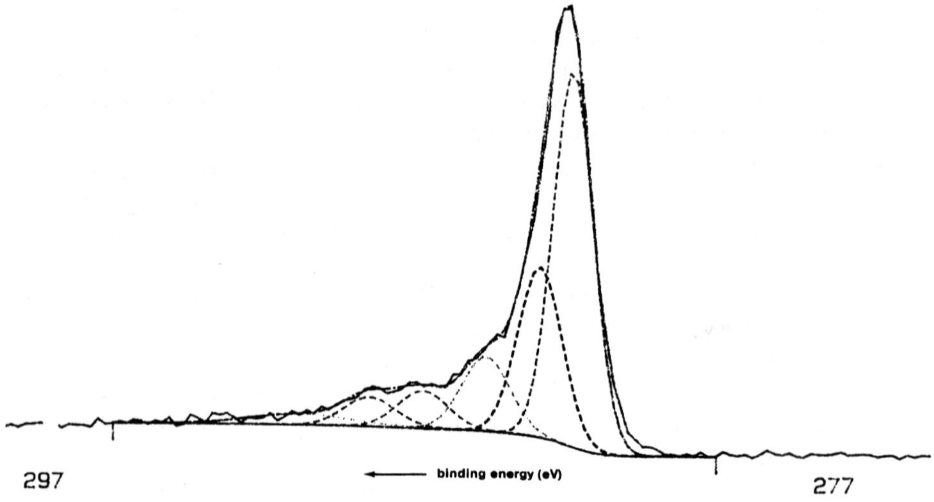

Figure 6. High resolution C_{1s} spectrum for unsized IM7.

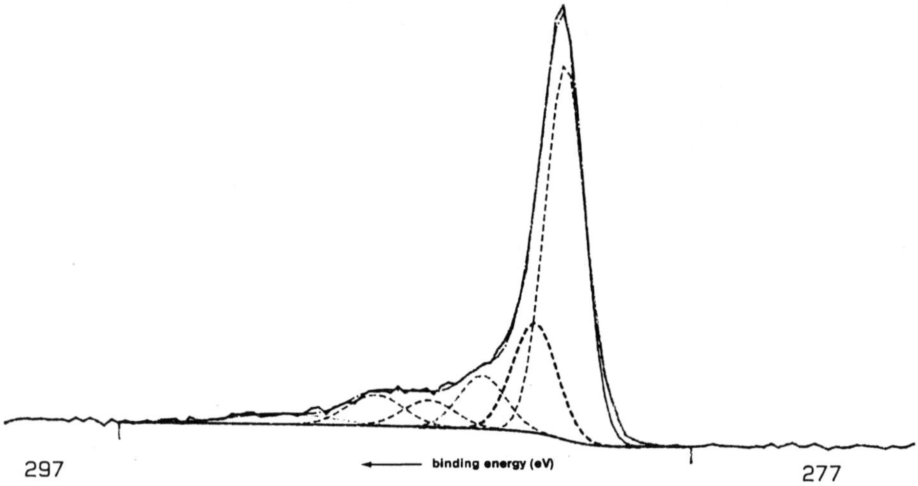

Figure 7. High resolution C_{1s} spectrum for CO_2 plasma treated IM6.

resolution peaks into component peaks. Component peak binding energy shifts were assigned based upon known literature values [37]. These assignments are given in Table 9 along with the component peak values. For purposes of discussion, the total C_{ls} and O_{ls} is normalized to 100 percent for both fibers.

Examination of the O_{ls} results given in Table 9 shows that the majority of the increased oxygen on the IM7 control surface is in the form of ketone or amide groups (binding energy = 531.6 eV). A similar conclusion is drawn from the C_{ls} data. IM6 shows 4.9 percent of the doubly bound carbon-oxygen, while the IM7 has 9.4 percent. Such groups would be expected to increase the fiber surface energy but not be available to form strong interfacial bonds. These results show that the electrochemical oxidation treatments used by most fiber manufacturers yield a mixture of surface species that is weighted toward carbonyls. An advantage of the glow discharge plasma treatment is that the surface species can be weighted towards acids, hydroxyls, etc. by proper selection of gases and operating parameters.

CO_2 plasma treatment increases the surface oxygen concentration by augmenting the surface oxygen species present at crystallite edges and defects, and possibly by attacking the graphite crystallite basal planes. Components of high resolution C_{ls} and O_{ls} envelopes are presented in Table 9. On the unsized IM6 fiber, the substantial peak at 290.7 eV is partially attributed to carbonate functionality, which also contributes to the peak at 286.4 eV. The appearance of a peak component at 288.7 eV on unsized IM6 shows that oxidation during fiber manufacture implants some carboxyl functionality. The component at 288.7 eV is substantially larger on plasma treated surfaces; this observation, together with the appearance of an O_{ls} peak component at 533.8 eV, shows that additional oxygen incorporated by plasma treatment is present primarily as carboxyl groups. The peak at 290.7 eV is substantially diminished, but the component at 286.4 eV is unchanged, suggesting that CO_2 plasma converts carbonate structures to hydroxyl as well as carboxyl functionalities.

Table 9.
Components of C_{1s} and O_{1s} ESCA Photopeaks

Binding Energy, eV	Species	Unsized IM7	Unsized IM6	Unsized IM6, CO_2 plasma (120 sec)
C_{1s}:		percent of C_{1s} photopeak		
284.6	Primary carbon	49.7	59.9	56.5
285.3	ß-carbon in conjugated systems	24.7	16.6	18.4
286.4	hydroxyl, ether, ester, single bond	11.6	10.9	9.4
287.6	ketone, amide	5.9	2.7	5.2
288.7	carboxyl, ester	4.8	2.6	6.5
290.7	aromatic shakeup, carbonate	3.3	7.3	4.0
O_{1s}:		percent of O_{1s} photopeak		
531.6	ketone, amide	60.3	52.0	38.8
532.8	carbonxyl, ester double bond; hydroxyl, ether single bond	19.8	48.0 nr*	45.5
533.8	carboxyl, ester single bond	16.5	48.0 nr*	15.7
535.2	carbonate	3.3	48.0 nr*	nd**

*not resolved
**not detected.

3.5. Fiber Bulk Properties

Single fiber tensile tests were conducted on as-received and CO_2 plasma treated IM6 and IM7 carbon fibers to evaluate tensile strength and tensile modulus. Weibull plots of tensile strength versus failure probability were constructed for each fiber tested. The data fell close to a straight line in Weibull coordinates, which indicates that they can be described by a Weibull distribution. Weibull analysis results are given in Table 10.

The data in Table 10 show that the tensile strength of IM7 control is 12.5% higher than the IM6 control, and the tensile modulus is 19% higher. The IM7 has much more uniform properties as shown by the decrease in the range of tensile strengths measured and the increase in shape parameter from 3.8 for the IM6 to 7.5 for the IM7. The modulus coefficient of variation is also much lower for IM7 than IM6.

CO_2 plasma treatment of the IM7 changed the properties only slightly: the tensile strength increased by 2.2% and modulus decreased by 7.8%. The treated IM7 still had a 19% higher tensile strength and a 9.5% higher modulus than the CO_2 plasma treated IM6, however. The measured values agree with the values reported by Hexcel (Magna, UT) within approximately 10%. Considering the limited number of single fiber tests conducted, this varation constitutes excellent agreement and allows the conclusion that the CO_2 plasmas examined did not degrade the fiber longitudinal properties.

Table 10.
Weibull Scale and Shape Factors and Tensile Modulus

Fiber	Scale Factor	Shape Factor	Tensile Modulus
IM6, as received	4137 MPa (600 kpsi)	3.80	268 GPa (38.9 Mpsi)
IM6, CO_2 plasma (120 sec)	3999 MPa (580 kpsi)	5.90	269 GPa (39.0 Mpsi)
IM7, as received	4650 MPa (675 kpsi)	7.50	319 GPa (46.3 Mpsi)
IM7, CO_2 plasma (120 sec)	4757 MPa (690 kpsi)	4.50	294 GPa (42.7 Mpsi)

3.6. Polycarbonate Characterization

Surface energy components of the thermoplastic matrix material are necessary to evaluate wetting behavior on the carbon fibers and to predict acid/base interactions. Initially, samples of the resin in fiber form were drawn so that resin surface energy could be measured using the Wilhelmy technique. The first attempt was to spin or draw fibers from 10%–20% by weight solutions of the resins. These fibers, however, did not have the consistent diameters required to measure surface energy accurately.

A technique that proved successful for measurement of polymer surface energy used precut fibers of glass or carbon (IM7) coated with the polymer. Polymer coatings were produced by dipping the fibers into 5% polycarbonate solutions in methylene chloride.

The same mathematical treatments as used on wetting data from carbon fibers were applied to the polycarbonate wetting data. In Table 11, dispersive energy components from methylene iodide and tricresyl phosphate wetting data are compared. The two values show reasonably consistent agreement. Using the energy components from the methylene iodide wetting, the calculated dispersive and non-dispersive wetting components for formamide and ethylene glycol are presented in Table 12.

The wetting values show that the primary component of surface energy for the polycarbonate is dispersive, i.e., the acid/base wetting components for the wetting liquids are essentially zero. This result means that the approach to improving bonding between the fibers and resin should concentrate on the formation of covalent bonds, rather than acid/base attractions. Because the CO_2 plasma treatment opens bonding sites on the graphite crystallite basal planes, it is the first choice for improving interfacial bonding.

Table 11.
Dispersive Work of Adhesion Measurements on Polycarbonate[(a)]

(CH_2I_2)		TCP	
W_{adv} (mJ/m^2)	γ^d (mJ/m^2)	W_{adv} (mJ/m^2)	γ^d (mJ/m^2)
90.5	40.8	77.1	36.3

[(a)] Resin coated on glass fiber
W^{adv} - advancing work of adhesion
γ^d - dispersive component of surface energy

Table 12.
Work of Adhesion Components of Polycarbonate[(a)]

Formamide($HCONH_2$)			Ethylene Glycol ($HOCH_2CH_2OH$)		
W_{adv} (mJ/m^2)	W^d (mJ/m^2)	$W^{a/b}$ (mJ/m^2)	W_{adv} (mJ/m^2)	W^d (mJ/m^2)	$W^{a/b}$ (mJ/m^2)
78.3	80.3	-2.0	68.8	69.2	-0.4

[(a)] Resin coated on glass fiber

3.7. Composite Mechanical Properties

The effects of plasma modification on fiber-matrix adhesion were measured with two interface sensitive stress states on unidirectional laminates: transverse tension and transverse flexure. Before testing, each laminate was characterized with optical microscopy to ensure quality. Only plates with low void content and uniform fiber distribution were selected for testing. Initial mechanical test results are given in Tables 13 and 14.

An analysis of the test results in Tables 13 and 14 reveals several important points. Bending test values given in Table 13 show a nominal 29% improvement in properties for laminates fabricated with CO_2 plasma treated fibers and a 16% improvement for laminates fabricated with NH_3 treated fibers. Coefficients of variation for all the flexural data in Table 13 are unusually high, even for composites. The transverse tensile test results given in Table 14 do not indicate significantly improved properties over the control, however. This disagreement is believed to be related to the consistent failures near the grips. Transverse tensile tests are sensitive to small misalignments that can impose significant bending moments on the samples.

The results presented in Table 13 are somewhat surprising for several reasons. The CO_2 plasma was expected to provide improved adhesion from acid/base interactions and, perhaps, some chemical bond formation through an ester interchange type of mechanism. Ammonia plasma treated fiber should have a basic surface. An aminated surface should not promote any of the adhesion mechanisms one would expect with the acidic surfaces. These observations indicate that other phenomena are occurring besides the expected bonding mechanisms discussed above.

In an attempt to further increase adhesion, a series of plasmas designed to incorporate surface moieties with increasing acidic strength were applied to the IM7 fibers before fabricating into polycarbonate matrix composites. Transverse flexural results as a function of acidic strength are presented in Table 15.

The results of Table 15 clearly show that, rather than promoting bonding, stronger acidic treatments degrade interfacial adhesion in this system. It appears that the stronger acids are desorbing during composite processing and causing extensive chain scission of the polycarbonate, forming a low molecular weight layer at the interface that acts as a weak boundary layer. Both acid and base catalyzed hydrolysis of esters and polycarbonate specifically are well known [44]. These reactions have not been reported for solid state interactions; however, the high processing temperatures for polycarbonate would facilitate their occurence. Thus, it appears that both the acidic

Table 13.
Transverse Flexural Strength of Unidirectional IM7/Polycarbonate Laminates

Sample Type	$\bar{S}_{22}$ MPa (kpsi)	Std. Dev. MPa (kpsi)	C.V.[a] %	Range MPa (kpsi)	N[b]
Controls:					
1	91.0 (13.2)	18 (2.6)	20	63.4-108.0 (9.2-15.7)	14
2	103.0 (15.0)	22 (3.2)	22	64.1-140.0 (9.3-20.3)	11
3	77.1 (11.1)	20 (2.9)	26	63.9-120.9 (9.2-17.4)	8
Avg.	93.3 (13.5)	23 (3.3)	24	63.2-140.0 (9.2-20.3)	33
CO_2 Plasma (120 sec):					
1	126.0 (18.2)	21 (3.1)	17	91.7-148.0 (13.3-21.4)	11
2	108.4 (15.6)	18 (2.6)	17	78.5-123.0 (11.3-17.7)	5
Avg.	119.8 (17.4)	22 (3.2)	17	77.7-147.7 (11.2-21.4)	16
NH_3 Plasma (60 sec)					
1	127.0 (18.4)	14 (2.0)	11	109.0-148.0 (15.8-21.5)	10
2	84.8 (12.2)	10 (1.5)	13	71.6- 95.9 (10.3-13.8)	8
3	111.2 (16.0)	33 (4.8)	30	81.3-165.4 (11.7-23.8)	5
Avg.	108.1 (15.7)	26 (3.8)	24	71.3-164.3 (10.3-23.8)	23

[a] Coefficient of Variation
[b] Number of Specimens

Table 14.
Transverse Tensile Strength of Unidirectional IM7/Polycarbonate Laminates

Sample Type	$\bar{S}_{22}$ MPa (kpsi)	Std. Dev. MPa (kpsi)	C.V.[a] %	Range MPa (kpsi)	N[b]
Controls:					
1	62.7 (9.1)	9.0 (1.3)	14	51.7-73.1 (7.5-10.6)	4
2	63.8 (9.3)	1.3 (0.2)	2	57.1-65.5 (9.1- 9.5)	4
Plasma Treated:					
CO_2 (120 sec)	65.9 (9.6)	6.1 (0.9)	9	58.6-73.1 (8.5-10.6)	4
NH_3 (60 sec)	63.1 (9.2)	4.2 (0.6)	7	57.2-66.2 (8.3- 9.6)	4

[a] Coefficient of Variation
[b] Number of Specimens

and basic fiber surface treatments are degrading the polycarbonate chains near the interface by chain scission at the ester linkages. A portion of these scissioned chains will chemically bond to the fiber surface at acidic or basic sites. This mechanism provides a means to improve adhesion if the hydrolysis reaction can be controlled.

The proposed hydrolysis mechanism may account for the observations in the literature of varying adhesion between polycarbonate and different types of carbon fibers. Hinkley [1] has previously observed that widely different fracture toughnesses resulted between polycarbonate composites reinforced with carbon fibers from different manufacturers. He also reported that the polycarbonate matrix was degraded to lower molecular weight after contact with XAS carbon fibers (Hexcel, Magna, UT) as measured by gel permeation chromatography. Similar observations have been noted with glass filled and unfilled polycarbonate injection molding compounds. It was not shown that hydrolysis was occurring in those systems, but it was shown that the polymer degraded in mechanical properties and molecular weight.

These observations then lead to an approach for improving adhesion between polycarbonate and carbon fibers. The fiber surface should be functionalized with acidic or basic chemical moieties. Reaction between the polycarbonate and the fiber surface sites can then be facilitated at lower temperatures by the use of an appropriate esterification catalyst. To evaluate such an approach, a set of fibers was treated with the CO_2 plasma. A ferric acetyl acetonate catalyst was added to the polycarbonate-methylene chloride prepreg solution at a concentration of 0.01 wt %. Composite laminates were then prepared following the identical process used for previous laminates. Transverse flexural results for the acidic surface, catalyzed system are given in Table 16.

The combination of catalyst and acidic carbon surface proved successful for improving interfacial adhesion. As seen in Table 16, the transverse flexural strength rose to 140 MPa, a 50% increase over the average control value. More importantly, the coefficient of variation dropped to 6%. (Control and acidic or ammonia plasma treated coefficients of variation without catalyst are 3–4 times higher.) The catalyzed CO_2 plasma treatment thus results in a more uniform interface with higher adhesion than

Table 15.
Transverse Flexural Strength of IM7/Polycarbonate Laminates with Acidic Forming Surface Plasma Treatments

Treatment*	$\bar{S}_{22}$ MPa (kpsi)	Std. Dev. MPa (kpsi)	C.V.[a] %	Range MPa (kpsi)	N[b]
ATI-2	78.6 (11.4)	20 (2.9)	25	55.8-121.0 (8.1-17.5)	11
ATI-3	67.0 (9.6)	17 (2.4)	25	39.6- 82.0 (0.7-11.8)	7
ATI-4	53.1 (7.7)	11 (1.6)	21	34.5- 68.3 (5.0- 9.9)	11

*Increasing acidic strength (2 to 4).
[a] Coefficient of Variation
[b] Number of Specimens

Table 16.
Transverse Flexural Strength of IM7/Polycarbonate with CO_2 Plasma Treatment and Catalyst

Sample Type	$\bar{S}_{22}$ MPa (kpsi)	Std. Dev. MPa (kpsi)	C.V.[a] %	Range MPa (kpsi)	N[b]
CO_2 plasma (120 sec) w/catalyst	140.4 (20.2)	9 (1.2)	6	127.8-147.5 (18.4-21.2)	6

[a] Coefficient of Variation
[b] Number of Specimens

the other surfaces examined. One can hypothesize that acidic surface sites are reacting with polycarbonate at lower temperatures in the presence of the catalyst during composite processing before the acidic surface sites would have desorbed at higher temperatures in the absence of reaction with the polycarbonate.

Alternative mechanisms can also be hypothesized for the observed increase in interfacial adhesion. Polycarbonates generally have a polydispersity greater than 2 and contain numerous low molecular weight additives. Diffusion of these low molecular weight species to the carbon/polycarbonate interface can also adversely affect interfacial adhesion [6]. It is possible that the transesterification catalyst functioned to transform low molecular weight species into higher molecular weight polycarbonate chains and the CO_2 plasma treated carbon fiber surface, thus reducing the interfacial weak boundary layer formed during composite processing.

Fracture surfaces were examined using scanning electron microscopy. Such an analysis is often useful in determining failure modes. Photomicrographs of the control and catalyzed CO_2 treated laminates are shown in Figures 8 and 9. At low magnification (50×), the treated material has fewer loose fibers on the surface and shows evidence of fiber bundle breakage. At higher magnification (1000×), the treated material shows more resin adhering to the fibers or failure in the polycarbonate. All samples showed a mix of polycarbonate coated fibers and bare fibers; however, the catalyzed CO_2 plasma treated samples had very few bare appearing fibers. Failure in those samples occurred predominately by cohesive failure in the polycarbonate.

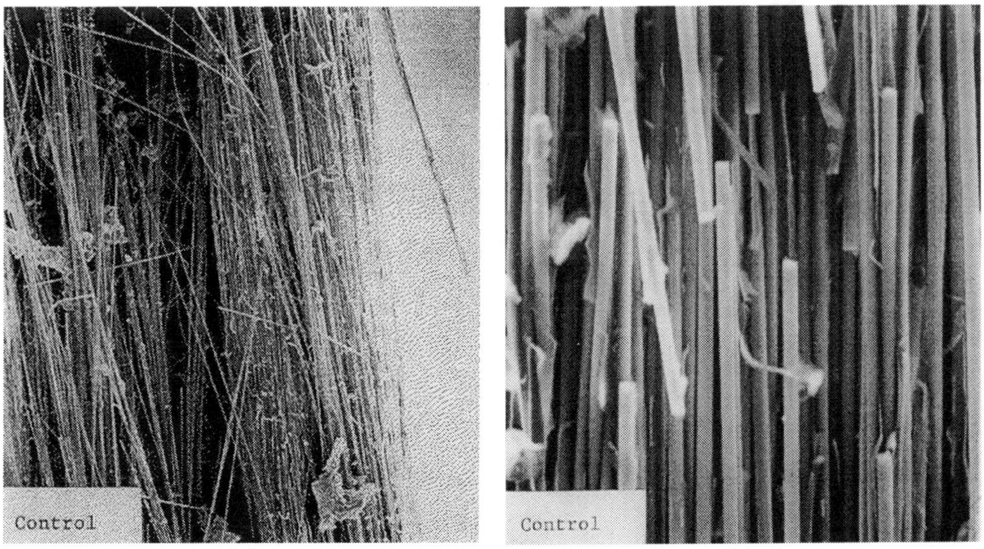

(a) (b)

Figure 8. Transverse flexure fracture surfaces of control IM7/polycarbonate laminates; (a) 40×, (b) 500×.

(a) (b)

Figure 9. Transverse flexure fracture surfaces of CO_2 plasma treated, catalyzed IM7/polycarbonate laminates; (a) 40×, (b) 500×.

4. CONCLUSIONS

This investigation shows that radio frequency glow discharge plasmas are an effective method for manipulating intermediate modulus carbon fiber surface chemistry. Such fiber surface modification provides a means to control interfacial adhesion in reinforced composite materials. Ion exchange, single filament wetting, inverse gas chromatography, ESCA and titration measurements were used to quantify the effects of ammonia (basic) and CO_2 (acidic) plasma treatments. Results demonstrate that fiber surface chemistry and surface energy can be varied over a wide range through the incorporation of specific chemical moieties into the surface structure. Surface reactions with the plasma are rapid, often reaching steady state in 30 s. Concentrations of specific surface groups are a strong function of plasma reactor configuration and operating parameters. Fiber bulk properties are not degraded by plasma exposure under the conditions examined.

CO_2 plasmas are particularly effective in modifying carbon fibers. Surface oxygen content, surface energy, and acidic sites are increased by a factor of two over as-produced IM6 fibers. High energy sites as seen in inverse gas chromatography measurements are increased by 50%. These results indicate that the CO_2 plasma treatment is functionalizing the graphite crystallite basal planes to some extent.

Both ammonia and CO_2 plasmas treated fibers are effective in increasing adhesion with polycarbonate resin, but show large coefficients of variation in test data as do unsized control fibers (24%). Plasma treatments incorporating stronger acidic moieties reduced interfacial adhesion. We hypothesize that the acid/base character of the fiber surface under the high temperature processing conditions necessary for composite fabrication acts to scission polycarbonate chains near the interface. Such a mechanism forms chemical bonds to the fiber surface at the hydrolysis site but also can form a low molecular weight weak boundary layer in the interphase region at high degrees of chain scission. The acidic moieties are also labile and may be desorbing at the high processing temperatures, which would cause the formation of a weak low molecular weight boundary layer by scissioning polycarbonate chains in the interphase region. This type of behavior explains the variations in interfacial bond strength with different fiber types that have been noted in the literature and also explains the scatter in interfacial strength measurements seen in this study.

An approach to controlled interfacial adhesion between polycarbonate and carbon fibers was implemented whereby the carbon fiber surface was rendered acidic by plasma modification and a catalyst added to promote esterification of the polycarbonate. These laminates were 50% higher in transverse flexural strength than the control laminates and scatter in the test data was reduced by a factor of 4. The catalyst appears to promote transesterification with surface groups before they desorb, or by transforming low molecular weight species into higher molecular weight polycarbonate chains in the interphase.

The primary achievement of the surface treatments was to improve adhesion by increasing formal covalent bond density rather than acid/base adduct formation at the interface. The principle of promoting electron donor/acceptor interaction across the interface guided the development of the surface treatments, however, and was very useful for developing diagnostic procedures such as wetting and IGC.

Acknowledgments

The authors are indebted to Dr. Jeffrey Hinkley at NASA Langley Research Center for his insights and helpful comments during the course of this research. His observations greatly increased the quality of the research and this document.

REFERENCES

1. J. A. Hinkley, in: *Composite Applications: The Role of Matrix, Fiber, and Interface*, T. Vigo and B. Kinzig (Eds), p. 267. VCH Publishers, New York, 1992.
2. J. A. Hinkley, *J. Reinf. Plastics Composites* **9**, 470 (1990).
3. R. A. Bucher and J. A. Hinkley, *J. Thermoplastic Composite Mater.* **5**, 2 (1992).
4. S. Clemans and T. Hartness, *SAMPE Q.* 38–42 (July 1989).
5. R. L. Brady, and R. S. Porter, *J. Appl. Polym. Sci.* **39**, 1873–1885 (1990).
6. V. K. Raghavendran, M. C. Waterbury, V. Rao, and L. T. Drzal, *J. Adhesion Sci. Technol.* **11**, 1501–1512 (1997).
7. E. Fitzer, K.-H. Geigl, W. Hüttinger, and R. Weiss, *Carbon* **18**, 389–393 (1980).
8. F. M. Fowkes, *J. Adhesion* **4**, 155 (1972).
9. F. M. Fowkes, *J. Adhesion Sci. Technol.* **1**, 7 (1987).
10. K. L. Mittal and H. R. Anderson, Jr. (Eds), *Acid-Base Interactions: Relevance to Adhesion Science and Technology*. VSP, Utrecht, 1991.
11. D. M. Riggs, R. J. Shuford, and R. W. Lewis, in: *Handbook of Composites*, G. Lubin (Ed.). Van Nostrand Reinhold, New York, 1982.
12. E. Fitzer and R. Weiss, *Carbon* **25**, 455–467 (1987).
13. A. Ishitani, *Carbon* **19**, 269–275 (1981).

14. M. Hudis, in: *Techniques and Applications of Plasma Chemistry*, J. R. Hollahan and A. T. Bell (Eds), pp. 113–147. John Wiley, New York, 1974, and references cited therein.
15. B. N. Chapman, *Glow Discharge Processes.* John Wiley, New York, 1980.
16. M. Venugopalan (Ed.), *Reactions under Plasma Conditions*, Vols. 1 and 2. John Wiley, New York, 1971.
17. R. F. Baddour and R. S. Timmins (Eds), *The Application of Plasmas to Chemical Processing.* Pergamon Press, 1967.
18. H. V. Boenig, *Plasma Science and Technology.* Cornell University Press, Ithaca, NY, 1982.
19. K. L. Mittal and A. Pizzi (Eds), *Adhesion Promotion Techniques: Technological Applications.* Marcel Dekker, New York, 1999.
20. J. C. Goan, "Method for Improving Graphite Fibers for Plastic Reinforcement and Products Thereof," U.S. Patent 3,634,220 (1972).
21. K. C. Hou, "Surface Modification of Carbon Fibers," U.S. Patent 3,745,104, July 10, 1973.
22. J. C. Goan, "Aminated Carbon Fibers," U.S. Patent 3,776,829 (1973).
23. H. Fujimaki, "Process for the Surface Treatment of Carbon Fibers," U.S. Patent 4,009,305 (1977).
24. S. Ueno and H. Kamata, "Method for Imparting Improved Surface Properties to Carbon Fibers and Composite," U.S. Patent 4,487,880 (1984).
25. J. F. Evans and T. Kuwana, *Anal. Chem.* **51**, 3 (1979).
26. J. B. Donnet, M. Brendle, T. L. Dhami, and O. P. Bahl, *Carbon* **24** (6), 757–770 (1986).
27. I.-H. Loh, PhD Thesis, Massachusetts Institute of Technology, 1985.
28. R. E. Allred, E. W. Merrill, and D. K. Roylance, in: *Molecular Characterization of Composite Interfaces*, H. Ishida and G. Kumar (Eds), pp. 333–375. Plenum Press, New York, 1985.
29. H. H. Madden and R. E. Allred, *J. Vacuum Sci. Technol.* **A4**, 1705–1707 (1986).
30. S. P. Wesson and R. E. Allred, in: *Inverse Gas Chromatography*, D. R. Lloyd, T. C. Ward, and H. P. Schreiber (Eds), pp. 204–216, ACS Symp. Ser. No. 391, ACS, Washington, DC, 1989.
31. R. E. Allred and L. A. Harrah, *Proc. 34th SAMPE Int. Sym. & Exhib.*, Covina, California (1989).
32. J. A. Hinkley, W. D. Bascom and R. E. Allred, in: *Interfaces in Composite Materials*, C. G. Pantano and E. J. H. Chen (Eds), pp. 351-356. Materials Research Society, Pittsburgh, PA, 1990.
33. S. P. Wesson and R. E. Allred, *J. Adhesion Sci. Technol.* **4**, 277-301 (1990).
34. R. E. Allred and L. A. Harrah, *Proc. 25th Int. SAMPE Tech. Conf.*, pp. 475–489 (1993).
35. R. E. Allred and W. C. Schimpf, *J. Adhesion Sci. Technol.* **8** (4), 383–394 (1994).
36. K. C. Kapur and L. R. Lamberson, *Reliability in Engineering Design.* John Wiley, New York, 1977.
37. D. T. Clark, in: *Characterization of Metal and Polymer Surfaces*, L.-H. Lee (Ed.), Vol. 2, p. 5. Academic Press, New York, 1977.
38. S. P. Wesson and J. S. Jen, *Proc. 16th National SAMPE Tech. Conf.*, p. 375, 1984.
39. A. T. James and A. J. P. Martin, *J. Biochem.* **50**, 679 (1952).
40. M. M. Davis and H. B. Hetzer, *J. Res. Natl. Bur. Std.* **54,** 309–320 (1955).
41. H. A. Benesi, *J. Phys. Chem.* **61,** 970–973 (1957).
42. O. Johnson, *J. Phys. Chem.* **59**, 827–831 (1955).
43. J. L. Kardos, F. S. Cheng, and T. L. Tolbert, *Polym. Eng. Sci.* **13**, 455–461 (1973).
44. W. L. Hawkins, *Polymer Stabilization.* Wiley-Interscience, New York, 1972.

Acid-Base Interactions, Vol. 2, pp. 581–592
Ed. K.L. Mittal

Polymer orientation at solid surfaces: Acid–base interaction origins and some property responses

A. LEPPE, J. LARA and H. P. SCHREIBER

CRASP, Department of Chemical Engineering, Ecole Polytechnique, PO Box 6079, Stn Centre Ville, Montreal, Quebec H3C 3A7, Canada

Abstract—An ethylene-acrylic acid graft copolymer, shown by inverse gas chromato-graphy to be acidic, has been adsorbed on versions of rutile (TiO_2) pigments with acidic and basic surfaces. Surface energy and rheological measurements on dispersions of the pigments in solutions of the copolymer lead to the conclusion that the copolymer adopts different chain configurations at the interface of the two pigments. In the case where the basic rutile is involved, strong acid–base interactions promote chain orientations leading to isotropic interphases. When the acidic pigment is used, the interphase becomes non-isotropic. Mechanical property data show that on aging at elevated temperature and high relative humidity the durability of compounds with isotropic interphases far exceeds that of compounds where the interphase is thought to be non-isotropic. The reasoned selection of components for polymer compounds, on the basis of their acid–base interaction capabilities, is vital to the optimization of their properties.

Keywords: Acid–base interaction; interphases; surface energies; dispersion viscosity; tensile strength; ductility.

1. INTRODUCTION

The past several years have witnessed a great rise in the attention paid to interfacial phenomena in multi-component polymer systems. Such vital properties as their mechanical strength and durability, their rheological characteristics, wetting and adhesion properties, all may be said to depend critically on events occurring at contacts among the system components. A rational research target has been the quantification of interactions, an objective which has used to advantage the division of interactions into their dispersion and non-dispersion components. Particular attention has been given to the measurement of non-dispersion interactions: These arise from a variety of sources, including polar, dipolar, hydrogen bond, ionic interactions, and the like. A number of workers have opted to sum these as acid–base interactions. Acid–base theory, of course, is still in a state of some flux, as outlined in recent publications [1, 2]. Not surprisingly, a variety of approaches have been formulated to an acid-base interpretation of non-dispersion interactions. For example, Good and coworkers have advanced an acid-base theory founded on the concept of molecular orbitals in hydrogen bonding. Their approach has been frequently reviewed [3, 4]. Fowkes and coworkers [5–7] adapted the Drago modification [8] of Lewis theory, using calorimetric enthalpy of interaction data to assign Drago E and C parameters to some polymers and solids involved in multi-component systems. Respectively, E and C represent the electrostatic and covalent contribution to the net acid-base interaction. Much of Fowkes' work may be regarded as seminal to the furthering of acid–base concepts in polymers, but the choice of Drago theory results in materials being designated as either acid or base, but not both. Clearly, however, many macromolecules possess both electron donor and acceptor capabilities, rendering the E and C descriptors too limiting.

In our laboratories, quantitative evaluations of acid-base interactions have made use of the Gutmann theory of molecular interactions [9], and of its application to polymers as developed by Papirer, Schultz and their associates [10–12]. The method of inverse gas chromatography](IGC) [13], has been our experimental tool of choice, making possible the designation of acid–base interaction parameters to polymers and to a wide variety of solids used in polymer systems. The matter is developed further later in this communication.

An additional element relevant to the present work is the remarkable orientational degree of freedom open to some polymers. Consider a polymer such as a two-component, phase-separated polyurethane. On solidification, the thermodynamic requirement to minimize the surface free energy favors a surface excess of the appropriate component, in the present example the soft polyether or polyester segment. When placed in contact with a polar medium however, the minimization of interfacial tension would favor a surface excess of the polar, hard segment of the polymer. The polymer surface may, therefore, be expected to reorganize in response to the thermodynamic demands made on it. Surface restructuring of the kind has been documented not only for polyurethanes as in this example [14, 15], but for many other polymers and copolymers [16–18]. The orientational degrees of freedom open to more complex macromolecules have recently spawned much activity under the catchword of 'smart' polymers [19].

In the present paper we report an apparent link between the orientational state of an ethylene-acrylic acid graft copolymer (E-g-AA) and the acid-base interaction between the polymer and pigmenting rutile (TiO_2) solids. Further, by way of relating these fundamental aspects of polymer behavior to practical considerations, we note the consequences of changes in the degree of acid–base interaction and in chain orientation on some mechanical and rheological properties of compositions involving the E-g-AA polymer.

2. EXPERIMENTAL

2.1. Materials

The E-g-AA copolymer, obtained from AT Plastics, Inc. (Brampton,On., Canada) was a melt flow index= 4.5 specimen, with a mole ratio of E/AA = 98/2. Its density was reported to be 0.921 $g \cdot cm^{-3}$.

Two rutile samples were used as substrates for the adsorption of E-g-AA. Both were commercially coated by the supplier, Tioxide Canada, Inc. Sample R-1 had a BET surface area of 9.7 m^2/g; R-2 of 9.3 m^2/g. The solids were essentially non-porous with mean particle diameters, respectively, of 0.226 and 0.230 μm. Both had densities of 4.11 g/cm^3.

2.2. Procedures: IGC

The techniques of IGC were used to obtain values of dispersion surface energies (γ^d) and of acid–base interaction constants, K_a and K_d for pure polymer, pigments and for combinations of polymer adsorbed in varying quantities on the pigment solids. Details of IGC experimentation and of the procedures used to obtain the characterizing parameters have been documented frequently [10–14] and are noted here only briefly. A Varian 3400 gas chromatograph, with both ionization flame and hot wire detectors, was used throughout this work. Previously degreased, washed and dried stainless steel columns were used to house the stationary phases. In the case of E-g-AA, the polymer was deposited from hot xylene solutions on Chromosorb AW (60/100 mesh) support (with a surface area of 1.1 m^2/g), using Rotovap apparatus. Conventional ashing procedures [13] showed that 8.3 wt% of E-g-AA had been deposited. Samples of rutile and of rutile with adsorbed E-g-AA (see below) were packed directly in columns, the weights of solids falling in the range 4.6–5.2 g.

IGC data were collected in the temperature range 40–70°C, using as vapors the n-alkanes from C_6–C_{10}, to represent probes capable of interacting with the solids only through dispersion forces, and chloroform, benzene, tetrahydrofuran (THF), diethyl ether and acetone as probes capable of acid–base interaction. The latter were chosen on the basis of Gutmann's classification [9]: chloroform and benzene being acids, THF and the ether bases and acetone a typical amphoteric vapor. Vapors were injected at very high dilution in at least triplicate, with variations from mean retention times not exceeding 4 %. In order to obtain γ^d values, the retention volumes, V_n, of n-alkane probes were plotted against T_b, the probes' normal boiling point, following the procedures of Sawyer and Brookman [20]. The slopes of the functions yield the required datum. As noted in an earlier paper [21], this is a reliable representation of experimental data which yields values of γ^d, well within experimental error of evaluations based on the equation stated by Schultz and co-workers [11, 12], but avoids the difficulties associated with the widely used approach of Papirer, Schultz and co-workers [10–12] where an estimate must be made of the cross-sectional area of adsorbed probe molecules. The invariably linear plots of RT ln V_n vs. T_b served as a reference for the data of polar probes, with the acid-base contribution to free energies of desorption given by the distance of the relevant datum from a point on the alkane line at the corresponding value of T_b. Thus,

$$-G^{ab} = (V_n)^{ab}/(V_n)_{ref}. \tag{1}$$

Availability of G^{ab} data over the specified (30°C) temperature range allowed for the calculation of H^{ab} and, again following well established precedent [10–14], the acid-base interaction parameters for the various stationary phases were then obtained from

$$-H^{ab} = K_a(DN) + K_d(AN^*), \tag{2}$$

where AN and DN are the acceptor (acid) and donor (base) numbers of the vapor probes as given by Gutmann. The correction of Riddle and Fowkes [22] has been used to obtain values of AN*.

2.2.1. Adsorption of Polymer on TiO_2. In order to place varying, controlled amounts of E-g-AA onto rutile surfaces, a standard 1 wt.% solution of the polymer in p-xylene was prepared under reflux. The solution could be maintained turbidity-free to near 50°C, and 60°C was therefore chosen as the working temperature for adsorption. To 50 ml. aliquots of the solution at that temperature were added carefully weighed amounts of rutiles R-1 and R-2. The suspensions were stirred for 5–6 h, whereupon the solids were filtered and washed at room temperature with copious amounts of xylene. The materials were dried to constant weight in a vacuum oven at 75°C, a process requiring some 16 h. Gravimetric analyses were used to establish the quantity of adsorbed polymer, removal of the adsorbed mass being effected by ashing at 600°C under a flow of nitrogen. Samples of pure R-1 and R-2 were subjected to the same procedure in order to establish a correction due to weight loss from the uncoated substrate.

2.2.2. Rheological Measurements. The main objective here was an estimation of the configuration adopted by the E-g-AA when adsorbed on the rutile solids. The relative viscosities of pigment dispersions prepared by the above method were used for the purpose. To effect corresponding states of dispersion, carefully weighed amounts of pigment were introduced into 200 ml. of E-g-AA solutions at a polymer concentration of 5 wt%, contained in an earthenware jar. This also contained 50 g of stainless steel balls, used as a grinding medium. The jar was placed on a roll-mill, and agitated for 24 h (roll rotation = 50 rpm). The steel balls were then removed and aliquots of the dispersion placed in a Bohlin VOR Rheometer were scanned over more than 3 decades of shear. All measurements were at 30°C, and reported results are averages of at least 2 independent determinations. A mean deviation of ± 4% applies to the data.

2.2.3. Mechanical Properties. Samples of solid E-g-AA containing rutiles R-1 or R-2 in quantities of up to 5 wt.% were prepared by mixing in a Brabender Plasticorder. The temperature was 170°C, the rotational speed 50 rpm and mixing was continued until a steady torque reading was obtained, generally within a period of 2–3 min. Mixtures were then compression molded for 2 min in a Carver press at 170°C, under 1820 kg of applied pressure. Dog-bone specimens, cut from the sheets, were conditioned for 24 h under ambient conditions, and then tested for initial mechanical responses on an Instron table model tester. The jaw separation speed was 10 mm/min. The stress-strain curves were the source of tensile strength and ductility values, and of the combined tensile -elongation product. An aging procedure was applied to specimens of these compositions. These were placed in a controlled humidity chamber, set at 60°C and 50% relative humidity (R.H.) and aged for up to 360 h. Measurements of mechanical responses were repeated at selected aging times. All mechanical property results are averaged from at least triplicate measurements. In all reported cases the standard deviation was less than ± 6%.

3. RESULTS AND DISCUSSION

3.1. IGC Results

The data obtained from IGC measurements on pure starting materials are given in Table 1. The dispersion surface energies of the two rutiles are similar and clearly greater than that of the copolymer. Dispersion forces, therefore, should favor wetting and adsorption processes. The pigments are different in their acid-base properties. Although both have electron donor and acceptor capabilities, acidity is dominant in the case of R-1, while the R-2 surface is dominated by basic sites. The copolymer also ranks as a mild acid. The combination of γ^d, Ka and K_d, however, suggests that very little of the AA component is actually found in the surface layer. The γ^d is essentially equivalent to values for pure polyethylenes [23]; although a value for pure PAA is not available, it may be expected to exceed 40 mJ/m^2 [24]. The AA moiety thus appears not to contribute significantly to the surface energy of the copolymer, *as prepared for the IGC measurement.* The polar groups are oriented into the bulk of the polymer, away from the air interface. The K_a, K_d data support this contention. Polyethylene surfaces usually are mildly oxidized, leading to low but finite values of K_a and K_d. Much higher K_a and K_d indexes would be expected for the AA group. The strong prevalence of pure hydrocarbon linkages in the surface region of the copolymer is in line with demands for the minimization of the total surface free energy[1].

Turning now to the properties of pigments with adsorbed E-g-AA, a striking difference between systems using R-1 and R-2 may be suggested immediately. In the case of R-2, strong acid-base coupling between substrate and polymer should favor a polymer configuration allowing it to approach the adsorbing surface to molecular distances. The preferred orientation at the pigment/polymer interface would place the acidic AA groups in close contact with the basic surface.The orientational state of adsorbed copolymer chains therefore would be similar to those at the air interface. The intervening interphase may therefore be thought of as isotropic (or isomorphic). With R-1 as the substrate, the dominant force at close approach of the adsorbing macromolecule would be of the acid-acid type. Minimization of interfacial tension would then promote placement of the E component at the pigment/polymer interface, with the grafted AA groups oriented toward the bulk of the adsorbed layer. This would be the mirror image of orientation at the polymer/air interface, rendering the interphase non-isotropic. These descriptions imply orientational degrees of freedom in E-g-AA, and this is supported by earlier literature results [11, 12].

The experimental data for adsorbed systems are reported in Table 2. They support the expectations stated above. In the case of R-1 substrate, noteworthy is the systematic change in γ^d and in K_a as the quantity of adsorbed E-g-AA increases and as, consequently, the interphase thickens. At very low masses of adsorbed copolymer, both parameters differ substantially from the results for pure E-g-AA in Table 1, but approach those values as the interphase thickens. At the highest surface coverage, both

[1] The support Chromosorb was found, in calibration work, to be weakly interactive, with K_a and K_d values near 0.3. It is possible therefore that AA groups of chains in immediate contact with the support may also orient preferentially. However, the deposited copolymer layer is too thick for the support to exert influence on the orientation of chains at the air interface.

Table 1.
Interaction parameters for starting materials

Material	γ^d (mJ/m^2)	K_a	K_d
E-g-AA	31.7	2.8	1.1
Rutile R-1	40.8	4.7	1.6
Rutile R-2	41.1	1.9	3.0

Table 2.
Surface characteristics of rutiles with adsorbed copolymer

Substrate R-1				Substrate R-2			
E-g-AA (g/m^2)	γ^d (mJ/m^2)	K_a	K_d	E-g-AA (g/m^2)	γ^d (mJ/m^2)	K_a	K_d
0.014	37.7	3.9	1.4	0.011	32.6	3.0	1.1
0.092	36.1	3.3	1.4	0.078	32.0	2.9	1.5
0.188	34.9	3.0	1.4	0.171	32.2	3.0	1.3
0.251	33.5	2.8	1.3	0.266	32.2	2.8	1.2
0.665	32.6	2.5	1.1	0.509	32.1	2.9	1.1
0.803	32.3	2.6	1.0	–	–	–	–

γ^d and K_a are within experimental uncertainty of the earlier values. The K_d data are unresponsive to variations in adsorbed quantity, as indeed is the case for R-2 systems. When R-2 is the adsorbing solid, none of the tabled parameters shows significant dependence on adsorbed amounts of copolymer. Even at the lowest surface coverage, K_a and K_d match the values in Table 1 for E-g-AA. A very slight rise in γ^d may be due to minor perturbation of the polymer chains in immediate contact with the substrate.

The availability of copolymer density and adsorbed mass data permits a rough estimate to be made of the adsorbed layer thickness. In so doing, we assume that the surface area of the support solid has not been affected by the adsorption process and that surface coverage is uniform, the latter an admittedly tenuous proposition, given a probable heterogeneity in the energies of adsorption sites. The representation in Figure 1, using γ^d as the comparison datum, results from the calculation. This suggests that the non-isotropic interphase of the layer adsorbed on R-1 has a thickness of about 0.50- 0.60 μm. With R-2, the constancy of γ^d points to an isotropic interphase, where the chain orientation is similar throughout the adsorbed polymer layer.

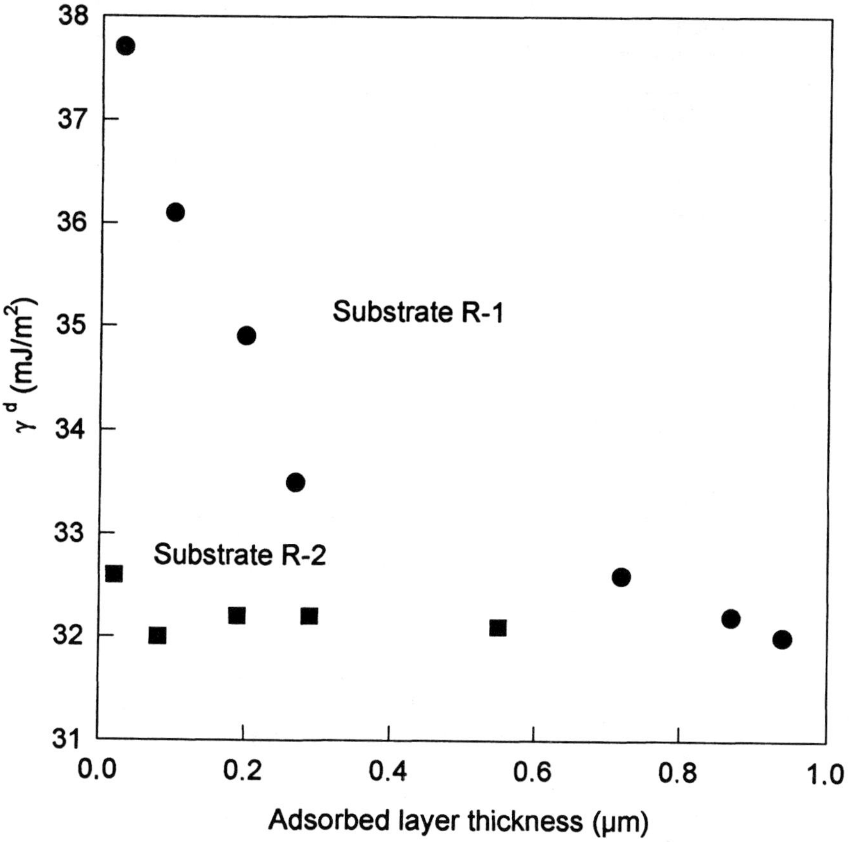

Figure 1. Variation of dispersion surface energy, γ_s^d, with adsorbed layer thickness in E-g-AA/rutile combinations. Data at T = 40°C.

3.2. Rheological Data

The aim of this part of the study was to inquire further into the configuration of polymer chains close to the pigment surface. Attention was focused on η_r, the relative viscosity of dispersions, this being the ratio of viscosities of pigment dispersions and of the unpigmented polymer solution. The theory of solids dispersions relates η_r with the solids volume fraction, φ_s, and many mathematical variations of the relationship have been proposed [25]. Relating to our case, when a macromolecule adsorbs on the solid surface the average particle dimension increases, and thus the true φ_s will exceed the apparent quantity calculated on the basis of the mass of solids introduced. A comparison of η_r for dispersions of R-1 and R-2 in E-g-AA solutions at equivalent values of the apparent solids volume fraction should then provide some insight into the shape of the adsorbed polymer.

Dispersion viscosities were found to be mildly shear-thinning, and as shown in Figure 2, the relative viscosities of R-1 and R-2 dispersions were distinct, most particularly at high φ_s. No attempt was made to fit the data to one of the expressions

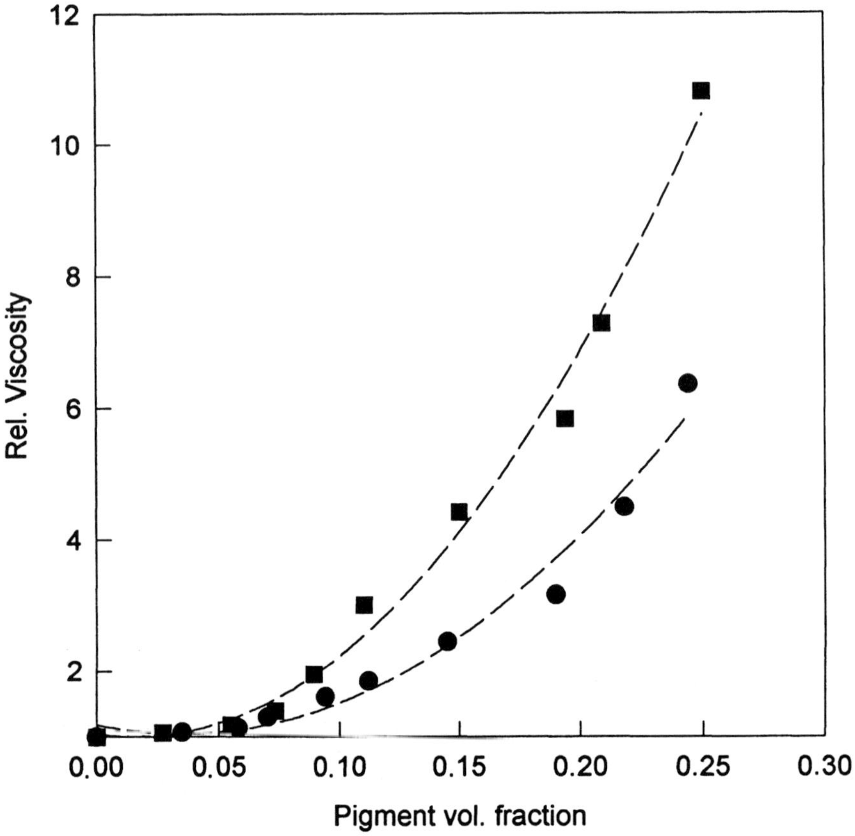

Figure 2. Relative viscosities of rutile dispersions in p-xylene solutions of E-g-AA copolymer. T=30°C. (■) Rutile R-1 substrate, (●) Rutile R-2 substrate.

drawn from theories referred to above. The difference in the behavior of the two sets of materials was sufficient for current needs. Since the quantity of available polymer was constant in the preparation of dispersions, the amount of adsorbed polymer must vary inversely with φ_s. Evidently, the adsorbed polymer has a noticeably greater effect on the η_r of dispersions of the acidic R-1. This may be viewed as indicating that layers of E-g-AA closest to the pigment surface are more strongly bound to the R-2 surface than to R-1, in the latter case extending further into the solvent carrier medium. Acid-base interactions again may be cited as the origin of the closer association between R-2 and the copolymer.

3.3. Mechanical Properties

The initial values of mechanical property parameters are listed in Table 3. Although the differences in behavior between systems with R-1 and R-2 pigments are very slight, the tensile strength in systems using pigment R-2 does appear to be somewhat

Table 3.
Initial mechanical properties of rutile-filled E-g-AA copolymers (Data recorded following 24 h under ambient conditions)

Wt.% rutile added	Tensile strength (Mpa)		Ductility (%)	
	R-1	R-2	R-1	R-2
0.05	24.1	25.0	140	133
0.10	25.5	25.7	118	110
0.20	26.3	27.7	92	90
0.50	28.0	29.9	79	70
1.00	28.8	30.8	68	64
3.00	29.3	31.8	55	48
5.00	30.7	34.0	42	32

higher than that in the corresponding R-1 compounds; the latter, however, have slightly greater ductilities. These results may be viewed as added independent evidence for the stronger bond at the polymer/R-2 interface. The difference in mechanical properties between these systems becomes far more apparent upon accelerated aging. We have expressed these in terms of the product, (TE), of tensile strength and elongation. Comparisons are made in Figs. 3 and 4, the former showing TE vs aging time for compounds with 1 wt.% pigment, while in Fig. 4 are given data for compounds with 5 wt.% particulates. The durability of mechanical properties is far greater in compounds with R-2, where interfaces are marked by strong acid-base interactions, and interphases are believed to be isotropic. The loss of mechanical strength and ductility in compounds with R-1 pigment becomes apparent in the first few hours of exposure to the 60°C/50% RH environment, about 15% of the initial TE value being lost in that time. At the conclusion of the 360 h aging period, R-1 compounds have lost about 50% of their initial property, while the loss in systems with R-2 is only of the order of 10%. The most apparent reason for the difference in behavior is in the interfacial strength of the two series and in the morphology of their interphases. It may be proposed that in the weakly bonded R-1 compounds, the intrusion and condensation of water in the interface results in substantial loss of mechanical strength and ductility. The effect of acid-base interaction in systems with R-2, leading to a closer approach of E-g-AA chains to the adsorbing surface, and to the resulting isotropic interphases, is considered to restrict the access of intruding water. The mechanical property data in particular stress the need to select components judiciously if property optimization is sought for multi-component polymer systems. The consideration of acid-base effects appears to be an important element in the selection process.

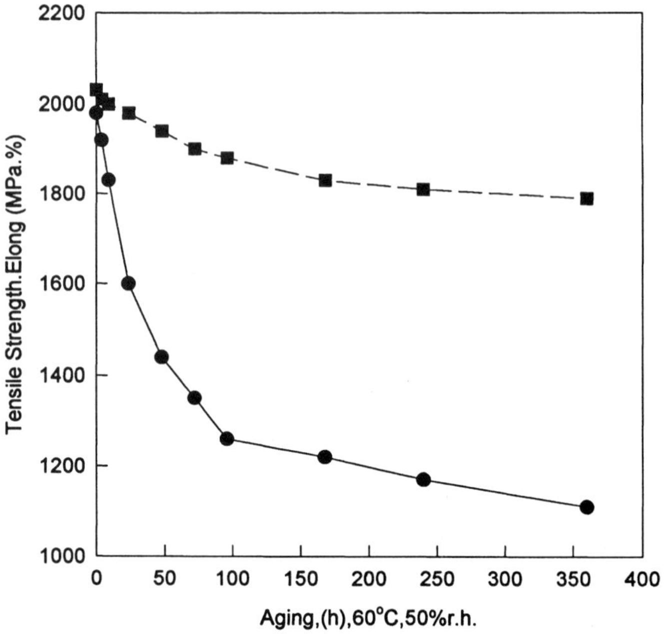

Figure 3. Variation of the (tensile strength.elongation) product of rutile-filled E-g-AA copolymer blends with accelerated aging time. Rutile concentration = 1 wt.%. (●) R-1, (■) R-2.

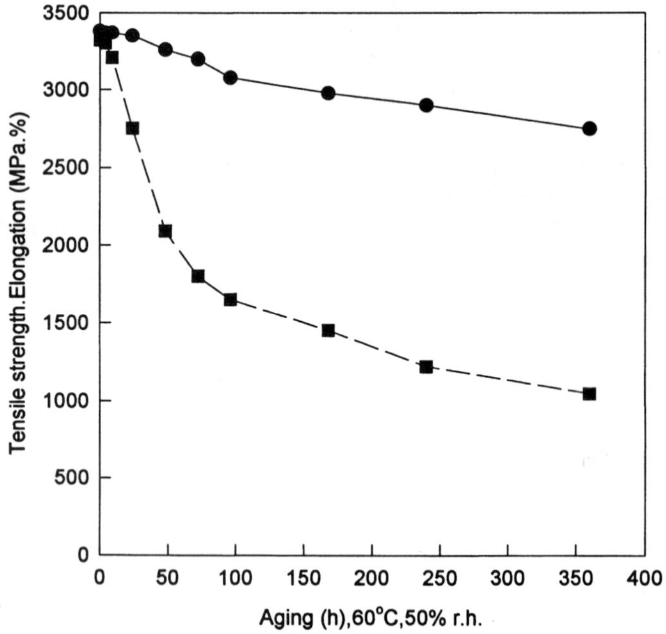

Figure 4. Variation of the (tensile strength.elongation) product of rutile-filled E-g-AA copolymer blends with accelerated aging time. Rutile concentration = 5 wt.%. (●) R-1, (■) R-2.

4. CONCLUSION

Though indirect, the evidence presented strongly supports the notion that the organization of an E-g-AA macromolecule at the surfaces of specifically chosen (rutile) substrates is determined by the acid-base forces acting at the interface. Where acid-base forces are appreciable, the organization of the copolymer at the air and substrate interfaces is similar, thus creating an interphase region in which the configuration of the polymer remains constant, and independent of the interphase thickness. That region then may be thought of as being isotropic or isomorphic. In contrast, when acid-base forces are not a dominant feature of the interface, a non-isotropic or non-isomorphic interphase is created, due to the different organization of the polymer at substrate and air interfaces. The nonisotropic region is restricted to the vicinity of the substrate, in the present study to a thickness somewhat greater than 50 μm. The mechanical properties of copolymer filled with alternate versions of the rutile demonstrate the superiority of the compositions in which acid-base interactions are favored. Acid-base considerations are to be considered in formulating complex polymer systems for optimum performance.

Acknowledgment

Financial support for this work was obtained from Operating Grants of the Natural Sciences and Engineering Research Council, Canada.

REFERENCES

1. P. Mukhopadhyay and H. P. Schreiber, *Colloids Surfaces A* **100**, 47 (1995).
2. K. L. Mittal and H. Anderson, Jr. (Eds), *Acid-Base Interactions: Relevance to Adhesion Science and Technology*. VSP, Utrecht, The Netherlands (1991).
3. R. J. Good and A. K. Hawa, *J. Adhesion* **63**, 5 (1997).
4. R. J. Good, *J. Adhesion Sci. Technol.* **6**, 1269 (1992).
5. F. M. Fowkes, *J. Adhesion Sci. Technol.* **4**, 669 (1990).
6. F. M. Fowkes and M. A. Mostafa, *Ind. Eng. Chem. Prod. Res. Dev.* **17**, 3 (1978).
7. F. M. Fowkes, in: *Physicochemical Aspects of Polymer Surfaces*, K. L. Mittal (Ed.), vol. 2, pp. 583–603. Plenum Press, New York (1983).
8. R. S. Drago, N. M. Wong and D. C. Ferris, *J. Amer. Chem. Soc.* **114**, 91 (1992).
9. V. Gutmann, *The Donor-Acceptor Approach to Molecular Interactions*. Plenum Press, New York (1978).
10. C. Saint-Flour and E. Papirer, *J. Colloid Interface Sci.* **91**, 63 (1983).
11. J. Schultz and L. Lavielle, in: *Inverse Gas Chromatography*, D. R. Lloyd, T. C. Ward and H. P. Schreiber (Eds), ACS Symp. Series No. 391, Ch.14. Amer. Chem. Soc. Washington, DC (1989).
12. M. Nardin and J. Schultz, *Composite Interfaces* **1**, 177 (1993).
13. D. R. Lloyd, T. C. Ward and H. P. Schreiber (Eds.), *Inverse Gas Chromatography*, ACS Symp. Series No. 391. Amer. Chem. Soc., Washington, DC (1989).
14. A. Sengupta and H. P. Schreiber, *J. Adhesion Sci. Technol.* **5**, 947 (1991).
15. Z. Deng and H. P. Schreiber, *J. Adhesion* **36**, 71 (1991).
16. J. Schultz, *J. Adhesion* **37**, 73 (1992).
17. J. H. Chen and E. Ruckenstein, *J. Colloid Interface Sci.* **135**, 496 (1990).
18. J. D. Andrade, in: *Polymer Surface Dynamics*, J. D. Andrade (Ed.), Ch. 1. Plenum Press, New York (1988).

19. J. T. Koberstein, D. E. Duch, W. Hu, T. J. Lenk, R. Bhatia, H. R. Brown, J.-P. Lingelser and Y. Gallot, *J. Adhesion* **66**, 229 (1998).
20. D. T. Sawyer and D. J. Brookman, *Anal. Chem.* **40**, 1847 (1968).
21. U. Panzer and H. P. Schreiber, *Macromolecules* **25**, 3633 (1992).
22. F. L. Riddle, Jr. and F. M. Fowkes, *J. Amer. Chem. Soc.* **112**, 3259 (1990).
23. H. P. Schreiber, Y. B. Tewari and D. Patterson, *J. Polym. Sci., Phys. Ed.* **11**, 15 (1973).
24. S. Wu, *Polymer Interface and Adhesion*, p. 89. Marcel Dekker, New York (1982).
25. A. B. Metzner, *J. Rheology* **29**, 739 (1985).

Acid-Base Interactions, Vol. 2, pp. 593–600
Ed. K.L. Mittal

Acid–base interactions in the measurements of surface energies of textile fibers and finish liquids

Y. K. KAMATH* and C. J. DANSIZER

TRI/Princeton, PO Box 625, Princeton, NJ 08542, USA

Abstract—Surface energies of organic textile fibers have been determined, using their interaction with liquids of appropriate acid-base or dispersive character. Work of adhesion calculated from the fiber-liquid wetting force determined by the Wilhelmy method has been used as a measure of thermodynamic interaction between the fiber and the liquid surfaces. Conventional methods of calculating surface energy from these measurements, assuming that the molecules at an interface can express both dispersive and nondispersive interactions simultaneously, lead to erroneous values of surface energy, especially in the case of liquids which can strongly hydrogen bond with the solid surface. In such cases assuming that the entire work of adhesion arises from nondipersive (acid–base) interactions leads to more realistic data for the surface energies of fibers and finish liquids. This shows that the surface energies of fibers determined by this method depend on the nature of the liquid used, and the assumption that the molecules at an interface can express both dispersive and nondispersive interactions simultaneously needs to be re-examined. The method has also been used to evaluate surface energies of liquids which are used as lubricants in spin finishes.

Keywords: Fibers; finishes; surface energy; dispersive component; Lewis acid–base; wetting force; work of adhesion; Wilhelmy principle; solid–liquid wetting.

1. INTRODUCTION AND BACKGROUND

Surface energies of solids are determined by their interaction with other substances in the form of liquids or gases. These interactions occur between the molecules at the interface, and in the case of a solid-liquid interface the intensity of this interaction is given by the work of adhesion, W_{Adh}. In the case of a simple liquid such as a hydrocarbon, the work of adhesion is essentially dispersive and its magnitude is given by [1]:

$$W_{Adh} = W^{d} = 2(\sigma_s^{d}\,\sigma_l^{d})^{1/2} \qquad (1)$$

where, σ's are the dispersive (superscript d) components of the surface energy of the solid and the liquid, specified by the subscripts s and l, respectively. These interactions consist of 'dispersion' (London), dipole–dipole (Keesom) and dipole-induced dipole (Debye) interactions. Good and Chaudhury [2] have referred to these interactions as Lifshitz–van der Waals interactions, σ^{LW}.

* To whom correspondence should be addressed. E-mail: ykamath@triprinceton.org

In the case of more complicated molecules, significant interactions can arise at the interface from Lewis acid-base interactions involving specific functional groups. Fowkes [3], assuming the additivity of such interactions, proposed that the work of adhesion at the interface be considered as the sum of the dispersive and the non-dispersive (acid–base) contributions, so that,

$$W_{Adh} = W^d + W^{ab} \tag{2}$$

and W^{ab} is given by

$$W^{ab} = nf\,(E_aE_b + C_aC_b). \tag{3}$$

In Equation (3) n is the interaction site density; $f = 4.18 \cdot 10^{10}$ (kcal-to-erg conversion factor); and E's and C's are the Drago coefficients [4] for the acid and base components in units of $(\text{kcal/mol})^{1/2}$.

Good *et al.* [5] gave the acid–base component of the work of adhesion by

$$W^{ab} = 2\,[(\sigma_l^+\sigma_s^-)^{1/2} + (\sigma_l^-\sigma_s^+)^{1/2}] \tag{4}$$

Where σ's are the acidic (+) and the basic (-) components of the surface energies of the solid and the liquid (subscripts s and l). Therefore, the total work of adhesion is given by

$$W_{Adh} = 2\,[\,(\sigma_s^d\,\sigma_l^d\,)^{1/2} + (\sigma_l^+\sigma_s^-)^{1/2} + (\sigma_l^-\sigma_s^+)^{1/2}] \tag{5}$$

Equation (5) has three solid surface energy components, which can be calculated by solving the three simultaneous equations obtained by making work of adhesion measurements in three different liquids, the surface energy components of which are known (see Table 1). Good *et al.* [5] have calculated the acid-base components of a series of liquids from interfacial tensions, assuming water to be truly amphoteric, so that the non-dispersive component of its surface tension is divided equally between acidic and basic components, (i.e. $\sigma_l^+ = \sigma_l^- = 25.5$ mN/m). The surface energy components of liquids used in this study are given in Table 1 [5].

Recently, Tate *et al.* [6] showed that the surface energy measurements of nylon 66 (N-66) fibers using 3 liquids (methylene iodide, water and formamide) and applying Equation (5) did not reflect the true nature of the polyamide surface, which is characterized by both acidic and basic sites. More reasonable values were obtained by assuming the entire work of adhesion in water and formamide to be acid–base in character, neglecting the dispersion component of these molecules. In the present work this analysis has been extended to poly(ethylene terephthalate) (PET) and

Table 1.
Surface energy components of liquids (mJ/m^2)

Liquid	σ_l	σ_l^d	σ_l^+	σ_l^-
Methylene iodide	50.8	50.8	-	-
Formamide	58.0	39.0	2.3	39.6
Water	72.8	21.8	25.5	25.5

poly(propylene) (PP) fibers with acidic and dispersive characteristics, respectively. We have also included finish liquids in this study which are random or block copolyethers of ethylene oxide and propylene oxide (EO/PO), and are essentially basic in their surface properties.

2. EXPERIMENTAL

2.1. Materials

The textile fibers used in this study, i.e. N-66, PET and PP were specially prepared without the application of any finishes. The diameters of these fibers were in the range of 20–30 μm. The fibers were exhaustively rinsed in demineralized water and were dried in a conditioned environment (65%RH at 21°C).

Methylene iodide and formamide were obtained from Aldrich Chemical Co.and were used as received. Demineralized water was prepared in the laboratory using ion exchange columns. The fluorosurfactant (FC 129) used for perimeter measurement was a 3M product and was used as received.

2.2. Methods

In this work surface energies were determined from wetting force measurements of liquids on single fibers, using the TRI/Scan Surface Force Analyzer. This apparatus is essentially a computer controlled Cahn recording microbalance used for making wetting force measurements by the well-known Wilhelmy principle [7]. Liquid surface tensions were measured by the platinum wire detachment method using the same apparatus. Fiber perimeters were also determined by wetting force measurements in a fluorosurfactant solution (FC 129, 1%) with a surface tension of 16 mN/m. Under these conditions the contact angle can be assumed to be 0° and the wetting force (F_w) is given by $F_w = P\,\sigma_l$, from which the perimeter P can be calculated.

The fiber-liquid wetting force measured by this method is given by

$$F_w = P\,\sigma_l\,\mathrm{Cos}\,\theta \tag{6}$$

where P is the fiber perimeter, σ_l is the liquid surface tension and θ is the contact angle. We would like to refer to Cosθ as the fiber-liquid interaction parameter, given by the ratio of the fiber-liquid adhesion tension and the surface tension of the liquid ($\sigma_l\,\mathrm{Cos}\theta\,/\sigma_l$). This makes Cos$\theta$ more of a physical rather than mathematical quantity.

The work of adhesion, which is given by

$$W_{Adh} = \sigma_l\,(1 + \mathrm{Cos}\theta), \tag{7}$$

is the parameter of choice in relating surface energies to thermodynamic interactions at the interface.

3. RESULTS AND DISCUSSION

3.1. Fiber Surface Energies

The work of adhesion of fibers in probe liquids is given in Table 2. Each entry is an average of measurements on 6 specimens.

The surface energy components of fibers calculated from the data in Table 2 using Equation (5) are listed in Table 3.

In Table 3 the data for PP fibers show that they are predominantly dispersive with negligible acidic or basic character. However, the surface acidity (σ_s^+) of both N-66 and PET fibers is not explained by these data. Dye adsorption experiments on these fibers clearly show the presence of surface carboxyl groups.

Alternatively, if the surface energy components are calculated from the data of Table 2, assuming that the entire work of adhesion in water and formamide is acid–base in character (for each solid we get two simultaneous equations of type 4 which can be solved for the surface energy components), more reasonable values are obtained for N-66 and PET, but not for PP. This can be seen from the data in Table 4. High value for the acid component in the case of PP is obviously wrong, which indicates that the assumption is not applicable to PP. Polyamide N-66 and PET fibers, as expected, show both acidic and basic characteristics.

Table 2.
Work of adhesion (mJ/m^2) of fibers in probe liquids

Fibers =>	N-66	PET	PP
Liquids			
Meth. iodide	82.1(2.5)	83.7 (3.0)	68.0 (1.5)
Formamide	89.2 (2.4)	87.4 (4.2)	60.7 (2.6)
Water	96.7 (2.1)	82.7 (1.4)	59.2 (3.2)

Note: Numbers in parentheses are Standard deviations

Table 3.
Surface energy (mJ/m^2) components of fibers (Using Equation (5))

Fiber	σ_s^d	σ_s^+	σ_s^-
N-66	33.9	0.16	10.7
PET	35.2	0.29	4.7
PP	23.3	0.16	3.2

Table 4.
Surface energy components of fibers, assuming acid-base interactions only (mJ/m^2)

Fiber	σ_s^+	Σ^-
N-66	39.7	14.4
PET	43.0	2.7
PP	20.1	1.9

This exercise clearly shows that the values of surface energy components obtained by this analysis depend on the solid-liquid systems used and the assumption regarding the concurrent expression of their dispersive and acid-base characters. This aspect will be discussed further in a separate section.

3.2. Surface Energy Components of Finish Liquids

Spin finishes are applied to melt spun synthetic fibers as lubricants to protect them during processing. The principal component of these finishes are random or block co-polymers of ethylene oxide and propylene oxide (EO/PO) of different molecular weights. Strong interaction of these liquids with surfaces of fibers is important in spreading and uniform distribution on the surfaces of fibers. Liquid surface energies in combination with the surface energies of the solid play an important role in this interaction.

We have seen that Equation (5) can be used to calculate surface energy components of solids, knowing the surface energy components of three probe liquids, by making specific assumptions regarding the nature of interactions at the interface. Conversely, we should be able to evaluate the surface energy components of liquids of interest, knowing the surface energy components of three solids. Fibers used in this study are suitable for this purpose. PP fibers are almost entirely dispersive, and PET and N-66 fibers are acidic and amphoteric, respectively. Surface energy components of finish liquids determined by this method are reported in Table 5.

It should be emphasized that in Table 5 the dispersive component is calculated from the work of adhesion with PP, and the acidic and the basic components were calculated from the work of adhesion with N-66 and PET fibers, assuming that the entire work of adhesion is acid-base in character. This analysis clearly shows that these liquids are basic. They are electron donors on account of the ether oxygen with lone pairs of electrons.

Table 5.
Surface energy components of spin finish lubricant liquids

	Work of Adhesion (mJ/m^2)			Surface (mJ/m^2)	Energy	Component
Fiber =>	N-66	PET	PP	σ_l^d	σ_l^+	σ_l^-
Lubricant						
UCON 50 HB 100	64.7	63.8	55.1	32.6	1.0	21.3
EO/PO L-31	61.9	61.2	56.0	33.6	0.8	19.7
EO/PO 10/90	65.3	64.3	60.6	39.4	1.1	21.5
EO/PO L-81	65.5	70.0	59.7	38.2	0.3	29.8

3.3. Molecular Interactions at Solid–liquid Interfaces

The dispersive and non-dispersive components of surface energies obtained from this study are summarized in Table 6.

The data presented in this work show that total surface energy ($\sigma^d + \sigma^{AB}$) of a solid is unlikely to be expressed completely in any measurement. The values obtained for surface energy components of solids and liquids depend very much on the nature of the conjugate molecules used in the measurement. This has been observed by other authors also [8, 9]. Strong acid-base interactions at the solid-liquid interface can lead to specific orientations, which can suppress dispersive interactions completely, and vice versa. In those cases where strong interactions are not involved, it is possible that both dispersive and acid–base interactions are partially expressed, i.e. ($m\sigma^d + n\sigma^{AB}$), where, $0 < m,n < 1$. However, it will be difficult to determine the values of m and n in such cases. In most of the work reported here either m or n has been assumed to be zero, depending on whether the dispersive or the acid-base effect predominates.

Further support for this concept comes from the dependence of intermolecular interactions on intermolecular distances as shown in Figure 1 [10]. Stronger attractions at the interface due to hydrogen bonds and metallic, covalent and ionic bonds reduce the influence of van der Waals (dispersive) interactions.

Table 6.
Surface energy components of fibers and finish liquids (mJ/m^2)

Fiber or Liquid	σ^d	σ^{AB}
N-66	33.9	47.8
PET	35.2	21.5
PP	23.3	1.5
UCON 50 HB 100	32.6	9.2
EO/PO L-31	33.6	7.9
EO/PO 10/90	39.4	9.7
EO/PO L-81	38.2	6.0

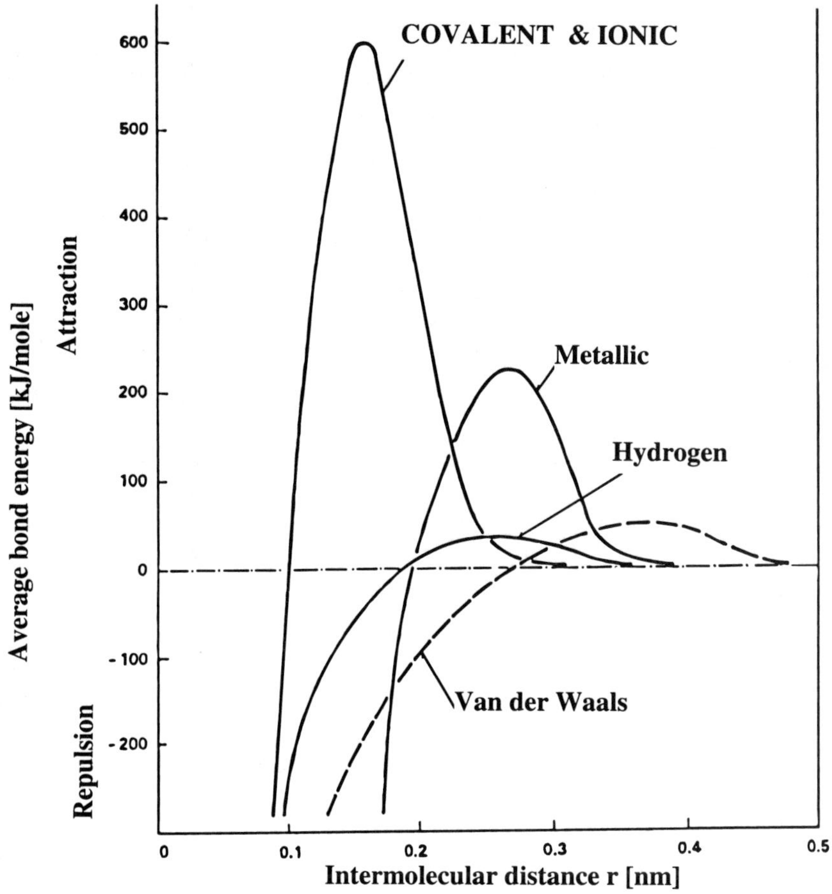

Figure 1. Intermolecular interactions at an interface as a function of the distance between the molecules, from Gutowski [10].

4. CONCLUSIONS

The wetting force data and the surface energy components calculated therefrom for organic textile fibers and finish liquids show that the conventional approach based on concurrent expression of dispersive and acid-base interactions lead to values which do not reflect the true chemical nature of the surface. On the other hand, reasonable values are obtained by assuming either dispersive or acid-base interactions exclusively, depending on the nature of the solid and the liquid involved. The reason for this behavior is the dependence of intermolecular distances on the strength of the interaction and the elimination of weak interactions in favor of strong ones, when possibility exists for both in a given molecule. Specific orientation of the molecules of the more mobile phase (liquid) in the vicinity of the less mobile phase (solid) can contribute significantly to the supression of dispersive interactions in favor of acid-base (donor-acceptor) interactions.

REFERENCES

1. L. A. Girfalco and R. J. Good, *J. Phys. Chem.* **61**, 904 (1957).
2. R. J. Good and M. K. Chaudhury, in: *Fundamentals of Adhesion*, L. H. Lee (Ed.), pp. 137–152. Plenum, New York (1991).
3. F. M. Fowkes, *J. Phys. Chem.* **66**, 382 (1962).
4. R. S. Drago, G. C. Vogel and T. E. Needham, *J. Am. Chem. Soc.* **93**, 6014 (1960).
5. R. J. Good, M. K. Chaudhury and C. J. van Oss, in: *Fundamentals of Adhesion*, L. H. Lee (Ed.), pp. 153–172. Plenum, New York (1991).
6. M. L. Tate, Y. K. Kamath, S. P. Wesson and S. B. Ruetsch, *J. Colloid Interface Sci.* **177**, 579–588 (1996).
7. B. Miller, and R. A. Young, *Textile Res. J.* **45**, 359 (1975).
8. D. Y. Kwok, R. Lin, M. Mui and A. W. Neumann, *Colloids Surfaces A* **116**, 63–77 (1996).
9. D. Y. Kwok, T. Giezelt, K. Grundke, H.-J. Jacobasch and A. W. Neumann, *Langmuir* **13**, 2880–2894 (1997).
10. W. Gutowski, in: *Fundamentals of Adhesion*, L. H. Lee (Ed.), pp. 87–136. Plenum, New York (1991).

Acid-Base Interactions, Vol. 2, pp. 601–614
Ed. K.L. Mittal

Modifying acid–base surface chemistry on E-glass fibers by acid leaching processes

LEANNE G. BRITCHER, SUSANNE GRASER and JANIS G. MATISONS*

Polymer Science Group, Ian Wark Research Institute, University of South Australia, Mawson Lakes 5095, Australia

Abstract—This study examined the attachment of two different amines onto E-glass fibers that had been treated with hydrochloric acid for 2 or 24 hours. XPS and DRIFTS techniques were used to evaluate the changes in acid-base surface parameters. Acid treatment did remove the alkali metals ions such as Ca and Al, therefore making the surface not only more silica-like, but also more acidic. Methylamine and butylamine were then adsorbed onto the untreated, the 2 hour and 24 hour acid treated E-glass fibers and their different adsorption behaviours were compared. The XPS studies revealed that after 2 hour acid treatment, the glass surface became highly active towards adsorption of either methanol (from the rinsing solution) or organic amines (by subsequent treatment). Changes in surface microporosity affect methylamine adsorption, much more than butylamine adsorption.

Keywords: E-glass fibers; surface modification; leaching; amine; XPS; DRIFTS; acid–base.

1. INTRODUCTION

The ability to tailor oxide surface properties has become a very important area for a variety of applications such as biomaterials [1, 2], chromatography [3–5], catalysis [6, 7], and electronics [8]. Glass fibers are often used to reinforce polymer resins. A key factor in the successful employment of such materials is the interaction of various coupling agents at the glass–fiber interface. Recent work by the Polymer Science Group has shown that siloxane oligomers bearing aminohydroxy and amino acid functional groups may, in fact, produce better sizes for composite manufacture, than conventional silane coupling agents [9, 10]. These siloxanes cannot attach to glass surfaces by the hydrolysis mechanisms used by silane coupling agents. Instead the OH, CO_2H and NH groups of the functionalized siloxane oligomers must interact (via hydrogen bonding) with the glass surface. If the acid–base character of the glass fiber surface is altered, even further improvements in adhesion can be gained.

E-glass fibers contain a mixture of acidic (SiO_2, B_2O_3) and basic (Al_2O_3, CaO and Na_2O) oxides which allow the adsorption of a host of reagents, either having basic or acidic properties. Modifying the glass, by making the surface more acidic, should improve the adhesion of organic bases such as amines. Acid leaching can alter the nature

* To whom correspondence should be addressed. E-mail: Polymer@unisa.edu.au

of the glass fiber surfaces. Both borosilicate and aluminosilicate glasses had improved chemical activity after a short immersion in either hydrofluoric or nitric acid [11]. Such mineral acids leach metal ions from the glass surface, replacing the Ca^{2+} and Al^{3+} ions with H^+ [12, 13] according to reaction (1):

$$\overline{M^{n+}} + nH^+ \Leftrightarrow M^{n+} + \overline{nH^+} \tag{1}$$

(the bar denotes association of the metal ion with the glass phase).

The aim of this study was to focus on improving the adhesion of amines to E-glass fibers by pretreating such fibers with a strong mineral acid for various time periods. The elements leaching from the surface could be monitored by inductively coupled plasma (ICP) spectrscocopy. To understand the bonding mechanisms and changes on the E-glass fiber surface, Diffuse Reflectance Infrared Fourier Transform Spectroscopy (DRIFTS) and X-ray Photoelectron Spectroscopy (XPS) were used.

2. EXPERIMENTAL

2.1. Materials

E-glass fibers (water sized) supplied by Ahlstrom Glass Fiber Pty. Ltd. were used as supplied. The average composition of the E-glass fibers as supplied by the manufacturer is as follows: 55% SiO_2; 22% CaO; 15% Al_2O_3+Fe_2O_3; 6% B_2O_3; <1% alkali content; <1% other.

All reagents were of at least laboratory grade. Hydrochloric acid (37%, ACE Chemicals, South Australia), was used as supplied. Butylamine was dried over potassium hydroxide for 16 h, refluxed over potassium hydroxide for 2–3 h, then fractionally distilled from calcium oxide (b.p. 78°C). Methylamine was fractionally distilled (b.p. 48°C). Methanol (ACE Chemicals, South Australia) was dried by distilling over magnesium turnings and iodine using the method of Armarego and Perrin [14]. Toluene (ACE Chemicals, South Australia), was distilled over sodium wire and benzophenone until a violet colour had developed, indicating the formation of the benzophenone ketyl radical.

The glass fibers (2g) were placed in polyethylene vessels containing concentrated hydrochloric acid for 1 h then washed with methanol (4×40 ml) and dried at 110°C for 3 h. The procedure was repeated with different batches of glass fibers staying in the HCl solution for 1, 2, 3, 4, 5 and 24 h. After treatment the HCl solutions were analyzed by Inductively Coupled Plasma (ICP) spectroscopy for evidence of any leached ions.

The adsorption of the amines was achieved by immersing the treated fibers (1g) in toluene (80 ml) containing the appropriate amine (0.8 ml). The solutions were then allowed to stand at room temperature for 3 h, decanted from the glass fiber and the fibers washed with toluene (4×40 ml), and dried (110°C, 3h). The untreated, 2 h leached and 24 h leached fibers were analyzed in detail.

2.2. Diffuse Reflectance Infrared Fourier Transform Spectroscopy (DRIFTS)

The glass fibers were analyzed with a single-beam Nicolet Magna Model 750 spectrometer in the wavenumber region 4000-650cm^{-1}, with the use of an MCT-A liquid nitrogen cooled detector. The interferogram was apodized using the boxcar method installed in the FTIR software (OMNIC FTIR software version 2.0). Signal-to-noise ratio was generally better than 15 000 to 1, using 256 scans. The fibers were mounted parallel to each other in a specially constructed sample holder, which was placed in a Spectra Tech diffuse reflectance apparatus. Each glass fiber sample was scanned 256 times with the fibers at an angle of 90°, with respect to the direction of the infrared beam, for maximum signal-to-noise ratio. The background spectrum was taken from high purity (IR) grade KBr powder (Merck), placed in a sample cup, and levelled to the top of the cup using a spatula. No pressure was applied to the KBr powder in the cup, when packing or levelling.

2.3. X-ray Photoelectron Spectroscopy (XPS)

XPS data were obtained using a Perkin-Elmer PHI 5600 XPS system with a concentric hemispherical analyser and Mg K_α X-ray source operating at 300 W. Pass energies of 89 and 18 eV were used for survey and elemental spectra, respectively. The angle between the sample surface and the analyzer was 45°. The pressure during analysis ranged from 10^{-8} to 10^{-9} Torr. A five min Argon ion etch was applied after the intitial XPS analysis; this sputter rate being calibrated for the complete removal of a Ta_2O_5 film (5 nm thick) on Tantalum. All binding energies (BEs) were referenced to the C1s neutral carbon peak at 284.6eV, to compensate for the effect of surface charging.

2.4. Inductively Coupled Plasma (ICP) Spectroscopy

To determine which elements were removed from the glass fibers after treatment with concentrated HCl, ICP analysis of the HCl solutions after the glass fibers treatment was conducted. A sample of the HCl solution (5ml) was diluted with water (100 ml) and the solution was analyzed for calcium, aluminium, silicon, sodium, potassium and boron on a Spectroflame ICP spectrometer (Germany). The standards of all of the elements except silicon were 5% in nitric acid. The silicon standard was 20% in hydrochloric acid.

2.5. Streaming Potential Analysis (SPA)

The untreated E-glass fibers were sent to the Institute of Polymer Research in Dresden, Germany for SPA on a PAAR EKA Electrokinetic Analyzer using a cylindrical measuring cell. Streaming potential measurements were recorded in an aqueous solution of 3×10^{-4} M potassium chloride. Streaming potentials were then recalculated to zeta potentials and plotted versus pH to give information on the surface charge of the glass fiber samples.

3. RESULTS AND DISCUSSION

E-glass fibers allow a wide range of reagents to adsorb or interact with the surface, due to the presence of both acidic and basic groups. The surface chemistry of the glass is determined by each component that introduces either basic or acidic sites to the glass surface. We were interested in increasing the number of acidic sites on the glass surface, to promote rapid adsorption of basic organic amine compounds. By removing some of the basic alkali metals such as Al and Ca with concentrated mineral acid, the glass surface approaches that of pure porous silica gel [11]. Streaming potential measurements on untreated E-glass fibers (Figure 1) show an isoelectric point (IEP) at pH=4.7, suggesting an acidic surface. The surface would already promote the interaction of organic amines with surface active sites. Such surface interactions with an organic base have been defined [15] by Equation (2):

$$\Delta_B = pK_{A(B)} - IEP \quad (2)$$

where Δ_B = surface interaction with an organic base, $pK_{A(B)}$ = ionization constant of an organic base, and IEP= isoelectric point.

Table 1 shows the Δ_B interaction values for two amines, methyl and n-butylamine, with E-glass fiber, alumina and silica. After the E-glass fibers have been treated with mineral acids, the IEP should lower from the pH=4.7 of a glass surface, towards a silica surface (IEP at pH 2.0) and therefore the acid-base forces between organic amines and such a treated surface should be even stronger. We have tried to systematically increase the surface acidity of E-glass fibers by immersing them in hydrochloric acid for various times, which will make the fibers to have a more silica-like surface, as basic surface species such as Ca^{2+} and Al^{3+} are removed.

E-glass fibers were treated with concentrated HCl for periods of 1, 2, 3, 4, 5 and 24 h, then washed with methanol and dried for 3 h at 110°C. ICP analysis of the HCl treatment solutions shows that calcium was leached from the surface the most followed by aluminium and boron (Figure 2). The levels of potassium and sodium were below the detection limits of the instrument. A linear regression best fit line has been drawn for each element as its concentration increases in the treatment solution with time, however, over longer periods of time (up to 20 days), leaching of these cations is exponential [16]. ICP analysis has provided information on what has been removed from the surface into solution, but to determine the changes at the surface of the glass, DRIFTS and XPS techniques were used.

Table 1.
Acid–base interactions for organic amines with different surfaces [15]

		$\Delta_B = pK_{A(B)} - IEP$		
Amine	$pK_{A(B)}$	Glass (IEP 4.7)	SiO_2 (IEP 2.0)	Al_2O_3 (IEP 8.0)
methylamine	10.6	5.9	8.6	2.6
n-butylamine	10.8	6.1	8.8	2.8

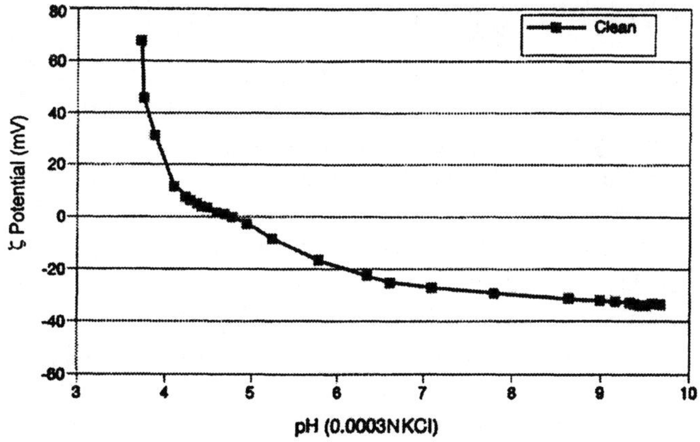

Figure 1. Streaming potential measurements on untreated E-glass fibers.

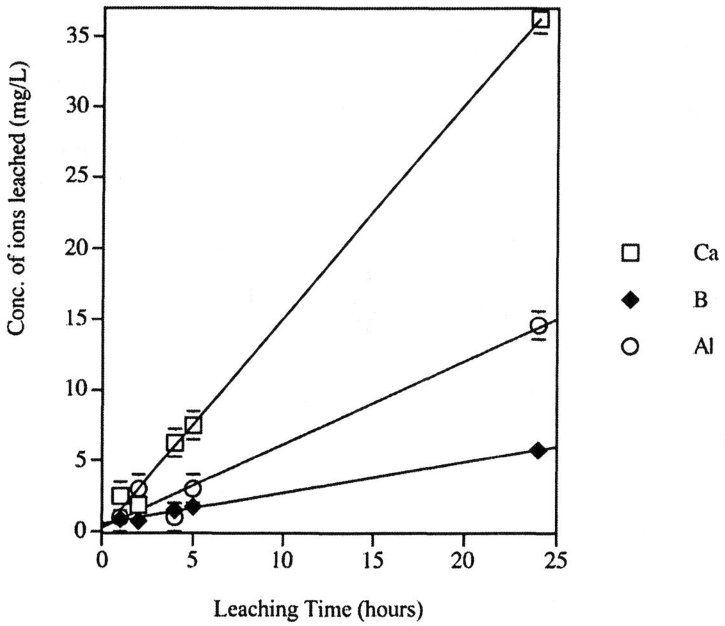

Figure 2. Leaching of cations from E-glass fibers in concentrated HCl solution.

The DRIFTS spectra of the untreated, the 2 h acid treated and the 24 h acid treated E-glass fibers are shown in Figure 3. A small peak at 1634 cm^{-1} is now appearing, which is assigned as an angle deformation mode of silica. The relative area of the broad OH stretching region between 3000–3600 cm^{-1} has increased, with increasing treatment time in acid. There are three possible processes occurring that could result in such an increase:

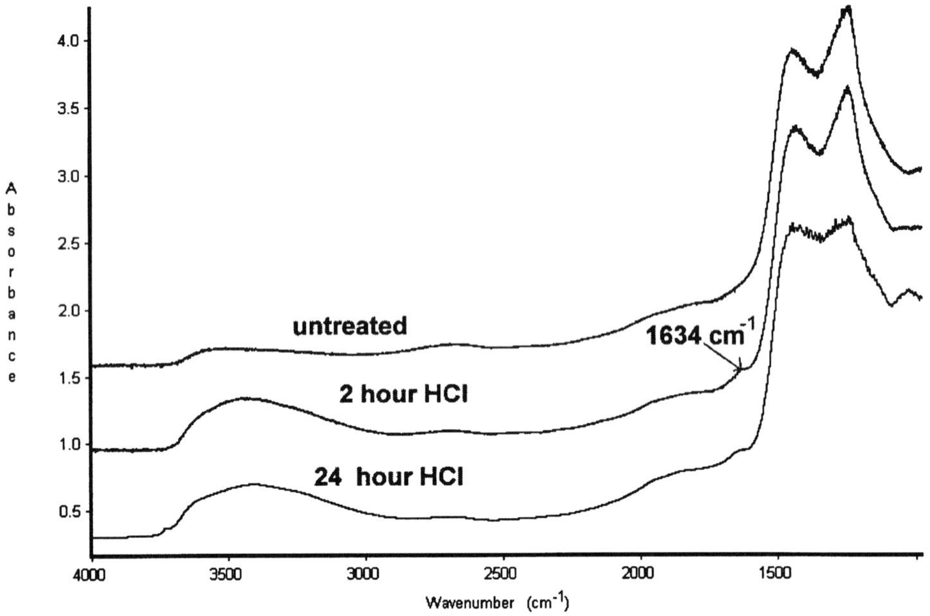

Figure 3. DRIFT spectra of the untreated, the 2 h acid treated and 24 h acid treated E-glass fibers.

1. As acid leaching occurs more SiOH surface groups will form, which will result in stronger hydrogen bonding to the water in the hydrochloric acid solution, and so cause the increase in hydrogen bonded water seen by DRIFTS (3000-3400 cm^{-1}).
2. Water resident in the bulk of the glass diffuses to the surface after the leaching, facilitated by the 'pores' left by departing Ca or Al ions.
3. Methanol used to rinse the fibers has adsorbed onto the fiber surface after the acid treatment.

It is uncertain whether one or all of these mechanisms are contributing to the increase in the OH stretching vibrations and whether the same mechanisms are operating in each of the acid leached samples, as the XPS data (*vide infra*) suggest that there are differences in the adsorption by varying times of acid treatment. After 24 h treatment a shoulder starts to appear at 3760 cm^{-1} in the DRIFTS spectrum, indicating an increase in the number of free terminal OH groups. Boron oxide is still on the glass surface after 24 h as the broad overtone at 2700 cm^{-1} remains visible. The DRIFTS spectra do show changes in the glass surface after acid leaching for 2 or 24 h, but XPS data should give more information about changes in the chemical species occurring on the fiber surface.

Figures 4 (a) and (b) show the comparative XPS data for the untreated and acid treated glass fibers before and after a 5-nm argon ion etch, respectively. After 2 h of leaching, the calcium and boron levels have decreased, while aluminium has increased slightly, but chlorine from the HCl has adsorbed (7.4%). However, chlorine is not present in the 24 h leached sample and here the glass surface consists mainly of a silica-like surface containing only small amounts of the minor glass elements. Clearly the surface structure and adsorption sites are different in the 2 hour and 24 hour leached samples. Argon ion etching of the fiber shows that the acid treatment has not just leached the elements in top surface layers of the glass, as there are differences between the leached and untreated surfaces (Figure 4(b)).

(a)

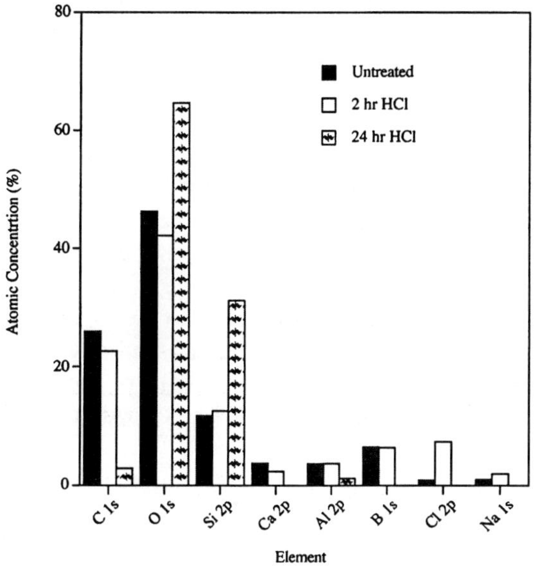

(b)

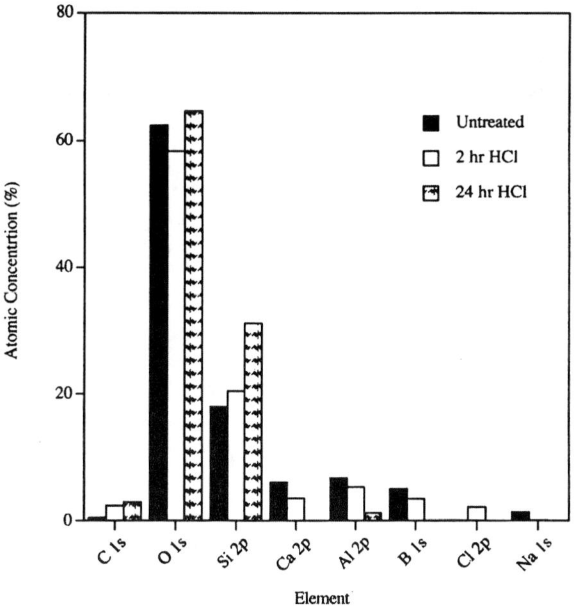

Figure 4. Comparative XPS data for the untreated, the 2 h acid treated and the 24 h acid treated E-glass fibers; (a) no argon ion etch and (b) after a 5 nm argon ion etch.

The high level of carbon on the surface, especially after 2 h acid treatment, is surprising, as the acid should have a cleaning effect on the surface. A closer look at the C (1s) multiplex (Figure 5), reveals that there are C=O species present (B.E.=287.5 eV). Methanol used to wash the fibers after acid treatment appears to have adsorbed onto the fibers and some oxidation has occurred, which is known to happen on silica surfaces under similar conditions [17, 18]. After the 24 h acid treatment there is still a small amount of C=O (7.5% of the now greatly reduced carbon signal present). Although, only a small amount of the more basic metal ions was removed after 2 h acid treatment, the surface appears to be highly active as both methanol and chlorine are adsorbing. The 'surface activity' then disappears after 24 hours as the chlorine is subsequently removed by the extended acid treatment, and smaller amounts of methanol remain. Tanaka *et al.* [19] found that the pore volume of E-glass fibers initially increased after short periods of acid treatment, but then decreased due to the collapse of micropores on prolonged acid treatment. The stage at which the glass surface is effectively microporous (2 h) is likely to contribute to the increased adsorption capacity or higher surface activity of the E-glass fibers. Such porosity will assist the adsorption of organic amine molecules, but whether the strength of increased acid–base interactions will continue to assist such adsorption upon collapse of the micropores remains to be determined (*vide infra*).

The information about the changes in the glass surface structure can also be gained by comparing the oxygen and silicon multiplexes after each treatment. The FWHM (Full Width at Half Maximum) of the oxygen multiplex narrows after 2 h acid treatment from 3.5cV (untreated) to 2.5eV (treated). After 2 h the O (ls) binding energy shifts from 531.5eV to 532.6eV, consistent with a more silica-like surface. Further acid treatment (24 h) does not result in any further changes in the oxygen multiplex. The silicon multiplex similarly narrows and the binding energy shifts from 102.2 eV (untreated) to 103 eV (2 h) and then to 103.5 eV (24 h), again consistent with a silica-like surface. The binding energies of the minor glass elements, if still present, remain unchanged. In summary, the DRIFTS and XPS analyses have shown that acid treatment gradually causes a silica-like layer to form on the glass surface, which does however have varying surface activity depending on the residence time of the glass surface in acid solution.

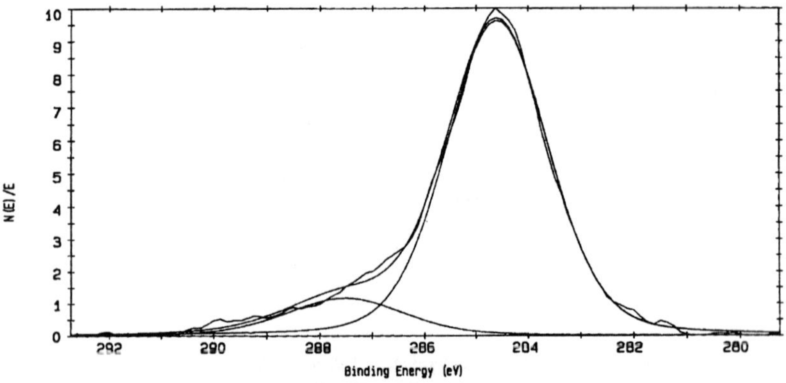

Figure 5. The carbon multiplex for the 2 h acid treated E-glass fibers.

Butylamine and methylamine were adsorbed onto the untreated and the acid treated glass fibers from toluene solution. The DRIFTS spectra for both amines adsorbed onto the fibers are shown in Figures 6 (a) and (b). Both the unleached glass fiber spectra show no C–H absorbances between 2800-3000 cm^{-1} after either amine treatment. In spite of the DRIFTS results, XPS data (*vide infra*), do show some amine

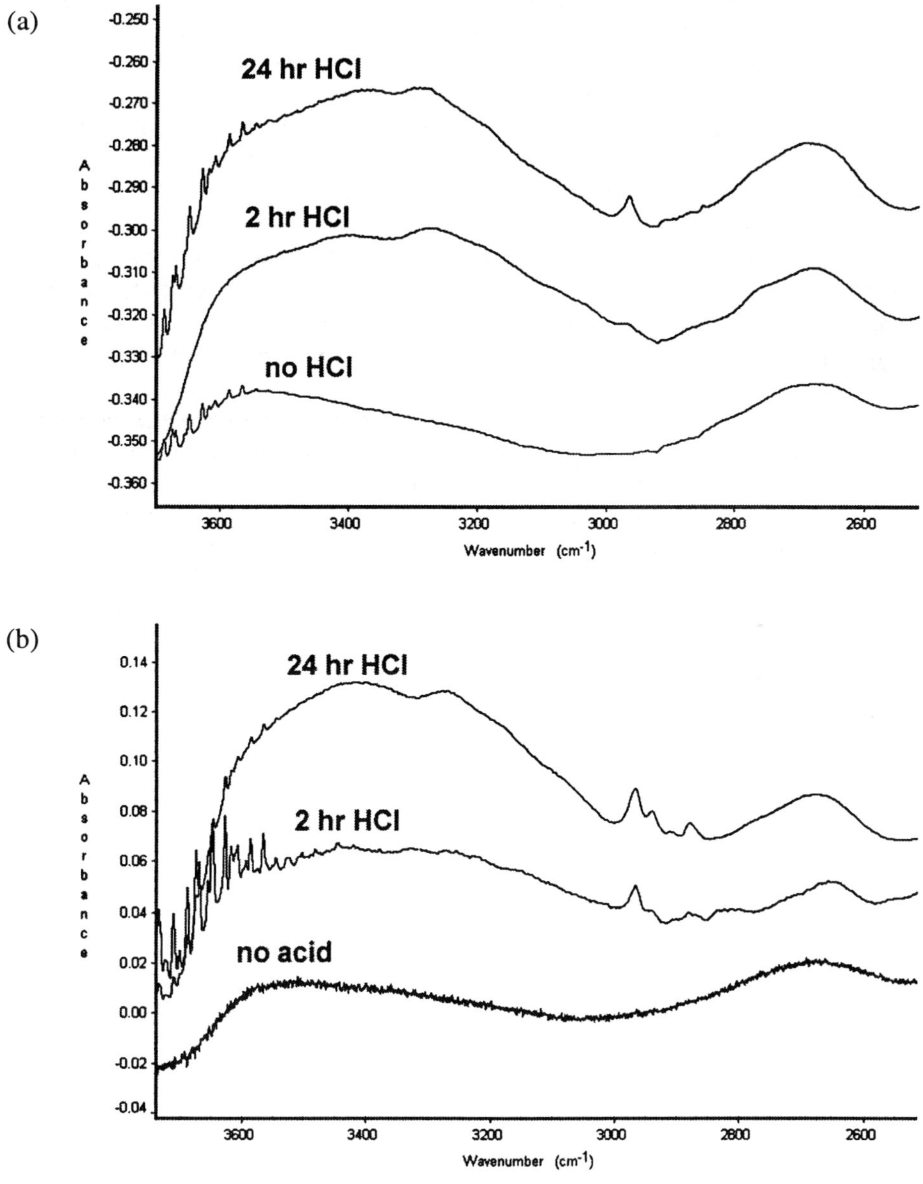

Figure 6. DRIFT spectra of untreated, the 2 h acid treated and 24 h acid treated E-glass fibers; (a) with methylamine adsorbed and (b) with butylamine adsorbed.

on the glass surface, indicating that the DRIFTS technique is not as surface specific or sensitive as XPS. The 2 h acid treated fibers do show signs of amine adsorption, with broad N–H stretching vibrations at 3250 cm^{-1} and 3375 cm^{-1} (from N–H or NH_3^+). Asymmetric and symmetric CH_2 (2850 cm^{-1} and 2925 cm^{-1}) and CH_3 (2900 cm^{-1} and 2975 cm^{-1}) stretching vibrations of the adsorbed butylamine (Figure 6(b)) are seen using the 2 h acid treated glass samples, and these become more prominent using the 24 hour acid treated glass samples. The CH_3 stretching vibrations have the highest peak intensities suggesting that the organic amines are pointing with the amino groups toward the surface, as expected (only for the 24 h acid treated fibers, are the methylamine CH_3 stretching vibrations clearly seen). The amines are likely to be adsorbing into micropores after 2 hs and are not detected by the DRIFTS. After 24 h the surface area of the glass should be increased, therefore more amine can be detected. The DRIFT spectra suggest that more amine is on the glass treated longer in acidic solutions, but the XPS data reveal a slightly different picture.

Tables 2 and 3 provide the XPS data for the methylamine and butylamine treated glass fibers, respectively. Both amines have adsorbed onto the untreated and the acid treated surfaces, with the highest nitrogen content seen for the 2 h HCl treated glass fibers (methylamine adsorbed). In all of the samples the concentration of Ca and Al decreased, while the carbon content increased, after amine adsorption. The chlorine level, which was quite high (7%) in the 2 h acid treated fibers (without amine), has considerably decreased (0.67% methylamine) and after argon etching the chlorine level is still low, indicating each amine displaces chloride ions from the surface. It is not possible to detect by XPS whether methanol still adsorbs on the 2 h acid treated/amine adsorbed fibers, as the C–N and C–O groups occur at the same binding energy in the carbon multiplex. Nitrogen levels for butylamine adsorbed onto the 2 h and the 24 h acid treated samples actually increase after argon etching, whereas methylamine adsorbed fibers show a decrease in nitrogen levels. The carbon level is higher in the butylamine treated fibers because of its inherent higher carbon content. There is no evidence for C=O species prior to argon etching, so it is unlikely that methanol is still adsorbing onto these surfaces. The (CH_3) carbon from butylamine would be at the outer surface, as amino groups interact with the surface silica groups, so the C–N species are likely to be revealed after etching. Amine adsorption has increased with acid treatment of the surface, but it also appears that the 2 h acid treatment has highly activated the glass surface, so that very high levels of methylamine are adsorbed. The high nitrogen content before and after argon etching suggests that there are multiple layers of methylamine self-associating outwards from the glass surface. Butylamine surface concentration does not increase with acid treatment, instead it appears that the amine groups interact with the M–OH groups on the glass; this is consistent with the longer carbon chain disrupting the self associating hydrogen bonding networks of the organic amine. Before acid treatment, the amine groups may associate with each other via hydrogen bonding, and the strength and number of surface acidic sites alone causes association with the glass surface (Figure 7(a)). The increased microporosity of the surface (after 2 h) may help increase adsorption of the smaller methylamine, where the longer alkyl chain prevents further association of more butylamine in the confined space (Figure 7(b)). Therefore, it appears that surface acid-base interactions are not the sole driving force causing methylamine adsorption, but surface porosity also plays a significant role. Acid-base interactions appear to drive the amines like butylamine to the acidic surface sites, but hydrogen bonding associations may well determine molecular associations beyond the glass surface sites.

The nitrogen multiplexes show differences in the ratio of acidic (NH_3^+, B.E.= 401.8 eV) and neutral (NH_2, B.E. = 399.3 eV) amine species for every sample. Figure 8 shows a comparison of acidic and neutral amine species for the methylamine treated surfaces. It appears that there is a limit to the amount of acidic species formed on the surface (Figure 8). The untreated glass with butylamine adsorbed showed only the neutral amine species, and the nitrogen concentration of the 2 h and 24 h HCl treatments was very low. After argon etching, curve fitting was not done as the etching

Table 2.
XPS data for methylamine adsorbed onto variously treated and untreated E-glass fibers

Sample	C 1s	O 1s	Si 2p	Ca 2p	Al 2p	B 1s	Cl 2p	Na 1s	N 1s
0 h HCl	22.35	47.78	15.60	1.96	4.44	2.60	0.75	0.93	3.50
0 h HCl after Ar ion etch	1.89	61.14	19.96	5.41	6.66	3.48	0.21	0.28	0.96
2 h HCl	34.12	38.79	16.35	0.41	0.50	0.61	0.67	0.25	8.29
2 h HCl after Ar ion etch	7.49	56.69	28.70	0.77	0.61	<0.05	0.82	0.52	4.41
24 hour	19.78	52.69	23.84	0.28	0.61	0.41	<0.05	<0.05	2.40
24 h HCl after Ar ion etch	4.09	62.66	29.18	0.24	1.85	<0.05	<0.05	<0.05	1.98

Table 3.
XPS data for butylamine adsorbed onto variously treated and untreated E-glass fibers

Sample	C 1s	O 1s	Si 2p	Ca 2p	Al 2p	B 1s	Cl 2p	Na 1s	N 1s
0 h HCl	29.30	43.87	11.42	3.16	3.25	4.47	1.07	1.16	2.28
0 h HCl after Ar ion etch	3.02	61.69	18.60	5.53	5.62	3.90	0.31	0.50	0.81
2 h HCl	34.63	39.15	24.68	0.26	0.37	<0.05	<0.05	<0.05	0.91
2 h HCl after Ar ion etch	13.00	52.92	31.07	0.17	0.57	<0.05	<0.05	<0.05	2.28
24 h HCl	29.74	43.40	26.30	<0.05	<0.05	<0.05	<0.05	<0.05	0.56
24 h HCl after Ar ion etch	6.88	58.77	30.29	0.28	1.31	<0.05	<0.05	<0.05	2.52

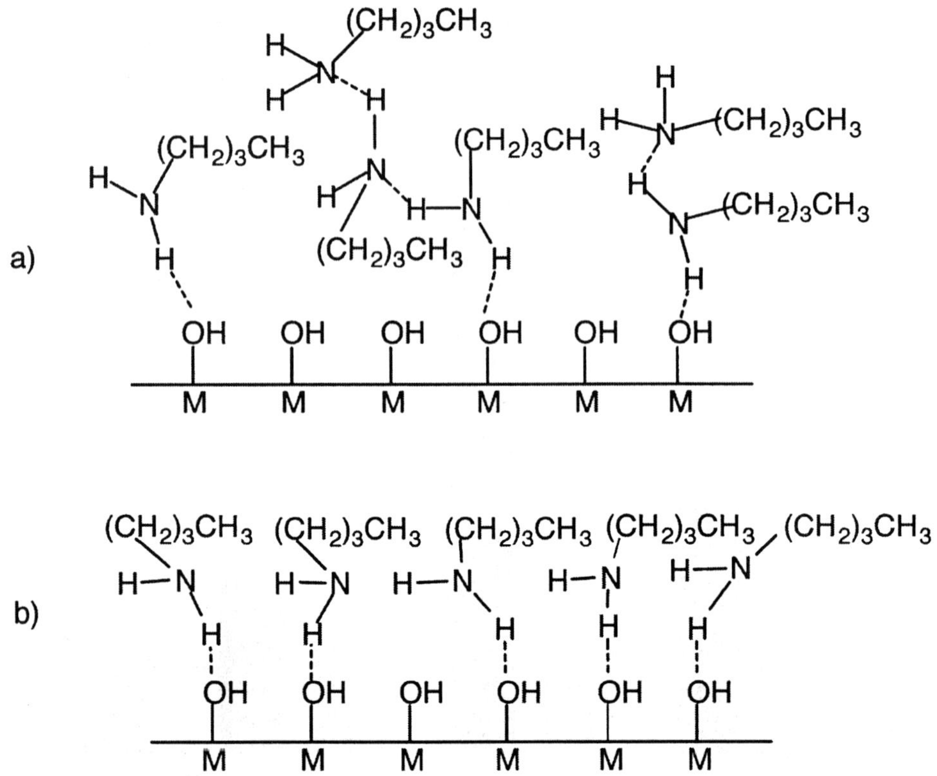

Figure 7. Proposed adsorption mechanism for butylamine adsorbed onto (a) untreated E-glass fibers and (b) acid treated E-glass fibers.

process may cause changes in the nitrogen surface species and their levels. Fowkes [20] has shown by depth profiling with XPS that the acidic nitrogen species are located at the glass interface. The concentration of acidic nitrogen groups did decrease when more methylamine was resident on the surface, further suggesting that multilayers of hydrogen bonded methylamines self associate above the amines interacting directly with the glass surface. Interestingly, when less acidic nitrogen species are detected, the carbon multiplex concurrently reveals some oxidation of the amines on the surface (C=O, B.E.=288 eV). The observed carbonyl groups resulting from oxidation could either interact with the surface M-OH groups or hydrogen bond to surface amino groups. The amount of acidic species detected by XPS is really dependent on how much their signals are attenuated by other surface groups, and an angle resolved depth profile is really needed to determine if the ratio of acidic to neutral amino species does change with treatment.

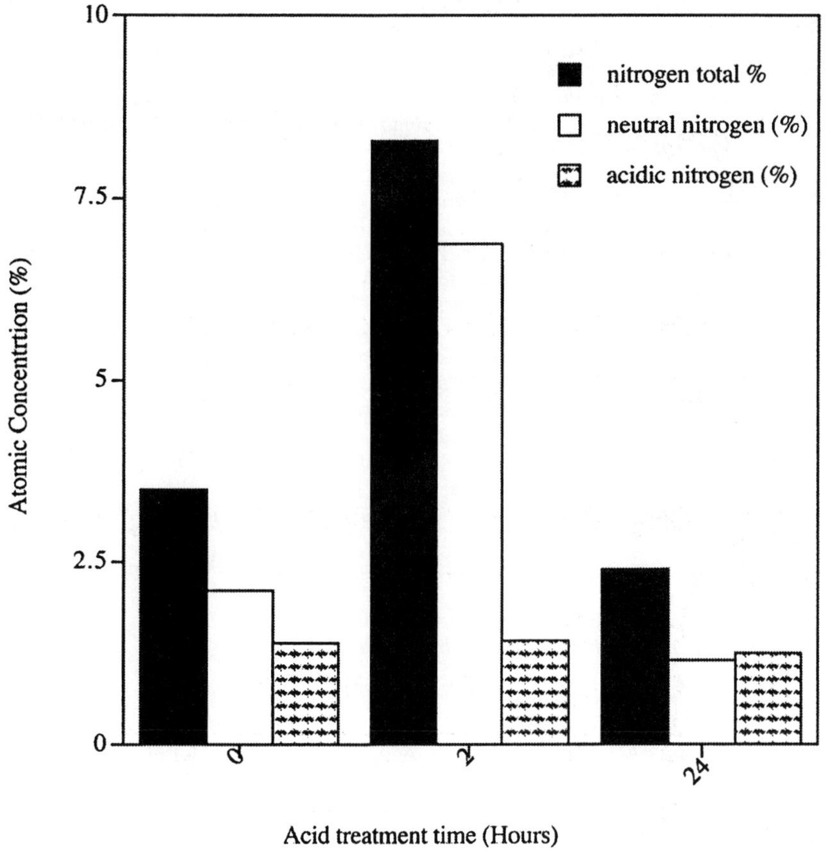

Figure 8. Comparison of the atomic concentration of acidic and neutral nitrogen species for methylamine adsorbed onto untreated and acid treated E-glass fibers.

4. CONCLUSIONS

The treatment of E-glass fibers with hydrochloric acid creates a more acidic type silica surface. DRIFTS and XPS studies revealed that acid treatment resulted in a highly activated surface towards adsorption, after just 2 h. The adsorption of methylamine on the glass surface increased dramatically after 2 h acid treatment of the glass fibers, but then decreased after 24 h. Acid–base interactions and an increase in the microporosity of the glass surface both affect the adsorption of organic amines on glass fiber surfaces. The concentration of butylamine on the glass surface did not increase with time of acid treatment, given the butylamine adsorbs with the amino groups interacting directly with surface hydroxyl groups. Therefore, increasing the acidity of the E-glass fiber surface can allow more amine groups to adsorb at the interface; but with smaller molecules, the microporosity will also affect the adsorption behaviour.

REFERENCES

1. C. Ziegler and W. Gopel, *Curr. Opin. Chem. Biol.* **2**, 585–591 (1998).
2. B. Kasemo, *Curr. Opin. Solid State Mater. Sci.* **3**, 451–459 (1998).
3. A. Gambero, L. T. Kuboto, Y. Gushikem, C. Airoldi, J. M. Granjero, E. M. Taga and E. F. C. Alcantara, *J. Colloid. Interface. Sci.* **185**, 313–316 (1997).
4. M. J. Wirth and H. O. Fatunmbi, in: *Chemically Modified Surfaces*, J. J. Pesek and I. E. Leigh (Eds), pp. 203–209. Royal Society of Chemistry, Cambridge (1994).
5. S. Heron and A. Tchapala, *Analysis* **21**, 327–347 (1993).
6. Y. Iwasawa, in: *Tailored Metal Catalysts*, Y. Iwasawa (Ed.), pp. 1–74. D. Reidel, Dordrecht (1986).
7. M. Lenarda, L. Storaro and R. Ganzerla, *J. Mol. Catalysis A- Chem.* **111**, 203–237 (1996).
8. C. T. Campbell, *J. Chem. Soc. Faraday Trans.* **92**, 1435–1445 (1996).
9. P. S. Arora, J. G. Matisons, A. Provatas and R. St. C. Smart, *Langmuir* **11**, 2009–2017 (1995).
10. A. Provatas, J. G. Matisons and R. St. C. Smart, *Langmuir* **14**, 1656–1663 (1998).
11. M. L. Hair, in: *Clean Surfaces: Their Preparation and Characterization for Interfacial Studies*, G. Goldfinger (Ed.), pp. 269–283. Marcel Dekker, New York (1970).
12. A. G. Metcalfe and G. K. Schmitz, *Glass Technol.* **13**, 5 (1972).
13. H. Hojo, and K. Tsuda, *Proc. 34 th Ann Conf. on Reinforced Plastics/Composites*, Institute of the Society of Plastics Industry, New Orleans, Paper 3-D (1979).
14. W. L. Armarego and D. D. Perrin, *Purification of Laboratory Chemicals*, 3rd Edn. Pergamon, Oxford, UK (1988).
15. J. C. Bolger, in: *Adhesion Aspects of Polymeric Coatings*, K. L. Mittal (Ed.), pp. 3–18. Plenum Press, New York (1983).
16. R. L. Jones and H. D. Chandler, *J. Mater. Sci.* **20**, 3320 (1985).
17. P. Kohli, *Characterization and Properties of Treated Fumed Silica*, Masters Thesis, University of South Australia (1995).
18. W. Hill, H. Miessner and G. Ohlmann, *J. Chem. Soc. Faraday Trans.* **85**, 691 (1989).
19. H. Tanaka, K. Kuraoka, H. Yamanaka and T. Yazawa, *J. Noncryst. Solids* **215**, 262–270 (1997).
20. F. M. Fowkes, in: *Acid–Base Interactions: Relevance to Adhesion Science and Technology*, K. L. Mittal and H. R. Anderson (Eds), pp. 93–115. VSP, Utrecht, The Netherlands (1991).

Acid-Base Interactions, Vol. 2, pp. 615–624
Ed. K.L. Mittal

Effect of PMMA coating on the acid-base interactions of carbonate rock

EMIL CHIBOWSKI* and LUCYNA HOŁYSZ
Department of Interfacial Phenomena, Faculty of Chemistry, Maria Curie-Skłodowska University, 20-031 Lublin, Poland

Abstract—Wicking and contact angle experiments with probe liquids, n-octane, diiodomethane, water and formamide, were conducted to determine the surface free energy components of carbonate rock originating from Lublin's region (Poland). This rock is a relatively soft and porous material which can be used for inside and outside building walls finishing. Because the material is porous and easily wettable by water, so for practical purposes its surface must be hydrophobized. Also, if this material were coloured that would be more attractive for the application. The study reported here also has a more general aspect relative to monuments protection against erosion. As the hydrophobizing agent PMMA solution in chloroform was used, and for coloring effects Methylene Blue, "Ultra Extra Blue" (Remix, Poland) and Yellow Dye (Inchem., Poland) were selected. The PMMA treated surface showed drastically reduced electron-donor component, from 57.3 mJ/m^2 (untreated surface) to 13.5 mJ/m^2, which is practically equal to the value for the pure PMMA surface. The electron-acceptor component increased slightly after the treatment (from 1.02 to 1.64 mJ/m^2) and the apolar component did not change. As a consequence, the work of water spreading on such rock surface became much more negative, -40.4 mJ/m^2 (-4.0 mJ/m^2 for untreated surface), which means that the surface became highly hydrophobic. The dyeing of the surface without the polymer deposition caused a slightly less hydrophobization of the surface, but in all cases the work of water spreading was in the range -34 to -25.5 mJ/m^2. It means that the hydrophobization was effective in case of dyeing also. So it is clear that both PMMA deposition as well as dyeing can be used for practical purposes where hydrophobization of a surface is required.

Keywords: Carbonate rock; hydrophobization; PMMA treatment; dyeing; wetting; acid-base components.

1. INTRODUCTION

In the region of Lublin (Poland) there are many carbonate rock deposits which, among other applications, can be used as an attractive material for finishing outside or inside walls of buildings. The rocks are relatively soft and, therefore, they can be easily cut in the shape of bricks or plates. As they are porous they can also be colored with some organic or even inorganic dyes. However, a serious disadvantage in using these rocks is their porous and hydrophilic character. As a result, they suck moisture and easily undergo erosion. Therefore, for any practical purpose they have to be hydrophobized. However, porosity in the hydrophobized material is often needed for ventilation purpose.

* To whom correspondence should be addressed. E-mail: emil@hermes.umcs.lublin.pl

The purpose of this study was some fundamental investigations of one of the carbonate rocks with respect to modification of its surface to import hydrophobic character. First, to evaluate qualitatively the changes in the surface properties, penetration rates of water into bars of the modified rock were measured. The surface was pretreated with cationic or anionic surfactant, as well as n-dodecane. However, for a quantitative evaluation of the hydrophilic/hydrophobic changes in the surface properties, the surface free energy components were determined by the wicking method [1–4] or contact angle method. Here PMMA solution in chloroform was used as the hydrophobizing agent. This study has also a more general aspect, for example, to protect historical monuments against erosion [5], especially because of increasing acidity of rain and environmental moisture.

To determine the surface free energy components the approach of van Oss *et al.* [6–8] was applied. In this approach, the surface free energy components are considered to be apolar Lifshitz-van der Waals, γ^{LW}, and polar electron-donor, γ^{-}, and electron-acceptor, γ^{+}. The work of adhesion between a solid (s) and a liquid (l) can then be expressed as:

$$W_a = 2\left[\left(\gamma_s^{LW}\gamma_l^{LW}\right)^{1/2} + \left(\gamma_s^{-}\gamma_l^{+}\right)^{1/2} + \left(\gamma_s^{+}\gamma_l^{-}\right)^{1/2}\right] \qquad (1)$$

From the Young equation the following relationship between contact angle and W_a results:

$$\gamma_l\left(1+\cos\theta\right) = W_a = 2\left[\left(\gamma_s^{LW}\gamma_l^{LW}\right)^{1/2} + \left(\gamma_s^{-}\gamma_l^{+}\right)^{1/2} + \left(\gamma_s^{+}\gamma_l^{-}\right)^{1/2}\right] \qquad (2)$$

If contact angles of three probe liquids (for example, diiodomethane, water and formamide) are measured (for which their surface tension components are known) then simultaneous solution of three Eq. (2) allows calculation of three unknown components of the surface free energy of the solid. In the case of the wicking method to determine these components, the wicking rates of three probe liquids have to be measured for bare and rock bars equilibrated (precontacted) with the liquid vapor, but the procedure is more complicated and the details are described elsewhere [1–4].

2. EXPERIMENTAL

The composition of the carbonate rocks and some of the characteristics are presented in Table 1 [9]. These characteristics vary over a range depending on the deposit origin. The principal components of the rocks are CaO and SiO_2. The storage of the rocks in an open area leads to some changes in their properties, and this is shown in Fig. 1a–d for the rocks originating from three different deposits. As can be seen the aging processes occur during first 5–6 months [9]. The porosity, permeability and toughness of the rocks increase, except for porosity of one rock (Fig. 1a). The observed changes must be due to chemical interactions of the rock with water and carbon dioxide, as well as, some physicochemical interaction with water vapor. It may be concluded that after the rock comes in contact with atmosphere, first, "fast" erosion

Table 1.
Composition of the carbonate rocks and some of their characteristics

Composition / Parameter	Value
CaO, %	20 – 45
MgO, %	0.2 – 1.6
SiO_2, %	14 – 33
Fe_2O_3, %	0.8 – 4
Al_2O_3, %	1 – 5
$CaCO_3 + MgCO_3$, %	53 – 71
Density, g/cm^3	2.2 – 2.6
Permeability, % w/w	10 – 37
Porosity, % v/v	35 – 51
Toughness (dry), MPa	2.3 – 38
Toughness (wet), MPa	1.5 – 18

processes take place (for about 6 months) and then any changes in the rock structure are much slower. However, any longer contact with, for example, acidic rain would cause a slow dissolution of the carbonates present in the rocks.

To determine the surface free energy components from the wicking and/or contact angles method, bars of $25 \times 5 \times 5$ mm in size were cut from the rock plate. Then they were dried at 110 °C for 90 min and kept in a desiccator. The details about experimental procedure and theoretical background of the wicking method were published elsewhere [1–4].

To modify the external surface as well as the inside of pores the bars were dipped into 5 ml of PMMA solution in chloroform (0.6g/100 ml) for 30 min, then dried at room temperature for 15 min and kept in a desiccator. Other bars, instead of dipping them into the PMMA solution, were wetted with the above solution (0.2 ml on each bar wall) and then were treated in the same way as above.

The effect of dyeing of the rock on its surface free energy components was also studied. For this purpose Methylene Blue, "Ultra Extra Blue" (Remix, Poland) and a Yellow Dye (Inchem, Poland) were used. In case of Methylene Blue the bars were dipped into 20 ml of 0.003 M aqueous solution for 16 h, dried at 60 °C for 90 min and then treated with PMMA by dipping the bars into the solution or were wetted by the PMMA solution as described above. The same procedure was applied in the case of "Ultra Extra Blue" dye, except the concentration was 0.5g/20 ml H_2O. The bars were similarly colored with the yellow liquid dye (2 ml + 18 ml H_2O) by keeping them for 16 h in the solution and then drying at 60 °C for 90 min, but after cooling the bars to 20 °C the PMMA solution (0.2 ml per wall) was spread over because chloroform rinsed out the dye, if the bar was dipped into the solution.

Thus prepared bars were used for wicking experiments or contact angle measurements if polar probe liquid did not wick the bar. The wicking [1–4] was performed at 20 ± 1 °C for a distance of 2 cm and repeated 3–4 times for each sample. The average penetration time for a particular distance was taken for determination of the surface free energy components. Similarly, contact angles were measured for 7–10 droplets of

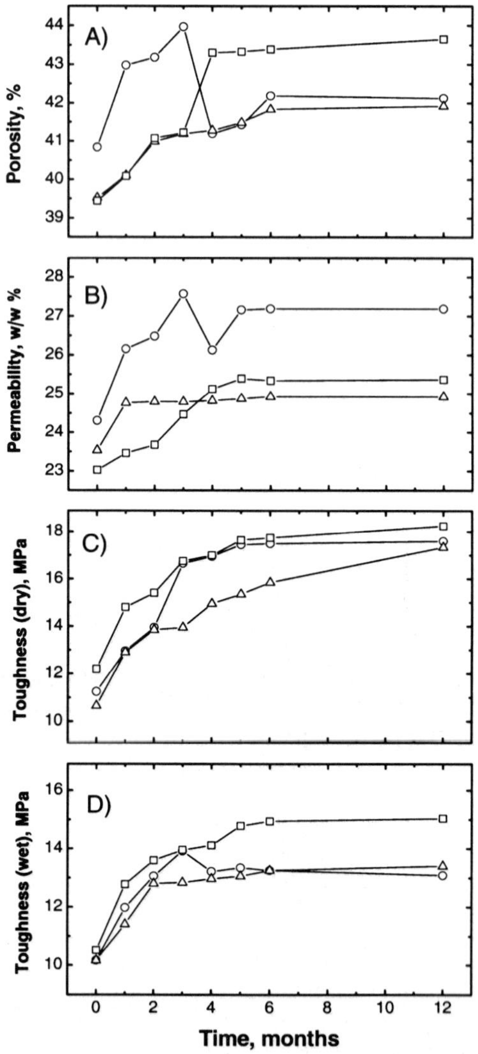

Figure 1. Changes in some parameters of carbonate rock of three different origins, as a function of the storage time: A) porosity, B) permeability, C) toughness of dry rock, and D) toughness of wet rock.

each probe liquid on both sides (advancing contact angles) and the reproducibility was within ±2 deg. For wicking experiments, n-octane, water and formamide were used as the probe liquids and for contact angles instead of n-octane diiodomethane was employed. The schematic of the wicking chamber is shown in Fig. 2.

Prior to the above described experiments, to learn about the hydrophobization effect of some surfactants and oil, the bars were also dipped in 5 ml of 10^{-3} M methanolic or aqueous solution of sodium dodecyl sulfate (SDS) or cetyltrimethyl ammonium bromide (CTMABr), 10 ml of n-dodecane, or 10 ml of paraffin oil solution in n-heptane (5 ml + 5 ml), for 16 h. Then the bars were dried at 50 °C for 1 h and used for wicking of water.

Wicking chamber

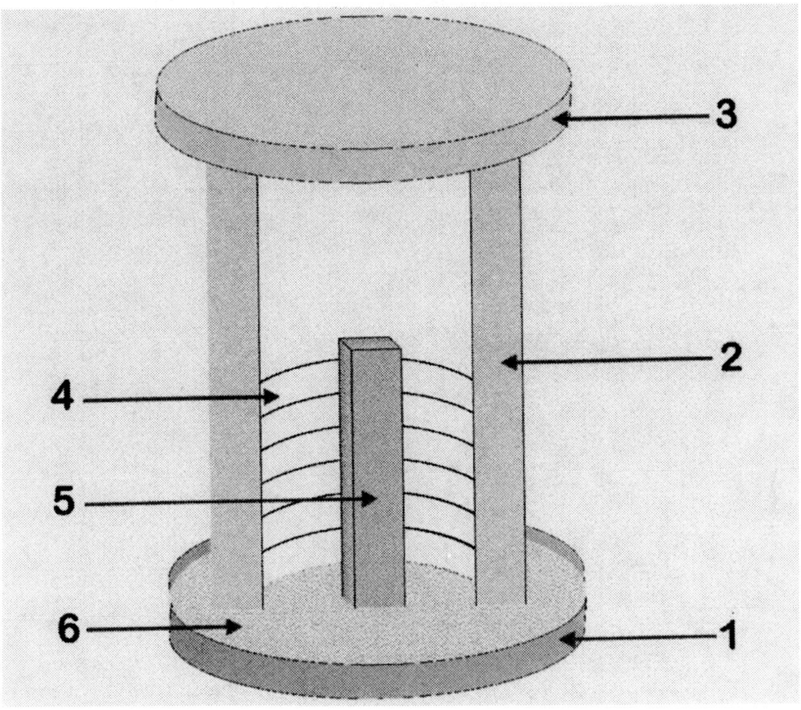

1 – liquid container
2 – glass cylinder
3 – cap
4 – ruler
5 – rock bar
6 – probe liquid

Figure 2. Schematic representation of the wicking chamber.

3. RESULTS AND DISCUSSION

The qualitative hydrophobization effects of the carbonate rock surface with different substances are shown in Fig. 3 as the penetration times of water into modified bars of the rock with that for the untreated bar. According to Washburn's equation [1–4], the relation between squared penetrated distance x versus time t should be a linear function:

$$x^2 = \frac{R\,t}{2\eta} \Delta G \quad (3)$$

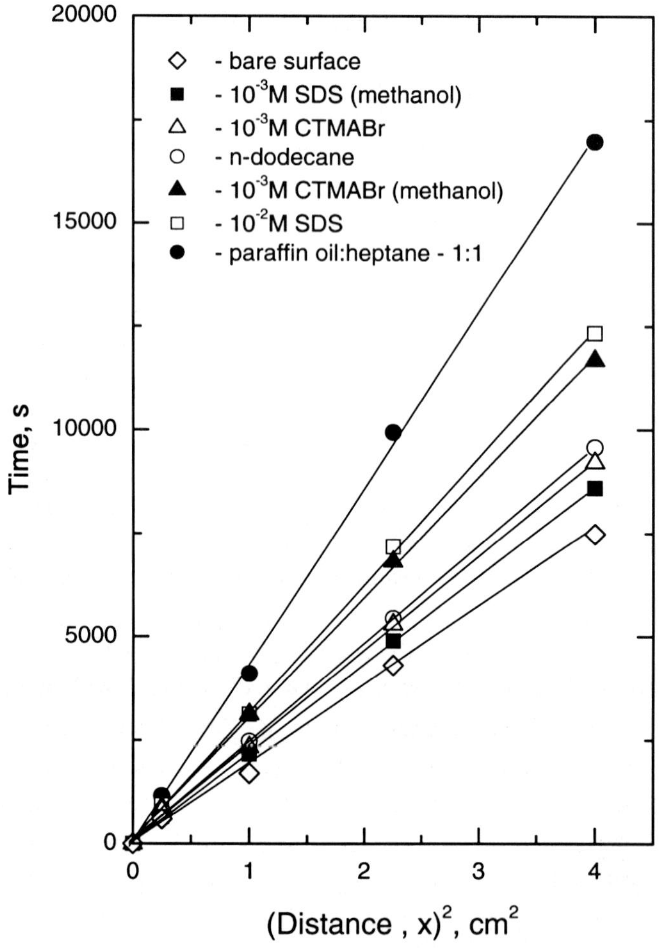

Figure 3. Wicking times of water into the carbonate rock bars treated with different substances as indicated vs. squared distances.

Where R is the effective radius of the capillaries, η is the liquid viscosity and ΔG is the change in the specific free energy accompanying the wicking process [1–4]. As can be seen from Fig. 3 all penetration results fit Washburn's relationship and the best hydrophobization effect (the longest penetration times) is obtained for the surface treated with paraffin oil, followed by 10^{-2} M aqueous solution of SDS, and 10^{-3} M methanolic solution of CTMABr. It should be stressed that water did not penetrate spontaneously into the rock bars treated with PMMA solution. One would expect that paraffin oil should hydrophobize the surface to such an extent that water could not penetrate the pores. As this is not the case, it may be concluded that the amount applied was not enough to form a uniform film inside the pores or the water molecules are to big to penetrate them. On the contrary, as after the PMMA treatment water does not penetrate the rock, one would conclude that the pores are sealed off. But, n-octane penetrated easily into the PMMA treated samples, although a decrease in the effective radius R was observed relative to the untreated sample. The R values determined with n-octane are shown in Table 2 together with Lifshitz-van der Waals component of surface free energy for the samples. The polymer treatment reduces the pore size by ca. 30% and dyeing with Methylene Blue and then PMMA treatment causes a further decrease in the pore size by 46%. Nevertheless, the apolar component is practically the same independent of the surface treatment (Table 2).

As mentioned above, to determine the apolar and polar components of the solid surface free energy from the wicking experiments three probe liquids have to be used and the wicking has to be conducted for bare and the liquid vapors equilibrated (pre-contacted) bars of the rock [1–4]. The results of such wicking experiments are presented in Fig. 4 for the unmodified bars of the rock. Distances of 2 cm were wicked with n-octane, water, and formamide. The straight-line plots were obtained. A good reproducibility of the results was obtained for 3–4 repeated experiments using a new rock bar each time. In case of the PMMA treated samples, where water did not wick the bars, contact angles were measured and the rock surface free energy components were calculated. The results of such determinations are collected in Table 3. The apolar component γ_s^{LW} determined from wicking is practically the same for untreated and PMMA treated rock, i.e., 34.6 and 34.9 mJ/m^2, respectively (Table 3). The value obtained from the wicking method is lower than that calculated from contact angles of diiodomethane (43–46 mJ/m^2) for sample "painted" with PMMA solution in chloroform. These values are slightly higher than the γ_s^{LW} of PMMA surface (41.8 mJ/m^2, Table 3) (also obtained from contact angle), but agree very well with the value obtained by van Oss *et al.* [10], 40.6 mJ/m^2 . The reason for the obtained differences in the γ_s^{LW} component of the PMMA, is probably due to different energetic states of the molecules in the film deposited on the rock surface (as determined from contact angles) and those inside the pores (determined from wicking). Also, a nonuniform coverage of the rock surface is possible. However, it worth noting that 1–2 mJ/m^2 differences in the component are, in fact, within the accuracy of such types of experiments.

As to polar electron-donor γ_s^- and electron-acceptor γ_s^+ components, the rock is predominantly exposing the electron-donor component, which is connected with the presence of the surface –OH groups and adsorbed water molecules. The PMMA

Table 2.
Apparent radius R and the Lifshitz-van der Waals component, γ_S^{LW}, of the carbonate rock variously treated

Carbonate Rock Treatment	R, μm	γ_s^{LW}, mJ/m^2
Untreated	1.175	34.6
PMMA (dipped)	0.807	34.9
Methylene Blue/ PMMA (dipped)	0.632	34.1
Ultra Extra Blue/ PMMA (dipped)	0.877	35.6

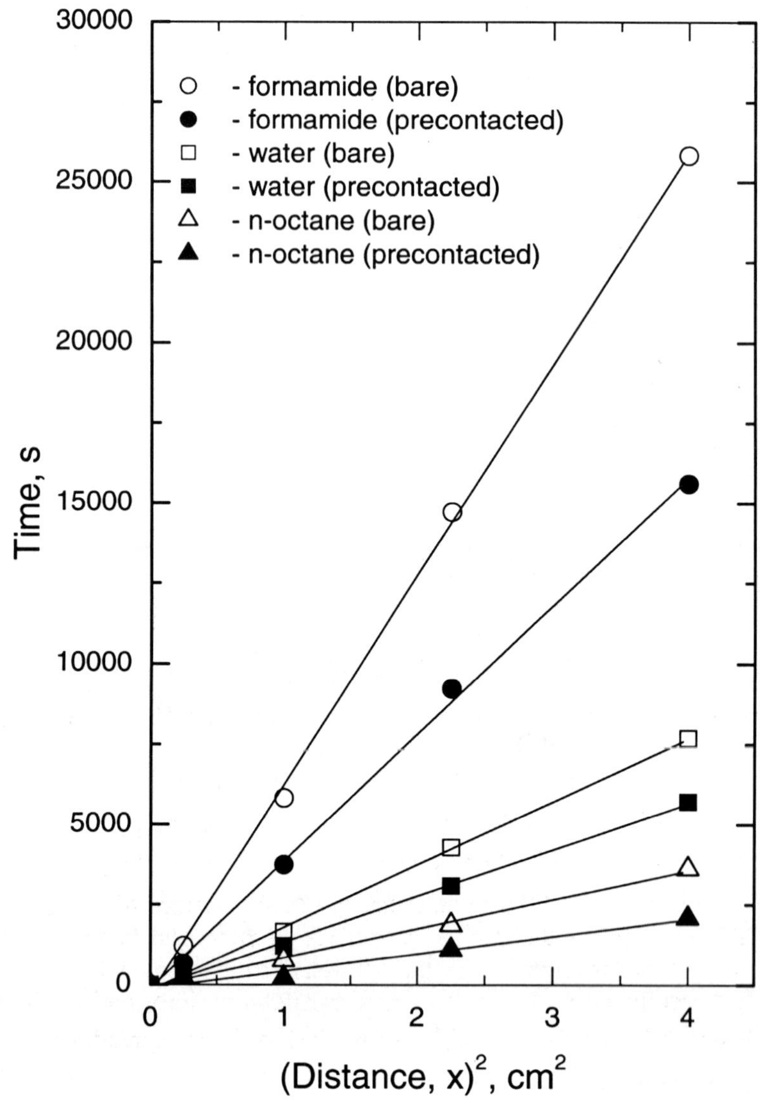

Figure 4. Penetration times of the probe liquids, n-octane, water and formamide, into bare and rock bars precontacted with the liquid vapors vs. squared distances.

treatment by dipping the rock sample in the chloroform solution causes drastic reduction in this component, practically to the value for pure polymer, i.e. from 57.3 to 13.5 mJ/m^2 (Table 3). However, after wetting the surface ("painting") with the same solution the γ_s^- interaction is reduced less, i.e., to 21.1 mJ/m^2. The reduction of electron-donor component by PMMA depends also on the kind of the dye used for the sample dyeing and it varies from 15.8 for Methylene Blue, to 18.9 and 31.1 mJ/m^2 for "Ultra Extra Blue" and Yellow Dye, respectively. In Table 3 are also shown values of

Table 3.
Surface free energy components of the original carbonate rock investigated and after treatment, and work of spreading W_s, in mJ/m^2

Treatment of the rock	γ_s^{LW}	γ_s^+	γ_s^-	γ_s^{AB}	γ_s	W_s
Untreated rock	34.6 (TLW)*	1.02	57.3	15.3	49.9	–4.0
Dipped in chloroform solution of PMMA	34.9 (TLW)*	1.64	13.5	9.42	44.3	–40.4
"Painted" with chloroform solution of PMMA	43.3	0.12	21.1	3.2	46.5	–34.2
Methylene Blue/PMMA, dipped	34.1 (TLW)*	1.44	15.8	9.5	43.7	–38.7
Ultra Extra Blue/PMMA, dipped	45.0	0.06	18.9	2.13	49.7	–34.7
Yellow Dye/PMMA, the rock "painted" with the solution	46.1	0.002	31.1	0.5	46.6	–25.5
PMMA, bulk polymer surface	41.8	0.02	13.7	1.05	42.8	–46.4

* TLW - determined from thin layer wicking method.

acid-base component (γ_s^{AB}) and the total surface free energy of the rock (γ_s). The values were calculated based on the van Oss *et al.*'s approach [6–8)], i.e.:

$$\gamma_s = \gamma_s^{LW} + \gamma_s^{AB} = \gamma_s^{LW} + 2\left(\gamma_s^+ \gamma_s^-\right)^{1/2} \quad (4)$$

It is clearly seen that the total value of the solid surface free energy is less informative than the knowledge of individual components [6, 7]. For example, the γ_s for the untreated rock sample is 49.9 mJ/m^2, while that of PMMA surface 42.8 mJ/m^2, which is not a big difference, while the surface properties of these two materials are significantly different. More informative is the acid-base component, which are 15.3 and 1.05 mJ/m^2, respectively. However, to obtain quantitative information about the hydrophobic/hydrophilic state of the sample the work of water spreading, W_s, should be calculated as follows:

$$W_s = W_a - W_c = 2\left[\left(\gamma_s^{LW}\gamma_l^{LW}\right)^{1/2} + \left(\gamma_s^- \gamma_s^+\right)^{1/2} + \left(\gamma_s^+ \gamma_s^-\right)^{1/2}\right] - 2\gamma_l \quad (5)$$

where the subscript *s* refers to the rock and *l* to water. Thus calculated work of spreading, using the surface free energy components from Table 3, are also listed in the same table. Now it is seen that the untreated carbonate rock sample is only slightly hydrophobic, because the work of water spreading is slightly negative. Because of the Laplace pressure in the pores, water can easily fill the pores. However, after impregnation with PMMA it becomes highly hydrophobic with the work of water spreading –40.4 mJ/m^2, which is only 6.6 mJ/m^2 less than the work of water spreading on the surface of the bulk polymer. It means that hydrophobization of the surface with PMMA is efficient and can be considered for use for practical purposes.

4. SUMMARY

The results presented show that the wicking method can deliver a valuable information on the hydrophobic/hydrophilic state of the pore surface, e.g., of the carbonate rock. On the other hand, contact angle measurements are more informative about the outside surface of the rock, especially after PMMA treatment. Although the surface free energy components are relative values, the calculated works of water spreading are the absolute ones [8]. Such experiments have practical aspects, for example, to protect monuments against erosion, or to improve properties of carbonate rocks, which may be used as a lining material for building walls.

REFERENCES

1. E. Chibowski and L. Hołysz, *Langmuir* **8**, 710 (1992).
2. E. Chibowski and F. González-Caballero, *Langmuir* **9**, 1069 (1993).
3. E. Chibowski and L. Hołysz, *J. Adhesion Sci. Technol.* **11**, 1289 (1997).
4. E. Chibowski in: *Acid-Base Interactions: Relevance to Adhesion Science and Technology*, Vol. 2, K. L. Mittal (Ed.). VSP, Utrecht, The Netherlands, in press.
5. C. Della Volpe, A. Penati, R. Peruzzi, S. Siboni, L. Toniolo and C. Colombo, *J. Adhesion Sci. Technol.*, in press.
6. C. J. van Oss, M. K. Chaudhury and R. J. Good, *Chem. Rev.* **88**, 927 (1988).
7. C. J. van Oss, R. J. Good and M. K. Chaudhury, *Langmuir* **4**, 884 (1988).
8. C. J. van Oss, *Interfacial Forces in Aqueous Media.* Marcel Dekker, New York (1994).
9. S. Kozłowski, *Mineral Resources of Middle–East Poland.* Geological Press, Warsaw (1985) (in Polish).
10. C. J. van Oss, R. F. Giese and W. Wu, *J. Dispersion Sci. Technol.* **19**, 1221 (1998).